# 16 ISCMP: Progress in Solid State and Molecular Electronics, Ionics and Photonics



Journal of Physics: Conference Series
Volume 253

Varna, Bulgaria
29 August - 3 September 2010

ISBN: 978-1-61782-128-8
ISSN: 1742-6588

**Printed from e-media with permission by:**

Curran Associates, Inc.
57 Morehouse Lane
Red Hook, NY 12571

**Some format issues inherent in the e-media version may also appear in this print version.**



Printed by Curran Associates, Inc. (2010)

For permission requests, please contact the Institute of Physics
at the address below.

Institute of Physics
Dirac House, Temple Back
Bristol BS1 6BE UK

Phone: 44 1 17 929 7481
Fax: 44 1 17 920 0979

techtracking@iop.org

**Additional copies of this publication are available from:**

Curran Associates, Inc.
57 Morehouse Lane
Red Hook, NY 12571 USA
Phone: 845-758-0400
Fax: 845-758-2634
Email: curran@proceedings.com
Web: www.proceedings.com

# TABLE OF CONTENTS

**IN MEMORIAM**

Joseph M Marshall graduated as BSc Hons. Physics, Sheffield, UK in 1965 and got a PhD degree in Edinburgh in 1970. In the same year he became Research Fellow in the University of Edinburgh, UK, then Lecturer (1974), Senior Lecturer (1981) and Reader (1984) in the University of Abertay Dundee. In 1987 he took the position of Professor of Electronic Materials in the University of Wales Swansea, UK and in 2002 became Emeritus Professor in the same University.

His scientific papers (over 180) span three areas: Experimental and theoretical studies of the conductive, photoconductive and other properties of crystalline, polycrystalline, amorphous and organic semiconductors; Computer based numerical modelling of carrier transport and other properties of semiconducting materials and electronic devices; Design, fabrication and characterisation of thin-film sensor devices and systems, for applications in information technology, medicine, automotive industry, etc.

He was Member of the Organising Committee of the annual Chelsea Conferences on Amorphous and Liquid Semiconductors (London, UK), Scientific Chairman of the Organising Committee of the 1990–2008 Biennial International Schools on Condensed Matter Physics (Varna, Bulgaria), Finance Chairman of the 1993 International Conference on Amorphous Semiconductors (Cambridge, UK), Co-Director of the 2001 NATO Advanced Study Institute 'Photovoltaic and Photoactive Materials – Properties, Technology and Applications' (Sozopol, Bulgaria) and co-editor of their Proceedings.

He was a recipient of the Marin Drinov Medal of the Bulgarian Academy of Sciences (1998) and the Georgi Nadjakov Sign of Honour of the Institute of Solid State Physics (2005).

# COMMITTEES

## *ORGANISING COMMITTEE*

***Chairman: Alexander G Petrov*** Director, Institute of Solid State Physics, Bulgarian Academy of Sciences, Sofia, Bulgaria
***Chairman Emeritus: Nikolay Kirov*** Institute of Solid State Physics, Bulgarian Academy of Sciences, Sofia, Bulgaria

### Local Organising Committee:

**Chairman:** Alexander G Petrov
**Members:** Diana Nesheva, Doriana Dimova-Malinovska, Eleonora Popova, Lyubomila Dedinska, Christo Popov, Vasil Lovchinov, Marina Primatarowa, Emilia Vlaikova, Irina Velkova, Hassan Chamati

## *PROGRAMME COMMITTEE*

**Chairman:** A G Petrov
**Members:** D Alexandrov (Thunder Bay), V Celebonovic (Belgrade), D Dimova-Malinovska (Sofia), B Dulmet (Besancon), A Grechnikov (Moscow), M Gunes (Mugla), C Main (Dundee), D Nesheva (Sofia), M C Petty (Durham), M Popescu (Bucharest), S Reynolds (Dundee), K Shimakawa (Gifu), J Singh (Darwin), N Starbov (Sofia), M Tomilin (St Petersburg), Ph Vanderbemden (Liege), A Vaseashta (Washington)

## *LOCAL SCIENTIFIC COUNCIL*

**Chairman:** A G Petrov
**Members:**
*Bulgarian Academy of Sciences* N Sabotinov (President)
I Nedkov (Scientific Secretary, Physics)
*Institute of Solid State Physics* N Kirov, D Nesheva, V Lovchinov, St Andreev, M Primatarowa
*Institute of Electronics* R Enikov (Director)
*Central Lab. of Solar Energy and New Energy Sources* D Dimova-Malinovska
*Institute of Optical Materials and Technologies* V Saynov, N Starbov
*Central Lab. of Applied Physics* R Kakanakov (Director)
*Sofia University – Faculty of Physics* A Andreeva, S Russev (Heads of Departments)

**PREFACE**

We are pleased to introduce the Proceedings of the 16 ISCMP, organized by the Institute of Solid State Physics of the Bulgarian Academy of Sciences. The Chairman of the School was Professor Alexander G Petrov. The School was dedicated to the late Professor Joe Marshall, who served for a long time as Chairman and Honorary Chairman and left us just after having completed the proceedings of the previous School. Like previous events, the School took place in the beautiful Black Sea resort of Saint Constantine and Elena near Varna, going back to the renewed facilities of the Panorama hotel. Participants from 19 different countries delivered 34 invited lecturers and 75 posters, contributing to three sessions of poster presentations. Papers submitted to the Proceedings were refereed according to the high standards of the *Journal of Physics: Conference Series* and the articles published in this volume illustrate the diversity and the high level of the contributions.

Not the least significant factor in the success of the 16 ISCMP was the social program, both the organized events (Welcome and Farewell Parties) and the variety of pleasant local restaurants and beaches.

These Proceedings are published for the first time in *Journal of Physics: Conference Series*. We are grateful to the Journal's staff for encouraging this idea.

The Scientific Committee of the ISCMP dedicates this volume of the Proceedings to the living memory of Professor Joe Marshall, Honorary Chairman of the ISCMP. The Committee decided that the next event will take place again in Saint Constantine and Elena, in September 2012. It will be entitled: Open Problems in Condensed Matter Physics, Biomedical Physics and their Applications.

**Doriana Dimova-Malinovska, Diana Nesheva, Alexander G Petrov and Marina T Primatarowa**

# Measurement and modelling of transport in amorphous semiconductors

**C Main and S Reynolds**

University of Dundee, School of Engineering, Physics and Mathematics, Small's Wynd, Dundee DD1 4HN, UK.

E-mail: c.main@dundee.ac.uk

**Abstract**. Experimental transient photoconductivity in a-Si:H and μc-Si:H is analysed by a Fourier Transform technique to reveal a relative density of states. It is often found that a deep-lying structure probably associated with defects, at around 0.5 – 0.6 eV depth exhibits an apparent shift against the background of band-tail states as the temperature is varied. We investigate by numerical modelling whether this shift arises from differential attempt-to-escape frequencies in a multi-trapping context, or from a temperature - dependent hopping transport energy.

## 1. Introduction

Joe Marshall was a mentor, colleague and personal friend for over 40 years. In his scientific work he was scrupulously honest, valuing truth and possessing a healthy disrespect for authority. His philosophy could be encapsulated in the motto of the Royal Society - "*Nullius in Verba*" which translates as 'on the word of no one'. He was also keen to promote the interests and development of younger scientists, in his teaching and in support of their attendance at schools and conferences. All who knew Joe as a friend or colleague, will miss him.

The transient photocurrent response (TPC) following excitation by a short laser pulse is determined by a combination of carrier transport processes. The non-equilibrium distribution of excess carriers created by such a pulse will diffuse and drift while relaxing in energy toward thermal equilibrium. If the immediate 'post-pulse' excess carriers initially exist in extended states, their relaxation and transport will involve the participation of localised states either by 'multi-trapping' (MT) processes in which transitions occur only between localised and extended states, or by direct inter-site tunnelling or 'hopping'. In the former case, any measured current arises from trap-limited band transport, whilst in the latter case the current arises from hopping transport.

It is possible to model numerically the TPC in a distribution of localized states (DOS) mediated by either of the above processes. Marshall [1,2] has studied, by computer modelling applicable to an arbitrary DOS, the case of transient photocurrents in the case of hopping transport, employing Monte-Carlo techniques. The present authors have further developed a finite difference numerical model based on Marshall's work which includes transport by either of the above processes, or even a simultaneous combination of both MT and hopping [3,4].

To complement such work, Fourier Transform (FT) techniques have been developed by the present authors to reveal, in a *spectroscopic* fashion, information on the energy distribution of localised states (DOS) from *measurements* of the transient photocurrent decay [5,6], assuming in the first instance that

Journal of Physics: Conference Series **253** (2010) 012001 doi:10.1088/1742-6596/253/1/012001

MT processes dominate transport. It its basic form, as described below, this method can only give a relative DOS scale and a relative energy scale as determined by a fixed value of attempt-to-escape frequency ν, although there are ways to alleviate this difficulty.

The motivation for this work arises from studies over a range of temperatures, of TPC in undoped a-Si:H and μc-Si:H, the DOS being obtained by MT-FT analysis. These studies, while revealing such features as band-tails and deep defect structure, often show an apparent systematic shift of the defects to deeper relative energy, against the tail state 'background' as temperature is increased in the range 100K – 300K. A typical set of results is shown in figure 1 illustrating the TPC in an a-Si:H sample over the temperature range 150K – 330K, and figure 2 showing the FT derived DOS for these data, using an attempt-to-escape frequency $\nu = 10^{12}$ Hz. There is an apparent shift of the defects of about 0.05 eV over this temperature range. One possible explanation for this is simply that the defect states have a much smaller attempt-to-escape frequency than the tail states. It turns out that a value of $10^{10}$ Hz would 'fix' the defects relative to the tail states. However, there is also an indication that this shifting ceases above room temperature, which would be difficult to explain in this way, since the 'shift phenomenon' would continue to apply at elevated temperatures.

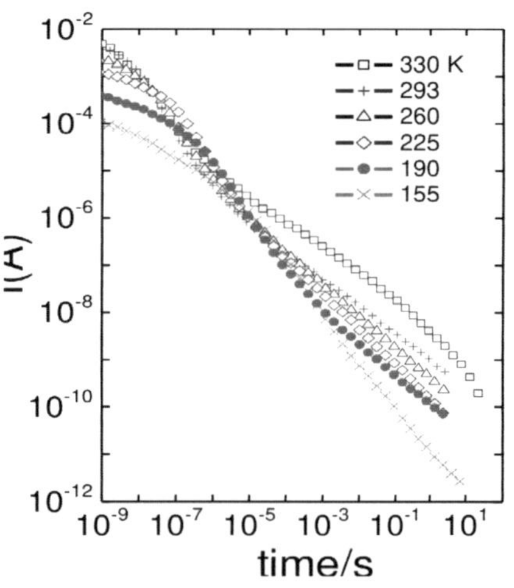


**Figure 1.** Transient Photocurrents (TPC) for a-Si:H coplanar thin film sample, over the temperature range 155K – 330 K.

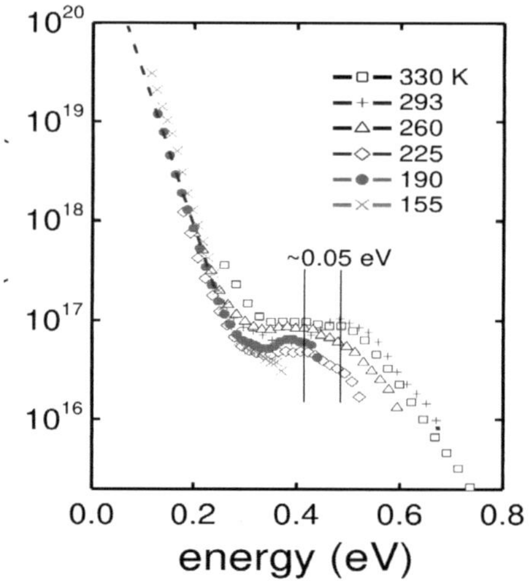


**Figure 2.** FT-DOS for the data of figure 1 computed using ν = 1012 Hz. Between 190K and 300K, there is an apparent shift of defect energy of about 0.05 eV.

An alternative explanation might invoke hopping transport, where it has been shown [7,8] that in an exponential band-tail, there is an energy within the band tail, at which hopping transport is maximum – the so-called 'transport energy'. This energy is temperature dependent, approaching the conduction band edge as temperature is increased. In such a situation, any deeper lying levels which have a significant influence on the effective drift of excess carriers would appear to lie at successively deeper energies below the transport energy as the temperature rises.

The purpose of the present work is therefore to evaluate whether the energy shift arises as a numerical artefact of the FT analysis or as a physical consequence of changes in the dominant energy location of hopping transport with temperature.

## 2. Numerical methods

### *2.1. Modelling hopping and multi-trapping transport*

Following Marshall [1], we represent the continuous distribution *g(E)* of localised states by a 'ladder' of state grouped into slices of energy with $\Delta E$. Thus, the level '$i$' of this discretised distribution has an approximate density $g(E_i)\Delta E$ cm$^{-3}$. Firstly, we write the probability of hops to iso-energetic or deeper sites at energy $E_j$ from a site at energy $E_i$ as

$$p_{i,j} = \nu_0 \left( g\left(E_j\right)\Delta E / N_{i,deep} \right) \exp\left(-r_{i,deep}/2R_0\right), \tag{1}$$

where $\nu_0$ is an attempt-to-hop frequency, $N_{i,deep} = \int_{E_i}^{\infty} g(E)dE \approx \sum_k g(E_k)\Delta E \quad (k \geq i)$ is the total number of iso-energetic and deeper sites, with corresponding mean separation $r_{i,deep} = \left(N_{i,deep}\right)^{-1/3}$, and $R_0$ is a localisation parameter for the sites. In this paper, we restrict $R_0$ to a typical value of around $7\times10^{-8}$ cm. The ratio $g\left(E_j\right)\Delta E / N_{i,deep}$ represents the fraction of all such sites in slice '$j$', since the probability of hopping to a given slice is proportional to the fractional concentration of sites it contains. For hops upward in energy, from a site in a deeper slice at energy $E_j$ up to $E_i$, detailed balance gives the inverse hop probability

$$p_{j,i} = p_{i,j}\left( g(E_i)/g(E_j) \right)\exp\left(-\left(E_j - E_i\right)/kT\right), \tag{2}$$

where $kT$ is the thermal energy.

Similarly, multi-trapping transition probabilities may be written, for trapping of a carrier in the conduction band by a trap in slice '$i$',

$$p_{c,i} = \left(\nu / N_C\right) g\left(E_i\right)\Delta E, \tag{3}$$

while for the inverse process, thermal re-emission, the expression is

$$p_{i,c} = \nu \exp\left(-\left(E_C - E_i\right)/kT\right), \tag{4}$$

where $\nu$ is an attempt-to escape frequency, $N_C$ is the effective density of states in the conduction band, and $E_C$ is the mobility-edge energy. We assume that the occupancy of states is always low, so that the system may be assumed linear.

The various probabilities above may be incorporated into a set of rate equations describing the time dependence of the densities of electrons in both localised and extended states. With appropriate initial conditions, such as the introduction of excess free or localised carriers, these will describe the progress of carrier relaxation in for example, a transient photocurrent experiment. The rate equations are as follows:

$$\frac{dn}{dt} = -n\sum_i p_{c,i} + \sum_i n_{ti} p_{i,c} \tag{5}$$

$$\frac{dn_{ti}}{dt} = np_{c,i} - n_{ti} p_{i,c} + \sum_j n_{tj} p_{j,i} - n_{ti} \sum_j p_{i,j}\ , \tag{6}$$

where $n$ is the instantaneous density of free electrons, and $n_{ti}$ is the density of electrons on sites in slice '$i$'. In the following, we will assume that there are '$M$' levels, including the conduction band.

To perform a discrete 'time-stepping' numerical solution, we can use these equations to compute updated values of each of the variables $n$, $n_{ti}$,... as time is incremented by a step $\Delta t$, using previous values of the variables. Re-casting the above equations in matrix-vector form, we obtain

$$\mathbf{n}(t+\Delta t)=\mathbf{A}.\mathbf{n}(t), \tag{7}$$

where **n**(*t*) is the vector of the instantaneous free and localised carrier densities, and **A** is an $M\times M$ 'stochastic' transition matrix with elements

$$A_{1,1}=1-\sum_{i=2}^{M}p_{c,i}\Delta t, \qquad A_{i,1}=p_{c,i}\Delta t, \qquad A_{1,i}=p_{i,c}\Delta t, \tag{8a,b,c}$$

$$A_{i,i}=1-\left(p_{i,c}+\sum_{j=2}^{M}p_{i,j}-p_{i,i}\right)\Delta t \quad, \qquad A_{i,j}=p_{j,i}\Delta t \quad (j\neq i). \tag{8d,e}$$

The system of equations (5) and (6) is 'stiff'. In practice this means that time step $\Delta t$ cannot exceed the shortest time-constant of the system, in this case, the overall trapping time of free carriers into the ensemble of traps $\tau_{trap}=\left(\sum_i p_{c,i}\right)^{-1}$. It is possible to deal with such systems using stable 'implicit' numerical methods, and the authors have done so in previous work on MT simulation [9]. However, in the present case, there is a simple artifice, which allows the explicit formulation described above, to be used to follow the time evolution of the system over very long time-spans. Using a suitably small value for $\Delta t$ of $10^{-14}$ s, we can, by continued squaring of the matrix **A**, generate very long effective time steps, *viz*-

$$\mathbf{n}(t+2^{m}\Delta t)=\mathbf{A}^{2m}.\mathbf{n}(t). \tag{9}$$

In this way, we may extend the simulation time-scale from $10^{-14}$s to $10^{+6}$s, in only 72 steps, although extended-precision arithmetic is required to minimise truncation errors. Initial conditions for the simulation of TPC, with MT-only transport or MT+hopping, are determined by setting the leading element of vector **n**(0) to a normalised value of *e.g.*, 1.0, and all other elements to zero.

Computation of the transient current for free carriers, uses the instantaneous extended state conductivity $\sigma_{ext}=nq\mu_{ext}$, where $q$ is the electronic charge and $\mu_{ext}$ is the free carrier mobility. For hopping carriers, and for low fields, we compute the effective diffusivity of carriers in each slice, as determined by the hop rate from sites in a given slice, and some 'average' hop distance, here – the distance from slice '*i*' to slice '*j*' taken as the mean separation of sites in slice '*j*', *viz*., $\bar{r}_j=\left(g(E_j)\Delta E\right)^{-1/3}$.

The contribution $\sigma_i$ of each level to the hopping conductivity, and the total hopping conductivity $\sigma_{hop}$ may then be found, after application of the Einstein relation giving,

$$\sigma_i=n_{ti}q^2\sum_j p_{i,j}\,\bar{r}_j^{\,2}/6kT, \tag{10}$$

$$\sigma_{hop}=\sum_i n_{ti}q^2\sum_j p_{i,j}\,\bar{r}_j^{\,2}/6kT. \tag{11}$$

It is useful to note here the analytical approximate result obtained by Monroe [7], which for an exponential tail produces a 'transport energy' $E_t$ at which the hopping transport peaks, *viz*

$$E_t = 3E_0 \ln\left(\frac{2kT}{3E_0 R_0 (4\pi N_0/3)^{1/3}}\right), \tag{12}$$

where $E_0$ is the characteristic tailing energy and $N_0$ is the integrated DOS of the tail. The temperature dependence of this energy appears in the logarithmic factor.

*2.2. Computing the DOS (MT only)*

Applying a Fourier transform to equations (5) and (6), omitting the hopping terms, gives in the frequency domain,

$$\hat{n}(\omega) = \int_0^\infty n(t)[\cos(\omega t) - j\sin(\omega t)]dt = \frac{N_0}{(A_\omega + jB_\omega)}, \tag{13}$$

where, treating the density of states (DOS) as continuous in energy,

$$A_\omega = \omega_R + \int_0^{E_f} \nu\sigma g(E)\left(1+\exp\left(-\frac{2(E_\omega - E)}{kT}\right)\right)^{-1} dE \tag{14}$$

and

$$B_\omega = \frac{i(0)\sin(\varphi(\omega))}{|\hat{I}(\omega)|} = \omega + \int_0^{E_f} \frac{\nu\sigma}{2} g(E)\operatorname{sech}\left(\frac{E_\omega - E}{kT}\right)dE\,. \tag{15}$$

In the above, $\hat{I}(\omega)$ represents the frequency domain photocurrent associated with $\hat{n}(\omega)$, *i(0)* is the initial value of the transient photocurrent – *i.e.* when all $N_0$ excess carriers are free, and $E_f$ is the Fermi energy. The term $k$ is Boltzmann's constant, and $T$ is the absolute temperature.

Equation (15) is a Fredholm integral equation of the first kind, and is 'ill-conditioned'. A simple *approximate* solution [6] replaces the peaked sech function with a delta function, to give

$$g(E_\omega) \approx \frac{N_c 2}{\nu\pi kT}\left(\frac{i(0)\sin(\varphi(\omega))}{|\hat{I}(\omega)|} - \omega\right). \tag{16}$$

with energy scale

$$E_\omega = kT \ln\left(\frac{\nu}{\omega}\right). \tag{17}$$

We note that if the value of the attempt-to escape frequency ν is known then a given feature in the DOS will produce a feature in the computed $g(E_\omega)$ at a specific frequency $\omega = \nu\exp\left(-\frac{E_\omega}{kT}\right)$ which varies with temperature so that the computed $E_\omega$ remains fixed at all temperatures. Otherwise, an incorrect value of ν will result in a 'shifting' feature.

## 3. Simulation results and discussion

The simulation method described in section 2 gives a typical result for hopping transport as shown in figure 3. Here the progress of the peak of the charge density and the mean hopping energy are shown. The results agree with the Monroe model. Parameters used for this and subsequent computations are $E_0$= 30meV, $R_0$=7.3×10$^{-8}$ cm, $g(E_c) = 4\times10^{21}$ cm$^{-3}$eV$^{-1}$, and $\nu = 10^{12}$ Hz. We find that the temperature

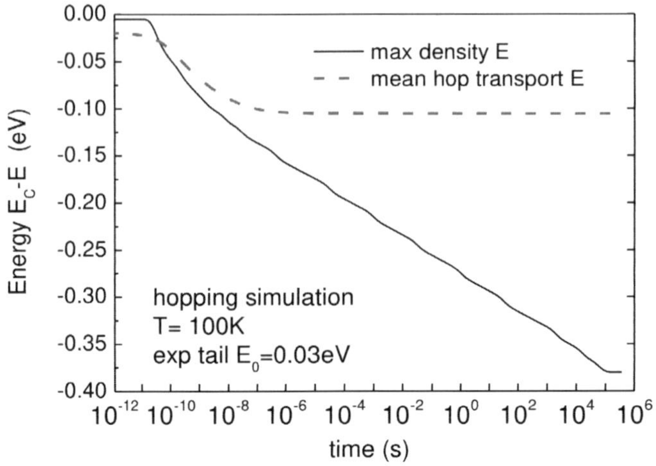


**Figure 3.** Numerical simulation of hopping transport at 100K in an exponential tail of characteristic energy 0.03eV. The progress of the peak of the charge density and the mean hopping energy are shown. The results agree with the Monroe model.

dependence of the transport energy also agrees with the Monroe model, equation (12). Note however, that we justify use of the numerical simulation on grounds that it may be applied to situations with an arbitrary DOS.

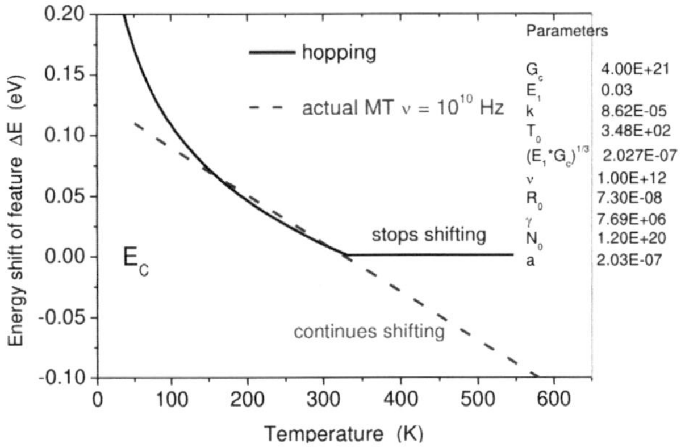


**Figure 4.** Numerical simulation of the Temperature dependence of the apparent energy shift for a deep feature in the computed DOS for MT transport, using $\nu=10^{12}$ Hz when the correct value was $10^{10}$ Hz, and for hopping transport which gives a shift of the transport energy. Exponential tail of characteristic energy 0.03eV.

We wish to test whether this upward movement of the transport energy with temperature effectively results in an increase in the apparent depth of the underlying defects comparable with that observed. Figure 4 shows the result of this computation with the hopping parameters given above. It does indeed seem possible that this is a reasonable explanation, even to the extent that the shifting ceases at elevated temperatures, as observed.

On the same figure we illustrate the alternative explanation – that the chosen value of ν may be correct for bandtail states, but incorrect for the defect states. Here we find that the required 0.05 eV shift between 200K and 300K may be obtained by assuming that the correct ν for the defect states is much lower, *viz* ~ $10^{10}$ Hz. While this does give the required shift, it can be seen that this continues as the temperature rises above 300K, in disagreement with the observation.

However, it may be pointed out that the FT-DOS method applies only when we ignore the possibility of hopping and that the above argument, though plausible, may thus be invalid. To partly answer such an objection, we computed the TPC for a DOS containing an exponential bandtail and a Gaussian defect feature, allowing the possibility of simultaneous MT and hopping processes, using the same parameters as before. The defect feature is centred at 0.5 eV below Ec, has characteristic energy 0.06 eV, and peak density $10^{17}$ $cm^{-3}$ $eV^{-1}$

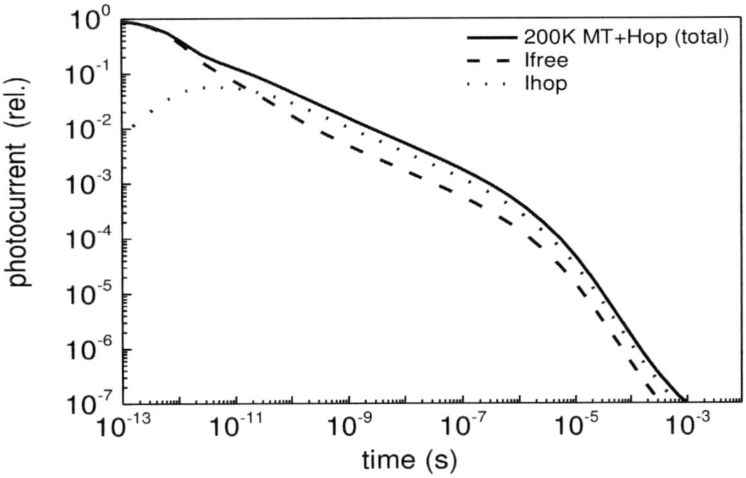


**Figure 5.** Numerical simulation of TPC for simultaneous MT and hopping transport at 200K in a DOS with exponential tail of characteristic energy 0.03eV and a Gaussian defect centred at 0.5 eV, width 0.06 eV. The relative free carrier and hopping transport currents are shown, and the total current. At 200K, the hopping current is slightly larger than the free carrier current.

Figure 5 shows the computed relative TPC currents (using a free carrier mobility of 10 $cm^2V^{-1}s^{-1}$) for the hopping and extended state transport components, at 200K. At this temperature, with the chosen parameters, the hopping component is slightly higher than the free carrier component over most of the time range. Figure 6 shows the same computation, now at 300K. Here the free carrier current is larger than the hopping current by a factor of about 4. Perhaps surprisingly, using quite reasonable values for hopping parameters, even at room temperature the hopping current is not insignificant.

In figure 7 we show the FT-computed DOS for the simulated free + hopping TPC transients. Although the existence of hopping current may again invalidate the analysis, the result is directly comparable to that obtained experimentally. The shift of peak defect energy matches the experimental result, and the cessation of shifting above 300K is also evident. The figure also demonstrates that the inclusion of hopping into the simulation has a significant effect even at 300K. One of the DOS curves for T = 300K (filled square symbol) was obtained from a simulation including simultaneous MT and hopping processes; the other (open circle symbol) from an 'MT- only' simulation.

## 4. Conclusions

We have compared alternative explanations of the apparent shifting in energy, of features in the FT-derived DOS in a-Si:H as obtained from TPC measurement. The alternatives are firstly – that the shift

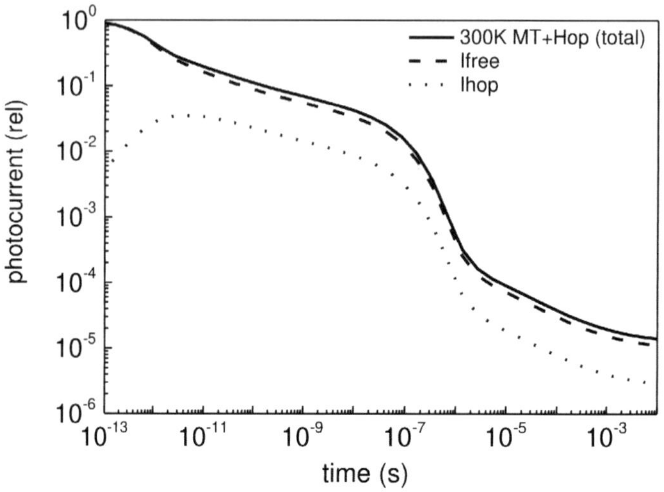


**Figure 6.** Numerical simulation of TPC for simultaneous MT and hopping transport at 300K for the same DOS as in Figure 5. The relative free carrier and hopping transport currents are shown, and the total current. At 300K, the free carrier current is larger than the hopping current by a factor of about 4.

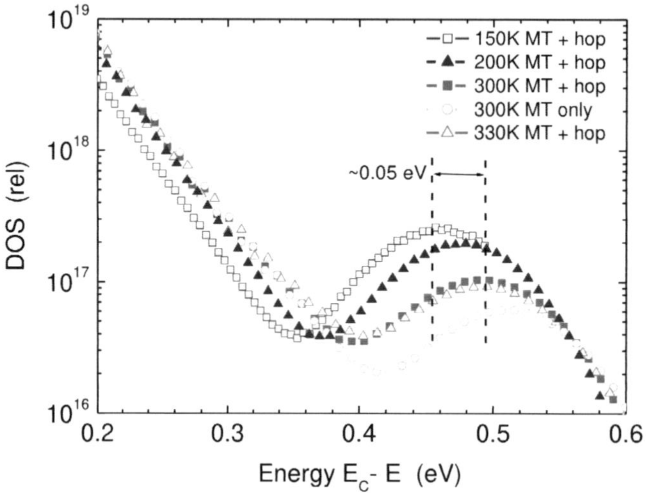


**Figure 7.** Computed FT – DOS from numerical simulations of TPC for simultaneous MT and hopping transport for a DOS with tail and defect states, at several temperatures. The apparent shift of defect states is a result of the shift of the hopping transport energy.

is merely an artefact of the FT calculation, involving an incorrect value of attempt to escape frequency, and secondly – that it arises from the physical shift in the transport energy when hopping conduction predominates. A definitive answer is not yet possible, as the FT analysis assumes only transport of free carriers with multiple trapping. However the latter explanation seems to be more plausible at the time of writing. Further development of the FT analysis, to include hopping, would be useful.

**Acknowledgments**

The authors are most grateful to the Organizing Committee of the 16th International School of Condensed Matter Physics, for their invitation to present the Joe Marshall Memorial Lecture, and further, to acknowledge the encouragement received over many years from Joe Marshall, whose ideas continue to provide inspiration for work in this field.

**References**

[1] Marshall J M 2000 *Philos. Mag. Lett.* **80** 691
[2] Marshall J M 2003 *J. Mat. Sci. Materials in Electronics* **14** 611
[3] Main C, Marshall J M, Reynolds S and Rose M J 2005 *Mat. Res. Soc. Symp. Proc.* **862** A15.3
[4] Main C, Marshall J M and Reynolds S 2005 *J. Opt. Adv. Mat.* **7** 107
[5] Main C 2002 *J. Non-Cryst. Solids* **299-302** 525-530
[6] Main C and Reynolds S 2000 *Appl. Phys. Lett.* **76** 3085
[7] Monroe D 1985 *Phys. Rev. Lett.* **54** 146
[8] Baranovskii S D, Rubel O and Thomas P 2005 *Thin Solid Films* **487** 2
[9] Main C and Brüggemann R 1993 *in Electronic and Optoelectronic Materials for the 21st Century,* ed N Kirov and A Vavrek (World Scientific Press) pp 270-278

# Carrier mobility, band tails and defects in microcrystalline silicon

**Steve Reynolds**

School of Engineering, Physics and Mathematics, University of Dundee, Nethergate, Dundee DD1 4HN, United Kingdom

E-mail: s.z.reynolds@dundee.ac.uk

**Abstract**. The development of microcrystalline silicon thin films and devices is briefly reviewed. Transport mechanisms, and the attendant key parameters of carrier mobility, band-tail width and defect density, are linked to film structure and composition. In particular we discuss the wide (but systematic) variations in time-of-flight mobility and its unusual field-dependence. While microcrystalline silicon remains an inferior semiconductor to single-crystal silicon, we propose, and support by means of a computer model, that present device-grade material may be of sufficient quality to justify re-examining whether useful thin-film bipolar devices might be developed. These could find application as more sensitive photo-detectors, and as current drivers in organic LED displays and logic circuits.

## 1. Introduction

Although hydrogenated microcrystalline silicon (μcSi:H) thin films [1] have been studied for at least as long as their amorphous silicon (aSi:H) counterparts [2], their large-scale exploitation in devices has a somewhat shorter history. Until the mid-1980s it was generally believed that the microcrystalline silicon films with the best electronic properties would contain a high crystalline volume fraction, consistent with the observation that crystalline semiconductors generally possess more desirable electrical properties than amorphous semiconductors – high-mobility transport in delocalised states, and a low defect density. However, at that time, highly-crystalline (> 90%) silicon films were found to contain a high density of electrically-active mid-gap defects [3], some $10^{18}$ $cm^{-3}$ compared with below $10^{16}$ $cm^{-3}$ in as-prepared high-quality aSi:H. Recombination and restricted movement of the Fermi level due to these defects cast doubt over the suitability of microcrystalline silicon for use in solar cells and thin-film transistors, which developed as important technological applications of amorphous silicon [4].

Following more detailed work on the relationships between deposition parameters and film structure in the 1990s, a number of groups developed *pin* solar cells containing microcrystalline silicon absorber layers [5-7]. Most depositions were made using plasma-enhanced chemical vapour deposition (PECVD), although it was demonstrated that hot-wire (HW) CVD material could perform equally well [8]. The development of single-junction μcSi:H and the 'micromorph' μcSi:H/aSi:H tandem solar cells has been reviewed [9].

It is important to distinguish between microcrystalline silicon μcSi:H, which forms the subject of the present work, and other forms of crystalline thin-film silicon such as polycrystalline [10], protocrystalline and polymorphous films [11]. μcSi:H is mixed-phase in nature, with crystalline grains

20 nm or less in size often aggregated into columns embedded in more disordered material. Significant hydrogen is present, as in amorphous silicon. A key feature is that the cracks and voids present in the highly crystalline films, frequently prepared under high RF powers from a process gas containing around 1% of silane in hydrogen, are diminished or absent in more compact films of good electrical quality, deposited at lower powers and higher silane concentrations. The variation in structure and composition that may be obtained by adjustment of deposition parameters is illustrated schematically in figure 1. Diagnostic tools such as electron spin resonance indicate that defect density reaches a minimum at the transition between growth regimes [12, 13], at a phase composition of roughly equal quantities of microcrystalline and amorphous material. It is commonly believed that this corresponds to the point at which the crystalline column boundaries are suitably passivated, giving rise to optimum performance of both solar cells [6] and thin-film transistors [14].

From an applications point of view, there are a number of important differences between aSi:H and μcSi:H. Most fundamentally, their band gaps are 1.8 eV and 1.1 eV respectively, exerting a strong influence on their optical absorption spectra and on dark current magnitudes. Furthermore, aSi:H behaves as a 'classical' disordered semiconductor with a largely exponential Urbach tail reflecting the joint density of states, whereas the absorption spectrum of μcSi:H bears a stronger similarity to that of crystalline silicon, as illustrated in figure 2, and is constrained by momentum-conservation, characteristic of an indirect semiconductor, as well as energetic considerations. The photovoltaic efficiency of a stabilized aSi:H solar cell is similar to its μcSi:H counterpart, taking into account that the former suffers more markedly from light-induced degradation, but the deposition rate of good-quality microcrystalline films is lower. This is exacerbated by the fact that a rather thicker μcSi:H absorber is desirable to compensate for its lower optical absorption coefficient in the visible region of the spectrum. Thus it is currently somewhat more economically favourable to produce amorphous silicon modules, although improvements in deposition rate [15] and light-trapping [16] (to maximise the optical path length and reduce the thickness required) may influence this in future. The so-called 'micromorph' tandem solar cell, in which the bottom cell is μcSi:H and the top cell aSi:H, has been developed intensively [9] and is now in commercial production. As well as improving module efficiency over a single junction by 2 to 3% the total absorber layer thickness may be reduced, which reduces total deposition time, and the thinner aSi:H layer combined with the more stable μcSi:H cell improves overall stability.

Silicon thin-film transistors (TFTs) [17] have proven highly successful commercially as switching elements in active-matrix liquid crystal (LC) displays. An important parameter in this application is the off: on channel resistance ratio, which can exceed $10^6$ in the case of amorphous silicon. Microcrystalline silicon TFTs look unlikely to rival the amorphous device in LC display applications unless stability is a major issue. However the μcSi:H TFT has a lower on-resistance, which means that in principle it can drive more current and switch more quickly than its amorphous counterpart, and is

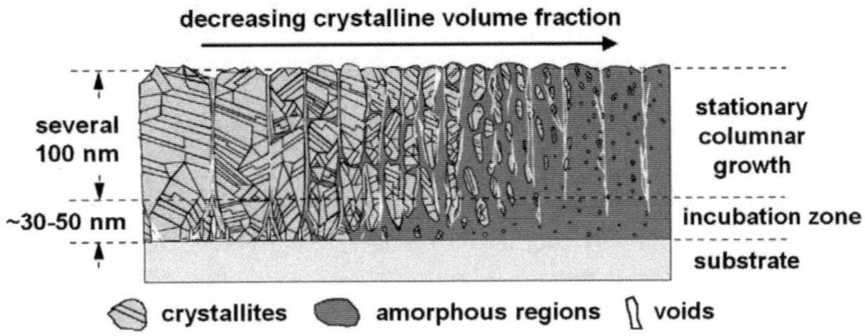


**Figure 1.** Schematic representation of a silicon thin film, after Vetterl et al [6].

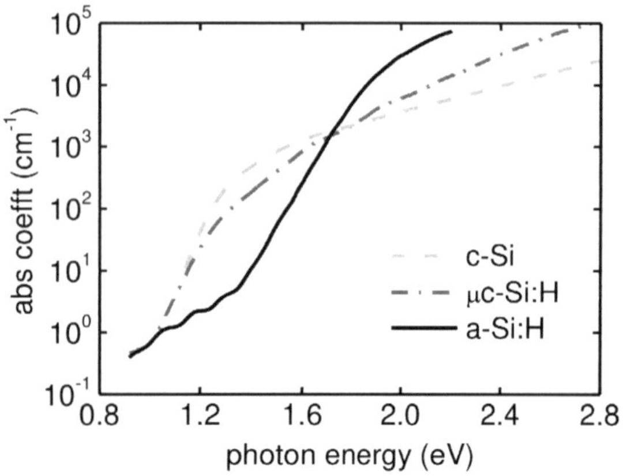


**Figure 2.** Constant-photocurrent spectra of crystalline, microcrystalline (Raman crystalline volume fraction approximately 60%) and amorphous silicon.

therefore more suited to applications in photoemissive displays such as organic LEDs, and in digital logic. The lower channel resistance is primarily a result of increased carrier mobilities, discussed in more detail later, and particularly so in the case of holes where an improvement of over 100 has enabled p-channel [18] and ambipolar TFTs [19], and CMOS logic gates [20], to be developed.

In the following sections, we consider carrier transport in microcrystalline silicon, building on the models developed previously for disordered materials such as amorphous silicon, and augmenting them where appropriate. In particular, the multiple-trapping model will be applied in the interpretation of photoconductivity data. Central to this model is the idea that carrier transport takes place in extended states, but the free carrier density is strongly influenced by the presence of localized states, in the form of band tails and defects, and by recombination centres. By studying both transient and steady-state photoconductivity it is possible to develop a consistent picture of the microscopic conduction properties, and to map this to film structure and composition. In optimised solar cells, carrier extraction appears to be substantially improved over amorphous silicon, which suggests that bipolar transistor action might be achievable. Although appropriate devices have not yet been prepared, possible designs and test procedures are considered, and a computer simulation of the characteristics of an aSi:H/μcSi:H heterojunction bipolar photo-transistor is presented.

## 2. Relationship between structure, composition and transport

### *2.1. Structural and compositional elements*

As discussed above, mixed-phase μcSi:H films contain proportions of both crystalline and amorphous materials. Based on evidence from structural studies [21, 22] we consider the topology of the crystalline fraction somewhat crudely, as small cubic grains of side $g$ that may coalesce into crystalline rectangular prism-shaped columns of square base width $w$ and height (in the growth direction) $h$. This is shown schematically in figure 3(a). We assume the columns are disposed some distance apart, and may be separated by voids, by disordered material, or by a combination of the two. In this model, each grain has six faces, which may be considered internal or external to the column, as shown in figure 3(b). At the small grain boundaries, bonds may be unperturbed (figure 3(c)), as within a perfect crystal, deformed by changes in bond length and/or angle (figure 3(d)), or broken (figure 3(e)), resulting in silicon dangling bonds. Furthermore, the 'space' between columns may be a void, or may be occupied by some disordered form of silicon, such as 'standard' amorphous silicon or a form intermediate between amorphous and crystalline, as illustrated in figures 4(a), 4(b).

We may now examine whether such a model stands up to quantitative considerations, and what predictions might therefore be made from it. Firstly, the model must explain the large difference in defect density observed between highly-crystalline porous material and optimal mixed-phase material.

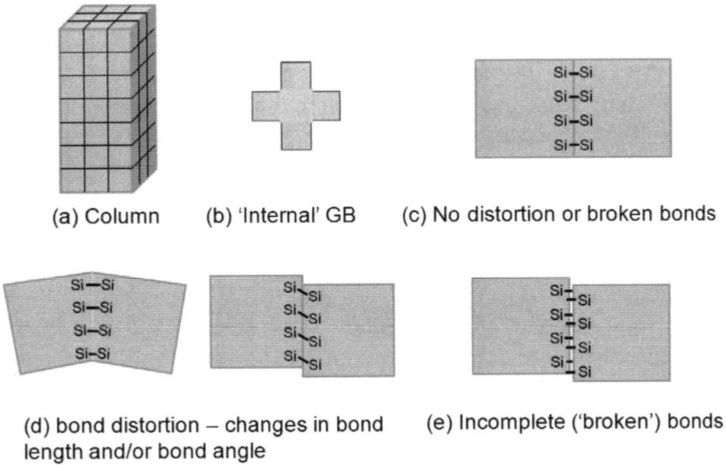


**Figure 3.** Highly simplified representation of microcrystalline silicon grains and columns.

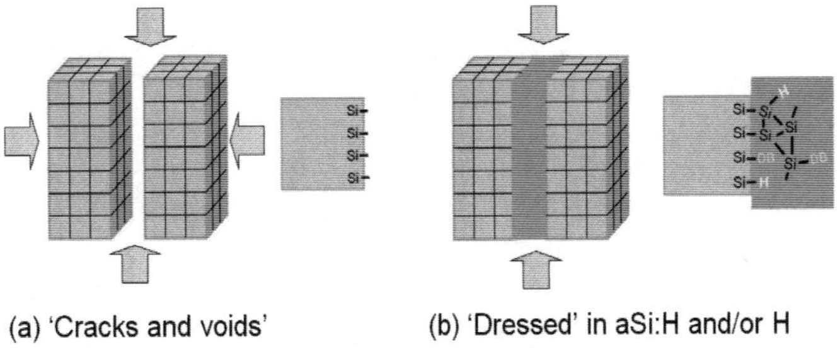


**Figure 4.** Column boundaries – possible configurations.

If closely-spaced columns separated by thin cracks form the bulk, then for a column volume $w^2h$ there is an associated surface area $4wh$ (ignoring the top and bottom surfaces). If the effective bulk defect density is due entirely to the exposed column surfaces, then it may be expressed as

$$N_B = (4wh)N_S/(w^2h) = 4N_S/w \quad (1)$$

For a bare silicon surface, the total surface state density $N_S$ lies in the range $10^{10}$ – $10^{12}$ cm$^{-2}$ depending on morphology and chemical treatment [23]. Given a typical column width of order 50 nm [21], the calculation suggests that the effective bulk defect density could approach $10^{18}$ cm$^{-3}$, in keeping with spin density [12, 13] and field-effect [3] measurements. The passivating influence of the material between columns thus appears quite strong, reducing the surface density to around 1% of this figure in optimised μcSi:H. Whether this is due to termination of what would otherwise be dangling bonds, or an internal field-effect retarding carrier motion directed towards the column boundaries, is not clear.

This model may also be used to assess whether the internal boundaries, between small grains within columns, contribute to the overall defect density, and to speculate on the nature of this boundary. The atomic density of silicon is around $5\times10^{22}$ cm$^{-3}$, and since the measured defect density in optimised μcSi:H can be as low as $5\times10^{15}$ cm$^{-3}$ as few as one in ten million atoms may contribute a dangling bond. Taking a cubic grain of side $g$ = 10 nm, we find that 16% of the atoms lie at the surface; evidently very few dangling bonds are concealed within columns. TEM and XRD evidence suggests there is also little disordered material within columns [21] and so the internal grain boundaries cannot be preserved in the same way as the column boundaries. A typical sample contains only 3 to 5% hydrogen which, even if all the hydrogen was effective, would leave some 10% of atoms at internal surfaces un-passivated. It thus appears that the grains must undergo some specific structural accommodation that enables the bond order to be preserved across boundaries. This may occur through 'twinning', such that adjacent grains share atoms on common lattice sites [22].

### *2.2. Transport models*

Experiments such as transient [24, 25] and modulated [26] photoconductivity and time-of-flight spectroscopy (TOF) [27-30] may be interpreted in terms of a multiple-trapping model [31], and generally support the view that μcSi:H possesses quite extensive conduction and valence band-tails arising from bond distortions in column material. ESR studies [12, 13] reveal significant spin densities, ascribed to dangling bonds at column boundaries. Possible contributions from the amorphous regions outside columns should not be disregarded, though conductive AFM studies [32] indicate, at least for transport through the film, that in more highly-crystalline films amorphous regions generally support only a small fraction of the total transport current. Kocka *et al* [33] present a transport model which combines reasoning similar to section 2.1 with a proposed density of states (DOS) and electrostatic barriers between crystalline and amorphous regions.

Values of electron and hole mobility, band-tail slopes and integrated defect densities are composition-dependent, and the nature and energetic position of the defects is still open to question. There is no single definitive 'microcrystalline silicon' material, owing to the wide range of possible compositions and structures depending on deposition conditions, and thus it is highly unlikely that a single transport model is appropriate for all films in all cases. For example, in amorphous silicon it is accepted that while transport in extended states mediated by multiple-trapping takes place at higher temperatures, band-tail hopping presides at temperatures below 100-150 K. μcSi:H is evidently structurally inhomogeneous which suggests that transport may be similarly anisotropic [33, 34]. There are regions of 'conducting' material (crystalline) embedded in a host of 'less conducting' material (amorphous), that have led to interest in percolation models of conduction [35], and the effective medium approach [36]. Both have proved successful in predicting certain aspects of transport such as the conductivity-composition transition and temperature-dependence of Hall effect.

As most of the above aspects of transport have already been addressed at some length in the literature, we confine ourselves here to a discussion of a new topic, the composition-dependence of transport in microcrystalline silicon, measured by TOF.

## 3. Transport measurements

### *3.1. Composition-dependence of time-of-flight mobility*

The variation of electron and hole TOF mobility at 300 K with crystalline volume fraction as determined using Raman spectroscopy ($I_{CRS}$) is shown in figure 5. In all cases, samples were *pin* structures, with transparent electrodes on each side. Typical i-layer thickness was 4 μm. Preparation conditions are as described previously [15]. Mobility was measured as $\mu = L / Ft_D$, where $L$ is the i-layer thickness, $F = V / L$ is the electric field for a given voltage bias $V$ and $t_D$ is the transit time, measured here by the construction illustrated in figure 6(a).

Samples containing more than 30% crystalline material exhibit electron mobilities of 3-4 $cm^2/Vs$, with a corresponding figure of 1-2 $cm^2/Vs$ for holes. In comparison with the values indicated for amorphous silicon, electron mobility is a factor of 2-3 higher, whereas the figure for holes is around 300 times higher. For $I_{CRS}$ in the region of 20% both mobilities fall rapidly, which is broadly in keeping with the conductivity transition threshold predicted by percolation theory. However, the mobility values do *not* attain typical values for aSi:H, as might be expected from percolation theory if transport became controlled by current flow through the 'host' material. The electron mobility falls

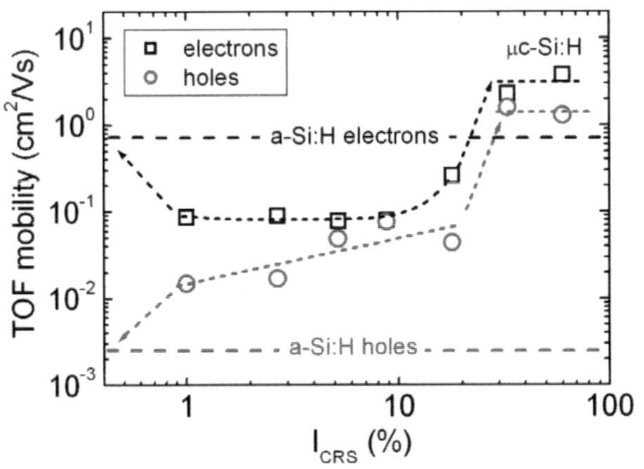


**Figure 5.** Plots of electron and hole TOF mobility at 300 K *vs.* Raman crystalline volume fraction. Dotted lines indicate the trends of the data. Dashed lines show electron and hole mobilities for fully amorphous samples. Each value is normalised to a sample length to electric field ratio of $10^{-7}$ cm$^2$/V.

*below* that of aSi:H, whereas the hole mobility remains substantially above that of aSi:H. This situation persists with diminishing $I_{CRS}$, to the limit of detection by Raman spectroscopy. In our PECVD system under the deposition conditions used, this corresponds typically to a process gas concentration *SC* of around 6.5% silane in hydrogen. Further increase in *SC* to 10% results in 'fully amorphous' material, with the mobilities indicated.

*3.2. Dispersion parameters*

Multiple-trapping theory applied to a simple exponential distribution of trap states [31] predicts that at a temperature $T$ the pre-transit current $I(t) \propto t^{-(1-\alpha)}$, where $\alpha = T / T_0$ is the dispersion parameter for a given band tail slope $T_0$. Additionally, in the same scenario, $\mu(F) \propto (L / F)^{1-1/\alpha}$. Both relationships yield values for $\alpha$, which should be equal. However this is not the case; in this example the pre-transit slope at $T = 300$ K gives $\alpha = 0.50$, whereas the field-dependence of mobility gives $\alpha = 0.87$. The latter case therefore implies a much steeper band tail. Data are presented over a more extensive temperature range in figure 7. It can be seen that while there is good agreement between both methods for both electron and hole TOF on amorphous silicon, the microcrystalline samples yield band-tail slopes that are consistently steeper when measured through the field-dependence of mobility.

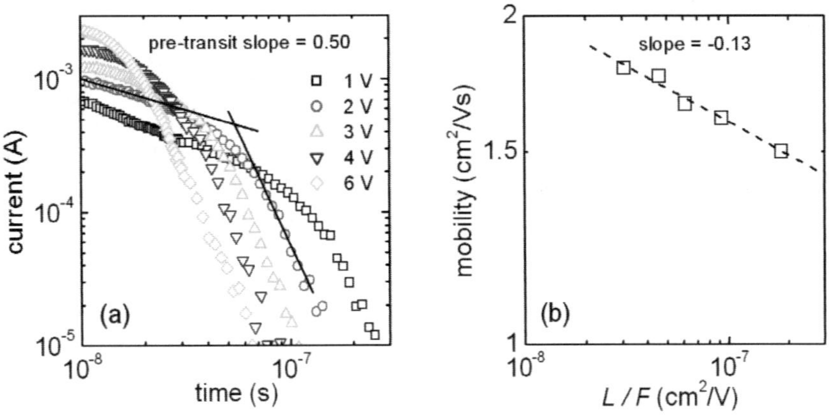


**Figure 6.** (a) Hole TOF transients for a highly-crystalline sample at 300 K, over a range of bias voltages. Pre-transit slope yields $\alpha = 0.50$. (b) Mobilities calculated from transit times plotted *vs.* $L$ / $F$ ratio (log-log scale). Note only weak field-dependence, yielding $\alpha = 0.87$.

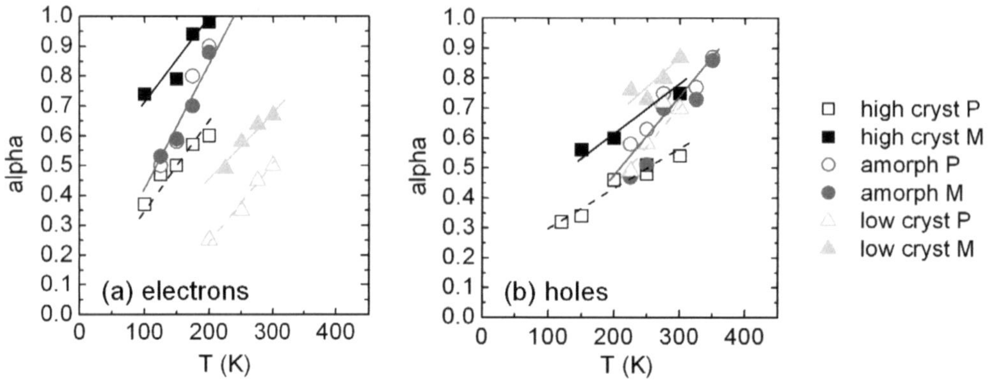


**Figure 7.** Variation of dispersion parameter $\alpha$ with temperature, for (a) electron TOF, (b) hole TOF. The highly crystalline material corresponds to $I_{CRS}$ = 60%, the amorphous material to $SC$ = 10% and the low crystalline material to $I_{CRS}$ ~ 1%. The letters P and M denote measurement of $\alpha$ from pre-transit $I(t)$ slope and from variation in mobility with applied electric field, respectively.

### *3.3. Discussion*

TOF data on more crystalline ($I_{CRS}$ > 30%) films are in general agreement with the predictions of multiple-trapping in a disordered semiconductor, *i.e.* the transits show dispersive behaviour, consistent with the presence of states extending from the band-edge into the mobility gap, and quite low extended state mobilities, somewhat higher than those in amorphous silicon especially in the case of holes, but much lower than in crystalline silicon. However, the unexpected field-dependence of mobility and the strange effects in low-crystalline material suggest the need for a closer look.

Previous studies [29, 30] have tended to concentrate more on lower temperature measurements where a larger portion of the pre-transit region is available for fitting or analysing data parametric in temperature at a specific electric field. This is understandable given the greater difficulty in accurately recording the shorter pre-transit region (compared with amorphous silicon) due to system bandwidth limitations. Brinza *et al* [37] however, made a comprehensive study of both temperature- and field-dependence of TOF transients for silicon films prepared using the expanding thermal plasma (ETP) technique. They discovered something rather similar to the above, though more extreme - that certain samples exhibited *field-independent* electron and hole mobilities over a wide temperature range, despite their typical 'dispersive' $I(t)$ signatures. They also noted a somewhat enhanced room temperature hole mobility, around 0.01 $cm^2/Vs$. The field-independence was accounted for in a computer model by adding an additional shallow Gaussian density of states to the exponential band tail. The rationale for this was that the ETP material was found to contain nanocrystalline inclusions, and that the states associated with these are suggested to lie in a narrow range within the amorphous silicon bandgap, positioned according to the band offsets.

This idea is interesting, and seems worth pursuing here. From our studies we can identify three aspects of transport in microcrystalline silicon that differ from amorphous silicon, and thus require independent explanation: (a) Hole mobility commensurate with electron mobility for material with crystallinity greater than 30%; (b) Relatively low electron mobility for low-crystallinity (or Raman amorphous) material; (c) Relatively high hole mobility for low-crystallinity material. We proceed as follows. Above the percolation threshold, both electron and hole currents circulate in predominantly crystalline channels, which offer enhanced hole mobility through reduced disorder, though transport is still limited by trapping and emission. Possibly the extended-state hole mobility is also increased. This could be the case for optimum phase mixture material. As the crystalline content is reduced below the percolation threshold, transport must take place through the lower-mobility 'amorphous' material. There would appear to be two additional factors – (i) that barriers to transport may exist between crystalline and amorphous regions, which control current flow, (ii) that the material may be an 'effective medium' in which both crystalline and amorphous densities of states (DOS) co-exist. Taking case (i), it seems reasonable to expect an effective mobility lower than in either sub-phase, due to electrostatic barriers. This could account for the reduced electron mobility, though it is generally believed that the energy offset at the conduction band (CB) is small compared with that at the valence band (VB). This raises the question of why hole transport is not adversely affected – holes appear to be *more* mobile than in amorphous material. We might imagine case (ii) as a superposition of the amorphous and the microcrystalline DOS. At low concentration, the crystalline element may introduce *localised* states superimposed on the amorphous band tails, reducing the free carrier density in the amorphous material extended states and hence the TOF mobility. At higher concentration, a new set of *extended* states, rather like impurity bands in a crystalline semiconductor, may result. Depending on their energetic location and capture properties, it seems possible that both could occur simultaneously, *i.e.* holes propagate in new extended states, but electrons are impeded by new localised states. It may be possible to simulate both the carrier distribution and the effective mobility for such a system.

## 4. Bipolar thin-film transistor

### 4.1. Bipolar versus Field-effect

It is interesting to consider the use of BJTs, or a combination of bipolar and field-effect transistors, in certain applications such as photoemissive displays or digital logic, where improved current drive or switching speed may be desirable. As mentioned above, carrier mobilities obtained from curve-fitting to the transfer function of μcSi:H field-effect TFTs exceed 100 $cm^2/Vs$ for electrons and 10 $cm^2/Vs$ for holes [38]. While the exact figures depend on fabrication details, and in particular on oxygen content of the films, it is clear that they are far in excess of the equivalent values for aSi:H. TFTs are majority-carrier devices however, and cannot provide a direct indication of the prospects for thin-film bipolar junction transistors (BJTs), which rely on transport properties of both carrier types.

In a BJT, minority carriers are injected by a forward-biased emitter-base junction and normally must diffuse across the neutral base region to be swept out at the reverse-biased base-collector junction. Current amplification, defined as the ratio $\beta$ between collector and base currents may be achieved provided (i) the current injected from emitter to base significantly exceeds that from base to emitter, and (ii) only a small proportion of the current injected into the base (minority carriers) is lost to recombination [39]. There have been few reports of BJT action in amorphous silicon [40, 41], with only modest values of $\beta$ (2-3) being achieved. To the author's knowledge, there have been no reports of BJTs fabricated from optimum-phase mixture μcSi:H. However, the broad area of non-crystalline bipolar TFTs is the subject of a US Patent [42], which may have deterred any recent commercial interest.

Based on the steady-state photocarrier grating method, ambipolar diffusion lengths ranging from 100 to 200 nm are obtained for both aSi:H and μcSi:H [43]. These figures can be taken as an indicator of minority-carrier transport efficiency through the neutral base region. In contrast, much improved carrier extraction properties are observed in device-quality μcSi:H *pin* solar cell absorber layers than in equivalent aSi:H devices, as shown in figure 8. This is especially true in the case of n-side illumination, where hole collection might be expected to be the rate-limiting step. Improved transport properties of both carrier types in μcSi:H would reduce both recombination loss and space-charge accumulation in the BJT. However it should be recognised that the base must be doped in order to obtain transistor action. In aSi:H, minority carrier lifetimes are diminished by doping due to the simultaneous creation of additional defects, although more optimistically a trend of increasing diffusion length with increasing conductivity is shown in μcSi:H [43]. Further, if μcSi:H (bandgap 1.1 eV) is used as base, with an aSi:H emitter (bandgap 1.7 eV), a heterojunction with much improved injection properties is expected. The successful use of aSi:H as a hetero-emitter in high-frequency crystalline silicon BJTs has already been demonstrated [44].

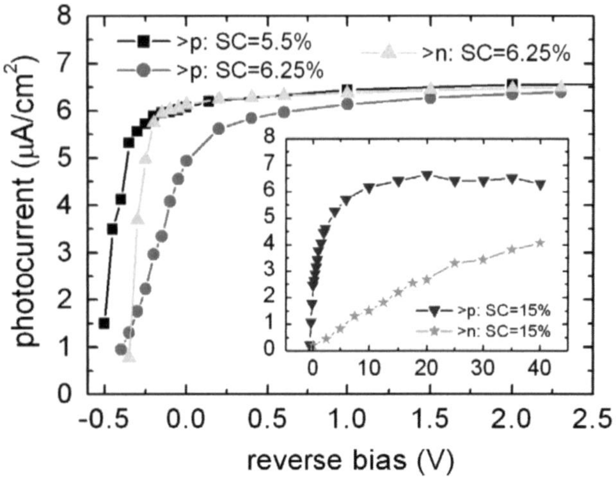


**Figure 8.** Photocurrent *vs.* reverse bias voltage for microcrystalline (main figure) and amorphous silicon (inset) *pin* solar cells. All cells have an i-layer thickness of 4 μm. The symbols >p and >n refer to illumination by 470 nm light (flux $5.1\times10^{13}$ cm$^{-2}$s$^{-1}$) from p and n sides, respectively.

*4.2. Computer model*

As a first attempt, we have modelled an *npn* thin-film BJT, consisting of $n^+$ aSi:H emitter and collector regions 30 nm wide, identical to those normally used in solar cell simulations, and a *p*-type μcSi:H base 500 nm wide. Detailed parameters are given in Table 1. To test the BJT, an input signal in the form of a base current is required. The simulator SC-Simul [45] is designed for two-terminal applications, and provides no independent control of the potential of intermediate layers. Therefore an optical signal, uniformly absorbed throughout the base region was applied, giving a photo-transistor structure. The output characteristics of this photo-BJT are shown in figure 9. The effective base current density due to the optical signal is $J_B = eG_BW$, where $G_B$ is the generation rate and $W$ the base width, and the collector current density is $J_C$. A current gain $\beta = J_C / J_B$, of around 6 is obtained with a neutral defect density of $10^{15}$ cm$^{-3}$. The variation of $\beta$ with defect density is shown in figure 10.

*4.3. Summary*

Given the large uncertainty in the input parameters, it may be concluded from this model that design of a silicon thin-film based bipolar device with potentially useful levels of current gain may be possible. However, the maximum tolerable defect density stretches the limits of present materials, and the base width and doping level require careful adjustment to obtain a compromise between suitable current gain whilst avoiding 'punch-through' breakdown [38]. By making use of microdoping techniques [46] it may be possible to deposit a graded doping profile in the base, providing a built-in field to assist the transport of minority carriers.

**Table 1.** Model parameters used in photo-BJT simulation.

| | $E_G$ (eV) | $\mu_{0e}$ (cm²/Vs) | $\mu_{0h}$ (cm²/Vs) | $N_C$ (cm$^{-3}$) | $N_V$ (cm$^{-3}$) | $kT_C$ (eV) [2] | $kT_V$ (eV) [2] | $E_D$ (eV) | $N_D$ (cm$^{-3}$) [3] | $U$ (eV) |
|---|---|---|---|---|---|---|---|---|---|---|
| n-type aSi:H [1] | 1.70 | 1.0 | 0.1 | $10^{20}$ | $10^{20}$ | 0.1 | 0.15 | 1.1 | $10^{19}$ | 0.25 |
| p-type μcSi:H [1] | 1.12 | 50 | 20 | $10^{20}$ | $10^{20}$ | 0.02 | 0.034 | 0.56 | varied | 0.2 |

**Notes:**

Key: $E_G$ energy bandgap; $\mu_{0e}$ free electron mobility; $\mu_{0h}$ free hole mobility; $N_C$ effective density of CB states; $N_V$ effective density of VB states; $kT_C$ CB tail slope; $kT_V$ VB tail slope; $E_D$ defect peak energy; $N_D$ defect density; $U$ correlation energy. All energies measured relative to $E_V = 0$

[1] Fermi level positions: n-type aSi:H 1.5 eV, p-type μcSi:H 0.28 eV

[2] Electron and hole capture coefficients into CB and VB tail states: $10^{-9}$ cm$^3$s$^{-1}$

[3] Electron and hole capture coefficients into (i) neutral defects: $5\times10^{-9}$ cm$^3$s$^{-1}$, (ii) charged defects: $5\times10^{-8}$ cm$^3$s$^{-1}$

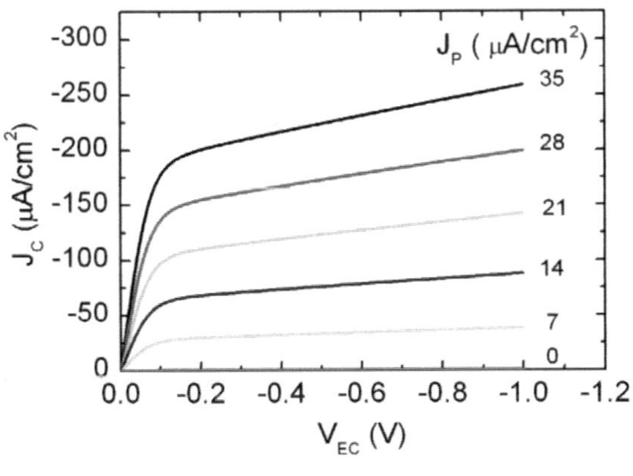


**Figure 9.** Simulated output characteristics of bipolar μcSi:H TFT, for $N_D = 10^{15}$ cm$^{-3}$.

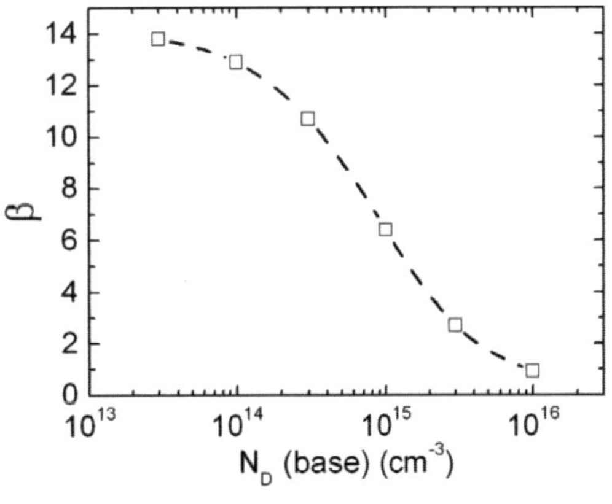


**Figure 10.** Simulated variation in current gain $\beta$ vs. base defect density $N_D$, at $V_{CE} = 0.5$ V.

## 5. Conclusions

Microcrystalline silicon thin films of mixed-phase composition show good optoelectronic properties, which are largely resistant to light-induced degradation. This enables them to be used in single-layer and tandem solar cells of comparatively high stable efficiencies. Improvements in deposition rate have been achieved without seriously compromising material quality, although presently this remains a disadvantage in comparison with amorphous silicon. Field-effect transistors show improved bias stability, current-handling and switching speeds, enabling a wider range of potential applications such as drive elements for organic LED photoemissive displays. The enhanced performance of p-channel devices due to the much higher hole mobility offers the potential to develop CMOS logic applications.

Transport landscapes based around the structure of the device-grade mixed-phase material, as examined using Raman spectroscopy, electron microscopy and x-ray diffraction, have been developed. Optoelectronic properties through the film can be explained semi-quantitatively in terms of transport via columnar structures consisting of crystalline grains, in which the column boundaries are passivated by a combination of disordered material and hydrogen. Remaining unsatisfied bonds give rise to mid-gap defects as in amorphous silicon, and band tails result from bond distortions within the columns.

A multiple-trapping model based on this scenario appears to satisfy most aspects of the measured transport behaviour, at least in the case of transient photoconductivity. In general, the conduction band tail has a similar or slightly steeper slope than in amorphous silicon, around 20 to 25 meV, though the valence band tail is noticeably steeper, at 30-35 meV compared with 45-50 meV. Electron time-of-flight mobility at 300 K for comparable sample geometries is two or three times higher in microcrystalline than amorphous material, approaching 5 cm$^2$/Vs at 300 K. Hole mobility is dramatically larger, by a factor of several hundred, and can reach 1.5 cm$^2$/Vs. In order to fit the multiple-trapping theory a lower attempt-to-escape frequency for holes trapped in the valence band is required. Other measurements such as Hall-effect suggest that transport is sensitive to potential barriers at column boundaries. The Hall mobility represents an average value, and hence is lower than in the above case and is Fermi-level position (doping) dependent. However, the trend towards increasing mobility in comparison to amorphous silicon remains in evidence. Field-effect measurements, particularly on top-gate TFTs, suggest much higher mobility values than are obtained from time-of-flight, due possibly to the higher crystalline content at the tops of the columns. To what extent transport in these materials is anisotropic remains an open question, but there are a number of direct and indirect measurements suggesting this is the case. Given the highly asymmetric structure development, it is possible that the effective medium parameters are quite different along the film, compared with through the film, which may also be anticipated to modify the percolation threshold.

Our understanding of transport in lower crystalline fraction material is incomplete. Based on time-of-flight measurements, hole mobility persists at a greatly enhanced level down to some 30% Raman crystalline volume fraction. Below this, both the hole and the electron mobilities appear to fall, and poor electronic quality material is obtained, in which the electron mobility falls below that of amorphous silicon. However, the spin density and CPM defect shoulder are not significantly greater, suggesting that it is primarily the transport path that is subject to additional disruption. Certain aspects of the transport behaviour are difficult to account for satisfactorily with a simple exponential band-tail model, and lead to inconsistencies in electric-field and temperature dependencies similar to those reported for thin-film silicon prepared by the expanding thermal plasma technique. It could be speculated that in such materials the density of states is influenced by additional compositional effects, which might for example involve sub-bands or the localisation of carriers on small islands of crystalline material, from which there is an additional energy cost for one carrier type (the electron) to escape. This is in contrast to the improved medium-range order proposed as a beneficial element found in the structure of protocrystalline silicon thin films.

Although optimum-phase microcrystalline silicon currently does not deliver a game-changing improvement in optoelectronic properties, it is demonstrably better than amorphous silicon in a number of respects. Improved transport properties, especially in the case of holes, opens up the prospect of designing bipolar devices. Models based around the parameters obtained from transport measurements predict a best-case scenario in which a bipolar thin-film TFT with a useable current gain could be realised. While a detailed study of potential benefits has not yet been carried out, there is the prospect that if such devices could be successfully fabricated they might find use in peripheral current drive applications, similar to BiCMOS crystalline silicon technology, and in thin-film digital logic where current amplification may be beneficial at certain nodes, for example to improve fan-out.

## Acknowledgments

The author is most grateful to the Organizing Committee of the 16th International School of Condensed Matter Physics, for their invitation to present the Milko Borissov Memorial Lecture [47]. Sincere thanks are extended to my co-workers, in particular Charlie Main, Rudi Brüggemann, Vlad Smirnov, Reinhard Carius and Friedhelm Finger for their invaluable scientific contributions, and to technical staff at IEF-5, Forschungszentrum Jülich, for their diligent preparation and characterization of samples.

## References

[1] Vepřek S and Mareček V 1968 *Solid St. Electr.* **11** 683
[2] Chittick R C, Alexander J H and Sterling H F 1969 *J. Electrochem. Soc.* **116** 77
[3] Le Comber P G, Willeke G and Spear W E 1983 *J. Non-cryst. Solids* **59&60** 795
[4] Wagner S 2010 *Phys. Stat. Sol.* A **207** 501
[5] Yamamoto K, Yoshimi M, Tawada Y, Okamoto Y, Nakajima A and Igari S 1999 *Appl. Phys.* A **69** 179
[6] Vetterl O, Finger F, Carius R, Hapke P, Houben L, Kluth O, Lambertz A, Mück A, Rech B and Wagner H 2000 *Solar Energy Materials & Solar Cells* **62** 97
[7] Meier J, Flückiger R, Keppner H and Shah A 1994 *Appl. Phys. Lett.* **65** 860
[8] Klein S, Repmann T and Brammer T 2004 *Solar Energy* **77** 893
[9] Shah A V, Schade H, Vanecek M, Meier J, Vallat-Sauvain E, Wyrsch N, Kroll U, Droz C and Bailat J 2004 *Prog. Photovolt: Res. Appl.* **12** 113
[10] Seto JYW 1975 *J. Appl. Phys.* **46** 5247
[11] Fontcuberta i Morral A and Roca i Cabarrocas P 2001 *Thin Solid Films* **383** 161
[12] Finger F, Baia Neto L, Carius R, Dylla T and Klein S 2004 *Phys. Stat. Sol* (c) **1** 1248
[13] Astakhov O, Carius R, Finger F, Petrusenko Y, Borysenko V and Barankov D 2009 *Phys. Rev.* B **79** 104205
[14] Chan K-Y, Knipp D, Gordijn A and Stiebig H 2008 *J. Appl. Phys.* **104** 054506

[15] Mai Y, Klein S, Carius R, Wolff J, Lambertz A, Finger F and Geng X 2005 *J. Appl. Phys.* **97** 114913
[16] Müller J, Rech B, Springer J and Vanecek M 2004 *Solar Energy* **77** 917
[17] Wagner S, Gleskova H, Cheng I-C and Wu M 2003 *Thin Solid Films* **430** 15
[18] Cheng I-C Allen S and Wagner S 2004 *J. Non Cryst. Solids* **338** 720
[19] Chan K-Y, Knipp D, Kirchhoff J, Gordijn A and Stiebig H 2009 *Sol. St. Electron.* **53** 635
[20] Chan K-Y, Gordijn A, Stiebig H and Knipp D 2009 *IEEE Trans. Electron Dev.* **56** 1924
[21] Houben L, Luysberg M, Hapke P, Carius R, Finger F and Wagner H 1998 *Philos. Mag* A **77** 1447
[22] Houben L, Luysberg M and Carius R 2003 *Phys. Rev.* B **67** 45312
[23] Henrion W, Rebien M, Angermann H and Röseler A 2002 *Appl. Surf. Sci.* **202** 199
[24] Smirnov V, Reynolds S and Main C 2004 *J. Non Cryst. Solids* 338 **421**
[25] Brüggemann R, Main C and Reynolds S 2002 *J. Phys. Condens. Matter* **14** 6909
[26] Brüggemann R, Kleider J P and Longeaud C 2000 *Proc. 16th European Photovoltaic Solar Energy Conference*, ed. H Scheer (London: James and James) p 11
[27] Fejfar A, Beck N, Stuchlíková H, Wyrsch N, Torres P, Meier J, Shah A and Kočka J 1998 *J. Non-Cryst. Solids* **227-230** 1006
[28] Serin M, Harder N and Carius R 2003 *J. Mat. Sci.: Mat. El.* **14** 733
[29] Dylla T, Finger F and Schiff E A 2005 *Appl. Phys. Lett.* **87** 032103
[30] Dylla T, Reynolds S, Carius R and Finger F 2006 *J. Non-Cryst. Solids* **352** 1888
[31] Tiedje T 1984 *Hydrogenated Amorphous Silicon II* eds. J D Joannopoulos and G Lucovsky (New York: Springer-Verlag) p 261
[32] Fejfar A, Vetushka A, Kalusová V, Čertik O, Ledinský M, Rezek B, Stuchlik J and Kočka J 2010 *Phys. Stat. Sol.* A **207** 582
[33] Kočka J, Fejfar A, Stuchlíková H, Stuchlík J, Fojtík P, Mates T, Rezek B, Švrček V and Pelant I 2003 *Solar Energy Materials & Solar Cells* **78** 493
[34] Unold T, Brüggemann R, Kleider J P and Longeaud C 2000 *J. Non-Cryst. Sol.* **266-269** 325
[35] Overhof H and Otte M 1996 *Future Directions in Thin Film Science and Technology* eds. J M Marshall, N Kirov, A Vavrek and J M Maud (Singapore: World Scientific) 23
[36] Shimakawa K 2000 *J. Non-Cryst. Solids* **266–269** 223
[37] Brinza M, Emelianova E V and Adriaenssens G J 2005 *Phys. Rev* B **71** 115209
[38] Sze S M 2002 *Semiconductor Devices – Physics and Technology* (New York: John Wiley and Sons) Ch. 5
[39] Lee C-H, Sazonov A, Nathan A and Robertson J 2006 *Appl. Phys. Lett.* **89** 252101
[40] Kuo Y, Lei Y and Nominanda H 2004 *Mat. Res. Soc. Symp. Proc.* **808** A4.7.1
[41] Nara Y and Matsumura M 1984 *Jpn. J. Appl. Phys.* **23** L714
[42] Kuo Y 2002 *Bipolar thin-film transistors and method for forming* US Patent 6500721
[43] Brüggemann R, Kleider J P, Longeaud C, Mencaraglia D, Guillet J, Bourée J E and Niikura C 2000 *J. Non Cryst. Solids* **266-269** 258
[44] Garner D M and Amaratunga G A J 1999 *Solid State Electronics* **43** 1973
[45] Rösch M *SC-Simul* (University of Oldenburg, Germany)
Available for download at: http://www.greco.uni-oldenburg.de/index.html
[46] Voz C, Peiro D, Fonrodona M, Soler D, Bertomeu J and Andreu J 2000 *Solar Energy Materials and Solar Cells* **63** 237
[47] Kamisheva G 2009 *7th International Conference of the Balkan Physical Union* **1203** 1336

# Steady-state photocarrier grating technique for the minority-carrier characterisation of thin-film semiconductors

**R Brüggemann**
Institut für Physik, Carl von Ossietzky Universität Oldenburg, 26111 Oldenburg, Germany

E-mail: rudi.brueggemann@uni-oldenburg.de

**Abstract.** The steady-state photocarrier grating technique allows the determination of the ambipolar diffusion length in low-mobility semiconductors and provides access to the minority-carrier properties in terms of the mobility-lifetime product. The technique probes the excess carrier distribution under the presence of a sinusoidally modulated photogeneration rate which is achieved by the superposition of two coherent laser beams. The relatively simple technique has been established in many laboratories and has been successfully applied to a variety of thin-film semiconductors. The basic theory of the method, experimental set-ups, variants of the experimental realisation in terms of the grating-period or the electric-field variation and some of the various applications of the steady-state photocarrier grating method on thin-film silicon, chalcopyrite and other semiconductors are presented and reviewed in this contribution. Worthwhile information can be obtained on the recombination and the density-of-states in the band-gap of the semiconductor from experiments with the steady-state photocarrier grating method in combination with dark- and photoconductivity measurements.

## 1. Introduction

The steady-state mobility-lifetime products of electrons and holes represent the photoelectronic properties of the majority and minority carriers under stationary illumination, in contrast to the mobility-lifetime products from transient photocurrent experiments like time-of-flight. Both mobility and lifetime are important properties which determine transport and recombination. In contrast to the majority-carrier properties which can be determined rather straightforwardly by stationary photocurrent measurements under the assumption that the minority-carrier contribution can be neglected, the minority-carrier properties can only be revealed by more sophisticated methods. In this respect the steady-state photocarrier grating (SSPG) technique has had an enormous impact since it was suggested by Ritter, Zeldov and Weiser named RZW hereafter, in 1986 [1].

RZW outlined in their paper on the diffusion length measurement in photoconductive insulators the basic features of the SSPG technique which is based on the carrier diffusion under the presence of a spatial sinusoidal modulation in the photogeneration rate achieved by the interference of two coherent laser beams – technically achieved by splitting the beam from a laser by a beam splitter. The resulting modulation in the photogeneration rate in the semiconductor bulk induces a so-called photocarrier grating with a spatial modulation in the excess carrier densities. From photocurrent measurements at different grating periods the ambipolar diffusion length can be determined by the RZW analysis that assumes ambipolar transport and charge neutrality.

The proposal by Ritter *et al* [1], following papers on the analysis [2,3] and the relatively simple set-up triggered the wide-spread application of the SSPG method. It was quickly adopted in other laboratories [4-8].

In parallel, critical and more in-depth accounts on the underlying theory were given to put the technique on a firm ground or to describe its limits when applied to semiconductors with traps, in extension to the rather intuitive theoretical approach in the early publication. In a later analysis Ritter *et al* [9] numerically solved the transport equations in the small-signal approximation and confirmed the validity of the approach in the "lifetime regime", in which the dielectric relaxation time under illumination, determined from the relaxation time with photoconductivity, is much shorter than the carrier lifetimes – in contrast to the "relaxation time regime". Theoretical accounts were also given by Balberg [10,11], Li [12], Sauvain *et al* [13], Abel *et al* [14] and Hattori *et al* [15]. The latter authors showed that under the conditions in the "lifetime regime" the RZW evaluation procedure of the SSPG technique gives correct results.

In this article, the SSPG method will be introduced, the experimental measurement procedure will be outlined and details of the data analysis will be described. An overview on the applications of SSPG on a variety of semiconductors will then be given which illustrate that SSPG has become a widespread technique that is established in many laboratories.

## 2. Description of the SSPG concept

Figure 1 illustrates the beams $L_1$ and $L_2$ which represent plane waves with wavelength $\lambda$ which impinge on the semiconductor where they suffer refraction according to Snell's law so that the angle between the two beams changes from $\theta$ to $\theta'$. The analysis of the plane-wave interference for weak absorption in $z$ direction results in a photogeneration-rate pattern $G(x,y,z)$ according to [1]

$$G(x,y,z) = G(x) = (G_1 + G_2) + 2\gamma_0 \left(G_1 G_2\right)^{0.5} \cos\left(2\pi x / \Lambda\right) \tag{1}$$

where $G_1$ and $G_2$ refer to the photogeneration rate resulting from beam $L_1$ and $L_2$ alone. A grating quality factor $\gamma_0$ is introduced here with values close to 1 for a good grating quality.

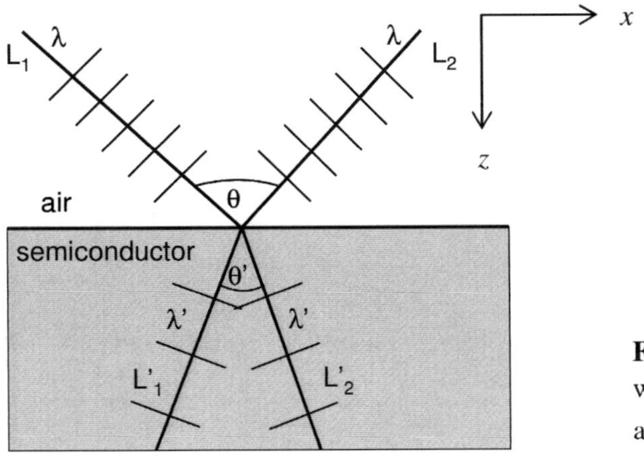


**Figure 1.** Sketch of the interference experiment with two planes waves from the laser beams $L_1$ and $L_2$.

Under the typical conditions with coplanar electrodes on a thin-film semiconductor with a gap in the range of say 0.3 mm to 2 mm, the number of grating periods amounts to a few hundred at typical grating periods in the range of 1 μm.

The grating period in equation (1) is given by

$$\Lambda = \lambda / \left[ 2\sin\left(\theta / 2\right) \right] \tag{2}$$

for which it is noted that it is only the angle $\theta$ between the two beams in air that enters the equation. For weak absorption in which the generation rate is almost constant in the $z$ direction there is only an $x$ dependence in $G(x)$.

The RZW analysis suggests to exploit two conditions of the laser-beam arrangement at a given $\Lambda$: one measurement under coherent conditions with current density $j_{\text{coh}}(\Lambda, L_1 + L_2)$ and one measurement under incoherent conditions with current density $j_{\text{inc}}(\Lambda, L_1 + L_2)$. RZW also suggested that a lock-in technique may be used with relative measurement to the current density with one beam only, with $j_1(\Lambda, L_1)$. A parameter $\beta(\Lambda)$ is defined given by

$$\beta(\Lambda) = \frac{j_{\text{coh}}(\Lambda, L_1 + L_2) - j_1(\Lambda, L_1)}{j_{\text{inc}}(\Lambda, L_1 + L_2) - j_1(\Lambda, L_1)} = \frac{U_{\text{coh}}(\Lambda)}{U_{\text{inc}}(\Lambda)}, \tag{3}$$

where $U_{\text{coh}}(\Lambda)$ and $U_{\text{inc}}(\Lambda)$ are lock-in amplifier measurements under coherent and incoherent conditions. For a number of position that relate to different grating periods $\Lambda$, the parameter $\beta(\Lambda)$ is determined by measuring $U_{\text{coh}}(\Lambda)$ and $U_{\text{inc}}(\Lambda)$.

RZW related this $\beta(\Lambda)$ parameter with the ambipolar diffusion length $L$ according to

$$\beta(\Lambda) = 1 - \frac{2Z}{\left[1 + (2\pi L / \Lambda)^2\right]^2} \tag{4}$$

From a measured set of $\beta$ values at a number of $\Lambda$ positions a fit can be made with the two fit parameters $L$ and $Z$. While L is the diffusion length that one wants to determine Z is a fit parameter, typically between 0.5 and 1, which contains information on the grating quality, the intensity dependence of the photoconductivity and the ratio between dark and total current under illumination [1].

Equation (4) can be changed to a form

$$(1-\beta)^{-1/2} = \left[(2Z)^{-1/2}(1/L)^2\right](2\pi/\Lambda)^2 + (2Z)^{-1/2}, \quad L = \left(\frac{a}{b}\right)^{1/2}, \quad Z = (2b^2)^{-1} \tag{5}$$

in which measured $\beta$ values are processed to be plotted as ordinate values or alternatively to [7

$$1/\Lambda^2 = \left[Z^{1/2}/(2\pi L)^2\right]\left[2/(1-\beta)\right]^{1/2} - (2\pi L)^{-2}, \quad L = \left(\frac{1}{-4\pi^2 b}\right)^{1/2}, \quad Z = \frac{a^2}{b^2}. \tag{6}$$

Here, $L$ and $Z$ are determined from linear fits.

The relation to the minority-carrier properties can be shown to be given by [14]

$$L = \sqrt{2 \frac{e}{kT} \frac{(\mu\tau)_{\text{min}} (\mu\tau)_{\text{maj}}}{(\mu\tau)_{\text{min}} + (\mu\tau)_{\text{maj}}}}, \tag{7}$$

where the minority carrier mobility-lifetime $(\mu\tau)_{\text{min}}$ product is determined by the product of the free carrier mobilities $\mu_{\text{min}}$ and the free carrier recombination lifetime $\tau_{\text{min}}$. For the majority carriers holds, $(\mu\tau)_{\text{maj}} = \mu_{\text{maj}}\tau_{\text{maj}}$.

Complementary, the steady-state photoconductivity $\sigma_{\text{ph}}$ is given by

$$\sigma_{\text{ph}} = eG\left[(\mu\tau)_{\text{min}} + (\mu\tau)_{\text{maj}}\right]. \tag{8}$$

These two equations with the two unknowns allow the $(\mu\tau)_{\text{maj}}$ and $(\mu\tau)_{\text{min}}$ determination but no straightforward correlation on whether the electrons or the holes are the majority/minority carriers. In the typical case that one carrier type dominates the two equations (7) and (8) reduce to

$$L = \sqrt{2 \frac{e}{kT} (\mu\tau)_{\text{min}}}, \tag{9}$$

and

$$\sigma_{\text{ph}} = eG(\mu\tau)_{\text{maj}}. \tag{10}$$

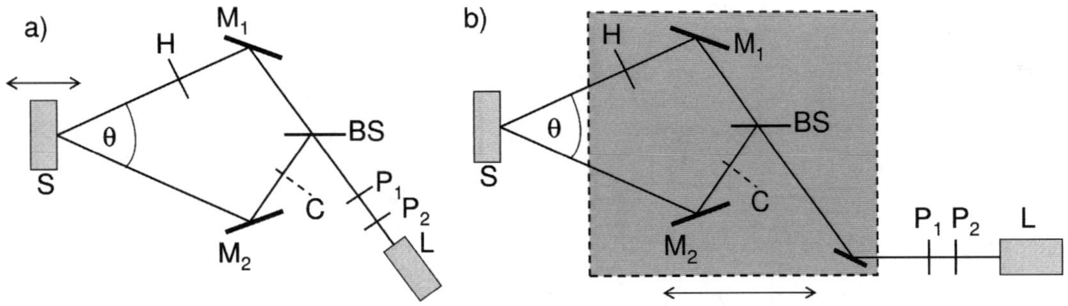


**Figure 2.** Schematic of the experimental setups that may be used for the SSPG method. The double arrow indicates which element is moved linearly in order to change the angle between the two beams that originate from one laser beam of a laser L that is split by a beam splitter (BS). In (a) the sample S is moved; in (b) an optical table, indicated by the shaded area, is moved. Such a set-up has been arranged by C. Longeaud and R. Brüggemann at LGEP, Paris. H denotes a half-wave plate and C a chopper. P1 and P2 are polarisation filters, M1 and M2 are mirrors.

## 3. Experimental set-up

A variety of experimental set-ups is possible for the SSPG technique. A common feature is the splitting of a laser beam which is then guided symmetrically onto the sample in order to induce an interference grating. For example, the two beams from the laser L, split by the beam splitter BS, hit the sample S in the gap between the electrodes in figure 2(a) which is drawn according to the original proposal by RZW. The sample is moved on a linear stage to change the angle between the two beams. The beam splitter BS separates the photon flux of the two beams such that the ratio is in the range of

10 or less. The half-wave retardation plate H is positioned in the strong beam, in order to rotate the polarisation, i.e., rotation of H by 45° changes the polarisation of the beam by 90°. An additional feature, not contained in [1] are the polarisation filters which define the polarisation and can also be used for a variation of the photon flux by rotation of the polarizer P2, keeping P1 fixed.

Figure 2(b) is different in that most of the optical elements are arranged on an optical table that is moved along a linear stage. The sample S is fixed. The advantage here is that the length of the light path is the same for every grating period while in arrangement (a) the length of the path changes and the laser spot may widen at longer distances. Depending on the cryostat that is available for temperature-dependent measurements, the arrangement in (b) can work with a fixed cryostat while in (a) the cryostat must be flexible enough to be moved over distances in the range of typically at least 1 meter.

## 4. Results

This chapter discusses results on the photoconductive properties on a number of thin-film semiconductors like hydrogenated amorphous silicon and its alloys in relation to the SSPG method, i.e., the minority-carrier properties in the steady state. As these are linked with the majority-carrier properties the latter will also be discussed in comparison when appropriate.

### *4.1. Hydrogenated amorphous silicon*

Hydrogenated amorphous silicon (a-Si:H) has been in the focus of SSPG applications since the early work by Ritter *et al* [1] in which the method was demonstrated on a-Si:H. Figure 3 shows typical experimental data from our laboratory of a device quality a-Si:H in a $\beta$ vs. $\Lambda$ plot (a) (equation 4) and in a linear plot (b) (equation 5). The very good agreement between the two evaluation schemes with $L$ = 154 nm indicates the quality of the measurements.

Figure 4 shows SSPG results for a high-defect a-Si:H sample. In contrast to the previous figure the $\beta$ values are almost constant. Note also that the variation in the ordinate values in figure 4(b) is much less than in figure 3(b). The good agreement between the two evaluations is a good indicator that the measurement is reliable even though the diffusion length is short.

It is noted that it is not possible to achieve such a good agreement for all samples under all the different conditions like low temperature or low generation rate. Any deviations between the values of the two analyses can be inspected and in some cases exempting individual data points may results in better agreement. The experimental error in the linear plot may become larger for larger $\beta$ because of the 1 - $\beta$ term in the denominator in equations (5) or (6).

*4.1.1. Temperature and generation-rate dependence.* In figure 5(a), the experimentally determined monotonous decrease of $(\mu\tau)_p$ with decreasing temperature is a signature often exhibited by a-Si:H samples. The experimental results in figure 5(a) can be roughly reproduced by numerical simulation of the photoelectronic properties of a-Si:H in figure 5(b) [16,17]. The increase in the splitting of the quasi-Fermi levels with decreasing temperature is related to an increase in the density of effective recombination centres for the excess carriers. These typically recombine through the electronic states between the quasi-Fermi levels leading to the observed decrease in lifetime or $(\mu\tau)_p$ with decreasing temperature.

At a given temperature, the mobility-lifetime products in figure 5 drop with increasing photogeneration rate $G$ which is related to the above-mentioned increase in quasi-Fermi level splitting. There has been some controversy in the literature about the consequences of the $G$ dependence with respect to density-of-states models in the band-gap of a-Si:H. Balberg *et al* [18] suggested that one correlated dangling bond level is not sufficient to explain the experimental data and he suggested that there is evidence for a defect-pool model based defect distribution with more than one defect peak.

The Neuchatel group pointed out that a proper balancing of charge in one correlated dangling bond and in the band-tail states can also explain the experimental findings [13].

*4.1.2. Fermi-level position.* For *n*-type samples there is a well-known increase in the photoconductivity with increasing dark conductivity. The change of the occupation of dangling bonds becoming negatively charged and thus more ineffective for electron capture, results in the increase in the lifetime despite a possible increase of the absolute dangling bond density upon doping. In terms of mobility-lifetime products, the situation in figure 6(a) has been established [19]. It shows $(\mu\tau)_{\text{min}}$ and $(\mu\tau)_{\text{maj}}$ for quite a number of samples plotted vs. Fermi level position $E_c - E_f$. The whole set of data points illustrates that for *n*-type samples with smaller $E_c - E_f$ the electrons are the majority carriers and here $(\mu\tau)_{\text{maj}} >> (\mu\tau)_{\text{min}}$. The value for $(\mu\tau)_{\text{maj}}$ increases with higher doping until at the highest doping levels the dangling bond density has become too large so that $(\mu\tau)_{\text{maj}}$ then decreases with further shift of $E_f$ towards $E_c$.

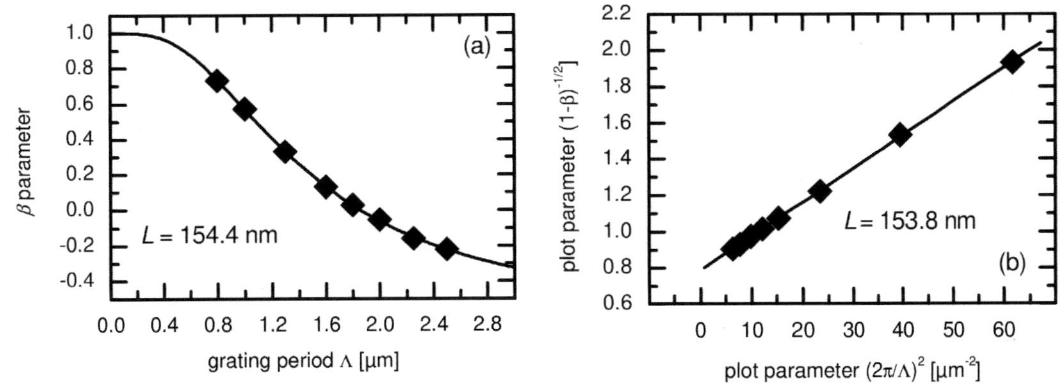


**Figure 3.** The symbols represent the experimental data points for an undoped a-Si:H sample. The full lines are fits. Both types of analysis give good agreement in the value for *L*.

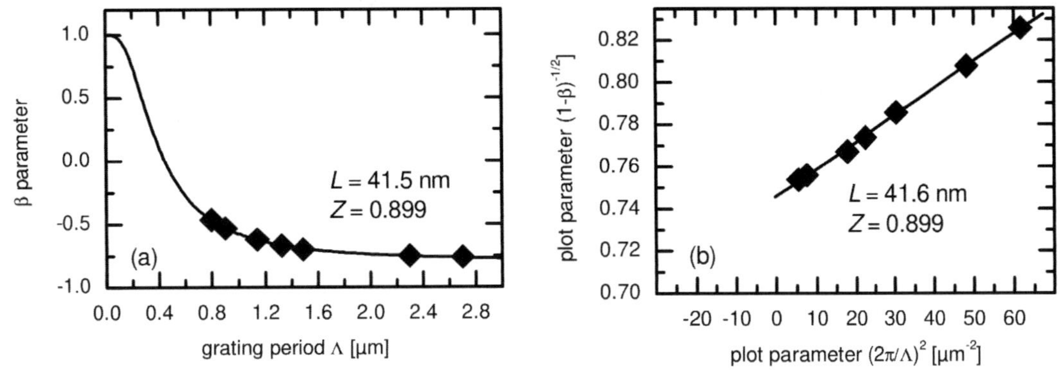


**Figure 4.** The symbols represent the experimental data points for a high-defect density undoped a-Si:H sample. The full lines are fits. Both types of analysis give good agreement in the value for *L*.

For $p$-type samples the role of the electrons and the holes changes. Here, again $(\mu\tau)_{maj}$ >> $(\mu\tau)_{min}$ and $(\mu\tau)_{maj}$ = $(\mu\tau)_p$. There is no full symmetry in the figure because the underlying density-of-states distribution in the band-gap is not symmetric. The lines are results from numerical simulation with a full model for charge neutrality, with dangling bonds and recombination which reproduces the experimental tendency quite well [19].

The ratio $b=(\mu\tau)_n/(\mu\tau)_p$ in figure 6(b) covers quite a wide range. For most of the nominally undoped samples with $E_c - E_f$ in the range 0.65 eV to 0.75 eV, $b$ is in the range of 20 to 300, indicating that electrons are the majority carriers in undoped a-Si:H. The presentation also indicates that there is no one-to-one relation between $b$ and $E_c - E_f$ but that the ordinate values show some scatter e.g. for $n$-type samples.

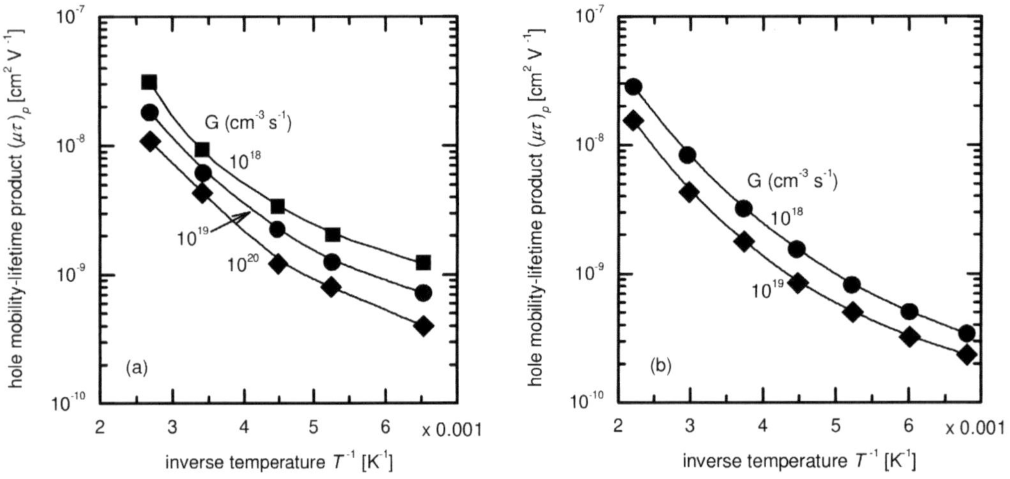


**Figure 5.** Temperature dependence of the $(\mu\tau)_p$ product of an undoped a-Si:H sample at different photogeneration rates (a), after [17]. Numerical-simulation results with a free-hole mobility of 1 cm$^2$ V$^{-1}$ s$^{-1}$ which reproduce the monotonous $(\mu\tau)_p$ with decreasing temperature (b), after [18].

Shortly after the SSPG proposal, the properties of intentionally doped films were studied by Yang *et al* [6]. Low-concentration boron doping was shown to lead to an increase in the diffusion length, in agreement with figure 6(a). With higher doping level the diffusion length dropped drastically. These results were interpreted as a change in the minority-carrier type. For the higher doping levels and the shift of the Fermi level towards the valence-band edge, the electrons become the minority carriers and the increasing defect density upon doping then leads to the reduction in the diffusion length. An important observation was a benefit in the diffusion length at small shifts of $E_f$ towards mid-gap which was thought to be beneficial for the net $i$-layer carrier collection in a-Si:H based *pin* solar cells.

*4.1.3. Electric-field effects.* In the early publications on SSPG there has been concern with respect to the electric-field limitations in two ways: electric-field modulation due to space charge and phase shift in the modulation of the electron and hole distributions because the electric field $E$ pulls these apart.

The analysis by Abel *et al* [14] takes into account electric-field effects and can been applied to study the $\beta$ dependence on grating period Λ and electric field $E$. An example is given in figure 7 [20].

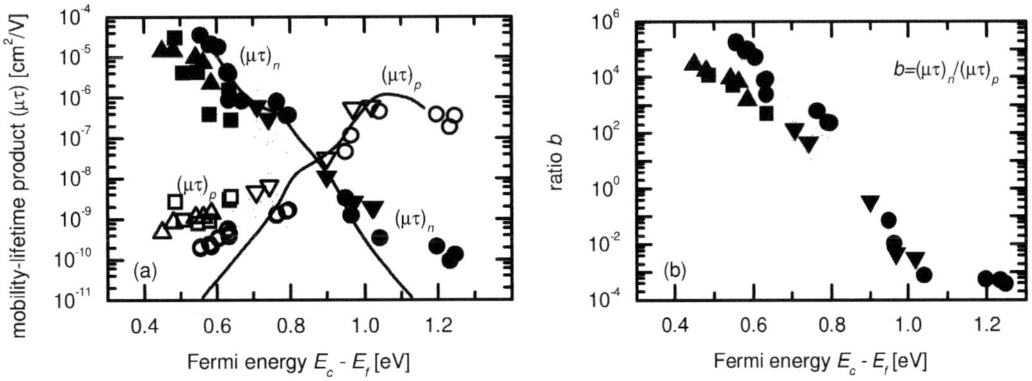


**Figure 6.** Variation of the mobility-lifetime products (a) and the ratio $b$ (b) for the variation of the Fermi level by doping. Different symbols denote different sample series. The full line is from numerical simulation to model the behaviour of the data represented by the circles, after [19].

Figure 7(a) is a check that the incoherent contributions maintain Ohmic character even at the high voltages, represented by a slope of 1 in the log-log plot. For the coherent contribution there is some deviation so that a voltage-dependent $\beta$ is the result. The full lines in Figure 7(b) are fits according to the theory by Abel *et al* [14] which includes both the effect of space charge and external electric field. It is noted that, depending on the measurement position, i.e., Λ value, there is quite a variation in the range of the measured $\beta$ values when the voltage is increased. Good agreement could be achieved between the experiment and the fit functions for the different grating periods. Accounts on the effective dielectric relaxation time could thus be given [20].

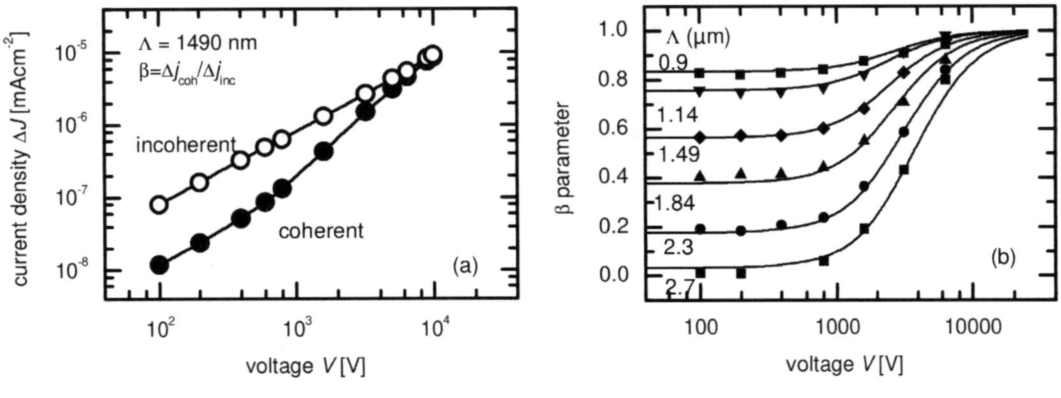


**Figure 7.** Experimental $E$ dependence of the SSPG data of a sample with electrode gap 0.5 mm. The open symbols in (a) demonstrate the Ohmic character under incoherent conditions. The full data points vary linearly with voltage only up to 8 kV/cm (400 V). In (b), for the coherent case, the drop in photocurrent is no longer maintained at high $E$. The voltage dependence of $\beta$ was measured at the different grating periods, indicated in (b), after [20].

*4.1.4. Light-soaking and degradation.* The time-dependent decrease of the photocurrent is a signature of many a-Si:H films, the details of which are analysed in terms of defect-creation models. Any increase in the defect density should also have an influence on the minority-carrier properties so that SSPG measurements in combination with light-induced degradation may reveal the underlying recombination physics that is determined by the time-dependent increase in the dangling-bond density upon illumination. In view of the Fermi-level related discussion above it is illustrative to point out the possibility that the diffusion length may increase upon light-soaking, as illustrated by the SSPG results in figure 8. The figure shows that for a nominally undoped a-Si:H sample the Fermi level shifted towards mid-gap upon light-soaking leading to an increase in the diffusion length although more defects have been created. Upon the Fermi-level shift, the defects become less negative and thus less attractive for hole capture, so that the minority carrier properties in terms of $L$ can be improved.

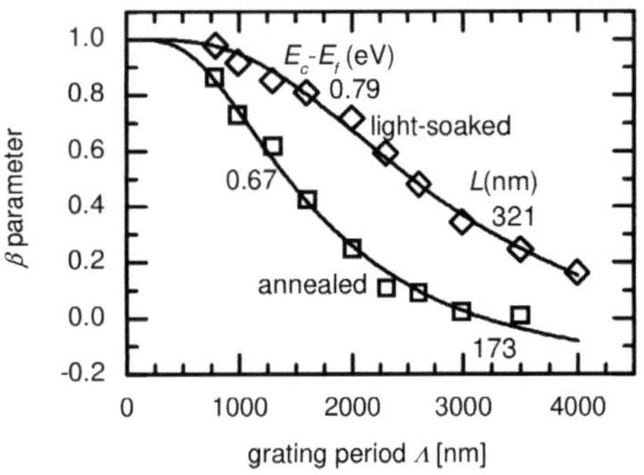


**Figure 8.** Increase of the diffusion length upon light-soaking of a-Si:H, after [21].

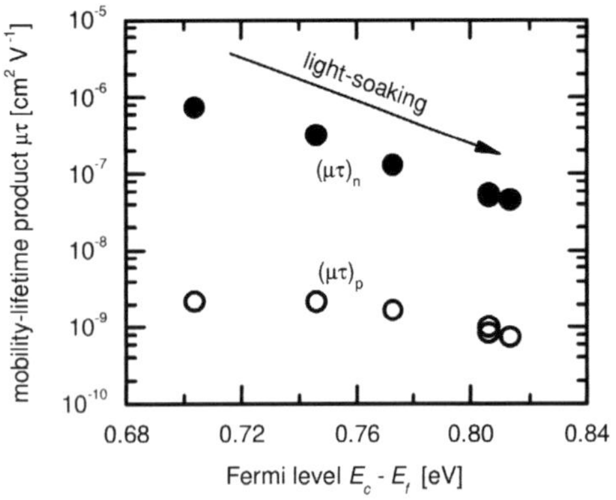


**Figure 9.** Effect of light-soaking on the Fermi level and the mobility-lifetime products. Compilation after Morgado [22].

With respect to the influence of the mobility-lifetime products on the Fermi level and the reported change of the Fermi level upon light-soaking, it appears very helpful to always monitor the dark current upon degradation. It is then interesting to discuss the light-soaking results by Morgado [22], which are compiled in figure 9. These changes give an indication of new defect creation as the Fermi level shifts but $L$ or $(\mu\tau)_p$ does not increase in contrast to figure 8. Upon longer light-soaking both mobility-lifetime products show a decrease which Morgado also related to the variation in the sub-gap absorption and defect density.

The combination of dark current, photocurrent and SSPG measurements has the potential to further elucidate the details of light-induced recombination and may, together with numerical modelling of the experiments, shed light on the question under which condition the main recombination channel is either via dangling bonds or via the band-tails and how and if there is a transition between the two channels by changing the temperature or the photogeneration rate

### *4.2. Hydrogenated silicon germanium*

One of the first reports on the application of the SSPG method was given by Bauer *et al* [4] on hydrogenated amorphous silicon germanium (a-$Si_{1-x}Ge_x$:H) alloys. The focus of this research and later publications on these alloys was to characterise the photoelectronic properties for samples with increasing x and to study the influence of Ge incorporation and its influence on the defect density and the band tails. Typically, both the photoconductivity as a signature of the majority carriers and the diffusion length as a signature of the minority carriers decrease with increasing x but they are affected in a different way by the concomitant shift of the Fermi energy towards mid-gap [23].

Figure 10 summarises the hole mobility-lifetime products for quite a number of a-$Si_{1-x}Ge_x$:H samples as a function of Ge content. For $x < 0.43$, the values of the $(\mu\tau)_p$ product stay $> 10^{-9}$ cm$^2$ V$^{-1}$, i.e., diffusion lengths of about 100 nm. Fölsch *et al* [24] report slightly larger values in this x-range. These authors report a strong drop in $(\mu\tau)_p$ for x larger than 0.5 at which the combination of increasing dark conductivity and decreasing photoconductivity makes the diffusion length very difficult to measure.

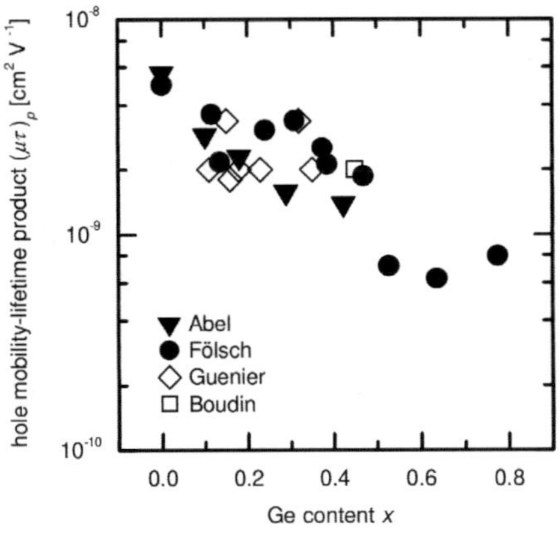


**Figure 10.** Hole mobility-lifetime products of a- a-$Si_{1-x}Ge_x$:H and pm-$Si_{1-x}Ge_x$:H thin films. Optimised films maintain a diffusion length in the range of 100 nm. At x = 0.45 there is a strong drop in the mobility-lifetime products. More details are in the text.

The data by Gueunier *et al* [25] on so-called polymorphous silicon germanium alloys (pm- $Si_{1-x}Ge_x$:H) are added in figure 10 in the range up to x=0.35 at which the optical gap $E_{04}$, i.e., the energy at which the absorption coefficient is $10^4$ cm$^{-1}$, was 1.5 eV. The reported *L* values of the annealed samples scatter in the range between 100 nm and 130 nm, values as also reported by Bhaduri *et al* [26] in the range x = 0.4 – 0.5 and represented by the symbol at x = 0.45 in figure 10. The selected results above show that it is possible to deposit hydrogenated amorphous silicon-germanium alloys in a Ge-content range $x < 0.45$ for which the deterioration in the photoelectronic properties is not too severe.

### *4.3. Hydrogenated amorphous silicon carbide*

The diffusion length from SSPG was determined by Li *et al* [27] for a corresponding characterisation and optimisation of hydrogenated amorphous silicon carbide films (a-$Si_{1-x}C_x$:H) of different band gaps and different feed stock gases. These authors find a correlation between the increase of the diffusion length and better solar cell properties. They make it clear though that, as mentioned above, the sole value of the diffusion length may not be the only indicator of the defect density of a sample. They deduce from the complementary photo- and dark conductivity results that Fermi level related occupation of defects may have lead to higher values of the diffusion length of some of their a-$Si_{1-x}C_x$:H samples.

Mohring *et al* [16,17] reported on the temperature-dependent and generation-rate dependent mobility-lifetime products of majority- and minority carriers in optimised a-$Si_{1-x}C_x$:H with an optical band gap up to 1.95 eV. For the alloys, the $(\mu\tau)_p$ product decreases with decreasing temperature, similar to the findings in figure 5 for a-Si:H. At any temperature $(\mu\tau)_p$ decreases with increasing x which has been related by Mohring *et al* to the increased dangling bond density with increasing x. For a-$Si_{1-x}C_x$:H with a band gap of 1.94 eV, a value of (1-2) × $10^{-9}$ $cm^2$ $V^{-1}$ for $(\mu\tau)_p$ was determined. and with $(\mu\tau)_n$ about a factor of ten higher.

### *4.4. Hydrogenated microcrystalline silicon*

The terms polycrystalline silicon, because of some large-grain materials, or nanocrystalline silicon, because of the nanocrystallites in a microcrystalline structure, were temporarily fashionable for what is typically subsumed today under hydrogenated microcrystalline silicon (μc-Si:H). The application of the SSPG method to microcrystalline silicon is rather straightforward on the one hand as, for example, sample geometries and absorption coefficients are somewhat but not significantly different. On the other hand, there are some obstacles involved because of the typically higher dark conductivity and some possible effect of surface roughness, which reduces the grating quality.

Droz *et al* [28] expand on the similarity in the transport between amorphous and microcrystalline silicon. Especially, both the majority and minority carrier properties as a function of doping level were shown to be similar to a-Si:H. Goerlitzer *et al* [29], Brüggemann [30] and Okur *et al* [31] report that the fitting parameter *Z* takes small values in μc-Si:H because of the small ratio of photo- to dark conductivity. Another obstacle is the reduction in the grating quality factor because of optical scattering which deteriorates the grating. The *Z* parameter was shown to increase after polishing the rough surface of μc-Si:H samples [30].

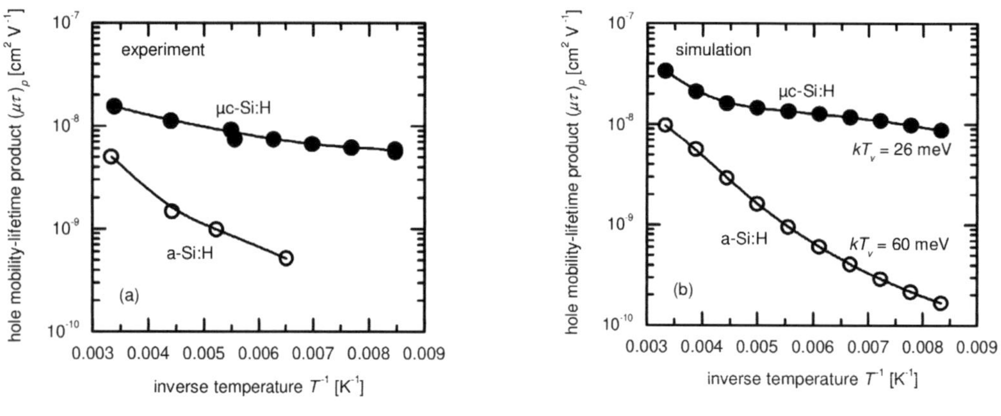


**Figure 11.** Temperature dependence of the majority and minority carrier mobility-lifetime products of μc-Si:H with a comparison of the experimental (a) and the simulation (b) results, after [32]. The valence-band tail is described by an exponential exp[-*E*/($kT_v$)].

The temperature dependence of $(\mu\tau)_p$ with a nominal crystalline volume fraction from Raman spectra of 0.66 was studied by Brüggemann and Kunz [32], as shown in figure 11. In contrast with a-Si:H, the experimental temperature-dependent decrease of $(\mu\tau)_p$ is much less than in a-Si:H. Application of numerical modelling, with which the small decrease can be reproduced (figure 11(b)) indicates that an introduction of a steeper valence-band tail in the low-band gap μc-Si:H leads to the different behaviour in comparison with a-Si:H.

Balberg [33] also reported on the temperature dependence of the two mobility-lifetime products of μc-Si:H and their dependence on the photogeneration rate. In addition to the experimental data they also performed numerical simulations in order to correlate the experimental and simulation results. Close inspection of this comparison shows that the agreement is not always good. Some of the input parameters in the model for the simulation have rather extreme values so that here is still a challenge here to provide more experimental evidence for this proposed model.

A relation between the diffusion length and the deep defect density was presented by Brüggemann [34]. It was shown that $L$ is correlated with the optical absorption coefficient at 0.8 eV. A value of 0.8 eV was chosen in order to represent an energy that is out width a band-tail region and deeper in the band-gap. The experimental data from quite a number of samples are shown in figure 12. The two open symbols represent one μc-Si:H sample in which the deep defect density was increased by proton irradiation [35]. For this sample the increase in the absorption coefficient upon irradiation correlates almost inversely linearly with a decrease in $L$. For the other samples there is substantial scatter. It may thus be possible that for low absorption coefficients another recombination channel limits the minority-carrier lifetime. Similar experiments on relating sub-gap absorption with the diffusion lengths in microcrystalline silicon were reported by Günes *et al* [36].

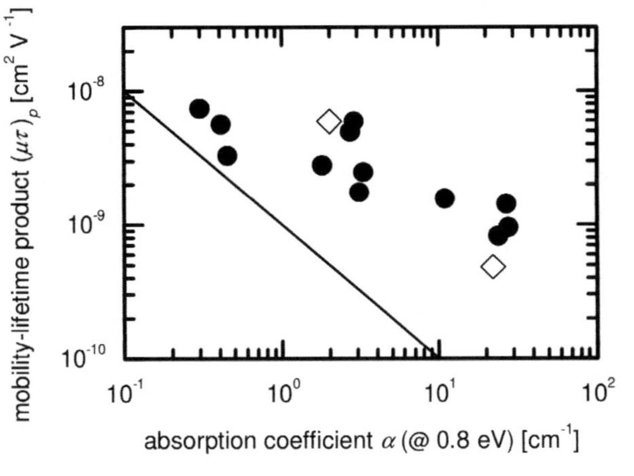


**Figure 12.** Variation of the room-temperature diffusion length with the absorption coefficient at 0.8 eV of a number of microcrystalline silicon samples, after [34].

There are indications that transport is inhomogeneous in μc-Si:H with respect to the transport path parallel or perpendicular to the film-growth direction. The SSPG experiment with coplanar electrode geometry and transport perpendicularly was complemented by the Prague group with surface-photovoltage (SPV) measurements for the diffusion length determination parallel to the growth direction to which the reader is referred [37]. For the SPV evaluation one must be careful as drift contributions were alleged to result in longer SPV diffusion lengths than determined by SSPG in a-Si:H.

*4.5. Hydrogenated microcrystalline germanium*

Badran *et al* [38] reported on the determination of the minority-carrier diffusion length in microcrystalline germanium (μc-Ge:H). Because of the high dark conductivity it was necessary to perform the SSPG experiments at lower temperature for the photocurrent to be measurable. The measured diffusion lengths were quite short. The conclusion by Badran *et al* was that μc-Ge:H appears suitable for diode applications as a sensor, as was demonstrated previously, under reverse bias but not as a solar cell because the diffusion lengths were too short.

### *4.6. Other semiconductors*

In one of their first papers on the SSPG method, Ritter and Weiser [3] also demonstrated its applicability to semi-insulating GaAs. Because of the longer diffusion length in this semiconductor, measurements at grating periods up to 70 μm were performed. The determined diffusion length was 10 μm, in agreement with alternative diffusion length measurements.

For high-band gap GaN and AlGaN/GaN double layers, Niehus and Schwarz [39,40] gave accounts on the application of SSPG. They report diffusion lengths up to 3 μm and find some relation between the Al doping and the excess carrier properties. The authors also raise some general questions concerning to applicability of SSPG and the meaning of the SSPG parameters in these experiments with high-band gap samples.

Thin-film chalcopyrite semiconductors were characterised soon after the SSPG method was proposed. Balberg *et al* [5] determined the majority and minority carrier properties of $CuGaSe_2$ thin films. The photo- to dark conductivity ratio was about 100 under their experimental conditions so that the experiments were not made impossible because of the background doping in these films. Here, electrons are the minority carriers for which the authors determined a diffusion length of 115 nm. From the intensity dependence of the mobility-lifetime products it was concluded that there is a sharp drop in the density-of-states in the band-gap close to the band edges – in contrast to a-Si:H.

Menner *et al* [41] report SSPG results on $CuGaSe_2$ thin films and determined the highest diffusion lengths for a Cu/Ga ratio at 0.95, i.e., close to the stochiometric point. SSPG-measurements in the Cu-rich region were impossible because of the high dark conductivity. Menner *et al* also aimed at revealing inhomogeneity in the growth process by application of a laser with a short absorption depth and by illumination from either the air/film and the substrate/film side. They concluded that a better crystal structure close to the substrate exists. These authors also raised the issue that transport may be different between in the direction of growth and perpendicular to it. A comparison with diffusion-length measurements in the direction of growth by spectral response suffered from the limited knowledge of the absorption coefficient of the CuGaSe films but indications were given that these diffusion lengths were a bit longer than those from SSPG.

Lubianiker *et al* [42] applied SSPG and photoconductivity for complementary measurements of $CuInS_2$ films. In this comprehensive study, the authors investigated both n-type and p-type samples by illumination from the air-side and the substrate-side. For the deduced recombination model and density-of-states distribution the reader is referred to the literature.

Belevich and Makovetskii [43] give an account of diffusion length measurements in polycrystalline p-type $CuInSe_2$. In their SSPG measurements in the range of grating periods of 4 μm to 9 μm with different electric field they determine photogeneration-rate dependent minority-electron diffusion lengths between 360 to 420 nm, decreasing with increasing photon flux. They relate the values of the diffusion length to the dimension of the crystallites in the polycrystalline film.

Generally, it is noted that one should not be too optimistic with respect to the application of SSPG to the $Cu(In,Ga)Se_2$ system. The experiment may be hampered by the large dark conductivity in $Cu(In,Ga)Se_2$, as was reported by Zweigart *et al* [44]. These authors have achieved to determine diffusion lengths for a variety of chemical compositions. But they also point out the difficulties in the SSPG application and evaluation. This is not surprising because the Fermi level is shifted towards the valence-band edge. If $E_f$ is a few tens of meV above $E_v$ the dark current will be too high. Higher photon flux for illumination may help or measuring at lower temperatures. Because the dark-current activation energy is also low, the decrease of the temperature will only help if the photocurrent does not decrease too much as well.

### *4.7. Further measurement modes*

Schmidt and Longeaud [45] developed an analysis of the SSPG method with emphasis on the density-of-states determination at the quasi-Fermi energy. Several photocurrent measurements are needed at a specific grating period. From an RZW fit the additional parameter Z can be determined. From a variation of the temperature or the generation rate the quasi-Fermi energy can be shifted so that a scan can be performed. The application of the method was demonstrated on a-Si:H [45] and μc-Si:H [46]. In addition to the measurements performed in coplanar electrode configuration, other configurations have been studied. With respect to sandwich configurations, Bernhard *et al* [47] applied the SSPG method to amorphous semiconductor multilayer structures.

## 5. Summary

The evaluation of the experimental data on the basis of the RZW approach provides the basis of determining the ambipolar diffusion length in thin-film semiconductors when the experiment is performed under the necessary condition of small modulation depth, short enough dielectric relaxation time and sufficiently high photocurrent values with distinct differences between the minority-carrier and the majority-carrier mobility-lifetime product. A simple relationship between the diffusion length and the minority-carrier mobility-lifetime product can then be applied.

The overview over the range of applications introduced into the temperature, generation-rate and Fermi level dependence of the mobility-lifetime products in a-Si:H and other silicon-based thin-film semiconductors. Results on chalcopyrite and other semiconductors were briefly presented which show that SSPG can be a method of choice not only in thin-film silicon.

## References

[1] Ritter D, Zeldov E and Weiser K 1986 Steady-state photocarrier grating technique for diffusion length measurement in photoconductive insulators *Applied Physics Letters* **49** 791

[2] Ritter D and Weiser K 1986 Ambipolar drift-length measurement in amorphous hydrogenated silicon using the steady-state photocarrier grating technique *Physical Review* B **34** 9031

[3] Ritter D, Weiser K and Zeldov E 1987 Steady-state photocarrier grating technique for diffusion-length measurement in semiconductors - theory and experimental results for amorphous-silicon and semiinsulating GaAs *Journal of Applied Physics* **62** 4563

[4] Bauer G H, Nebel C E, Mohring H-D 1988 Diffusion lengths in a-SiGe:H and a-SiC:H alloys from optical grating technique *Mater. Res. Soc. Symp. Proc.* **118** 679

[5] Balberg I, Albin D and Noufi R 1989 Mobility-lifetime products in $CuGaSe_2$ *Applied Physics Letters* **54** 1244

[6] Yang L, Catalano A, Arya R R and Balberg I 1990 Effect of low-level boron doping and its implication to the nature of gap states in hydrogenated amorphous-silicon *Applied Physics Letters* **57** 908

[7] Balberg I, Delahoy A E and Weakliem H A 1988 Self-consistency and self-sufficiency of the photocarrier grating technique *Applied Physics Letters* **53** 992

[8] Liu J Z, Li X, Roca i Cabarrocas P, Conde J P, Maruyama A, Park H and Wagner S 1990 Ambipolar diffusion length in a-Si-H(F) and a-Si,Ge-H,F measured with the steady-state photocarrier grating technique *Conf. record of the twenty first IEEE Photovoltaic Specialists Conf. (Kissimimee, USA, 21 – 25 May 1990)* (Piscataway: IEEE) p 1606

[9] Ritter D, Zeldov E and Weiser K 1988 Ambipolar transport in amorphous-semiconductors in the lifetime and relaxation-time regimes investigated by the steady-state photocarrier grating technique *Physical Review* B **38** 8296

[10] Balberg I 1990 The theory of the photoconductance under the presence of a small photocarrier grating *Journal of Applied Physics* **67** 6329

[11] Balberg I 1991 Theory of the small photocarrier grating under the application of an electric-field *Physical Review* B **44** 1628

[12] Li Y M 1990 Phototransport under the presence of a small steady-state photocarrier grating *Physical Review* B **42** 9025

[13] Sauvain E, Hubin J, Shah A and Pipoz P 1991 Effect of the dangling-bond charge on the ambipolar diffusion length in a-Si-H *Philosophical Magazine Letters* **63** 327

[14] Abel C-D, Bauer G H and Bloss W H 1995 Generalized theory for analytical simulation of the steady-state photocarrier grating technique *Philosophical Magazine* B **72** 551

[15] Hattori K, Okamoto H and Hamakawa Y 1992 Theory of the steady-state-photocarrier-grating technique for obtaining accurate diffusion-length measurements in amorphous-silicon *Physical Review* B **45** 1126

[16] Mohring H D and Abel C D, Brüggemann R and Bauer G H 1991 Characterization of high electronic quality a-SiC-H films by mu-tau products for electrons and holes *Journal of Non-Crystalline Solids* **137** 847

[17] Mohring H D, Abel C D, Brüggemann R and Bauer G H 1997 Characterization of high electronic quality a-SiC:H films by mu tau-products for electrons and holes *Journal of Non-Crystalline Solids* **210** 36 (Erratum to [16])

[18] Brüggemann R 1993 *PhD thesis* Philipps-Universität Marburg

[19] Brüggemann R, Abel C-D and Bauer G H 1992 Implications of the defect pool model for the simulation of a-Si:H solar cells *Proc. of the 11th E.C. Photovoltaics Solar Energy Conf. (Montreux, Switzerland, 12-16 October 1992)* ed L Guimaraes *et al* (Chur: Harwood Academic Publishers) p 676

[20] Brüggemann R and Badran R I 2004 Electric-field dependence of photocarrier properties in the steady-state photocarrier grating experiment *MRS Symp. Proc.* **808** 133

[21] Morgado E 2002 Light-soaking and annealing kinetics of majority and minority carrier mobility-lifetime products in a-Si:H *Journal of Non-Crystalline Solids* **299** 471

[22] Morgado E 2004 Influence of light-soaking and annealing on electron and hole mobility-lifetime products in a-Si:H *Journal of Non-Crystalline Solids* **338** 386

[23] Abel C D and Bauer G H 1993 Evaluation of the steady-state photocarrier grating technique with respect to a-Si:H and its application to a-SiGe:H alloys *Progress in Photovoltaics* **1** 269

[24] Fölsch J, Finger F, Kulessa T, Siebke F, Beyer W and Wagner H 1995 Improved ambipolar diffusion length in a-$Si_{1-x}Ge_x$:H alloys for multi-junction solar cells *MRS Symp. Proc.* **377** 517

[25] Gueunier M E, Kleider J P, Brüggemann R, Lebib S, Cabarrocas P R I, Meaudre R and Canut B 2002 Properties of polymorphous silicon-germanium alloys deposited under high hydrogen dilution and at high pressure *Journal of Applied Physics* **92** 4959

[26] Bhaduri A, Chaudhuri P, Williamson DL, Vignoli S, Ray P P and Longeaud C 2008 Structural and optoelectronic properties of silicon germanium alloy thin films deposited by pulsed radio frequency plasma enhanced chemical vapor deposition *Journal of Applied Physics* **104** 063709

[27] Li Y M, Fieselmann B F and Catalano A 1991 *Conf. Record 22nd IEEE Photovoltaic Specialists Conf. (Las Vegas, USA, 7 – 11 October 1991)* (Piscataway: IEEE) p 1231

[28] Droz C, Goerlitzer M, Wyrsch N and Shah, A 2000 Electronic transport in hydrogenated microcrystalline silicon: similarities with amorphous silicon *Journal of Non-Crystalline Solids* **266** 319

[29] Goerlitzer M, Beck N, Torres P, Meier J, Wyrsch N and Shah A 1996 Ambipolar diffusion length and photoconductivity measurements on "midgap" hydrogenated microcrystalline silicon *Journal of Applied Physics* **80** 5111

[30] Brüggemann R 1998 Improved steady-state photocarrier grating in nanocrystalline thin films after surface-roughness reduction by mechanical polishing *Applied Physics Letters* **73** 499

[31] Okur S, Gunes M, Finger F and Carius R 2006 Diffusion length measurements of micro crystalline silicon thin films prepared by hot-wire/catalytic chemical vapor deposition (HWCVD) *Thin Solid Films* **501** 137

[32] Brüggemann R and Kunz O 2002 Temperature dependence of the minority-carrier mobility-lifetime product for probing band-tail states in microcrystalline silicon *Physica Status Solidi* B **234** R16

[33] Balberg I 2002 A simultaneous experimental determination of the distribution and character of the two band tails in disordered semiconductors *Journal of Optoelectronics and Advanced Materials* **4** 437

[34] Brüggemann R 2005 Mobility-lifetime products in microcrystalline silicon *Journal of Optoelectronics and Advanced Materials* **7** 495

[35] Brüggemann R, Brehme S, Kleider, J P, Gueuneier, M E and Bronner W 2004 Effects of proton irradiation on the photoelectronic properties of microcrystalline silicon *Journal of Non-crystalline Solids* **338-340** 77

[36] Günes M, Göktas O, Okur S, Isik N, Carius R, Klomfass J and Finger F 2005 Sub-bandgap absorption spectroscopy and minority carrier transport of hydrogenated microcrystalline silicon films *Journal of Optoelectronics and Advanced Materials* **7** 161

[37] Svrcek V, Fejfar A, Fojtik P, Mates T, Poruba A, Stuchlikova H, Pelant I, Kocka J, Nasuno Y, Kondo M and Matsuda A 2002 Importance of the transport isotropy in mu c-Si : H thin films for solar cells deposited at low substrate temperatures *Journal of Non-Crystalline Solids* **299** 395

[38] Badran R I, Brüggemann R and Carius R 2009 Minority-carrier properties of microcrystalline germanium *Journal of Optoelectronics and Advanced Materials* **11** 1464

[39] Niehus M and Schwarz R 2006 New results on diffusion lengths measurements in wide bandgap semiconductors, obtained from steady state photocarrier gratings (SSPG) *Superlattices and Microstructures* **40** 350

[40] Niehus M and Schwarz R 2006 Diffusion lengths in GaN obtained from steady state photocarrier gratings (SSPG) *Physica Status Solidi* C **3** 2103

[41] Menner R, Zweigart S, Klenk R and Schock H W 1993 Ambipolar diffusion length in $CuGaSe_2$ thin-films for solar-cell applications measured by steady-state photocarrier grating technique *Japanese Journal of Applied Physics* - Part 1 **32** 45

[42] Lubianiker Y, Biton G, Balberg I, Walter T, Schock H W, Resto O and Weisz SZ 1996 Dependence of the phototransport properties on the position of the Fermi level in polycrystalline $CuInS_2$ films *Journal of Applied Physics* **79** 876

[43] Belevich N N and Makovetskii G I 1994 Ambipolar diffusion and ambipolar carrier drift in $CuInSe_2$ films *Semiconductors* **28** 988

[44] Zweigart S, Menner R, Klenk R and Schock H W 1995 Application of the steady-state photocarrier grating technique for determination of ambipolar diffusion lengths in Cu(In,Ga)(S,Se)2-thin films for solar cells *Materials Science Forum* **173** 337

[45] Schmidt JA and Longeaud C 2005 Analysis of the steady-state photocarrier grating method for the determination of the density of states in semiconductors *Physical Review* B **71** 125208

[46] Souffi N, Bauer G H and Brüggemann R 2006 Density-of-states in microcrystalline silicon from thermally-stimulated conductivity *Journal of Non-Crystalline Solids* 352 1109

[47] Bernhard N and Bauer G H 1995 Physical-properties of a-Si:H based compositional periodic multilayers *Physical Review* B **52** 8829

Journal of Physics: Conference Series **253** (2010) 012003 doi:10.1088/1742-6596/253/1/012003

# The application of NLC for detecting solid crystals surface homogeneity

**M G Tomilin**

St.-Petersburg State University of Information Technologies, Mechanics and Optics, Kronverksky pr. 49, St.-Petersburg, 197101, Russia

E-mail address: mgtomilin@mail.ru

**Abstract**. The goal of the paper is to demonstrate wide fields of new polarizing microscope application based on nematic liquid crystals (LCs) for detecting different solid crystals surface inhomogeneities. Optical polarizing microscope (OPM) based on LCs makes possible to observe the invisible physical fields' distribution on the objects' surfaces. The OPM novelty consists in LC spatial light modulator (SLM) introduction in optical scheme to detect local deformations in real time. LC SLM applied as recording media has to be in direct contact with the surface under investigation. It gives the possibility to detect the invisible physical fields on the surface: intermolecular interactions, electrical, magnetic fields, etc.

The theory of LC layer deformations was developed to find the relation between real size of structural defect D and the size of its image D' visualized with NLC layer. OPM method was used for detecting different aspects of surface inhomogeneities for different types of solid crystals. The new results were obtained for twinning boundaries in piezo quartz resonator having industrial application. LC SLM demonstrates non destructive method and better accuracy in comparison with etching, and the simplicity in comparison with x-radiation detecting. LC SLM may be combined with interference, phase-contrast and even near-field microscopes.

## 1. Introduction

The optical microscope operations are limited with illumination distribution detecting on the object surface in reflective regime and absorption object parameters detecting in transparent regime. The functions are increased by observation the objects in polarizing light. The inside tension and optical activity in transparent materials may be additionally observed. Optical polarizing microscopy is very powerful tool for investigation in many fields of science and technology. But it is helpless in detecting invisible physical fields' distribution on the object surface. The combination of optical polarizing microscope (OPM) with liquid crystal spatial light modulator (LC SLM) in contact with objects' surface makes possible to increase the microscopes' power and functions. The propagation of LC SLM application to other microscopes types with super high resolution is also discussed.

## 2. The optical polarizing microscope scheme and principle of operation

The novelty of OPM consists in LC SLM introduction in optical scheme to observe LC layer local deformations in real time [1]. LC SLM applied as recording media has to be in contact with the surface under investigation (figure 1). In this case LC SLM gives the possibility to detect the invisible physical fields on the object's surface: intermolecular interactions, electrical, magnetic fields, etc.

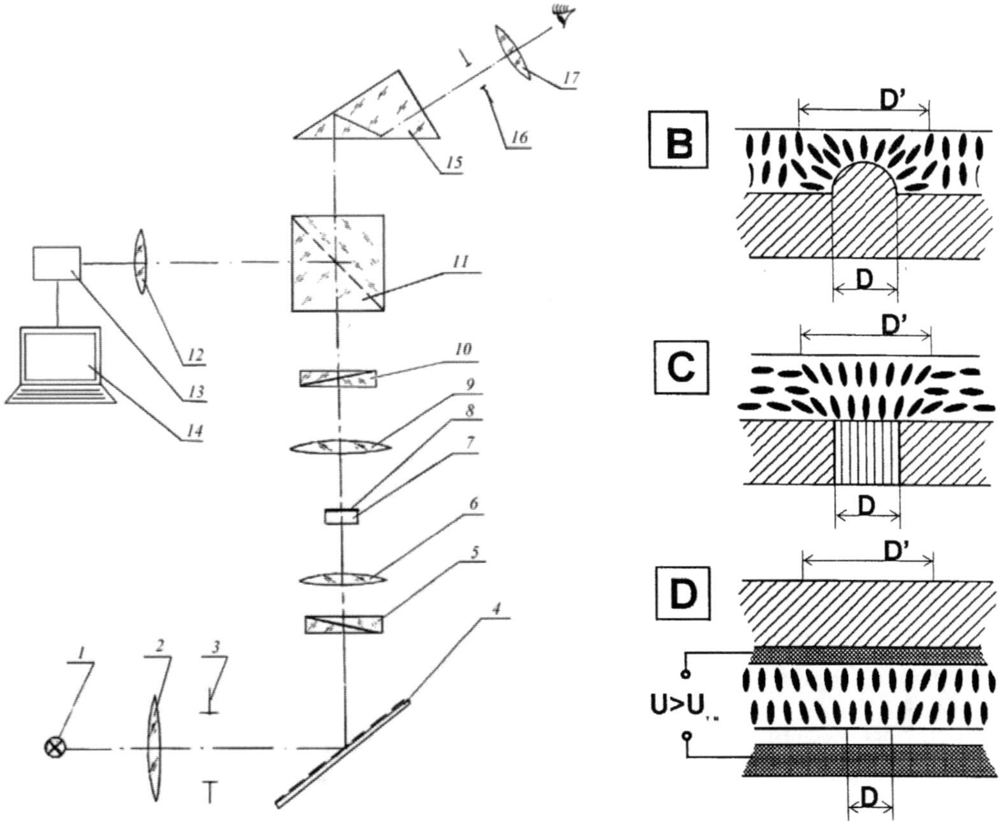


**Figure 1.** Left – optical scheme: 1-light source, 2-collector, 3-aperture, 4-mirror, 5-polarizer, 6-condenser lens, 7-object, 8-LC SLM, 9-microscope lens, 10-analyzer, 11-prism-cube, 12-ocular 1, 13-CCD-sensor, 14-computer, 15-prism, 16-aperture, 17-ocular. Right – B, C, D – defects types. B – Microrelief defect; C – structural inhomogeneities; D – the distribution of non uniform electrical or magnetic field. D - Defect's size. D' - defect's image size in NLC.

As it is shown on figure 1 B, C and D the possibility arises to visualize microrelief and structural defects. It also opens the horizons for bio analytic applications using LC flow in micro fluidic channels for detecting macromolecules local defects.

Nematic LCs are more sensitive to physical fields detecting in comparison with sensitivity of cholesteric and smectic LCs having super molecular structure. The principle of operation is shown on figure 2. In polarized light we observe not the invisible structural defect D, but only the deformations in LC layer D' induced by defect (left) (figure 2). In non polarized light we observe the invisible structural defect D through transparent LC layer and see nothing (right). It is the reason to study the deformations of NLC layer, induced by initial fields.

## 3. The theory of NLC layer deformations

The light intensity over NLC layer $I(x, y)$ modulated by deformed NLC structure is described by equation:

$$I(x, y) = I_0 \, Sin^2[\delta(x,y)/2] \tag{1}$$

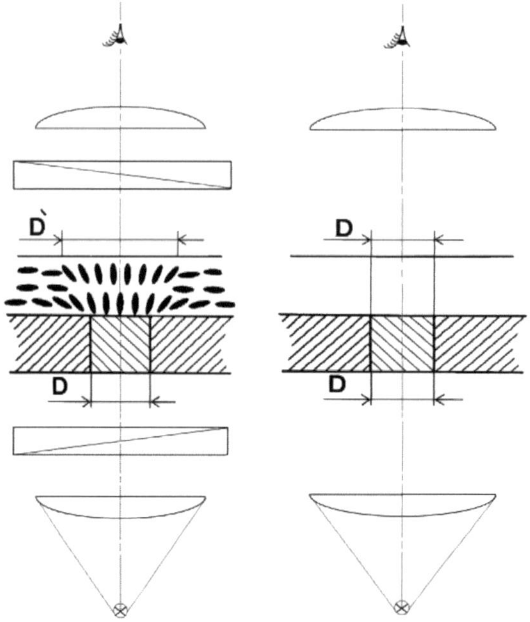


**Figure 2**. OPM with LC SLM operation principle of [1].

The phase delay *δ(x, y)* caused by the NLC birefringence:

$$\delta(x,y) = {}^{2\pi}\!/_{\lambda}\left[-n_0 \cdot H + \int_0^H n(x,y,z)dz\right] \quad (2)$$

*H* is the thickness of NLC; *n (x, y)* is film reflective index of deformed zone; $n_0$ is non-deformed layer the refractive index. If the orientation has no twist deformation then only orientation bending occurs, hence:

$$n(x,y,z)=[n_e^{-2}Sin^2\varphi(x,y,z) + n_0^{-2}Cos^2\varphi(x,y,z)]^{-1/2} \quad (3)$$

*φ(x,y,z)* is the deflection angle of the long axis of the molecules; $n_o$, $n_e$ are the refractive indices of NLC layer for ordinary and extraordinary polarization. The high value of NLC optical anisotropy permits to use very thin layers to obtain sufficient value of phase delay.

The theory of LC layer deformations near the structural defects or obtained by magnetic or electrical fields is developed and described in [2, 3].

The most interesting and practically important case is the detecting structural inhomogeneities in different materials (for example, on the solid crystals surface), shown on fig. 1C. The theory gives the distribution of LC molecules orientation near the structural defect in the case of strong anchoring conditions (figure 3).

The NLC molecules orientation induced by structural inhomogeneities is described by equation:

$$\pi\frac{\varphi(x,y)-\varphi_a}{\varphi_d-\varphi_a} = arctg\frac{sh(D-X)ctg\frac{Y}{2}}{ch(D-X)+1} + arctg\frac{Sh(D+X)ctg\frac{Y}{2}}{ch(D-X)+1}; \quad (4)$$

$$X=\frac{\pi x}{H},\quad Y=\frac{\pi y}{H},\quad D=\frac{\pi d}{H}.$$

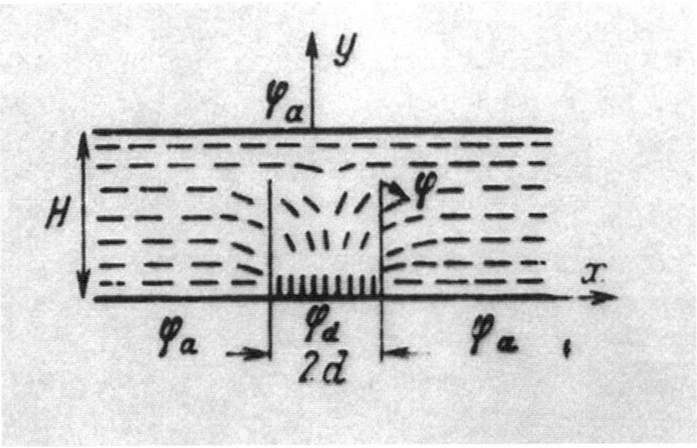


**Figure 3.** Solitary structural defect and local orientation of LC molecules induced by it.

The discussed theory was used for explanation the results of experimental examination the structural defects on the surface of solid crystals and other materials. It gives possibilities to calculate parameters of external fields.

## 4. The application of OPM with LC SLM to detecting solid crystals surface homogeneity

OPM with LC SLM may be effectively used for detecting the NLC orientation on the surface of solid crystals, twinning boundaries detecting, phase transition application for enlarging the contrast of twinning boundary image, structural blocks and defects detecting in crystals and semiconductors, crystalline phase detecting in vitreous materials and domains detecting in ferroelectric crystals.

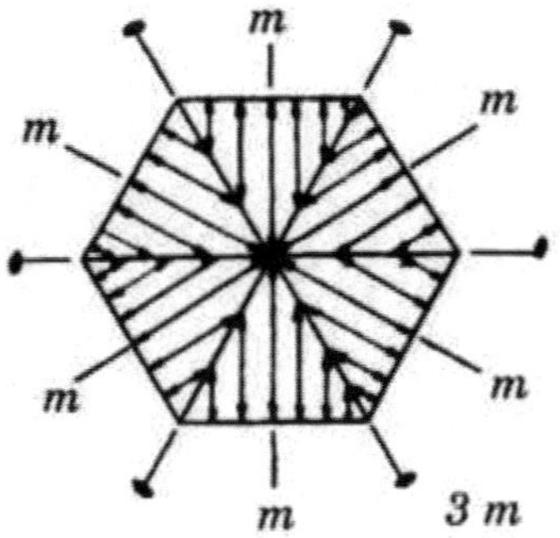


**Figure 4.** Plans of symmetry detecting in guanidine aluminum sulfate gexagidrate with application of E-field to tolans mixture (left) and the scheme of molecules orientation on mono crystalline surface (right).

*4.1. NLC orientation on the surface of solid crystals*

At the beginning of LC materials study solid crystals were used by Ch. Mogen (1913), F. Grandjean (1916), P. Gobert (1938) and other scientists to study the behavior of LCs on solid crystals [4]. The goal was to achieve the stable and repeatable orientation of LC layers. It was demonstrated that easy LC molecules orientation has the relation with properties of crystal symmetry At 1916 Grandjean showed that LC molecules on the surface of solid crystals are spontaneously oriented along crystallographic directions. At 1970 P.G. de Gennes established that the number of easy orientation directions is elated to the symmetry of the substrate. At 1980 N. Tikhomirova & A. Ginsberg investigated surface symmetry properties of crystals NaCl, KBr, LiF, GASH, etc. [5]. At now days LC mostly used to detect the planes of crystal symmetry and solid crystals structural defects (figures 4, 5a, b). LCs gives the possibility to detect the ideal symmetry changes in crystals induced by inclusions (figures 5c and d).

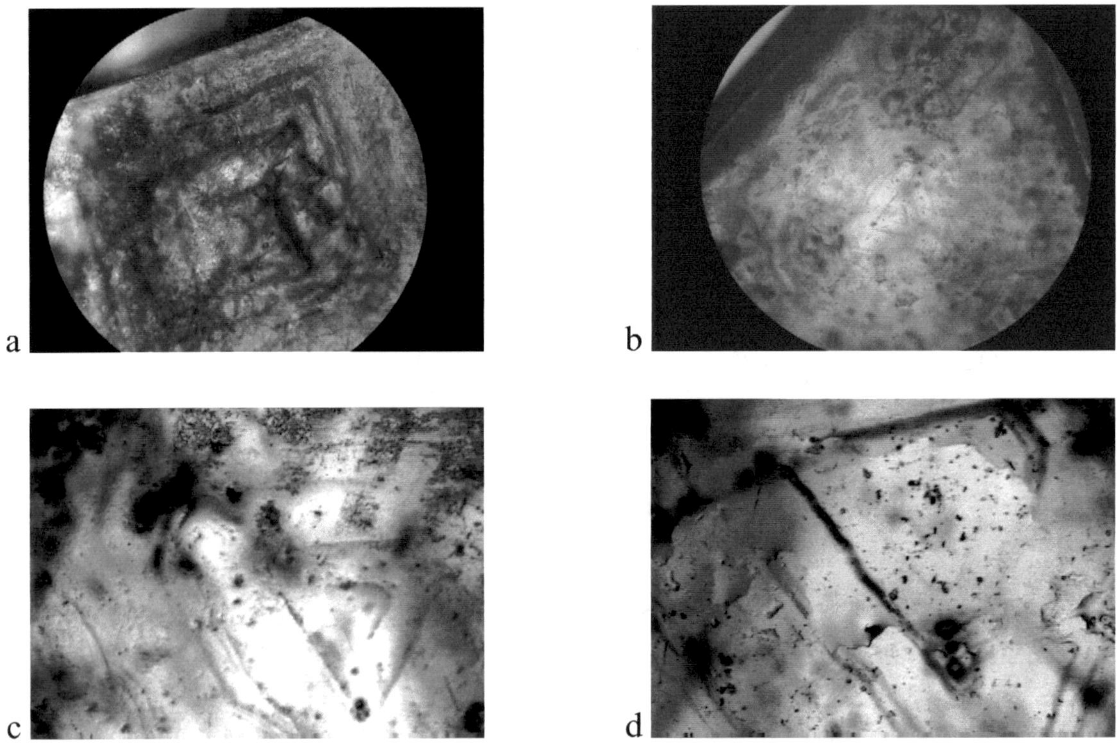



**Figure 5.** Planes of symmetry on the surface of crystals KCl (a) and alums (b) visualized with NLC (MBBA+EBBA). Inclusions disturb the ideal symmetry of alums (c, d). Magnification 50x.

*4.2. Twinning boundary detecting*

The optical application of crystals with twinning defects is forbidden almost in all cases. Twinning defects usually are visualized by etching. Unfortunately it dramatically changes the crystal surface quality. It is possible to submit etching by non destructive technique based on LCs. The example of LCs application to twinning boundary detecting in Island spar is shown on figure 6.

In comparison with etching the application of LCs on the Island spar surface doesn't change the quality of polished crystal surface. In addition NLCs visualize very small size structural defects that are not detected by etching (figure 6, right).

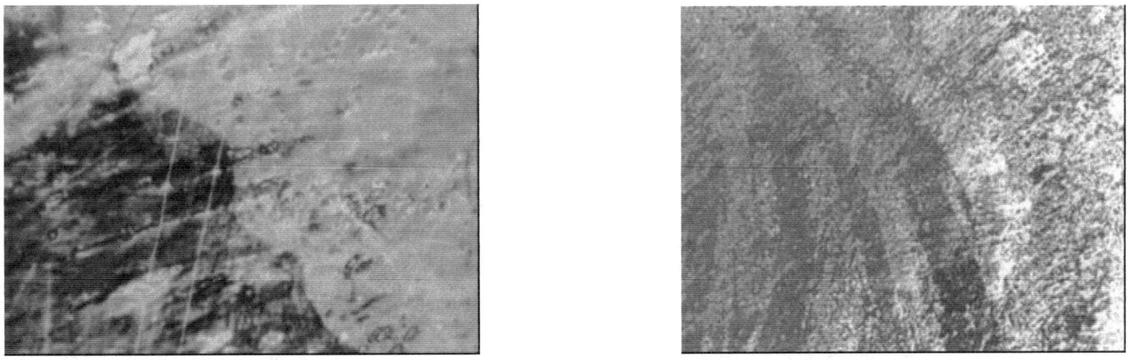

**Figure 6.** Twinning boundary in Island spar visualized with MBBA:EBBA, 50x (left) and visualized boundary details, 250x (right) [6].

*4.3. Phase transition application for enlarging the contrast of twinning boundary image*

The image of twinning boundaries, visualized in Island spar at room temperature, has a low contrast (figure 7a). Contrast may be increased by heating LCs up to the cleaning temperature to isotropic phase at one part of the boundary with smaller anchoring energy (figure 7b).

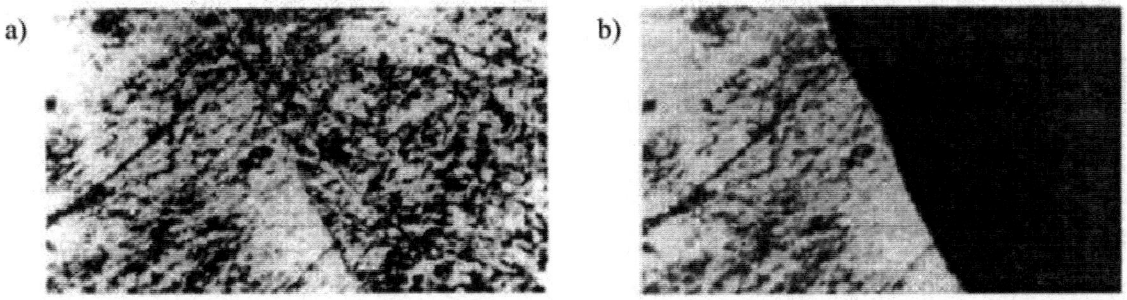


**Figure 7.** Phase transition application in LCs for enlarging visual contrast of twinning boundary in Island spar: MBBA:EBBA, *ta*=20°C; *tb*=53,4°C; 50x [7].

Structural defects in crystalline quartz resonators also may have natural reason or appear in the fabrication process, changing frequency parameters of resonator (figure 8a, b). We suggested the use of NLCs in industry for detecting structural defects with more accuracy than etching (figure 8c) and much easy than X-ray technique [8].

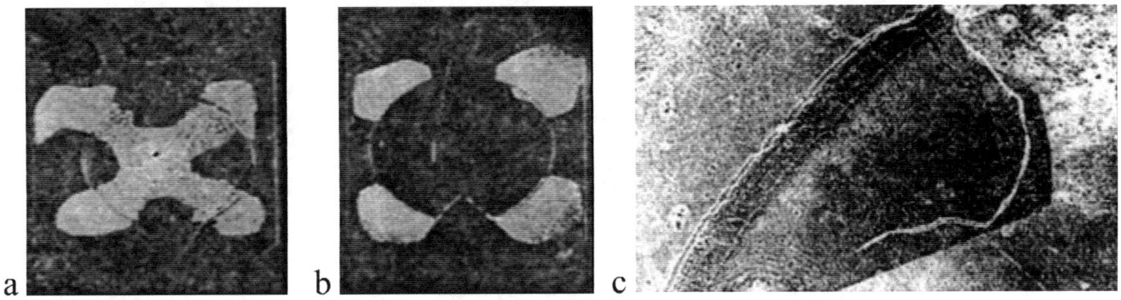


**Figure 8.** LCs application to structural defects detecting in industrial crystalline quartz resonator, MBBA:EBBA, 50x.

*4.4. Structural blocks detecting in Mica*

NLCs may efficiently visualize structural inhomogeneities in mica that characterize the insulator properties of the material (figure 9, left and right).

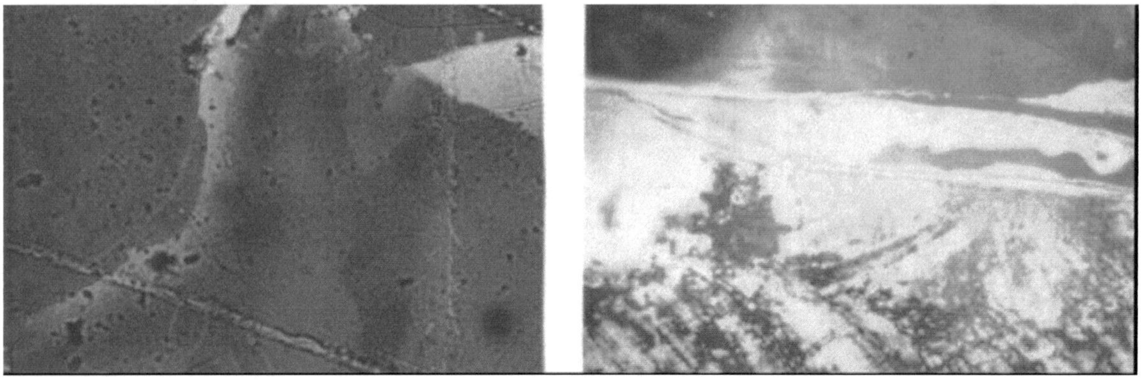

**Figure 9.** Blocks in {KAl2[AlSi3O10](OH,F)2} visualized with NLCs. MBBA, $t = 20°C$. 100x.

*4.5. Structural defects in laser crystals detecting*

Non linear laser crystals optical characteristics depend on their structural uniformity. Structural defects dramatically decrease the efficiency of frequency transformation. It was demonstrated that NLC method gives more information and higher spatial resolution on structural uniformity in comparison with second harmonic generation. The efficiency of NLC method was confirmed by series of experiments with KTP laser crystals [9]. The example of detecting structural defects on KTP crystal surface is shown on figure 10.

*4.6. The crystalline phase detecting in vitreous materials*

The invisible solid crystals appear in vitreous materials under the influence of UV radiation and heating. It is possible to visualize their appearance using NLCs (figure 11) [10].

*4.7. Structural defects in semiconductors*

NLCs may be used for detecting lattice defects and dislocations accumulation on different types of Si-surface (figures 12 and 13).

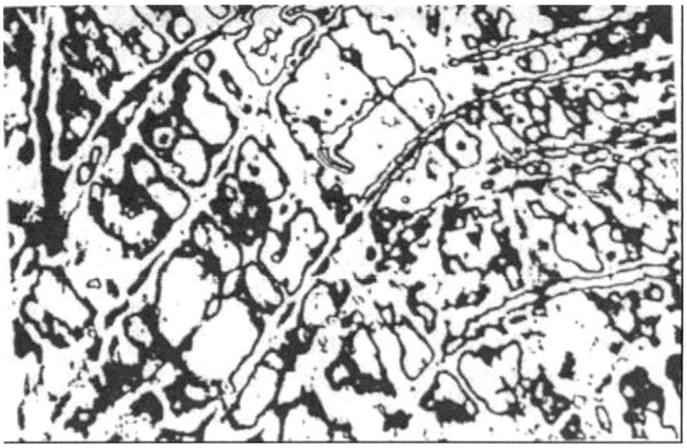

**Figure 10.** Structural defects in KTP laser crystal decreasing the efficiency of frequency transformation visualized with MBBA:EBBA; 100x.

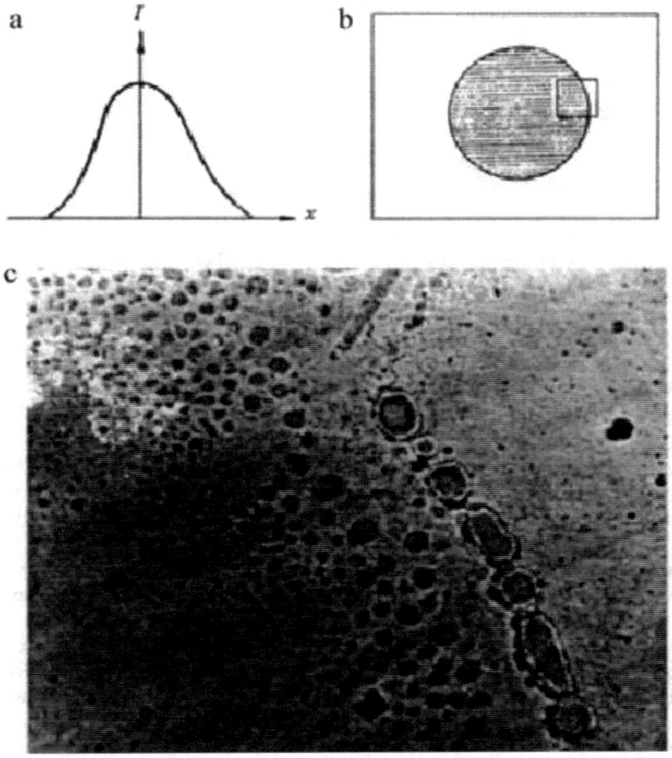


**Figure 11.** The distribution of light intensity in laser beam section (*a*), the object fragment subjected to laser radiation (*b*). Solid crystals in vitreous materials visualized with NLCs; 1300x (*c*). MBBA:EBBA [10].

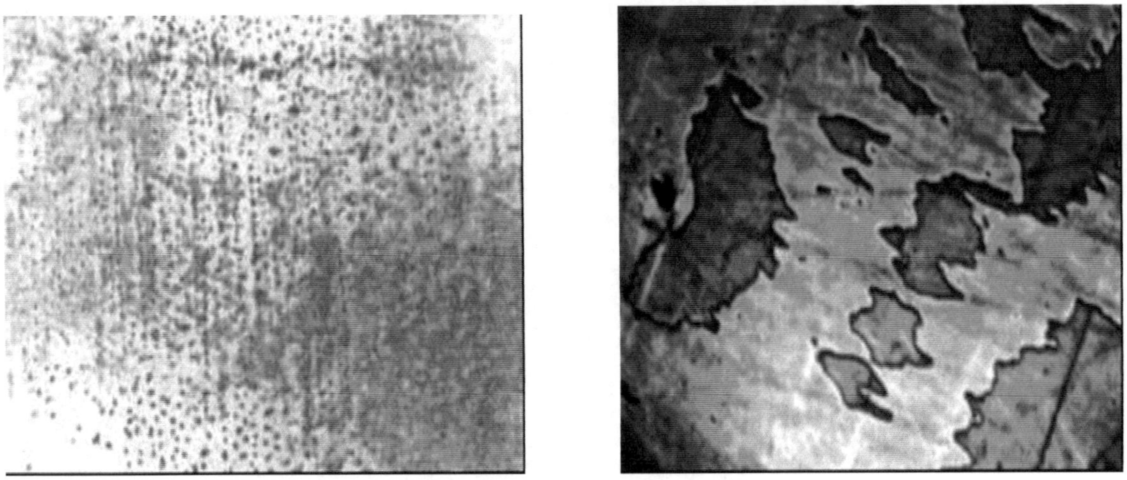

**Figure 12.** Crystalline lattice defects and dislocations accumulation visualized on Si surface (left). Structural defects on the surface of α-Si:H layer (right). 500x. MBBA:EBBA. $t = 20°C$.

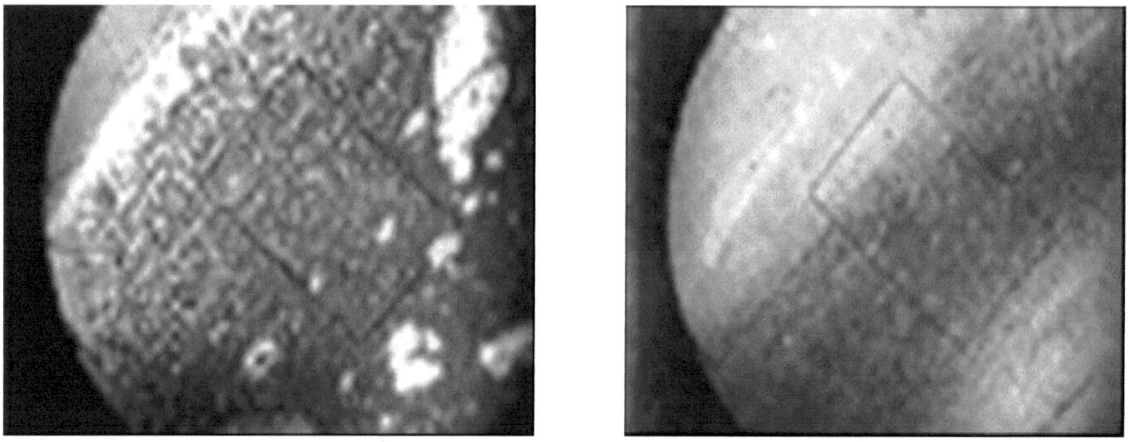

**Figure 13.** Dislocations accumulation visualized on $\alpha$-Si/Fe layer before (left) and after (right) annealing (t=800°C, 2 hours). MBBA:EBBA. 200x. $t = 20$°C [11].

*4.8. Domains detecting in ferroelectric crystals TGS and X-radiation defects*

Longitudinal domains are not visualized on the surface of ferroelectric crystal in optical microscope. The application of NLCs gives the possibility to visualize in polarizing microscope not only the longitudinal domains but also domain wall movement with change of applied electric field. X-radiation defects were easily visualized on the surface of ferroelectric crystal TGS (Cu-radiation; D – 0,2 mrad); TGS thickness – 1 mm (figure 14) [12].

The idea to make visible the distribution of physical fields (defects) on the material surface may be extended on other types of microscopes, for example near field microscope. Such idea may be used in visualizing defects at nanoscale level. The first application of LCs in super resolution microscopy was described in paper [13].

## 5. Summary

OPM with LC SLM opens wide possibilities in visualizing structural inhomogeneities of solid crystals in material science and high technologies. OPM with LC SLM application may be spread on interference, phase-contrast, confocal and near-field microscopy and may be considered as new

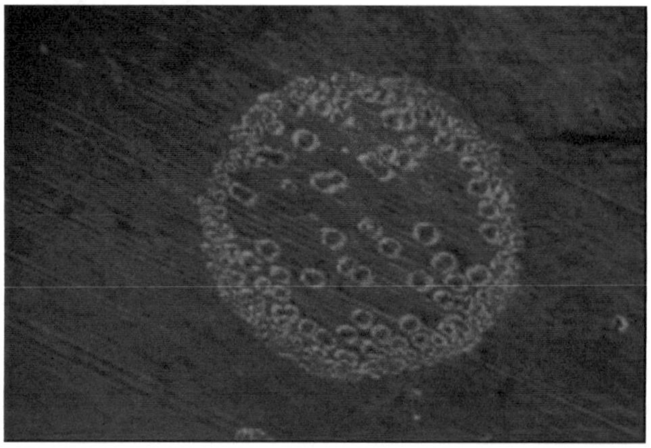

**Figure 14.** Longitudinal domains and X-radiation defects visualized on the surface of ferroelectric crystal TGS; MBBA:EBBA, 50 **x.**

universal instrument to detect invisible physical fields on objects' surface with high sensitivity and spatial resolution

**References**

[1] Tomilin M G and Barsukov O A 2010 *Opt. and Spectr.* **1** 122
[2] Aero E L and Tomilin M G 1987 *Sov. J. Opt. Techn.* **8** 50
[3] Tomilin M G 2001 *The interaction of liquid crystals with the* surface (St.-Petersburg: Polytechnika) p 325
[4] Sonin A *The surface physics of LCs* 1995 (Amsterdam: Gordon and Breach Publ.) pp 118-122
[5] Tikhomirova N A and Ginsberg A V 1980 (*Oxford-Budapest*: *Pergamon Press)* p 651
[6] Tomilin M G 1990 *Mol. Cryst. Liq. Cryst.* **193** 7
[7] Tomilin M G 1993 *Optical information processing. SPIE* **2051** 286
[8] Pasinkova O V, Ruban Yu V, Tomilin M G and Chudor I V 1991 *Sov. J. Opt. Techn.* **7** 69
[9] Kalintzev A G, Rachmanova A R, Tomilin M G and Yakobson V E 1995 *Sov. J. Opt. Techn.* **11** 86
[10] Glebov L B, Muchamadeeva A R, Nikanorov N V and Tomilin M G 1994 *Sov. J. Opt. Techn.* **7** 63
[11] Kerpeleva S Yu, Kudoyarova V Ch, Tomilin M G, Terukov E I 2003 *Sov. J. Opt. Techn.* **7** 71
[12] Tikhomirova N A, Dontsova L I and Pikin S A 1983 *Proc. of V-th ICSC on LC (Odessa, USSR, 2-9 Sept 1983)* **1**. Part 2 p189
[13] De Luca A, Barna V, Atherton T J, Carbone and Rosenblatt C 2007 *Abstracts of 12$^{th}$ Intern. Topical Meet. on Optics of LCs*. 2007 *(Puebla City, Mexico, 1-5 Oct 2007)* O-65

# The two dimensional Hubbard model: a theoretical tool for molecular electronics

**Vladan Celebonovic**

Institute of Physics, Pregrevica 118,11080 Zemun-Belgrade,Serbia

E-mail: vladan@ipb.ac.rs

**Abstract.** When speaking about molecular electronics, the obvious question which occurs is how does one study it theoretically. The simplest theoretical model suitable for application in molecular electronics is the two dimensional Hubbard model. The aim of the present paper is to introduce this model, and give some examples of the systems which it can describe. After a short mathematically oriented discussion, it will be shown how to calculate the electrical conductivity of a particular planar system: a rectangular lattice with mutually independent conductivities along the two axes,but without using the $2D$ Hamiltonian. This system could find applications in high Tc studies. It will finally be shown that the electrical conductivity of graphene can be determined not by using the full formalism of the $2D$ Hubbard model, but by a slight reformulation of the Hamiltonian of the $1D$ Hubbard model.

## 1. Introduction

The electron was discovered in 1897. It was quickly shown that metals contained electrons which were free to move,while insulators did not. In this way the famous problem of metal $\rightarrow$ insulator transitions ($MI$ transitions for short) entered modern physics. Some sucess in analyzing the $MI$ transition was achieved in the early days of quantum mechanics ([1] and references given there). A more general theory of $MI$ transitions was formulated towards the middle of the last century [2].

Attempts to develop a microscopic theory of the $MI$ transition were crucial in the development of the so called Hubbard model $HM$ ([3] and later work) . In its mathematically simplest form, the main ingredient of the $HM$ is a collection of atomic orbitals, and the main physical idea is the idea of *tight binding.* Expressed in common language, this means that the wave function of an electron is centered on the lattice site of an ion, and that any electron can "hop" one lattice spacing at a time.

In statistical physics, a model is specified by its Hamiltonian. The Hamiltonian of the HM is given as the sum of two terms, the "free" kinetic term $H_0$ and the "interaction" term $H_I$

$$H = H_0 + H_I \tag{1}$$

For an example of a recent discussion on the formulation of the Hubbard model see [4]. In two dimensions, expressed within the second quantisation formalism, the Hamiltonian of the Hubbard model has the following form:

$$H = -\sum_{ij}\sum_{\sigma} t_{ij} c_{i\sigma}^{+} c_{j\sigma} + \frac{1}{2}\sum_{ijkl}\sum_{\sigma\sigma'} < ij|v|kl > c_{i\sigma}^{+} c_{j\sigma'}^{+} c_{l\sigma'} c_{k\sigma} \tag{2}$$

In this expression,latin indices denote sites on a lattice, $\sigma$ is the spin index, and symbols of the form $c_{i\sigma}^{+}$ denote second quantisation operators creating an electron with spin $\sigma$ at a lattice site $i$. The hopping integral between sites $i$ and $j$ is denoted by $t_{ij}$. Finally, $v$ is the Coulomb potential.

The Hubbard model may seem relatively simple, judging by the form of its Hamiltonian. However, it is only *apparently* simple. Although nearly 60 years have passed since it was proposed, this model has been exactly solved only for the 1D case [5]. In spite of that, the Hubbard model and its applications are still provoking scientific interest. A recent search ( beginning of August 2010 ) with the key words "Hubbard model" gave 1043 papers published in the journal Physica B. A similar search on http://prola.aps.org gave 1613 papers in Phys Rev B and Phys Rev Lett with the same key words in 2000.-2010.

The relevance of the $2D$ Hubbard model to modern condensed matter physics follows from the fact that both in "standard" high temperature superconductors, as well as in those containing $Fe$, planar sheets of charges play important roles.

Calculations are usually rendered easier by introducing the following two approximations in the Hamiltonian:

$$t_{ij} = \begin{cases} t & (i,j) \quad nn \\ 0 & otherwise \end{cases}$$

and

$$< ij\,|v|\,kl >= \begin{cases} U & (i=j=k=l) \\ 0 & (otherwise) \end{cases}$$

where $nn$ denotes nearest neighbors.

## 2. Some mathematical details

With these approximations the Hamiltonian of the two dimensional Hubbard model takes the form

$$H = -t \sum_{<i,j>} \sum_{\sigma} (c_{i,\sigma}^{+} c_{j,\sigma} + h.c) + U \sum_{i} n_{i,\uparrow} n_{i,\downarrow} \tag{3}$$

where $h.c$ denotes hermitian conjugation.

A logical first step in using the $2D$ Hubbard model in theoretical studies of molecular electronics would be finding its general solution. Even with the two approximations mentioned above the general solution of the $2D$ Hubbard model has not been found. The $2D$ Hubbard model can be solved in two particular limiting cases: these are the *band limit* and the *atomic limit.*

The *band limit* corresponds to the case $U = 0$, and the Hamiltonian then has the form

$$H_{U=0} = \sum_{\vec{k},\sigma} \epsilon(\vec{k}) c_{\vec{k},\sigma}^{+} c_{\vec{k},\sigma} \tag{4}$$

and

$$\epsilon(\vec{k}) = -\sum_{i0} t_{i0} \exp(i\vec{k} \cdot \vec{R}_i) \tag{5}$$

where $\vec{k}$ belongs to the Brillouin zone. In the case of a $2D$ square lattice with lattice step $a$, it can be shown that

$$\epsilon(\vec{k}) = -2t[\cos(k_x a) + \cos(k_y a)] \tag{6}$$

The *atomic limit* of the Hubbard model corresponds to the opposite case: $t = 0$ but $U \neq 0$. The Hamiltonian then has the form

$$H_{t=0} = U \sum_i n_{i,\uparrow} n_{i,\downarrow} \tag{7}$$

As the hopping energy is $t = 0$ this limit has no interest for studies of the transport properties of the Hubbard model. The Hubbard model can be solved in each of these two limiting cases separately. Complications arise when attempting to solve it in "intermediary" cases, when both $t \neq 0$ and $U \neq 0$.

A related problem, with more practical applicability is the calculation of the electrical conductivity of the Hubbard model and,in broader terms, of any physical system whose Hamiltonian is known. These calculations have applications in material science, in studies of materials as different as Q1D organic conductors, high $T_c$ superconductors, graphene,...

*2.1. Calculating the conductivity*

Imagine that the Hamiltonian H of a many-body system is at some sufficiently remote moment in the past perturbed by a time-dependent external field $h(t)$ . In real experiments, this external field can be high external pressure, which is at some moment turned on and increases with time. At time $t$ this Hamiltonian can be expressed as $H(t) = H + V(t)$, where $V(t) = h(t)A$, and the symbol $A$ denotes the parameter of the system with which the perturbing external field $h(t)$ is coupled. A common example of $A$ is the particle number density.

As a consequence of the existence of the perturbation $h(t)$, the system under study is not "isolated" any more. This implies that the average value of the observable represented by the operator $A$ depends on the details of the perturbing field; solving such a problem is a complicated task. This problem can be reduced to the Linear Response Theory if the perturbation $h$ is small enough.

This idea is the basis of the statistical mechanical theory of ireversible processes, proposed by Kubo [7]. The aim of this theory is to develop a scheme for the calculation of the kinetic coefficients for quantities such as the electrical and thermal conductivity. Kubo has shown that this calculation can be performed as a calculation of time correlation functions in equilibrium. From the viewpoint of pure theory, Kubo's theory solves the problem - it gives formal expressions for the required physical quantities. However, the expressions it gives are far too complex for application to real materials.

Another method applicable to the calculation of the kinetic coefficients is the so called "memory function" method, recently reviewed in [8]. This method is a logical continuation of the work by Kubo. It was practically developed in the 1970s, and within this method the electrical conductivity can be calculated as follows:

$$\chi_{AB}(z) = << A; B >> = -i \int_0^\infty \exp izt < [A(t), B(0)] > dt \tag{8}$$

where $\mathbf{A} = \mathbf{B} = [\mathbf{j}, \mathbf{H}]$, $\mathbf{j}$ denotes the current operator, and

$$\sigma(z) = i \frac{\omega_P^2}{4\pi z} [1 - \frac{\chi_z}{\chi_0}] \tag{9}$$

The symbol $\omega_P$ denotes the plasma frequency, which is given by $\omega_P^2 = 4\pi e^2 n/m$, and where $e, n, m$ are the electron charge, number density and mass respectvely. The current operator is defined as follows [9]:

$$\mathbf{j} = \frac{\partial}{\partial t} \mathbf{P} = i[\mathbf{H}, \mathbf{P}] \tag{10}$$

where

$$\mathbf{P} = \sum_i \mathbf{R_i} n_i \tag{11}$$

is the polarisation operator. In expression (11) $\mathbf{R_i}$ denotes the position of lattice site $i$, and $n_i$ is the number of particles at lattice site $i$. Calculating the electrical conductivity of a $2D$ Hubbard model by the memory function method is, in principle, possible but it would be tedious. Just as an illustration, the current operator would be given by an expression of the form:

$$\mathbf{j} = i[\mathbf{H}, \mathbf{P}] = [\mathbf{H_0} + \mathbf{H_I}, \sum_{i,j} \mathbf{R_{i,j}} n_{i,j}] \tag{12}$$

in which the free part and the interaction part of the Hamiltonian would have to be inserted from the Hamiltonian of the 2D HM. Note that in the last expression $R$ and $n$ are labelled by two indexes because of the dimensionality of the lattice.

The following section of this paper contains a solution to the simplified problem: the calculation of the conductivity of a $2D$ square lattice, with mutually independent conductivities along the axes.

## 3. The conductivity of a square lattice

Imagine a $2D$ square lattice of side length $a$. If the lattice sides are denoted by $x$ and $y$, the electric current flowing through such a system can be expressed as

$$\vec{j} = j_x \vec{e_x} + j_y \vec{e_y} \tag{13}$$

where $\vec{e_x}$ and $\vec{e_y}$ are unitary vectors of the two lattice axes. The total current is given by

$$j = [j_x^2 + j_y^2]^{1/2} \tag{14}$$

As by definition $\vec{j} = \sigma \vec{E}$ where $\sigma$ is the electrical conductivity and $\vec{E}$ the electric field, and assuming that $E_x = E_y = E$, it finally follows that

$$\sigma = [\sigma_x^2 + \sigma_y^2]^{1/2} \tag{15}$$

It is assumed here that the conductivities along the two axes are mutually independent, which means that the conductivity of this system can be exressed using known results for the conductivity of $1D$ correlated electron systems [10]. Full details of the calculation are given in [10]. The electrical conductivity is given by:

$$\sigma_R(\omega_0) = (1/2\chi_0)(\omega_P^2/\pi)[\omega_0^2 - (bt)^2]^{-1}(Ut/N^2)^2 \times S = p_F \times S \tag{16}$$

and the symbol $S$ denotes the following function

$$\begin{aligned} S = 42.49916 \times (1 + \exp(\beta(-\mu - 2t)))^{-2} + 78.2557 \times \\ (1 + \exp(\beta(-\mu + 2t\cos(1+\pi))))^{-2} + (bt/(\omega_0 + bt)) \times \\ (4.53316 \times (1 + \exp(\beta(-\mu - 2t)))^{-2} + \\ 24.6448(1 + \exp(\beta(-\mu + 2t\cos(1+\pi)))))^{-2}) \end{aligned} \tag{17}$$

where $\mu$ denotes the chemical potential of the electron gas, determined in [10] as:

$$\mu = \frac{(\beta t)^6 (ns - 1)\,|t|}{1.1029 + .1694(\beta t)^2 + .0654(\beta t)^4} \tag{18}$$

In equations (15)-(17), $\omega_0$ denotes the real part of the frequency, $\beta$ is the inverse temperature, $s$ the lattice constant, $n$ the mean number of electrons per lattice site and $U$ is the on-site Coulomb repulsion. The number of lattice sites is denoted by $N$. The static limit of the susceptibility is denoted by $\chi_0$. Equations (15)-(17) give the possibility for investigating the dependence of the conductivity on the temperature $T$, band filling $n$, frequency $\omega_0$ and hopping parameter $t$. The symbol $b$ denotes a numerical constant, $b = -1.83879$. Applying expressions (15)-(17) is straightforward. For the sake of simplicity, it was assumed that the parameters have the same values along the two axes, meaning that $t_x = t_y$ etc. The numerical values were chosen as follows: $N = 150$, $\chi_0 = 1/3$, $U = 4t$, $\omega_0 = 3t$, $\omega_P = 12t$, $\beta = 11600/T$, and the conductivity was normalized to 1 at $n = 0.7$,$t = 0.01eV$, $t = 100K$. Examples of results of such calculations are presented in the figures; figures 1 and 2 correspond to $n = 0.7$, figure 3 corresponds to $n = 0.8$ and figure 4 to $n = 1.2$. The figures show that the values of the hopping energy and doping have a drastic influence on the shape of the teperature dependence of the conductivity, as well as on the intensity of the maximum. The existence of such an influence is known in experiments on high $T_c$ materials and $Q1D$ organic conductors; see for example [11] and references given there.

An interesting extension of the results of the present calculation could be obtained by redefining the total current as

$$\vec{j} = mj_x\vec{e_x} + nj_y\vec{e_y} \tag{19}$$

with $m$ and $n$ being any real numbers.

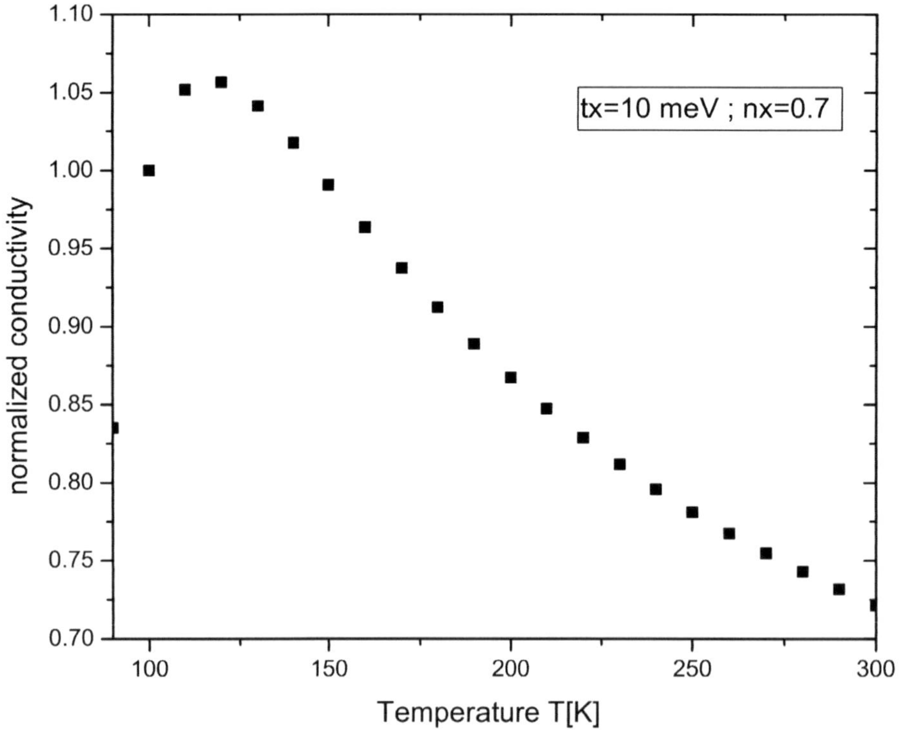


**Figure 1.** Normalized conductivity of a square lattice for $t_x = 10$ meV and $n = 0.7$.

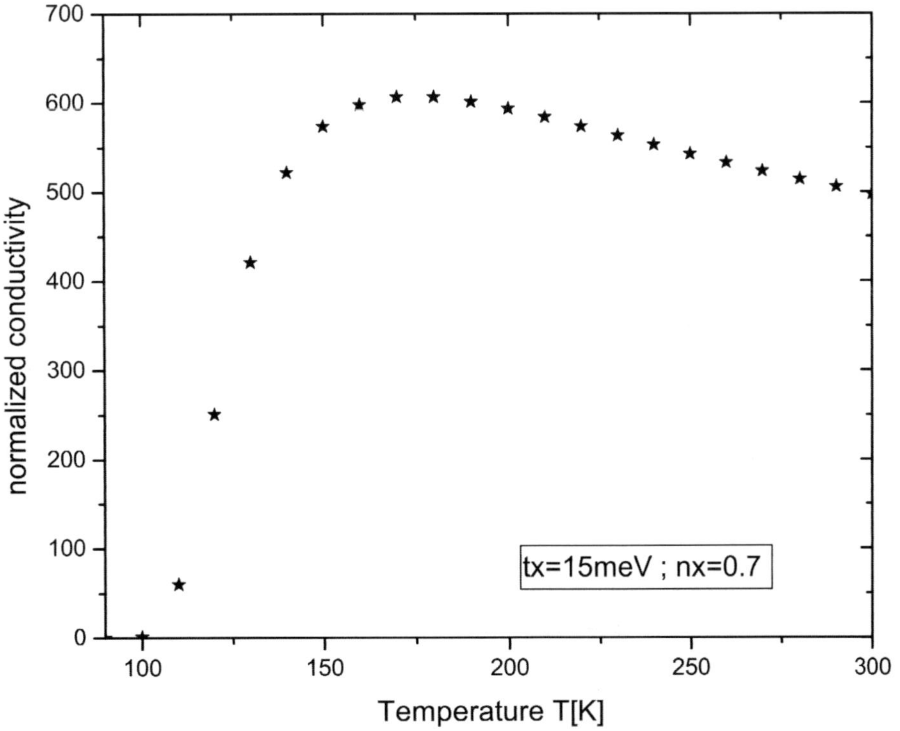


**Figure 2.** Normalized conductivity of a square lattice for $t_x = 15$ meV and $n = 0.7$.

## 4. Graphene: is the 2D Hubbard model applicable?

It was a "certainty" in statistical physics for more than 70 years that 2D materials cannot exist. It was argued by people of no less scientific stature than Landau or Peierls (quoted in [12]) that such materials would be therodynamically unstable, because a divergent contribution of thermal fluctuations in low dimensional crystal lattices would lead to atomic displacements comparable to atomic spacings. Much to the astonishement to the scientfic community, a $2D$ allotrope of carbon, later named graphene, was discovered in 2004 [13]. Graphene is a monolayer of carbon atoms packed in a honeycomb lattice [14]. Experiments have shown that graphene is an extremely interesting material. For example, it is the thinnest and strongest known material, and it can support current densities six times greater than copper [15]. It has been shown that electron transport in graphene is governed by the Dirac equation, that its conductivity never becomes zero, and that the charge cariers in graphene are massless Dirac fermions [14]. Graphene being planar, the obvious question is whether or not its transport properties can be studied within the Hubbard model, using, for example, the memory function method. Performing such a calculation would be possible by insertion of expression (3) into equations (8)-(10), but it could easily become complex. There is however a seemingly simpler way to complete such a calculation: use the Hamiltonian of the $1D$ Hubbard model, and simply change the physical meanings of various variables which occur in it. The Hamiltonian of the $1D$ Hubbard model has

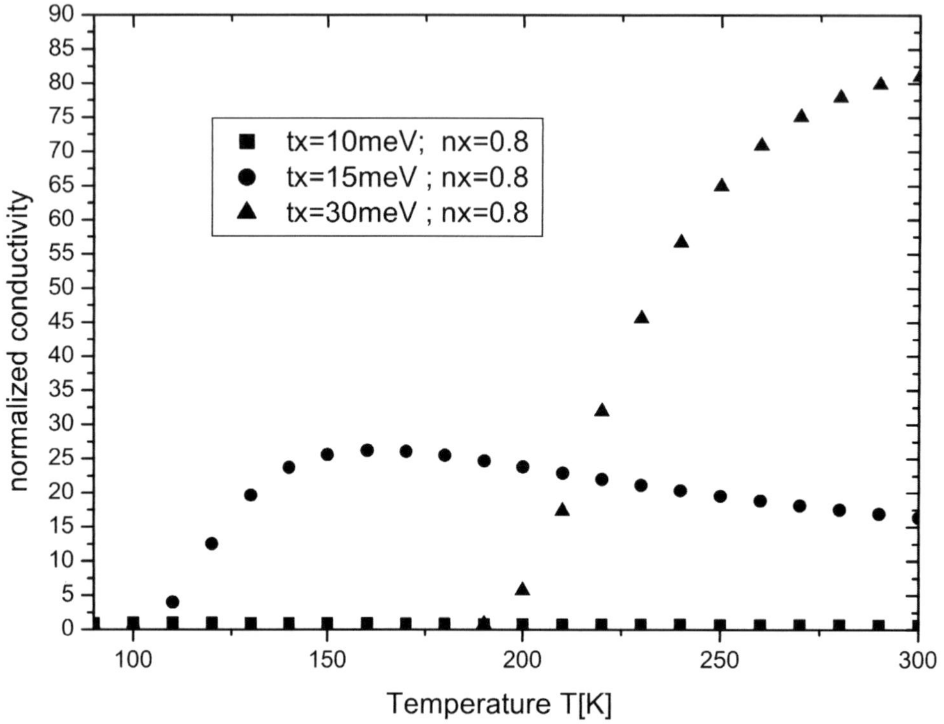


**Figure 3.** Normalized conductivity of a square lattice for 3 values of $t_x$ and $n = 0.8$.

the following mathematical form:

$$H = -t\sum_{i,\sigma}(c^+_{i+1\sigma}c_{i\sigma} + c^+_{i\sigma}c_{i+1\sigma}) + U\sum_{i} n_{i,\uparrow}n_{i,\downarrow} \quad (20)$$

Symbols of the form $c^+_{i\sigma}$ denote second quantisation operators creating an electron with spin $\sigma$ at a lattice site $i$, while $n_{i,\uparrow}$ is the number operator for the number of electrons at an ion on lattice site $i$ having spin up. The calculation of the electrical conductivity using this Hamiltonian and the "memory function" method is discussed in detail in [10]. With application to graphene in view, the Hamiltonian given by equation (19) can be reformulated in the following form

$$H = t(\sum_{j,\Delta,\sigma} c^+_{j,\sigma}c_{j+\Delta,\sigma} + h.c) + U\sum_{j} c^+_{j,\sigma}c_{j,\sigma}c^+_{j,-\sigma}c_{j,-\sigma} \quad (21)$$

where $c^+_{j\sigma}$ and $c_{j\sigma}$ are the creation and anihilator operators of electrons in a graphene unit number $j$ and spin $\sigma$, $t$ is the overlap integral between adjacent graphene units, which is determined by overlapping of the wave functions of graphene electrons, $\Delta$ is the vector relating two adjacent units in the hexagonal and $U$ is the energy of Coulomb repulsion of the electrons in the same graphene unit.

From the point of view of mathematical form, equations (19) and (20) are almost identical; this implies that the results of any calculations in which they are used are also going to be of

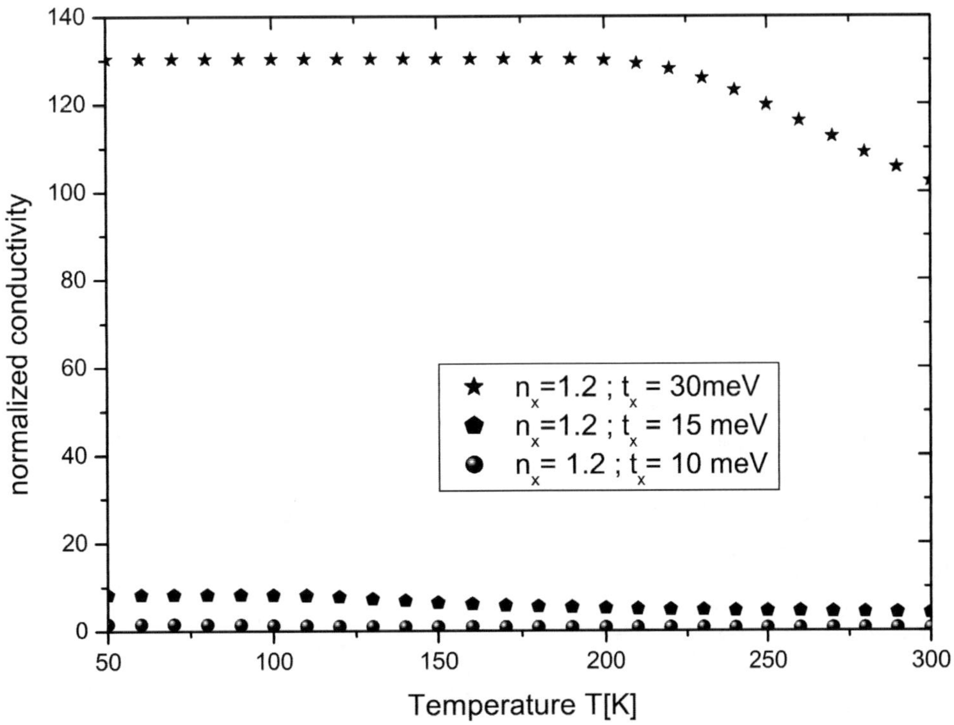


**Figure 4.** Normalized conductivity of a square lattice for $t_x$ meV and $n = 1.2$.

similar form. There exist, however, differences in the physical nature of the quantities which occur in them. For example, in a $1D$ problem, electrons "hop" between lattice sites, while in the case of graphene charge carriers hop between graphene units.

At the time of this writing (mid August 2010) attempting to apply the $2D$ Hubbard model to the conductivity of graphene is an ongoing calculation. The commutator $[j, H]$ has been calculated and it has the same mathematical form as the corresponding commutator in the $1D$ case. Details are slowly "coming in" and will be discussed elsewhere.

## 5. Conclusions

In this paper the Hubbard model in $2D$ was introduced and discussed to some extent. No attempt was made of a general discussion of the model, but attention was focused on aspects relevant for its transport properties. This aspect is useful for theoretical studies in molecular electronics, which is the topic of this meeting. A method capable of providing a full general solution for the electrical conductivity, called the memory function method, was briefly discussed. The general expression for the conductivity was not discussed.

However, the conductivity of a particular system: a square lattice with mutually independent conductivities along the axes was discussed to some extent. The temperature dependence, as well as the dependence on doping, were shown on several numerical examples, calculated for a set of arbitrarily chosen lattice parameters. Possibilities for rendering this calculation more complex, and therefore more physically interesting, were also indicated.

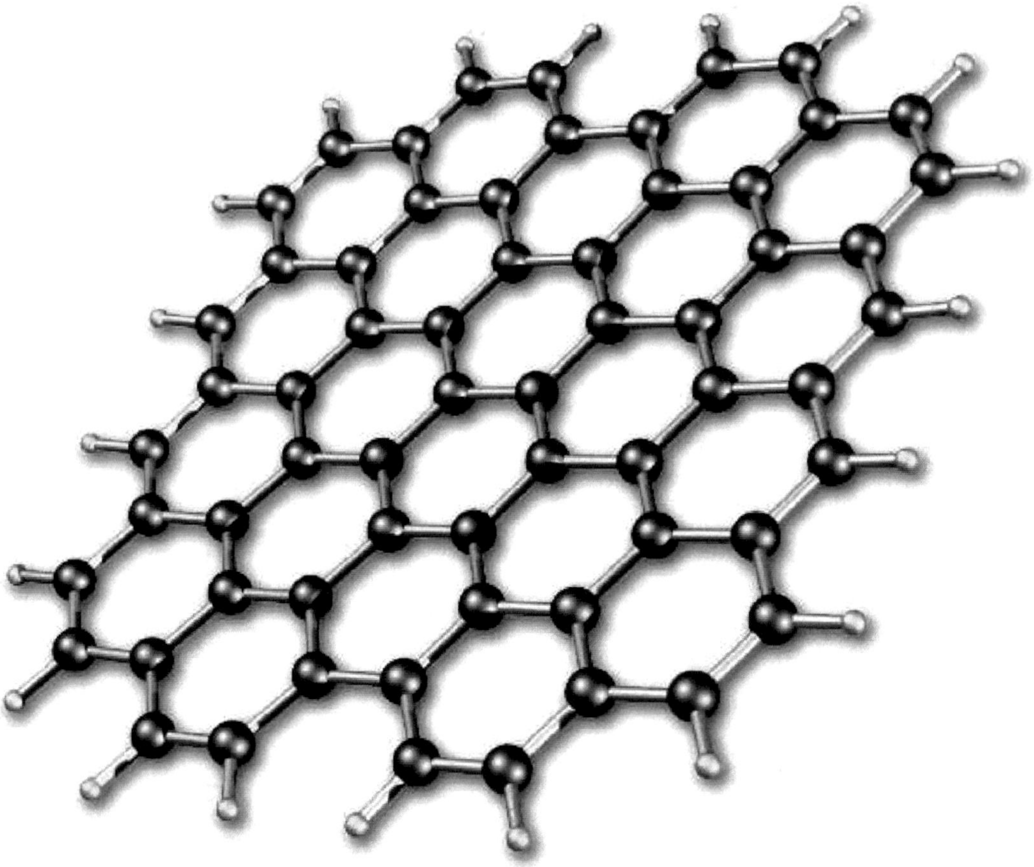

**Figure 5.** The structure of graphene

The paper ends with a brief discussion of graphene and a possibility of using the Hubbard model in the calculation of its conductivity. It is pointed out that by slightly reformulating the physical meaning of various parameters in the $1D$ Hubbard model, it should be possible to use it in calculating the conductivity of graphene. Some details of this calculation have been performed, the results are encouraging, but the full calculation is far from being finished.

An interesting detail has emerged after this paper was completed. Namely, the idea of trying to use the $2D$ Hubbard model was here proposed by purely theoretical reasoning: this model gives good results for the conductivity of $1D$ correlated electron systems, so it seemed a logical extension to try it in the $2D$ case. Some experiments studying the influence of electron-electron interaction on the low temperature conductivity of graphene have just been published, which can be interpreted as an argument in favour of the purely theoretical proposal made in the present paper [16].

**Acknowledgments**
The preparation of this paper was financed through the research project 141007 funded by the Ministry of Science and Technological Development of Serbia.

## References

[1] Mott N 1982 *Proc.R.Soc.London* A **382** 1
[2] Mott N 1968 *Rev.Mod.Phys.* **40** 677
[3] Hubbard J 1963 *Proc.R.Soc.London* A **276** 238
[4] Celebonovic V 2009 *J.of Optoelectronics and Advanced Materials* **11** 1135
[5] Lieb E H and Wu F Y 1968 *Phys.Rev.Lett.* **20** 1445
[6] Tasaki H 1998 *J.Physics Condens.Matter* **10** 4353
[7] Kubo R 1957 *J.Phys.Soc.Japan* **12** 570
[8] Balucani U, Lee M H and Tognetti V 2003 *Phys.Rep.* **373** 409
[9] Mahan G D 2000 *Many-Particle Physics,third edition* (New York:Kluwer Academic/Plenum Publishers)p 24
[10] Celebonovic in 2006 *Trends in Material Science* ed.B M Caruta (New York:Nova Science Publishers)p 241
[11] Carlson E, Emery V J, Kivelson S A and Orgad D 2002 *Preprint* cond-mat/0206217
[12] Geim A K and Novoselov K S 2007 *Nature Materials* **6** 183
[13] Novoselov K S, Geim A K, Morozov S V, Jiang D, Zhang Y, Dubonos S V, Grigorieva I V and Firsov A A 2004 *Science* **306** 666
[14] Novoselov K S, Geim A K, Morozov S V, Jiang D,Katsnelson M I, Grigorieva I V, Dubonos S V and Firsov A A 2005 *Nature* **438** 197
[15] Geim A K 2009 *Science* **324** 1530
[16] Kozikov A A, Savchenko A K, Narozhny B N and Shytov A V 2010 *Phys. Rev.* B **82** 075424

# Charge-transfer in some physical processes

**S Nešpůrek[1,2], J Nožár[1,3], D Rais[1], S Pochekaylov[1], J Šebera[1] and A Kochalska[1]**

[1] Institute of Macromolecular Chemistry, Academy of Sciences of the Czech Republic, Heyrovsky Sq. 2, 162 06 Prague 6, Czech Republic

[2] Faculty of Chemistry, Brno University of Technology, Purkyňova 118, 612 00 Brno, Czech Republic

[3] Charles University in Prague, Faculty of Mathematics and Physics, Ke Karlovu 3, 121 16 Prague 2, Czech Republic

E-mail: nespurek@imc.cas.cz

**Abstract**. The background of the intra- and inter-molecular electron transfer and some properties of charge transfer states (excitons) are mentioned. On the example of gas sensors the importance of the charge transfer in the dark is discussed. The utilization of the photoinduced charge transfer in electronic processes is demonstrated on the examples of photoconductivity, polymer photodegradation and molecular orientation.

## 1. Introduction

Dark and photoinduced electron transfer is an important process which influences many physical phenomena. The transfer of electron from a donor to acceptor is a quite obvious process observable even in dark. As a result is the formation of charge transfer complex with a characteristic optical absorption in (very often) visible spectral region. Acceptor-gas sensors work on this principle. In the optically excited state, there is a much stronger charge transfer, caused by the transfer of an electron from the highest occupied molecular orbital of donor to the lowest unoccupied molecular orbital of acceptor. As a result a charge-transfer exciton (ion-pair of the positive and negative species) is formed. It concerns an excited, neutral and polar state which influences many electronic processes, among others photoconductivity, polymer photodegradation and molecular orientation. The paper summarizes some facts included with the invited lecture at the 16$^{th}$ International School on Condensed Matter Physics: Progress in Solid State and Molecular Electronics, Ionics and Photonics.

## 2. Intramolecular charge transfer (CT)

We will demonstrate the electron transfer on the $H_2CO$ model of the benzaldehyde molecule (**1**, scheme 1). Each bond corresponds to a pair of bonding/antibonding orbitals, single bond being filled $\sigma$ orbitals and the double bond consisting of a filled $\sigma$ orbital and a filled $\pi$ orbital of higher energy. The lone pairs of the oxygen atom are filled non-bonding orbitals. The set of energy states is shown in figure 1 [1]. The n-$\pi$* excitation removes on electron from the O atom, this electron is transfered into the $\pi$* orbital localized half on the C atom and half on the O atom. When orbitals are localized in different zones of the molecule, the transition implies a charge transfer (CT). It is usually accompanied by a change in dipole moment, so long as the molecule has no centre of symmetry. The molecule usually has two parts which can be distinguished as "electron donor" and "electron acceptor".

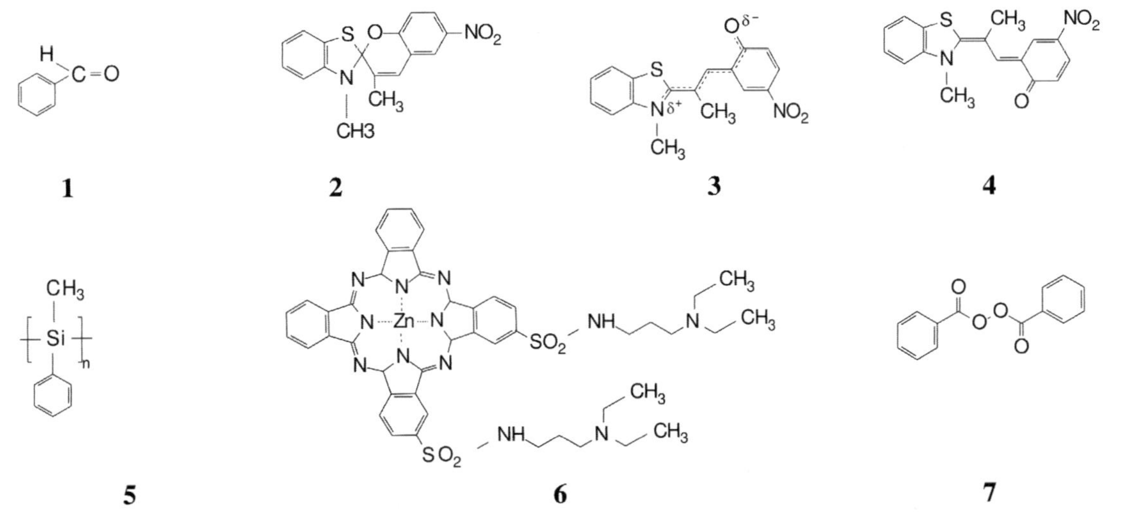


**Scheme 1.** Chemical structures of materials mentioned in this paper:
**1**: Benzaldehyde, **2**: 3',3'-dimethyl-6-nitro-2,3-dihydrospiro[benzothiazole-2,2'-[2*H*]chromene, **3**: 1,3,3-trimethyl-2[2-(5-nitro-2-oxidophenyl)vinyl]-3*H*-indol-1-ium, **4**: CTT conformer of **3**, **5**: Poly[methyl(phenyl)silylene, **6**: 3-diethylamino-1-propylsulphonamide substituted Zn-phthalocyanine, **7**: Dibenzoyl peroxide.

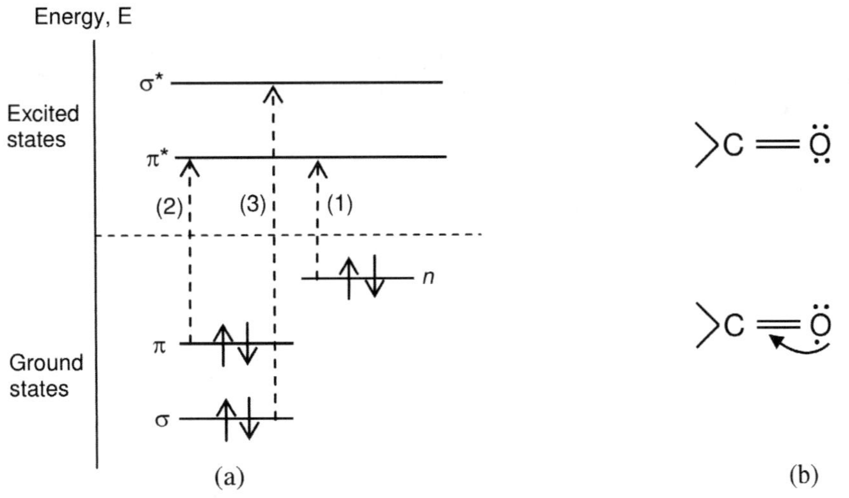


**Figure 1.** (a) Electronic excitations in benzaldehyde molecule. Transitions: (1) - $n \rightarrow \pi^*$, (2) - $\pi \rightarrow \pi^*$, (3) - $\sigma \rightarrow \sigma^*$. (b) Charge transfer during the $n \rightarrow \pi^*$ excitation.

*2.1. Charge transfer in photochromic compounds*
The process mentioned above is often observable in photochromic molecules. The photochromic process can be schematically presented as a photoreversible reaction which, in its simplest form, can be written as:

$$X \underset{h\nu_2,\ \text{heat}}{\overset{h\nu_1}{\rightleftarrows}} Y$$

In general, $h\upsilon_1$, the activating radiation (associated with electromagnetic energies from the ultraviolet and visible regions), causes a molecule, X, (with an absorption maximum at a wavelength $\lambda_1$) to convert to a product, Y (absorbing at $\lambda_2$). The reaction is not necessarily complete; the concentrations of X and Y depend on the thermodynamics of the system and kinetics of processes taking place under given conditions. As a typical example can be mentioned the molecule of 3',3'-dimethyl-6-nitro-2,3-dihydrospiro[benzothiazole-2,2'-[2*H*]chromene (**2**, scheme 1) which undergoes a reversible photochemically driven reaction, resulting in production of highly polar merocyanine, 1,3,3-trimethyl-2[2-(5-nitro-2-oxidophenyl)vinyl]-3*H*-indol-1-ium (**3**, scheme 1) [2,3]. The overall mechanism responsible for the photochromism may be described as a heterolytic bond cleavage and ring opening, but there are also cis-trans bond conversions and triplet-triplet contributions to the mechanism. In the mentioned example several conformers with different dipole moments can be formed (e.g., structure **4**). The dipole moments are: the basic compound **2** – 5.9 D, the conformer **3** – 12.8 D, **4** – 12.0 D [4].

### *2.2. Charge transfer during polaron transport*

This effect will be demonstrated on a semiconducting polymer, poly[methyl(phenyl)silylene] (PMPSi, **7**, scheme 1). Polysilanes with their uninterrupted chains of silicon atoms and with significant electron delocalization along the polymer chain [5] are interesting photosensitive materials. They also possess promising photoconductive properties [6-8]. PMPSi chain, with methyl and phenyl groups on each silicon atom, behaves as molecular wire with weak intermolecular interactions. Slowly moving charge carrier induces a deformation of PMPSi soft molecule; atoms in charged molecule (or its part) are shifted to new positions to preserve minimal potential energy. The new conformation and charge redistribution result in the change of frequencies of vibration modes. All these variances move along with the charge carrier (together they form a so-called polaron quasiparticle) and create a potential well around it. This self-trapping process results in the decrease of charge carrier mobility. Thus, the charge carrier is localized on a monomer unit (or on several of them) and further transport needs a charge transfer to the neighbor monomer unit (or to the chain consisting of several of them). Similar situation arises if two chain sequences are separated by some irregularity, e.g., all-trans chain conformation or by chain cis-conformation by branching points. In this case CT plays also an important role.

On molecular level, three factors, the electronic coupling (electron transfer integral) $t$ between the individual parts of the molecule, the reorganization energy $\lambda$ during the charge transport, and the effective length of the electron transfer $L$, are important for the charge movement. It is assumed that charge carrier localized on a molecular segment and its transport – hopping can be described by Marcus theory [9,10]. The rate of the transfer can be then expressed as

$$k = \frac{4\pi^2}{h}\frac{1}{\sqrt{4\pi\beta kT}}t^2\exp\left(-\frac{\beta}{4kT}\right) \qquad (1)$$

where $h$ is the Planck constant, $t$ is the transfer integral that reflects the strength of electronic interactions between two adjacent species, $T$ is the temperature, $k$ is the Boltzmann constant, and $\beta$ is the total reorganization energy. The reorganization energy $\beta$ consists of two terms $\beta_1$ and $\beta_2$ (equations 2-4).

$$\beta = \beta_1 + \beta_2 \qquad (2)$$

$$\beta_1 = E_+(Q_N) - E_+(Q_+) \qquad (3)$$

$$\beta_2 = E_+(Q_+) - E_+(Q_N). \tag{4}$$

Here, $E_+(Q_+)$ is the total electronic energy of the charged state with the geometry of a neutral molecule (segment), $E_+(Q_+)$ is the total energy of the charged state with the geometry of a charged molecule, $E_N(Q_+)$ is the total energy of the neutral state with the geometry of a charged molecule (segment), and $E_N(Q_N)$ is the total energy of the neutral state with the geometry of a neutral molecule (segment). β is frequently called as reorganization energy. $\beta_2$ as polaron binding energy. Electron transfer integral of the charge $t$ can be calculated according to Ref. [11] as

$$t = \frac{E_{HOMO} - E_{HOMO-1}}{2} \tag{5}$$

i.e., as the energies spliting of HOMO and HOMO–1 levels which are assumed to be energy levels responsible for charge carrier (here positive charge, hole) transport. The meaning of the parameters $E_+(Q_N)$ and $E_+(Q_+)$ is the following:

At the beginning of the charge transfer process the charge carrier is localized on the molecule (molecular segment) – the localization energy is $\beta_2$. After the charge hop on another molecular segment its geometry starts to change and self-trapping process begins. The final localization energy is $\beta_2$. Simultaneously, the back conformation charge of the first molecular segment proceeds – the energy needed is $\beta_1$. Note, the both energies, $\beta_1$ and $\beta_2$, are comparable.

*2.3. Photoinduced charge transfer in PMPSi and molecular electret network formation*

It has been shown that the first long-wavelength absorption band in PMPSi with the maximum at $\lambda_{max}$ = 325 nm is formed mainly by the delocalized $\sigma - \sigma^*$ transitions in the Si backbone. The dipole moment of this transition is directed along the polymer backbone. The $\pi - \pi^*$ aryl-like excitations are responsible for the short wavelength absorption with the maximum at 276 nm ($\alpha = 5.2 \times 10^4$ cm$^{-1}$).

During the ($\sigma$–$\sigma^*$) excitations, an electron is promoted from the bonding to an antibonding orbital. This process is followed by electron transfer along the polymer chain. Thus, an electron-hole pair is formed. However, the pairs generated in the same chain segment very often recombine geminately with a very fast decay rate, due to the electron delocalization on the Si backbone. An electron transfer to phenyl group stabilizes the pair. The intrachain electron transfer is possible as it follows from electrochemical experiments [12,13]. Thus, quite strong dipole of the type Si backbone-phenyl is formed (see figure 2). The exposure with the linearly polarized light results in the angularly dependent excitation of the main chain segments which are oriented parallel to the electric field vector of the exciting light. Figure 3 shows the polar diagram of the absorbance at 333 nm.

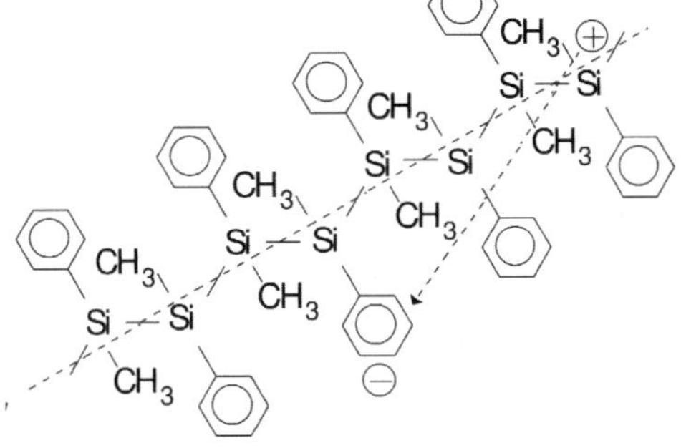

**Figure 2.** Schematic representation of the formation of an induced dipole in PMPSi chain.

This dipole is quite stable because of the formation of metastable electronic states. It was mentioned above that during the polaron formation [14] (and similarly during the energy transport [15] the deformation of the silicon chain occur. Positive polaron is characterized by Si–Si bond stretching; at the same time, the Si–Si–Si bond angles damp out. The stretching of the bonds and the shrinking of the chain conformation leads to the formation of metastable [16] electronic states in the HOMO-LUMO gap (band-edge states) with quite long lifetime.

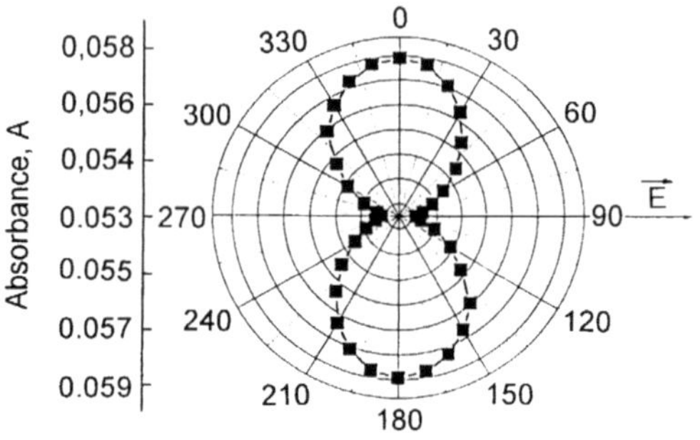


**Figure 3.** Polar diagram of the absorbance at 333 nm after the irradiation with linearly polarized light of 333 nm (HBO 100 W, 7 min). Electric field vector visualizes the electric field vector of the incident light..

Due to the long lifetime of ion-pairs, the detection of dipoles by means of liquid crystal alignment experiments was possible. The details of the experiment can be found in Ref. [17]. The measurements of the photoinduced dichroism in the PMPSi films and the azimuthal anchoring energy of the liquid crystal deposited on these films revealed the absence of correlation in their exposure-dose kinetics. The azimuthal anchoring energy of the liquid crystal reached its maximum just at the beginning of the irradiation when the dichroism was still negligible (cf. figure 4). On further irradiation the dichroism increased but the azimuthal anchoring energy of liquid crystal dropped to zero. Because the liquid crystal – polymer surface interactions include dispersive (van der Waals), polar, and steric terms we ascribe the result shown in figure 5 to the polar species and steric effects [18].

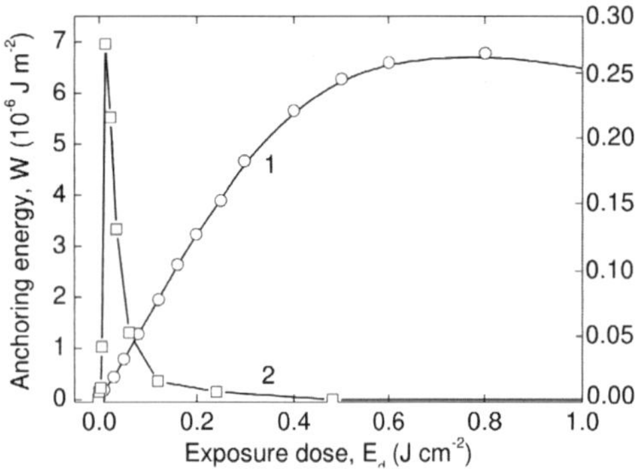


**Figure 4.** Changes of azimuthal anchoring energy of liquid crystal (Merck, LC 2 LT-4801) (1) and dichroism calculated at 333 nm (2) on irradiation of PMPSi film with linearly polarized light of He-Cd laser ($\lambda$ = 325 nm, P = 10 mVcm$^{-2}$).

The "nano-electret" network mentioned above can be used for the fabrication of polarized light emitting diodes. If on the light-treated surface an electroluminescent material is deposited molecules trend to be oriented due to the interactions with the dipoles and polarized elkectroluminescence is observed [19].

## 3. Intermolecular charge transfer

In case of the crystal which consists from molecules of D type, under the excitation the following reaction can take place:

$$D + D^* \rightarrow D^{+\delta} + D^{-\delta}$$

Here, D is the molecule in the ground state, D* is the molecule in the excited state and δ is the amount of charge transfered from one to another molecule. As the result an ionic-pair or charge-transfer exciton is formed. In organic materials the term charge-transfer exciton is to denote an excited, neutral and polar state, where the exciton radius is only one or two times the nearest-neighbor intermolecular distance [20].

The charge-transfer (CT) energy $E_{CT}$, as dependent on distance $r_{CT}$ molecules D and D*, may be described by a Coulomb-type law [21]

$$E_{CT}(r_{CT}) = E_{g-opt} - \frac{e^2}{4\pi\varepsilon\varepsilon_0 r_{CT}} \tag{6}$$

where $E_{g\text{-}opt}$ is the optical energy gap (separation energy between nonrelaxed negative and positive polaron levels), $e$ is the unit charge and $\varepsilon\varepsilon_0$ is the electric permittivity. Note, that conduction levels of relaxed states of positive and negative molecular polarons are separated by the adiabatic energy $E_{g\text{-}ad}$. Expression (6) shows that $E_{CT}$ ($r_{CT}$) dependence has a convergence limit $r_{CT} \to \infty$, which is determined by the width of optical energy gap. The situation is demonstration in figure 5. Charge-transfer states are formed after the autoionization and thermalization of excited molecule. The dissociation of charge-transfer (ion-pair) state leads to the free charge carrier generation.

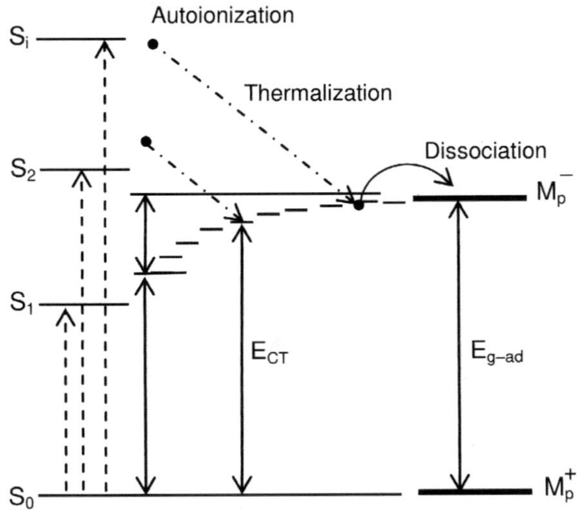


**Figure 5.** Schematic energy diagram in electronic states of molecular material. $S_0$ is the neutral ground singlet state, $S_i$ are excited singlet states. $E_{CT}$ is the energy of charge-transfer state $U_{CT}$ is the Coulombic energy of CT state.

The presence of acceptor (A) molecules supports electron transfer. The transfer can be realized even in the dark and charge transfer complex can be formed. In this case the relation (6) is substituted by the relation (8)

$$D + A \to D^{+\delta} + A^{-\delta}.$$

As a result of this reaction is the formation of charge-transfer complex (δ < 1) or paramagnetic radical salt (δ = 1). The transition energy is usually found in the visible spectral region.

In the optically excited state, there is a much stronger CT interaction, caused by the transfer of an electron from the highest occupied molecular orbital of donor D to the lowest unoccupied molecular orbital of acceptor A. The energy of this transition depends on the ionization energy of D, the electron affinity of A, and the Coulombic and polarization energy gained by creating two oppositely charged ions in a polarizable material. Since optical transitions take place in times of $\leq 10^{-15}$ s, the contributions of molecular rearrangement are not great; the energy of the transition may be written as

$$h\nu_{CT} = I_D - E_A - C - P \quad (7)$$

where $C$ is the Coulombic energy of the two charges, $P$ is the polarization energy of the CT dipole, $I_D$ is the ionization potential of the donor, and $A_C$ is the electron affinity of the acceptor. Very often the transition energy is found in the visible spectral region.

The photoinduced electron transfer from a donor to an acceptor is very important process in charge carrier photogeneration. Then, the photoconductivity is mainly influenced by an extrinsic process via a nonrelaxed exciplex formed between singlet excitons of the donor and acceptor-like dopands or photooxidation products A. The following mechanism is proposed:

$$D^* + A \rightarrow (D^* \cdots A) \rightarrow (D^+ \cdots A^-)^* \xrightarrow{1} D^+ \overset{r_{CT}}{\cdots} A^- \xrightarrow{2} D^+ + A^-$$

Here, $(D^* \cdots A)$ represents an encounter complex, $(D^+ \cdots A^-)^*$ a nonrelaxed exciplex state, $(D^+ \overset{r_{CT}}{\cdots} A^-)$ the CT state and $D^+ + A^-$ free charge carriers. The process $\xrightarrow{1}$ is a thermalization, the process $\xrightarrow{2}$ represents free charge carrier generation via electric field induced dissociation CT state. The electric field dependence of the charge photogeneration efficiency is strong and can be described by the Onsager theory of geminate recombination of ion pairs [22]. Measurements performed with PMPSi and pyrromellitic dianhydride acceptor indicate that both the lifetime and concentration of the ion-pairs increase in the presence of acceptor.

Note, that in polymers pendant groups can act as acceptor centres. In this case the electron transfer from the backbone to the pendant side group leads to the formation of intramolecular CT state. From the point of energy structure, these CT states are localized below the conduction level.

### *3.1. Acceptor-gas sensors*

The change of the absorption spectra of CT complexes can be utilized in the construction of sensors for acceptor-like gases. Here, also CT process between the sensing material (donor) and the detection medium (acceptor gas) plays an important role. As an example we will present $NO_2$ sensor based on 3-diethylamino-1-propylsulphonamide substituted Zn-phthalocyanine (ZnPcSu, see scheme 1, **6**).

Capobianchi et al. [23,24] proposed two-step mechanism of $NO_2$ interaction with phthalocyanines (Pcs). First step consists of transferring one electron from Pc to $NO_2$. During the second step, another electron is transferred from Pc to the gas molecule, thus forming bication of Pc complexed with two $NO_2$ molecules:

$$MePc + NO_2^{\cdot} \leftrightarrow (MePc)^{+} \cdots NO_2^{-}$$

$$(MePc)^{+\cdot} \cdots NO_2^{\cdot} \leftrightarrow (MePc)^{2+} \cdots 2NO_2^{-}$$

The equilibrium of the first step is shifted in the direction of reaction products. The second step reaction is equilibrated closer to the initial reagents. This explains the partial reversibility of $NO_2$ sensors reported in literature.

Absorption spectra of Pc consist of two main bands: Q-band at 600 – 700 nm and Soret band with the maximum at about 350 nm. It was found, that the spectrum of ZnPcSu is slightly changed in the presence of $NO_2$ – see figure 6. A new band at about 500 nm appears (decrease of the differential transmittance $\Delta T$). Intensity at the same time transmittance in the range of Q-band increases. From this result we suspect that CT complex is formed between ZnPcSu (donor) and $NO_2$ (acceptor). It supports the mechanism proposed by Capobianchi et al. Note, that a typical optical absorption of phthalocyanine CT complexes appears near 500 nm [25].

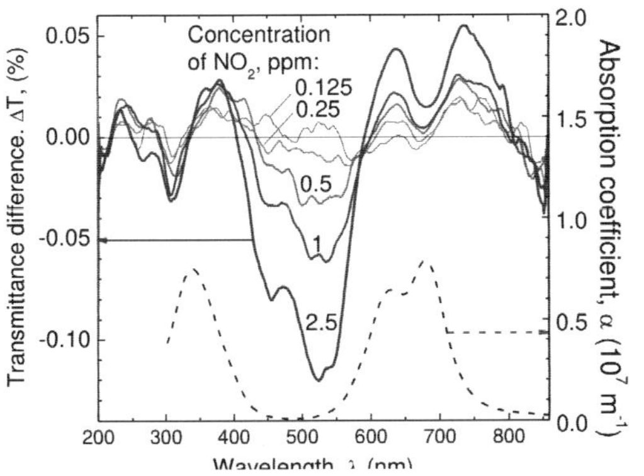


**Figure 6.** Differential optical transmittance spectra of ZnPcSu film exposed to $NO_2$. Optical absorption spectrum of ZnPcSu film (dashed line), is given for comparison [26].

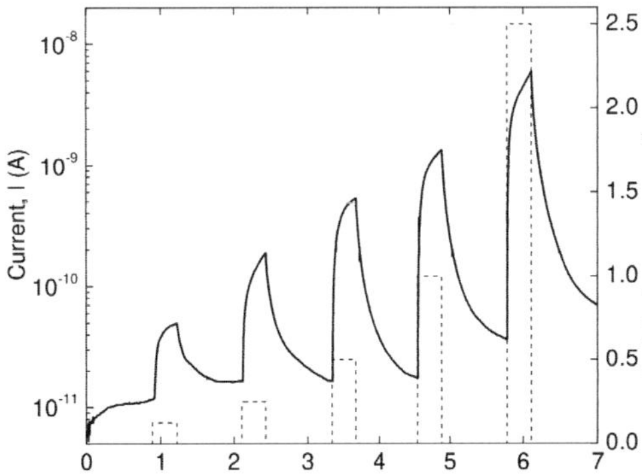


**Figure 7.** Kinetics of electrical current of ZnPcSu thin film during $NO_2$ exposures (full curve). $NO_2$ concentrations are indicated by dashed lines.

The charge-transfer complex formation usually increases free charge carrier concentration. Nitrogen dioxide causes increase the conductivity of ZnPcSu thin film as it is shown in figure 7. It also confirms the hypothesis of CT complex [26].

According to quantum chemical calculations the interaction of $NO_2$ with central Zn atom is very probable. Interactions of $NO_2$ with Pc conjugated skeleton and/or sulphonamide substituents are very weak if any. Oxygen from $NO_2$ forms the coordination bond Zn···O=N=O. Electron density of HOMO is shifted to $NO_2$, confirming the CT complex formation. Theoretical spectral are in agreemental observations.

### *3.2. Polymer photodegradation: The improve of UV degradation efficiency by means of CT process*

UV exposure of poly(organylsilanes) results in photobleaching and Si-Si bond scission, producing low-molecular-weight fragments. Thus, the interesting use of polysilanes is in area of photolithography due to the fact that polysilanes can be easily changed to materials having completely disparate properties when exposed to radiation. The contribution of polysilanes to large scale integrated circuit manufacturing technology is that silicon atoms included in the polymer chain are not recognized as impurities in silicon devices. Other metal atoms can cause contamination problems. For efficient photolithography process materials with high speed of photostimulated reactions are needed. Even though the photodegradation of polysilanes is quite fast, an enhancement is eligible. We will present a possibility concerning the increase of the photodegradation efficiency in PMPSi by electron acceptor additive utilizing the charge transfer process. As mentioned above the formation of intramolecular charge-transfer exciton chain-phenyl result in an 36 μs which follows from flash photolysis measurements) the electron can move back to the chain and recombine with delocalized hole (cation radical) which results in charge equilibrium and chain stabilization. If strong electron acceptor additive is present, the electron can be transferred to the additive and in the case of anion radical unstability the back electron transfer to chain is limited, quantum efficiency of photodegradation increases. For experiments, dibenzoyl peroxide (BP, **9**, scheme 1) was used.

The degradation of PMPSi chain under UV illumination is accompanied by two changes in absorption spectrum: Absorbance decreases and its maximum is shifted to higher energies. It means that bonds scission of polysilane chain takes place. Figure 8 shows the dependence of the maximum shifts on the UV illumination time for PMPSi (curve 1) and PMPSi+BP (curve 2).

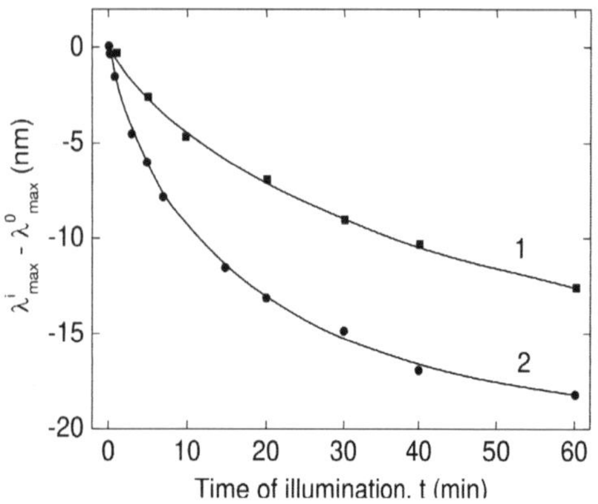


**Figure 8.** Photodegradation of PMPSi with various additives characterized by the wavelength shift of the band maximum. 1 – neat PMPSi, 2 – PMPSi + benzoyl peroxide.

Two possible reactions were tested to understand the interactions of PMPSi chain and the additive which is assumed to be in anion radical form.

(a) In the first case, the electron is localized on the additive molecule for a significant time. During the localization silicon bonds in the main chain is weakened and chain can undergo a bond scission. However, the electron relaxation back to polymer limits the photodegradation efficiency.

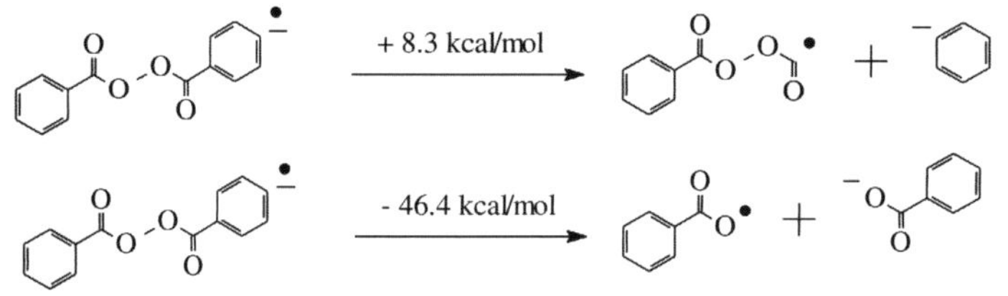


**Scheme 2.** Two ways of the spliting of dibenzoyl peroxide anion radical.

(b) The second case requires the presence of a non-stable bond in anion radical of the additive [27].

Thus, a radical and a anion (nucleophilic) group is expected to be formed (see scheme 2). The later can subsequently attack positive charged silicon chain. The additive molecule must satisfy the condition that its energy in a separate state (nucleophilic anion and radical) must be lower than the energy in a joint state (anion-radical).

## Conclusion

In some physical and chemical processes an electron transfer is important both in the dark and excited state. Especially the charge transfer between an optically excited donor and an acceptor can improve quantum efficiency of charge carrier photogeneration, polymer degradation and liquid crystal alignment. The knowledge of the charge transfer mechanism is also important to understand polaron transport and functionality of some gas sensors.

## Acknowledgements

The research was supported by the Grant Agency of the Academy of Sciences of the Czech Republic (grant No. KAN401770651).

## References

[1] Suppan P 1994 *Chemistry and Light* (Cambridge, The Royal Society of Chemistry)
[2] Tamci N and Miyasaka H 2000 *Chem. Rev.* **100** 1875
[3] Abe Y, Nakamo R, Horii T, Satoski O and Irie M 1996 *J. Photochem.Photobiol.* **A 95** 209

[4] Toman P, Bartkowiak W, Nešpůrek S, Sworakowski J and Zalesny R 2005 *Chem. Phys.* **316** 267
[5] Miller R D and Michl J 1989 *Chem. Rev.* **89** 1359
[6] Nešpůrek S 1999 *Mater. Sci. Eng. C* **8-9** 319
[7] Abkowitz M and Stolka M 1991 *J. Inorg. Organomet. Polym.* **1** 487
[8] Nešpůrek S and Eckhardt A 2001 *Polym. Adv. Technol.* **12** 427
[9] Marcus R A 1993 *Rev. Mod. Phys.* **65** 599
[10] Farazdel A, Dupuis M, Clementi E and Arivam A 1990 *J. Am. Chem. Soc.* **112** 4206
[11] McCrary V R, Sette F, Chen C T, Lovinger A J, Robin M B, Stohr J and Zeigler J M 1988 *J. Chem. Phys.* **88** 5925
[12] Pitt C G, Carey R N and Toren E C Jr 1972 *J. Am. Chem. Soc.* **94** 3806
[13] West R 1971 *Proc. Symp.Organosilicon, Pittsburgh*
[15] Mintmire J W 1989 *Phys. Rev.B* **39** 13350
[14] Arkhipov V I, Emelianova E V, Kadashchuk A, Blowsky I, Nešpůrek S, Weiss D S and Baessler H 2002 *Phys. Rev.B* **65** 165218
[16] Takeda K, Shiraishi K, Fujiki M, Kondo M and Marigaki K 1994 *Phys. Rev. B* **50** 5174
[17] Nešpůrek S, Zakrevsky Y, Stumpe J, Sapich B and Kadashchuk A 2006 *Macromolecules* **39** 690
[18] Yaroshchuk O and Kadashchuk A 2000 *A Appl. Surf. Sci.* **158** 357
[19] Nešpůrek S, Wang G, Rais D, Rakušan J, Kratochvílová I, Stumpe J and Zakrevskyy Y 2006 *Mol. Cryst. Liq. Cryst.* **448** 1
[20] Pope M, Swenberg Ch 1999 *Electronic Processes in Organic Crystals and Polymers* (New York: Oxford University Press)
[21] Sebastian L, Weiser G, Peter G and Baessler H 1983 *Chem. Phys.* **75** 103
[22] Onsager L 1938 *Phys. Rev.* **54** 554
[23] Baldini F, Capobianchi A, Falai A and Pennesi G 1998 *Sens. Actuators B* **51** 176
[24] Capobianchi A, Paoletti A M, Pennesi G and Rossi G 1998 *Sens. Actuators B* **48** 333
[25] Bortchagovsky E G, Kazantseva Z I, Koshets I A, Nešpůrek S and Jastrabik L 2004 *Thin Solid Films* **460** 269
[26] Lee Y L, Chang C H 2006 *Sens. Actuators B* **119** 174
[27] Kani R, Nakano Y, Yoshida H and Hayase S 1998 *Macromolecules* **31** 8794

# Measurement of dielectric properties at low temperatures: application to the study of magnetoresistive manganite/ insulating oxide bulk composites

**P Vanderbemden[1], B Rivas-Murias[1,2], V Lovchinov[3] and B Vertruyen[1]**

[1] University of Liège, SUPRATECS research group, Departments of Electrical Engineering & Computer science (B28) and Chemistry (B6), Sart-Tilman, B-4000 Liège, Belgium
[2] University of Santiago de Compostela, Departamento de Química-Física, 15.782 Santiago de Compostela, Spain
[3] Institute of Solid State Physics, Bulgarian Academy of Sciences, 72 Tzarigradsko Chaussee Blvd., 1784 Sofia, Bulgaria

E-mail : Philippe.Vanderbemden@ulg.ac.be

**Abstract**. In this paper, we report low temperature dielectric measurements of bulk composite electroceramic samples containing a colossal magnetoresistive (CMR) manganite phase ($La_{0.7}Ca_{0.3}MnO_3$ [abbreviated LCMO]) and an insulating phase ($Mn_3O_4$). Details of the experimental system are given and possible experimental artefacts due to moisture are outlined. For a LCMO volume fraction of ~ 16%, the permittivity of the LCMO/ $Mn_3O_4$ composite at $T = 50$ K is found to be much higher than that of pure $Mn_3O_4$ and magnetic field dependent. This effect is related to an extrinsic space charge polarization mechanism between the insulating phase ($Mn_3O_4$) and the conducting magnetoresistive phase (LCMO).

## 1. Introduction

In the past few years there has been a significant interest in functional materials which exhibit a coupling between ferroelectric and magnetic properties [1]. These materials had already been characterised in the 1960s [2], but the interest was renewed recently, due to the discovery of new compounds with remarkable properties [3-6] as well as the perspective of engineering applications in which the electric polarization can be tuned magnetically. The general term "magnetodielectric" can be used to describe materials whose permittivity is sensitive to an external magnetic field ("magnetocapacitance"), whereas "multiferroic" refers specifically to compounds where ferroelectric and ferromagnetic orders can coexist simultaneously [1]; the latter include e.g. several transition metal perovskites [7] or some rare-earth manganites [3-9]. In addition to such "intrinsic" magnetoelectric materials, it has also been shown that magnetocapacitance properties can be observed due to an "extrinsic" effect caused by the combination of magnetoresistance and Maxwell-Wagner effect at the electrode/ceramic interface or at grain boundaries, e.g. in composites [8]. Such interfacial magnetocapacitive effects are usually associated to a strong frequency dependence of the parameters of the parallel $R \parallel C$ equivalent circuit of the material [7,8]. Although such properties are not directly related to a "true" magnetoelectric coupling, they are nevertheless of definite technological interest.

In the present study, we report the dielectric properties of bulk composite samples containing a (conducting) Colossal Magneto Resistive (CMR) manganite phase [10-12] and an (insulating) oxide secondary phase. We focus on the $La_{0.7}Ca_{0.3}MnO_3$/ $Mn_3O_4$ composite system, where $La_{0.7}Ca_{0.3}MnO_3$ (LCMO) is the manganite phase and $Mn_3O_4$ the insulating phase. In our previous works, we have shown that such composite materials can be prepared with almost constant manganite compositions throughout the series of samples with different proportions of the two phases [13]. We have shown that the electrical properties of this system exhibit a clear percolation threshold for a volume fraction of the conducting phase ($f_{LCMO}$) of ~ 0.19, i.e. for $f_{LCMO} > 0.19$, the composite material is conducting (resistivity $\rho < 10^3$ Ω.cm) whereas for $f_{LCMO} < 0.19$, the resistivity exceeds $10^6$ Ω.cm. The fundamental reason for studying such composites is that, in the conducting regime ($f_{LCMO} > 0.19$), the presence of the insulating $Mn_3O_4$ phase forces the electric current to meander through the LCMO grains, thereby increasing significantly the colossal magnetoresistance effect at low magnetic fields [14,15]. Due to the fact that the insulating phase ($Mn_3O_4$) is ferrimagnetic for $T < 42$ K, these materials are characterised by unusual magnetoresistive properties at low temperature [16]. In addition to affecting the electrical properties, grain boundaries in these materials were also shown to affect the thermal conductivity of the composite [17]. In this paper, we concentrate on the insulating regime ($f_{LCMO} < 0.19$) and consider the low temperature dielectric properties of such composites and investigate possible magnetic field dependence of the permittivity. We first describe the experimental system used for characterising the dielectric properties at low temperature and then compare the behaviour of a LCMO/$Mn_3O_4$ composite sample to that of pure $Mn_3O_4$.

## 2. Experiment

The composite $La_{0.7}Ca_{0.3}MnO_3/Mn_3O_4$ materials were synthesized from precursor powders obtained by the spray drying technique [18]. This consists in spraying an aqueous solution of metallic cations into droplets, which are dried by a hot air flow. The detailed synthesis process has been described previously (see [13] and references therein). The samples were characterised by several techniques. Powder X-ray diffraction patterns were collected with a Siemens D5000 diffractometer (Cu $K_\alpha$ radiation). The morphology of the samples was studied by scanning electron microscopy using a Philips XL30 FEG-ESEM. The cationic composition was checked by an Energy Dispersive X-ray Analysis (EDAX system), coupled to the electron microscope. The density was measured using the Archimedes method in order to calculate the volume fraction of each phase, i.e. the LCMO phase, the $Mn_3O_4$ phase and the total (open + closed) porosity.

The dielectric properties were measured as a function of temperature and magnetic field in a home-made experimental set-up based on a Quantum Design PPMS (Physical Property Measurement System) [19-22]. Our experimental arrangement is shown schematically in figure 1. Either silver or gold electrodes are first deposited on both sides of the ceramic sample in order to realize a parallel plate capacitor. A circular guard ring is also deposited around the low voltage (V-) electrode and connected to earth. The sample holder is based on the sample puck of the PPMS (figure 1): the low voltage electrode is thermally anchored to the puck itself through silver paste and a small brass disk, but isolated electrically using a thin cigarette paper and 7031 GE varnish. A brass cap around the system, acting as electric shield, is connected to earth. In view of minimizing thermal conduction through the wires within the experimental chamber, no additional coaxial cable is used; connexions between the sample and the measuring instruments are achieved with the existing wires along the external face of the sample chamber of the PPMS. The sample chamber itself is connected to earth and parasitic capacitance between the three wires (shown in red, blue and green colours in figure 1) is minimized by choosing the wires that are positioned at approx. 120° from each other. The dielectric properties are measured using a Perkin Elmer 7260 Lock-In amplifier acting both as a voltage source and high sensitivity ammeter. For each measuring frequency (in the range 5 Hz – 250 kHz), phase correction is achieved by measuring a Solartron calibrated reference capacitor (11.92 pF).

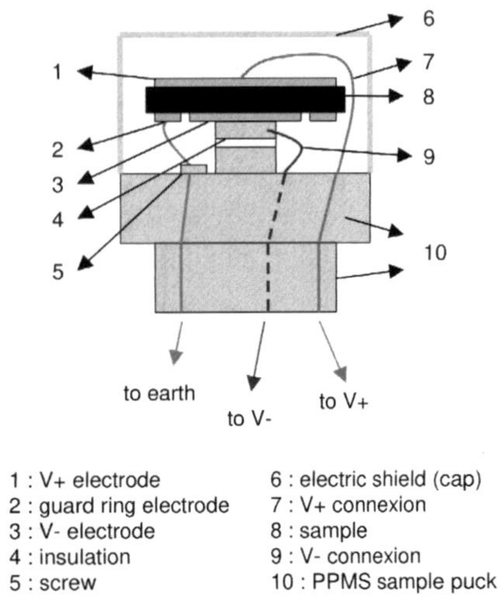


**Figure 1.** Schematic illustration of the experimental system for dielectric measurements

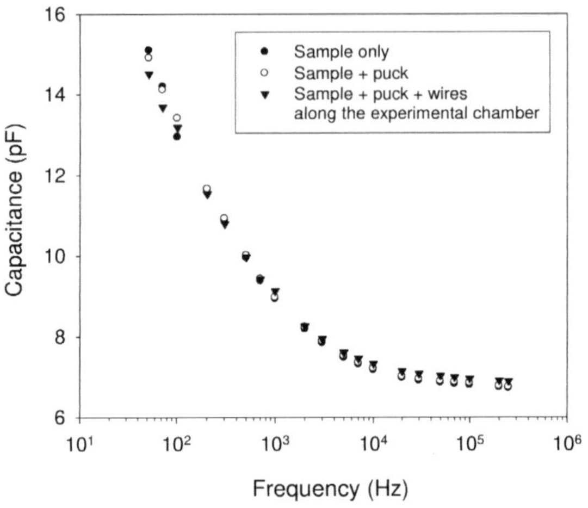


**Figure 2.** Room-temperature frequency dependence of the capacitance of a composite sample (LCMO / $Mn_3O_4$) having a LCMO volume fraction equal to ~ 0.16 for 3 different experimental arrangements.

## 3. Results and discussion

### *3.1. Experimental system*

Figure 2 shows the room-temperature frequency dependence of a composite sample (LCMO / $Mn_3O_4$) for which the dimensionless LCMO volume fraction is equal to 0.16. The volume fraction of the pores in this sample is evaluated to 0.24; the volume fraction of $Mn_3O_4$ is 0.60. Three different experimental arrangements are used. The first is the system described in the paragraph above, where the sample is placed on the PPMS puck, inserted in the sample chamber. In order to evaluate possible sources of parasitic capacitances, the sample was also measured (i) placed on the PPMS puck, but inserted in an electric shield or (ii) removed from the puck holder ("sample only" in figure 2) and placed in an electric shield. As can be seen in figure 2, the capacitance measured by the three methods leads to results that are close to each other; the agreement is found to be better than 4% at 10 Hz and better than 1% at 10 kHz. The parasitic capacitance between the connecting wires along the sample chamber was evaluated to 0.3 pF, almost frequency independent. The result shown in figure 2 shows that the simple experimental system described in figure 1 is perfectly appropriate for measuring the dielectric properties of the composite ceramic samples.

In the experimental chamber, the temperature of the sample can be varied using the temperature control system of the PPMS. The temperature dependence of parameters $R$ and $C$ of the parallel equivalent circuit of the material are shown in figure 3. If no particular precaution is taken (blue triangle symbols) the $R(T)$ and $C(T)$ curves are found to exhibit a clear anomaly at $T \sim 238$ K: a significant jump is observed in the $C(T)$ plot whereas the resistance displays a sharp (negative) peak. The particular temperature at which the anomaly appears is not reproducible and is found to vary within a ~ 40 K window from one experimental run to another. These features are related to the presence of moisture in the pores of the ceramic sample and / or in the experimental chamber.

After 72 h of conditioning the samples in dry atmosphere and turbomolecular pumping the sample chamber for 24 h, both the jump in $C(T)$ and the peak in $R(T)$ disappear (white symbols in figure 3). These peaks are related to the change of dielectric properties of water at solid-vapor or solid-liquid

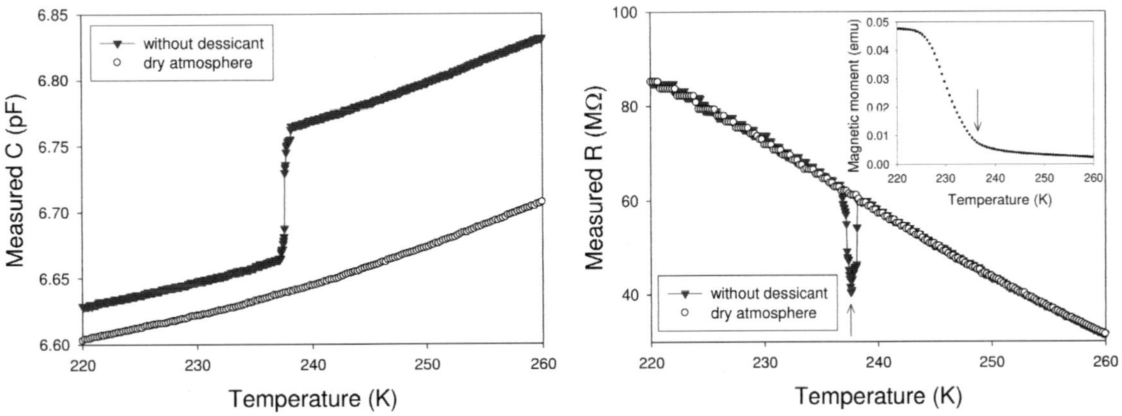


**Figure 3.** Measured temperature dependence of the capacitance $C$ and parallel resistance $R$ of a composite sample (LCMO / $Mn_3O_4$) with a LCMO volume fraction equal to ~ 0.16. The data shows that a significant experimental artefact can be observed due to moisture in the pores of the sample and / or in the experimental chamber. The inset of in the left graph shows the temperature dependence of the DC magnetic moment of the same material measured under a 0.1 T applied magnetic induction.

phase transitions [23], arising at temperatures that are strongly pressure dependent. The data shown in figure 3 give thus experimental evidence that spurious experimental artefacts can be observed in measuring the dielectric properties of bulk porous ceramics that have not been kept in dry atmosphere. It can be noted that the jump and peaks observed in figure 3 might be particularly misleading since they occur at a temperature (indicated by an arrow) close to the ferromagnetic – paramagnetic transition temperature of the LCMO phase in our composite samples (cf. inset in figure 3).

### *3.2. Dielectric properties of composite samples*

At T = 300 K, the capacitance of a bulk composite sample (LCMO / $Mn_3O_4$) with $f_{LCMO} = 0.16$ is a decreasing function of frequency, as shown in figure 2. At 1 kHz, the absolute value of in-phase permittivity of the bulk composite is found to be approximately 2.7 times that of bulk $Mn_3O_4$. The presence of interfaces with different densities of charge carriers, e.g. at dielectric – electrode interfaces or at grain boundaries can be responsible for the behaviour shown in figure 2. In the particular case of the composite sample, the manganite phase produces an extrinsic interfacial or space-charge polarization mechanism, resulting in a noticeable increase in the real part of the permittivity with respect to that of $Mn_3O_4$ [24-26].

Figure 4 shows the magnetic field dependence of the in-phase permittivity of the composite sample at $T = 50$ K, for a magnetic field ramped from 0 to 4 teslas and then cycled between two symmetric values. The data for the composite material (red curve) are compared to the behaviour of $Mn_3O_4$ (black curve). As can be seen, the LCMO / $Mn_3O_4$ sample displays a perceptible magnetocapacitance (relative variation ~ 0.4 % in the magnetic field range investigated) and the $C(H)$ plot displays a hysteretic behaviour. Conversely, the permittivity of the $Mn_3O_4$ sample at 50 K is almost magnetic field independent. More precisely the in-phase permittivity decreases slightly with the applied field but the relative variation of permittivity is smaller than 0.02 % in the magnetic field range investigated.

The behaviour of $Mn_3O_4$ depicted in figure 4 is in agreement with recent literature data putting into evidence a magnetodielectric coupling associated with specific magnetic structures and understood with a spin-phonon coupling model [27,28]. These intrinsic magnetodielectric effects are linked to complex magnetic phase transitions arising at $T$ =34, 40 and 42 K but the permittivity decreases very slightly with magnetic field in the paramagnetic phase ($T > 42$ K). By contrast, the permittivity of our

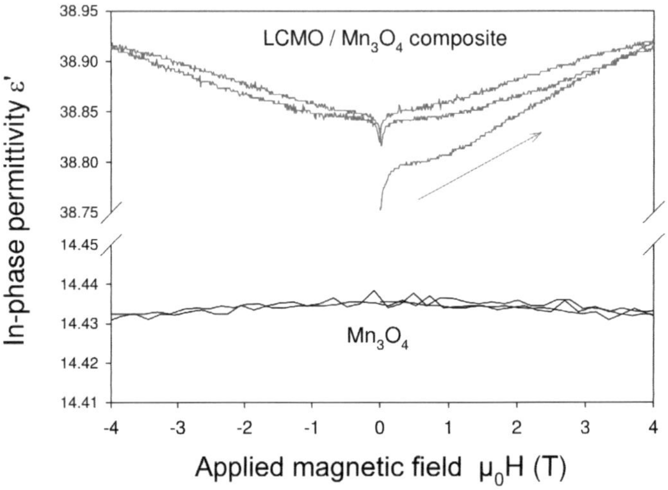


**Figure 4.** Comparison of the magnetic field dependence of the in-phase permittivity measured at T = 50 K for pure $Mn_3O_4$ and for a composite (LCMO / $Mn_3O_4$) ceramic.

LCMO / $Mn_3O_4$ composite is found to increase with magnetic field. Combined with the pronounced frequency dependence (figure 2), this magnetodielectric effect is linked to the heterogeneous nature of the composite and can be understood through a combination of the magnetoresistive phase (LCMO) and the Maxwell-Wagner effect at interfaces [8]. As can be seen in figure 4, the magnetodielectric effect is more pronounced around zero. The intrinsic low-field magnetoresistance of the conducting LCMO phase [15] below its transition temperature (~230 K, cf. inset in figure 3) is likely to be responsible for this effect. Similarly, the hysteretic behaviour shown in figure 4 is reminiscent of the hysteretic resistance vs. magnetic field properties of the composite in the vicinity of the percolation threshold ($f_{LCMO} \sim 0.19$).

## 4. Conclusions

We have measured the dielectric properties of a bulk manganite/insulator polycrystalline composite (LCMO / $Mn_3O_4$) at relatively low manganite content (16%), so that the whole material is insulating. We have described the experimental set-up for measurements at low temperature and emphasized artefacts that can be observed if the bulk ceramic is not kept in dry atmosphere. The permittivity of the composite is found to be frequency dependent and much higher than that of pure $Mn_3O_4$. A clear magnetic field dependence of the permittivity is observed at $T = 50$ K. At this temperature, intrinsic magnetodielectric effects of $Mn_3O_4$ are very small. The behaviour observed in the composite is therefore ascribed to Maxwell-Wagner effects arising at the interfaces between the insulating and the conducting magnetoresistive phase.

**Acknowledgments**
The authors would like to thank J.-F. Fagnard for carrying out careful electrical measurements and M. Ausloos for fruitful scientific discussions. Electron microscopy was performed in the Catμ microscopy centre (ULg, Belgium). We thank ULg for research grants, under references SFRC 09-47 and D09-09. This work is part of a collaboration programme financed by Wallonie Bruxelles International (WBI, Belgium) and the Bulgarian Academy of Sciences (BAS, Bulgaria).

## References

[1] Khomskii D I 2006 *J. Magn. Magn. Mater.* **306** 1
[2] Smolenskii G A et al 1971 *Segnetoelectrics and antisegnetoelectrics* (in Russian) (Leningrad: Nauka)
[3] Kimura T, Goto T, Shintani H, Ishikaza K, Arima T and Tokura Y, 2003 *Nature* **426** 55
[4] Lawes G et al 2005 *Phys. Rev. Lett.* **95** 087205
[5] Kimura T, Lawes G, Goto T, Tokura Y and Ramirez A P 2005 *Phys. Rev. B* **71** 224425
[6] Hur N, Park S, Sharma P A, Guha S and Cheong S W 2004 *Phys. Rev. Lett.* **93** 107207
[7] Lawes G, Kimura T, Varma C M, Subramanian M A, Rogado N, Cava R J and Ramirez A P 2009 *Prog. Solid State Ch.* **37** 40
[8] Catalan G 2006 *Appl. Phys. Lett.* **88** 102902
[9] Radulov I, Nizhankovskii V I, Lovchinov V, Dimitrov D and Apostolov V 2006 *Eur. Phys. J. B* **52** 361
[10] Siwach P K, Singh H K and Srivastava O 2008 *J. Phys.: Condens. Matter* **20** 273201
[11] Coey J M D, Viret M, and von Molnar S 1999 *Adv. Phys.* **48** 167
[12] Ziese M 2002 *Rep. Prog. Phys.* **65** 143
[13] Vertruyen B, Cloots R, Ausloos M, Fagnard J F and Vanderbemden P 2007 *Phys. Rev. B* **75** 165112
[14] Mathur N D, Burnell G, Isaac S P, Jackson T J, Teo B S, MacManus-Driscoll J L, Cohen L F, Evetts J E and Blamire M G 1997 *Nature* **387** 266
[15] Vertruyen B, Cloots R, Rulmont A, Dhalenne G Ausloos M and Vanderbemden P 2001 *J. Appl. Phys.* **90** 5692
[16] Vertruyen B, Cloots R, Ausloos M, Fagnard J F and Vanderbemden P 2007 *Appl. Phys. Lett.* **91** 062514
[17] Mucha J, Vertruyen B, Misiorek H, Ausloos M, Durczewski K and Vanderbemden P 2009 *J. Appl. Phys.* **105** 063501
[18] Vertruyen B, Rulmont A, Cloots R, Fagnard J F, Ausloos M, Vandriessche I and Hoste S 2005 *J. Mater. Sci.* **40** 117
[19] Bellido N, Martin C, Simon C and Maignan A 2007 *J. Phys.: Condens. Matter* **19** 056001
[20] Ray S K, Buroker L, Williamsen M S, Zou Y, Sen S and Guptasarma P 2010 *Mater. Res. Soc. Symp. Proc.* **1199** F06-28
[21] Kundys B, Martin C and Simon C 2009 *Phys. Rev. B* **80** 172103
[22] Muralidharan R, Jang T H, Yang C H, Jeong Y H, and Koo T Y 2007 *Appl. Phys. Lett.* **90** 012506
[23] Hasted J B 1961 *Progress in Dielectrics* (London: Heywood and Company)
[24] von Hippel A 1954 *Dielectrics and Waves* (London: Artech House)
[25] Rivas J, Mira J, Rivas-Murias B, Fondado A, Dec J, Kleeman W and Señaris-Rodríguez M A 2006 *Appl. Phys. Lett.* **88** 242906
[26] Freitas R S, Mitchell J F and Schiffer P 2005 *Phys. Rev.* B **72** 144429
[27] Tackett R, Lawes G, Melot B C, Grossman M, Toberer E S and Seshadri R 2007 *Phys. Rev.* B **76** 024409
[28] Suzuki T, Adachi K and Katsufuji T 2006 *J. Phys.:Conf. Ser.* **31** 235

# The state-of-the-art and future development of the photovoltaic technologies – the route from crystalline to nanostructured and new emerging materials

**D Dimova-Malinovska**
Central Laboratory of Solar Energy and New Energy Sources, Bulgarian Academy of Sciences, 72 Tzarigradsko Chaussee, 1784 Sofia, Bulgaria

E-mail address:doriana@phys.bas.bg

**Abstract**. Photovoltaic technology has received raising attention as one of the most promising approach to environment friendly energy production. The main efforts in the PV technologies are focussed on increasing of the efficiency and decreasing of the price of the solar cells and modules. Solar cells made applying nanostructured and new emerging materials are being investigated to address this challenge. In this paper an overview of the present state of knowledge of material aspects of photovoltaic cells is given, and new materials, including nanomaterials, with the potential for using in photovoltaic devices are presented.

## 1. Introduction

Electricity from the Sun (Photovoltaics - PV) is considered as one of the most prospective and clear source of energy. Because of universal availability of solar energy and immense potential of direct conversion of solar radiation into electricity to use, it was predicted that photovoltaic will become a very serious alternative for fossil fuels in the nearest future. PV-generated electricity is totally environmental-friendly. As mankind will probably have to face energy crisis, factors such as energy independence, energy security, stability of energy supply and variety of energy sources become much more vital these days.

Different semiconducting inorganic and organic materials are used to prepare solar cells and modules. In this work, an overview of the present state of knowledge of the materials aspects of photovoltaic cells will be given, and new materials, namely nanomaterials, with the potential for application in photovoltaic devices will be introduced.

## 2. World's PV market

From the first space applications to the GW planned systems, more than 50 years have passed. The last decade has seen PV technology emerging as a potentially major technology for power generation in the World (figure 1). By the end of 2008, the World cumulative PV power installed was approaching 16 GW and today, almost 23 GW are installed globally which produce about 25 TWh of electricity on a yearly basis. Europe is leading the way with almost 16 GW of installed capacity in 2009, representing about 70% of the World cumulative PV power installed at the end of 2009 while Japan (2.6 GW) and the USA(1.6 GW) are following behind. The 2010 World PV market could reach up between 8.2 and 12.7 GW of new installations under the Moderate scenario and the Policy-Driven scenario, respectively [1].In a policy Driven scenarios, the World annual PV market could reach up to

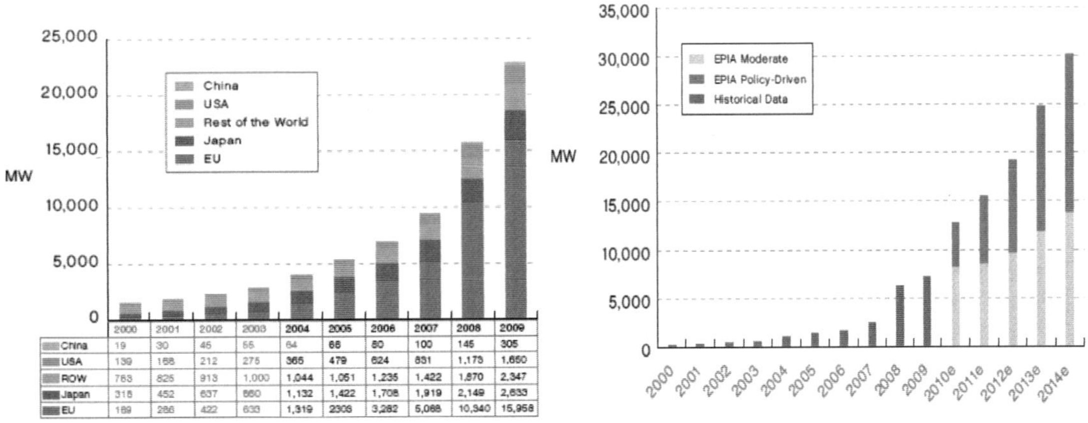


| | 2000 | 2001 | 2002 | 2003 | 2004 | 2005 | 2006 | 2007 | 2008 | 2009 |
|---|---|---|---|---|---|---|---|---|---|---|
| China | 19 | 30 | 45 | 55 | 64 | 68 | 80 | 100 | 145 | 305 |
| USA | 139 | 168 | 212 | 275 | 365 | 479 | 624 | 831 | 1,173 | 1,650 |
| ROW | 763 | 825 | 913 | 1,000 | 1,044 | 1,051 | 1,235 | 1,422 | 1,870 | 2,347 |
| Japan | 318 | 452 | 637 | 860 | 1,132 | 1,422 | 1,708 | 1,919 | 2,149 | 2,633 |
| EU | 189 | 286 | 422 | 633 | 1,319 | 2306 | 3,282 | 5,068 | 10,340 | 15,958 |

**Figure 1.** Historical development of World cumulative PV power installed in main geographies till 2009 (left) and the market forecasts until 2014 according different scenarios (right) [1].

30 GW in 2014 based, of course, on favorable conditions established by policy makers, regulators and the energy sector at large. The announced world-wide PV production capacity would also be sufficient to cover the expected evolution of the market in the coming five years. The PV market forecasts demonstrate a bright future for the PV market and industry in the coming 5 years.

## 3. Photovoltaic materials - state-of the-art

Solar cells are in fact large area semiconductor diodes (figure 2). Due to photovoltaic effect energy of light (energy of photons) converts into electrical current. At p-n junction, an electric field is built up which leads to the separation of the charge carriers (electrons and holes). At incidence of photon stream onto semiconductor material the electrons and holes are generated, if the energy of photons is sufficient. Contact to a solar cell is realized by metal electrodes. If the circuit is closed, meaning an electrical load is connected, and then direct current flows.

A number of properties are required for candidate PV materials and device structures. The most essential ones concern optical and electrical conditions:

i) Strong light absorption over a large spectral range. This property implies that a tunable band gap is desirable. The peak of absorption should be at 1.4–1.5 eV, for optimal efficiency.

ii) Good carrier collection properties for both - minority and majority carriers, a low carrier recombination loss (in the bulk, at grain boundaries and at the front and back surfaces).

iii) Low cost, so that the thin film structures are preferable.

iv) Stability as functions of both time and illumination conditions (stable active materials, stable metal contacts, resistance to corrosion);

v) High abundance of the source materials (for large-scale production);

vi) Environment friendly technology;

PV technologies presently fall into three categories:

i) **Crystalline silicon flat plates** (single crystalline and multicrystalline silicon wafers) - 1$^{st}$ generation solar cells;

ii) **Si-based thin films** (polycrystalline silicon, amorphous silicon and its tetrahedral alloys, protocrystalline and microcrystalline silicon) and **polycrystalline films of CdTe, CuIn(Ga)Se$_2$** -2$^{nd}$ generation solar cells, and

iii) **GaAs** and a multijunction approach using **III-V alloy** material combinations; number of concentrator approaches (including Si, GaAs, GaAlAs) - 1$^{st}$ generation solar cells. This type of solar cells are suitable for installation in the regions with a high intensity of the solar radiation, under sunlight concentration and for space application, they will not be object of discussion in this review.

***Crystalline (c-Si) and multicrystalline (mc-Si) silicon*** based technologies dominate today's PV production. Si is presently the most mature and best studied candidate for terrestrial PV applications. It

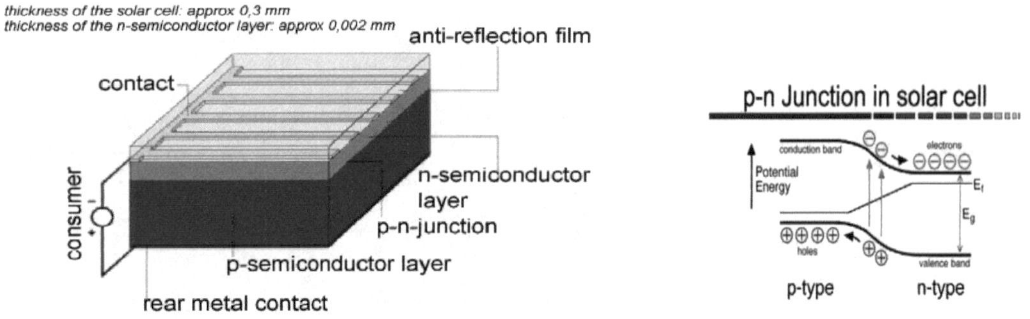


**Figure 2.** Scheme of the solar cell and p-n junction.

offers several advantages over other PV materials, including abundance, an established technology base, high material quality and stability, and good surface passivation characteristics. The obvious disadvantages are the indirect band gap and the currently high processing cost for silicon material and devices. Hence, research continues both in improving silicon material and in studying and developing other Si-based technologies (e.g. thinner Si wafer -100-50 μm, metallurgical silicon, etc.).

Today, the most efficient c-Si solar cell is a "passivated emitter and rear locally diffused" (PERL) structure (figure 3), showing an efficiency of about 24.4% [2,3]. The main features of this cell are: front surface reflection loss reduced by inverted pyramids, high quality emitter-diffusion profile and high quality passivating thermal oxide on the front and back surfaces to reduce surface recombination losses, small front-contact fingers and localized rear p+ diffusion to reduce the contact contribution to total recombination losses. A similar approach has been applied for a multicrystalline silicon solar cell, where "honeycomb" front surface texturing is performed. As a result, multicrystalline Si solar cells with 19.8% efficiency have been produced [4].

***Amorphous Silicon solar cells.*** a-Si:H is a non-crystalline solid, lacking long-range periodic ordering of its constituent atoms. However, it does have local order on an atomic scale. This short-range order is directly responsible for the observation of semiconductor properties such an optical absorption edge and an activated electrical conductivity. A typical amorphous silicon cell employs a p-i-n design where an intrinsic layer (i-layer) is sandwiched between a p-layer and a n-layer (figure 4). The flexible solar cells are based on a sophisticated multi-layer amorphous silicon thin-film solar cell have been developed.

This spectrum-splitting cell, shown schematically in figure 5, is constructed of separate p-i-n type - a-Si:H solar sub-cells, each with a different spectral response characteristic. This allows the cell to convert the different visible and near infrared wavelengths of sunlight with optimal efficiency. Amorphous cells with different light absorption properties and deposited continuously, one on top of another, to capture the broad solar spectrum more effectively. This increases the energy conversion efficiency of the multi-cell device and improves performance stability. The multi-junction approach has resulted in world record stabilised efficiencies for these cells of 15%.

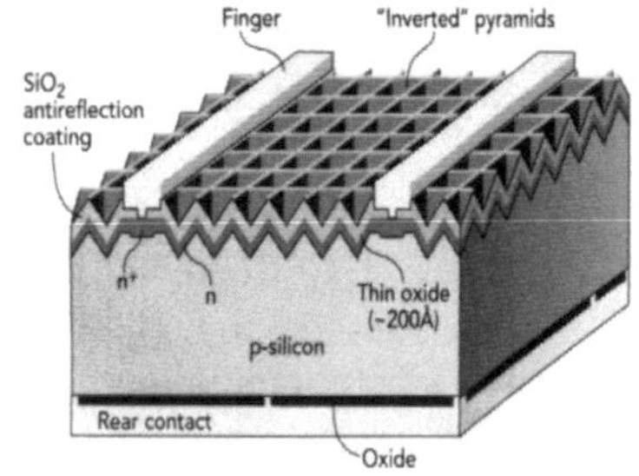


**Figure 3.** Structure of PERL cell: c-Si with a pyramidal textured surface [4].

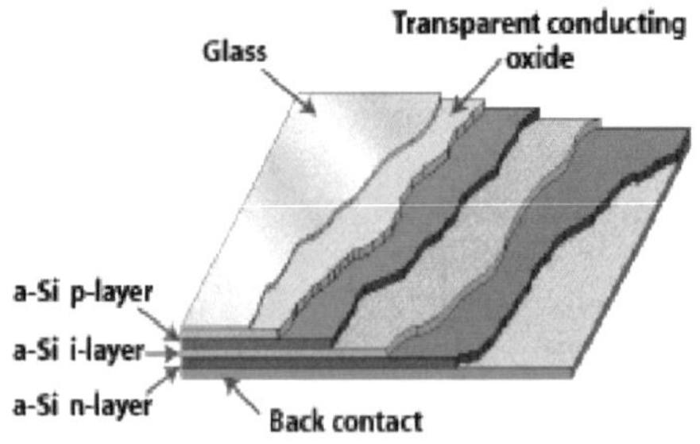


**Figure 4.** A typical a-Si:H solar cell employs a p-i-n design, in which an intrinsic layer (i-layer) is sandwiched between a p-land an n-layer.

Journal of Physics: Conference Series **253** (2010) 012007 doi:10.1088/1742-6596/253/1/012007

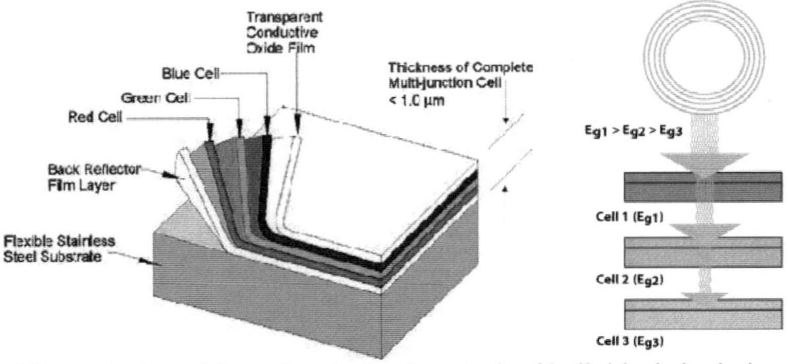


**Figure 5.** A multijunction device is a stack of individual single-junction cells in descending order of bandgap ($E_g$).

***Chalcopyrite solar cells (figure 6).*** Ternary chalcopyrite $CuInSe_2$ (CIS) and its modification Cu(In,Ga)Se (CIGS) are the first thin film materials to achieve 10% efficient solar cells. The device stability, together with improvements of the cell efficiency, has attracted the interest of researchers and industry in CIS-based PV technology. The maximum efficiency reported to date is 19.19% 50 x 50 mm test cells [5]. For modules on flexible Ti foil substrate, the efficiency of 17.9 % is reported [6] and 13.9% (for 90 $cm^2$) [7].

***Cadmium telluride solar cells (figure 7).*** Many of the basic properties of CdTe make it an ideal material for use in thin film solar cells. For example: i) its energy gap is direct and its value of 1.45eV supports maximal conversion efficiency, ii) it may be doped n- or p-type, and the preparation of the stoichiometric compound may be achieved easily in the production process, iii) CdTe thin films can be deposited at high rates by various methods: sublimation/condensation, close spaced sublimation, chemical spraying, electro-deposition, screen printing, chemical vapour deposition and sputtering. A disadvantage of CdTe solar cells concerns environmental and safety problems, since the product contains Cd, a toxic element. It has been shown, however, that modules can be produced under normal industrial risk conditions, and suitable techniques for waste treatment and recycling of modules are available [8]. Maximal efficiency of solar modules with CdTe thin films is achieved of 18% in the laboratory conditions and 11.1% by industry [9].

## 4. Solar spectrum

Solar radiative energy has its origin in a nuclear fusion reaction in the sun. The resulting energy is emitted mainly as electromagnetic radiation in the spectral range 0.2–3 µm (figure 8). The spectral distribution of the solar radiation which reaches the earth can be approximated by that of a black body at a temperature of 5800K. However, there is a sufficient departure from this idealized spectrum and hence it is desirable to use more exact data.

The sunlight is absorbed and scattered when it passes through the atmosphere on its way to the earth's surface. Basically, three sources of atmospheric absorption and scattering are important: atmospheric gases ($O_2$, $N_2$, etc.), aqueous vapour, and dust. The solar spectrum under different conditions is shown in figure 8 (left) [10]. The solar spectrum is very broad and contains significant power density

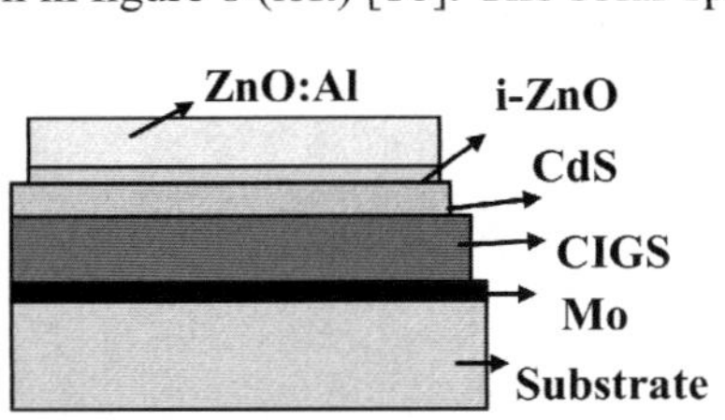


**Figure 6.** Scheme of a typical CIS solar cell.

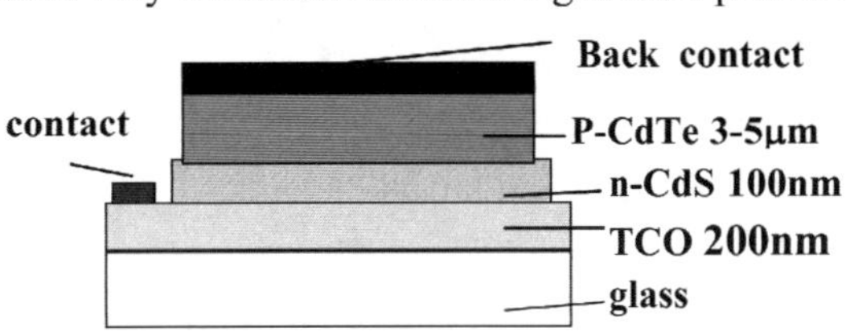


**Figure 7.** Scheme of a typical CdTe-/CdS solar cell.

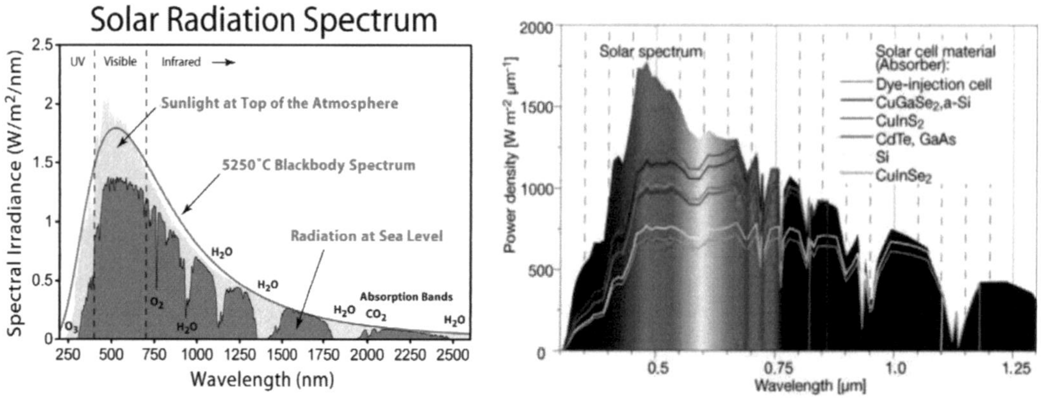


**Figure 8.** Solar spectrum on the earth surface (left) and the sensitivity of different type solar cells (right) [10].

both below and above the band gap of semiconductors applied for solar cells (figure 8). Figure 8 (right) illustrates the spectral sensitivity of different types of solar cells on the background of the solar spectrum at the earth surface. It is seen that different materials used for solar cells fabrication could transfer only part of the Sun spectrum into electricity.

Increasing the efficiency and decreasing the production cost of terrestrial photovoltaic devices is important for their widespread acceptance as a major power generation technology for the future. One of the main factor limiting the efficiency of solar cells is the spectral mismatch between the band energy (absorption spectrum) of the absorber material and the incident solar light. The solar spectrum is very broad and contains significant power density both below and above the band gap of the materials applied. Therefore the conversion efficiency limit, e.g. for a single-junction Si solar cell, according to the detailed balance analysis, yield a maximum of 31% under standard AM 1.5 solar spectrum radiation[11].

Several approaches have recently been proposed **to increase** the efficiency of single-junction solar cells above the theoretical limit. Within the category of high-efficiency approaches, a distinction is made between approaches that tailor the properties of the active layer to better match the solar spectrum and approaches that modify the incoming solar spectrum and function of the periphery of the active device, without fundamentally modifying the active layer properties. Advances in nanotechnology and nanomaterial are relevant to both approaches.

Another main factor, which limits large-scale use of solar energy, is the cost, which is still too high. The most advanced solar cell processes make extensive use of high temperature processes, vacuum processes, pattern definition steps, and/or expensive materials (including the silicon wafers). Therefore, not only the efficiency enhancement but also **the cost reduction** of solar cells has to be a target of innovative developments.

An analysis of current trends in the PV field shows that both **improved efficiency** and **lowered cost** can be obtained by implementing nanomaterials and nanotechnology to existing solar cells. Solar cells entirely based on nanostructured materials can ultimately be an alternative to the dominant bulk crystalline Si technology ($1^{st}$ generation) and thin film based PV technology ($2^{nd}$ generation) Thus, incorporation of nanostructured and new materials in $1^{st}$ and $2^{nd}$ generation solar cells to enhance conversion efficiency and lower cost per unit Wat, is required and will provide an evaluation process towards creation of new generation of advanced solar cells, **the $3^{d}$ generation,** which can produce efficiency greater than 20% yet at the same cost levels as thin film solar cells.

## 5. Nanomaterials and emerging technolologies for PV application

Nanomaterials such as nanowires and nanoparticles have unique advantages that make them usefull in photovoltaic devices. The nanometer size objects have very large surface areas per unit volume which offers the possibility to form very large interfacial area. The quantum confinement effects due to the

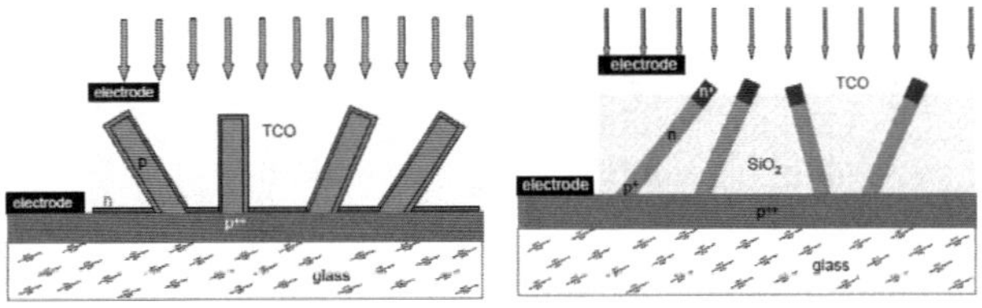


**Figure 9.** Scheme of nanowire solar cells with radial and axial p-n-junction.

nanosize effect results in increasing of the band gap of materials that help to design materials with different band gap. Nanomaterial can be produced by non vacuum technologies on large area substrate. All these advantages offer the possibility to increase the efficiency of solar cells and modules and the reducing their price.

### *5.1. Silicon nanowires*

While the 3$^{d}$ generation imply use of new band structure concepts that allow for limiting efficiencies greater than 31% Shockley-Queisser limit, a first step taken toward such device is development of thin film-based concept with using Si nanowires [11,12]. Silicon nanowire carpets can be produced by Vapor Liquid Solid method (VLS) using metal (Au or Al) nanotemplate [12], galvanic wet etching process [13], Plasma Enhance Chemical Vapour Deposition (PECVD) [14] and arc discharge [15]. There are several potential advantages of silicon nanowire-based solar cells compare to conventional thin film. One of the key features of nanowire devises is the ability to design the structure such that optical absorption is effectively de-coupled from charge transport consideration. In the bulk Si solar cell it is necessary that the cell be greater than ~120 μm in thickness to absorb the light, which in turn may lead to recombination losses as minority carriers diffuse to the top of the p-n junction. In the nanowire device it is possible to build a so-called conformal p-n junction that is only 100's of nanometers to a few microns width. Nanowire arrays possess excellent anti-reflection and light trapping properties that enable broadband absorption enhancement compared to solid thin films of the same thickness [14]. Nanowires also provide a direct path for charge transport resulting in lower series resistivity and lower probability of non-radiative recombination. Independent of the nanowire preparation method two designs of nanowire solar cells are under consideration with the p-n junction either radial or axial (figure 9).

Figure 10 shows a higher resolution SEM image, indicating the metal droplets on top of each wire deposited by VLS method. The optical reflection, transmission and absorption of a Si nanowire carpet are shown in figure 10 b. Typical is the very low reflectivity of about 5% up to a wavelenghth of 800 nm which demonstrates perfect antireflection properties. The I-V curves of the cell n-doped nanowires grown on p-type wafer are demonstrated in figure 10 c. The open circuit voltage is rather low. However the doping of the nanowire is not optimised. This result is one of the first demonstrations of possible application of the Si nanowires in a solar cell.

### *5.2. Silicon quantum dot*

The radiative efficiency of the bulk Si solar cells is limited theoretically to 31% due to the incomplete utilization of high energy photons and transmission of the photons with less energy than the Si band gap. One way to enhance the efficiency is to use stack of solar cells in which each cell has band gap that is optimized for the absorption of a certain spectral region. The theoretical efficiency of tandem solar cells with a bulk Si bottom cell increases to 42.5 % when one additional cell with 1.8 eV band is used and to 47,5% with two other solar cells with band gaps of 1.5 and 2 eV placed on top of the bulk cell [16]. Promising materials for the fabrication of Si based high band gap solar cells are Si quantum dots (Si QD) in dielectrics. By restricting the dimensions of Si to less than Bohr radius of bulk crystalline Si (~5 μm), quantum confinement causes its effective bandgap to increase. Heteroface

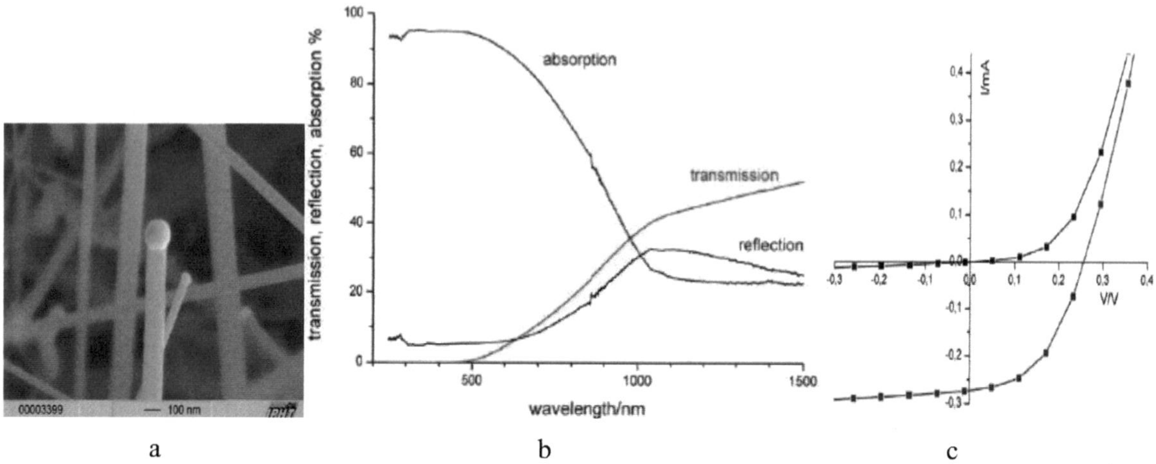


**Figure 10.** SEM images of Si nanowire grown by the VLS method with Au droplet on top (a) optical transmission, reflection and absorption of a nanowire carpet (b) and I-V curve of a cell consisting of n-doped nanowires grown by the VLS method onto a p-doped wafer [12].

structures composed of Si QD deposited on p-i-n homojunction crystalline silicon are fabricated and studied [17]. Figure 11 demonstrates the scheme of the devices with Si QD, the dependence of the open circuit voltage of the device and the spectral response of internal quantum efficiency (IQE) of the cell with various sizes of p-type Si QD and n-type c-Si heteroface devices. The IQE in the case of 3 nm Si QD size shows higher values in the entire range wavelength. For the short wavelength response, the IQE is around 96% at 500 nm for 3 nm Si QD heteroface device and the efficiency of this solar cell is 13.01 %. This result is an encouraging step towards application of Si QD in tandem Si solar cells. Further device optimization is required to improve photovoltaic efficiency, including not only process parameters (substrate resistivity, QD deposition conditions and thickness, etc.) but also device structure such as front metallization, base doping concentration, back-surface field, etc.

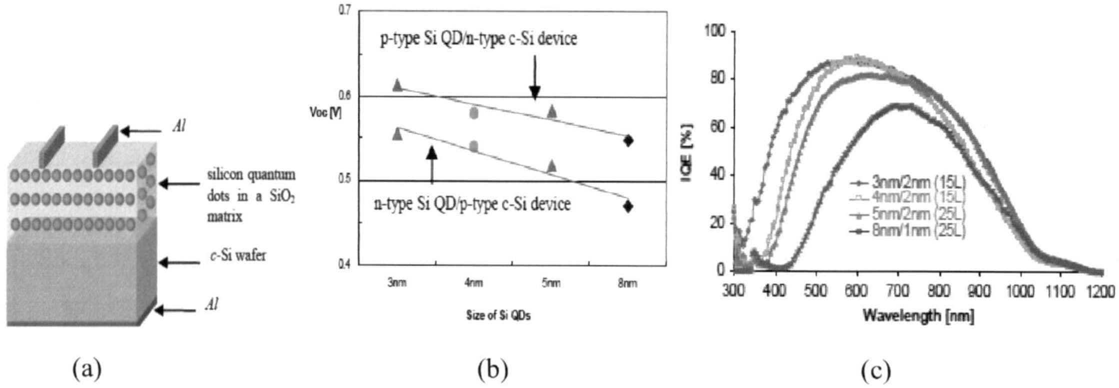


**Figure 11.** Scheme of the Si QD on c-Si substrate heteroface device (a), dependence of the $V_{oc}$ on the Si QD size (b) and internal quantum efficiency of the cell with different size of the SiQD(c) [17].

### *5.3. Nanowires of $A_2B_6$ semiconductors*

Increased optical absorption and stability of thin film silicon solar cells is the main goal of all thin film silicon solar cell developments. This is possible to do by increasing the light scattering and subsequent light trapping due to the nano-rough interfaces.

The photovoltaic (PV) devices based on inorganic $A_2B_6$ semiconductor with vertically oriented nanowires and nanorods structure have been extensively investigated, such as ZnO [15,18-26], CdS [27], covered with a-Si:H heterostructures or filled with organic photo-active materials [35],

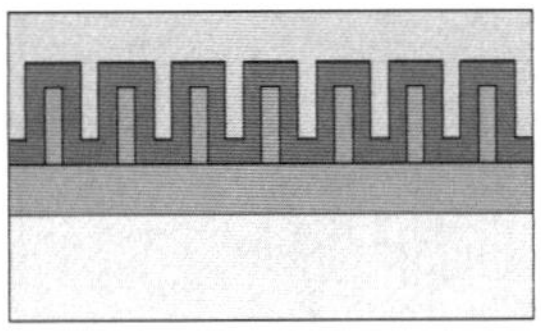

**Figure 12.** Schematic sketch of 3-dimensional a-Si:H (red color) solar cells on glass (ligh blue)/TCO nanocolumns (blue), covered by ZnO (green), serving as the second electrode [18].

**Figure 13.** SEM of ZnO nanowires deposited by the method of arc discharge [15].

fullerences [26], etc. These solar cells with ordered or well-aligned nanowires or nanorods, in comparison with the bilayer structure, have the advantages of very large donor–acceptor (D–A) interface area, continuous and minimized carrier conduction path ways which can significantly increase the power conversion efficiency of cells. Light scattering and subsequent light trapping due to nano-rough interfaces of the TCO is generally applied, too. Figure 12 shows the scheme of 3-dementional a-Si:H solar cells on superstrate galss/TCO/nanocolumns, covered by ZnO serving as electrode [18].

Single-crystal ZnO nanowire arrays have attracted research interest because they may be used as building blocks for a new generation of devices. In the framework of photovoltaic applications, they are an interesting option to be applied as n-type wide band material in nanostructures solar cells. They show quantum confinement and direct path for electrons promoting higher electron diffusion coefficient of ZnO nanowired areas with respect to nanoparticles. ZnO nanowire could be prepared by different techniques such as chemical and electrochemical deposition, sol-gel processing, spray pyrolysis, Vapor Liquid Solid (VLS), arc discharge, etc.[15, 20-24].

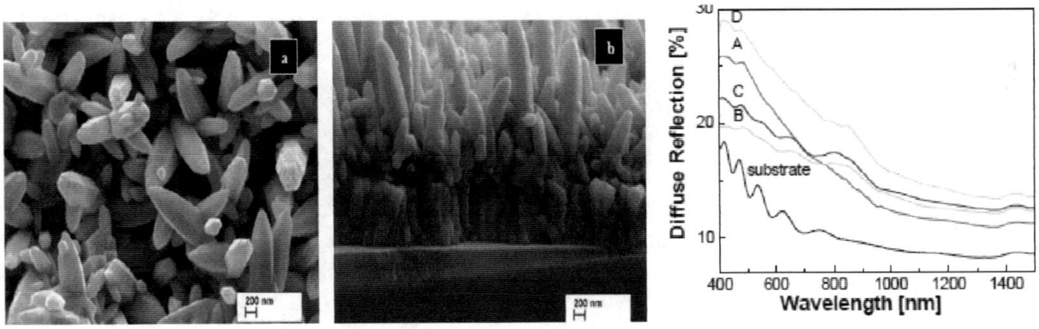


a b c

**Figure14.** SEM image of the surface (a) and the cross section (b), and diffuse reflection spectra of ZnO films electrochemicaly deposited on $SnO_2$:F coated glass substrate (c) [24].

Figure 13, 14a and 14b show the SEM images of ZnO nanostructures prepared by the method of arc discharge and electrochemical deposition, respectively. The high value of defuse reflection (figure 14 c) of ZnO nanostructured films demonstrate the possibility to be applied as a back contact for increased light trapping in thin film solar cells.

The ZnO nanowired films deposited on the glass/ITO electrode have been used in ZnO/organic hybrid solar cells [25]. It was found that fill factor (FF) was remarkably improved from 50% to 65% along with a small change in short circuit current and open voltage. The ZnO nanorods work as the electron collectors by shortening the average electron diffusion distance in the polymer PCBM network in the bulk heterojunction. The ZnO nanorods work also as direct paths to transport charge to the ITO electrode, resulting in a decreased charge recombination. Therefore, a large FF is observed.

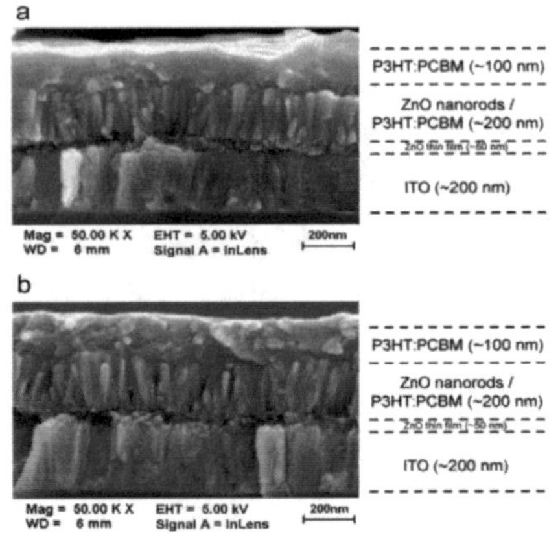


**Figure 15.** Cross-sectional SEM images of of polymer/ZnO hybrid solar cells (a) with and (b) without the solution-processed fullerene interlayer [26].

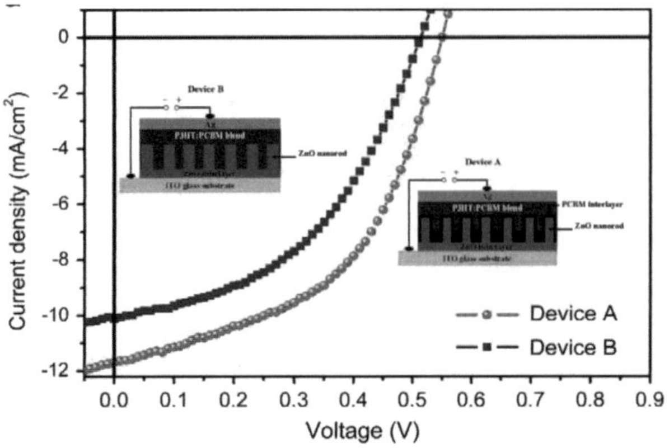


**Figure 16.** I-V characteristics of the PV devices under 100mW/cm$^2$ AM 1.5 irradiation. Insets: Device A represents the hybrid solar cells with the solution-processed fullerene interlayer. Device B denotes the hybrid solar cells without the fullerene interlayer [26].

Applying fullerene interlayer between the ZnO nanorod arrays of polymer/ZnO hybrid solar cells an enhancement in the efficiency, short-circuit current density, open circuit voltage and fill factor compare to the devices without the interlayer has been achieved (figure 15 and 16). This enhancement is due to the improved infiltration of the photoactive layer into ZnO nanorod spacing [26].

CdS has been proved to be an excellent photoactive and charge transport material in PVdevices [25–28], and its direct band-gap of 2.4 eV is appropriate to be an acceptor in solar cells. Recently, the synthesis and application of CdS nanowire arrays or nanorod arrays have become an attractive research field. Hybrid solar cells based on one-step fabricated vertically aligned CdS nanorod arrays and poly(2-methoxy-5-(20-ethylhexyloxy)-1,4-phenylenevinylene)(MEH-PPV) have been fabricated. The vertically aligned CdS nanorods provide the minimal electron transport path to reduce the charge carrier recombination at the heterojunction interface [27]. It is believed that the device performance of polymer/CdS nanorod composites can be improved further by improving the CdS nanorod arrays structure and introducing the higher mobility polymers.

To explore the potency of the proposed strategy, highly ordered, single-crystalline nanopillars of n-CdS directly on aluminium substrate and embedded them in a thin film of p-CdTe as the optical absorption material (figure 18) have been synthesized [28]. Conventional thin-film photovoltaics rely

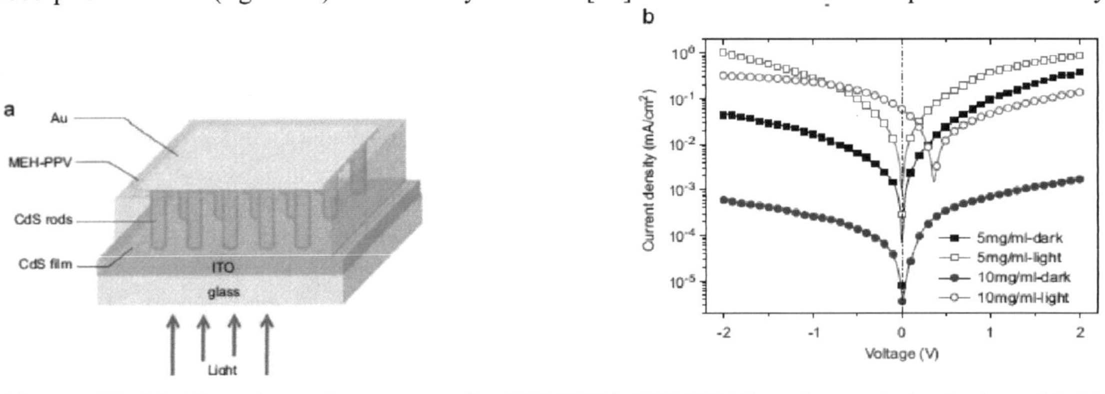


**Figure 17**. (a) The schematic structure for ITO/CdS/MEH-PPV/Au photovoltaic device; (b) I-V characteristics for ITO/CdS nanorod arrays/ MEH-PPV (dip-coating)/Au devices dip-coated from the MEH-PPV solution with two different concentrations of in the dark and under illumination at an AM 1.5 global solar irradiation (100mW/cm$^2$, 1 Sun) [27].

on the optical generation and separation of electron-hole pairs with an internal electric field, as shown in figure 2. Among different factors, the absorption efficiency of the material and the minority carrier lifetime often determine the energy conversion efficiency [29].

In this regard, simulation studies have previously shown the advantages of three-dimensional (3D) cell structures, such as those using coaxially doped vertical nanopillar arrays, in improving the photocarriers' separation and collection by orthogonalizing the direction of light absorption and electrons-holes separation (Figure 18 b). This type of structure is particularly advantageous when the thickness of the device is comparable to the optical absorption depth and the bulk minority carrier lifetimes are relatively short. Under such circumstances, the optical generation of carriers is significant in the entire device thickness and the 3D structure facilitates the efficient electron-hole separation and collection [28]. Figure 19 shows the I-V characteristics of a typical cell under different illumination intensities, P, ranging from 17 to 100 mWcm$^{-2}$ (AM 1.5G). Specifically, an efficiency $\eta$= 6% is obtained with an open circuit voltage $V_{oc}$ = 0.62 V, short circuit current density $J_{sc}$= 21mAcm-$^{2}$ and fill factor FF= 0.43 under AM1.5G illumination. The dependency of the performance characteristics on the illumination intensity is shown in figure 19 (down). As expected, $J_{sc}$ exhibits a near-linear dependency on the intensity because in this regime the photocurrent is proportional to the photon flux with a constant minority carrier lifetime. On the other hand, $V_{oc}$ increases only slightly from 0.55 to 0.62 V with a linear increase of $J_{sc}$, which we attribute to a slight thermal heating of the device. As the efficiency of a solar cell is expressed as $\eta = V_{oc}J_{sc}FF/P$ and FF slightly decreases with light intensity, the efficiency is about 6% and shows minimal dependence on the illumination intensity as shown in figure19 (above). It should be noted that this modest efficiency is obtained without the use of an antireflective surface coating or concentrators.

## 6. Photon management concept

Photon management concept employed to the solar cells is based on the phenomena as i) by surface

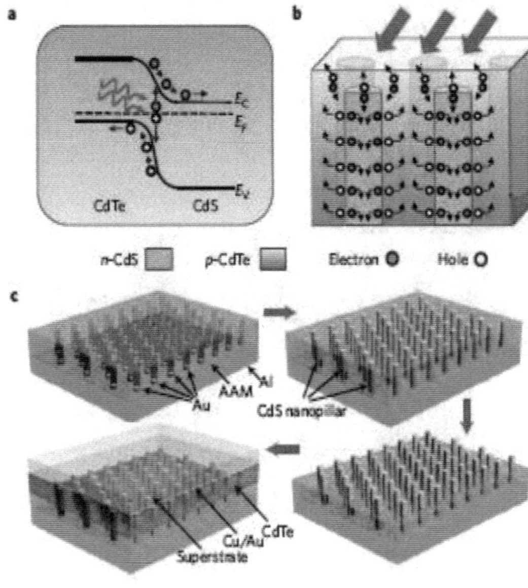


**Figure 18**. CdS/CdTe solar cell with nanoparticle: a Energy band diagram of CdTe/CdS heterojunction, b) cross-sectional schematic diagram, illustration the enchanced carrier collection efficiency, c) fabrication processes flow [13].

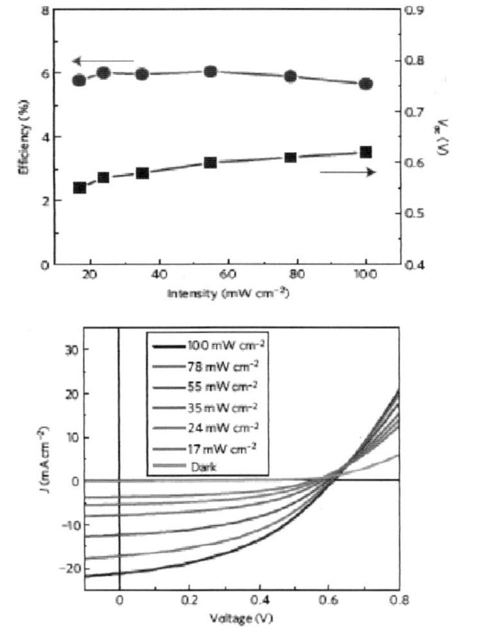


**Figure 18.** I-V characteristics (down), and dependence of the efficiency and $V_{oc}$ on the light intensity (above) of n-CdS/CdTe solar cell with CdS nanopillar [13].

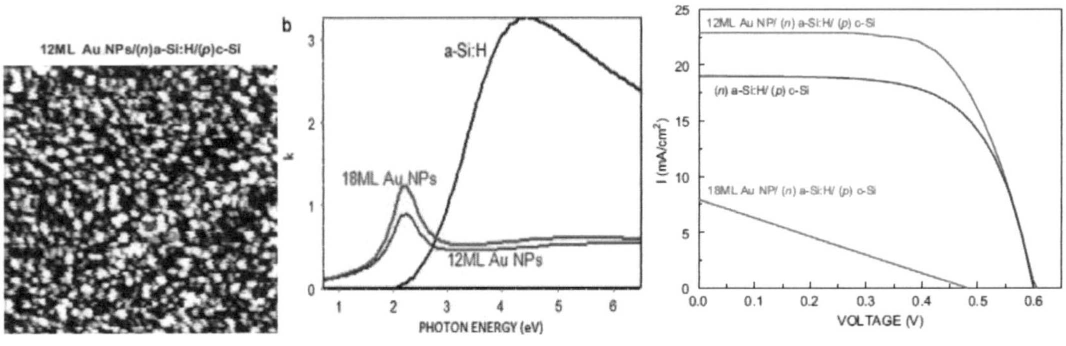


**a** **b** **c**

**Figure 20**. AFM image of 12 monolayers Au nanoparticles on n-type a-Si:H/ p-type c-Si heterojunction,(a), spectrum of extinction coefficient of two nanoparticle Au layers and of a-Si:H thin film (b) and I-V characteristics of heterojunction without and with Au nanoparticle with 12 Mls (20 nm) and 18 Mls (30 nm) (c) [30].

Plasmon Resonance (SPR) effect and ii) spectral modification of the incident light - up- and down-conversion.

### *6.1. Surface plasmon resonance*

SPR from noble metal nanoparticles (NPs) is a promising way to improve the efficiencies of silicon based photodiodes and solar cells. Exploitation of the localized absorption of metal Au nanoparticles via SPR and improvement of the characteristics of solar cells is demonstrated [30]. Methods of electroless plating, sputtering or thermal evaporation with post thermal treatment or ion irradiation, deposition from colloidal solution are used to prepare Ag or Au nanoparticles [30-32].It is shown that a density of 1.3 x $10^{11}$ $cm^{-2}$ of Au nanoparticles with an approximately 20 nm diameter (12 monolayers (Mls)) deposited by sputtering on the surface of n-type a-Si:H/p-type c-Si heterojunctions enhance performance (short circuit current) of solar cell due to the improved absorption of light by the SPR of Au NPs (figure 20)[30].

### *6.2. Up- and down- conversion*

**Up-down conversion** can be regarded as one of the approach for development of the Third Generation Photovoltaics - to use the energy incident on a solar cell more effectively. Most of the attempts to improve of the solar cell efficiency to date have aimed at modifying the structure of the solar cells itself, by fitting its spectroscopic sensitivity to the fixed sunlight spectrum. A reverse concept, however, is to fit the sunlight spectrum to the solar cell sensitivity via a wavelength-shifting operation. This can be done e.g. by the inclusion of layers containing rare earth ions. Rare earth ion complexes absorb light of short (or long) wavelength, and then emit light with longer-down conversion, (or shorter – up-conversion) wavelengths. If the wavelength shifts for rare earth ion are applied to solar cells, the higher (or lower) photon energy component of sunlight can be shifted to the longer (shorter) wavelength, at which cells can produce electrical power with high efficiency. Candidate rare earth ions for improvement of the conversion efficiency are $Ce^{3+}$, $Sm^{3+}$, $Eu^{3+}$, $Yb^{3+}$, $Nb^{3+}$, $Tb^{3+}$, $Er^{3+}$,$Dy^{3+}$ embedded in different glass matrices.

The efficiency of thin film solar cells can be improved with the addition of a photon down-conversion top layer. This layer converts incident ultraviolet light of the solar spectrum to visible light, which transmits through the glass and is efficiently absorbed by the active layer of the solar cell.

In the case of **down-conversion** doping of the transparent antireflection coating of the solar cell with suitable rare earth elements, which have absorption bands in the UV region and emission bands in the visible region (in the best case at the spectral sensitivity peak of the solar cells), would results in increasing of the conversion efficiency. Not all rare-earth ions, however, are suitable for increasing the

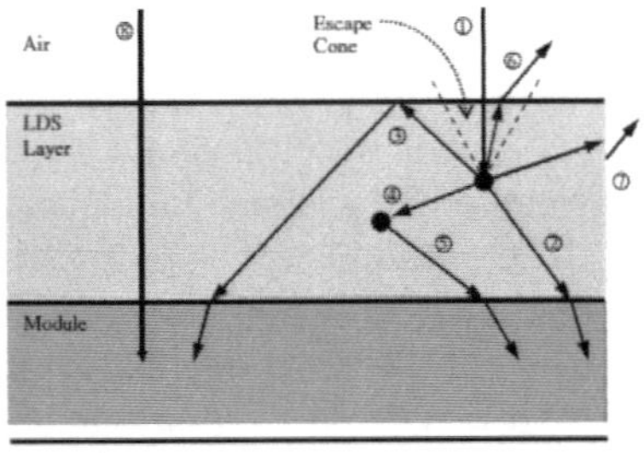


**Figure 21.** A diagram of a PV module with an LDS layer on top. Light having λ within the absorption band of the luminescent species 1 will be absorbed and re-emitted at longer λ.

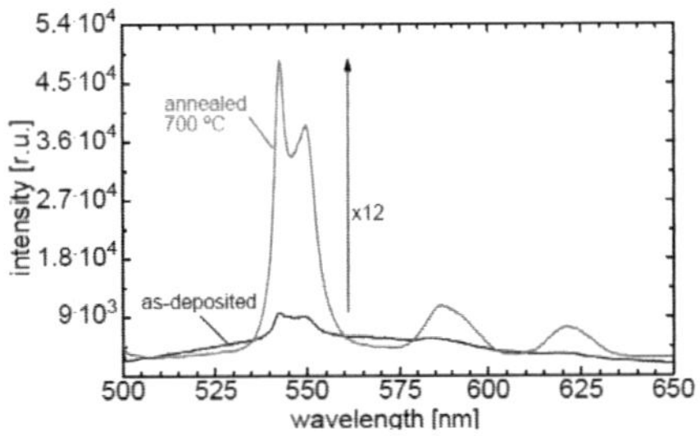


**Figure 22.** PL spectra of an as-deposited 260 nm thin $SiO_2$:Tb film on a Si substrate with a Tb concentration of 4.61 at.% and annealed at 700 °C PL is exited by 488 nm line of $Ar^+$ line of Ar laser[37].

efficiency via up-conversion. Several rare-earth ions satisfy these conditions: $Ce^{3+}$, $Sm^{3+}$, $Eu^{2+}$, $Eu^{3+}$, $Yb^{3+}$, $Nb^{3+}$, $Tb^{3}$ [33]. It has to be considered that the position of the emission band of the rare earth, however, depends on the host material and the ambient temperature. The rare-earth ions can be involved in the antireflection coating by different methods: ion implantation, magnetron co-sputtering of the ion and AR coating, sol-gel [33-37]. A diagram of a PV module with a luminescencent down-shifting (LDS) layer on top is presented in figure 21. An example of PL spectrum of a LDS material– a $SiO_2$ thin film with embedded $Tb^{3+}$ ions, deposited by magnetron co-sputtering on glass substrate is given in figure 22. The photoluminescence spectrum is exited by 488 line of $Ar^+$ line. Three characteristic bands were observed and identified as the following 4f-4f transitions: $^5D_4 - {}^7F_5$ (~ 546 nm), $^5D_4 - {}^7F_4$ (~ 586 nm) and $^5D_4 - {}^7F_3$ (~ 622 nm). The first transition was much stronger than the others because of the highest value of the corresponding transition matrix element according to the Judd-Ofelt theory [39]. It had a structure with two maxima at 542 nm and 549 nm, which was ascribed to Stark splitting and has been also observed for $Tb^{3+}$ in some glasses [40]. Such layer deposited on the front side of the glass substrate with solar cells on the back side will transmit the emitted light through the glass which will be efficiently absorbed by the active layer of the solar cell.

Application of the **up-converters** to the solar cells and modules could reduce the losses of the incident power. E.g. silicon solar cells lose about 35% of the incident power because of the thermalisation losses and 20% because photons with energy below the band gap are transmitted straight through the device. The theoretical efficiency limit is raised from near 30% [11] up to about 40% [40] for Si solar cells with upconverter illuminated by non-concentrated light.

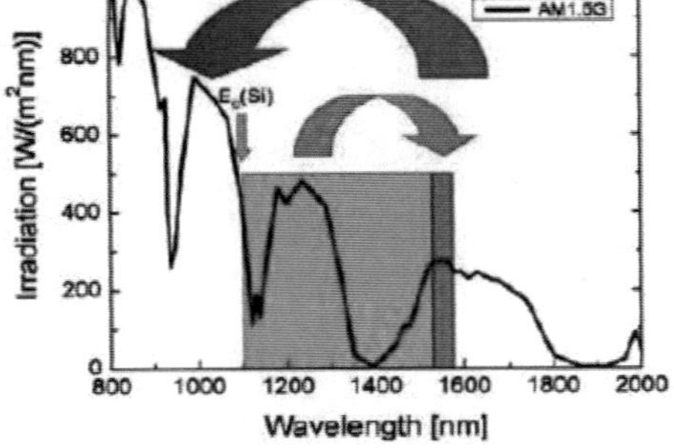


**Figure 23.** Spectral concentration and upconversion. The fluorescent material absorbs in a broad spectral range (green) and emits in the absorption range of the upconverter (red). The upconverter emits photons with energies above the bandgap $E_g$ of the silicon solar cell.

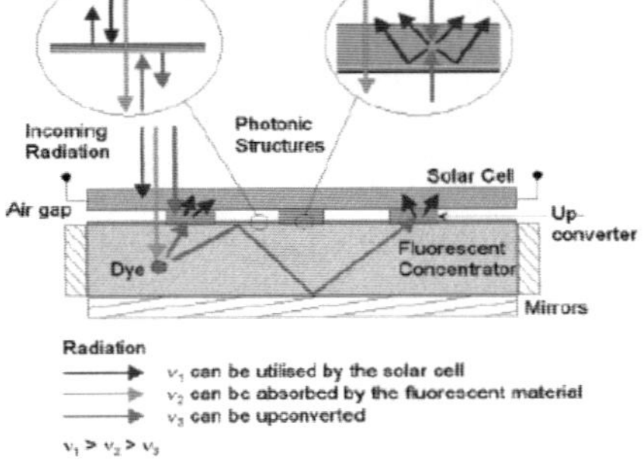


**Figure 24.** Potential setup of an advanced upconverter system.

Unfortunately, promising up-converting materials, such as erbium-doped $NaYF_4$, have only a narrow absorption range. More photons can be converted, when the upconverter is combined with a fluorescent material. The fluorescent material should absorb over a wide spectral range and emits in the absorption range of the upconverter (figure 23). PbSe/PbS core-shell nanocrystal quantum dots area promising fluorescence material for this application. An advanced upconverting system layout, which includes the combination of up-converter and fluorescent material, and the application of photonic structures for spectral management has been proposed [42, 43]. The experimental data demonstrate the effect of an UC material on the long-wave-length response of amorphous Si solar cells (figure 25 [ 42]). The I-V characteristics of a-Si:H solar cell with UC phosphor is a factor of three larger (figure 26 [42].Up today the enhancement of the quantum efficiency of the real c-Si solar cells with $\beta$-$NaYF_4$:20% $Er^{3+}$ is very low (0.011%) [43].

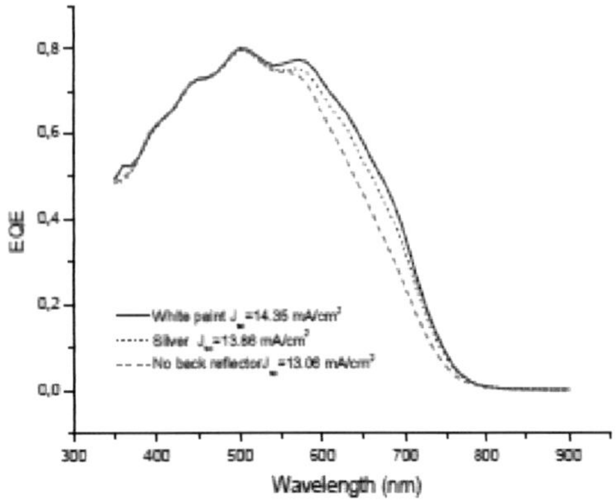


**Figure 25.** Spectral response measurements for solar cells with1 µm ZnO:Al 0.5% layer as the back contact and different back reflectors [42].

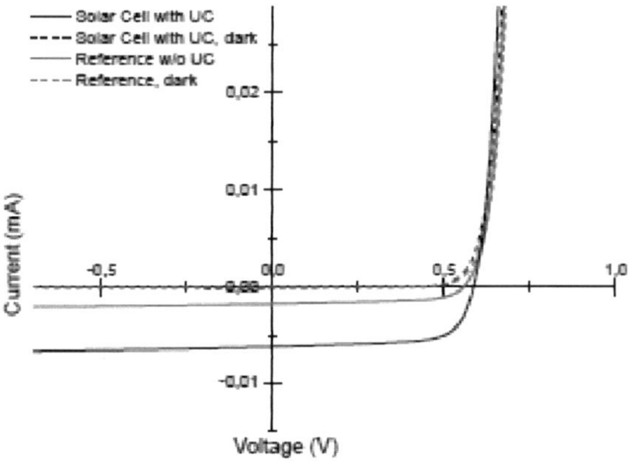


**Figure 26**. I-V curves of a-Si:H solar cells with and without UC layer attached at the back. The current in the cell with the UC phosphor is a factor of three larger [42].

## 7. Conclusion

The third generation of solar cells as low cost and high efficiency revolutionary technologies needs to be developed to make photovoltaic energy economically competitive. Nanostructured materials are likely to play important roles both in reducing the cost of second generation thin film solar cells and in the development of third generation photovoltaic devices because they offer some significant and unique advantages for solar-to-electric energy conversion. Proposed novel solar cell architectures that make use of nanomaterials have already been demonstrated but these are still not efficient and reliable enough to compete with first and second generation solar cells in the near future for generating electricity. Compared to the traditional solar cells the new technologies based on nanomaterials are still in their infancy. In some areas, progress from fundamental research to prototypes has been achieved. Future advances and the rate at which we can achieve them will depend entirely on the amount of resources the society chooses to invest in photovoltaics.

**Acknowledgement**
The work is supported by the project NanoPV (EC 7FP) and PV NMS NET (IEE EC Programme).

## References

[1] Global market outlook for photovoltaics until 2014. 2010 (Brussels European Photovoltaic Industry Association) www.epia.com
[2] Benick J, Hoex B, Dingemans G, Kessels W M M, Richter A, Hermle M and Glunz S W 2009 *Proc. 24th EPVSEC(Hamburg, Germany 21-25 September 2009)* pp 863-866

[3] Zhao J, Wang A, Altermatt P P, Wenham S and Green M A 1996 *Sol. Energy Mat. Sol. Cells* **41/42** 87

[4] Zhao J, Wang A, Green M and Ferrazza F 1998 *2nd World Conference on PV Solar Energy Conversion (Vienna, Austria 6-10 July 1998)* pp 1681-1684

[5] Windeln J, Schock H W, Felse C, Metzner H, Mannstadt W, Klenk R, Gruhn T, Kaufmann C, Rudigier-Voigt E, Speit B, Haag M, Rissom T, Kieven D, Ludwig C, Haarstrich J and Oertel M 2009 *24th EPVSEC (Hamburg, Germany21-25 Sept 2009)* pp 2443-2446

[6] Nakada T, Yagioka T, Horiguchi K, Kuraishi T and Mise T 2009 *24th EPVSEC (Hamburg, Germany 21-25 Sept 2009)* pp 2425-2428

[7] Dimller, B, Gross, E, Hariscos, D, Kesler, F., Lotter, E., Powalla, M., Springer, J, Stein U, Voorwinden., G, Gaen, M, and Schaffler R 1998 *2nd World Conference on PV Solar Energy Conversion (Vienna, Austria 6-10 July 1998)* pp 419-412

[8] Bohland J, Dapku T, Kamm K and Smigielski K 1998 *Proc.2nd World Conference on PV Solar Energy Conversion (Vienna, Austria 6-10 July 1998)* pp 716-719

[9] Sinke W C, Ballif C, Bett A, Dimmler B, Dimova-Malinovska D, Fath P, Mason N, Ferrazza F, Gabler H, Hall M, Martí A, Mellikov E, Milner A, Mogensen P, Panhuber C, Pearsall N, Poortmans J, Protogeropoulos C, Sarre G, Sarti D, Strauß P, Topič M, van Zolingen R and Żdanowicz T, Working Group 3: Science, Technology and Applications EU PV Technology Platform, "A strategy research agenda for photovoltaic solar energy technology" 2007 *Proc. 22nd EPVSEC (Milano, Italy 3-7 Sept 2007)* pp 3344-3349

[10] Thekaekra M P 1974 *Suppl. Proc. 20th Ann. Meet. Inst. Environ. Sci.* pp 21- 30

[11] Schockley W and Queisser H J 1961 *J.Appl.Phys.* **32** 510

[12] Andrea G, Pietsch M, Sivakov V, Stelzner Th, Gawlik A, Christiansen S and Falk F 2008 *Proc. 23th EPVSEC Valencia, Spain 1-5 Sept 2008)* pp 163-166

[13] Peng K Q, Huang Z P and Zhu J 2004 *Adv. Mater.* **16** 73

[14] Tsakalakos L, Balch J, Frontheiser J, Korevaar B, Sulima O and Rand J 2008 *Proc.23thEPVSEC (Valencia, Spain 1-5 Sept 2008)* pp11-16

[15] Koprinarov N , Marinov M, Dimova-Malinovska D, Starbova I, Konstantinova M ,Nichev H, and Tzacheva Tz 2008 *J. Optoelectronics &Advanced Materials* **11** 1304

[16] Conibeer G, Green M, Cho E-C, Konig D, Cho Y H, Fangsuwannarak T, Scardera G, Pink E, Huang Y and Puzzer T 2008 *Thin Solid Films* **516** 6748

[17] Park S, Cho E, Hao X, Conibeer G and Green M 2008 *Proc.23th EPVSEC, (Valencia Spain 1-5 Sept 2008)* pp 189-192

[18] Vanecek M, Poruba A, Remes Z, Holovsky J, Purkrt A, Babchenko O, Hruska K, Meier J and Kroll U 2009 *Proc.24th EPVSEC (Hambourg, Germany 21-25 Sept. 2009)* pp 2286-2289

[19] Pauzauskie P J and Yang P 2006 *Nanowire photonics, Materials Today* **9** 36

[20] Stelzner T, Balch J, Fronheisner B A, Korevaar B A, Sulima O and Rand J 2007 *Appl. Phys. Lett.* **91** 233117

[21] Kong X, Ding Y and Wang Z 2004 *J. Phys. Chem. B* **108** 570

[22] Dai Y, Zhang Y, Li Q K and Nan C W 2002 *Chem. Phys. Lett.* **358** 83

[23] Lim Y S, Park J W, Hong S T and Kim 2006 *Mat. Sci. Engineering* B **129** 100

[24] Dimova-Malinovska D, Andreev P, Sendova-Vassileva M, Nichev H and Mikli V 2009 *Proc.24th EPVSEC (Hambourg, Germany 21-25 Sept. 2009)* pp 391-394

[25] Takanezawa K, Tajiama and Hashimoto K 2008 *Appl. Phys. Lett* **93** 063308-1-3

[26] Huang J S, Chou C Y and Lin C F 2010 *Sol Energy Mat Sol Cells* **94** 182

[27] Jiang X, Chen F, Xu H, Yang L, Qui W, Shi M, Wang H and Chen H 2010 *Sol. Energy Mat. Sol. Cell* **94** 338

[28] Three Z, Fan H, Razavi J-W, Do Moriwaki A, Ergen O, Chueh Y-L, Leu P W, Ho J C, Takahashi T, Reichertz L A, Neale S, Yu K, Wu M, Ager J W and Javey A 2009 *Nature Materials* **8** 648

[29] Andrä G, Bergmann J, Falk F and Ose E 2004 *Proc.19th EPVSEC* pp 872-875

[30] Lasuro M, Giangregorio M, Bianco V, Sacchetti A, Carezzuto P and Bruno G 2009 *Sol. Energy Mat. Sol. Cells* **93** 1749
[31] Blackwood D J and Khoo S M 2010 *Sol. Energy Mat. Sol. Cells* **94** 120
[32] Pivin J C, Sendova-Vassileva M, Nikolaeva M, Dimova-Malinovska D and Martucci A 2002 *Appl. Phys* A**75** 401
[33] Kawano K, Sado T, Nihikawa M and Nakata R 1998 *Proc. 2nd Word Conference on PV Solar Energy Conversion (Vienna, Austria 6-10 July 1998)* pp 334-337
[34] Sendova-Vassileva M, Vuchkov A, Angelov O and Dimova-Malinovska D 2003 *J. Mater. Sci: Mater. in Electronics* **14** 853
[35] Sendova-Vassileva M, Nikolaeva M, Angelov O, Vuchkov , Dimova-Malinovska D and Pivin J C 2002 Photovoltaic and Photoactive Materials - Properties, Technology and Application: *Proc. NATO ASI Scientific series, II Mathematic, Physics and Chemistry* vol 80) ed. J. M. Marshall and D. Dimova-Malinovska (Dordrecht: Kluwer Academic Publisher) pp 329-332
[36] de Graaf D, Stelwage, S J, Hintzen H T and de With G 2003 *J. Non-Cryst. Solids* **325** 29
[37] Baumgartner K, Ahrens B, Angelov O, Sendova-Vassileva M, Dimova-Malinovska D, Holländerd B, Schweizer S and Carius R 2010 *Proc. of SPIE* vol. 7725 77250T-1
[39] van Dongen M. A 1992 *J. Non-Cryst. Solids* **139** 271
[40] Trupke T, Shalava A, Richardsa B S, Wurfel P and Green M A 2006 *Sol. Energy Mat. Sol. Cells* **90** 3327
[41] Fischer S, Goldschmidt J C, Löper P, Janz S, Peter M, Glunz S W, Kige A, Lifshitz E, Krämer K, Biner D, Bauer G H and Brüggemann R 2008 *Proc.23th EPVSEC (Valencia, Spain 1-5 Sept 2008)* pp 620-623
[42] de Wild J, Meijerink A, Rath J K, van Sark W G J H M and Schropp R E I 2009 *Proc.24th EPVSEC (Hambourg, Germany 21-25 Sept. 2009)* pp 350-353.
[43] Loper P, Goldschmidt J C, Fischer S, Peters M, Meijrnik A, Biner D, Kramer K, Schultz-Wittmann O, Glunz S W and Luther J 2008 *Proc.23th EPVSEC (Valencia, Spain 1-5 Sept 2008)* pp 173-180

# Computational modelling and optimisation of the fabrication of nano-structures using focused ion beam and imprint forming technologies

**S Stoyanov[1], C Bailey[1], Y K Tang[1], S Marson[2], A Dyer[2], D Allen[2] and M Desmulliez[3]**

[1] Computational Mechanics and Reliability Group, School of Computing and Mathematical Sciences, University of Greenwich, London SE10 9LS, UK
[2] Precision Engineering Centre, Cranfield University, Bedfordshire, MK43 0AL, UK
[3] MicroSystems Engineering Centre, Heriot-Watt University, Edinburgh EH14 4AS, Scotland, UK

E-mail: s.stoyanov@gre.ac.uk

**Abstract.** Focused Ion Beam (FIB) and Nano-Imprint Forming (NIF) have gained recently major interest because of their potential to enable the fabrication of precision engineering parts and to deliver high resolution, low-cost and high-throughput production of fine sub-micrometre structures respectively. Using computational modelling and simulation becomes increasingly important in assessing capabilities and risks of defects with respect to product manufacturability, quality, reliability and performance, as well as controlling and optimising the process parameters. A computational model that predicts the milling depth as function of the ion beam dwell times and a number of process parameters in the case of FIB milling is investigated and experimentally validated. The focus in the NIF study is on modelling the material deformation and the filling of the pattern grooves during the mould pressing using non-linear large deformation finite element analysis with hyperelastic non-compressive material behaviour. Simulation results are used to understand the risk of imperfections in the pattern replication and to identify the optimal process parameters and their interaction.

## 1. Introduction

Recent demands for advanced nano- and micro-engineering have been the driving force in researching, developing and promoting the use of a number of processing technologies that have the potential for producing the required technical specifications and quality at an acceptable cost. In particular, there is a major interest and demand for the fabrication of various precision engineering parts and micro-tools, and for the production of fine small scale structures. The key issues relate to the ability to realise processes that can offer high resolution, high yield and low cost fabrication routes. These technologies are required in a wide range of applications, from the manufacturing of miniaturised components for various heterogeneous systems, to micro-fluidics, bio-medical, MEMS, and embedded test devices.

The development of nano- and micro-engineering structures, parts and components imposes a number of unique challenges [1]. Most commonly the challenges are associated with the need to use completely new, or little known manufacturing processes that may have significant uncertainty in terms of dimensional resolution and accuracy. The material behaviour at the micro-scale could also be very different to the behaviour observed in the macro-scale world. Finally, the design performance and quality could be very uncertain as a result of the miniaturised extremely small scale nature of these products. Nowadays, the use of computational modelling and software plays an important role in studying, generating new knowledge and understanding the physical behaviour, expected quality and the risk of defects in these products, or in identifying the optimal manufacturing process control.

The computational modelling and process optimisation of two particular nano- and micro-fabrication processes, focused ion beam (FIB) and nano-imprint forming (NIF), are presented and discussed. The computational analyses focus mainly on issues related to either the optimal process control or to the risks of defects in the fabricated structures, and aim to produce results that can help to formulate design rules for improved quality, reliability and performance.

The most attractive feature of the FIB that has helped the development of this technology is its potential for direct milling and material removal while in the same time having almost no material restrictions [2,3]. In FIB milling, ion beam with certain intensity is directed to substrate material and the energised ions, when hitting the target surface, cause the transfer of the energy to the target atoms on the surface. This leads to a collision cascade and atoms from the target surface are sputtered. By moving the ion beam in a controlled manner different nano- and micro- features can be realised. The ability to control the depth variation is the most important aspect of the FIB milling process. A computational model that can be used to predict the required ion beam dwell times for a predefined shape is investigated. The emphasis is on a study for validating this model by comparing the model predictions with experimental data for milling a trench-shaped feature.

Nano-Imprint Forming process on the other side allows large scale pattering and fabrication of fine structures at very high resolution in a very cost effective manner compared with other existing processes [4]. The key concerns in this technology relate to the process control and the subsequent quality of the fabricated nano- and micro- structures. The main objective is to realise the process in a way so that the risk of defects and imperfections is minimised. In particular, a process phenomenon of significant importance for the quality of the patterns is the material deformation and the filling of the pattern grooves during the mould pressing step. We demonstrate the use of non-linear large deformation finite element analysis with hyperelastic non-compressive material behavior to study the effect of process parameters (imprint pressure and temperature) and pattern related parameters (the aspect ratio of the grooves) on the quality of the fabricated structures. The modeling results identify design rules that can help the optimal process set up and the control of the process parameters.

## 2. Computational Modelling of FIB

The principle of operation for FIB is bombardment of a target surface through high energy gallium $Ga^{+}$ (or other) ions. As a result of this, small amount of material sputters in the form of secondary ions, natural atoms and secondary electrons. The ion beam is moved in a raster scan manner which is digitally controlled over a pixel matrix defined on the target surface area. The size of a pixel is determined by the step of the beam movement. The time the ion beam spends on a pixel is called dwell time. The sputtering yield is the measure for the efficiency of the material removal and depends on a number of parameters such as the ion energy, angle of incidence of the ions to the target surface, ion flux and mass of the ions and target material atoms.

### *2.1. Computational Model for FIB sputtering*

The first models for simulation of ion beam milling have been limited in the sense that they tried to provide predictions for the final geometry based on a particular set of FIB parameters. The requirements for the FIB process at present are much more demanding than its original ion milling abilities. In particular, a critical control required for FIB milling is the one of the depth variation.

A mathematical model that can be used to predict the etched shape or to calculate the dwell times required to achieve a predefined shape is investigated [5]. The model relates the milling depth and the ion beam dwell times at the pixels placed over the target surface using a number of process parameters. The governing equation for the sputtering process is then discredited over each element of the pixel matrix so that ultimately a system of linear equations that relates the dwell times $t_{ij}$ with the sputtering depth $H_{ij}$ at any pixel $(i,j)$ is constructed.

If $(x_i, y_j)$ denotes the centre of the pixel $(i,j)$, then the sputtering at this pixel in terms of depth as a result of the material removal at that pixel can be expressed as

$$H_{ij} = \iint \frac{\Phi(x,y)}{\eta} f_{x,y}(x_i, y_j)\, Y(E_0, \alpha_{x_i,y_j})\, t_{x,y}\, dx\, dy \tag{1}$$

where $H_{ij}$ is the sputtering depth at the point $(x_i, y_j)$, $\Phi(x,y)$ is the ion flux at point $(x,y)$ on the target surface ($cm^{-2}s^{-2}$), $\eta$ is the atomic density of the target material (atoms/$cm^3$), $Y(E_0, \alpha_{x_i,y_j})$ is the sputtering yield (atoms per incident ion at point $(x_i, y_j)$), $t_{x,y}$ is the dwell time of the ion beam at point $(x_i, y_j)$ (seconds), $f_{x,y}(x_i, y_j)$ is the ion beam density distribution function in two dimensions.

The sputtering yield in Equation (1) is function of the angle of incidence $\alpha_{x_i,y_j}$ of the ion beam at point $(x_i, y_j)$ and the ion energy $E_0$ as well as the type of ion source and target material. Generally, the yield increases from perpendicular ion beam incidence to a maximum at angle 60° to 85°, and then rapidly decreases due to the strong reflection at grazing incidence. The sputtering yield can be obtained experimentally. A software package, SRIM (The Stopping and Range of Ions in Matter), can be also used to predict the sputtering rate for particular ion source and target material under specified energy conditions by Monte Carlo binary collision simulation [6].

In FIB milling, the ion flux is not uniform and has a distribution associated with it. Typically, it is assumed that this distribution is Gaussian. The standard deviation of the distribution can be estimated by assuming that the beam spot diameter is equal to full width at half maximum (FWHM). The ion beam geometry in terms of density distribution $f_{x,y}(x_i, y_j)$, as required in the FIB model above, is taken to be Gaussian bi-variate density function:

$$f_{x,y}(x_i, y_j) = \left( \frac{1}{\sqrt{2\pi}\sigma} \right)^2 e^{-r^2/(2\sigma^2)} \tag{2}$$

where $r^2 = (x_i - x)^2 + (y_j - y)^2$ is the radial coordinate for an ion beam focused at $(x,y)$.

### 2.2. *FIB Experiment and Model Validation*

Two FIB experiments using FEI 200 FIB machine with gallium ion source and silicon wafer as the target surface are conducted. The results from the first experiment are intended to provide data on the actual sputtering yield, and the second test is undertaken to examine the predictive capability of the model and for model validation.

*2.2.1. Experimental derivation of the sputtering yield.* The FEI FIB 200 system is used to set the milling of a simple boxed shaped feature. The objective is to estimate the amount of sputtered material and knowing the process parameters to obtain experimentally a value for the sputtering yield for the particular set of FIB process conditions. The target material used in the study is p-Silicon (B doped)-(100) orientation, and the scanned area on the target surface is selected with dimensions 5x10 μm. The ion beam is set with current 350 pA (corresponding to beam diameter 55 nm). The beam is set to raster with 50% overlap, angle of incidence 0 degrees (perpendicular to the target surface), ion energy 30 keV and at dwell times of 1 μs. The FIB current of 350 pA with overlap 50% is associated with ion flux of 1.44e24 ions/sec $m^2$ at the target surface. The total time of milling for this experiment is set to 460 sec. This results in 13,890 beam scans over the target area.

The cross sectional analysis shows some variation in the depth of the produced shape ranging from 0.883 μm to 1.09 μm (figure 1). From the analysis profiles we calculate that the sputtered volume is 48.176 $\mu m^3$. For total time of milling equal to 460 sec and ion beam current of 350 pA, the total amount of ions delivered to the sample over this milling period of time is 1.00488e12 ions. With known atomic density of the target material (5e10 atoms/$\mu m^3$), we can now evaluate the average sputtering yield. The calculated sputtering yield value is 2.397 atoms/ion. This experimentally derived value is well in the range that can be found in the literature and very similar to the SRIM software prediction.

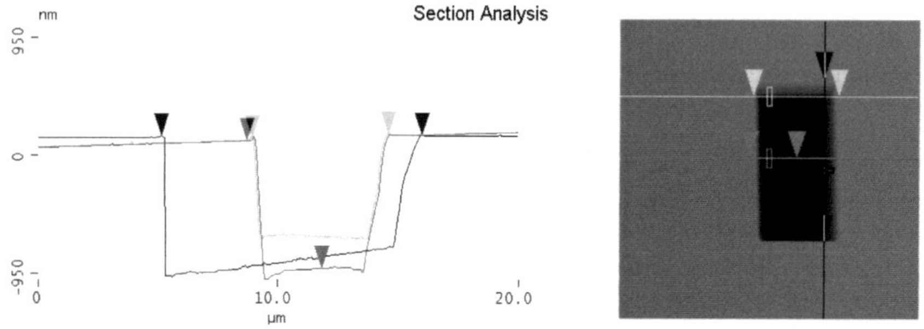


**Figure 1.** Cross sectional analysis of the milled shape (sputtering yield case study).

*2.2.2. Model Validation.* The validation example is designed with experiment of milling a step-wise parabolic trench using the same FIB system (FEI 200) and the same FIB process settings as used in the previous study. The desired trench has depth at the centre line 1.94 μm. The target area on the silicon surface that defines the trench is 10 x 10 μm.

The parabolic step-wise cross-section of the targeted shape for the trench, as shown in figure 2, features five patterns. The top most (at the surface) is pattern 1, the bottom most is pattern 5. The width of each subsequent pattern is reduced by 2 μm of the width of the pattern above. The width of pattern 1 is 10 μm. Note that in figure 2 the horizontal and the vertical axis are not equally scaled. The depth at each of the five patterns that realise the shape of interest are detailed in table 1.

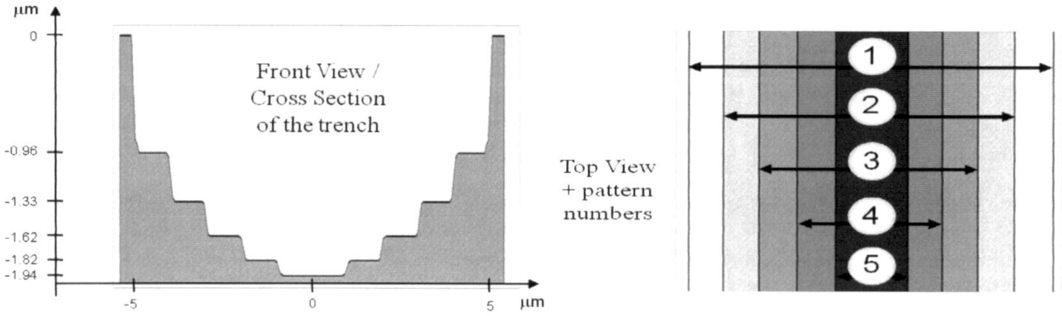


**Figure 2.** Shape dimension of the trench feature used in model validation study

**Table 1.** Definition of the parabolic shape depth variation.

| Pattern n | 1 | 2 | 3 | 4 | 5 |
|---|---|---|---|---|---|
| Set depth values from FIB model (□m) | 0.96 | 1.33 | 1.62 | 1.81 | 1.94 |

The computational model discussed in the previous section is employed to make predictions for the required pixel dwell times to mill this predefined shape. The model parameters are detailed in table 2 and based on process milling as used in the experiment for verifying the sputtering yield values.

**Table 2.** FIB model parameters.

| Parameter | Value |
|---|---|
| Target material atomic density | $5 \times 10^{22}$ atoms/cm$^3$ |
| Beam Spot diameter | 55 nm (Full Width at Half Maximum, FWHM) |
| Ion beam flux | 1.446e12 ions / mm$^2$ sec |
| Beam overlapping | 50% (pixel size : dx = 55 nm, dy = 27.5 nm) |
| Ion beam distribution | Gaussian bi-normal, with standard deviation 0.02786 μm |
| Sputtering rate | 2.39 atoms/ion (experimentally derived) |

The computational model for FIB is discretised over the pixel matrix with individual pixel dimension 55 x 27.5 nm. The required depth at each pixel is specified so that the sputtered shape corresponds to the five patterns step-wise shaped trench targeted with this experiment. The system of equations is solved for the unknown dwell times. The solution is detailed in the first two columns of table 3. The modelling results indicate the dwell times if the entire shape is milled in a single scan.

In order to apply the modelling predictions for the dwell times to a real experiment with the FEI 200 system, the model outputs must be transformed into an appropriate input data for the machine. In particular, the milling of such a shape would always be attempted not in a single beam scan but ratter using a large number of scans. For a feature with variable depth such as the five level step-wise trench considered in this study, the FIB system can be set in one of the following two regimes. First option is to execute a fixed number of iterations across all pixels using different pixel dwell times to achieve the shape. Alternatively, the FIB can be run into mode where the dwell time is fixed for all pixels but different pixels are scanned different number of times (so called pattern mode). Both settings are equivalent as there is a linear direct relationship between the pixel dwell time and the number of scans at that pixel. The FEI 200 can be set with fixed dwell time for all pixels and allows defining patterns that will be subjected to different number of scans. The model predictions are transformed into an equivalent FIB set up in terms of number of scans on each pattern, respectively into total time for scanning each particular pattern. The data is shown in the last two columns of table 3.

**Table 3.** FIB model predictions for dwell times and the eequivalent FEI FIB 200 set up.

| Pattern number | Model predictions for pixel dwell time, final shape in one scan | Equivalent FEI FIB 200 set up (fixed dwell time 1 μs and different pattern scanning times, respectively number of scans | |
|---|---|---|---|
| | Dwell time (μs) | Number of beam scans over the respective pattern | Total (cumulative) time for scanning the pattern area (sec) |
| 1 | 13 706 | 13 706 | 908 |
| 2 | 19 257 | 19 257 | 1 203 |
| 3 | 23 403 | 23 403 | 1 369 |
| 4 | 26 227 | 26 227 | 1 443 |
| 5 | 27 905 | 27 905 | 1 465 |

The milling of the trench is undertaken in five steps that correspond to milling each pattern area, in a sequence, for the identified respective total time for that pattern. Pixel dwell time is predefined at 1 μs. Because the last column in table 3 specifies the cumulative time for milling the respective pattern area, in reality the total scanning time at any pattern is the incremental time from the previous milling time. For example, we start with milling the area of pattern 1 for time 908 sec, and then we continue with milling the area which corresponds to pattern 2 for time 295 sec (the increment from pattern 1), etc.

Figure 3 shows the output data and graphs from a cross section analysis undertaken to obtain information for the actual shape and the trench depth variation. Several cross sectional planes are examined. Table 4 details the depths as defined and used in the model, the actual measured depths for the experimentally sputtered shape, and also provides the percentage errors between the two values.

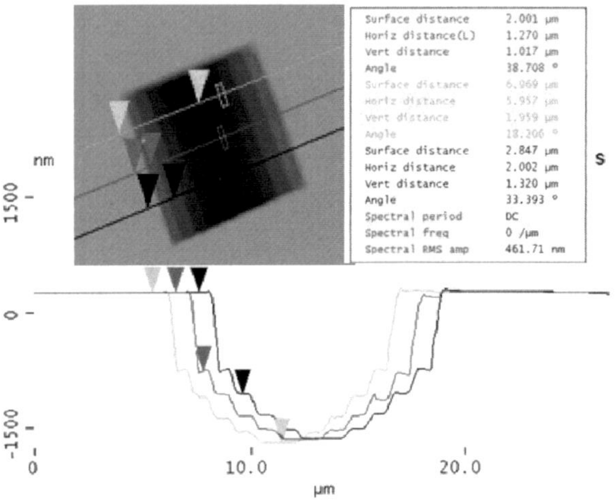


**Figure 3.** Cross sectional analysis of the sputtered trench (model validation case study).

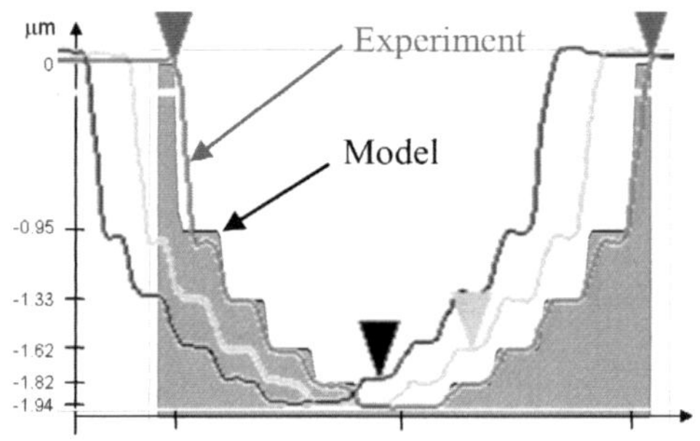


**Figure 4.** Model vs. experimental cross-sectional profiles.

**Table 4.** Comparison between model and experimental data for depth variation.

| Pattern number | 1 | 2 | 3 | 4 | 5 |
|---|---|---|---|---|---|
| Set Z values from FIB model (μm) | 0.96 | 1.33 | 1.62 | 1.81 | 1.94 |
| Z measured (μm) ± 0.03 from sectional analysis | 1.01 | 1.32 | 1.61 | 1.80 | 1.96 |
| Difference in model and experimental cross-sectional depths (%) | 5.0 | 0.8 | 0.6 | 0.5 | 1.0 |

The data in table 4 shows that there is a very good agreement between the intended shape as analysed with the model and the actual experiment set with the model predictions (see also Fig. 4). Only at the level of the first top most step the difference in depth between the model and the experiment is larger at 5 %. The depth at all other subsequent patterns is extremely well predicted, with errors less than one percent. At the centre line of the trench the model predictions is for depth 1.94 μm versus the experimental measurement of 1.96 μm. The model validation study has shown that the FIB computational model can be used as a powerful tool to predict how to set the FIB system in order to accurately achieve the desired shape without following a trial-and-error approach. However, it is critical to use accurate model input data related to the process parameters, in particular the sputtering yield and also to account for variation in the FIB process parameters.

## 3. Computational Modelling and Simulation of Nano-Imprint Forming

The principle of the generic NIF process involves several steps: (A) softening (above Tg) of a thin film of formable material deposited on a substrate, (B) pressing a rigid mould which has the required (negative) pattern of features onto the formable material, (C) curing the formable material by cooling down to temperatures below Tg, and finally (D) releasing the mould. The sequence of these process steps results in a negative replication of the mould pattern onto the formable material with process ability for very high resolution [7]. Figure 5 outlines schematically the imprint forming process steps.

Whereas conventional techniques use convection heating, the NIF approach considered in this paper uses focused microwave radiation to soften the material. This method has already been investigated for a different application by using advanced multi-physics modelling techniques to simulate the physical phenomena and process issues related to microwave heating [8, 9]. The focus in this work is on modelling the mould pressing step only (step B in figure 5).

### *3.1. Modelling NIF Mould Pressing Step*

Most applications in NIF are based on polymeric materials that have suitable properties for the particular method of softening the material. In this investigation Poly-Methyl MethaAcrylate (PMMA) with glass transition temperature (Tg) 105°C is studied. Above Tg, the target PMMA material is soft and the pressing of the mould into the polymer promotes the flow of the formable material into the mould pattern grooves. The actual elevated temperature to which the polymer is taken and then deformed is an important process parameter. The imprint temperature is not an independent process parameter. For example, higher imprint temperature is associated with lower imprint pressure at the expense of longer process steps.

The imprint pressure is another process parameter that has a major impact on the quality of the replication. An insufficient pressure will result in incomplete filling of the pattern grooves and may lead subsequently to shape defects. Conversely high imprint pressures should also be avoided as this will cause high residual stresses in the cured polymer during the subsequent process steps. Therefore one of the most important parameters to be optimised is the imprint pressure.

The deformation process and the process parameters are also influenced by the shape of the pattern and the aspect ratio (AR) of the grooves. Important relationships between the process parameters are identified and design rules for accurate and defect-free replications are formulated using the modelling results. In this work the effects of the step dwell times are not analysed.

Above Tg the PMMA is assumed to behave as non-compressive rubber-like hyper-elastic body. The Money-Rivlin hyper-elastic model is used to model the large deformations during the deformation process. The constants at different temperatures required by the Money-Rivlin material model used to simulate the visco-elastic behavior of the PMMA above Tg are discussed and reported in reference [10]. In all simulations that follow the mould is modelled as a rigid body. The analysis is also set as a large strain and large deformation analysis. Finally, contact boundary and interfaces between the mould and the polymer are included. No friction is assumed between the mould and the polymer. The simulations are carried out using ANSYS software. Hirai et al. have validated through experiments the above modelling approach in the simulation of the dynamics of the deformation process for thermal nanoimprint lithography [11].

The finite element analysis model of the pressing step is two-dimensional and comprises a single groove section of the mould pattern and the associated part of the formable material. A nominal case is designed to simulate the deformation process at imprint temperature of 170°C and for mould pattern where the groove has height and width of 300 nm resulting in aspect ratio AR=1. The thickness of the PMMA is also 300 nm. The nominal model is illustrated in figure 6.

### *3.2. Simulation Results*

*3.2.1. Effect of Imprint Pressure.* To understand the effect of imprint pressure, the nominal case is run with several different pressures and the deformation profile is observed. This parametric study is used to understand what percentage of the groove is filled as a result of applying various in value imprint pressures P and ultimately to indicate the nominal pressure $P_0$ required. The nominal pressure $P_0$ is the pressure at which the groove is completely filled with material. We also define a quantity called relative imprint pressure, $P_R$. The relative imprint pressure is the ratio of the applied pressure and the nominal pressure, $P_R=P/P_0$, and provides a more generic interpretation of the simulation results in terms of relative quantities. Figure 7 shows the simulated polymer deformation profile at four different relative imprint pressures and details the percentage, in volume, of the groove which is successfully filled. This quantity is defined as the filling ratio. The relationship between the relative imprint pressure and the filling ratio is non-linear. For example, at 1/3 of the nominal imprint pressure, about half of the groove is penetrated by the polymer and at imprint pressure of half the nominal pressure about 76% of the groove is successfully filled. The existing trend is illustrated in figure 8.

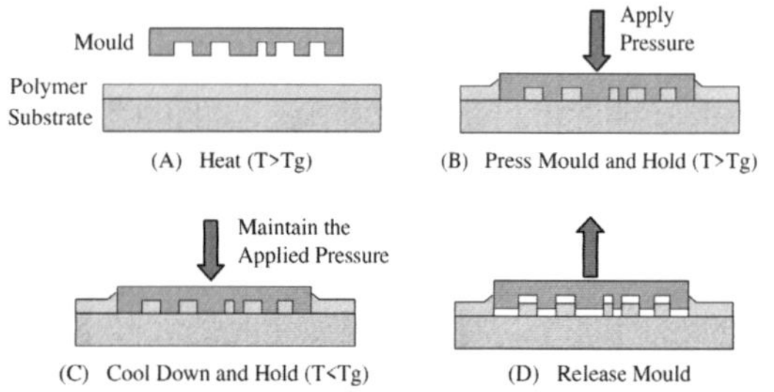


**Figure 5.** Schematic outline of imprint forming generic process steps.

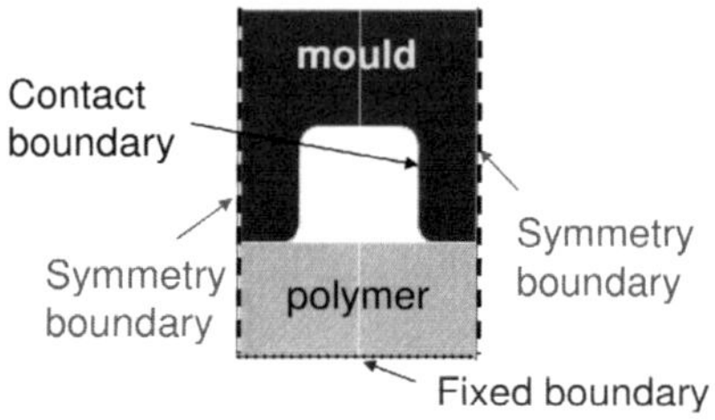


**Figure 6.** Imprint forming model and boundary conditions.

The simulation results of the nominal case show that the absolute value of the required nominal pressure $P_0$ that can promote a complete filling of the pattern is 2.28 MPa. It is useful to consider the nominal imprint pressure in relation with the polymer elastic modulus, $E_0$, at the relevant imprint temperature. The required nominal imprint pressure $P_0$ for the nominal case at imprint temperature 170°C is approximately twice the elastic modulus, $P_0=1.9E_0$. The ratio $P_0/E_0$ defines the normalised imprint pressure $P_N$ that can be used as a relative process parameter to indicate the imprint pressure relations to the formable material behavior and properties.

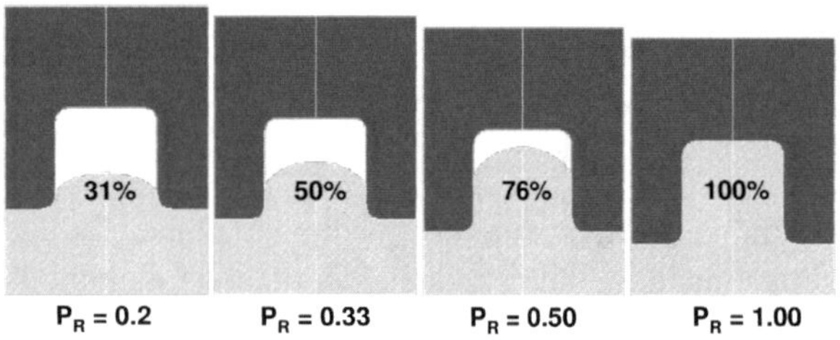


**Figure 7.** Deformation profiles from simulations at different relative imprint pressures vs. filling ratio.

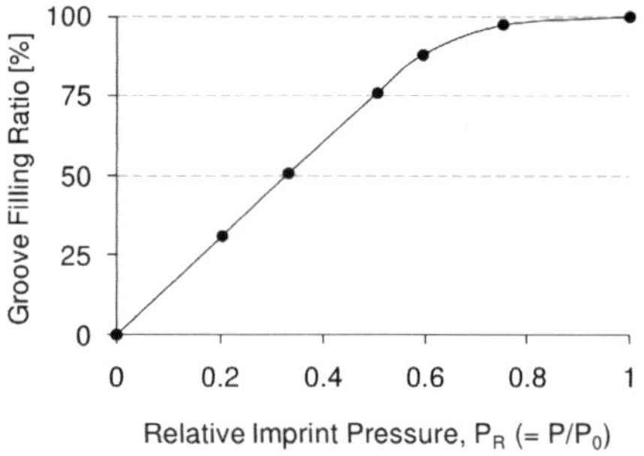


**Figure 8.** Relative imprint pressure vs. filling ratio.

*3.2.2. Effect of the Aspect Ratio of the Grooves in the Pattern.* The pattern aspect ratio (AR) is defined as the ratio of groove height over groove width. In order to fill in completely patterns with different aspect ratios, different nominal imprint pressures are required. To gain knowledge on the pressure trends, three different aspect ratio cases (0.5, 1, and 1.5) are modelled and analysed as illustrated in figure 9.

The different aspect ratios are achieved by varying the width of the grooves and by keeping the height constant at 300 nm as in the nominal case. The thickness of the PMMA layer is also unchanged at 300 nm. The three aspect ratio cases are simulated at 170°C.

The results show that, for higher aspect ratio, higher imprint pressure is required to achieve a complete filling of the pattern. Figure 10 illustrates the simulated data in terms of aspect ratio vs. normalised imprint pressure $P_N$. Simulation results indicate non-linear relationship between AR and the imprint pressure. With exactly the same process conditions and materials as used for the analysis of the nominal case, increasing only the pattern aspect ratio from 0.5 to 1.5 would require 45% higher imprint pressure to ensure accurate shape replication.

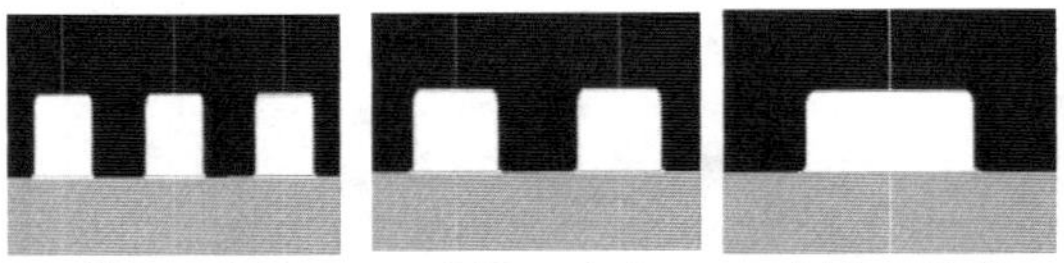


**Figure 9.** Aspect ratio cases.

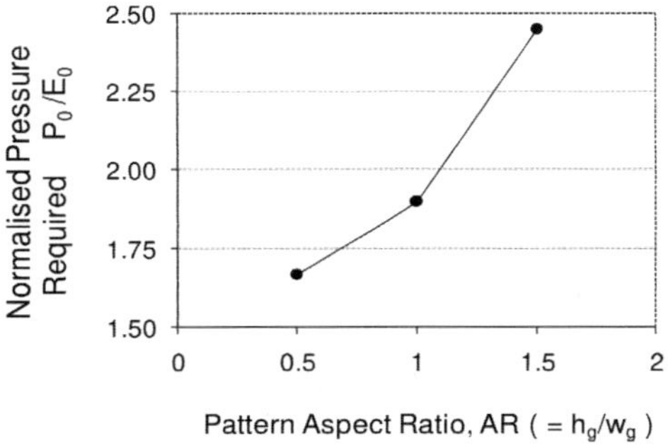


**Figure 10.** Aspect ratio vs. imprint pressure.

*3.2.3. Effect of Imprint Temperature.* The PMMA shows different material behavior at different imprint temperatures because the viscous flow properties change. Therefore, the imprint temperature as a process parameter has influence on the other process parameters. The aim of this part of the study is to evaluate the requirements for imprint pressure when the imprint forming is undertaken at different temperatures. To model this effect, the data for PMMA visco-elastic behaviour at temperatures 150, 170 and 190°C is used [10]. The simulations undertaken to evaluate the relationship between the imprint temperature and the imprint pressure are based on the nominal case pattern scenario.

As expected, because at higher temperatures the polymer becomes less viscous and easy to deform, lower imprint pressure need to be applied. As figure 11 shows, the required imprint pressure lowers when the imprint temperature increases. For the PMMA investigated in this work and in terms of absolute values, it is predicted that the imprint pressure at temperature 150°C is 3.3 MPa and at temperature 190°C the imprint pressure is over two times lower (1.4 MPa).

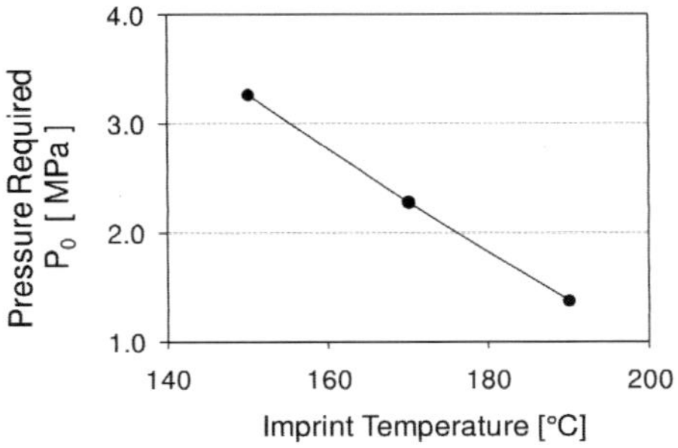


**Figure 11.** Imprint temperature vs. imprint pressure.

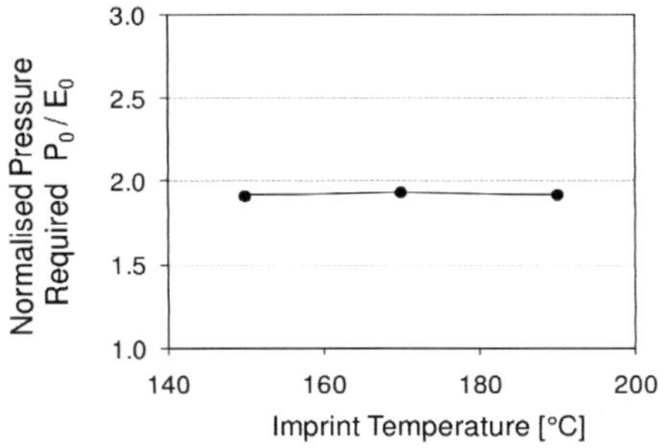


**Figure 12.** Imprint temperature vs. normalised imprint pressure.

The relationship between the imprint temperature and the normalised imprint pressure $P_N$ (=$P_0/E_0$) provides further insight into the existing trends. Elastic modulus $E_0$, here, is different at different temperatures. The graph in figure 12 reveals an interesting result. The normalised imprint pressure is virtually independent of the imprint temperature and, for the nominal case of the pattern geometry and the PMMA being analysed, it is close to constant 1.9.

## 4. Conclusions

Computational approach to study two nano- and micro-manufacturing processes, FIB and NIF, is researched and presented. A computational model for estimating FIB milling dwell times to achieve predefined shapes is validated. The analysis has shown that the model accuracy is very good. In the case study reported the difference in the depth of a parabolic step-wise trench predicted with the model and observed in the experiment was 1%. Clearly, variations do exists in the process and there is also some uncertainty associated with the process parameters. Quantifying the effect of these uncertainties and incorporating their effect in the model as part of a risk analysis assessment is where we focus our interest at present. Additional computations that account for re-deposition effects are also considered though more relevant in scenarios of milling high aspect ratio features.

The simulation results on the NIF process have provided insights into the potential risks associated with the technology and have shown trends that can help minimising commonly observed defects. For the particular PMMA polymer studied in this work the optimal imprint pressure was identified. The imprint pressure is the most important parameter that can result in shape imperfections if not optimised. This parameter depends also on other process parameters such as the imprint temperature and the aspect ratio of the grooves. Future work will address the effect of other process parameters such as the dwell times of the process steps and issues that are specific to the way in which the formable material is softened. An effort will be placed also on gathering relevant material and process data required by simulation models.

**Acknowledgments**

The authors acknowledge the UK funding council - EPSRC - for supporting the work detailed in this paper and in particular the 3D-MINTEGRATION project under grant EP/C534212/1.

**References**

[1] Desmulliez M and Topham D 2008 3D-Mintegration: the design and manufacture of 3D miniaturised integrated products, *Proc. 2nd Electronics System-Integration Technology Conference* (London, UK, 1-4 Sept 2008) pp 737-741
[2] Gamo K 1993 Focused Ion Beam technology, *Semicond. Sci. Technol.* **8** 1118
[3] Tseng A A 2004 *J. micromechanics and microengineering*, **14** 15
[4] Chou S Y and Krauss P R 1997 *Microelectronics Engineering* **35** 237
[5] Nassar R, Vasile M and Zhang W 1998 *J. Vac. Sci. Technol.* B **16** 109
[6] SRIM2010, http://www.srim.org/
[7] Chou S Y and Krauss P *Microelectronics Engineering* **35** 237
[8] Tilford T, Sinclair K, Bailey C, Desmulliez M, Goussettis G, Parrott A and Sangster A 2007 *J. Soldering and Surface Mount Technology* **19** 26
[9] Sinclair K, Desmulliez M and Sangster A 2006 A Novel RF-Curing Technology for Microelectronics and Optoelectronics Packaging, *Proc. 1st Electronics System-Integration Technology Conference* (Dresden, Germany, 5-7 Sept 2006) pp 1149-1157
[10] Stoyanov S, Amalou F, Sinclair K, Bailey C and Desmulliez M 2008 Modelling the nano-imprint forming process for the production of miniaturised 3D structures, *Proc. 9th EuroSIME* (Freiburg-im-Breisgau, Germany, 21-23 April 2008) pp 497-504
[11] Hirai Y, Fujiwara M, Okuno T and Tanaka Y 2001 *J. Vac. Sci. Technol.* B **19** 2811

16 ISCMP: Progress in Solid State and Molecular Electronics, Ionics and Photonics IOP Publishing
Journal of Physics: Conference Series **253** (2010) 012009 doi:10.1088/1742-6596/253/1/012009

# Electroreflectance investigations of quantum confined Stark effect in GaN quantum wells

**A Drabinska, K Pakula, J M Baranowski and A Wysmolek**

Institute of Experimental Physics, University of Warsaw, Hoza 69, Warsaw, Poland

e-mail: Aneta.Drabinska@fuw.edu.pl

**Abstract**. In this paper we present room temperature electroreflectance studies of GaN quantum wells (QWs) with different well width. The electroreflectance measurements were performed with external voltage applied to the structure therefore it was possible to tune the electric field inside QW up to its completely screening and furthermore even reversing it. The analysis of QW spectral lines showed the Stark shift dependence on applied voltage and well width reaching about 35 meV for highest voltage and widest well width. It was possible to obtain the condition of zero electric field in QW. Both broadening and amplitude of QW lines are minimal for zero electric field and increases for increasing electric field in QW. The energy transition is maximum for zero electric field and for increasing electric field it decreases due to Stark effect. Neither amplitude and broadening parameter nor energy transition does not depend on the direction of electric field. Only parameter that depends on the direction of electric field in QW is phase of the signal. The analysis of Franz-Keldysh oscillations (FKOs) from AlGaN barriers allowed to calculate the real electric field dependence on applied voltage and therefore to obtain the Stark shift dependence on electric field. The Stark shift reached from -12 meV to -35 meV for 450 kV/cm depending on the well width. This conditions were established for highest forward voltages therefore this is the value of electric field and Stark shift caused only by the intrinsic polarization of nitrides.

## 1. Introduction

Due to potential application for many optoelectronic systems [1-4], nitrides for several years are under advanced studies focused on the development and improvement of their optical properties. It has been predicted by theory and confirmed by experiment that nitrides in their wurtzite crystal structure are pyroelectric materials [5-8]. Large values of spontaneous polarization which are present without any external electric field (0.034 $Cm^{-2}$ for GaN and 0.09 $Cm^{-2}$ for AlN) are characteristic for these crystals. The orientation of this spontaneous polarization is always along the [000-1] direction and cannot be inverted by external electric fields. The orientation of the spontaneous polarization relative to the c-axis can only be controlled by adjusting the polarity of the material. The piezoelectric constants of group III nitrides are a factor of 5–20 larger than those of InAs, GaAs and AlAs. The macroscopic nonlinear pyroelectric polarization of wurtzite nitrides (both spontaneous and piezoelectric polarization) dramatically affects the optical and electrical properties of multilayered AlGaN/GaN hetero-, nanostructures and devices, due to the huge built-in electrostatic fields and bound interface charges caused by gradients in polarization at surfaces and heterointerfaces. The realization of undoped AlGaN/GaN heterostructures leads to the existence of a gradient in internal polarization and therefore it is relatively easy to obtain two-dimensional electron gases (2DEGs) in GaN/AlGaN structures [9]. Observations of Franz–Keldysh oscillations and Stark effects in InGaN/GaN and

AlGaN/GaN QW structures provide direct evidence for the presence of strong electric fields caused by gradients in polarization [10-13]. Controlling the electric field in QWs by external bias is crucial for the investigation of optical properties of quantum wells. Therefore the principle of these properties is the change of the absorption coefficient as a function of an applied voltage [12, 13]. For bulk material, this is due to the Franz-Keldysh effect (FKE) [14]. In the case of quasi-two-dimensional semiconductor structures, i.e., quantum wells, the corresponding phenomenon is the quantum-confined Stark effect (QCSE) [15]. Although both effects have been extensively studied in GaAs-based structures [16], much less work has been done using GaN-based alloys.

## 2. Samples

The investigated structures were grown on the sapphire substrate in a horizontal, law pressure Metal-Organic Vapor Phase Epitaxy (MOVPE) reactor. On the 20 nm nucleation layer there were grown sequentially: $Al_{0.09}Ga_{0.91}N$ layer and highly silicon doped $Al_{0.09}Ga_{0.91}N$ layer, both 600 nm thickness. Next on the n-type layer there were grown: 10 nm back $Al_{0.09}Ga_{0.91}N$ barrier, GaN QW and 20 nm $Al_{0.09}Ga_{0.91}N$ top barrier. The structures were covered with very thin cap layer of low-temperature GaN. We present measurements of three structures with different QW thickness: 3 nm, 6 nm and 9 nm, marked in this paper as sample A, B and C respectively. The nominal concentration of dopants was $2 \times 10^{18}$ $cm^{-3}$ for narrower QWs and $2 \times 10^{18}$ $cm^{-3}$ for widest QW.

In order to decrease the number of screw dislocations, ensure high resistivity of undoped layers and ensure high quality and flatness of interfaces, two of the structures were grown on modified low-temperature AlN nucleation layer. It has been shown that the 3D or 2D growth mode of AlGaN layers depends predominantly on the growth conditions of the low temperature (LT) nucleation layer [17]. Commonly used 3D growth mode is achieved on LT GaN or AlN nucleation layer. Growth of a successive layer begins from separated sites, where individual 3D crystallites are formed. Threading dislocations present in crystallites bend on their facets, which reduces the quantity of dislocations. However, slight crystallographic misorientations between crystallites lead to the creation of new dislocations during coalescence of the crystallites. As a result, edge and mix dislocations appear at similar densities of about $10^9$ $cm^{-2}$. Modification of growth conditions of LT AlN nucleation layer, especially reduction of their growth rate, leads to drastic changes in properties of the layer. Growth of successive AlGaN layer at high temperature starts evenly on whole surface retaining atomic flatness. Thus growth at high temperature occurs only by 2D mode. Therefore, it is possible to grow a very thin AlGaN layers directly on top of LT nucleation layer. Such layers contain large number ($10^{10}$ $cm^{-2}$) of edge dislocations, and relatively small number (less than $10^8$ $cm^{-2}$) of mix dislocations. The widest QW was grown on the standard GaN buffer.

## 3. Experimental technique

Modulation spectroscopy is the branch of optical spectroscopy that deals with the measurement and interpretation of changes in optical spectra caused by modification of measurements condition. The general concept of modulation spectroscopy is that an optical spectrum that contains of many broad lines can be reduced to a series of derivative-like lines [18-21]. In this work we present only one particular modulation i.e. electric field modulation. In the case of electro-modulation reflectivity (EMR) the measured parameter is a periodic variation in reflectivity caused by small periodic changes of electric field in the sample. The benefits of this technique is that it gives very sharp, sensitive to build-in electric field, lines even at room temperature.

Electroreflectance (ER) is one of the method of implementing EMR. It requires at least two contacts on the structure and modulation voltage is applied between them so the internal electric field is modulated. In this paper we used Schottky barrier electroreflectance (SBER), in which one of the contacts is a semitransparent gold Schottky contact evaporated on the structure surface. Melting indium into a side of the sample made ohmic contacts to the n-doped $Al_{0.09}Ga_{0.91}N$. In this case the electric field parallel to the growth direction is modulated. The ER is measured in the depleted region which extends from the surface through the QW and barriers to the n-doped $Al_{0.09}Ga_{0.91}N$.

The reflectivity was measured with a monochromatic light obtained from a Xenon lamp. The light of intensity $I_0$ was focused on the semitransparent Schottky electrode and the reflected light was detected by silicon detector. The intensity signal consists of dc value $I_0R(E)$ and ac value $I_0\Delta R(E)$, measured by a voltmeter and lock-in amplifier synchronized with the modulation source, respectively. Dividing both signals eliminates dependence on the intensity of the light source, and yields to the quantity $\Delta R/R(E)$. Except the modulation voltage it is also possible to apply constant voltage in this method. This allows to change the electric field in depletion layer and furthermore to measure the dependence of optical properties of QW and barriers on electric field.

## 4. Results and discussion

The electroreflectance maps (figure 1) were measured at room temperature for energy range 3.3 – 3.8 eV for sample A and 3.3 – 4.0 eV for samples B and C and voltage range -1.0 – 0.7 V, -1.5 – 0.8 V and -1.5 – 1.0 V for samples A, B and C respectively. For energies higher than 3.6 eV for all samples characteristic vanishing oscillations are clearly observed. These are Franz-Keldysh oscillations from AlGaN barriers. For all three samples the period of this oscillations increases with increasing reverse bias. For energies lower than 3.6 eV transitions in QW are observed. The position of these lines changes with applied voltage. This lines vanishes from ER spectrum for voltage -1.1 and -0.7 V for sample B and C respectively.

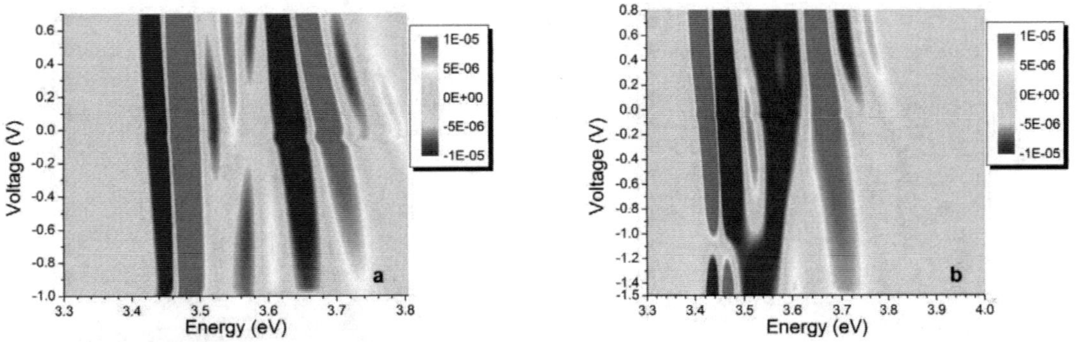


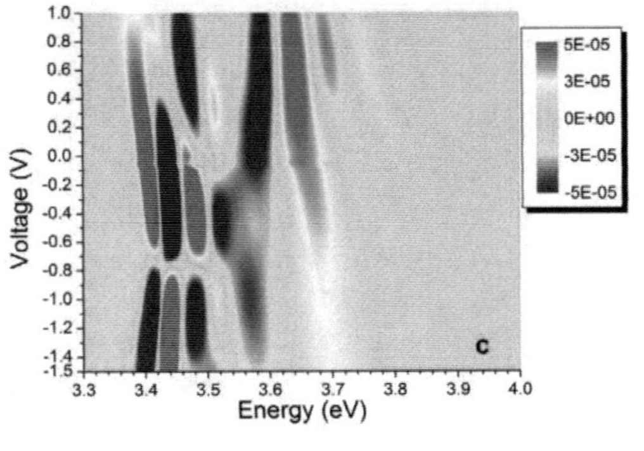


**Figure 1.** Electroreflectance signal as a function of energy and applied voltage of sample A – (a) , sample B – (b) and sample C – (c).

### *4.1. Fitting procedure*

The perturbation, due to the modulation of electric field, destroys the translational symmetry and hence the free charge carriers in the crystal can be accelerated. The lineshape of ER signal from layers can be expressed as a function of broadening parameter ($\Gamma$), transition energy ($E_g$), electric field ($F$) and effective mass in field direction ($\mu_\parallel$) using following formula [22, 23]:

$$\frac{\Delta R}{R} = \mathrm{Re}\left[ Ae^{i\varphi} \frac{H(z)}{(E - i\Gamma)^2} \right], \qquad (1)$$

Journal of Physics: Conference Series **253** (2010) 012009 doi:10.1088/1742-6596/253/1/012009

where $H(z)=2\pi\left\{e^{-i\pi/3}Ai'(z)Ai'\left[ze^{-2i\pi/3}\right]+ze^{-2i\pi/3}Ai(z)Ai\left[ze^{-2i\pi/3}\right]\right\}+i\sqrt{z}$,

$z=\frac{E_g-E+i\Gamma}{\hbar\theta}$ and $(\hbar\theta)^3=\frac{(eF\hbar)^2}{2\mu_\parallel}$.

For carriers in bound states, like quantum wells, the effective mass along the field direction is infinitive and electro-optic energy ($\hbar\theta$) vanishes. These types of particles are confined in space and the electric field is not able to accelerate them. In this case the effect of an electric field applied along direction of confinement is very different that it was described previously. For example, considering the square quantum well, the effect of the electric field along the confinement direction adds a linear potential, which tilts the quantum well and the barriers, thus changing the shape of the potential well. The wave functions are mixed in such a way, that electrons and holes are shifted in opposite directions, but they are still confined. In addition, their energy of the state is changed as well. In high electric field the quantum well may be perturbed in a way that tunneling may take place. In this case, the measured ER signal hardly depends on electric field and can be approximated by the formula:

$$\frac{\Delta R}{R}=\mathrm{Re}\left[Ae^{i\varphi}\left(E-E_g+i\Gamma\right)^{-2}\right] \tag{2}$$

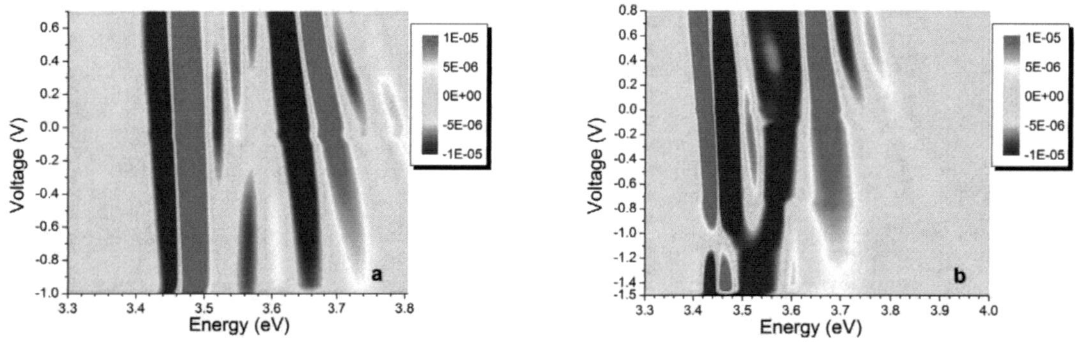


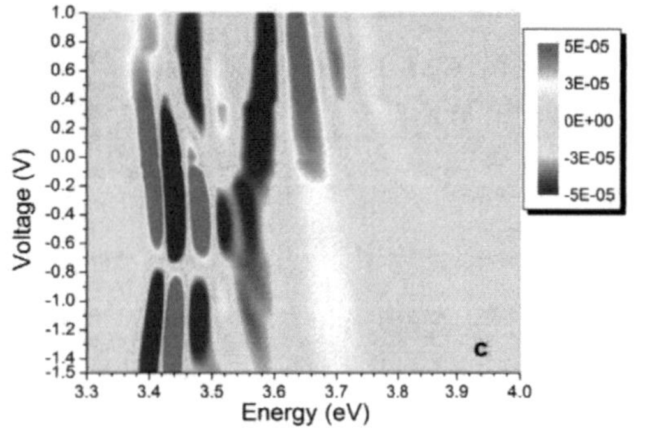


**Figure 2.** Fit of electroreflectance signal as a function of energy and applied voltage of sample A – (a), sample B – (b) and sample C – (c). QW lines were fitted using equation (2) and FKO using equation (1). Fitting procedure used one, two and three QW spectral lines for samples A, B and C respectively.

For all three samples for each voltage there was fitted theoretical lineshapes (from one to three QW lines for energy lower than 3.6 eV and FKOs for energies higher 3.6 eV). Since the lineshape of each transition is complicated, depends on several parameters (4 for QW transitions and 5 for transition in barriers) and individual lines overlap with each other, the fitting procedure is tricky. First, fitting was performed separately for FKO and QW. For samples B and C, were there was more than one QW lines, fit was performed step by step, adding individual QW lines starting from the lowest energies. Finally with well established initial parameters, there was performed the final fit of spectrum for each voltage. The results of the fitting are presented in figure 2.

### *4.2. QW lines*

Due to the spontaneous and piezoelectric polarization positive and negative polarization charge is present on both AlGaN/GaN interfaces. Although the direction of the electric field in the outer barrier is established by the surfaces states the internal electric field in QW is opposite to the one in the inner barrier and to the electric field build by the applied bias. Although internal polarization can not be changed by applying external bias, it is possible to add external electric field that can compensate this internal electric field. Furthermore by applying high external bias it is possible to obtain the net electric field opposite to the polarization induced field.

For low reverse bias the electric field in QW is opposite to the one in barriers (figure 3a), therefore QW lines shift into blue with increasing reverse bias and reaches maximum for bias -1.0 V and -0.7 V for samples B and C respectively (figure 1b and c). For sample A it is not possible to establish voltage for maximum QW transition energy, since it was not reached in experiment (figure 1a), but it is greater than maximum reverse voltage (-1.0 V). This voltage is sufficient enough to completely compensate electric field in QW (figure 3b) The decrease of screening voltage with increasing well width is due to increasing depletion layer width and different voltage-electric field dependence for each sample. For zero electric field, for samples B and C the electro-reflectance signal from QWs vanishes. Higher reverse biases builds opposite electric field to the QWs so the lines appear again in spectrum but with opposite phase. Increasing reverse bias increases electric field in the QW therefore lines shift back into red (figure 1b and c). This time electric field is in the same direction than the one in barriers (figure 3c).

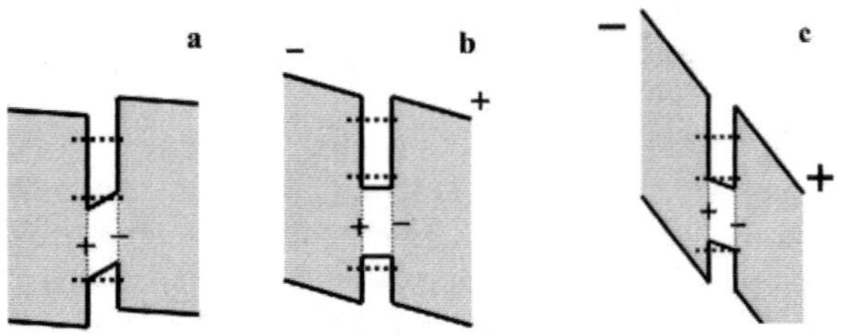


**Figure 3.** The schematic band structure for III-nitride QW (considering its pyroelectricity i.e. spontaneous and piezoelectric polarization) without external electric field (a) and in external electric field opposite to the pyroelectric field (b) and (c). (b) presents the case when external electric field has the same value that fields induced by pyroelectric polarization therefore the net electric field in QW is zero.

For all measured samples for energies below 3.6 V, using equation (2), there were fitted QW lineshapes. Fitting one QW transition for sample A, two for sample B and three for sample C gave the best fitting results. The fitted energy positions, broadening parameters and amplitude of the sample B are presented in figure 4 a, b and c respectively. For screening voltage (zero electric field in QW), for both transitions, the amplitude and broadening parameter are minimal. Increasing electric field in barriers causes an increase of both: the amplitude and broadening parameter of fitted lines. During the change of direction of electric field in QW the modulation of voltage has not changed. This results in the inversion of phase of ER signal after the inversion of the electric field. For sample A and C the dependence is very similar. Square dependence for each QW transition was fitted to obtain the value of screening voltage ($U_0$) and the energy transition ($E_0$) for zero electric field for B and C samples. Fitted screening voltage are -1.0 V and -0.7 V for sample B and C respectively. Transition energies for sample B are 3.432 eV, 3.465 V and for sample C – 3.413 eV, 3.441 eV, 3.482 eV. For sample A in ER measurement the conditions of screening electric field wasn't reached therefore we can only say that screening voltage is over -1 V of reverse voltage and transition energy is larger than 3.464 eV. However the energy dependence on bias suggests that these parameters are very close to the "zero electric field" case.

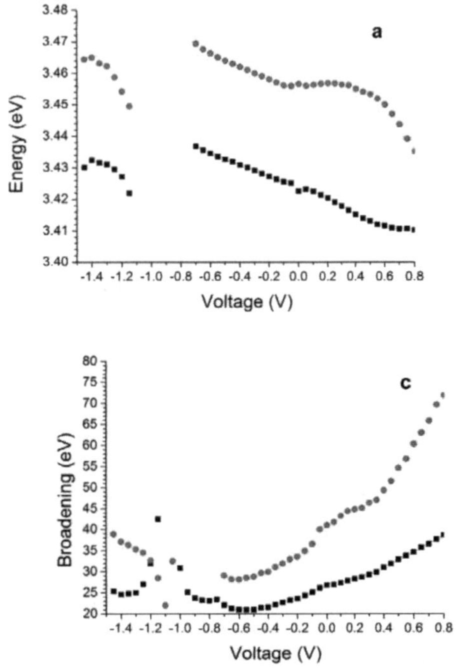


**Figure 4.** Fitted energy, amplitude and broadening of two QW transitions for sample B. The ground and exited states are marked with black squares and red circles respectively.

To obtain the Stark shift of each transition there was subtracted energy transition for "zero electric field" ($E_0$) from the fitted energy transition for each voltage. Since the value of Stark shift does not depend on the direction of electric field the Stark shift was presented (figure 5) as a function of absolute value of the difference between the actual voltage and screening voltage ($U_0$). The value of Stark shift depends on the well width and for 1.6 V reaches -14 meV, -20meV and -35 meV for sample A, B and C respectively.

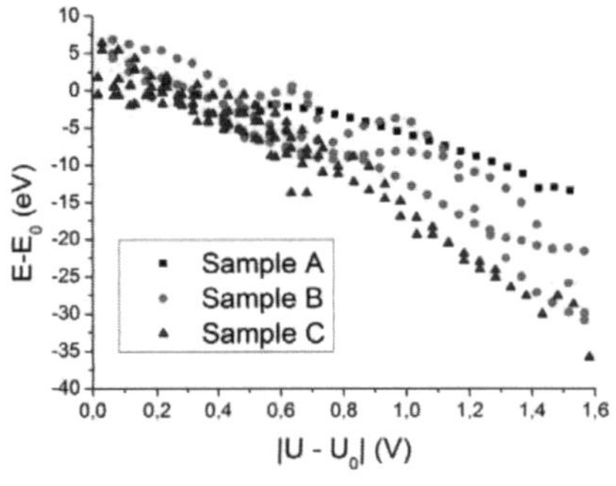


**Figure 5.** Stark shift dependence on voltage applied to QW for sample A (black squares), B (red circles) and C (blue triangles).

*4.3. Franz-Keldysh oscillations from barriers*

Although the exact expression describing Franz-Keldysh oscillations is complicated (equation (1)) it is relatively easy to evaluate the value of electric field without full analyze of the spectrum. Due to periodic dependence of ER signal as a function of $(E\text{-}E_g)^{3/2}$ it is possible to calculate the electric field ($F$) from the period ($T$) of FKOs using the following expression:

$$T = \frac{3\pi e F \hbar}{\sqrt{8\mu_{\parallel}}}$$

Unfortunately for the higher reverse voltage amplitude of FKOs from barriers is very small and only two-three oscillations are observed, therefore obtained value of electric field is not accurate. However it is clearly visible for all samples that the electric field in barriers, the same as the period of oscillations, increases with increasing reverse voltage. Analysis of FKOs in all samples showed that electric field increases linearly with the applied voltage. Linear dependence of electric field on voltage can be understood assuming he constant width of depletion layer. The best fit to the formula:

$$F = -\frac{U - U_B}{d},$$

where $U_B$ is Schottky barrier height and $d$ is depletion layer width, gives the depletion layer width 33 nm, 36 nm and 39 nm for samples A, B and C respectively. This agrees with nominal widths of depletion layer for each sample. Obtained Schottky barrier height was equal 1.95 V.

The presence of intrinsic polarization (spontaneous and piezoelectric) results in electric field jump on each interface in heterostructures. On the first AlGaN/GaN interface, the electric field in QW is reduced compering to the one in barrier. From analysis of QW lines it was possible to obtain value of voltage for which the electric field in QW is zero. Therefore, for that voltage, the value of electric field in barriers gives the difference of intrinsic polarization between AlGaN and GaN layers. Obtained the drop of polarization was 0.0059 C/m$^2$ for sample B and 0.0052 C/m$^2$ for sample C. The spontaneous polarization in GaN (0.034 C/m$^2$) and AlGaN (0.039 C/m$^2$) is the same for both samples, therefore the difference in these values (0.0007 C/m$^2$) is caused by the piezoelectric polarization. The difference in piezoelectric polarization between individual samples is most probably caused by the different growth mode (2D mode for sample B and 3D for sample C) which produced different stress distribution.

More accurate analyses of Franz-Keldysh oscillations are possible after fitting data with equation (1). First of all the fitting procedure showed that for forward and low reverse voltage the best fit gives fitting with two sets of FKOs with the same energy transition and different electric field (figure 6a), whereas for higher reverse voltage it was enough to fit only one set of FKO (figure 6b).

**Figure 6.** Fit of sample B FKO region using equation (1) for 0.8 V (a) and -0.2 V (b). Points represent experimental data, solid red line is the best fit (two set of FKOs for 0.8 V and one set of FKOs for -0.2 V). For 0.8 V with blue line is presented the best fit using only one set of FKOs.

Both barriers have the same Al content, therefore the spontaneous polarization in back and top barrier is the same. Assuming the same stress in both layers (QW is too thin to produce significant stress comparing to $Al_2O_3$ substrate and 1200 nm of AlGaN layers), the value of electric field in both barriers should be the same. However if the quantum state is occupied and 2DEG is present in QW, it produces additional electric field opposite in top and back barrier. Therefore for forward and low reverse voltage QW is occupied. This results in two values of electric field in barriers. For higher reverse voltage the electric field in back barriers is large enough to produce sharp triangular potential barrier (figure 3c) and allow the tunneling of carriers from the QW. For higher reverse voltage the tunneling is so effective that QW is not longer occupied. This is confirmed also in luminescence measurement, in which the intensity of the PL peak dramatically decreases when sufficient reverse voltage is applied.

### *4.4. Quantum Confined Stark effect*

Analysis of Franz-Keldysh oscillations showed that electric field in barriers depends linearly on the voltage for all investigated samples. Since electric field in QW is shifted by intrinsic polarization, but the dependence on bias is the same, it was possible to translate the voltage dependence of the Stark shift in QWs (figure 5) to its electric field dependence (figure 7). Maximum electric field applied to the QW was about 450 kV/cm and for this electric field the Stark shift was -14 meV, -20meV and -35 meV for sample A, B and C respectively.

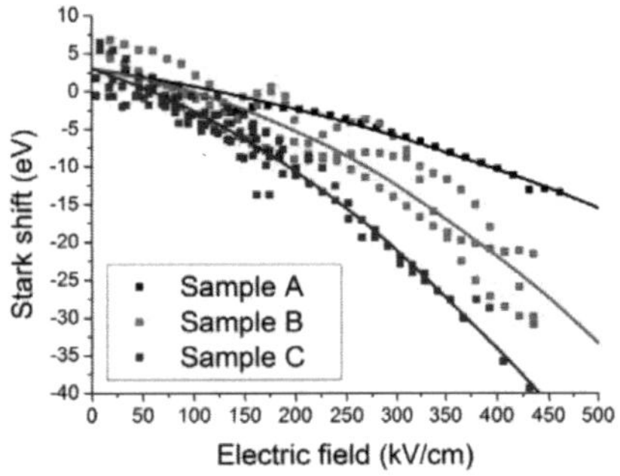


**Figure 7.** Stark shift dependence on electric field in QW for sample A (black squares), B (red circles) and C (blue triangles). Fitted with equation (3) parabola dependencies are shown as a black, red and blue lines for sample A, B and C respectively.

For electron confined in a QW, drift in an electric field $F$ is impossible. Instead its wave function $\psi(z)$ is moved by the field to the wall of the QW and squeezed. The Schrödinger equation should be completed with a potential $V = -eFz$. In the case of a rectangular QW of width $L$, with infinite barriers, the solution is a sum of the Airy functions [24]. The resulting energy eigenvalue ($W_{QW}$) is lowered by an energy of the order of $eFL$ (about 0.1 eV). The calculated dependence was nonlinear. In the range $eFL<10*W_{QW}$, it can be approximated with expression:

$$E - E_g - W_{QW} = -AF - BF^2, \tag{3}$$

where $A = \alpha \dfrac{eL}{W_{QW}}$, $B = \beta \left( \dfrac{eL}{W_{QW}} \right)^2$, $\alpha = 0.0090$ and $\beta = 0.0086$.

The above dependence was fitted to the Stark shift obtained from ER analysis (figure 7). The fitted coefficients are presented in Table 1.

**Table 1.** Fitted coefficients of Stark shift dependence on electric field (equation (3)).

| | $A\times10^{-29}$ [Cm] | $B\times10^{-55}$ [$C^2m^2$] | $B/A^2$ |
|---|---|---|---|
| SAMPLE A | 3.18±1.35 | 0.88±0.56 | 87±66 |
| SAMPLE B | 3.28±0.94 | 2.68±0.36 | 249±79 |
| SAMPLE C | 7.00±1.06 | 3.15±0.45 | 64±13 |

According to equation (3) quantity $B/A^2$ should be constant for all samples and equal to 122. Experimentally obtained values for sample A and C are lower and for sample B is bigger than the expectations. However both A and B samples agrees with this value. In measured QWs the highest electric fields are of the order of 450 kV/cm, for which for the widest quantum well $eFL \sim 10*W_{QW}$, therefore the limits of validity of equation (3) approximation is reached. This can explain the observed discrepancies.

## 5. Conclusions

We have presented results of the electroreflectance measurements on three $Al_{0.09}Ga_{0.91}N/GaN$ quantum wells with different well width (3, 6 and 9 nm). The electroreflectance measurements were performed at room temperature with external voltage applied to the structure. Analysis of Franz-Keldysh oscillations from barriers showed that applied voltage built over 800 kV/cm of electric field for highest reverse voltage. The electric field have been built in all layers in depletion region, which extended over QW and both barriers. The presence of intrinsic polarization (spontaneous and piezoelectric) results in electric field jump on each interface in heterostructures. On the first AlGaN/GaN interface, the electric field in QW is reduced compering to the one in barrier (for zero external electric field it is even inverted) and on the second GaN/AlGaN interface is again increased, so the electric field in both barriers is the same. Additional carriers in QW produces extra electric field which modifies the electric field in barriers (to one of the barriers its added to the other subtracted). This results in two set of Franz-Keldysh oscillations (of the same energy transition and different electric field) observed for forward and low reverse voltage. Screening voltage (voltage for which the electric field in QW is completely screened) in ER measurement was found to change from -0.7 V to over -1.0 V for different samples. The electric field value in barriers for screening voltage allowed to obtain the difference of total polarization between AlGaN and GaN layers equal 0.0059 C/m$^2$ and 0.0052 C/m$^2$ for samples B and C respectively. Screening voltage of QWs and electric field dependence on bias in barriers allowed to obtained the value of electric field in QWs for the whole voltage range.

Analysis of QWs lines dependence showed that both broadening and amplitude of lines is minimal for screening voltage and increases for increasing electric field in QW. The energy transition is maximum for screening voltage and for increasing electric field it decreases due to Stark effect. Neither amplitude and broadening parameter nor energy transition does not depends on the direction of electric field. Only parameter that depends on the direction of electric field in QW is phase of the signal. The Stark shift depends on the well width and electric field and reached from -12 meV to -35 meV for 450 kV/cm depending on the sample. This conditions were established for highest forward voltages therefore this is the value of electric field and stark shift caused only by the intrinsic polarization.

## References

[1] Razeghi M and Rogalski A 1996 *J. Appl. Phys.* **79** 7433
[2] Pulfrey D L, Kudek J J, Nener D B, Parish G, Mishra U K and Tarsa E J 1999 *phys. stat. sol. (a)* **176** 169
[3] Yang W, Nohava T, Krishnankutty S, Torreano R, McPherson S and Marsh H 1998 *Appl. Phys. Lett.* **73** 1086
[4] Teke A, Yun F, Dogan S, Reshchikov M A, Le H, Liu X Q, Morkoç H, Zhang S K, Wang W B and Alfano R R 2003 *Solid State Electronics* **47** 1401
[5] Bernardini F, Fiorentini V and Vanderbilt D 1997 *Phys. Rev. B* **56** R10024
[6] Zoroddu A, Bernardini F, Ruggerone P and Fiorentini V 2001 *Phys. Rev. B* **64** 45208
[7] Ambacher O 1998 *J. Phys. D: Appl. Phys.* **31** 2653
[8] Ambacher O, Majewski J, Miskys C, Link A, Hermann M, Eickhoff M, Stutzmann M, Bernardini F, Fiorentini V, Tilak V, Schaff B and Eastman L F 2002 *J. Phys.: Condens. Matter* **14** 3399
[9] Jena D, Heikman S, Speck J S, Gossard A, Mishra U K, Link A and Ambacher O 2003 *Phys. Rev. B* **67** 153306

[10] Wetzel C, Takeuchi T, Yamaguchi S, Katoh H, Amano H and Akasaki I 1998 *Appl. Phys. Lett.* **73** 1994

[11] Grandjean N, Massies J and Leroux M 199 *Appl. Phys. Lett.* **74** 2361

[12] Drabinska A, Baranowski J M, Pakula K, Caban P and Strupinski W 2009 *phys. stat. sol. (a)* **206** 816

[13] Drabinska A, Korona K P, Pakula K and Baranowski J M 2007 *phys. stat. sol. (a)* **204** 459

[14] Hosea T J C 2006 phys. *stat. sol. (b)* **189** 531

[15] Miller D A B, Chemla D S, Damen T C, Gossard A C, Wiegmann W, Wood T H and Burrus C A 1984 *Phys. Rev. Lett.* **53** 2173

[16] Miller D A B, Chemla D S and Schmitt-Rink S 1986 *Phys. Rev. B* **33** 6976

[17] Pakuła K, Baranowski J M and Borysiuk J 2007 *Crystal Research and Technology* **42** 1176

[18] Cardona M, Shaklee K L and Pollak F H 1967 *Phys. Rev.* **154** 696

[19] Aspnes D E and Studna A A 1973 *Phys. Rev. B* **7** 4605

[20] Gay J G and Klauder L T 1968 *Phys. Rev.* **172** 811

[21] Yin X and Pollak F H 1991 *Appl. Phys. Lett.* **59** 2305

[22] Aspnes D E 1967 *Phys. Rev.* **153** 972

[23] Hall D J, Hosea T J C and Lancefield D 1997 *J. Appl. Phys.* **82** 3092

[24] Miller D A B, Chemla D S, Damen T C, Gossard A C, Wiegmann W, Wood T H and Burrus C A 1985 *Phys. Rev. B* **32** 1043

Journal of Physics: Conference Series **253** (2010) 012010 doi:10.1088/1742-6596/253/1/012010

# Quadrature frequency resolved spectroscopy (QFRS) of radiative transitions of $Er^{3+}$ and $Nd^{3+}$ ions in chalcogenide glasses (ChGs)

**T Aoki[1], D Saitou[1], K Fujimoto[1], C Fujihashi[1], K Shimakawa[2], K Koughia[3] and S O Kasap[3]**

[1]Joint Research Centre of High-technology, Tokyo Polytechnic University, Iyama 1583, Atsugi, Japan
[2]Department of Electrical and Electronic Engineering, Gifu University, Gifu 501-1193, Japan
[3]Department of Electrical Engineering, University of Saskatchewan, 57 Campus Drive , Saskatoon, SK, S7N 5A9 Canada

E-mail: aoki@ee.t-kougei.ac.jp

**Abstract**. Wideband quadrature frequency resolved spectroscopy (QFRS) of photoluminescence (PL) lifetime distributions from 2 ns to 160 s is shown to be very effective in elucidating the characteristic features of radiative transitions of $Er^{3+}$ ions in GeGaSe and GeGaS chalcogenides glasses (ChGs). Undoped GeGaSe ChGs show triple-peak lifetime distributions of which two short-lifetimes are associated with singlet-triplet excitons and longest-lifetime, ~20 s, with radiative tunnelling (RT) of distant-pairs (DPs). Er-doped GeGaSe and GeGaS ChGs exhibit a double-peak lifetime distribution, consisting of a peak at ~3.3 and ~5.3 ms, respectively, a characteristic of the $Er^{3+}$ luminescence centre and another peak at ~20 s, similar to that of undoped GeGaSe ChGs. It is shown that the QFRS can separate and analyse two mixed radiative transitions of $Nd^{3+}$ ions, ${}^4F_{3/2}\rightarrow{}^4I_{15/2}$ and ${}^4F_{5/2},{}^2H_{9/2}\rightarrow{}^4I_{15/2}$ in GaLaS ChGs. From the QFRS results we can experimentally extract the branching ratio $\beta_J$ and lifetime $\tau \approx 77$ μs for 4 lasing transitions ${}^4F_{3/2}\rightarrow{}^4I_J$ ($J$ = 9/2, 11/2, 13/2, 15/2) of $Nd^{3+}$ ions in GaLaS ChGs, in particular, the weakest transition ${}^4F_{3/2}\rightarrow{}^4I_{15/2}$.

## 1.Introduction

Radiative transitions of rare earth (RE) ions such as $Er^{3+}$ and $Nd^{3+}$ embedded in a solid state matrix are of great interest for possible opto-electronic applications. The *f-f* transitions of RE ions are forbidden in the dipole approximation and become partly allowed owing to an admixture of other orbitals when they are incorporated in a solid matrix. This admixture crucially depends on the immediate vicinity of RE ions and its symmetry. A disordered material such as glass offers a distinct advantage because it lowers the symmetry of the RE environment and its short-range potential fluctuations facilitate the mixing of the wave-functions. Moreover, compared with crystalline matrices, glasses offer a variety of attractive properties. For example, they are optically isotropic, they allow size- and shape-scalability as well as a wide range of variations in their compositions. Glasses usually demonstrate high transparency, allowing operation at high optical power densities, and are less expensive and easier to fabricate than crystals. In particular, chalcogenide glasses (ChGs) possess lower phonon frequencies and higher refractive indices in comparison with traditionally used oxide and fluoride glasses. Most significantly, they allow effective doping with a large amount of RE ions, and corresponding oscillator strengths are largely enhanced especially in sulphide ChGs because of the high covalence of RE-sulphur bonds [1].

GeGaSe ChGs, in particular, have been shown to be excellent Er-host materials since they allow homogeneous Er-doping in large amounts (e.g., up to 2 at.% Er) with the subsequent formation of $Er^{3+}$ ions [1]. The photoluminescence (PL) of *f-f* transitions ($^4I_{13/2} \rightarrow {}^4I_{15/2}$) in $Er^{3+}$ centres in Er-doped GeGaSe ChGs with a near-stoichiometric composition and a high concentration of Ga features a single dominant peak in the PL spectrum at ~1.5 μm with a typical PL lifetime $\tau_{Er} \approx 3$ ms, which is almost independent of temperature *T*, and whose exact value only slightly depends on the composition of the host GeGaSe ChGs. The wavelength ~1.5 μm of $^4I_{13/2} \rightarrow {}^4I_{15/2}$ radiative transitions of $Er^{3+}$ ion matches the telecommunication window of the minimum optical loss of silica-based optical fibres. The relatively long lifetime of a few milliseconds and the ability to dissolve high concentrations of $Er^{3+}$ also make the Er-doped GeGaSe and GeGaS ChGs potential candidate materials for applications in optical amplifiers and lasers [2]. However, the PL mechanisms in these materials are still far from being understood and require further investigation for practical use as well as scientific interest [3–5].

Similarly, glasses doped with Nd are also popular materials widely used in solid-state lasers and optical amplifiers; here ChGs also are very successful as a host matrix for the same reason as Er-doped ChGs. One of the first ChGs lasers was demonstrated in 1996 by Schweizer in a $Nd^{3+}$ doped gallium lanthanum sulphide glass (GaLaS) and this was quickly followed by a GaLaS fibre laser. These lasers operated on the $^4F_{3/2} \rightarrow {}^4I_{11/2}$ radiative transition of $Nd^{3+}$ at wavelengths around 1080 nm in both pulsed and CW modes [6, 7]. In a previous paper we investigated the $^4I_{9/2} \leftrightarrow {}^4F_{3/2}$ optical transitions of $Nd^{3+}$:GaLaS ChGs using absorption and emission spectra at various temperatures. We determined the positions of Stark levels of $^4I_{9/2}$ and $^4F_{3/2}$ manifolds and the emission cross-section [8]. The radiative lifetime of the metastable $^4F_{3/2}$ manifold was determined independently from PL decay and quadrature frequency resolved spectroscopy (QFRS), both of which methods gave practically identical results. The observed lifetime was almost independent of temperature and agreed fairly well with that estimated from the Judd-Ofelt (JO) analysis. However, little is known about the $^4F_{3/2} \rightarrow {}^4I_{15/2}$ radiative transition giving rise to a PL band centred around 1840 nm. The branching ratio calculated using the JO analysis, $\beta_{15/2}$ corresponding to $^4F_{3/2} \rightarrow {}^4I_{15/2}$ transition, is about $2\times10^{-3}$, which is too small to be experimentally verified, though laser-oscillations were observed at the 1840 nm transition in Nd-doped Yttrium Aluminium Garnet (YAG) [9–11]. For practical applications of $Nd^{3+}$-doped GaLaS (for example in high-power glass lasers and optical amplifiers) the exact branching ratios $\beta_J$ ($J$ = 9/2, 11/2, 13/2 and 15/2) of all four abovementioned radiative transitions, or lasing channels should be known in order to suppress parasitic oscillations and undesirable spontaneous emission.

In addition to PL and PL excitation (PLE) spectroscopy, PL lifetime measurement is very important. Conventional measurement of PL lifetime obtained from PL decay is called time resolved spectroscopy (TRS). On measuring TRS of the PL decay consisting of multi-lifetime components, one must decrease the photocarrier generation rate *G* sufficiently to avoid the occurrence of non-linearity, and to ensure a geminate recombination condition, where a short-lived component appears in the decay immediately after the cessation of PL excitation. However, sometimes a long-lived component decreases to an undetectable level at a time corresponding to its own lifetime in the PL decay [12].

By contrast, QFRS yields detailed information on widespread lifetime distributions in amorphous semiconductors compared with the TRS and can discriminate a small lifetime component against the main lifetime component with sufficient sensitivity by virtue of the dual lock-in detection techniques used in this measurements system [12]. Nevertheless, QFRS has not been widely applied to the radiative transitions of RE ions as far as the authors know. One of the disadvantages of QFRS is that the minimum lifetime that can be resolved is limited to microseconds owing to the upper limit frequency of the conventional lock-in amplifiers. We have developed a dual-phase double lock-in (DPDL) QFRS system to measure nanoseconds lifetimes by using an RF lock-in amplifier and extended the longer lifetime limit to 160 s by using a digital lock-in amplifier in internal mode with synchronous filtering; therefore our wideband QFRS technique covers PL lifetimes of almost 11 decades from 2 ns to 160 s.

In the present paper, we first review previously published results to elucidate the PL mechanisms of undoped and Er-doped GeGaSe ChGs with the abovementioned wideband QFRS [13, 14], adding the QFRS of Er-doped GeGaS of sulphide based ChGs. We also present the results of the detailed analysis on the $^4F_{3/2}\rightarrow{}^4I_{15/2}$ radiative transition of $Nd^{3+}$in GaLaS ChGs with the QFRS improved in accuracy and the doubly-modulated PL spectroscopy at room temperature [15].

## 2.Experimental procedure

Er-doped GeGaSe ChGs of the actual composition $Ge_{28}Ga_6Se_{64.7}S_{0.8}Er_{0.5}$, which has been doped with $Er_2S_3$, and undoped ChGs of the actual composition $Ge_{27}Ga_9Se_{64}$ were prepared at University of Saskatchewan. The density of $Er^{3+}$ centres is estimated as $\sim 5\times 10^{19}$ cm$^{-3}$ for the 0.5 % Er-doped samples. The sample thickness was ~1.2 mm for the undoped and ~2.4 mm for the Er-doped samples. Er-doped GeGaS ChGs of the actual composition $Ge_{28}Ga_6S_{65.5}Er_{0.5}$ with an estimated $Er^{3+}$ density $\sim 2\times 10^{20}$ cm$^{-3}$ and a thickness of 1.23 mm were similarly synthesized at the same laboratory.

The samples were excited by various lasers but mostly at $E_x$ = 2.33 eV (532 nm), and PL was detected by a Hamamatsu 5509-72 infrared photomultiplier (PMT) at the photon energy ~0.73 eV (~1.7 μm) or greater with a long-pass filter to cut off the excitation laser light. Occasionally the PL was dispersed by a 10 cm-f/3.0 monochromator with a resolution of ~40 nm, which was installed between the sample and PMT. The PL lifetime distributions were measured by the aforementioned wideband QFRS system [12]. The newly detected slow component was analysed using the QFRS of the PL excited by bandgap light of 2.33 eV in superposition with sub-bandgap light of ~0.37 eV (3.39 μm) from an infrared (IR) laser.

Nd-doped GaLaS ChGs with the molar composition $(65Ga_2S_3)$:$(31.5La_2S_3)$:$(3La_2O_3)$:$(0.5Nd_2S_3)$ was prepared at the University of Southampton with an estimated concentration of $Nd^{3+}$ $\sim 8.9\times 10^{19}$ cm$^{-3}$ [15, 16]. A sample with a thickness ~3.3 mm was excited at room temperature by a laser operating at 2.33 eV with power density ~230 mWcm$^{-2}$. The PL was detected by an electronically cooled InGaAs PIN photodiode (Hamamatsu G5852-21) having the long-wavelength limit of 2050 nm and a cut-off frequency 40 MHz. The PL was dispersed by a 10 cm-f/3.0 monochromator with a resolution ~14nm. For the measurements of the 900 nm PL band, the InGaAs PIN diode was replaced by a Si photodiode to improve the responsibility of the detector. Both the spectral throughputs of the monochromator and the spectral sensitivity of the detector were calibrated with an incandescent lamp. We used the double lock-in technique to observe the weak PL in the range from 1400 to 2040 nm by doubly-modulating the laser light with an acousto-optic modulator and an optical chopper. Although the PL lifetime of an RE ion has been measured by TRS so far, we applied here the QFRS method because of its better sensitivity and time resolution [12, 15].

Each PL band of $Nd^{3+}$ belonging to $^4F_{3/2}\rightarrow{}^4I_J$ ($J$ = 9/2, 11/2, 13/2 and 15/2) was selected by a combination of short- and long-pass optical filters for the QFRS. The accuracy of the QFRS was improved as follows. By using the "ratio function" mode of the two lock-in amplifiers, the QFRS signal was normalised to excitation intensity to suppress the influence of the fluctuations of laser power. Compared to previous papers, we reduced the logarithmic lifetime interval of QFRS from 0.15 to ~0.002, which gave us an actual interval of ~0.4 μs in the vicinity of the peak of QFRS spectrum around ~80 μs. In order to recover a true lifetime distribution precisely, QFRS spectrum was deconvoluted with improved accuracy by using the statistics package software of *Mathematica*.

If two PL bands overlap in spectra but differ in lifetime $\tau$, they can be discriminated by modulating the PL emission. When a PL is modulated at an angular frequency $\omega$, the amplitude of PL with a lifetime $\tau$ is known to decrease as $[1+(\omega\tau)^2]^{-1/2}$ with increasing $\omega$ [12]. Therefore, a fast recombination process may be effectively separated from a slow process by measuring the frequency-dependent PL amplitude.

## 3. Results and discussion

*3.1 Undoped and Er-doped GeGaSe ChGs*

Figure 1 shows the QFRS spectra of undoped (host) and Er-doped GeGaSe glasses at 3.7 K and 294 K (room temperature) at the PL excitation energy $E_x$ = 2.33 eV and generation rate $G \approx 2\times10^{19}$ cm$^{-3}$s$^{-1}$, which corresponds to a power density ~20 mWcm$^{-2}$. The maxima of the data are normalised to unity. There are striking differences between the PL of the undoped and Er-doped samples: (1) the QFRS spectrum of the host glass reveals three lifetime peaks centred at $\tau_S \approx 20$ ns, $\tau_T \approx 80$ µs and $\tau_H \approx 20$ s, at

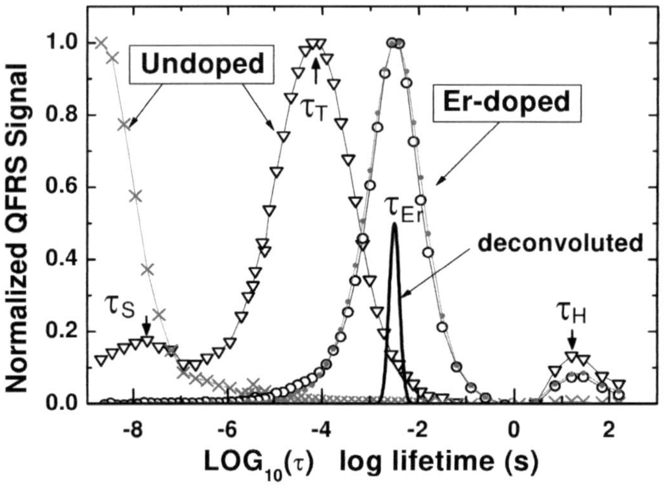


**Figure 1.** QFRS spectra of undoped (host) and Er-doped samples. ($\nabla$) shows QFRS spectrum of host measured at 3.7 K and ($\times$), at 294 K. ($\circ$) shows spectrum of Er-doped sample at 3.7 K and ($\bullet$), at 294 K. The sharp-pointed peak at $\tau_{Er} \approx 3.3$ ms shows the deconvoluted spectrum of the Er-doped sample with a FWHM ~1.9 ms at both the temperatures. All measured at a generation rate $G \approx 2\times10^{19}$ cm$^{-3}$s$^{-1}$ corresponding to a power density ~20 mWcm$^{-2}$ at $E_x$ = 2.33 eV.

3.7 K, whereas that of the Er-doped sample comprises of two peaks: a peak due to intrinsic $Er^{3+}$ radiative transition at $\tau_{Er} \approx 3.3$ ms and slow component at $\tau_H \approx 20$ s similar to that observed in the host: (2) a temperature rise from 3.7 K to 294 K decreases the magnitude of the intrinsic PL of the host by a factor of ~$4\times10^4$ accompanied by significant changes in the lifetime distribution. The PL of the Er-doped sample decreases only by ~10 over the same temperature rise without notable changes in the lifetime distribution, including the slow $\tau_H$ component; the deconvolution improved in accuracy by using the non-linear least square fit of *Mathematica* which gives a FWHM ~1.9 ms for the lifetime distribution of the $\tau_{Er}$-components at 3.7 and 294 K in stead of ~0.3 ms in the previous paper [13] (figure 1).

The lifetime peak of the $\tau_S$ component of the undoped sample falls short of the QFRS limit of 2 ns at 294 K, while the $\tau_T$ component in the middle of the lifetime distribution at 3.7 K exhibits a pronounced decrease accompanied by a shift of the lifetime peak from ~80 µs to 0.6 µs with temperature rise from 3.7 K to 294 K [14]. An activation energy of 78 meV estimated from the temperature dependence of ratio of the $\tau_S$ component to the $\tau_T$ component, coincides with the PL peak energy difference ~80 meV of the PL spectra of both the components obtained from PL emission-energy-evolved QFRS spectra [14]. Hence the $\tau_S$ and $\tau_T$ components are associated with singlet and triplet excitons, respectively, with a spin-exchange energy of ~80 meV, as previously reported for a-Si:H and a-Ge:H [17] and ChGs such as $As_2S_3$ glass and a-Se [18].

Figure 2 shows the quantum efficiencies (QEs) $\eta_S$ and $\eta_T$ of the corresponding $\tau_S$ and $\tau_T$ components as functions of temperature $T$ in addition to the lifetimes $\tau_S$ and $\tau_T$. Generally the QE is reduced by non-competing non-radiative recombination (NRR) and/or competing NRR in amorphous semiconductors. The non-competing NRR is introduced by defects and impurities, which rapidly capture photoexcited carriers normally within $10^{-12}$ s during the thermalisation of photocarriers from extended or higher tail states after the PL excitation. The non-competing NRR reduces $\eta_S$ and $\eta_T$ without shortening the measured PL lifetimes $\tau_S$ and $\tau_T$, whereas the competing NRR with a lifetime $\tau_{nr}$ reduces $\tau_S$ and $\tau_T$ since the composite or actual lifetime $\tau_m$ in competitive recombination is

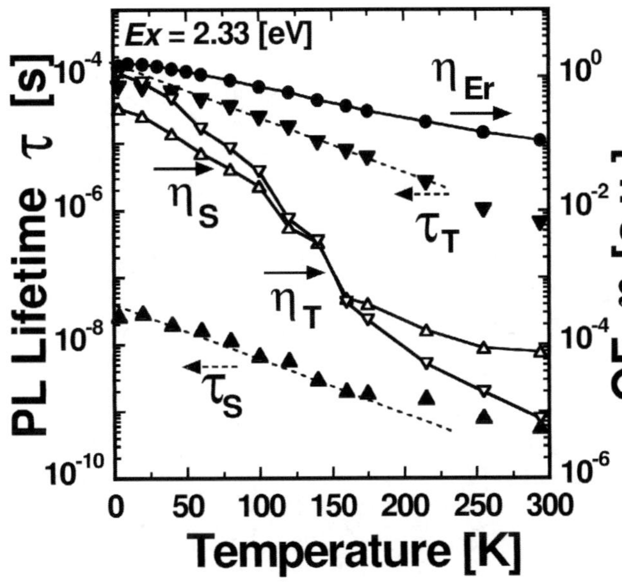


**Figure 2**. QEs (Δ) $\eta_S$, (∇) $\eta_T$ and lifetimes (▲) $\tau_S$, (▼) $\tau_T$ of the two excitonic components as functions of $T$ obtained for the undoped GeGaSe ChGs. QE (•) $\eta_{Er}$ of the $\tau_{Er}$ component of the Er-doped sample is also plotted.

expressed as $\tau_m^{-1} = \tau_r^{-1} + \tau_{nr}^{-1}$ with a radiative recombination time $\tau_r$ for each exciton, being usually independent of $T$ [14]. With increasing $T$ the QEs $\eta_S$ and $\eta_T$ decrease more strongly than the lifetimes $\tau_S$ and $\tau_T$ (figure 2). This difference indicates the contribution of the non-competing NRR to the decrease in both $\eta_S$ and $\eta_T$. Thus, the non-competing NRR contributes substantially to the thermal quenching of the excitons in undoped GeGaSe ChGs; namely as $T$ rises from 3.7 to 294 K, $\eta_S$ decreases by about two orders of magnitude greater than $\tau_S$ and $\eta_T$, by three orders of magnitude greater than $\tau_T$. The QEs, $\tau_S/\tau_r$ for the competitive recombination of $\tau_S$-component and $\tau_T/\tau_r$ for that of $\tau_T$-component show a similar $T$-dependence as $\exp(-T/T_0)$ with $T_0 \approx 53$ K estimated from the dotted lines in figure 2, and thus the characteristic temperature $T_0 \approx 53$ K for the competing NRR of both the $\tau_S$- and $\tau_T$-components, despite the difference of more than 3 orders of magnitude in $\tau_S$ and $\tau_T$. This is attributed to the difference in exciton capture cross sections, from which a characteristic energy of localised states $\varepsilon_0 \approx 20$ meV is obtained [14].

The observations are distinctly different for the Er-doped GeGaSe sample. It is shown in figure 2 that the QE $\eta_{Er}$ of the $\tau_{Er}$-component, i.e., the $^4I_{13/2} \rightarrow {}^4I_{15/2}$ emission decreases only by an order of magnitude for a $T$ increase from 3.7 to 294 K at $E_x$ = 2.33 eV, while the lifetime $\tau_{Er} \approx 3.3$ ms remains the same as shown in figure 1. This suggests the absence of a competing NRR with a lifetime $\tau_{nr}$ that would modify $\tau_{Er}$ through $\tau_m^{-1} = \tau_r^{-1} + \tau_{nr}^{-1}$. The $\tau_{Er}$-component is thus quenched by the non-competing NRR. In the Er-doped sample, the slow component with $\tau_H \approx 20$ s is still observable at $T$ = 294 K and $E_x$ = 2.33 eV without appreciable changes in the lifetime distribution obtained from the QFRS spectrum at 3.7 K, except for the magnitude (figure 1); the competing NRR is completely absent even in the slow $\tau_H$-emission process as well as the $^4I_{13/2} \rightarrow {}^4I_{15/2}$ emission of the $\tau_{Er}$-component. This implies that the $Er^{3+}$ luminescent centre in GeGaSe ChGs is extremely isolated from the competing NRR and possesses no back-transfer process [19].

Figure 3(a) depicts the excitonic recombination on the basis of Mott-Davis-Street (MDS) model [20]. When the exciting light creates an electron-hole (e-h) pair, the hole is captured by a nearby negatively-charged $D^-$ recombination centre, thereby changing the charge state to $D^0$. This causes a lattice relaxation around the centre and shifts the $D^0$ state deeper into the mid-gap. The excited electron in the conduction band is then bound to the hole captured by the $D^0$ centre, which forms a singlet (S) or triplet (T) bound-exciton. Alternatively, either singlet or triplet exciton bound to $D^0$ can be formed

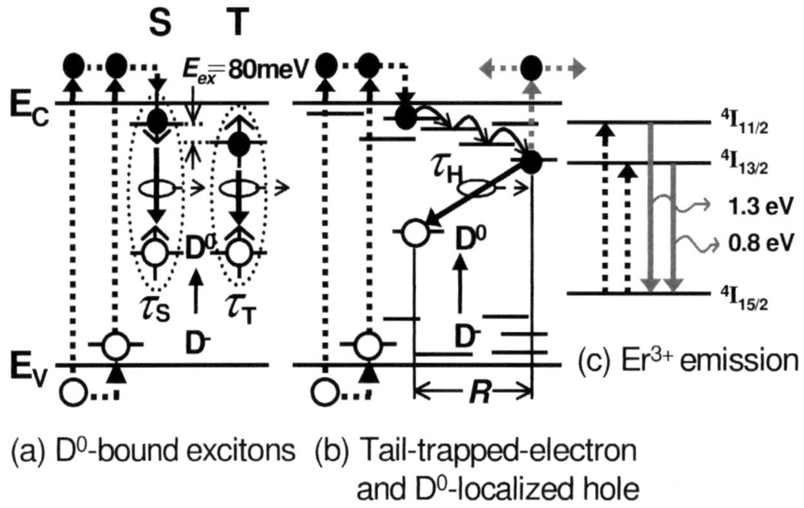


**Figure 3**. Possible excitation processes of the $Er^{3+}$ centre. Photoexcitation produces an electron and hole (e-h) pair, each of which is thermalised very fast into the respective tail states. One-type of charge carrier, e.g., hole is immediately captured by a charged defect $D^-$, which undergoes a consequent lattice relaxation and shift in energy to $D^0$. The process $D^- + h \rightarrow D^0$ also takes place when an electron of $D^-$ is photoexcited. (a) Photoexcited electron forms a bound singlet (S) or triplet (T) exciton with the hole captured at $D^0$ as a precursor, and (b) the photoexcited electron is trapped in tail-states, forming an e-h pair with the localised hole at $D^0$, which has an appreciable separation $R$ and subsequently recombines via tunnelling with a slow recombination rate. The complementary process involving the capture of an electron in $D^+$ is also possible. (c) Excitation and recombination at $Er^{3+}$ centre.

by exciting an electron of the $D^-$ to the conduction band. The complementary process involving a capture of an electron in $D^+$ is also possible. In the undoped sample, the excitons can then recombine and emit luminescence having different lifetimes $\tau_S$ and $\tau_T$ with appreciable Stokes shifts. In the Er-doped sample, on the other hand, the exciton recombines non-radiatively as a precursor, transferring its energy to a nearby $Er^{3+}$ ion, and excites it from the ground state to the excited state, followed by an $Er^{3+}$ luminescence emission as in figure 3(c).

The wideband QFRS technique revealed a third very-long-lived component almost universally in the PL of a-Si:H, a-Ge:H and ChGs, and associated it with the radiative tunnelling (RT) recombination of distant-pairs (DPs) by examining its dependences on the generation rate $G$, $E_x$ and the magnetic field. However, this is not the case for the slow $\tau_H$ component of the host, because it has little dependence on $G$ and $E_x$ to support DP recombination. Nevertheless the recombination rate $1/\tau_H$ is independent of PL emission energy $\hbar\omega$ whereas $1/\tau_S$ and $1/\tau_T$ are proportional to $(\hbar\omega)^3$ indicating excitonic recombination for the $\tau_S$ and $\tau_T$ components [14]. The independence of the recombination rate $1/\tau_H$ from $\hbar\omega$ suggests recombination of an uncorrelated e-h pair such as DPs [21]. Thus RT within the uncorrelated e-h pair cannot be discarded for the slow $\tau_H$ component. Impurities and defects in the structure open non-radiative channels to e-h pair recombination. The non-radiative tunnelling (NRT) channel, which is an exponential function of the distance between a tail-trapped carrier and a defect-state, competes with the RT channel in recombination events and thus the lifetime $\tau_H$ is a composite of the RT and the NRT recombination. Since the recombination rate is low for the long RT lifetime of the e-h pair appreciably separated by distance $R$ (figure 3(b)), $\tau_H$ is mostly determined by the impurity and defect densities. Such a situation was frequently encountered for glassy $As_2S_3$ [22] and defect-rich films of a-Si:H and a-SiN:H [23]. Generally, relatively large quantities of impurities and defects may be present in ChGs and thus they will play a rate-limiting role in the slow $\tau_H$-recombination.

In order to confirm the above concept, we superposed an infrared (IR) light of 0.37 eV (3.39 μm) and ~10 $mWcm^{-2}$ in the wideband QFRS excited at $E_x$ = 2.33 eV and $G \approx 10^{15}$ $cm^{-3}s^{-1}$ (~1 $\mu Wcm^{-2}$). Both the samples exhibit a close similarity; IR biasing induces a decrease of the $\tau_H$ component and a shortening of its lifetime peak $\tau_H$. The results are interpreted in terms of IR-induced release of the trapped carriers from localisation at the tail-states; the IR biasing activates a localised photocarrier, i.e., electron, which makes a hop-up to a higher tail-state or extended-state and then diffuses as shown by a red dashed arrow in figure 3(b). Thus, it may recombine either with a hole captured by a closer $D^0$ centre or with a hole at a closer non-radiative centre such as an impurity or defect, thereby shortening the lifetime $\tau_H$ and decreasing the amplitude of the $\tau_H$ component in both the undoped and doped samples [13, 14, 23].

Figure 4 shows the PL spectra of QFRS signals of the Er doped sample obtained by fixing QFRS frequencies correspondingly to $\tau_{Er} \approx 3$ ms and $\tau_H \approx 16$ s, together with the PL spectrum of the host obtained by DC measurements at 3.7 K. Surprisingly, even the slow $\tau_H$ component of the Er-doped sample exhibits sharp $Er^{3+}$emission lines at ~0.8 eV ($^4I_{13/2} \rightarrow {}^4I_{15/2}$) and ~1.27 eV ($^4I_{11/2} \rightarrow {}^4I_{15/2}$). However, the background spectrum of the $\tau_H$ component has close resemblance to the PL spectrum of the host, whereas the $\tau_{Er}$ component shows the pronounced $Er^{3+}$-emission lines with a quite different background compared with the $\tau_H$ component. This suggests a difference in the energy transfer mechanisms in the two components [13].

Deconvoluting the QFRS spectrum of the undoped sample at 3.7 K in figure 1, we estimated $\tau_S$, $\tau_T$ and $\tau_H$ components to be ~22, ~70, ~8 %, respectively. We postulate that the PL process occurs in cascade in the Er-doped sample as the precursor transfers its energy to the $Er^{3+}$centre, and hence the PL lifetime is determined by the longer-lived process. Thus, the faster excitonic precursors contribute mainly to the $\tau_{Er}$ component (figure 3(a)) and the precursor composed of a tail-trapped photocarrier and an oppositely-charged carrier captured to a $D^0$ centre contributes to the slow $\tau_H$ component

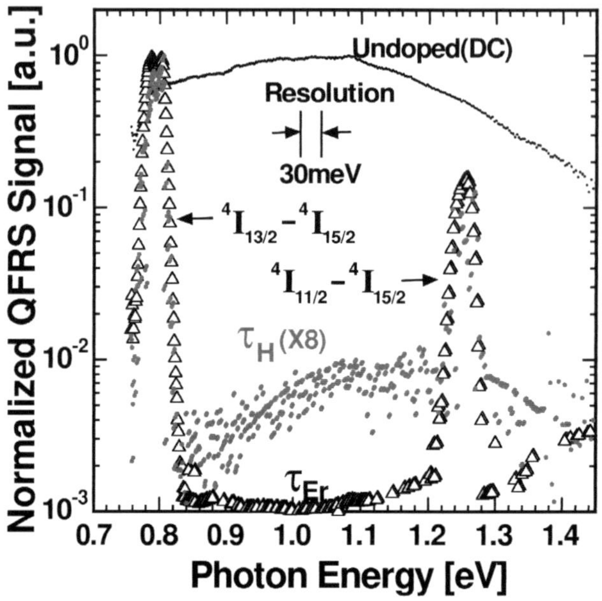


**Figure 4.** PL spectra of QFRS signals of Er-doped sample obtained at fixed frequencies corresponding to $\tau_{Er}$ (Δ) and $\tau_H$ (•). Peaks of the spectra for $\tau_{Er}$ and $\tau_H$ at 0.8 eV (~1.5 μm) are normalised to unity. PL spectrum of host obtained by DC measurement (•) is also shown. Vertical scale is logarithmic with arbitrary unit.

(figure3(b)). Deconvolution of the QFRS for the Er-doped sample at 3.7 K (figure 1) suggests that 51% of the whole excitonic component of the host is converted to the $\tau_{Er}$ component and 54% of the slow $\tau_H$ component of the host, to that of the Er-doped sample at low temperature [13].

The energy transfer mechanism of the main $Er^{3+}$ emission, i.e., the $\tau_{Er}$ component is of particular interest for photonic applications. Förster energy transfer is discarded, because this process, in principle, transfers energy between singlet excitons via dipole-dipole interactions whereas the triplet exciton component occupies ~70 % of all recombination events in the host. The Dexter energy transfer can take place between triplet excitons but its critical transfer distance is usually 0.3–1 nm. Since average distance between $Er^{3+}$ centres is estimated to be roughly ~3 nm for the 0.5 % Er-doped sample where density of $Er^{3+}$ centre amounts to ~$5\times10^{19}$ cm$^{-3}$, probably being much larger than that of defects (D-states), this process is also not relevant. Thus, we assert defect-related Auger excitation (DRAE) of the $Er^{3+}$centre to be the most probable energy transfer mechanism at least at low temperatures [5, 13, 24]. The energy transfer mechanism of the slow $\tau_H$ component is different from that of the $\tau_{Er}$ component, because the precursor consists of a tail-state trapped carrier and oppositely-charged carrier captured by a $D^0$ centre; and experimentally the background spectrum of PL differs from those of the $\tau_{Er}$ component. However, considering no involvement of excitons in the slow process, we propose the slow component to be controlled by DRAE as well [13].

### *3.2 Er-doped GeGaS ChGs*

Figure 5 shows quite similar QFRS spectra of Er-doped GeGaS, which were excited by the light of energy $E_x$ = 2.33 eV at generation rate $G \approx 1\times10^{16}$ and $1\times10^{18}$ cm$^{-3}$s$^{-1}$ at 3.7K and $G \approx 1\times10^{15}$ cm$^{-3}$s$^{-1}$ at

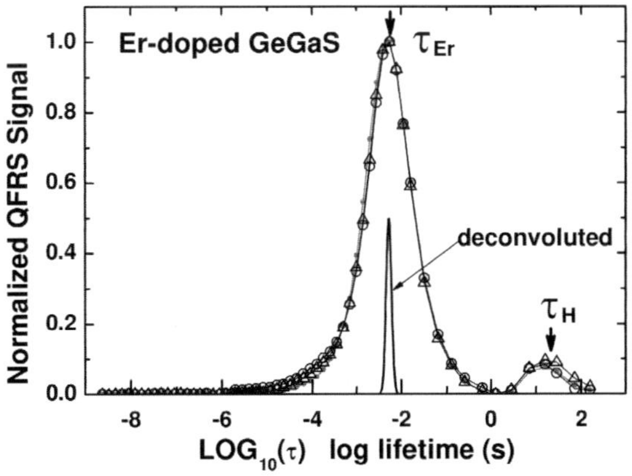


**Figure 5.** QFRS spectra of 0.5at. % Er-doped GeGaS ChGs. (○) G=$10^{16}$ and (Δ) G=$10^{18}$ $cm^{-3}s^{-1}$ at 3.7 K (•) G=$10^{15}$ $cm^{-3}s^{-1}$at 294 K. The deconvoluted spectrum of the data (Δ) has a peak at ~5.3 ms and a FWHM ~ 1.6 ms.

294 K (room temperature). All the maxima of the data are normalised to unity. In the Er-doped GeGaS sample the slow $\tau_H$ component is observable even at room temperature and its peak lifetime hardly depends on $G$ and $T$ similar to the behaviour of the Er-doped GeGaSe sample except $\tau_{Er} \approx 5.3$ ms differing from 3.3 ms, which is probably due to the difference in the host material (GeGaS). Incidentally, the QFRS on Er-doped GeGaS ChGs gave $\tau_{Er} \approx 4$ ms and no useful information about the slow $\tau_H$ component, while the TRS gave $\tau_{Er} \approx 5.2$ ms in Ref.[2].

### *3.3 Nd-doped GaLaS*

Figure 6 compares the PL spectra of Nd-doped GaLaS ChGs in the 1400–2040 nm spectral range measured at two sets of double-modulation frequencies of PL excitation light: low frequency set of 800 Hz and 15Hz (LF or closed symbols) and high frequency set of 15 kHz and 150 Hz (HF or open symbols); the double lock-in technique was introduced by doubly-modulating the excitation light to

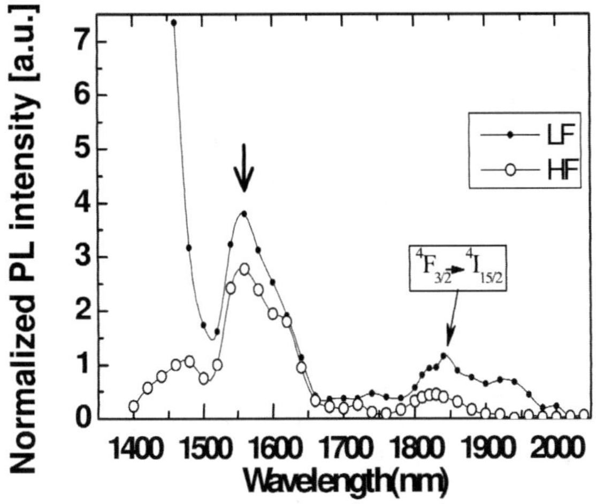


**Figure 6.** Normalised PL spectra of $Nd^{3+}$:GaLaS at room temperature from 1400 to 2040 nm. The spectra shown by solid symbols (•) are measured using a relatively low lock-in modulation frequency of 800 Hz (LF). The spectra shown by open circles (○) are measured using higher frequency of 15 kHz (HF). 2π×HF is higher and 2π×LF is lower than the inverse of 77 μs which is characteristic of the $^4F_{3/2} \rightarrow ^4I_J$ ($J$ = 9/2 ~ 15/2) transitions.

investigate this very weak spectral range in detail. The substantial reduction of peak at 1840 nm and the difference in the spectral shape detected for HF modulation indicates the dominance of a long-lifetime component, which is supported by figure 7(a) of the QFRS spectrum measured for 1840 nm PL band ranging from 1700 to 2040 nm. The deconvoluted QFRS spectrum reveals the presence of two components: dominant peak at 77 μs, which can be attributed to $^4F_{3/2}\rightarrow{}^4I_{15/2}$ transition, and a small peak at 3.8 μs, whose nature is discussed later. The relative contribution of both components to the 1840 nm band may be estimated by integrating the underlying areas. This procedure attributes ~83 % to the $^4F_{3/2}\rightarrow{}^4I_{15/2}$ transition with lifetime 77 μs and the remaining ~17 % to the short-lifetime component.

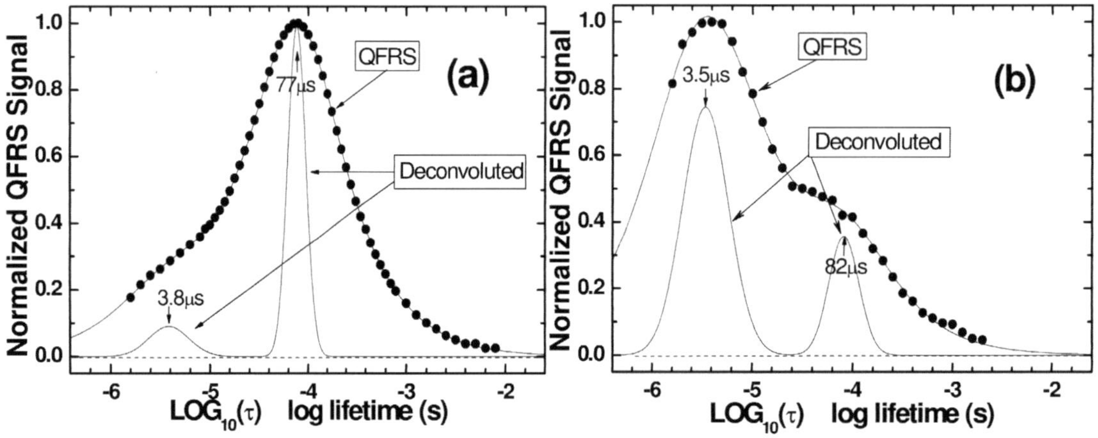


**Figure 7.** The experimental QFRS spectra are shown by solid circles for 1840 nm and 1560 nm emission bands in (a) and (b), respectively. Double peaked Gaussians are the results of deconvolution of the QFRS spectra with QFRS response function. The results of convolution of the double peaked Gaussians with QFRS response function fit the QFRS spectra well (solid curves)

The PL band between 1500 and 1650 nm with the peak at ~1560 nm in figure 6 (larger arrow) is less sensitive to the variation of modulation frequency from LF to HF in contrast to 1840 nm band, suggesting a higher content of the short lifetime component and thus deserves special discussion. Its overall intensity measured at LF is about 140% of PL band centred at 1840 nm. The deconvolved QFRS spectrum shows the presence of two Gaussian-like components centred at 3.5 and 82 μs and the dominant "fast" component controls about 76% of the overall spectrum (figure 7(b)). The huge difference between the dominant lifetime of 1560 nm band (3.5 μs) and the lifetime of $^4F_{3/2}\rightarrow{}^4I_{15/2}$ transition (~77 μs) suggests a different origin for this band. From the point of view of spectral position, this band may be assigned to radiative transitions originating from the doublet ($^4F_{5/2}$,$^2H_{9/2}$) to the $^4I_{15/2}$ manifold as illustrated by the dashed arrow in figure 8. These transitions would produce two unresolved lines peaking around ~1630 and ~1580 nm, which fall within the experimentally observed 1560 nm band.

However, the observed short lifetime of 3.5 μs requires further discussion for the transition probabilities of electric-dipole transitions $^4F_{5/2}$, $^2H_{9/2}\rightarrow{}^4I_J$ in $Nd^{3+}$:GaLaS ChGs, but unfortunately there is no direct information except that of fluoride glasses [25, 26], in which the total probability of the $^4F_{5/2}\rightarrow{}^4I_J$ transitions is only ~1.4 times larger than that of $^4F_{3/2}\rightarrow{}^4I_J$ transitions. Meanwhile the probability of $^2H_{9/2}\rightarrow{}^4I_J$ transitions is smaller by an order of magnitude. Therefore, the short lifetime of ~3.5 μs cannot be explained by the radiative transitions of $^4F_{5/2}$,$^2H_{9/2}\rightarrow{}^4I_J$ only. When the energy difference $\Delta E$ between two levels is relatively small, there is high probability of efficient multiphonon

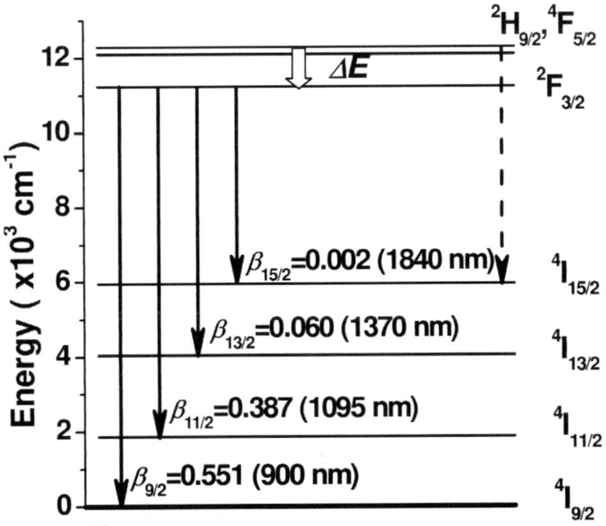


**Figure 8**. Possible radiative transition $^2H_{9/2}$,$^4F_{5/2}$ (doublet)$\rightarrow$$^4I_{15/2}$ (dashed arrow) and non-radiative transition $^2H_{9/2}$,$^4F_{5/2}$$\rightarrow$$^4F_{3/2}$ (open arrow) in $Nd^{3+}$ ion. Four solid arrows show the four lasing channels of $Nd^{3+}$ with the corresponding values of central wavelengths of emission bands as well as branching ratios $\beta_J$ ($J$ = 9/2, 11/2, 13/2 and 15/2) (Table 1).

relaxation from the upper energy level to the lower one by emitting phonons [27]. This may occur between the doublet ($^4F_{5/2}$,$^2H_{9/2}$) and the $^4F_{3/2}$ manifold where the energy gap $\Delta E \approx$850–1050 cm$^{-1}$ is close to typical phonon energies. Therefore, the radiative transitions compete with non-radiative multiphonon relaxation of $^4F_{5/2}$,$^2H_{9/2}$$\rightarrow$$^4F_{3/2}$ as shown by an open arrow in the energy scheme for $Nd^{3+}$ in figure 8. We estimated the multiphonon relaxation time to be about 9.0 μs at room temperature for $\Delta E$ = 950 cm$^{-1}$ which is comparable with the dominant lifetime ~3.5 μs of the 1560 nm band in figure 7 (b) [15]. Thus the PL spectrum and lifetime suggest that the 1560 nm band is caused by radiative $^4F_{5/2}$,$^2H_{9/2}$$\rightarrow$$^4I_{15/2}$ transitions competing with multiphonon relaxation. It is also worth noting that the 1560 nm band may be effectively smeared due to several factors such as the strong overlapping of Stark components in both transitions and the competing multiphonon relaxation. Therefore the PL of the 1840 nm band may be intermingled with the short-lived PL of the 1560 nm band.

Table 1 presents the branching ratios $\beta_J$ ($J$ = 9/2, 11/2, 13/2 and 15/2) corresponding to the four

**Table 1.** Central wavelength $\lambda$ (nm), branching ratio $\beta_J$ and lifetime $\tau$ (μs) obtained from Judd-Ofelt analysis and experiments. Four lifetimes obtained by JO analysis give the composite lifetime ~86 μ.

| Transition | Judd-Ofelt analysis | | | Experiment | | |
|---|---|---|---|---|---|---|
| $^4F_{3/2}\rightarrow$ | $\lambda$ (nm) | $\beta$ | $\tau$ (μs) | $\lambda$ (nm) | $\beta$ | $\tau$ (μs) |
| $^4I_{9/2}$ | 890 | 0.539 | 160 | 900 | 0.551 | 78.4 |
| $^4I_{11/2}$ | 1066 | 0.393 | 219 | 1095 | 0.387 | 77.3 |
| $^4I_{13/2}$ | 1390 | 0.067 | 1288 | 1370 | 0.060 | 78.4 |
| $^4I_{15/2}$ | 1893 | 0.002 | 47998 | 1840 | 0.002 | 76.6 |

Journal of Physics: Conference Series **253** (2010) 012010 doi:10.1088/1742-6596/253/1/012010

radiative transitions $^4F_{3/2}\rightarrow{}^4I_J$ in $Nd^{3+}$:GaLaS calculated by integrating the spectra of corresponding bands [15]. Prior to the integration, the spectra are properly calibrated. As for the band at 1840 nm, only 83% of overall integral are taken into account, which corresponds to the $^4F_{3/2}\rightarrow{}^4I_{15/2}$ transition only. It is noteworthy that these values are in a good agreement with previous estimations based on JO analysis. It should also be noted that according to QFRS, all four $^4F_{3/2}\rightarrow{}^4I_J$ transitions are competing with each other and therefore exhibit approximately the same lifetime of ~77 μs, with actual QFRS data scattered from 76.6 to 78.4 μs. The above lifetime is slightly different from that reported earlier in Ref. [8] (~90 μs) and might be due to the higher accuracy of the present data. In addition, the JO lifetimes in Table 1 give a composite lifetime ~86 μs.

## 4. Conclusion

The wideband QFRS discriminated three (triple-peak) PL components of undoped GeGaSe ChGs and two PL components of Er-doped GeGaSe ChGs at 3.7 K. Two components of undoped GeGaSe ChGs with lifetimes 20 ns and 80 μs originate from singlet and triplet excitons, respectively and are influenced by competing NRR with an identical characteristic temperature $T_0 \approx 53$ K as well as non-competing NRR; the latter more strongly. The Er-doped sample has the $Er^{3+}$ *f-f* emission of narrow lifetime distribution around the mean lifetime $\tau_{Er} \approx 3.3$ ms, being almost temperature independent. Both the undoped and Er-doped samples have slow emission of a lifetime of 20 s, but the latter does not show appreciable change in both lifetimes of 3.3 ms and 20 s even by elevating the temperature from 3.7 K to room temperature, suggesting little influence of competing NRR on Er luminescence centres. Similar QFRS results were obtained from Er-doped GeGaS ChGs except $\tau_{Er} \approx 5.3$ ms.

The undoped GeGaSe sample has broad and featureless PL spectra for all the three components but the Er doped sample has $Er^{3+}$ *f-f* emission spectra not only for the main $Er^{3+}$ emission of lifetime 3.3 ms, but also for the slow emission of lifetime 20 s. The precursor for the main $Er^{3+}$ emission is the exciton formation and that for the slow emission is a separately localised e-h pair, i.e., the tail-trapped one and the other captured by a $D^0$ centre.

In Nd-doped GaLaS ChGs, two overlapping PL bands in the wavelength range from 1400 to 2040 nm were separated by modulating the PL emission at two different frequencies at room temperature. One PL band possesses a peak wavelength of 1840 nm and a major lifetime 77 μs in common with the four lasing channels of $Nd^{3+}$. This band is attributed to the lowest energy $^4F_{3/2}\rightarrow{}^4I_{15/2}$ transition of the four radiative channels of $Nd^{3+}$. The other PL band with a peak wavelength of 1560 nm and a major lifetime 3.5 μs is attributed to the doublet $^2H_{9/2},{}^4F_{5/2}\rightarrow{}^4I_{15/2}$ radiative transitions. Its short-lived lifetime is interpreted by estimating the competing multiphonon relaxation. By suitably interpreting the two mixed radiative transitions $^4F_{3/2}\rightarrow{}^4I_{15/2}$ and $^2H_{9/2},{}^4F_{5/2}\rightarrow{}^4I_{15/2}$, we obtained the branching ratios of the four channels in Nd-doped GaLaS ChGs, agreeing with those from the Judd-Ofelt analysis.

**Acknowledgement**

Authors acknowledge I Ikeda, N Ohrui, S Kobayashi, M Munzar, R DeCorby, C Haugen, D W Hewak and D Tonchev for their contributions to the previous works which the present paper is based on. We thank NSERC and JPSPF for financial support.

## References

[1] Munzar M, Koughia K, Tonchev D, Maeda K, Ikari T, Haugen C, Decorby R, McMullin J N and Kasap S O 2005 *Opt. Mater.* **28** 225

[2] Koughia K, Saitou D, Aoki T, Munzar M and Kasap S O 2006 *J. Non-Cryst. Solids* **352** 2420

[3] Fuhs W, Ulber I, Weiser G, Bresler M S, Gusev O B, Kuznetsov A N, Kudoyarova V Kh, Terukov E I and Yassievich I N 1997 *Phys. Rev. B* **56** 9545

[4] Kühne H, Weiser G, Terukov E I, Kuznetsov A N and Kudoyarova V Kh 1999 *J. Appl. Phys.* **86** 896

[5] Bishop S G, Turnbull D A and Aitken B G 2000 *J. Non-Cryst. Solids* **266-269** 876

[6] Schweizer T, Hewak D W, Payne D N, Jensen T and Huber G 1996 *Electronics Letters* **32** 666
[7] Schweizer T, Samson B N, Moore R C, Hewak D W and Payne D N 1997 *Electronics Letters* **33** 414
[8] Koughia K, Soundararajan G, Kasap S O, Allen T, Haugen C, Decorby R, Ohrui N, Aoki T and Fujihashi C 2009 *J. Mater Sci: Mater Electron* **20** S19
[9] Feofilov P P, Timefeeva V A, Tolstoi M N and Belyaev L M 1965 *Optic. Spectroscopy* **19** 451
[10] Wallace R W 1971 *IEEE J.* QE **7** 203
[11] Kaminskii A A 1981 *Laser Crystals, Springer Series in Optical Sciences* vol 14 ed D L MacAdam (Berlin: Springer) pp244-325
[12] Aoki T 2006 *Optical Properties of Condensed Matter and Applications* ed J Singh (Chichester: John Wiley & Sons) p75
[13] Aoki T, Saitou D, Kobayashi S, Fujihashi C, Shimakawa K, Munzar M, Koughia K and Kasap S O 2007 *J. Mater Sci: Mater Electron* **18** S97
[14] Aoki T, Saitou D, Kobayashi S, Fujihashi C, Shimakawa K, Koughia K. Kasap S O 2007 *J. Optoelectron. Adv. Mater.* **9** 3143
[15] Aoki T, Koughia K, Fujimoto K, Fujihashi C, DeCorby R, Haugen C, Hewak D W, Tonchev D and Kasap S O 2009 *Phys. Status Solidi* C **6** 554
[16] Mairaj A K, Chardon A M, Shepherd D P and Hewak D W 2002 *IEEE J. Sel. Top. Quantum Electron.* **8** 1381
[17] Aoki T 2006 *J. Non-Cryst. Solids* **352** 1138
[18] Aoki T, Komedoori S, Kobayashi S, Shimizu T, Ganjoo A and Shimakawa K 2003 *J. Non-Cryst. Solids* **326-327** 27
[19] Kik P G, de Dood M J A, Kikoin K and Polman A 1997 *Appl. Phys. Lett.* **70** 1721
[20] Davis E 1979 *Amorphous Semiconductors* ed M H Brodsky (Berlin: Springer-Verlag) p 41
[21] Aoki T, Ikeda K, Kobayashi S and Shimakawa K 2006 *Philos. Mag. Lett.* **86** 137
[22] Aoki T, Saitou D, Ikeda K, Kobayashi S and Shimakawa K 2005 *J.Optoelectron. Adv. Mat.* **7** 1749
[23] Aoki T, Ikeda K, Ohrui N, Kobayashi S and Shimakawa K 2007 *J.Optoelectron. Adv. Mat.* **9** 70
[24] Yassievich I N, Bresler M S and Gusev O B 1998 *J. Non-Cryst. Solids* **226** 192
[25] Reisfeld R 1984 *Proc. Int. Symp. on Rare Earths Spectroscopy (Wrocław, Poland, 10-15 Sept.1984)* ed. B Jeżowska-Trzebiatowska, J Legendziewicz and W Stręk (Singapore: World Scientific) p 587
[26] Reisfeld R and Jørgensen C K 1987 *Handbook on the Physics and Chemistry of Rare Earths* vol 9 ed K A Gschneidner and L Eyring (Amsterdam: North-Holland) p 1
[27] Riseberg L A and Weber M J 1976 *Progress in Optics* **XIV** ed E Wolf (Amsterdam: North-Holland) p 89

Journal of Physics: Conference Series **253** (2010) 012011 doi:10.1088/1742-6596/253/1/012011

# Interaction anisotropy and random impurities effects on the critical behaviour of ferromagnets

**H Chamati [1] and S Romano [2]**

[1] Institute of Solid State Physics, Bulgarian Academy of Sciences, 72 Tzarigradsko Chaussée, 1784 Sofia, Bulgaria

E-mail: chamati@issp.bas.bg

[2] Dipartimento di Fisica "A. Volta", Università di Pavia, Via A. Bassi 6, I-27100 Pavia, Italy

E-mail: Silvano.Romano@pv.infn.it

**Abstract.** The theory of phase transitions is based on the consideration of "idealized" models, such as the Ising model: a system of magnetic moments living on a cubic lattice and having only two accessible states. For simplicity the interaction is supposed to be restricted to nearest–neighbour sites only. For these models, statistical physics gives a detailed description of the behaviour of various thermodynamic quantities in the vicinity of the transition temperature. These findings are confirmed by the most precise experiments. On the other hand, there exist other cases, where one must account for additional features, such as anisotropy, defects, dilution or any effect that may affect the nature and/or the range of the interaction. These features may have impact on the order of the phase transition in the ideal model or smear it out. Here we address two classes of models where the nature of the transition is altered by the presence of anisotropy or dilution.

## 1. Introduction

Materials in Nature can be found in qualitatively different phases having distinct properties. The change from one phase to another is the consequence of a variation of an intensive thermodynamic quantity, e.g., the temperature $T$, the pressure $P$, external electric or magnetic fields $E$ or $H$. Phase transitions are accompanied by abrupt changes in a number of macroscopic thermodynamic quantities. Some familiar examples of phase transitions include the gas–liquid transition (condensation), the liquid–solid transition (freezing), the normal–superconducting transition in conductors, the paramagnet–ferromagnet transition in magnetic materials, and the superfluid transition in liquid helium. Further examples are transitions involving amorphous or glassy structures, spin glasses, liquid crystals, charge–density waves, and spin–density waves.

In many cases, the two phases above and below the transition point, say $T_c$ for a temperature driven phase transition, may be discerned from each other in terms of some ordering that occurs in the phase below $T_c$. For example in the liquid–solid transition the molecules of the liquid get "ordered" in space when they form the solid phase. In a paramagnet, the magnetic moments of each atom can point in any random direction (in the absence of an external magnetic field), but in the ferromagnetic phase the moments are lined up in a particular direction of ordering. Thus in the high temperature phase (above $T_c$), the degree of ordering is smaller than in the low temperature phase (below $T_c$). To quantify the amount of ordering in a system one uses the so called *order parameter*, which is usually a vanishing quantity in the high temperature (disordered) phase.

The phase diagram shows regions within which homogeneous equilibrium states exist as a function of temperature and other thermodynamic variables like $P$, $E$, $H$. For some physical systems the chemical potential $\mu$ or composition variables are also involved. The different regions of the phase diagram are delimited by phase boundaries that mark conditions under which multiple phases can coexist at equilibrium. Phase transitions take place along phase boundaries marked by lines of equilibrium.

The theoretical framework that aims at describing phase transitions and related phenomena as a result from cooperative effects over macroscopic scales is a part of the realm of equilibrium statistical physics [1–3]. Statistical physics is based on probabilistic models of the interactions of microscopic entities forming large assemblages (macroscopic bodies). The probability a macroscopic system of volume $V$, having $N$ particles, in a state $\mathfrak{S}$ with energy $E$ at a temperature $T$, is given by

$$\mathscr{P}(\mathfrak{S}) = \frac{e^{-E(\mathfrak{S})/(k_B T)}}{\mathscr{Z}(T)}, \tag{1.1}$$

where $k_B$ is the Boltzmann constant, and $\mathscr{Z}(T)$ is the partition function

$$\mathscr{Z}(T) = \sum_{\mathfrak{S}} e^{-E(\mathfrak{S})/(k_B T)}, \tag{1.2}$$

relating microscopic degrees of freedom to macroscopic thermodynamic quantities. The function $\mathscr{Z}(T)$ depends upon any parameters that might affect the value of $E(\mathfrak{S})$. The expectation value of any statistical operator $\mathfrak{O}$ is defined via

$$\langle \mathfrak{O}(\mathfrak{S}) \rangle = \sum_{\mathfrak{S}} \mathfrak{O}(\mathfrak{S}) \mathscr{P}(\mathfrak{S}). \tag{1.3}$$

For example the internal energy $U$ is obtained by averaging $E(\mathfrak{S})$ over all the accessible states of the system. This is given by

$$U = \langle E(\mathfrak{S}) \rangle.$$

A fundamental thermodynamic quantity related to the partition function is the *free energy*

$$f(T) = -k_B T \ln \mathscr{Z}(T), \tag{1.4}$$

which contains all the information on the thermodynamics of the considered system.

According to the Ehrenfest classification scheme there are different kinds of phase transitions depending on the nature of the singularities of the thermodynamic quantities at the transition. Such transitions are categorized as first–, second–, or higher order transitions if the lowest derivative of the free energy that exhibit nonanalytic behaviour with a finite jump is the first, second or higher one. Within this classification the Berezinskiĭ–Kosterlitz–Thouless (BKT) [4, 5] may be considered as being of infinite order.

To describe universal features of phase transitions one uses relatively "simple" microscopic models such as the magnetic model involving interaction between magnetic degrees of freedom. In a *classical model* for a magnet, the spins (magnetic moments) may be represented by $n$–component unit vectors $\boldsymbol{s}_i$ located at sites $i$, with coordinates $\boldsymbol{x}_i$, belonging to a finite subset of a generic $d$–dimensional lattice $\Lambda_d$, i.e. $\mathfrak{L} \subset \Lambda_d$, with $|\mathfrak{L}| = N$. We are considering here saturated lattice models, where each lattice site hosts one spin. The simplest and probably most extensively studied cases considered in the literature assume a hypercubic lattice $\Lambda_d = \mathbb{Z}^d$ and isotropically interacting spins, thus the (nearest–neighbour) Hamiltonian

$$\mathscr{H}_{\mathfrak{L}} = -\frac{J}{2} \sum_{\langle i,j \rangle} \boldsymbol{s}_i \cdot \boldsymbol{s}_j - \boldsymbol{H} \sum_i \boldsymbol{s}_i, \tag{1.5}$$

where the coupling $J$ is restricted to nearest–neighbouring sites $i$ and $j$, with each distinct pair being counted once. $\boldsymbol{H}$ stands for an uniform magnetic field.

In the absence of the magnetic field, i.e. $\boldsymbol{H} = \mathbf{0}$, a ferromagnetic coupling, $J > 0$, favours a parallel orientation of the spins in the ground state, whereas thermal fluctuations tend to create an orientational disorder. On the other hand an interaction, $J < 0$, would be the precursor of an antiferromagnetic order at low temperatures. Actually, in the specific case considered here, i.e. nearest–neighbour coupling and bipartite lattice, and with $\boldsymbol{H} = \mathbf{0}$, the sign of the coupling constant is immaterial, i.e. models defined by $+J$ and $-J$ yield the same partition function, whereas the two correlation functions are connected by suitable sign factors.

For $n = 1, 2, 3$, the model corresponds to the Ising, planar rotator (PR) and Heisenberg (He), respectively. The Ising model is known to have a $Z_2$ discrete symmetry, while systems with $n \geq 2$, like PR and He, are said to possess a continuous $O(n)$ symmetry. It has become customary to refer to the number of component $n$ of the order parameter as symmetry index. The interaction with an external field breaks this symmetry and establishes a preferred direction for spin alignment. By reducing the external field to zero in the thermodynamic limit, the system may exhibit spontaneous magnetization pointing in the initial direction of the field. It has been shown [6] that the limit $n \to \infty$ leads to the spherical model [7] obtained by requiring the spins to be continuous variables subject to a global relaxed constraint $(\sum_{i=1}^{N} \boldsymbol{s}_i^2 = N)$ rather than forcing them to take unit lengths i.e. $|\boldsymbol{s}_i|^2 = 1$. The formal limit $n \to 0$ is relevant to the study of self–avoiding walk problem, which can be applied to polymers.

By now a number of rigorous results, assuming translational invariance, have been worked out, entailing existence or absence of a phase transition in the thermodynamic limit, depending on lattice dimensionality $d$ and number of spins components $n$ [8, 9].

Model (1.5) with different values of $n$ and $d$ is extensively studied in the literature via different methods and its behaviour as a function of the temperature is very well known. For a review with a rich list of references see [10]. At $H = 0$, it exhibits a second order phase transition,[1] for any $d > 1$ and with discrete spin variables, i.e. $n = 1$. Such a transition is characterized by a significant growth of the nearest–neighbour correlations for orientational fluctuations, and also the onset for long–range orientational correlations. On the other hand, according to the Mermin–Wegner theorem [11] in $O(n)$ symmetric models ($n \geq 2$) there can be no spontaneous symmetry breaking at finite temperatures for $d \leq 2$ meaning that the system remains orientationally disordered at any finite temperature. For $d > 2$ and $n \geq 2$ a second order phase transition takes place in the system. In the two–dimensional case $d = 2$ and $n = 2$ the system exhibits a BKT transition from a high temperature disordered phase to a low temperature phase with slow decay of the correlation function and an infinite magnetic susceptibility [12].

In the vicinity of a continuous phase transition, such as second order or BKT transition, there is only one dominating length scale related to the growth of fluctuations: The *correlation length*. Because of the diverging nature of the correlation length as the critical point is approached the microscopic details of the system becomes irrelevant. Thus the description of the singular behaviour of many thermodynamic observables requires a small number of universal variables: critical exponents, amplitudes and functions. This allows the arrangement of a great variety of different microscopic systems in universality classes of equivalent critical behaviour. A universality class depends upon the number of components of the order parameter and the dimensionality of the system. For more details the reader is invited to consult references [13, 14].

The model (1.5) can describe the properties of a wide variety of physical situations in the vicinity of their transition point, however, it may happen that the very experimental situation at hand requests the interaction model to be complicated (made physically richer and theoretically more challenging) in various ways. On the one hand, there exist different possible lattice types $\Lambda_d$, in addition to $\mathbb{Z}^d$. On the other hand, as for the orientational dependence, more elaborate potential models involve (in some combination or other) isotropic or anisotropic linear couplings between spin components, sometimes higher powers of scalar products among the interacting spins, multipolar (usually dipolar) interactions,

[1] Here and below the temperature will be measured in units of $J/k_B$.

Dzyaloshinski–Moriya terms, single–site anisotropy fields. Notice also that more distant neighbours are sometimes involved, or even, in principle, all neighbours may be coupled by long–range interactions. In some specific favourable cases one has even been able to match the model and its potential parameters to a specific experimental system [15]. The ferromagnetic ordering transition observed experimentally in the absence of an external field is more frequently second order, but first–order transitions are also known. They might, for example, result from doping by nonmagnetic impurities, anisotropy of interactions in spin space, or coupling to the lattice [16]. Other interaction models for different physical systems can be found in [17].

This review is devoted to the description and analysis of effects related to the nature of the interaction. This aims at gaining insights in the thermodynamics of some specific systems. The models considered here are some of the most popular models in the theory of phase transitions. They involve different kind of interactions that hopefully might be adapted to the characteristics of a given material. They describe ferromagnetism with anisotropic coupling, systems with random dilution and may to some extent be used to investigate fluids. Special attention is paid to the construction of the phase diagrams of these models that are determined from the investigation of different thermodynamic quantities such as: the free energy, the susceptibility, specific heat etc.

The review is organised as follows: In Section 2 we discuss the transitional behaviour of *generalised XY* models introduced in reference [18]. These are generalization of the XY model with a nontrivial coupling along the $z$ components of the spins. We construct the phase diagrams of the models in two and three dimensions and determine the effect of the nontrivial coupling on the nature and location of transition temperature. In Section 3 we review the effect of dilution of ferromagnets by the introduction of random impurities and present the phase diagram of the diluted Heisenberg in three dimensions and the diluted plane rotator in two dimensions. We conclude with Section 4, where we discuss the results and comment on other models sharing similar transitional behaviour.

## 2. Generalised XY models

Let us first consider interactions being anisotropic in spin space with nonvanishing and equal ferromagnetic couplings involving $m < n$ components of the partner spins only, while still keeping the interaction restricted to nearest neighbours. In this case the model reads

$$\mathscr{H}_{\text{anis.}} = -\frac{J}{2}\sum_{\langle i,j\rangle}\sum_{\alpha,\beta\leq m}\delta_{\alpha,\beta}s_i^{\alpha}s_j^{\beta}. \tag{2.1}$$

Among these models we may mention the *continuous* Ising ($m = 1$ and $n$ arbitrary) and the various versions of XY models ($m = 2$ and $n$ arbitrary), where the interaction is restricted to only one component and two components of the magnetic moments, respectively

In general the transitional behaviour of such models is analogous to that of their $O(m)$ isotropic counterparts: On the one hand, the universal critical behaviour of these models in the vicinity of their corresponding transition temperature is equivalent to their isotropic $O(m)$ counterparts i.e. the models share the same universal features (critical exponents and amplitudes). on the other hand, the transition temperature is a typical non–universal quantity, and is recognizably affected by the anisotropy.

A more general class of anisotropic spin models may be constructed by introducing some kind of extreme anisotropy coupling only a part of the spin components in some nontrivial manner in addition to the anisotropic interaction in the spin space. A model that will be discussed in this study is the so called "generalised" XY model introduced in reference [18]. This model involves 3–component unit vectors, and is defined by

$$\mathscr{H}_{XY}^{p} = -\frac{J}{2}\sum_{\langle i,j\rangle}(\sin\theta_i\sin\theta_j)^p\cos(\phi_i-\phi_j), \tag{2.2}$$

where $p \in \mathbb{N}$ is a parameter controlling the strength of anisotropy along the $z$–spin direction, and the spins are expressed in terms of the usual spherical coordinates $\theta_i$ and $\phi_i$ i.e. $\boldsymbol{s}_i =$

$(\sin\theta_i\cos\phi_i, \sin\theta_i\sin\phi_i, \cos\theta_i)$. Notice that by setting $p = 1$ ($p = 0$) we recover the familiar XY model (planar rotator). In this case of the planar rotator model, the $\theta$–dependence only survives in the free–spin measure. The Hamiltonian (2.2) can also be written in terms of spin components, in the more complicated form

$$\mathscr{H}_{XY}^{p} = -\frac{J}{2}\sum_{\langle i,j\rangle}\left[1-(s_i^z)^2-(s_j^z)^2+(s_i^z s_j^z)^2\right]^{(p-1)/2}\left(s_i^x s_j^x + s_i^y s_j^y\right). \tag{2.3}$$

As for the role of $p$ in Eq. (2.2), notice that it could be taken to be a real positive number, say ranging between 0 and 1 (and hence continuously interpolating between planar rotator and XY models). On the other hand, larger values of $p$ reinforce the out–of–plane fluctuations. This makes it possible to widely vary the anchoring of spins with respect to the horizontal plane which might have direct experimental relevance. As for the present model, this change of anchoring is ultimately reflected by the significant changes in transition behaviour.

The transitional behaviour of the model in two and three dimensions has been investigated by different approaches. First, it was proven rigorously [18], on the basis of the known behaviour of PR, that when $d = 2$ and for all values of $p$, the named potential models produce orientational disorder at all finite temperatures, and support a BKT–like transition. On the other hand, when $d = 3$, these models support ordering transitions taking place at finite temperatures. In both cases, the transition temperatures are bounded from above by the corresponding values for the PR counterpart. It was later proven [19], again rigorously, that the transition turns first–order for sufficiently large $p$. Notice that the threshold values had to be estimated by other means.

### 2.1. Two dimensions

Using Monte Carlo simulations, the thermodynamics of model (2.2) has been investigated in details in reference [20] for $d = 2$ and values of $p$ ranging from 2 to 5. Analysis of Monte Carlo data, showed that the model produces a BKT–(like) transition, possibly changing to a first–order transition for larger $p$, due to the large number of vortices and strong out–of–plane fluctuations. It has been found that the transition temperature is indeed decreasing with increasing $p$.

To gain insight into the transitional behaviour of the model for large values of $p$ we performed further Monte Carlo simulations at $p = 6$. Our analysis shows that the transition is most likely to have a weak first order nature. Unfortunately more simulations are required and different approaches are needed to be more conclusive.

In figure 1 we show the phase diagram of the two dimensional generalised XY model in the $(p, T)$ plane. It is seen that the model exhibits a phase transition from a paramagnetic phase to a BKT–like phase as the temperature decreases. The transition temperature is evaluated using Monte Carlo simulations according to reference [20]. The value of the transition temperature corresponding to $p = 1$ is taken from reference [21].

### 2.2. Three dimensions

For $d = 3$, different analytical approximations, such as a Mean Field (MF) approach, as well as its Two–Site Cluster (TSC) refinement have been used in reference [18] to estimate transition temperatures for $p = 2, 3, 4$, and then for higher values of $p$ in subsequent papers (see below). Notice also that $(\sin\theta)^p$, and hence the absolute value of the interaction potential, decreases with increasing $p$, and this aspect is reflected by the $p$–dependence of the estimated transition temperature. Transition temperatures have been estimated in reference [22] by self–consistent harmonic approximation, both for $d = 2$ and $d = 3$, and it was found that the transition temperature is decreasing against $p$. A study of the model in its continuum limit, carried out in reference [22] also showed that out–of–plane fluctuations, and consequently the magnon density, decrease with increasing $p$.

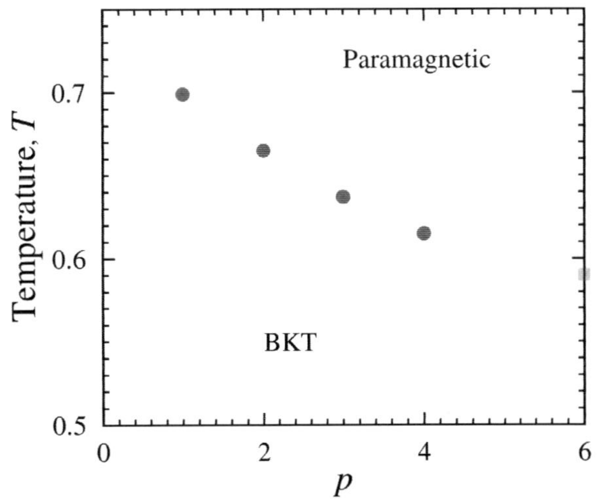


**Figure 1.** $(p, T)$ phase diagram of the two–dimensional generalised XY model. Full circles indicate a Berezinskiĭ–Kosterlitz–Thouless like phase transition, while the square is believed to corresponds to a weak first order phase transition.

We have also addressed the three dimensional generalised XY models for various values of $p$, by means of Monte Carlo (MC) simulation. We have investigated the transitional behaviour of various thermodynamic functions, such as the susceptibility and the specific heat, and made comparisons with MF and TSC predictions. MF yielded a tricritical behaviour with tricritical points having real (non–integer) values of the parameter $p$. As for simulation results, transitional behaviour characetristic of the XY model was found for $p = 2, 3, 4, 8$, the case $p = 12$ suggested tricritical behaviour, whereas evidence of first–order transitions was obtained for $p = 16, 20$.

In Figure 2 we present the $(p, T)$ phase diagram, where we can read off the behaviour of the transition temperature as a function of $p$ for $1 \leq p \leq 20$. MF and TSC data can be found in references [18, 19, 23, 24], while Monte carlo data are taken from references [23–25]. Notice that, for $p \leq 4$, TSC gives better estimates of $T_c$ than MF, then the two roles are exchanged for $p = 8, 12$, and finally the three methods give very similar answers when $p \geq 16$, i.e. where the transition has a pronounced first–order character.

## 3. Annealed dilution in ferromagnets

Another class of systems that has been extensively studied in the literature spans those with random impurities (disorder). For a review on diluted magnetism see reference [26]. The interest in these systems stems from the fact that in nature no system is really pure. Indeed, the presence of cavities, grain boundaries, lattice defects, chemical impurities (i.e. of other chemical individuals) or some other kind of disorder may affect the properties of the pure system, and result in different effects. To consider a simple but rather important example, in a system of magnetically coupled spins, some lattice sites may be occupied by nonmagnetic constituents (site–dilution). Both in experimental and theoretical terms, one can distinguish between annealed and quenched disorder.

In quenched systems, impurities are held frozen in randomly distributed fixed positions, without the possibility of overcoming potential barriers for diffusing into the host material: in this case the relaxation time is very long and thermal equilibrium between impurities and the constituents of the host is never reached. In annealed materials, impurities are allowed to diffuse randomly and to reach thermal equilibrium with the other constituents of the host material. One can also think of a two–component solution being very dilute with respect to one of the components. When the system is in the liquid phase,

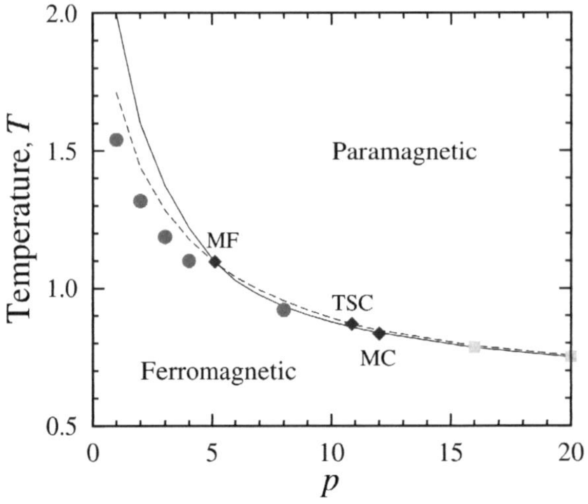


**Figure 2.** Phase diagram of the three dimensional generalised XY model. The solid and dashed lines stand for transition temperatures estimated via mean–field and two–site–cluster treatments, respectively. Full circles and squares indicate second and first order phase transitions temperatures evaluated via Monte Carlo simulations, respectively. Full diamonds shows the locations of tricritical points.

molecules of the two types can exchange their positions, and diffuse throughout the sample. Let then the system crystallize, at sufficiently low temperature. In the resulting solid, particles of the minority component are fixed in certain lattice sites only.

The simplest extension of equation (1.5) taking these situations into account reads

$$\mathscr{H}_{\text{dis.}} = -\frac{J}{2}\sum_{\langle i,j\rangle} \nu_i\nu_j\, \boldsymbol{s}_i \cdot \boldsymbol{s}_j. \tag{3.1}$$

where the occupation numbers $\nu_i$ equal "zero" for a site $i$ hosting a nonmagnetic impurity and "one" for a magnetic particle. The density of magnetic particles in the system is defined via

$$\rho = \frac{1}{N}\sum_{i}^{N}\langle \nu_i\rangle.$$

Within this notation the pure system, i.e. without impurities, corresponds to $\rho = 1$.

In the annealed case, Hamiltonian (3.1) can be interpreted as describing a two–component system consisting of interconverting "real" ($\nu_i = 1$) and "ghost", "virtual" or ideal–gas particles ($\nu_i = 0$). Both kinds of particles have the same kinetic energy, and the total number of particles equals the number of available lattice sites. In this case one works in the Grand–Canonical Ensemble where the probability (1.1) for a configuration, now involving the occupation numbers $\{\nu_i\}$, as well as the spins $\{\boldsymbol{s}_i\}$, is defined by

$$\mathscr{P} \propto \exp\left[-\beta\left(\mathscr{H}_{\text{dis.}} - \mu\sum_i \nu_i\right)\right], \tag{3.2}$$

where $\mu$ denotes the excess chemical potential of "real" particles over "ideal" ones. Of course, the interaction may be anisotropic in spin space, as in the cases outlined above, or involve more distant neighbours. The system remains translationally invariant on average. This model bears some similarity with the Blume–Emery–Griffiths model [27] for $^3$He impurities in superfluid $^4$He. More precisely, notice also that, starting from an assigned model, its lattice gas extensions can be written in general as

$$\mathscr{H}_{LG} = -\frac{J}{2}\sum_{\langle i,j\rangle} \nu_i\nu_j \, \boldsymbol{s}_i \cdot \boldsymbol{s}_j + \frac{\lambda}{2}\sum_{\langle i,j\rangle} \nu_i\nu_j \, , \tag{3.3}$$

where the purely positional $\lambda$ term only becomes immaterial in the pure limit $\mu \to +\infty$. In equation (3.1) we have chosen the simplest case $\lambda = 0$. Lattice gas models can be used to model adsorption, and, in general, the variable occupation numbers produce some fluidity of the system. One can also set the coupling $J$ term to zero, so that the resulting model becomes isomorph with an Ising model in external field. In general, the interplay between $J$ and $\lambda$ can produce a richer phase diagram: for example, when $\lambda > 0$ and $\mu$ is sufficiently large, the ground state may exhibit checkerboard positional order but no orientational one.

It is very well known that a small amount of annealed disorder does not affect the way the singularities take place in pure systems i.e. the phase transition remains a second order one. The $\rho$–dependent critical temperature is shifted towards $T_c(\rho) < T_c(1)$. If the chemical potential $\mu$ is held fixed, the properties of the phase transition are the same as those known for the pure system, corresponding to $\mu \to \infty$ and $\rho = 1$. If, however, the concentration $\rho$ is kept fixed during the transition, care must taken in characterising the transition. For details see reference [28].

A significant amount of impurities may alter the order of the phase transition or make it disappear. Investigations of the XY model, as a protype for $He^3$ – $He^4$ mixtures, in three dimensions, via high temperature series expansion of the partition function, show that by increasing the density of randomness the transition temperature decreases. At a certain value of $\rho$ the transition changes its nature and turns into a first order one [29]. The second order phase transition line ends at a tricritical point, which marks the begining of a line of fisrt–oder phase transition temperature that keeps decreasing as the concentration of impurities increases.

In the three–dimensional case, the topology of the phase diagram of model (3.1) had been investigated by MF and TSC approximations for the Ising [30], as well as PR cases [31] in the presence of a magnetic field, and for He at zero magnetic field [32]. These investigations were later extended [33] to the extremely anisotropic (Ising–like) two–dimensional model, and in the absence of a magnetic field, as well. The studied models were found to exhibit a tricritical behaviour i.e. the ordering transition turned out to be of first order for $\mu$ below an appropriate threshold, and of second order above it. When the transition is of first order, the orientationally ordered phase is also denser than the disordered one.

To check the predictions of the molecular–field–like treatments used to construct the phase diagrams, extensive Monte Carlo simulations has been performed [31–33] for particular values of the chemical potential. A number of thermodynamic and structural properties had been investigated. It had been found that there is a *second order* ferromagnetic phase transition manifested by a significant growth of magnetic and density fluctuations. The transition temperatures were found to be smaller than those of the corresponding values for the pure systems and the critical behaviour of the investigated models to be consistent with that of their pure counterparts. Furthermore it had been found that MF yields a qualitatively correct picture, and the quantitative agreement with simulation could be improved by TSC, which has the advantage of predicting two–site correlations. In general we found that simulations results are consistent with the molecular–field like treatments.

In figure 3 we report the phase diagrams in the chemical potential – temperature and density – temperature planes. We show the behaviour of the transition temperature $T_c$ versus chemical potential $\mu$. The plot also shows a fast, approximately linear, increase of $T_c$ with $\mu$ up to $\mu \approx 0$, and a slower one above this value. Moreover, MF and TSC results essentially coincide up to $\mu \approx 0$. In the phase diagram in the $(\rho, T)$ plane the existence of a first–order transition is reflected by a biphasic region. We found that the system exhibits a first–order phase transition from a dense BKT phase to a paramagnetic one. In the temperature–density phase diagram, both phases are expected to coexist over some range of densities and temperatures.

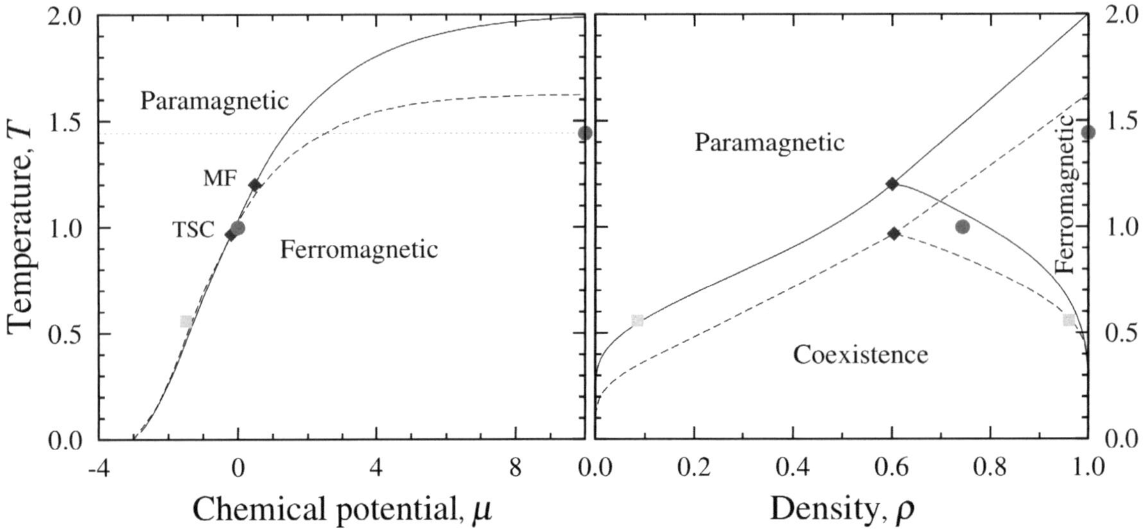


**Figure 3.** Phase diagrams of the three–dimensional Heisenberg model in the chemical potential – temperature and density – temperature planes. Solid and dashed lines correspond to mean–field and two–site–cluster results, respectively. Circles and squares stand, respectively, for estimates of the second and first order transition temperatures using Monte Carlo simulations. Diamonds mark the locations of tricritical points. The dotted line marks the transition temperature of the model in the absence of impurities.

Two–dimensional annealed lattice models were investigated [34, 35] as well, and the obtained results for $\mu = 0$ or a moderately negative $\mu$ were found to support a BKT phase transition with a transition temperature lower than that of the pure parent due to the presence of impurities. For negative and sufficiently large in magnitude $\mu$ we found evidence of a first order phase transition in agreement with renormalization group treatments [36, 37].

In figure 4 we show the phase diagram in the $(\rho, T)$ plane. The existence of a first–order transition is reflected here by a biphasic (binodal) region. The phase diagram obtained here is similar to the one resulting from the study of the diluted planar rotator model [31].

So far, we have discussed isotropic models. As we have mentioned, in addition to dilution one can consider anisotropic interactions as well. In reference [33] we have investigated two–dimensional continuous Ising spin models with two and tree component spins. The phase diagrams of these models were found to be topologically similar to those found for the Heisenberg model shown in figure 3. The main differences come from the locations of the phase boundaries.

## 4. Conclusion

The Statistical Mechanics of lattice systems is essentially based on the study of a number of relatively "simple" models which are gradually made more complicated. We have shown here some examples, involving generalized XY models and annealed magnetic systems, where rigorous results have been produced to prove existence and type of transition, supplemented by a variety of techniques (MF or TSC approximation and Monte Carlo simulation) for elucidating the resulting physical behaviour and estimating numerical values.

We obtained the phase diagram in two and three dimensions of a generalized version of the lattice XY model, where the out–of–plane fluctuations of the spins are controlled by a parameter $p$. Our investigation shows that the nature of the transition is highly affected by the strength of the out–of–plane fluctuations. For $p \geq 12$ at three dimensions and $p \gtrsim 6$ at two dimensions, it is first order, whereas for small values its

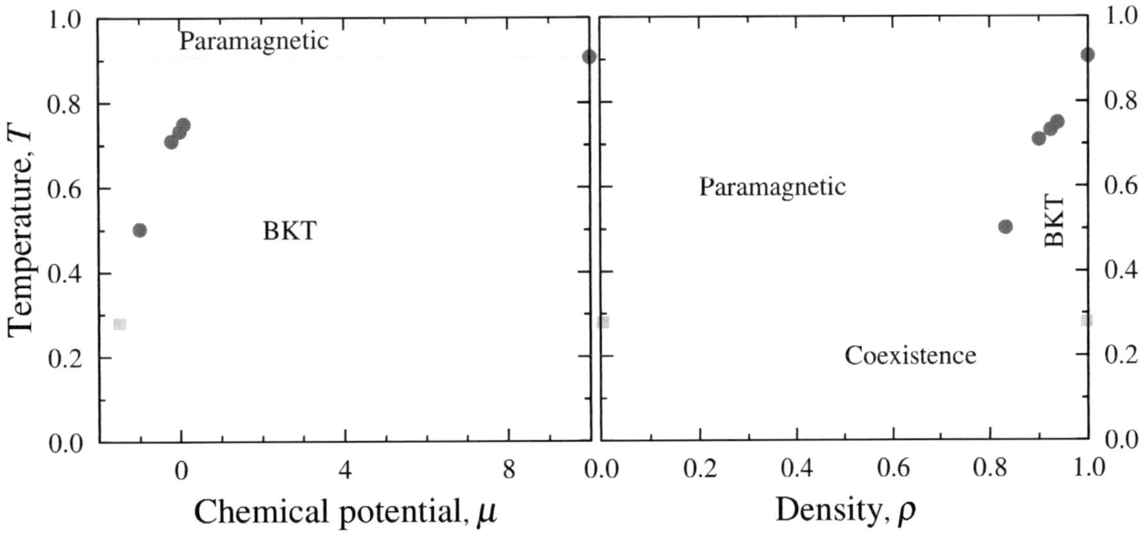


**Figure 4.** Phase diagrams of the two–dimensional planar rotator. Circles and squares stand, respectively, for estimates of the second and first order transition temperatures using Monte Carlo simulations. Diamonds mark the locations of tricritical points. The dotted line marks the transition temperature of the model in the absence of impurities.

transitional behaviour coincides with that of the original XY model. The transition temperature is found to decrease as $p$ increases at any dimension.

We constructed the phase diagrams of the annealed lattice plane rotator in two dimensions and the Heisenberg model in three dimensions. It is found that the transition changes from a second order at three dimensions and Berezinskiĭ–Kosterlitz–Thouless at two dimensions into a first order transition as the concentration of impurities is increased. In turn the transition temperature is found to decrease with increasing impurity density.

**Acknowledgments**
This work was supported by the Ministry of Education and Science of Bulgaria under Grant No Φ–1517.

**References**

[1] Yeomans J M 1992 *Statistical Mechanics of Phase Transitions* (New York: Oxford University Press)
[2] Pathria R K 1996 *Statistical mechanics* 2nd ed (Oxford, England: Butterworth–Heinemmann)
[3] Mazenko G F 2003 *Fluctuations, Order and Defects* (Hoboken: Wiley)
[4] Berezinskiĭ V L 1971 *Sov. Phys. JETP* **32** 493–500 [Zh. Eksp. Teor. Fiz. 59, 907–920]
[5] Kosterlitz J M and Thouless D J 1973 *J. Phys. C: Solid State Phys.* **6** 1181–1203
[6] Stanley H E 1968 *Phys. Rev.* **176** 718–722
[7] Berlin T H and Kac M 1952 *Phys. Rev.* **86** 821–835
[8] Georgii H O 1988 *Gibbs Measures and Phase Transitions* (*de Gruyter Studies in Mathematics* vol 9) (Berlin: Walter de Gruyter)
[9] Sinai Y G 1982 *Theory of Phase Transitions: Rigorous Results* (Oxford: Pergamon)
[10] Pelissetto A and Vicari E 2002 *Phys. Rep.* **368** 549–727
[11] Mermin N D and Wagner H 1966 *Phys. Rev. Lett.* **17** 1133–1136
[12] Gulácsi Z and Gulácsi M 1998 *Adv. Phys.* **47** 1–89
[13] Fisher M E 1998 *Rev. Mod. Phys.* **70** 653–681

[14] Stanley H E 1999 *Rev. Mod. Phys.* **71** S358–S366
[15] de Jongh L J and Miedema A R 2001 *Adv. Phys.* **50** 947–1170
[16] Binder K 1987 *Rep. Prog. Phys.* **50** 783–859
[17] Cardy J 1996 *Scaling and Renormalization in Statistical Physics* (*Cambridge Lecture Notes in Physics* vol 5) (Cambridge, England: Cambridge University Press)
[18] Romano S and Zagrebnov V 2002 *Phys. Lett. A* **301** 402–407
[19] van Enter A C D, Romano S and Zagrebnov V A 2006 *J. Phys. A: Math. Gen.* **39** L439–L445
[20] Mól L, Pereira A R, Chamati H and Romano S 2006 *Eur. Phys. J. B* **50** 541–548
[21] Evertz H G and Landau D P 1996 *Phys. Rev. B* **54** 12302–12317
[22] Mól L A S, Pereira A R and Moura-Melo W A 2003 *Phys. Lett. A* **319** 114–121
[23] Chamati H, Romano S, Mól L and Pereira A R 2005 *Eur. Phys. Lett.* **72** 62–68
[24] Chamati H and Romano S 2006 *Eur. Phys. J B* **54** 249–254
[25] Costa B V, Pereira A R and Pires A S T 1996 *Phys. Rev. B* **54** 3019–3021
[26] Stinchcombe R B 1983 Dilute Magnetism *Phase Transitions and Critical Phenomena* vol 7 ed Domb C and Lebowitz J L (London: Academic Press) chap 3, pp 151–280
[27] Blume M, Emery V J and Griffiths R B 1971 *Phys. Rev. A* **4** 1071–1077
[28] Fisher M E 1968 *Phys. Rev.* **176** 257–272
[29] Reeve J S 1976 *J. Phys. C: Solid State Phys.* **9** 2575–2587
[30] Sokolovskii R O 2000 *Phys. Rev. B* **61** 36–39
[31] Romano S and Sokolovskii R O 2000 *Phys. Rev. B* **61** 11379–11390
[32] Chamati H and Romano S 2005 *Phys. Rev. B* **72** 064424
[33] Chamati H and Romano S 2005 *Phys. Rev. B* **72** 064444
[34] Chamati H and Romano S 2006 *Phys. Rev. B* **73** 184424
[35] Chamati H and Romano S 2007 *Phys. Rev. B* **75** 184413
[36] Cardy J L and Scalapino D J 1979 *Phys. Rev. B* **19** 1428–1436
[37] Berker A N and Nelson D R 1979 *Phys. Rev. B* **19** 2488–2503

# Kinetics of electric field enhanced crystallisation in phase-change materials

**K Kohary and C D Wright**

College of Engineering, Mathematics and Physical Sciences, University of Exeter, Harrison Building, North Park Road, Exeter, EX4 4QF, UK

Email: k.i.kohary@exeter.ac.uk

**Abstract**. We study the effect of electric field on crystallisation kinetics in phase change materials within the framework of classical nucleation theory. We identify the parameter set where the electric field has a pronounced effect on the crystallisation of phase-change materials.

## 1. Introduction

Phase change materials have a unique combination of structural, electronic, and optical properties [1]. The amorphous and crystalline states are both stable at room temperature and exhibit a large difference in optical reflectivity and electrical resistance. At the same time the transition between the two phases can be achieved on very short timescales. These properties make phase-change materials suitable for a wide range of electrical applications including various optical and electrical memories. Well known phase-change materials are based on chalcogenide alloy systems, such as Ge-Sb-Te, Ag-In-Sb-Te, and In-Sb-Te [2]. In present phase-change materials are also considered for future emerging technologies, such as Probe-based Terabit Memory (ProTeM) [3] or for use in phase-change random access memory (PCRAM).

Crystallisation of phase-change materials is often modelled at the continuum level either using the Johnson-Mehl-Avrami-Kolmogorov (JMAK) approach or models based on separable nucleation and growth models. Unfortunately many of the assumptions on which JMAK formalism is based are violated in real switching events in phase-change devices. The approach based on nucleation and growth models deals with crystal clusters at or above the critical nucleus size. More sophisticated models based on rate equations should be used to study the role of sub-critical clusters [4]. Nevertheless, the simple nucleation and growth model has been successfully employed for a wide variety of problems [5,6,7]. An alternative to these continuum models is the cellular automata model developed by Wright et al [8], which uses the Gillespie algorithm to determine the outcome of a Monte Carlo simulation. More recently ab initio atomic scale models based on density functional theory have been used to model the crystallisation of phase-change materials [9].

In this paper, we use the nucleation switching model [7] to study the kinetics of crystallisation of phase-change materials. In the nucleation switching model the nucleation of amorphous materials is dependent on the applied electric field. Using the nucleation model we identify the parameter set where the electric field has a pronounced effect on the crystallisation of phase-change materials. The paper is organized as follows. In the next section we give a brief summary of the thermodynamics and kinetics of electric field enhanced crystallisation. In section 3, we present the crystallisation properties of phase-change material Ge-Sb-Te when the electric field is present. In particular, we discuss the crystallisation times at various temperatures and electric fields. The paper finishes with conclusions in section 4.

## 2. Electric field enhanced crystallisation

Crystallisation of phase-change materials enhanced by electric field has been recently studied by Karpov et al [7]. According to this work the nucleation rate is dependent of the applied voltage and the idea of field enhanced nucleation of a shunting crystalline cylinder has been put forward. In this model the initial crystalline clusters appear to be spherical and assumed to be highly conductive and therefore their induced dipole moment interact with the field, decreasing the system energy and thus facilitating nucleation. According to this scenario when the electric field is present the switching continues with the nucleation of a long ($h>>R$) conductive cylinder that acts as a "lightning rod" concentrating the electric field. The stronger electric field facilitates nucleation of an additional conductive particle at the cylinder end making it longer and hence further increasing the field.

A nucleated 'metal' particle changes the local electric field strength $E$ and decreases the electrostatic energy [7]

$$E = E_0/n \text{ and } W_E = -0.5\varepsilon E^2/n \quad (1)$$

where $\varepsilon = \varepsilon_r \varepsilon_0$ with $\varepsilon_r$ is the relative electric permittivity and $\varepsilon_0$ is permittivity of free space. The depolarising factor is $n = 1/3$ for a spherical particle, and for a long cylinder it takes the form of

$$n = (R/h)^2[\ln(2h/R) - 1]. \quad (2)$$

### 2.1. *Thermodynamics*

*2.1.1. Spherical particle.* The classical nucleation theory derived for condensed systems assumes that formation of a spherical cluster of new phase requires a free energy $G$ given by

$$G = A_{sph}\sigma - V_{sph}g \quad (3)$$

where $A_{sph} = 4\pi R^2$ and $V_{sph} = 4\pi R^3/3$ are the surface and volume of the spherical cluster with radius $R$, and $\sigma$ is the surface energy. The bulk free energy difference $g$ between the amorphous and crystalline phases is considered to be temperature dependent. In this paper we use the form established by Peng et al. [5]

$$g = g(T) = \begin{cases} H_2\left[1 - \dfrac{T}{T_g}\left(1 - \dfrac{H_1}{H_2}\dfrac{T_m - T_g}{T_m}\right)\right] & \text{if } T \le T_g \\ H_1\dfrac{T_m - T}{T_m} & \text{otherwise} \end{cases}, \quad (4)$$

but other slightly different forms are also known in the literature (see for example, [4,6]). The corresponding parameter values are listed in table 1. Minimising $G$ gives the critical radius $R_0$ and nucleation barrier $W_0$ of a spherical particle

$$R_0 = 2\sigma/g(T) \text{ and } W_0 = 16\pi\sigma^3/3[g(T)]^2, \quad (5)$$

and the temperature dependence of these parameters is illustrated in figure 1.

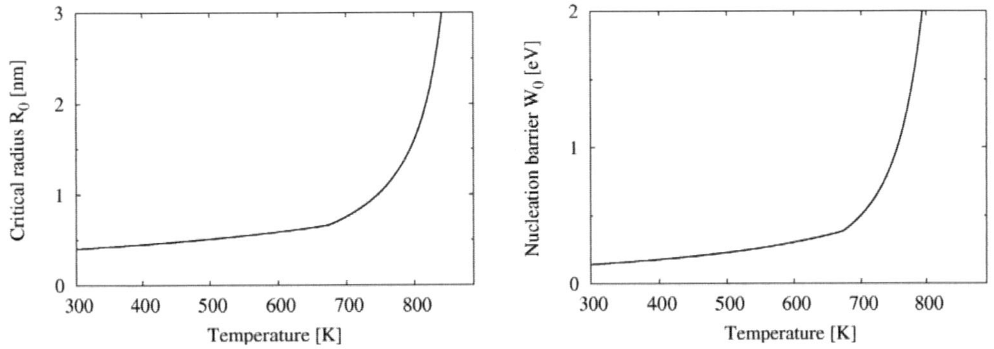


**Figure 1.** Critical nucleus radius and the corresponding nucleation barrier as function of the temperature according to Eq. (5).

The electric field modifies the free energy $G$ (see equation (3)) and in case of a spherical particle it obeys the form of

$$G_{sph} = A_{sph}\sigma - V_{sph}\left[g(T) + g_{sph}(E)\right], \tag{6}$$

with $g_{sph}(E) = 1.5\varepsilon E^2$ using $W_E$ in Eq. (1) and $n = 1/3$ for a spherical particle. Therefore, the electric field decreases both the critical radius and nucleation barrier when the field is large according to

$$R_{sph} = 2\sigma/\left[g(T) + g_{sph}(E)\right] \text{ and } W_{sph} = 16\pi\sigma^3/3\left[g(T) + g_{sph}(E)\right]^2. \tag{7}$$

This dependence is demonstrated in figure 2.

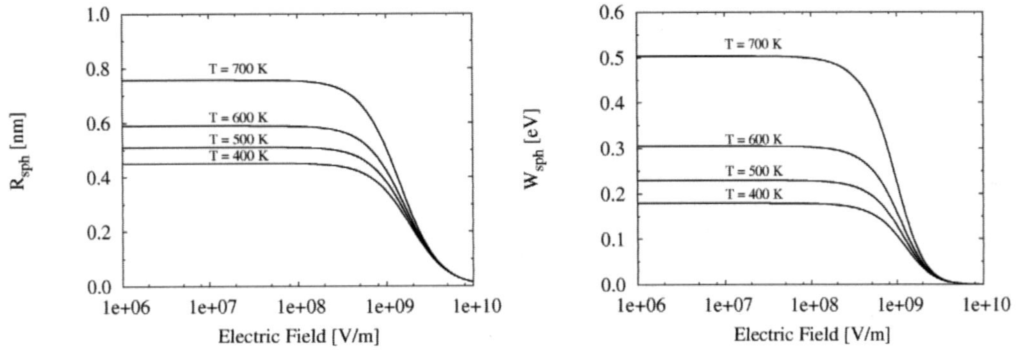


**Figure 2.** Electric field dependence of the critical radius and nucleation barrier according to Eq. (7).

*2.1.2. Cylindrical particle.* The free energy for a cylindrical particle is given by

$$G_{cyl} = A_{cyl}\sigma - V_{cyl}\left[g(T) + g_{cyl}(E)\right], \tag{8}$$

with $A_{cyl} = 2\pi Rh + 2\pi R^2$, $V_{cyl} = \pi R^2 h$ and $g_{cyl}(E) = 0.5\varepsilon E^2 (h/R)^2$, where following [7] we have neglected the logarithmic dependence in Eq.(2). The corresponding nucleation barrier normalised to $W_0$ as a function of cylinder radius $R$ and length $h$ (normalised to $R_0$) is shown in figure 3.

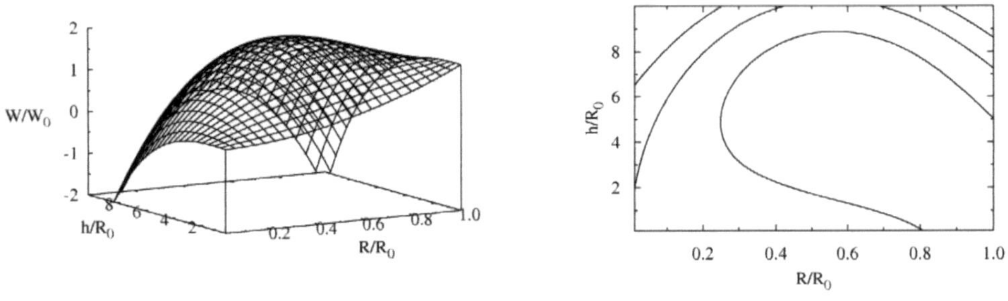


**Figure 3.** Surface and contour plots of the normalised nucleation barrier for cylindrical particles.

In nucleation theory the surface and volume energy terms have physical meanings if the cylindrical particle has a minimum radius $R_{\min} = \alpha R_0$. Following the work in reference [7], we are using $\alpha = 0.1$ throughout this paper. The nucleation barrier for a cylindrical shaped particle can be lower than the corresponding nucleation barrier of a spherical particle (the latter would approximately correspond to 'coordinates' of $(R = R_0, h = 2R_0)$ in figure 3). It can be shown that the relevant energy barrier for crystallisation along the line $R_{\min} = \alpha R_0$ is lowered compared to spherical particles according to

$$W_{cyl} = 1.5W_0\left[\alpha^2 + \left(2/3\sqrt{\pi}\right)\left(\alpha - \alpha^2\right)^{3/2}\left(E_0/E\right)\right], \tag{9}$$

as illustrated in figure 4. The characteristic field $E_0$ is defined to be equal to $E_0 = \sqrt{W_0/\varepsilon R_0^{\ 3}}$.

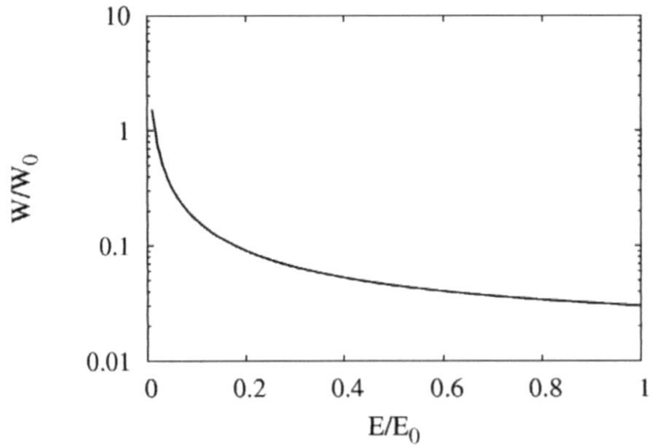


**Figure 4.** Reduction of the nucleation barrier for cylindrical particle with minimum radius $R_{\min} = \alpha R_0$ according to equation (9).

### *2.2. Kinetics*

The crystallisation kinetics of phase-change materials has been extensively studied by various groups for various applications [1]. It is common in all these calculations based on the nucleation theory, as described in section 2.1, that crystallisation occurs either by the nucleation of critical size particles and their subsequent growth within the untransformed region (homogeneous nucleation), or by the growth of crystallites at the boundary between the untransformed region and the surrounding crystalline matrix (heterogeneous nucleation). In this paper we only consider the first scenario as no specific geometry for the phase-change material is taken into account, although following [5,6], the phase-change layer is assumed to have been discretised into cells of $\Delta = 1\ \text{nm}^3$ in this work. The probability of an untransformed cell in the amorphous state to be become crystalline during the time interval $dt$ is given by

$$P = \nu \cdot dt \cdot \exp[-\beta(E_a + W)], \tag{10}$$

where $\nu$ is a frequency factor related to atomic vibrations, $\beta = 1/k_B T$ where $k_B$ is the Boltzmann constant, $E_a$ is the activation energy associated with nucleation and $W$ is the nucleation barrier.

**Table 1.** Parameters used in this study.

| Symbol | Value | Unit | Ref |
|---|---|---|---|
| $H_1$ | 418.9 | $J/cm^3$ | 5 |
| $H_2$ | 218.9 | $J/cm^3$ | 5 |
| $T_m$ | 889 | K | 5 |
| $T_g$ | 673 | K | 5 |
| $\sigma$ | 0.033 | $J/m^2$ | 10 |
| $\nu$ | 4E+25 | Hz | 5 |
| $E_a$ | 2.19 | eV | 5 |
| $\varepsilon_r$ | 10 | 10 | 7 |

## 3. Results and discussion

We follow the crystallisation of the amorphous material by calculating the time evolution of the crystalline fraction given by

$$\chi(t) = \chi(t-dt) + P[1 - \chi(t-dt)]. \tag{11}$$

Each simulation was performed at constant temperature and constant electric field starting from an initially amorphous sample (eg $\chi(t=0) = 0$).

### 3.1. *Spherical particles*

The crystallisation map depicts the required crystallisation times for the amorphous sample to fully crystallise (e.g. $\chi = 1$) at a given temperature and electric field. For spherical particles the effect of the electric field is negligible at low fields below $E < 10^7$ V/m, as seen in figure 5. At these low fields the temperature has the dominant effect and the optimal crystallisation time can be achieved for the temperature range between 650 and 700K according to figure 5. Below 400K and above 800K the crystallisation times can be orders of magnitude longer than for the optimised temperature range.

At larger electric fields (eg above $E > 10^8$ V/m), the field has a pronounced effect and reduces the crystallisation time. According to the results shown in figure 5 the electric field significantly enhances crystallisation speed by about two to three orders of magnitude (if keeping temperature constant but changing only the electric field).

Theoretical predictions for emerging probe storage applications [11] have shown that within these devices it is not uncommon to see high electric fields of up to $E \sim 10^8$ V/m. Therefore, in principle, crystallisation of spherical particles in probe storage devices can be enhanced by the electric field.

Nevertheless, we show below in section 3.2 that for cylindrical shaped particles the crystallisation time can be further reduced at even lower fields.

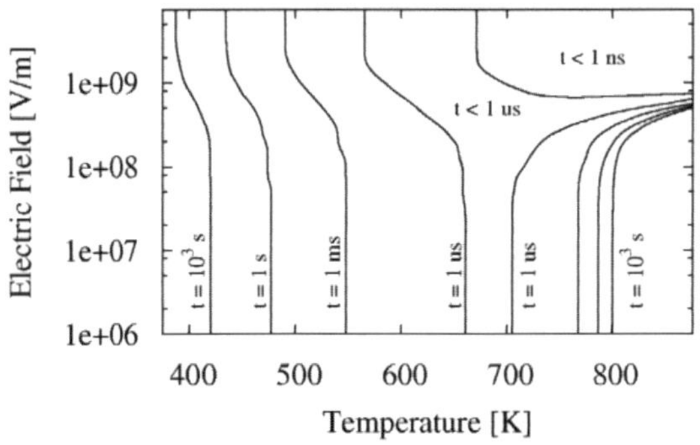


**Figure 5.** Crystallisation times for spherical particle at different temperatures and electric fields.

*3.2. Cylindrical particles*

The crystallisation times of cylindrical shaped particles may be smaller in comparison to spherical shaped particles due to lower nucleation barrier, as described in section 2.2. We have therefore repeated the calculations in section 3.1 using the free energy term in equation (8). We found that at low electric fields below $E < 10^7$ V/m the crystallisation times are much larger for cylindrical shaped particles than for the spherical particles, as shown in figure 6. On the other hand for large fields above $E > 10^8$ V/m the crystallisation times appear to be very similar both for spherical and cylindrical particles.

The crystallisation times for cylindrical particles is reduced in the 'transition regime' for fields between $10^7$ V/m $< E < 10^8$ V/m, as shown in figure 6. It can be seen that there is a much wider optimal temperature range for the crystallisation times. For example, it emerges from figure 5 that the optimal crystallisation times found within the temperature range between 650 and 700 K (below $E < 10^8$ V/m) for the spherical particles (figure 5) has a much broader temperature range from approximately 600 to 800 K for the cylindrical particles (figure 6).

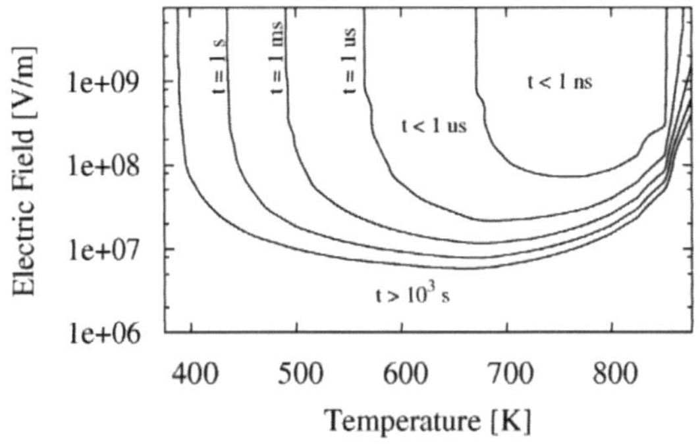


**Figure 6.** Crystallisation times for cylindrical particle at different temperatures and electric fields.

## 4. Summary
We studied the electric field enhanced crystallisation kinetics in phase-change materials using nucleation theory. The crystallisation times are reduced for spherical shaped particles at larger electric fields, but on the other hand crystallisation of cylindrical shaped particles can be further enhanced at even lower fields.

**Acknowledgments**
The EPSRC is gratefully acknowledged for support of this work via funding grant EP/F015046/1.

## References
[1] Raoux S and Wuttig M 2008 *Phase Change Materials: Science and Applications* (London:Springer)
[2] Ohta T and Ovshinsky S R 2003 Phase-Change Optical Data Storage Media *Photo-induced Metastability in Amorphous Semiconductors* ed A V Kolobov (New York: Wiley-VCH) pp 310-326
[3] http://www.protem-fp6.org/
[4] Senkander S and Wright C D 2004 *J. Appl. Phys.* **95** 504
[5] Peng C, Cheng L and Mansuripur M 1997 *Phys. Rev.* B **82** 4183
[6] Kim D H, Merget F, Foerst M and Kurz H 2007 *J. Appl. Phys.* **101** 064512
[7] Karpov V G, Kryukov Y A, Savransky S D and Karpov I V 2007 *Appl. Phys. Lett.* **90** 123504
[8] Wright C D, Blyuss K and Ashwin P 2007 *Appl. Phys. Lett.* **90** 063113
[9] Hegedus J and Elliott S R 2008 *Nature Mat.* **7** 399
[10] Surface energy value derived using relevant parameters in reference [5]
[11] Wang L and Kohary K [unpublished]

Journal of Physics: Conference Series **253** (2010) 012013 doi:10.1088/1742-6596/253/1/012013

# What is the origin of tail states in amorphous semiconductors?

**S Kugler**

Department of Theoretical Physics, Budapest University of Technology and Economics, H-1521 Budapest Pf. 91, Hungary and Joint Research Center for High-Technology, Tokyo Polytechnic University, 164-8678 Tokyo, Japan

E-mail: kugler@eik.bme.hu

**Abstract.** The tight-binding molecular dynamics simulations and reverse Monte Carlo structural modeling method were applied in order to investigate the existence of small bond angles (like those in triangles and squares) in amorphous silicon networks. The influence of small bond angles on the electronic density of states was analyzed. The presence of a number of smaller bond angles is necessary for a proper reproduction of the neutron diffraction data of amorphous silicon. Semi-empirical Hartree–Fock calculations show that these arrangements provide higher energy states in electronic density of states which are localized on these local structures. We consider that smaller bond angles are a new type of defects. These defect states are highly energetic and they are the origin of tail in amorphous semiconductors.

## 1. Introduction

In condensed matter physics, the electronic density of states (DOS) of a system describes the number of states at each electron energy level. The states can be occupied or unoccupied. A DOS of zero means that no states can be occupied at that energy level. These states form a so-called gap. The DOS is one of the most important properties of condensed materials considering electron transport, optical properties like transparency and related effects, etc. We must have information on it for electronic application as well. In crystalline case having translational symmetry plenty of different calculation methods exist to derive the DOS. They are usually based on Block electron theory. Huge number of such papers appeared containing electronic density of states calculations and results are usually displayed in inverse space (k space or reciprocal k-space), inside the first Brillouin zone of the lattice showing symmetry labels (see e.g.[1], page 204 for Si, page 207 for Ge and page 209 for GaAs). In the electronic DOS of crystalline semiconductors there is an energy interval where we can not find states at any k value. These gaps are about 1-2 eV in crystalline semiconductors. If the maximum energy of occupied and the minimum energy of unoccupied states belong to the same k than we say that this semiconductor has direct gap, opposite case this is an indirect gap semiconductor [1].

There is no periodicity in amorphous materials. Hence there is no reciprocal k-space. Block electron theory does not work! Should band gap occur in amorphous materials as it exists in crystalline case? The answer is yes but it is a little bit different. Only the number of states versus energy is the useful relationship. How to determine? The most common possibility is the cluster calculation which has two important steps. First, a good atomic configuration is needed so one must make a convenient computer modeling. Molecular Dynamics and Monte Carlo simulations are usually used for obtaining an atomic scale configuration. This structure allows us to carry out an electron energy calculation. Hartree-Fock ab initio or semi-empirical methods usually applied for deriving the DOS of noncrystalline semiconductors, but there are some others like density functional method or tight-binding model. So far the Hartree-Fock approximation seems to be the best method to calculate the electron energies although this method overestimates the gap.

In all handbooks we can find the following general description for the electronic density of states of pure amorphous semiconductors, see figure 1. The first surplice is that from the crystalline physics learned empty gap disappeared! Not empty gap called mobility gap (tail) can be found containing localized electron states which states do not exist in any pure crystalline semiconductors. The rest of states around gap are demoralized and the interface between localized and demoralized states is the mobility edge. A natural question: what is the origin of such states in the impurity-free amorphous semiconductors? The usual answer is “disorder”. That’s it no detailed description can be found in the literature.

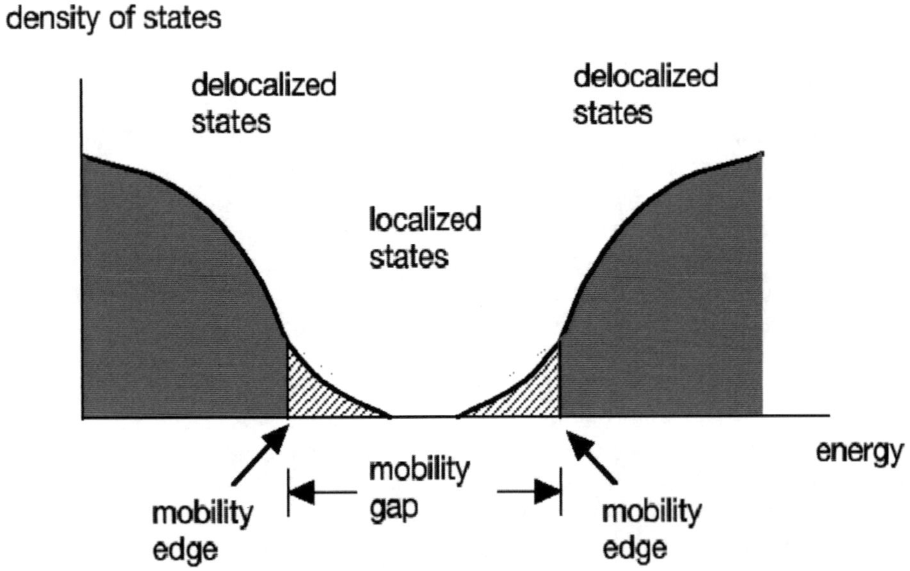


**Figure 1**. Traditional description of electronic density of states in amorphous semiconductors

There are several calculations on the band structure of amorphous silicon which is the model material of IVth column semiconductors. Most of cases Continuous Random Network model of Wooten et al. [2]. (WWW model) were applied for structural atomic scale configuration. Wooten’s model of a-Si having periodic boundary conditions contains bond angles in the internal between about $90^{o}$ and $150^{o}$. The classical empirical Keating potential was applied for the construction which interaction has a quadratic energy term of the difference between cosine bond angle and cosine ideal bond angle. This term avoids the large deviation from the canonic value of bond angle. The coordination defect-free network includes fivefold and sevenfold rings in addition to the six fold rings of the diamond-like structure. Since then, several computer generated models have been constructed using various classical empirical potentials or by applying different quantum mechanical methods [3-11]. Despite this effort the WWW model is still considered to be the best three-dimensional atomic scale representation of a-Si structure. We also used WWW model for the derivation of charge fluctuation in a-Si [12]. We observed that the angle deviation from the ideal bond angle plays the most important role in the charge accumulation on each silicon atoms and the bond length fluctuation is negligible in a-Si [12]. We got the similar conclusion for tetrahedrally bonded a-C [13]. (My/our all papers cited here can be downloaded from my homepage: http://www.phy.bme.hu/~kugler/pub2.html)

Journal of Physics: Conference Series **253** (2010) 012013 doi:10.1088/1742-6596/253/1/012013

## 2. Our amorphous silicon structure investigations

We have developed a tight binding molecular dynamics (TB–MD) computer code to simulate the real preparation procedure of amorphous structures, which is grown by atom-by-atom deposition on a substrate. Note that our differs from most other molecular dynamics (MD) studies where the amorphous networks are formed by rapid cooling from the liquid state. No laboratory where amorphous samples from the elements of column four could be prepared! Our MD method was successfully used for the description of the amorphous carbon growth first [14,15]. Later on the TB Hamiltonian of Kwon et al. [16] was used to describe the interaction between silicon atoms. All parameters and functions of the interatomic potential for silicon were fitted to the results of the local density functional calculations. The TB potential reproduces the energies of different cluster structures, the elastic constants, the formation energies of vacancies and interstitials in crystalline silicon. A surprising result was found in the ring statistics [17]. We defined a ring as a closed path, which starts from a given atom walking only on the first neighbor bonds. The size of a ring is the number of atoms in a closed path. The networks prepared by our models have a significant number of squares. Furthermore, triangles are also present in the atomic arrangements. Most of the theoretical models for a-Si do not contain such structural fractions.

Having our neutron diffraction data on evaporated amorphous silicon sample [18,19] RMC computer simulation study have been carried out. Potential independent reverse Monte Carlo (RMC) modeling has already been shown to be a powerful tool for interpreting amorphous structures. The unconstrained simulation contained a large number of bond angles of 60 degrees. At that time, we considered this result might be artifact. For the next case, a rather specific series of RMC calculations has been carried out. Our aim was to establish whether bond angles drastically different from the canonical value of 109.5 degrees are necessary to reproduce the neutron diffraction data of [20,21]. The RMC calculations were started from a 512 atom WWW model and only very small displacements (of the order of 0.002Å) were permitted during the Monte Carlo simulation. Bond angles was monitored regularly, after about each 50 000 accepted displacements. The computer program stopped after 500 000 accepted steps. The step development demonstrates that we can get better and better fit to the measured diffraction data if some bond angle get smaller and smaller. (At the beginning the smallest bond angle was about 90 degrees).

There is no direct experimental method for determination the atomic distributions in three dimensions. During the growth of amorphous structures it might be expected that covalently bonded atoms tend to form `natural configurations'. One way of ascertaining these natural configurations in a-Si is to consider the configuration of Si fragments embedded inside large molecules. The environment of silicon atoms might be considered as `white noise' around such fragments. A systematic analysis of structural data has been carried out using the Cambridge Structural Database (CSD) [22], this being the world's largest database of experimentally determined crystal structures containing the results of X-ray and neutron diffraction studies and three-dimensional atomic coordinates. We have collated from the CSD the experimentally determined structural data of molecules containing Si-Si-Si fragments. A search found more than 2000 targets. In the angle distribution of Si-Si-Si parts, two well-defined and unexpected regions have been found i.e. we observed triangles and squares as natural configurations of silicon atoms.

## 3. Density of states calculations

Quantum chemical cluster calculations at the AM1 level were carried out in order to find out whether the presence of triangles and/or squares cause variations in terms of the electronic properties [19 20, 21]. The electronic density of states (EDOS) of a part of WWW model and the modified WWW models containing triangles and squares were calculated. The first cluster (a part of the WWW model) contained about 100 fourfold coordinated Si atoms and a sufficient number of hydrogen's saturating the dangling bonds on the boundary of the cluster. This system can be taken as a reference system, as it contains no significant deviation from a locally nearly perfect tetrahedral order. Based on this network, we constructed other clusters by adding silicon (and hydrogen) atoms which formed one, two and three fused or individual triangles and squares. Significant differences were observed in terms of the electronic DOS: additional higher energy states appeared inside the gap, which are localized on the triangle(s) and square(s). There atomic scale arrangements construct the localized part of the mobility gap and form a shape of the electronic density of states having two important peaks.

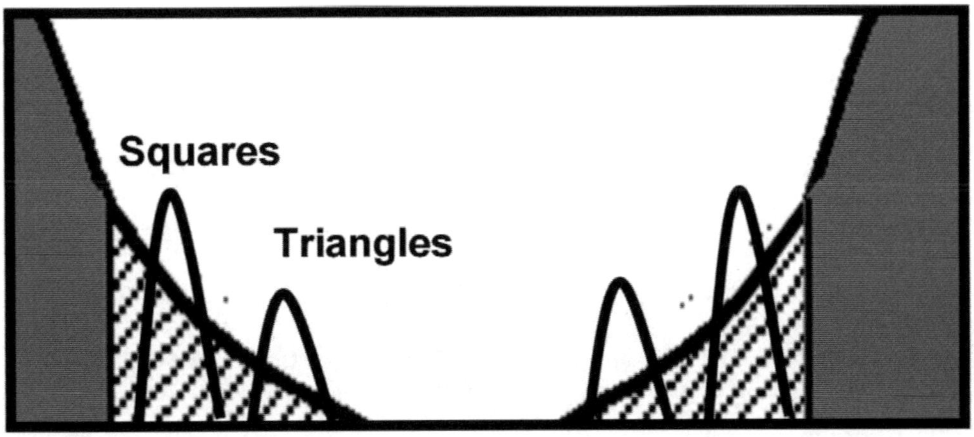


**Figure 2.** Electronic density of states of amorphous semiconductors containing two significant peaks belonging to the triangles and squares.

## 4. Conclusion

Our atomic scale structure investigations of tetrahedrally bonded amorphous semiconductors confirm that threefold and fourfold rings are also basic atomic arrangements of such materials. We consider triangles and squares as new types of defects. These defects are highly energetic. Accepting this result we must reconsider the electronic density of states of amorphous semiconductors. The localized mobility gap has structure i.e. two characteristic peaks can be found inside the tail. First larger peak belongs to the squares while the peak at larger energy is formed by triangles (figure 2). In the earlier investigation of electron transport, hopping conductivity, optical properties, etc. the tail was considered usually as an exponential or Gaussian decaying function. Triangles and/or squares have never been considered in any band structure calculations although they play important role in several phenomena.

## Acknowledgement

I am indebted to my coworkers: Laszlo Pusztai and Krisztina Kadas (Institute for Solid State Physics and Optics, Budapest, Hungary) and Krisztian Kohary (University of Exeter, UK). The computer simulations were partly carried out at the Tokyo Polytechnics University (Japan). I am indebted to Professor Takeshi Aoki (TPU) for providing us this possibility.

## References

[1] Sólyom J 2009 *Fundamentals of the Physics of Solids* Vol. II, Electronic Properties (Berlin:Springer-Verlag)
[2] Wooten F, Winer K and Weaire D 1985 *Phys. Rev. Lett.* **54** 1392
[3] Kelires P C and Tersoff J 1988 *Phys. Rev. Lett.* **61** 562
[4] Ishimaru M, Munetoh S and Motooka T 1997 *Phys. Rev.* B **56** 15133
[5] Vink R L C, Barkema G T and van der Weg W F and Mousseau N 2001 *J. Non-Cryst. Solids* **282** 248
[6] Stich I, Car R and Parrinello M 1991 *Phys. Rev.* B **44** 11092

[7] Toth G. Naray-Szabo G 1994 *J. Chem. Phys.* **100** 3742
[8] Hensel H, Klein P, Urbassek H M and Frauenheim T 1996 *Phys. Rev.* B **53** 16497
[9] Yang R and Singh J 1998 *J. Non-Cryst. Solids* **240** 29
[10] Herrero C P 2000 *J. Phys.: Condens. Matter* **12** 265
[11] Valladares A A, Alvarez F, Liu Z, Sticht J and Harris J 2001 *Eur. Phys. J.* B **22** 443
[12] Kugler S, Surjan P R and Naray-Szabo G 1988 *Phys. Rev.* B **37** 9069
[13] Kugler S and Naray-Szabo G 1991 *Jpn. J. Appl. Phys.* **30** L1149
[14] Kohary K and Kugler S 2001 *Phys. Rev.* B **63** 193404
[15] Kohary K 2001 "Computer simulations of structural and hopping conduction properties of disordered solids" *Ph.D. Thesis* (Marburg-Budapest)
[16] Kwon I, Biswas R, Wang C Z, Ho K M and Soukoulis C M 1994 *Phys. Rev.* B **49** 7242
[17] Kohary K and Kugler S 2004 *Molecular Simulation* **30** 17
[18] Kugler S, Molnar G, Peto G, Zsoldos E, Rosta L, Menelle A and Bellissent R 1989 *Phys. Rev.* B, **40** 8030
[19] Kugler S, Molnar G, Peto G, Zsoldos E, Rosta L, Menelle A and Bellissent R 1993 *Phys. Rev.* B **48** 7685
[20] Kugler S, Kohary K, Kadas K and Pusztai L 2003 *Solid State Comm.* **127**
[21] Kugler S, Kohary K, Kadas K and Pusztai L *2004 J. Non-Cryst. Solids* **338-340** 425
[22] Kugler S and Varallyay Z 2001 *Phil. Mag. Lett.* **81** 569

Journal of Physics: Conference Series **253** (2010) 012014 doi:10.1088/1742-6596/253/1/012014

# Switching mechanism in amorphous chalcogenides

**M Popescu and I D Şimăndan**

National Institute of Materials Physics, Bucharest – Magurele, P.O. Box Mg.7, Romania

E-mail address: mpopescu@infim.ro

**Abstract**. The electrical switching in solid compositions based on chalcogenide materials is discussed. Different mechanisms proposed for switching effect are reviewed. A new description of the switching phenomenon is done. The switching is regarded as due to formation and breaking of the links between the dendrites of crystalline nuclei in bulk materials, as a consequence of the energy pumped by an electrical field. This mechanism explains the very short switching time (<20ns), the possibility to get smart memories based on multisteps of resistivity and the high number of cycles supported by the cell ($10^{16}$).

## 1. Introduction

The memory effect in solid state has been a great challenge along the last decades. One of the leading class of materials where the switching phenomenon was discovered by Ovshinsky in 1968, is the class of chalcogenides. Various combinatioms of chemical elements including or not the chalcogenide ones have been prepared and tested regarding the switching effect. Multistep switching has been discovered recently and this opened the way toward intelligent memories.

In spite of large scale investigation of the switching effect in solids, the mechanism of switching is not fully understood. This is due to a variety of phenomena that are produced into a material when subjected to an electrical or optical field of excitation.

## 2. Resistance change versus charge storage

Traditional memory technologies are rapidly approaching miniaturization limits as the industry moves toward memory cells with 22-nm lateral features projected for 2016. The reason is that they are based on charge storage and it becomes increasingly difficult to reliably retain sufficient electrons in these shrinking cells. Magnetic and ferroelectric random-access memories, which are currently used in niche markets, share this struggle with scaling.

Nonvolatile memory concepts aimed at the horizon beyond 2013 are based on resistance change rather than charge storage.

The phase-change memory in chalcogenides, is based on change from crystalline to amorphous state and back: the crystalline state (on state, low resistivity state) is well known in many chalcogenides and for some systems, as e. g. Ge-Sb-Te, is characerized by very low (metallic) conductivity and high optical reflectivity. The amorphous state (off state, high resistivity state) is characterized by high resistivity and low optical reflectivity. The transition between these two states with very different properties is the basis of the phase-change memory used in modern switching applications.

### 3. Switching mechanism in electrolyte based memory cells

The first cell memory in solid electrolytes (programmable-metallization-PMC memory cell) has been obtained and investigated by Mitkova et all [1].

The basical material was 50 nm thick Ag doped Ge-Se film. The principle of working is the diffusion of silver in a chalcogenide material. During PMC cell operation it occurs the formation and removal of Ag-deposit through of an electrolyte amorphous medium. A sketch of the mechanism of switching stages is given in figure 1.

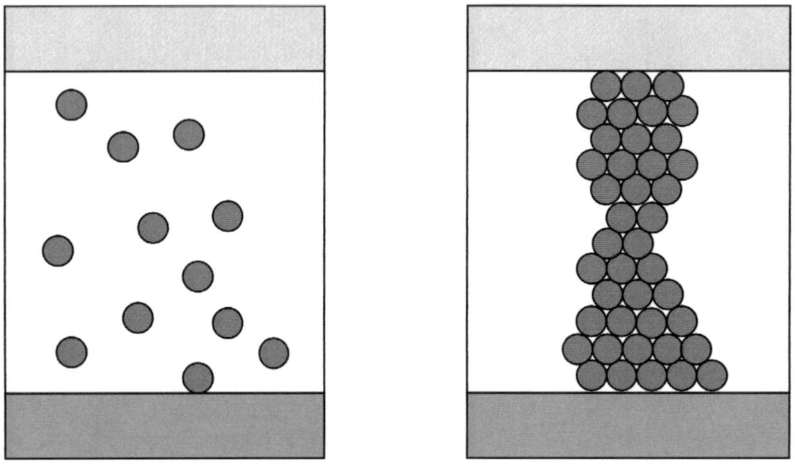

a) **WRITE**

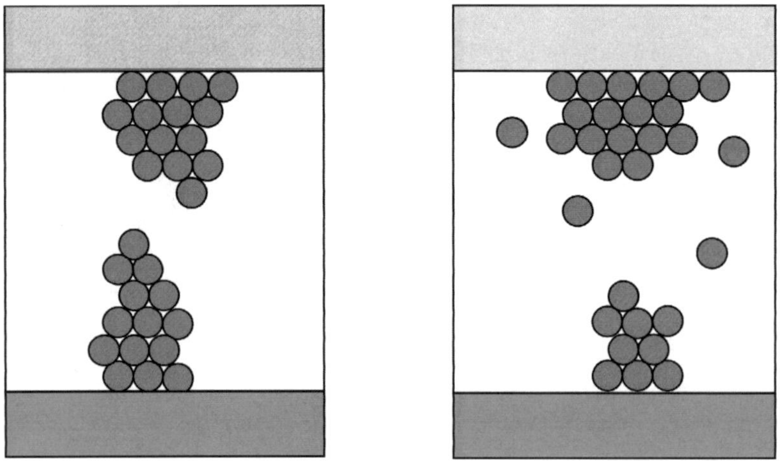

b) **ERASE**

**Figure 1.** The mechanism of switching in a memory cell based on the diffusion of Ag in a chalcogenide electrolyte (Ge-Se): a) writing step (A very small voltage rapidly injects excess silver ions into the electrolyte, to form silver dendrites on cathode); b) erasing step (A **very small** reverse voltage removes excess silver from the electrolyte, excess silver is **backed** on the silver electrode)

The usual materials (electrolytes) used by various companies in making memory cells based on the above described principle are:

- $CuS_2$ (NEC),
- GeTe (Samsung),
- Ge – S (Sony, Infineon, Micron, Adesto Technology).

There are also cells with resistivity-change memory based on transition-metal oxides. The cells work by the diffusion of metals into the oxide structures under the influence of an electrical field. The resistivity can be changed in a reversible way as a function of the electrical field polarity. Thus a reversible memory can be obtained. The systems investigated up to day are:

- Metal doped $WO_3$,
- Metal doped SiO.

## 4. The electrical switching mechanisms

### *4.1. The old views on the electrical switching mechanism in chalcogenides. The thermal switching and filamentary discharge.*

The observation of switching in thin semiconducting chalcogenide films goes back over a considerable period. Since Ovshinsky reported in 1968 [2], threshold processes in thin layers of chalcogenide glass alloys there was a lot of attempts to explain these observations. The problem is still challenging. Several scientists collected arguments to demonstrate that the switching is essentially thermal, the current in the ON state being carried by a hot filament. Others, as Adler et all [3], presented an electronic interpretation of the phenomenon.

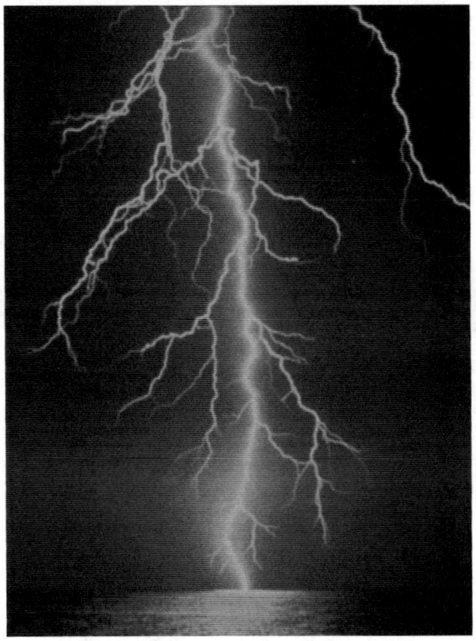

**Figure 2 a.** The electrical discharge into an insulating material (here in air)..

Thermal theory predicts ON-state "filament temperature" of the order of several hundreds degrees centigrade, depending on the assumptions made [4, 5], whereas the glass transition temperatures of the multicomponent chalcogenide alloys generally used for threshold switches are in the range 150 – 350 $^{o}$C. This leads to an expectation of gross instability, whereas the threshold switches are in fact highly stable (more than $10^{14}$ operations under pulse conditions). Step by step it has become clear that the ON-state channel in thin chalcogenide switches is not ordinarily hot and this is a strong evidence against a thermal model. Although the early theories attempted to reduce the switching effect to heating of the sample (as a whole or in the filament), later studies revealed that heating plays an insignificant role in reversible switching. It has been accepted that heating provides some

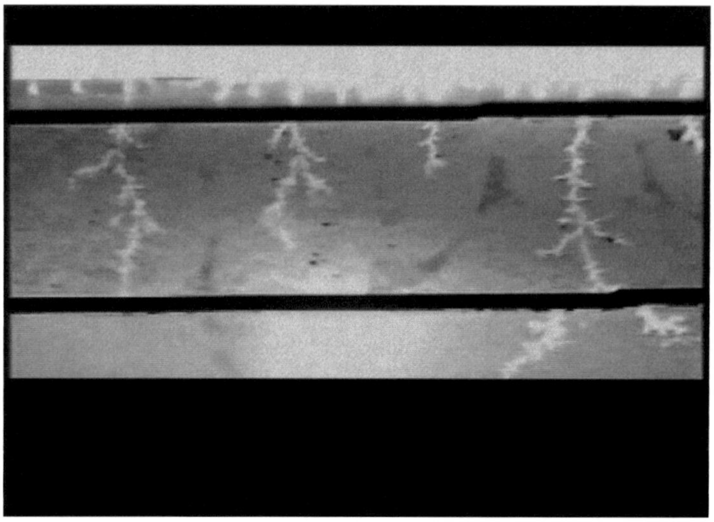

**Figure 2 b.** The filamentary structure of silver in PMC cells (courtesy of M. Mitkova).

contribution only to the switching delay time after the threshold field strength has been reached, as well as to the formation of memory and to the structural transformation of the material. Regarding the filamentary source of switching, the conducting channel is similar to the shape of an electrical discharge between two electrodes and through an insulating material (polymeric or not), (see figure 2a). In the PMC the Ag filaments exhibit a dendritic structure (figure 2b).

The application of sufficiently high electric fields to any material eventually results in deviations from linearity in the observed current – voltage characteristics. There are two general classes of explanations of such non-ohmic effects: thermal and electronic.

The thermal effects arise because the electrons accelerated by the field always emit phonons in an attempt to return to equilibrium. Electronic effecs are due to changes in the response of the charged carriers to high applied fields. In general both effects must be considered in any theoretical analysis [6]. As a conclusion, the switching in chalcogenide glasses has particularly desirable properties because of the flexibility resulting from the presence of lone-pair valence electrons while at the same time lead to a large density of valence alternation pairs

The switching phenomena follow from electronic effects, threshold switches from the field stripping of chalcogen lone-pair electrons and subsequent filled of lone-pair induced traps, while memory switching results from local structural rearrangements in less cross-linked chalcogenides under excitation conditions.

*4.2 The electronic mechanism. The two level system.*

The analysis of many papers [7] has shown that switching is produced by the following mechanism: the electric field generates free electrons and holes and this generation is most pronounced at the threshold field strength. These electrons and holes fill the traps. After all the traps have been filled, the resistance decreases sharply and switching occurs. Nevertheless, a high number of problems remain unclear: the nature of the traps, why the efficiency of trapping is maximum at the threshold field. No quantitative results are available. Different authors proposed different defects as traps: valence alternation pairs, dangling bonds, two-level systems.

Chaban [8] studied the glass transition and elaborated a model for this transition. According to Chaban's theory the glass transition point is the temperature at which ordered regions start to come in contact. This definition provides an explanation for the jumps of the heat capacity and other quantities at the glass transition point. The smallest change in heat capacity should correspond to the appearance and disappearance of one Frenkel hole in an ordered region. The related discreteness of levels should be observed at low temperatures. This idea has been used by Chaban [9] to built a model of low temperature anomalies in glasses. The ordered regions are characterized by a unique sensitivity to changes in the external conditions. They are sensitive to change of pressure, shear stress or electrical field strength, but also to the orientation of the nuclear spins, to the ejection of electrons from ordered regions by phonons, etc.

Chaban assumes that for adjacent electric field induced localized states of an electron of energy $E_e$ (pseudoexcitonic states) that have one electron each, it is energetically favourable to combine with the formation of a two-level system. The upper and lower levels of this system correspond to parallel and antiparallel spins, respectively. Chaban assumes that the energy difference between these levels is greater that $k_BT$ at room temperature. If the number of these two-level systems is larger than the percolation threshold, singlet electron pairs will be able to move throughout the sample in a resonant non-dissipative manner over the paths created by overlapping wave functions of these systems. Switching corresponds to the transition to a new type of conduction where the electrons propagate mostly through non-dissipative paths.

*4.3. The flip-flop mechanism.*
Kolobov et all[10] discussed this mechanism in the particular case of Ge-Sb-Te material, the structure of the crystalline phase is similar to the rock salt structure with a rather stable Te face-centered cubic (fcc) sublattice. Due to difference in covalent radii of the constituent species, Ge, and to a lesser extent Sb, are shifted from their corresponding fcc sublattice sites giving rise to a system of shorter bonds (stronger) and longer bonds (weaker). The shorter bonds are more rigid and thus provide a framework for the local structure. The germanium bonds (four) remain intact during transformation to amorphous state and thus the medium is fast and stable.

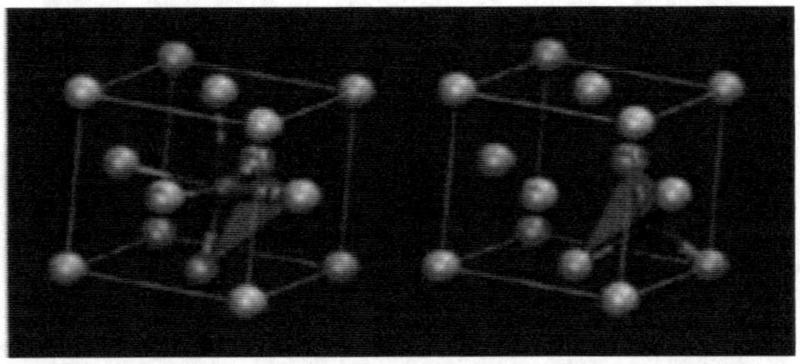

**Figure 3.** Fragments of the local structure of GST around Ge atoms in the crystalline (left) and amorphous (right) states. Stronger covalent bonds are shown as thicker lines, while weak interblock bonds are shown as thinner lines (after Kolobov [10]).

The rupture of the weaker bonds in GST (Ge-Te bonds) produces a net force on Ge atoms that makes it move along the [111] direction across the Te 111 plane. Provided the energy released by the bond relaxation is sufficient to overcome the energy barrier due to repulsion by Te atoms, Ge can flip across the plane in a tetrahedral site (figure 3).

The structural transformation consists in the switching of the Ge atoms between octahedral sites and tetrahedral sites due to the presence of strained Ge-Te bonds. Such mechanism seems to be common in binary Ge-Te and in quasi-binary $GeSb_2Te_4$ [10].

*4.4. Switching in multistage memory materials. Nucleation and growth.*
Phase change materials mainly include two families of materials: nucleation – dominated chalcogenide alloys and growth dominated chalcogenides alloys. In the first category enter GeTe –

$Sb_2Te_3$ alloys in which the crystallization is controlled by nucleation of crystallites. In the second category enter the In and Ag doped $Sb_2Te_3$ alloys in which crystallization occurs by motion of a glass-crystal interface [8].

*4.4.1. Growth of the nuclei without dendrites.* Existing models describing the crystallization behaviour of phase-change materials using optical and electrical memories are based on classical theory of nucleation. The values of the parameters: viscosity, diffusivity, fusion enthalpy must be estimated.

The normal memory effect is produced in the thin memory film by nucleation of crystalline germs and growth rapidly as dendrite development. The multistate must be related to the variable number of dendrite filaments across the chalcogenide material.

Crystallization often starts with the formation of small, unstable clusters of a new phase. Eventually some clusters reach a critical size beyond which they are stable, such that they can grow rather than dissolve. Homogeneous nucleation occurs at a random position in the original phase, while heterogeneous nucleation takes place at preferential sites as surfaces, interfaces and impurities.

As a function of the material type and preparation parameters the nuclei can grow without dendrites. In such a case the time of switching is considerably slower.

When the phase change is produced the nuclei must reorganize or develops in order to create percolation paths. In the case of non-dendritic development the percolation is considerable delayed. This is because the neighbouring nuclei are in a limited number and the probability of sticking and making crystalline paths is lower.

Figure 4 shows an image of $As_2S_3$ glass deposited. The round configurations grow without developing dendrites.

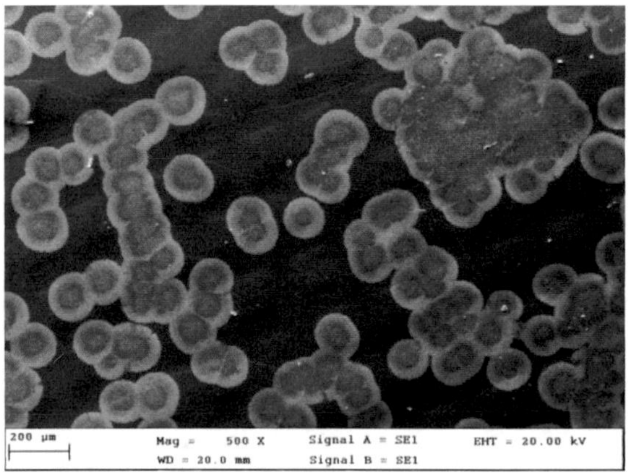


**Figure 4.** $As_2S_3$ deposit which grows in size without developing dendrites. The density of nuclei remains unchanged.

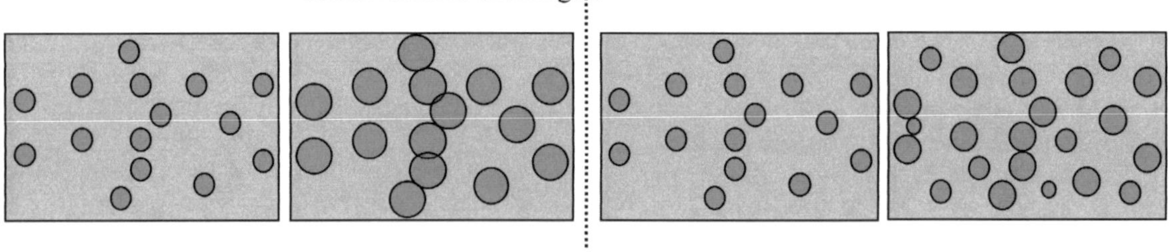

**Figure 5**. a) growth of the crystalline phase by size increase of crystallites, b) growth of the crystalline phase by multiplication of the crystallites and a moderate growth of crystallite size.

Figure 5 is a sketch of crystallization advancement in a amorphous material in two ways:

a) growth of the crystalline phase by size increase of initial nuclei,
b) growth of the crystalline phase by multiplication of the nuclei and a moderate increase in size.

The rigorous treatment of transient effects (e.g. transient nucleation rate case a) has been developed by Kashchiev et al. [13]. He has proposed the nucleation rate for isothermal annealing conditions to be:

$$I(t) = I^{ss} \left(\frac{4\pi\tau}{t}\right)^{1/2} * \exp\left(-\frac{\pi^2\tau}{4t}\right),$$

where I(t) is the nucleation rate ($cm^{-3}s^{-1}$) and t is the time of material annealing.

This function is plotted in figure 6 as a function of time for 250 °C, the incubation is ~ 2 ms much higher than the time of switching.

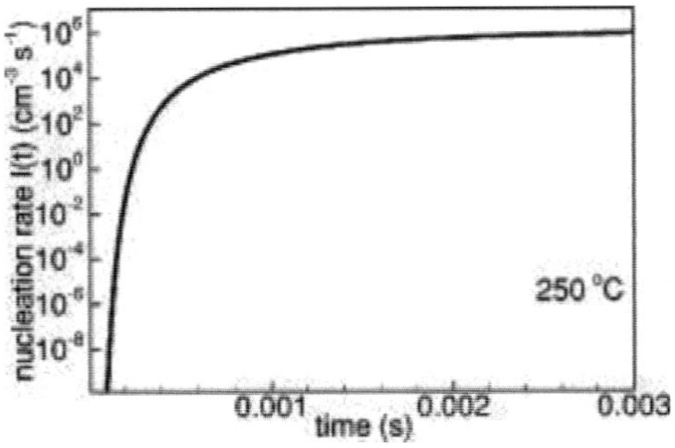


**Figure 6.** Transient nucleation rate of GST as a function of time for 250 °C [15] (case a).

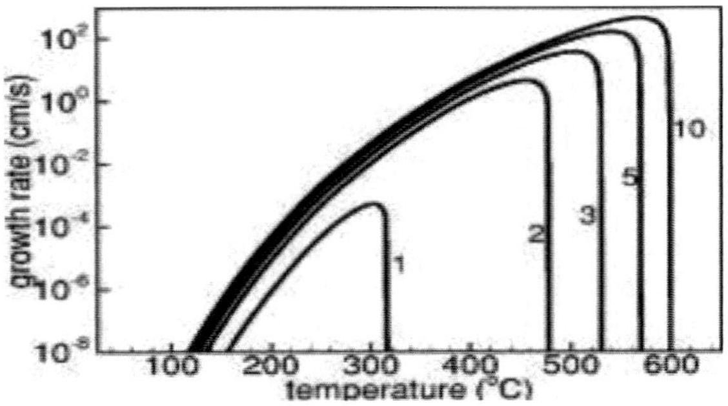


**Figure 7.** Growth rate of crystalline clusters as a function of temperature. Number next to each graph represents radius of cluster in nanometers [14] (case b).

The growth rate of crystalline clusters (case b) is plotted versus temperature in Fig.7. The process is described by the following equation:

$$\frac{dn}{dt} = g(n) - d(n)$$

where both the growth and dissolution rate is taken into account.

The growth rate is higher for higher temperatures but usually is situated in the range of $10^{-4} \sim 10^{-2}$ cm/s ($10^3 \frac{nm}{ns} \sim 10^5 \frac{nm}{ns}$). Usually, time of switching is 20 ns and the growth of nuclei is up to $2 \times 10^{-5} \sim 2 \times 10^{-3}$ nm.

In the case of memory switching, the ON state is triggered when the power is large enough to promote crystallization ($T>T_c$) but low enough to avoid melting of the conduction filament ($T<T_m$) at the end of the pulse, when power density is higher than that corresponding to $T_m$ , then a transition to OFF state occurs.

*4.4.2. Growth of the nuclei with dendrites.* After spontaneous formation of nuclei, further growth of these small crystalline clusters develops with more or less extension of the dendrite as a function of various parameters of the material. A typical material that shows dendrite crystallization is $Sb_2S_3$. As function of the deposition parameters many other amorphous materials develop crystallites with dendrites.

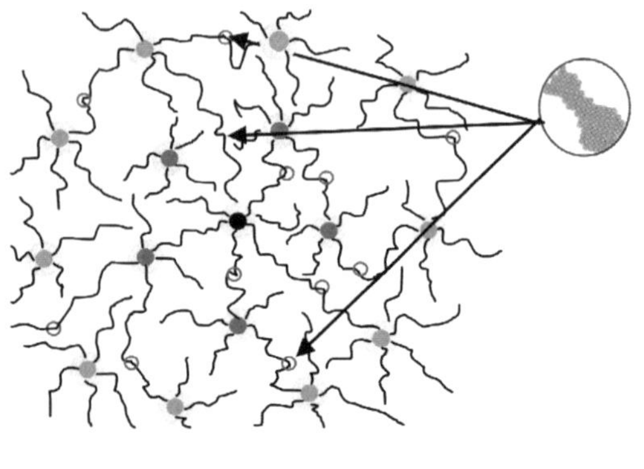

**Figure 8 a.** Switching model: state ON.

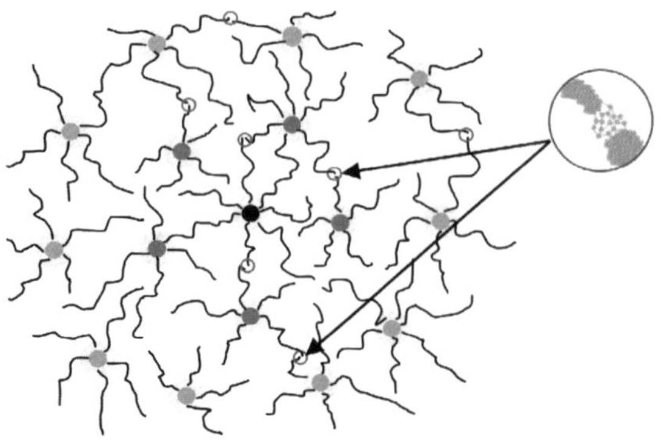

**Figure 8 b.** Switching model: state OFF.

Figure 8 shows the development of nuclei with long dendrites. These dendrites can occasionally meet and form thus a short conduction channel. This is a simple mechanism for the formation of a conducting channel into the material. The first step before switching under the action of an electrical field, in the amorphous materials are nucleation of small crystalline islands that grow and develop dendrites. The switching in material occurs when the dendrites become linked so that a percolation path appears (figure 8a - state ON). To go back a new pulse of the energy removes the inter-dendrite linkage and the percolation paths are destroyed. Thus state OFF is triggered (figure 8b).

Crystalline dendrites develop during the growth of many thin solid films, as e.g. $As_2S_3$. This is illustrated in figure 9.

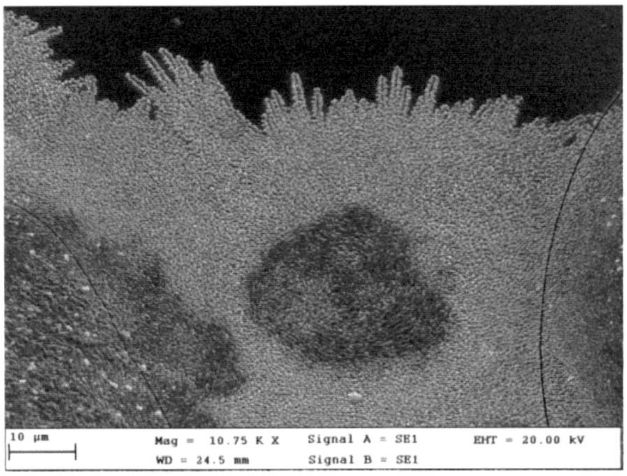


**Figure 9.** Crystalline dendrites are developed around a nucleus in thin film deposition ($As_2S_3$).

*4.5. Multi stage switching.*

The most prominent and widespread use of chalcogenide materials is in the rewritable phase-change optical memory disks and in rewritable digital versatile disks. Both optical and electrical data storage rely on the reversible phase transformation between amorphous and poly-crystalline states. On optical disks data recording is achieved by writing an amorphous dot on a crystalline film by local melting with a focused laser beam. When the laser is switched off, rapid quenching of the melt leads to amorphization of the dot. Since an amorphous dot has a lower reflectivity than the sourrunding crystalline background, the written bit can be read by a low power laser. Conversely, erasing the dot is achieved by heating the amorphous dot at temperatures less than melting point and allowing the amorphous dot to crystallize. Crystallization is much slower than amorphization.

Electrical memory uses the same two states of the chalcogenide material. A small volume of PC material is transformed by applying current pulses. Resistive heating are produced. Crystallization is initiated by a pulse heating. The state of the memory cell is read-out non-destructively by measuring the resistivity caused by the current pulse. The amorphous phase is obtained by a pulse sufficiently large to achieve melting. Subsequently, a rapid quenching allows the material to become amorphous. The crystalline phase is the low resistivity state (state ON). The amorphous phase is high resistivity state (state OFF). High speed crystallization is important. Therefore, phase separation is detrimental to short switching time. Moreover the atoms must not move on large distances. That is why only special materials as for example Ge-Sb-Te are recommended for switching.

The old materials based on Ge-Sb-Te-S suffer phase segregation [11] upon crystallization. The new materials based on Ge-Sb-Te exhibit congruent crystallization characterized by no phase segregation and rapid crystallization in ~ 20 ns. The life time of the switching is 10 years at 110 °C.

As a function of the material type and preparation parameters the nuclei can grow without dendrites. In such a case the time of switching is considerably higher. When the phase change is produced the nuclei must reorganize or develop in order to create percolation paths. In the case of non-dendritic development the percolation is considerable delayed. This is because the neighbouring nuclei are in a limited number and the probability of sticking and making crystalline paths is lower. When the nuclei develop by dendritic growth, the percolation threshold is reached more rapidly. When nuclei grow by dendrite extension, every nucleus becomes in contact with a higher number of neighbours.

As in the case of neuronal systems, a huge amount of conexions between cells determines a very good connectivity. Therefore the percolation can be attained easily. To block completely the crystalline continuity it is necessary to block the connection at random for a great amount of dendrites.

While the percolation limits in systems of lattice points of given symmetry are high and vary in broad range (from 0.20 to 0.43), in dendrite connected networks the percolation threshold can be considerably lower (down to 0.14 for taking into account the effect of the next nearest neighbours, or even 0.10 when next-next-nearest neighbours are considered).

The multistage memory makes use of new mechanisms in the deceptively simple, single, amorphous nano-dimensional spot in the low current operational regime. As current pulses are applied minute nanocrystalline region form and the volume fraction of such crystalline phases increases with each current pulse.. Crystallization can occur through nucleaton/growth upon the application of a current pulse. The microcrystallites generated by a sequence of pulses form a temporarily coherent sequence of states.

Figure 10 shows the multistage electrical switching in a chalcogenide material.

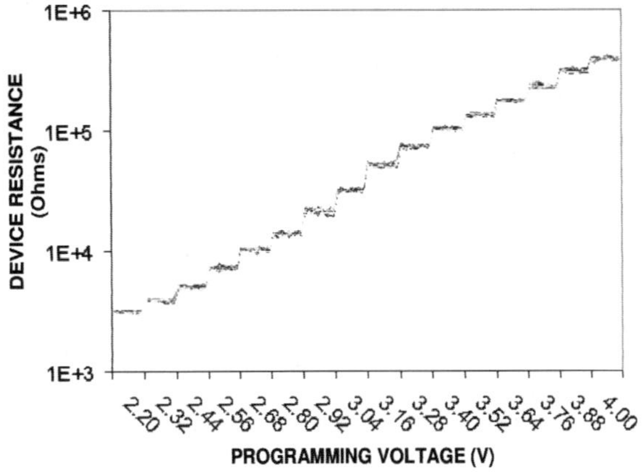


**Figure 10.** Programming energy changed by changing power at constant time during multistage switching.

The connection by dendrite explains the following properties of multiswitching:

a) the very low aging and the stability of the switching parameters in time,

Journal of Physics: Conference Series **253** (2010) 012014 doi:10.1088/1742-6596/253/1/012014

b) the very short time of switching.

The switching effect is flexible because the percolation during switching to stable ON state is easily reached. For dendrite growth of conducting nuclei the percolation threshold is much lower. For passing to the state OFF it is enough to melt only the dendrite thin contacts and not all the crystalline part. Thus the rate of switching to OFF state is higher. In the same time multi-state stable switching can be obtained.

The nano-crystallites are distributed randomly throughout the chalcogenide material. As they grow, a percolation path results, a continuous, high conductivity pathway across the material between the contacts. The increment of the crystallization that occurs upon application of a current pulse is dictated by the energy deposited by the pulse into the material. Once the energy is accumulated a new state occurs. The crystallites are developed up to a given dimension and spatial extension. This is in close analogy with neurosynaptic functionality in brain. Continued application of energy to a given structural state induces additional crystallization and further accumulation of energy until sufficient energy has been applied to reach the percolation transition. The energy required for percolation is the threshold from a high resistance state (state OFF) to a low resistance state (state ON). In the multistage memories based on optical mechanism, we must replace the resistance with reflectivity.

It seems the process is analogous with that produced in neuronal networks. Firstly, energy accumulates at a neuron through dendrites that link through synapses to other neurons. At a certain level of energy (a threshold energy) the neuron fires: This means a discharge is produced and the energy is transferred to other neurons. In the multi-stage device, the abrupt reduction in resistance that occurs at the percolation transition is analogous to the firing of a biological neuron.

A chalcogenide device with natural oxide on its electrodes exhibits some unique switching and memory behaviours. I-V characteristics showed a reversible bias-polarity-dependent switching behaviour via some intermediary states. Device resistance as a function of pulse height and decay behaviour from a low resistance state revealed that there were some intermediate states which were relatively stable in a certain voltage range and a certain period. These relatively stable intermediate resistance states make multi-state storage possible. These multi-states might be related to the number of dendrite filaments across the chalcogenide channel which could be induced by electrical pulses.

Regarding the importance of the material for multi-stage processes, it is remarkable the role played by $Sb_2Te_3$. A comparative study of crystallization processes in $Sb_2Te_3$ films using laser and thermal annealing techniques has been carried out by Reddy et all [12]. Usually the crystallization of this material is produced by developing long dendritic configuration. Adherent and pin-hole free amorphous $Sb_2Te_3$ thin films have been obtained by vacuum evaporation at substrate temperatures of $\leq 25$ °C. The films have been crystallized by thermal and laser annealing, and the crystallization processes monitored as a function of annealing temperature and laser scan speed. A comparative study of topography reveals disk-shaped crystallized areas in thermal crystallization and dendrite growth in the laser induced process. The crystallized films in both cases contain a single $Sb_2Te_3$ phase. Activation energy of 2 eV for crystallization, determined using differential scanning calorimetry indicates good room temperature stability of the amorphous states.

In Ge-Sb-Te the short range order is almost the same in the amorphous phase and in the corresponding crystalline one. The local order already present in the amorphous phase is likely responsible for the fast crystallization process (100 ns, while in GeTe is 300 ns). The phase change is completed without appreciable atomic movement. In binary system the local atomic order is different in crystalline and amorphous phase. In the amorphous state the germanium atoms are fourfold coordinated by 2 Te and 2 Ge atoms. When crystallization occurs the Ge-Ge bonds disappear and Ge atoms are uniquely surrounded by Te atoms. The phase transformation from amorphous to crystalline state requires the destruction of the Ge-Ge bonds and a significant movement of the Ge atoms to join the crystalline sites. These processes limit probably the crystallization speed and, therefore, the switching rate.

## 5. Mechanism of switching with fractal growth and dendrite interaction

The fractal growth of crystalline nuclei is the key factor of the rapid transition between the conducting and insulating state. In the same time the multistage memory can be easily explained. The crystalline nuclei are formed in the bulk amorphous matrix with various density and sizes.

As opposite to the old model with dendrite extending from one electrode to the other, the new model implies the formation of randomly distributed crystalline clusters that grow with dendrite development. These dendrites determine a higher probability to form percolation paths due to multiple possibilities to interconnect various developed nuclei. This is similar to the neuronal connections. The intricate connexions by synapses and long dendrites makes the connectivity much more complex than in the case of simple networks of atoms with simple geometry.

The mathematical percolation thresholds calculated for various lattices are given in table 1.

**Table 1**. Mathematical percolation thresholds in 3d lattices.

| **Lattice** | **Z** | **Site Percolation Threshold** | **Bond Percolation Threshold** |
|---|---|---|---|
| Ice | 4 | 0.433(11) | 0.388(10) |
| Diamond | 4 | 0.426(+0.08,-0.02) 0.4301(4) | 0.390(11), 0.3893(2) |
| Simple cubic | | 0.3116 | 0.2488126(5),0.2488125(25) |
| Icosahedral Penrose | 6 (average) | 0.285 | 0.225 |
| Penrose w/2 diagonals | 6.764 (average) | 0.27 | 0.207 |
| bcc | 8 | 0.2459615(10) | 0.1802875(10) |
| fcc | 12 | 0.1992365(10) | 0.1201635(10) |
| hcp | 12 | 0.1992555(10) | 0.1201640(10) |
| $La_{2-x}$ $Sr_x$ Cu $O_4$ | 12 | 0.19927(2) | |
| Penrose w/8 diagonals | 12.764 (average) | 0.188 | 0.111 |
| cubic with n.n.n. | 18 | 0.13735(5) | |
| simple cubic with short-length correlation | 6+ | 0.126(1) | |
| cubic with n.n.n.n. | 26 | 0.0976445(10) | |

n.n.n. = next-nearest neighbor, n.n.n.n. = next-next nearest neighbor

In the case of the spatial network of points the percolation threshold depends on the number of first neighbours. For spatial lattices with 4 neighbours (diamond) the percolation threshold is 0.43. For hcp lattice with 12 neighbours the percolation threshold is 0.20. for percolation if one takes into consideration the next nearest neighbours in the cubic lattice (18) the percolation threshold is 0.137. When one considers the percolation taking into account the next-next nearest neighbours, as in cubic lattice: 26 neighbours, the percolation threshold is 0.098.

In the new model the memory process occuring in a chalogenide film is as follows:

-two dendrites develop closely one to another.

-the dendrites can link at certain points and make a continuous channel of crystalline phase from one electrode to another (this is the ON-state).

-during a higher electrical pluse the dendrites are partially melted and the thinner zones of contact disappear (this is the OFF-state), the conductivity drops abruptly because the thin dendrite connections are destroyed.

-a new electrical pulse determines a new crystallization process with the formation of new bridges, and therefore new percolation paths.

What is important is the role played by the nuclei with dendrites. This is similar to the dendrite action in neuronal cells.

By accumulation of energy the nuclei can fire, as in the case of neurons, i.e. the energy is transferred primarily to dendrite who links one nucleus to another and can form or break the path between nuclei and thus the state ON or the state OFF of the circuit is triggered.

The dendrites are extending at long distances in-between the fundamental crystallites. Thus a crystallite can form channels to another crystallite situated not only in the first neighborhood but also in the next-next neighborhood and even farther. Thus the multiple possibilities to connect the crystalline nuclei, makes the percolation path to be achieved easier. Thus, as shown above, the percolation threshold can be very low and, therefore, the threshold condition is considerably easier to get and needs a lower amount of energy. Moreover, the energy supply needed for amorphization of the thin dendrite contacts is very low.

## 6. Discussion

Optical memory in the same class of chalcogenides is a big challenge for the electrical memory, because it is simple and can be used with simple devices (optical disks) already existent. The best materials belong to the class: Ge-Sb-Te. They are intended to replace the non-volatile flash memories.

Their main characteristics are:

- The commercial Optical Storage Media are based on Ge-Sb-Te chalcogenides,
- The materials are based on changes in structural configurations,
- The materials exhibit high Optical / Resistive Contrast,
- The switching characteristics can be very different. They depend strongly on the type of materials,

All these memories are in principle related to the electronic states in the chalcogenides

The multiswitching effect is based on configurational changes in the material: from amorphous state to intermediate states and crystalline states and back by careful operation.

Why the chalcogenides are the best materials for switching? Firstly chalcogenide materials are highly resistive despite the presence of defect and impurities. This is because the VAP defects pin the Fermi level near the gap center. This property gives a high resistive OFF state. This is totally different from the other semiconductors with comparable gaps. Secondly, the chalcogenides can easily and quickly undergo reversible structural transitions between amorphous and crystalline states. Thirdly, the chalcogenides are materials that undergo phase change with minimum motion of the constituent atoms. Thus the switching rate is high. Switching materials based on diffusion and phase separation are not appropriate because the switching is very slow due to low rate of processes implied.

## 7. Conclusions

The switching phenomenon is very important for modern optoelectronics. The chalcogenides are master materials for both electrical and optical switching.

Various switching mechanisms at the atomic scale have been proposed up to day.

The newest and the most subtile one is based on the formation of the pre percolation stage where the crystalline nuclei are grown with extended dendrites in an electric field or photo-excitation. More energy determines more spatial extention of the dendritic structure, accompanied by random linking of dendrites giving rise to the percolation path between electrodes (state ON). The OFF state appers when the excess of energy pumped in the system is enough to melt the narrow dendrite connections and to wipe out the percolation paths.

## References

[1] Mitkova M and Kozicki M N, 2004 *Non-Crystalline Materials for Optoelectronics* (Series: Optoelectronic Materials and Devices ) v **1** ch. 8 pp 211-258
[2] Ovshinsky S R 1968 *Phys. Rev. Lett.* **21** 1450
[3] Adler D, Henisch H K and Ovshinsky S R 1978 *Rev. Modern Physics* **50**
[4] Popescu C and Croitoru N 1972 *J. Non-Cryst. Solids* **8-10** 531
[5] Popescu C 1975 *Solid State Electron.* **18** 671

[6] Adler D, Shur M S, Silver M and Ovshinsky S R 1980 *J. Appl. Phys.* **51** 3289
[7] Madan M and Shaw M, 1988 *The Physics and Applications of Amorphous Semiconductors* (San Diego, Ca: Academic Press)
[8] Chaban I A 2007 *Physics of Solid State* **49** 420
[9] Chaban I A 1979 *Fiz. Tverd. Tela* **21** 1444
[10] Kolobov A V, Fans P and Taninaga J 2005 *J. Appl. Phys.* **44** 3345
[11] Njoroge W K and Wuttig M 2001 *J. Appl. Phys.* **90** 3816]
[12] Reddy G B, Dhar A, Malhotra L K and Sharmila E K 1992 *Thin Solid Films* **220** 111
[13] Kashchiev D 1969 *Surf. Sci.* **14** 209
[14] Senkader S and Wright C D 2004 *J. Appl. Phys.* **95** 504

Journal of Physics: Conference Series **253** (2010) 012015 doi:10.1088/1742-6596/253/1/012015

# Investigation of vibrational and electronic properties of oxide nanopowders by spectroscopic methods

**Maja Šćepanović, Mirjana Grujić-Brojčin, Zorana Dohčević-Mitrović and Zoran V. Popović**

Center for Solid State Physics and New Materials, Institute of Physics, University of Belgrade, Pregrevica 118, 11080 Belgrade, Serbia

E-mail: maja@ipb.ac.rs

**Abstract.** Raman spectroscopy technique has been successfully used to study the microscopic nature of structural and/or morphological properties of investigated nanopowders. The phonon confinement model has been applied for the estimation of nanocrystals dimensions ($TiO_2$ and $CeO_2$), as well as the correlation length (ZnO), from the frequency shift and asymmetrical broadening of Raman optical phonon modes. The particle size distribution in nanopowders has been estimated from the low frequency Raman spectra, using the fact that the acoustic phonon modes in nanosized $TiO_2$ and $CeO_2$ can be well described by the elastic continuum model. In addition, the appearance of surface optical phonon modes has been predicted theoretically by the dielectric functions approach and detected experimentally in the Raman spectra of ZnO nanopowders, whereas the resonant behaviour of the first and second order ZnO Raman modes has been used for estimation of the electron-phonon interaction in this kind of nanomaterial. The optical properties of oxide nanopowders have been investigated by spectroscopic ellipsometry and photoluminescence spectroscopy. It has been shown that bandgap energy, as well as the energies of the other interband electronic transitions in investigated nanopowders can be determined by spectroscopic ellipsometry. On the other side, the existence of the broad bands in visible region of the photoluminescence spectra of $TiO_2$ and ZnO nanopowders points out to the various electronic transitions mediated by defect levels within the bandgap.

## 1. Introduction

The dependence of structural and optical properties of oxide nanopowders on particle size, stoichiometry, microstrain, as well as on the impurities and defect concentration, opens possibilities for such materials to be widely used in optoelectronics, energy conversion, photocatalyses and many other applications. The investigation of several wide band-gap semiconductors ($TiO_2$, $CeO_2$ and ZnO nanopowders) has been demonstrated that the quantum size and confinement effects become important, and that the optical and electronic properties of nanomaterials differ significantly from their bulk counterparts, depending on length dimensions in comparison with the excitation wavelength and Bohr exciton radius [1-15]. Moreover, it has been shown that the structural, optical and electronic properties of oxide nanoparticles depend not only on crystal structure, bulk composition, and morphology of the oxide particles, but also on their defect structure [16]. Thus, controlling of defects, especially nonstoichiometry related ones, becomes one of the most important goals of contemporary technology of oxide nanomaterials. Therefore, main purpose of this study is to correlate the results of experimental investigation of vibrational and optoelectronic properties of nanopowders with the variation of the crystallite size, microstrain, nonstoichiometry, as well as the presence of intrinsic defects and impurities.

## 2. Vibrational properties of nanocrystalline powders

Raman scattering can give important information about the nature of the solid on a scale of a few lattice constants. Therefore, Raman spectroscopy technique can be used to study the correlation between the change in vibrational properties and structural and/or morphological changes in nanostructured material. At different length dimensions $L$ of nanostructure, compared to the excitation wavelength $\lambda$ and Bohr exciton radius $a_B$, new Raman features may occur [17]. In case of $a_B \approx L << \lambda$ the confinement effects become important for optical and acoustic phonons, and Raman spectra of nanostructure differ significantly from the spectra of its bulk counterpart. On the contrary, when $a_B << L << \lambda$ the confinement effect is not so relevant, but additional surface phonon modes may appear between the TO and LO modes in the Raman spectra of polar semiconductors.

### *2.1. Phonon confinement model and elastic sphere model*

It is well known that using optical techniques such as Raman spectroscopy, only zone-center optical phonons can be observed in ideal single crystals. However, this $q \approx 0$ selection rule is relaxed due to the interruption of the lattice periodicity in a nanocrystalline material. Hence, the behaviour of Raman modes in nanocrystalline materials due to crystallite size decrease can be simulated by phenomenological phonon confinement model (PCM) in an isolated nanoparticle. According to this model, the phonons in nanocrystal are confined in space and all the phonons over the entire Brillouin zone will contribute to the first-order Raman spectra. The weight of the off-center phonons increases as the crystal size decreases and the phonon dispersion causes an asymmetrical broadening and the shift of the Raman mode. According to Richter et al. [18] and Campbell and Fauchet [19], for crystallite size $L$ and Gaussian confinement function, the resulting Raman intensity $I(\omega)$ is presented as a superposition of weighted Lorentzian contributions over the whole Brillouin zone by the equation:

$$I(\omega) \propto \sum_{i=1}^{n} \int_{0}^{\infty} \rho(L) dL \int \frac{\exp\left(-q^2 L^2 / 8\beta\right) d^3 q}{\left(\omega - (\omega_i(q) + \Delta\omega)\right)^2 + \left(\Gamma / 2\right)^2}, \quad (1)$$

with $\rho(L)$ as distribution of crystallite size, $q$ - wave vector, $\beta$ - confinement strength and $\Gamma$ - the intrinsic mode linewidth. The sum is carried over $n$ dispersion curves $\omega_i$ $(q)$, depending on mode degeneration. The slope of dispersion curves away from the center of BZ determines the Raman line shift and asymmetrical broadening, as the crystalline size reduces: a negative slope, towards lower frequency, produces a redshifted (ZnO, $CeO_2$), while a positive slope results in blueshifted Raman peak ($TiO_2$). The peak centered at $\omega_i$ is shifted by the value of $\Delta\omega$ due to the strain [1, 2, 12].

Similar to the optical phonons, the acoustic phonons also get confined within the particles. In the elastic continuum limit, the confinement of long-wavelength acoustic phonons (sound waves) is followed by the appearance of discrete modes of particle. These modes depend on the elastic properties through the longitudinal and transverse sound velocities [20]. The frequencies of the spheroidal and torsional modes of the particle depend on the angular momentum associated with the vibration. Raman scattering from confined acoustic vibrations in nanoparticles has been observed in the low frequency part of spectra of various materials. The elastic sphere model (ESM) has been generally used to analyze the acoustic phonon confinement. This model describes the oscillations of homogeneous elastic sphere using Navier's equation and imposing spherical symmetry [13, 14]. Rigid boundary and stress-free surface condition can be applied as boundary conditions in solving that equation. The frequency of an acoustic Raman mode is related to the particle diameter $D$ as: $\omega^R{}_{ln} = \omega / 2\pi c = \beta_{ln}/D$ where $\beta_{ln}$ characterize the solutions of the Navier's equation, while $l$ is the angular momentum quantum number. According to the group theory analysis the spheroidal $l$=0 and $l$=2 modes are Raman active [21]. Thereby, the intensity of Raman scattering by nanoparticles can be expressed as [22]:

$$I(\omega) = A \frac{n(\omega)}{\omega} C(\omega) g(\omega), \quad (2)$$

where $n(\omega)$ is Bose-Einstein factor, $C(\omega)$ is the mode-radiation coupling factor, $g(\omega)$ is the phonon density of states and $A$ is the scaling constant. It had been shown that the mode-radiation coupling factor scales inversely with the vibrational frequency: $C(\omega)\sim 1/\omega$ for small spherical particles [22]. According to ESM, Raman spectrum would consist of discrete peaks if all the particles were of same dimensions. It can be assumed that the density of states reflects the particle size distribution, $N(D)$, through the relation: $g(\omega)=N(D)$ [23]. The most often size distributions encountered in different nanomaterials can be well approximated with Gaussian or log-normal distribution functions.

*2.1.1. Determination of particle size distribution in pure and La-doped $TiO_2$ nanopowders synthesized by sol-gel method.* Decreasing of nanocrystals size to a few tens of nanometers leads to the appearance of low frequency Raman modes and blueshift and asymmetrical broadening of dominant $E_g$ mode in the Raman spectra of anatase $TiO_2$ nanopowders, as shown in figure 1.

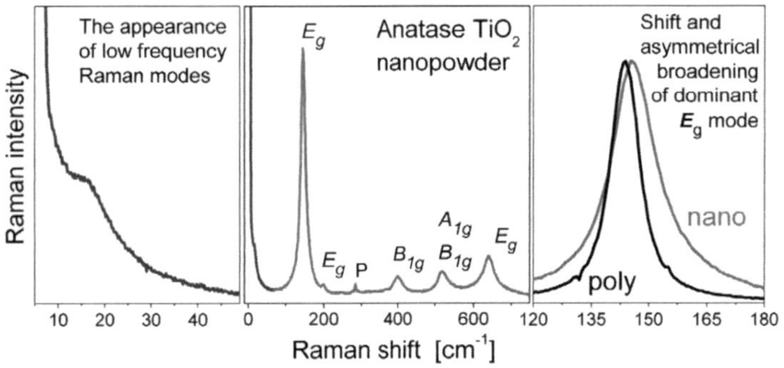


**Figure 1.** The effects of crystallite size decrease on Raman spectra of anatase $TiO_2$ nanopowder.

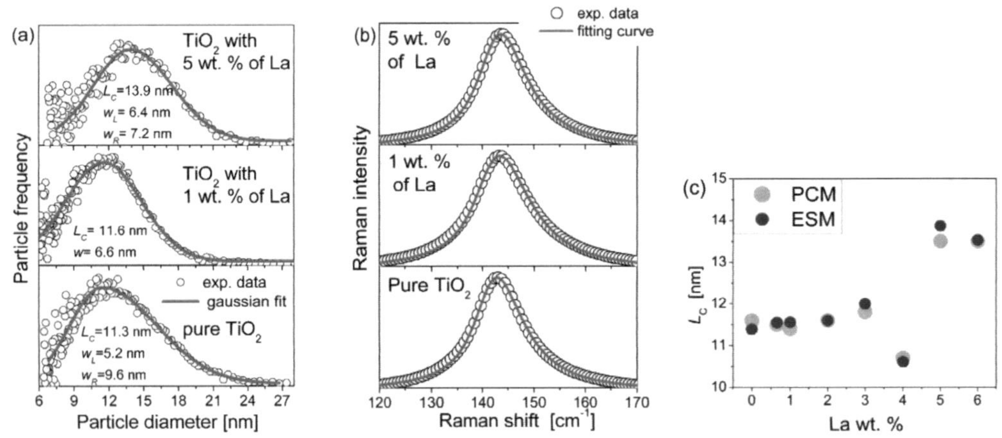


**Figure 2.** (a) Particle size distributions in pure and several La-doped $TiO_2$ nanopowders estimated from low-frequency Raman spectra ($w$ - the width of Gaussian distribution, $w_L$ and $w_R$ are left and right width of asymmetric Gaussian distribution). (b) PCM fits of anatase $E_{g(1)}$ Raman mode for pure and La-doped samples. (c) Most frequent particle size $L_C$ obtained by ESM (dark circles) and from PCM (light circles) in dependence of wt. % of La in $TiO_2$ nanopowders.

The particle size distribution in pure and La-doped $TiO_2$ nanopowders synthesized by sol-gel method has been estimated from the low frequency Raman spectra, using the fact that the phonon modes in nanosized $TiO_2$ observed in the low frequency region can be well described by the ESM, assuming that nanoparticles are of perfect spherical shape and isotropic [10]. The procedure proposed by Ivanda et al. [23] has been performed. Values of longitudinal and transversal sound velocities of rutile titania were used in calculations due to the lack of this data for anatase and value of $\beta_{\mathrm{ln}}$ was calculated. Inhomogeneous broadening of the spectrum as a consequence of the non-uniform particle size distribution of the sample was incorporated. The evaluated particle size distributions in pure and La-doped $TiO_2$ are shown in figure 2a together with the best fits of these distributions obtained by Gaussian or asymmetric Gaussian functions. The particle size distributions obtained by ESM were used as initial distributions for the best-fit calculations of the $E_{g(1)}$ mode by PCM (figure 2b) [1, 6, 10]. The dependence of the most frequent particle size $L_C$ on La content in $TiO_2$ nanopowders is shown in figure 2c. The values of $L_C$ in the distributions finally used in PCM are very close to the corresponding values of $L_C$ in the distributions obtained by ESM. Note that the effects of compressive strain in the calculations of $E_{g(1)}$ mode intensity in nanopowders with higher La content had to be included in PCM [10]. Obtained results have shown that the particle size in $TiO_2$ nanopowders doped with La up to 4 wt.%, is close to or smaller than in the pure $TiO_2$, whereas nanoparticles with higher La content are larger and strained. This study confirms that the combined use of ESM and PCM in the analysis of Raman spectra of $TiO_2$ nanopowders gives the information not only about particle size distribution but also about influence of dopants on microstrain in anatase nanostructures.

*2.1.2. Use of ESM and PCM in determination of particle size distribution in $CeO_2$ nanopowders synthesized by self-propagating room temperature method.* Decreasing of nanocrystals size has the similar effects on vibrational properties of $CeO_2$ nanopowders as in the case of $TiO_2$. The appearance of low frequency Raman modes together with large redshift and asymmetrical broadening of $F_{2g}$ mode in the Raman spectra of $CeO_2$ nanopowders, are evident from figure 3 [13].

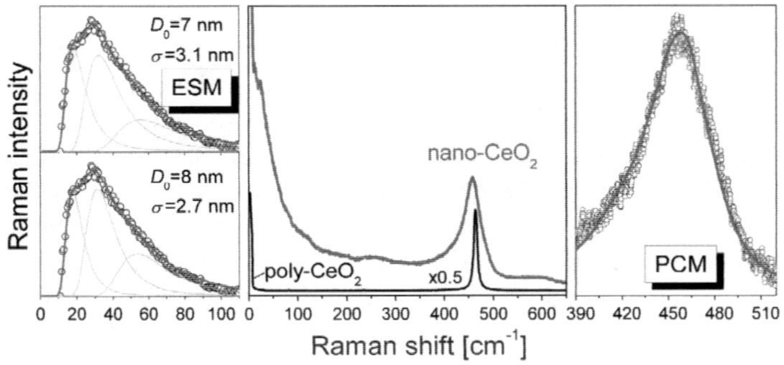


**Figure 3.** Effects of crystallite size decrease on Raman spectra of $CeO_2$ nanopowders

Raman intensity in the low-frequency range was calculated as the sum of three ESM modes intensities: ($l$=2, $n$=0), ($l$=0, $n$=0) and ($l$=0, $n$=1), taking into account the correspondence between Raman intensity and particle size distribution [14]. The best agreement between experimental and calculated spectrum was obtained for the rigid boundary case but not for stress-free boundary condition (figure 3, left). The reason can be found in high agglomeration of the nanoparticles or in microstrain effects that influence strongly the optical Raman mode in ceria nanopowders [14]. If the results from the fits (figure 3) are compared to the value of the mean particle size diameter $D_0$=8 nm, acquired from PCM (figure 3, right), it can be concluded that Gaussian size distribution with values of $D_0$ from the interval 7 – 8 nm describes best the particle size distribution of the investigated sample.

*2.2. Surface optical phonons*

Surface modes of small particles have been observed for particle sizes smaller than the wavelength of the exciting laser light inside the particle in many polar crystals [17]. Although for powders or nanocrystals embedded in homogeneous media 3D effective-medium models are successfully applied, it appears that 2D models are more appropriate to describe the morphology of polar porous semiconductors [24]. The effective dielectric function of the nanopowder thus is given by [25, 26]:

$$\varepsilon_{\parallel} = (1-f)\varepsilon^{air} + c\varepsilon_{\parallel}^{psc}, \tag{3}$$

(exactly) and

$$\varepsilon_{\perp} = \frac{(1+f)\varepsilon_{\perp}^{psc} + (1-f)\varepsilon^{air}}{(1-f)\varepsilon_{\perp}^{psc} + (1+f)\varepsilon^{air}}\varepsilon^{air}, \tag{4}$$

(in a two-dimensional Maxwell-Garnett approximation), where $\varepsilon_{\perp}^{psc}$ and $\varepsilon_{\parallel}^{psc}$ are the dielectric functions in normal and parallel polarization, whereas $f$ is the volume fraction of porous semiconductor (psc) and $\varepsilon^{air}$=1. Taking into consideration the plasmon contribution, the dielectric functions of semiconducting nanopowders with wurtzite structure can be written as [25]:

$$\varepsilon_{\parallel(\perp)} = \varepsilon_{A_1(E_1)} = \varepsilon_\infty \left( \frac{\omega^2_{A_1\mathrm{LO}(E_1\mathrm{LO})} - \omega^2 - i\omega\gamma}{\omega^2_{A_1\mathrm{TO}(E_1\mathrm{TO})} - \omega^2 - i\omega\gamma} - \frac{\omega_p^2}{\omega^2 + i\omega\Gamma} \right), \quad (5)$$

where $\omega_{A_1(E_1)}$ is the uncoupled phonon frequency of the $A_1(E_1)$ - type mode in bulk semiconductor and $\gamma$ is phonon damping rate. The plasmon frequency is given by $\omega_p=(Ne^2/\varepsilon_\infty m^*)^{1/2}$, whereas $\Gamma=e/\mu m^*$ is plasmon damping rate. $N$ is free carrier concentration, while $m^*$ is the effective mass of electron, $\mu$ - electron mobility, and $\varepsilon_\infty$ - static dielectric constant.

The Raman line shape of surface optical phonon modes is derived according to the assumption that the excitation of extraordinary phonons is dominant, and that the Raman intensities can be written as:

$$I_{eo} \sim \mathrm{Im}[-1/(\varepsilon_\parallel \cos^2\theta + \varepsilon_\perp \sin^2\theta), \quad (6)$$

where $\theta$ is the angle between the exciting light wave vector and the direction parallel to the $c$-axis. It is necessary to take into account multiple scattering by averaging $I_{eo}$ over $\theta$ with a weight $\exp(-\theta^2/\sigma^2)$, as well as inhomogeneity of the sample by averaging over $f$ with a weight $\exp(-(f-f_0)^2/\sigma_f^2)$ [25].

*2.2.1. Surface optical phonon modes in the Raman spectra of mechanically activated ZnO powders.* As a result of mechanical activation, the frequency and shape of Raman modes characteristic for wurtzite ZnO structure are changed due to the structural and stoichiometric variation, depending on the milling time and the choice of equipment [3, 12]. Observed redshift and asymmetrical broadening of the first order $E_2^{high}$ and $E_1$(LO) Raman modes (figure 4) are ascribed to decrease of correlation length, due to increase in intrinsic defect concentration induced by mechanical activation. Such behaviour is simulated by PCM, which includes Brillouin zone anisotropy and high sensitivity of $E_2^{high}$ mode to lattice strain [12]. The modes, which appear at about 510 and 550 cm$^{-1}$ in the Raman spectra of mechanically activated ZnO powders, are ascribed to the surface optical phonon (SOP) vibrations,

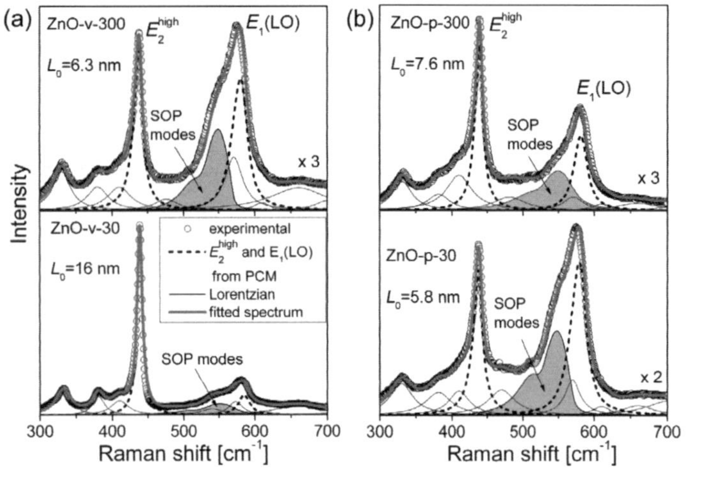


**Figure 4.** Raman spectra of ZnO powders activated in (a) vibro-mill (ZnO-v-30, ZnO-v-300) and (b) planetary mill (ZnO-p-30, ZnO-p-300) during 30 and 300 minutes. Fitted spectra is obtained as a sum of following modes: $E_2^{high}$ and $E_1$(LO) calculated by PCM, SOP modes calculated by dielectric function approach, and other Raman modes represented by Lorentzians.

which occur at grain boundaries between ultrafine nanocrystals and disordered regions in nanosize ZnO crystallites. With increasing disorder in activated ZnO powders, the SOP modes in experimental Raman spectra are significantly increased in intensity, whereas no measurable shift of these modes can be observed. The Raman line shape of SOP modes is efficiently simulated by dielectric function approach described in previous section. The great difference in the intensity and line shape of simulated SOP modes is mainly the result of variation in mean volume fraction $f_0$ of ZnO and phonon damping rates, as a consequence of vacancies induced porosity and high disorder in damaged surface layers. The higher relative SOP intensity and vacancies induced porosity is correlated to higher surface defects concentration, probably the oxygen vacancies.

### *2.3. Electron phonon interaction*

*2.3.1. Dependence of electron phonon coupling on correlation length in highly disordered ZnO nanopowders.* The electron-phonon coupling (EPC) is very important in understanding the optoelectronic properties of the crystalline semiconductors. The coupling strength between electrons and LO phonons in polar semiconductors is determined by the ratio of the second- to the first-order Raman scattering cross sections [27]. The intensity ratio of $E_1$(2LO) to $E_1$(LO) Raman modes, as an indicator of EPC strength, is shown in figure 5a for starting and several mechanically activated ZnO powders. From the dependence of the $E_1(2LO)/E_1(LO)$ ratio on incident photon energy it is obvious that the EPC strength has the highest value in the starting powder and decreases with activation. A great enhancement of the oscillator strength for the second-order resonance can only be understood if the relevant resonant intermediate state is a localized exciton, interacting strongly with the LO-phonon, forming a localized exciton-phonon complex [28]. The origin of great EPC strength, observed here in nonstoichiometric ZnO-0, may lie in the increasing concentration of oxygen vacancies ($V_O$) and interstitial atoms ($Zn_i$) followed by the presence of their electronic levels within the bandgap.

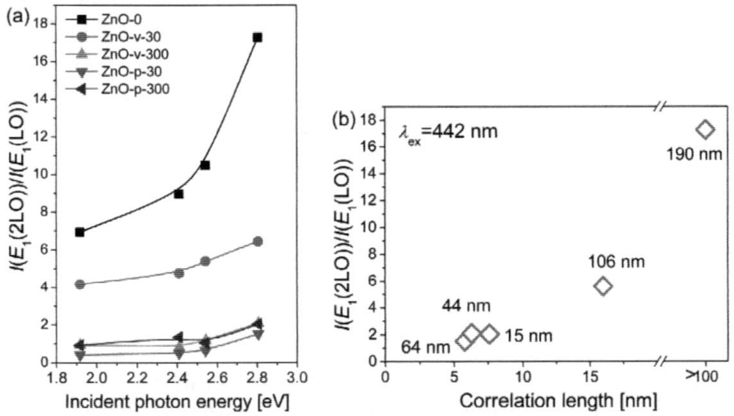


**Figure 5.** Dependence of the $E_1(2LO)/E_1(LO)$ ratio on (a) incident photon energy and (b) correlation length in starting and mechanically activated ZnO powders. Crystallite sizes determined by XRD analyses are denoted in (b) for all powders.

Weakening of the electron-LO-phonon coupling in the activated ZnO powders is a consequence of great sensitivity of second-order structures to atomic scale disorder [29] induced my mechanical activation. So, it can be concluded that low level of disorder, induced by nonstoichiometry of starting powder, strongly enhances EPC strength, whereas this strength become weaker in highly disordered activated powders. Note that the $E_1(2LO)/E_1(LO)$ ratio gradually decreases with activation time in the powders activated in vibro-mill. However, in the powders activated in planetary mill, the $E_1(2LO)/E_1(LO)$ ratio has minimal value after 30 min of activation, and slightly higher value after 300 min, though crystallite size continuously decreases with activation. These results have shown that electron phonon coupling strength in crystalline ZnO powders has the same trend as the correlation length, but not as crystallite size. Such conclusion is explicitly confirmed by figure 5b, where dependence of the $E_1(2LO)/E_1(LO)$ ratio on correlation length in starting and activated ZnO powders has been shown together with the crystallite size determined by XRD analysis.

*2.3.2. The influence of $Fe^{2+}/Fe^{3+}$ doping on vibrational properties of cerium dioxide nanopowders.* It can be seen from the Raman spectra of Fe-doped $CeO_2$, shown in figure 6, that changing of Fe valence state causes evident redshift and broadening of $F_{2g}$ Raman mode, followed by increased intensity of intrinsic oxygen vacancy mode at 600 cm$^{-1}$. This unexpected redshift and broadening of the $F_{2g}$ mode can be only a consequence of the coupling between charge located at $Ce^{3+}(Fe^{3+})$-$O^{2-}$ complex and $F_{2g}$ vibration arising from electron molecular vibration coupling [30]. The change in vibrational properties of Fe-doped $CeO_2$ nanopowders is followed by drastic variation of their magnetic properties as well [28].

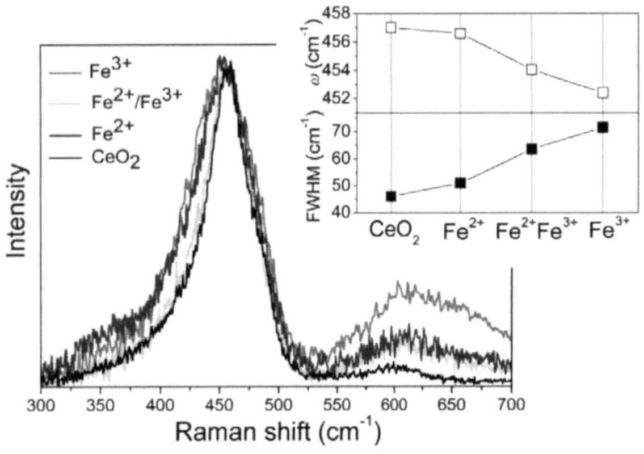


**Figure 6.** Raman spectra of undoped and Fe-doped $CeO_2$ samples. *Inset*: Dependence of full width at half maximum (FWHM) and frequency ($\omega$) of $F_{2g}$ Raman mode varying the valence state of dopant.

## 3. Optoelectronic properties of nanocrystalline powders

Optical and electronic properties of nanocrystalline materials strongly depend on their composition, structural and morphological characteristics, but also on the presence of defects. For crystallite size comparable to Bohr exciton radius of particular materials the quantum size effects become important, the bandgap grows, and the optical and electronic properties of nanomaterial differ significantly from its bulk counterpart. On the other hand, presence of intrinsic defects and impurities may introduce different electronic levels within the band-gap, which can cause the various electronic transitions mediated by these levels and variation of the optical bandgap energy as well.

### *3.1. Interband electronic transitions*

The dielectric function $\varepsilon(\omega)$ of $TiO_2$ nanopowders synthesized by laser-induced pyrolysis has been determined by spectroscopic ellipsometry in the energy range from 1.5 to 6 eV at room temperature. The features observed in $\varepsilon(\omega)$ have been fitted to analytical line shapes by using the second derivatives of experimental spectra (figure 7). In such a way, the energies corresponding to different interband electronic transitions have been determined. Obtained results point out that the energy of $E_1$ transition,

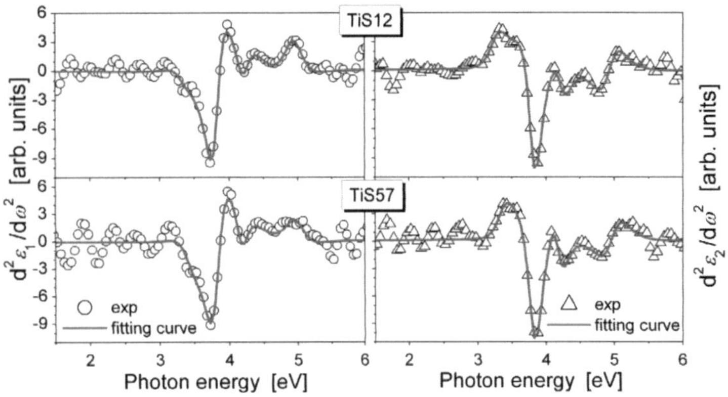


**Figure 7.** 2nd derivatives of $\varepsilon_1(\omega)$ (real) and $\varepsilon_2(\omega)$, (imaginary) components of dielectric functions obtained by SE measurements of anatase nanopowders, together with the fitting curves determined by using critical point (CP) analysis.

which can be related to the energy bandgap transition, is higher in nanopowder with smaller crystallite size, $d_c$. Namely, this energy reaches 3.384 eV in the powder with $d_c \approx 12$ nm (TiS57), whereas in the powder with $d_c \approx 17$ nm (TiS12) its value is 3.285 eV. Such increase (~0.1 eV with crystallite size decrease by ~5 nm) coincides with the behavior predicted for the energy bandgap in anatase nanostructures [11, 31, 32]. However, the energies related to the other interband electron transitions between valence and conduction band have similar values in both nanopowders - $E_1+\Delta E_1 \approx 3.8$ eV,

$B$≈4.2 eV and $E_2$≈4.9 eV. Possible explanation of such behaviour of the interband electron transitions may be the following: the last three transitions are considered to be direct, whereas the first one is can be assumed to be a kind of transition between direct and indirect.

### *3.2. Determination of band gap energy*

The values of the bandgap energy ($E_g$) in direct bandgap semiconductors can be obtained by extrapolating the straight-line portion of the $(\alpha E)^2$ vs. $E$ plots to the photon energy $E$ axis. Figure 8 shows those plots for nonactivated and mechanically activated ZnO powders, where the absorption coefficient have been obtained from measured SE data by using $\alpha = 4\pi k/\lambda$ ($\lambda$ is the incident light wavelength) [15]. It is obvious that $E_g$ for all samples is narrower than the bandgap of bulk ZnO (3.37 eV), with the lowest value in the most disordered powder ZnO-p-30. This is most likely a consequence of a drastic structural change in ZnO crystal, as some of the surface states, defects and dangling bonds induce formation of band tail and can cause the bandgap narrowing [33]. The shift of optical bandgap of ZnO to values smaller than the bulk $E_g$ can be also induced by electrons generated from $V_O$ and $Zn_i$ [34]. On the other side, the origin of the bandgap widening in the ZnO-p-300 powder is not related only to the lower concentration of defects but also to the raising zirconia content, due to prolonged milling in zirconia vessels. Namely, the presence of zirconia in these samples causes shifting the energy gap to the high energy, as the energy-gap values of $ZrO_2$ ranges 5.0–5.85 eV [35].

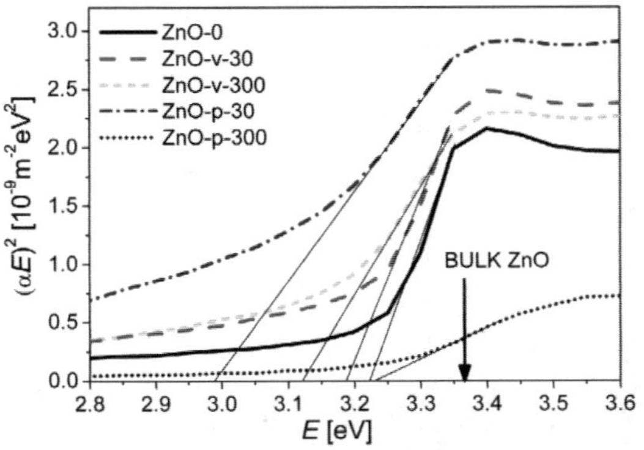


**Figure 8.** Plot of $(\alpha E)^2$ versus photon energy ($E$) for all samples. The values of the direct bandgap ($E_g$) are obtained by extrapolation of the linear portion of the curves to absorption $\alpha$ equal to zero.

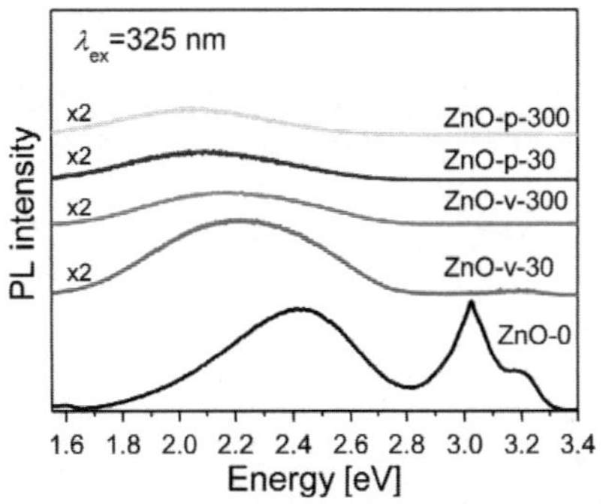


**Figure 9.** PL spectra of nonactivated and activated ZnO powders excited by 325 nm line of He-Cd laser. Some spectra are up-shifted for clarity.

### *3.3. Sub-bandgap electronic transitions in ZnO and $TiO_2$ nanopowders*

From the photoluminescence spectra of ZnO samples excited by 325 nm (3.81 eV) line of He-Cd laser shown in figure 9 it is obvious that a significant impact of defects on the electronic properties of the ZnO reflects not only on the bandgap energy variation (figure 8), but also on the changes in the PL spectra. The excitation with the photon energy greater than $E_g$ causes the appearance of the luminescence at different emission wavelengths in the visible range, due to different intrinsic or extrinsic defects in ZnO. Highly intensive violet emission (at ~3 eV) and green luminescence (at ~2.4 eV), whose origin were related to oxygen deficiency, dominate the PL spectrum of non-stoichiometric starting ZnO powder, where the most abundant bulk point defects are supposed to be $Zn_i$ and $V_O$ [15]. On the contrary, the PL in mechanically activated ZnO powders is readily quenched due to high defect concentration. Decrease in the PL intensity in mechanically activated ZnO powders can be also a consequence of an unlimited access of oxygen to the nanoparticles surface during activation in air and the fact that oxygen is efficient in electron collecting [15]. Moreover, increasing concentration of interstitial oxygen defects on the surface layers of ZnO nanoparticles is probably the main reason for the shift of the PL maximum to yellow and orange range in activated ZnO powders.

Under the laser irradiation with subband photon energy of 2.41 and 2.71 eV anatase $TiO_2$ nanocrystals also displayed strong visible light emission, even at excitation power as low as 0.05 W/cm$^2$ [11]. The line shape and position of this broad luminescence band vary with excitation energy and temperature, as can be seen from figure 10. The analysis of those variations suggests that the PL spectra of nanopowder TiS12, with larger crystallite size, are dominated by a radiative recombination of electrons via intrinsic surface states for all applied excitation energies and temperatures. The energy of these optical electron transitions is about 2.2 eV [11, 36]. However, excitation of the sample with smaller nanocrystallites, TiS57, by higher photon energies (close enough to the energy bandgap) activates self-trapped exciton luminescence with the energy of ~2.41 eV [37 and the references therein]. PL blueshift with sample heating also points out to the recombination of self-trapped excitons as dominant PL mechanisam in TiS57 nanopowder.

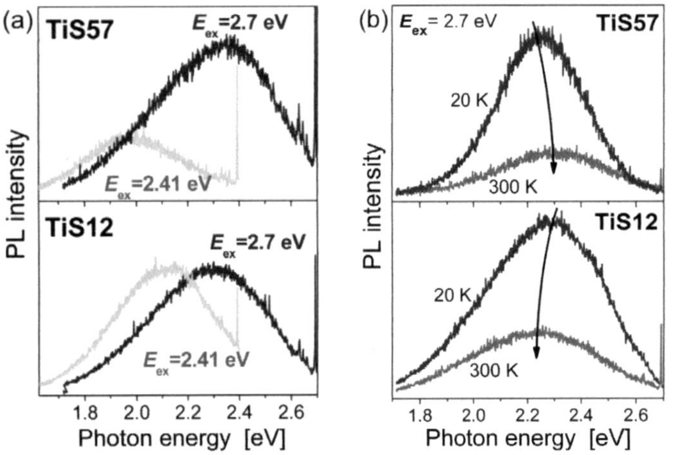


**Figure 10.** PL spectra of anatase nanopowders with variation of laser excitation energy (a) and temperature (b).

## 4. Conclusion

This study gives an overview of theoretical interpretation and experimental evidences of the strong correlation of vibrational and optoelectronic properties with composition, structure, morphology and presence of defects and impurities in nanocrystalline semiconductors. Raman scattering measurements have proven as a valuable tool for determining particle size distribution as well as surface and defect structure in nanosized materials, whereas spectroscopic ellipsometry and photoluminescence spectroscopy have shown their efficiency in estimation of the energies of optical bandgap, different interband electronic transition as well as electronic levels within the bandgap.

**Acknowledgments**

This work is supported by the Serbian Ministry of Science under Project No. 141047, SASA project F-134 and the Swiss National Science Foundation through grant IZ73Z0-128169.

**References**

[1] Šćepanović M J, Grujić-Brojčin M U, Dohčević-Mitrović Z D and Popović Z V 2006 *Mater. Sci. Forum* **518** 101
[2] Dohčević-Mitrović Z D, Šćepanović M J, Grujić-Brojčin M U, Popović Z V, Bošković S B, Matović B M, Zinkevich M V and Aldinger F 2006 *Solid State Commun.* **137** 387
[3] Šćepanović M, Srećković T, Vojisavljević K and Ristić M M 2006 *Sci. Sinter.* **38** 169
[4] Dohčević-Mitrović Z D, Grujić-Brojčin M, Šćepanović M, Popović Z V, Bošković S, Matović B, Zinkevich M and Aldinger F 2006 *J. Phys. - Condens. Mat.* **18** S2061
[5] Popović Z V, Dohčević-Mitrović Z, Konstantinović M J and Šćepanović M 2007 *J. Raman Spectrosc.* **38** 750

[6] Šćepanović M J, Grujić-Brojčin M, Dohčević-Mitrović Z D and Popović Z V 2007 *Appl. Phys.* A **86** 365
[7] Šćepanović M J, Grujić-Brojčin M, Dohčević-Mitrović Z and Popović Z V 2007 *J. Optoelectron. Adv. Mat* **9** 30
[8] Šćepanović M, Grujić-Brojčin M, Dohčević-Mitrović Z, Vojisavljević K, Srećković T and Popović Z V 2007 *Acta Phys. Pol.* A **112** 1013
[9] Aškrabić S, Dohčević-Mitrović Z D, Radović M, Šćepanović M and Popović Z V 2009 *J. Raman Spectrosc.* **40** 650
[10] Šćepanović M, Aškrabić S, Grujić-Brojčin M, Golubović A, Dohčević-Mitrović Z, Kremenović A and Popović Z V 2009 *Acta Phys. Pol.* A **116** 99
[11] Šćepanović M, Grujić-Brojčin M, Mirić M, Dohčević-Mitrović Z and Popović Z V 2009 *Acta Phys. Pol.* A **116** 333
[12] Šćepanović M, Grujić-Brojčin M, Vojisavljević K, Bernik S and Srećković T 2010 *J. Raman Spectrosc.* **41** 914
[13] Kostić R, Aškrabić S, Dohčević-Mitrović Z and Z V Popović 2008 *Appl. Phys.* A **90** 679
[14] Aškrabić S, Kostić R, Dohčević-Mitrović Z and Z V Popović 2007 *J Phys. Conf. Ser.* **92** 012042
[15] Šćepanović M, Grujić-Brojčin M, Vojisavljević K and Srećković T *J. Nanopart Res*, submitted
[16] Ischenko V, Polarz S, Grote D, Stavarachem V, Fink K and Driess M 2003 *Adv. Funct. Mater.* **15** 1945
[17] Irmer G 2007 *J. Raman Spectrosc.* **38** 634
[18] Richter H, Wang Z P and Ley L 1981 *Solid State Commun.* **39** 625
[19] Campbell I I and Fauchet P M 1986 *Solid State Commun.* **58** 739
[20] Arora A K, Rajalakshmi M and Ravindran T R 2004 *Encyclopedia of Nanoscience and Nanotechnology* vol 8, ed H S Nalwa (Los Angeles: American Scientific Publishers) p 499
[21] Duval E 1992 *Phys. Rev.* B **46** 5795
[22] Montagna M and Dusi R 1995 *Phys. Rev.* B **52** 10080
[23] Ivanda M *et al.* 2007 *J. Raman Spectrosc.* **38** 647
[24] Sarua A, Monecke J, Irmer G, Tiginyanu I M, Gartner G and Hartnagel H L 2001 *J. Phys. - Condens. Mat.* **13** 6687
[25] Ye J D, Tripathy S, Ren F-F, Sun X W, Lo G Q and Teo K L 2009 *Appl. Phys. Lett.* **94** 011913
[26] Tiginyanu I M, Sarua A, Irmer G, Monecke J, Hubbard S M, Pavlidis D and Valiaev V 2001 *Phys. Rev.* B **64** 233317
[27] Wang R P, Xu G and Jin P 2004 *Phys. Rev.* B **69** 113303
[28] Ramkumar C, Jain K P and Abbi S C 1996 *Phys. Rev.* B **54** 7921
[29] Du C L, Gu Z B, You Y M, Kasim J, Yu T, Shen Z X, Ni Z H, Ma Y, Cheng G X and Chen Y F 2008 *J. Appl. Phys.* **103** 023521
[30] Dohčević-Mitrović Z D, Paunović N, Radović M, Popović Z V, Matović B, Cekić B and Ivanovski V 2010 *Appl. Phys. Lett.* **96** 203104
[31] Reddy K M, Manorama S V and Reddy A R 2003 *Mater. Chem. Phys.* **78** 239
[32] Guisbiers G, Van Overschelde O and Wautelet M 2008 *Apl. Phys. Lett.* **92** 103121
[33] Shim E S, Kang H S, Pang S S, Kang J S, Yun I and Lee S Y 2003 *Mater. Sci. Eng.* B **102** 366
[34] Suchea M, Christoulakis S, Katsarakis N, Kitsopoulos T and Kiriakidis G 2007 *Thin Solid Films* **515** 6562
[35] Qadri S B, Kim H, Khan H R, Pique A, Horwitz J S, Chrisey D, Kim W J and Skelton E F 2000 *Thin Solid Films* **377-378** 750
[36] Zhang W F, Zhang M S, Yin Z and Chen Q 2000 *Appl. Phys.* B **70** 261
[37] Karunagaran B, Rajendra Kumar R T, Viswanathan C, Mangalaraj D, Narayandass Sa K and Mohan Rao G 2003 *Cryst. Res. Technol.* **38** 773

Journal of Physics: Conference Series **253** (2010) 012016 doi:10.1088/1742-6596/253/1/012016

# Role of excitons in the energy resolution of scintillators used for medical imaging

**Jai Singh**

School of Engineering and IT, B-purple-12, Faculty of EHS, Charles Darwin University, Darwin NT 0909, Australia

**Abstract**. Theoretical investigations suggest that the nonproportionality in a scintillator is caused by the high excitation density created within the track of an X-ray or γ ray photon entering in a scintillating crystal. In this paper an analytical expression for the scintillator yield is derived. For the case of $BaF_2$ scintillator the role of excitons created within the γ-ray track in the scintillator yield is studied. By comparing the results of two theories an analytical expression is also derived for an energy parameter which could otherwise only be determined by fitting the theoretical yield to the experimental data.

## 1. Introduction

The world today needs scintillators that have better energy resolution for improving diagnostic accuracy in medical imaging and identification of contraband nuclear materials in homeland and global security. Existing scintillators have poor energy resolution. Hence there is huge demand for inventing new scintillators with superior energy resolution for better human health and security. This is due to the fact that nuclear medical imaging systems, like the single photon emission computed tomography (SPECT) diseases and positron emission tomography (PET) are envisaged to be the most popular diagnostic systems of human health because they have faster patient throughput, greater patient comfort, diagnostic certainty and staging accuracy. This enables practitioners better treatment planning of diseases like cancers, tumours, heart diseases and Alzheimer's. In addition there are more than 49,000 airports and 926 seaports at present in the world which require detectors of contraband nuclear materials at each of their entrances for their country's border and homeland security. Each of these detecting systems also requires, now more than ever, scintillators with superior energy resolution.

A γ photon absorbed in a scintillator first creates a pair of very high energy electron (e) and hole (h) which relaxes in a cascade process and creates many secondary e-h pairs with energies close to the band gap; some of them recombine radiatively and some non-radiatively. The photons emitted due to the radiative recombination contribute to the light yield of a scintillator and is defined as the integrated light output per MeV of energy deposited in a scintillator. The light yield is not constant with the incident γ energy, typically in the range of 10 keV to 1 MeV, in most scintillators which is referred to as nonproportionality or nonlinearity [1-2] and it seriously affects their attainable energy resolution leading to poor quality and inaccurate imaging. A few theoretical attempts [3-4] have recently been made in understanding the nonproportionality in inorganic scintillators and accordingly it has been revealed that it is caused by the high excitation density created within the track of an X-ray or a $\gamma$ – ray photon entering in a scintillating crystal. In this paper, following our recent approach [4] the light yield is first derived as a function of the excitation density created within the track of a $\gamma$-photon. Then using the excitation density as derived by Jaffe [3], the yield is derived as a function of the total deposited energy. Finally the results are compared with those in [3].

## 2. Theoretical developments

Following [4] we consider a cylindrical $\gamma$-ray track of radius $r$ in a scintillating crystal and only excitons and pairs of electron and hole (e-h pairs) are generated in the track. Phonons are of course generated in relaxation of energetic carriers and self-trapping of excitons. The dynamical changes in exciton and e-h pair densities at any point, $x$, along the track can be expressed by the following two rate equations [4]:

$$-\frac{dn_{ex}(x)}{dt} = (R_{1x} + K_{1x})n_{ex}(x) + (R_{2x} + K_{2x})n_{ex}^2(x) - \gamma_{ex}n_{eh}(x) + \gamma_{xe}n_{ex}(x) + K_{3x}n_{ex}(x)^3 - f_x n(x)\delta(t), \tag{1}$$

$$-\frac{dn_{eh}(x)}{dt} = (R_{1eh} + K_{1eh})n_{eh}(x) + (R_{2eh} + K_{2eh})n_{eh}^2(x) - \gamma_{xe}n_{ex}(x) + \gamma_{ex}n_{eh}(x) + K_{3eh}n_{eh}^3(x) - (1 - f_x)n(x)\delta(t), \tag{2}$$

where $n_{ex}(x)$ is the excitonic concentration and $n_{eh}(x)$ is the concentration of excited e-h pairs not bound like excitons at any point $x$ on the track. $f_x$ is the fraction concentration of excitons , $(1 - f_x)$ is the fraction concentration of e-h pairs, and $n(x)\delta(t)$ represents the total number of excitations, $n(x) = n_{ex}(x) + n_{eh}(x)$, created by the incident energy at time $t = 0$ at any point $x$ along the track. Accordingly, $n(x)$ ($cm^{-3}$) is defined as:

$$n(x) = \frac{\left(-\frac{\partial E}{\partial x}\right)}{\pi r^2 E_{eh}}, \tag{3}$$

where $E$ is the total initial energy incident at any point $x$, $\pi r^2$ is the average area of cross section of the track and $E_{eh}$ is the average energy required to create an excitation in a scintillator, and here it is assumed to be three times the band gap energy ($E_g$) : $E_{eh} = 3E_g$ . The validity of the concept of average radius has been addressed in [4].

In equations (1) and (2), $R_{ix}$ and $K_{ix}$ denote the rates of radiative and non-radiative (quenching) recombination of excitons, respectively, and $i$ = 1, 2, denote through linear (1) and binary (2) processes. $R_{ieh}$ and $K_{ieh}$ ($i$ = 1, 2) are the corresponding rates of recombination for an e-h pair, and $K_{3x}$ and $K_{3eh}$ are rates of non-radiative Auger (ternary) recombination of excitons and an e-h pairs, respectively. It is assumed here that Auger processes do not contribute to any radiative recombination. $\gamma_{ex}$ and $\gamma_{xe}$ are rates of converting an e-h pair into an exciton and vice versa, respectively. It is important to consider both the possibilities for applying the theory at higher temperatures. According to Eqs. (1) and (2), we can classify scintillators in three categories: (i) excitonic with $f(x)$ = 1, (ii) non-excitonic with $f(x) = 0$, and (iii) mixed case $0< f(x) < 1$.

In the track all the radiative channels contribute to the emission of photons which are detected and hence the light yield or scintillator yield is defined as the total emitted photons divided by the total excitations created in the track. In this way following [4] the scintillator yield is obtained as (details of derivation are given in [4]):

$$Y = \frac{a_1 n + a_2 n^2}{a_3 n + a_4 n^2 + a_5 n^3} \quad , \tag{4}$$

where $n$ is the total number of excitations created in the track and using the Bethe expression for the stopping power [5] and following Jaffe [3] it is approximately obtained as:

$$n \approx [\frac{20 e^4 \kappa^2 N_e}{E_{eh} r^2}] E^{-1}, \tag{5}$$

where $e$ is the electronic charge, $\kappa = (4\pi\varepsilon_0)^{-1} = 8.9877\times10^9$, $N_e \approx 10^{30}$ m$^{-3}$ is the density of ground-state electrons in the track and $E$ is the total energy deposited in the track. Following [4] the other quantities appearing in the yield equation are defined as:

$$a_1 = R_{1x} A + R_{1eh} B \tag{6a}$$

$$A = \frac{[\gamma_{ex} + (R_{1eh} + K_{1eh}) f_x]}{(R_{1x} + K_{1x})(R_{1eh} + K_{1eh}) + \gamma_{ex}(R_{1x} + K_{1x}) + \gamma_{xe}(R_{1eh} + K_{1eh})} \tag{6b}$$

$$B = \frac{[(1 - f_x)(R_{1x} + K_{1x}) + \gamma_{xe}]}{(R_{1x} + K_{1x})(R_{1eh} + K_{1eh}) + \gamma_{ex}(R_{1x} + K_{1x}) + \gamma_{xe}(R_{1eh} + K_{1eh})} \tag{6c}$$

$$a_2 = \frac{R_{2x}}{2\tau_x} A^2 + \frac{R_{2eh}}{2\tau_{eh}} B^2; \quad \tau_x^{-1} = R_{1x} + K_{1x} \text{ and } \tau_{eh}^{-1} = R_{1eh} + K_{1eh} \tag{6d}$$

$$a_3 = (R_{1x} + K_{1x}) A + (R_{1eh} + K_{1eh}) B \tag{6e}$$

$$a_4 = \frac{(R_{2x} + K_{2x})}{2\tau_x} A^2 + \frac{(R_{2eh} + K_{2eh})}{2\tau_{eh}} B^2 \tag{6f}$$

$$a_5 == \frac{K_{3x}}{3\tau_x^2} A^3 + \frac{K_{3eh}}{3\tau_{eh}^2} B^3 \tag{6g}$$

Using Eq. (5) the yield in Eq. (4) can be written as:

$$Y = \frac{a_1 Q E^{-1} + a_2 Q^2 E^{-2}}{a_3 Q E^{-1} + a_4 Q^2 E^{-2} + a_5 Q^3 E^{-3}} \tag{7}$$

where
$$Q = \frac{20 e^4 \kappa^2 N_e}{E_{eh} r^2} \tag{8}$$

The yield obtained in equation (7) is of a similar form as obtained in [3] except that it takes into account up to third order processes. In [3] only two, one radiative and one non-radiative, processes are considered. The details of yield in equation (7) and associated analysis will be published elsewhere. Here onward we also consider only two channels so that the results can be compared with the previous theory [3]. We ignore the second order radiative and third order non-radiative processes, which means that neglect $a_2$ and $a_5$. This does not affect to any aspect of the mathematical formalism because the actual yield derivation is based on only first order processes [4]. Thus the yield in equation (7) can be expressed as:

$$Y = \frac{a1}{a_3} \frac{E^p}{E^p + E^q K^{p-q}} \quad ; \quad K = \frac{a_4}{a_3} Q \tag{9}$$

where $p = -1$ and $q = -2$. Except for the factor $a_1 / a_3$, the yield in equation (9) is the same as obtained in [3] by Jaffe, however here the energy parameter $K$ is not really a parameter but it is a function of the rates of radiative and non-radiative processes as it can be seen from equations (6) and (9). In [3] K has been determined by fitting the yield to the experimental data.

Experimentally the scintillator yield is calculated, as a customary or convention, in relation to the yield at $E$ = 662 keV. Here also if we divide the yield in equation (9) by its value at $E$ = 662 keV, we get an expression for the relative yield $Y_R$ identical to that obtained in [3] as:

$$Y_R = \frac{Y}{Y_{662}} = \frac{(662\text{keV})^{-\alpha} + K^{-\alpha}}{E^{-\alpha} + K^{-\alpha}} \tag{10}$$

where $\alpha = p - q$. For p = -1 and q = -2, $\alpha = 1$ . It is to be noted that in the relative yield $Y_R$ in equation (10) the factor $a_1 / a_3$ plays no role. This is based on the assumption that rates of radiative and non-radiative processes are independent of the incident energy, which is a reasonable assumption. In this approach the energy $K$ can be calculated from equation (9) if the required radiative and non-radiative rates in equation (6) are known. This will be carried out in the next section. The important point is that the energy $K$ is a function of the track radius, average energy of excitation and the rates of radiative and non radiative processes. Therefore $K$ comes out as the most important quantity in the yield that contains all information about the track as well as all the processes raking place within the track.

## 3. Results

The important aspect of this paper is that one can calculate $K$ theoretically from equation (9) and verify that it agrees or not with that obtained by fitting equation (10) to the experimental yields measured in some scintillators. Jaffe [3] has carried out a very comprehensive task of fitting equation (10) to the experimental data for seven scintillators, namely CsI, $CaF_2$:Eu, GSO:Ce; $K_2LaCl_5$:Ce; $Lu_2SiO_5$:Ce, $BaF_2$ and $RbGd_2Br_7$:Ce, and thus determined $\alpha$ and $K$. We have also calculated the yield for four scintillators, namely NaI:Tl, $BaF_2$, GSO:Ce and $LaCl_3$:Ce, in [4]. The common scintillators studied in [3] and [4] are $BaF_2$ and GSO:Ce. We will focus only on $BaF_2$ scintillator in this paper for comparison but a more comprehensive analysis involving other scintillators will be published elsewhere.

According to [4], $BaF_2$ is an excitonic scintillator which means that dominantly the created electron and holes form excitons. In this case $f_x = 1$ which gives $K$ from equation. (9) as [4]:

$$K = \frac{Q(R_{2x} + K_{2x})}{2(R_{1x} + K_{1x})} \tag{11}$$

For $BaF_2$ according to [4], $R_{1x} = 1.6x10^9\,\mathrm{s}^{-1}$, $K_{1x} = 4x10^8\,\mathrm{s}^{-1}$, $R_{2x} = 0$ and $K_{2x} = 2x10^{-17}\,\mathrm{m}^3\mathrm{s}^{-1}$. Taking $r$ = 2nm [4] and $E_{eh}$ = 15 eV [3], we get $Q = 6.8931\mathrm{x}10^{29}\,\mathrm{eV/m}^3$ and then from equation (11) we get $K$ = 3.44 keV. This is in fair agreement with 4.28 keV obtained by Jaffe [3] from fitting equation (10) to the experimental data. It also agrees quite well with the average value of 3 keV obtained in [3] for seven scintillators. This clearly demonstrates the significant role that rates of radiative and non-radiative processes play within the $\gamma$-ray track in a scintillator. The details of the yields in equations (7) and (10), including other scintillators, will be published elsewhere.

## 4. Discussions

The scintillator yield derived earlier in [4] following an analytical and phenomenological approach is used to evaluate the yield for a special case considered in [3], where only two processes, one radiative and other non-radiative, are considered to occur within the track of a $\gamma$-ray photon entering a scintillator. By comparing the analytical yield derived here in equation (10) with that obtained in parametric form one gets an analytical form of the parameter. This is a very distinct advantage of this comparative method because the parameter $K$ in equation (10) could only be determined earlier by fitting equation (10) to the experimental data. Now $K$ can be calculated without fitting if the rates and track parameters are known. The calculated $K$ here agrees quite well with its fitted value for $BaF_2$ scintillator. A detailed comparison for other scintillators, including the third order process through equation (7) will be carried out in future and published elsewhere.

It may be noted that here we have the value of $\alpha = 1$ in equation (10), which however varies for different scintillators in [3]. For, example, for $BaF_2$ $\alpha = 1.4$ and for CsI it is 1.29, etc. This may be attributed to the fact that Jaffe [3] has considered only two processes that quench excited carriers in the track, a radiative of order $p$ and a non-radiative of order $q$ without specifying $p$ and $q$. This is different from the present approach in equation (7) where three orders of processes are considered specifically and as the yield in equation (10) is a special case of equation (7) by neglecting the third order process, here $p$ = -1 and $q$ = -2. As the experimental data are representative of all processes occurring within the track, by fitting these only two processes, one is expected to get a value for $\alpha$ different from 1.

Here we have considered only the excitonic scintillator with $f_x = 1$. Apparently all the cases considered in [3] are close to this case; most are excitonic scintillators. The classic scintillator of sodium iodide doped with thallium (NaI:Tl) shows a different scintillator yield. In this case $f_x << 1$ [4]. The observed characteristics of the light yield show a rise with decreasing energy from the highest energy. Then a maximum occurs in the middle energy range which decreases again as energy decreases to the lower values. This feature of the yield in NaI:Tl or intrinsic NaI scintillators cannot be obtained by fitting data only to two processes. It is of course very desirable to study the yield in other well known scintillators used in SPECT and PET medical imaging systems, e. g., LYSO and YAP. However, here we present only some initial developments in the study of nonproportionality and as stated above a more comprehensive study will be published elsewhere.

## 5. Conclusions

An analytical expression for the scintillator yield is derived as a function of the $\gamma$-ray incident energy by incorporating the rates of first, second and third order processes. By comparing the yield derived in [3] and [4] an analytical expression is also obtained for the energy parameter $K$ which could otherwise be determined only by fitting the yield in equation (10) to the experimental data.

## References

[1] Dorenbos P, de Haas J T M and van Eijk C W E 1995 *IEEE Trans. Nucl. Sci.* **NS-42** 2190
[2] Moszynski M, Zalipska J, Balcerzyk M, Kapusta M, Mengesha W and Valentine J D 2002 *Nucl. Instr. And Meth.* A **484** 259
[3] Jaffe J E 2007 *Nucl. Instr. And Meth. In Phys. Res.* A **580** 1378
[4] Bizarri G E, Moses W W, Singh J, Vasile'v A N and Williams R T 2009 *J. Appl. Phys.* **105** 044507
[5] Bethe H A 1939 *Ann. Phys.* **5** 325

Journal of Physics: Conference Series **253** (2010) 012017 doi:10.1088/1742-6596/253/1/012017

# Towards real time 3D quantitative characterisation of *in situ* layer growth using white light interference microscopy

**PC Montgomery[1], F Anstotz[1], J Montagna[1], D Montaner[1], L Pramatarova[2] and E Pecheva[2]**

[1]*Institut d'Electronique du Solide et des Systèmes (InESS), Laboratoire commun UdS-CNRS, UMR 7163, 23 rue du Loess, 67037 Strasbourg, France.*

[2]*Institute of Solid State Physics, Bulgarian Academy of Sciences, 72 Tzarigradsko Chaussee Blvd., 1784 Sofia, Bulgaria.*

E-mail: paul.montgomery@iness.c-strasbourg.fr

**Abstract**. Quantitative 3D characterisation of layer growth or modification *in situ* in liquid systems is a challenge because of the changing morphology of the layer and the presence of the growth liquid. Because of the limited bandwidth of many surface profiling techniques, measurement of microscopic surface roughness is generally limited to that of static surfaces. The aim of the present work is to develop a new technique using high speed scanning interference microscopy combined with an adapted immersion head for use in liquid growth systems. For several years we have been developing a real time 3D surface analysis system based on a high speed camera and cabled logic processing, combined with continuous scanning white light interferometry. An optical measurement head is also being developed for use in liquid immersion conditions, with the view of measuring layer growth or modification in biomaterials. In this paper we report on the present status of the development of our second prototype 4D system and also of the optical immersion head for *in situ* measurements, describing the achievements made and the difficulties still to be overcome.

## 1. Introduction

Many new areas of materials development such as organic electronics, microsystems, photonic crystals and biomaterials [1] use some sort of layer growth and processing. Amongst the many characterisation techniques employed to study the different properties of the layers grown, 3D measurement of the surface morphology figures as a growing challenge, due to the high roughness and growing complexity of the layers. Knowledge of the roughness values, particle size, step heights, and structural shape of the layers plays a growing role in calibrating growth techniques and in understanding the functionality of the grown layers at different scales. Classical profilometry techniques such as stylus profiling and AFM generally require many seconds to many minutes or more to perform a single measurement of the 3D surface. Such a time constraint limits measurements to static samples, *ex situ*.

A key aspect of achieving real time quantitative microscopy is the requirement of a very high bandwidth in the measurement system. Real time microscopy of several to tens of 2D image slices per second has been demonstrated in the area of confocal microscopy for biological specimens [2] by increasing the scanning speed of the probe beam, but the 3D image acquisition time is several seconds or more due to the need to acquire several slices over the depth. By using probe tip resonance in AFM [3] and SNOM [4], 3D image acquisition can be increased to tens of images per second (i/s), but this is generally limited to small field sizes of 10 µm or less and with uncalibrated depth information. Several tens of i/s has been achieved in interference microscopy, but limited to the measurement of either small surface roughness (below λ/2) [5] or certain types of deeper sample structure [6]. More recently, in digital holography, a rate of 15 i/s has been demonstrated for measuring continuous smooth surfaces [7].

In interference microscopy, signal processing is generally carried out on the PC microprocessor, requiring a measurement time for a single 3D image of several seconds or minutes depending on the image size and scan depth [8]. A real time "smart pixel" CMOS camera with integrated dedicated hardware processing in pOCT (parallel optical coherence tomography) has demonstrated a 3D imaging rate of 14 i/s over a depth of 1.4 mm with a depth resolution of 4.5 µm [9]. While this technology is interesting for certain dedicated real time applications, it has the disadvantages of a small camera sensor size of 90x144 pixels, photodiodes with a very low fill factor of 10 % and lack of flexibility in terms of the choice of signal processing.

White light scanning interferometry (WLSI) is of great interest in quantitative real time microscopy because no physical scanning in the XY plane is necessary as in stylus profiling or near field microscopy, the measurement rate being limited only by the Z scan speed of the fringes, and the image acquisition and processing speeds. High depths can be measured (hundreds of µm) with an axial sensitivity of nm to tens of nm, depending on the surface roughness [8, 10]. For example, using a high speed CCD camera (250 i/s, 512x512 pixels full frame) synchronized with a continuously scanning piezo (10 Hz) and image processing carried out on a FPGA (Field Programmable Gate Array), we previously demonstrated a first prototype system for 3D measurement at a rate of 5 i/s over 3 µm [11]. In the present work we discuss the development of a second prototype based on a high speed CMOS camera (500i/s, 1280x1024 pixels full frame - 625 Mbytes/s) connected by a Cameralink bus to a PCI-X (64 bits, 66 Mhz) FPGA based vision processing board [12]. Results have been demonstrated on a laterally translated microchip.

A second difficulty of measuring layer growth is the physical growth conditions. Most of the previously mentioned techniques are generally used in dry, *ex situ* conditions. *In situ* measurement is made more complicated by the limited space available due to the growth chamber and the difficulty to approach the sample surface. In the case of growth in a liquid, the presence of the growth solution also limits the type of optics that can be used. A new optical measurement head is being developed that is suitable for use in liquid immersion conditions, with the view of measuring layer growth or modification in biomaterials [1]. In this paper we report on the present status of the development of our second prototype 4D system and also of the optical immersion head for *in situ* measurements, describing the achievements made and the difficulties still to be overcome.

## 2. Choice of algorithm for high speed processing with cabled logic

### 2.1. 3D measurement by fringe scanning

Optical interference microscopy makes use of the phase of the light reflected from the surface to be measured by comparing it with that of light reflected from a plane reference mirror. With the path lengths of each beam being nearly equal, a series of white light fringes are produced superimposed on an image of the surface in the plane of the camera [10]. Interference objectives typically used are of the Michelson or Mirau type that use a single objective for low to medium magnifications, or the Linnik type, that uses two separate objectives for higher magnifications [13]. The fringes are translated over the whole of the depth of the sample by varying the distance between the interference objective and the sample surface. A series of images is captured using a digital camera and the images processed to extract the fringe envelope at each pixel. The light intensity $I$ measured in a white light interferometer that is spatially incoherent has the following form:

$$I(x,y,z)=\sum_{\lambda_1}^{\lambda_2} I_0(x,y,z)[1+R\cdot\Gamma(\Delta L)\cos(\phi(x,y,z))]$$

The $x$ and $y$ coordinates correspond to the image coordinates and the $z$ coordinate indicates the axial location of the sample surface. The quantity $I_0$ is an offset related to the reference and object beam intensities. The interferogram envelope function $\Gamma$ is related to the spectral profile of the white light and $R$ is the effect of the reflection coefficient of the sample. Detection of the peak of the fringe intensity or the fringe envelope at a pixel over the depth scanned determines the position of the surface at that point. Many different signal processing techniques for determining the envelope peak have been proposed in the literature, varying in mathematical complexity, computational cost, processing time, robustness and axial measurement uncertainty. Some of the more computationally costly algorithms include demodulation, correlation, determination of the barycenter of the envelope, Fourier, Hilbert and wavelet techniques [13]. These techniques are suitable for static measurement (or post-processing after acquisition at high speed) where a processing time above one second is sufficient. On the contrary, for real time on-line measurements in which processing needs to be carried out "on-the-fly" at less than a second per 3D image, the use of complex mathematical functions in a microprocessor in a PC, a DSP or a GPU is not possible at present above a 2D image rate of around 90 i/s.

Processing images from a high speed camera above 500 i/s requires the use of cabled logic processing, for example using a FPGA The choice of the best algorithm is then mainly based on the possibility of integrating it into the FPGA, which reduces the availability of mathematical functions to basic operators such as adders, multipliers and comparators, together with memory requirement. Two of the simpler algorithms proposed in the literature, peak fringe scanning microscopy (PFSM) and the determination of the fringe visibility (FSA) are now described.

### *2.2. PFSM algorithm*

The PFSM algorithm [14] is based on the detection of the highest intensity of the fringe signal along the optical axis, which corresponds to the peak of the central zero-order fringe as long as the background intensity is always less than this value. This algorithm requires a first memory to record the position of the maximum for each pixel (altitude image) and a second memory for the current value of the maximum intensity (which gives the reflection image). The interest of this algorithm lies in its simplicity, requiring only one comparator, a limited number of images stored in on board memory and thus a considerable gain in processing time and storage space. It is well adapted for implementation on a FPGA. The axial resolution depends on the displacement step between images, with typical values being between 20 nm and 80 nm.

### *2.3. FSA algorithm*

The five sample adaptive (FSA) algorithm [15] is developed from a generalized form of the well known five-step phase-shifting algorithm used in monochromatic interferometry. Rather than measuring the phase, the modulation of the fringes is determined by measuring a series of five consecutive points along the Z axis at each pixel location. Based on the Hilbert approximation, an analytical expression of the modulation function can be established, given by the following expression:

$$M=C.\sqrt{\left((I_2-I_4)^2-(I_1-I_3)(I_3-I_5)\right)}$$

where C is a constant of standardization. This formula is exact when the phase shift between interferograms is 90°.

Using this relation, it is possible to measure the variation in fringe visibility, to detect the envelope of the interference fringes and to detect the peak value in order to determine the Z position. Although the FSA algorithm is inevitably more complex in terms of digital architecture than the PFSM algorithm, it is nonetheless far simpler than algorithms requiring complex mathematical functions. Like the PFSM algorithm, the FSA algorithm is well adapted to implementation in a FPGA. The axial resolution of the FSA algorithm in its simplest form is limited to the step height of 90°, typically around 80 nm.

## 3. High speed camera acquisition and processing system

### *3.1. Measurement system*

The real time measurement system (figure 1) consists of an adapted Leica DMR-X microscope equipped with a halogen lamp and a set of Michelson and Mirau interference objectives. Axial displacement of the sample is carried out by a piezo actuator (PI) controlled by a closed loop controller using a high accuracy capacitive position detector to achieve a sensitivity of 1 nm in the static mode. Image acquisition is performed with a high speed CMOS camera (BASLER A504k) and a CORECO "Anaconda" acquisition and processing board. An on-board FPGA is used for the high speed processing.

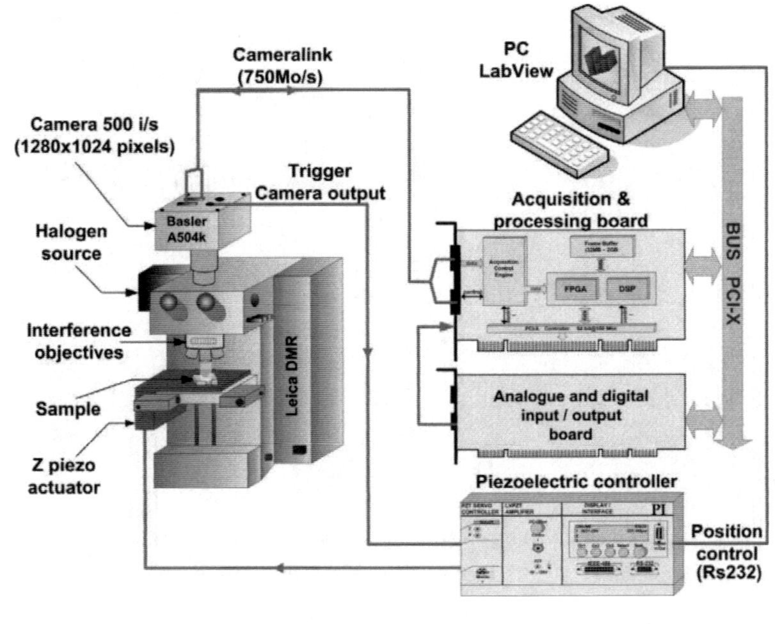


**Figure 1.** Layout of real time 3D measurement system

### *3.2. System synchronization*

Each element of the electronic system is synchronized by a clock signal originating from a function generator developed under LabView (figure 2). This clock signal is used to trigger the camera, to control the processing rate on the imaging board, to generate the ramp signal for the piezoelectric translator and to synchronize the image acquisition and processing on the PC. Complex low and high level software has been developed under LabView and $C^{++}$ to initialize and to configure the different parts of the system indicated by the dotted lines in figure 2. By this means it is thus possible to define the image size, the image acquisition rate, the exposure level, the piezo scanning rate and depth, the transfer rates to the imaging board, the type of processing on the FPGA, the 3D image rate, and the type of storage on the PC hard disc. The timing signals for synchronization of the camera, the piezo translator and the calculated 3D image are illustrated in figure 3. A symmetrical square wave is used for the timing signal and a stepped ramp signal to control the piezo translator. A single 2D image from the camera is acquired on the rising edge of the camera trigger signal.

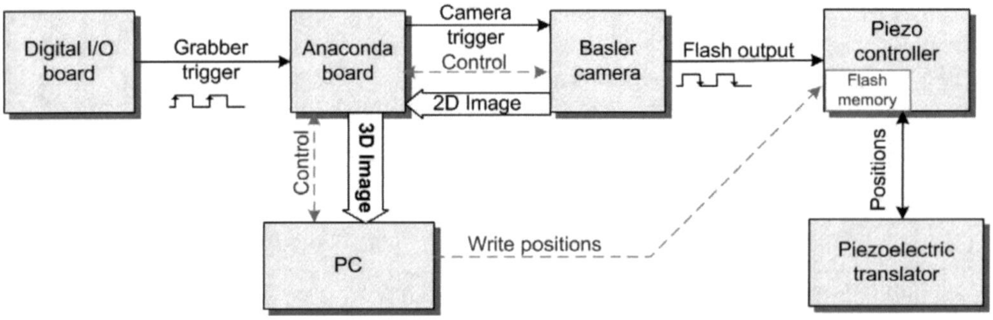


**Figure 2.** Block diagram of image acquisition, processing and synchronisation

The piezo translator is displaced one step on the falling edge of the camera trigger signal. In this way image acquisition occurs during the second half of the plateau of the stepped ramp signal to the piezo translator, minimizing axial movement of the translator and optimizing fringe contrast. The previously acquired image is sent to the imaging board on the falling edge of the camera trigger signal. These operations are repeated until a complete scan in one direction is completed (rising or falling).

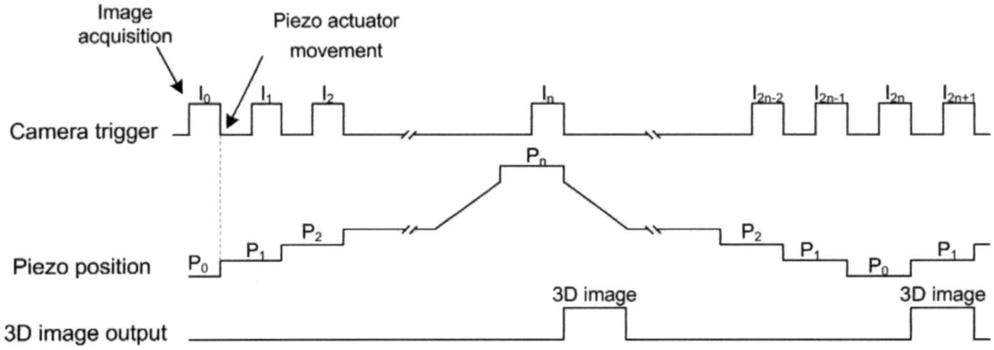


**Figure 3.** Timing signals for synchronizing different parts of the system

At the beginning of the scan in the opposite direction, the two images containing the height measurement data and the maximum intensities (reflection image) are transferred to the RAM of the PC via the PCI-X bus. Because the image data rate is far lower than the camera acquisition rate (by a factor of 50 to 200, depending on the number of images scanned along the optical axis), the PC software can be used to perform image filtering and construction of the pseudo 3D image. The image data can also be stored on hard disk. Such a system design makes it possible to perform real on-line measurements.

### *3.3. Implementation of algorithms in cabled logic*

The two fringe processing algorithms proposed were implemented in cabled logic, firstly by simulating them in VHDL (Very High Speed Integrated Circuit Hardware Description Language) and secondly by synthesizing the optimized VHDL code to the bit stream file specific to our FPGA circuit. Although the algorithms are fairly simple, much work was required to develop a suitable architecture for the cabled logic. Many difficulties had to be overcome in interfacing between the camera and the processing board via the CameraLink connection and between the FPGA processor and the PC via the PCI-X bus and the different connections used to control each part of the system. The development work for implementing the PFSM and FSA algorithms in cabled logic was initiated in the framework of the PhD by G. Johnson [12] and then taken further and optimized by J. Montagna during the present

Oséo project. A second option of using the minimum of the signal (black fringe) as well as the maximum (white fringe) has also been added. A « *ChipScope Pro* » tool was used on the FPGA chip to monitor directly the different signals and to optimize the algorithms developed.

### *3.4. System performance*

In this section we present the practical performance of the acquisition and processing part of the system in terms of the on-line 3D image rate as a function of the scanning depth. The calculation of the performance depends on many different parameters and is complex, but the main limitation is due to data transfer rates to the FPGA and its local environment of buffers and memories.

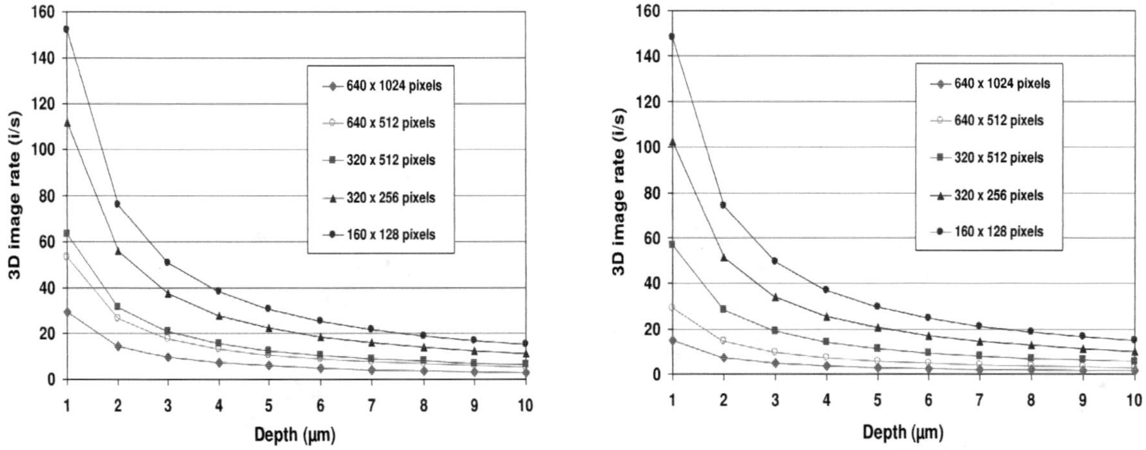


(a) PFSM algorithm (b) FSA algorithm

**Figure 4.** Comparison of performance of two algorithms in terms of 3D image rates as a function of scan depth and image size for a step size of 76 nm

The calculated results for the PFSM and FSA algorithms are given in figure 4 for different image sizes over an axial scan range of 1-10 µm and a step value of 76 nm between the acquired fringe images. Comparing the performance of the two algorithms, it can be noted that the 3D image rates are slightly higher for the PFSM algorithm for the same image sizes and scan depths. This improvement is due to the greater simplicity of the PFSM algorithm. For example, for a scan depth of 3 µm and an image size of 640x1024 pixels, the 3D image rate is 4.94 i/s for the FSA algorithm and 9.75 i/s for the PFSM algorithm. For the smaller image size of 160x128 pixels, the 3D image rate increases to 49.40 i/s for the FSA algorithm and 50.67 i/s for the PFSM algorithm. Whereas the step size is fixed for the FSA algorithm, for the PFSM algorithm it can be varied over a practical range of 20-80 nm, with a corresponding increase in processing time and reduction in 3D image rate for steps smaller than the example of 76 nm given.

## 4. Results of on-line measurements on laterally moving samples

Whilst the real time measurement system is being developed for measuring *in situ* layer growth, at this stage, it has been found that translating a microelectronic chip laterally using the motorized tables at different speeds is quite satisfactory for the purpose of testing the system under development. The chip used was a microfluxgate sensor, an integrated version of the fluxgate magnetic probe [16] fabricated using a process derived from a standard Bipolar process. The surface of the chip is covered with a 0.8 µm to 1.6 µm thick transparent passivation layer. The measurements in WLSI were performed on the buried interface consisting of the Silicon substrate and the metal connections.

### *4.1. Measurements with the PFSM algorithm*

Results of measuring the microfluxgate with the PFSM algorithm for large and small image sizes for an axial step size of 40 nm are given in figure 5.

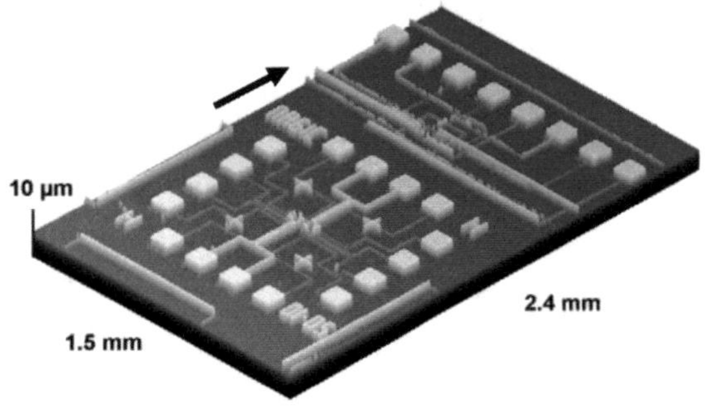


(i) $I_0$ t = 0 s, dx = 0 µm

(i) $I_0$ t = 0 s, dx = 0 µm

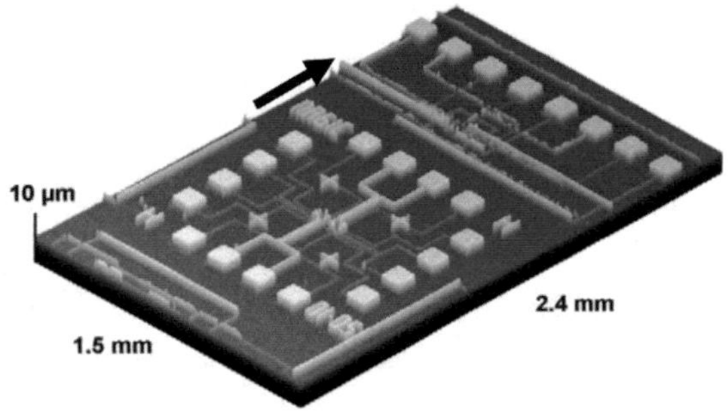


(ii) $I_8$ t = 4.2 s, dx = 84 µm
(a) Large image (640x1024 pixels, x5 Michelson objective), 40 nm step over 5 µm, lateral speed 20 µm/s, sampling rate 240 i/s and a 3D image rate of 1.9 i/s

(ii) $I_{17}$ t = 1.7 s, dx = 36 µm
(b) Small image (320x256 pixels, x10 Mirau objective), 40 nm step over 4 µm, a lateral speed of 20 µm/s, sampling rate of 1010 i/s and a 3D image rate of 10 i/s

**Figure 5.** Two sequences of measurements for different image sizes and sampling rates with the PFSM algorithm of the microfluxgate chip being displaced sideways at different speeds

Two images at different times are presented for each size, to show the advancing chip. The full results exist in the form of a video clip. The x5 Michelson objective was used for the large image size (640x1024 pixels, 1.5x2.4 mm) with a fringe image acquisition rate of 240 i/s over a scan range of 5 µm, giving a 3D image rate of 1.9 i/s. For the small image size (320x256 pixels, 375x300 µm), the x10 Mirau objective was used with a fringe image acquisition rate of 1010 i/s over a scan range of 4 µm, giving a 3D image rate of 10 i/s. The results contained much less noise and artifacts than the measurements made with the first prototype [5]. The most significant artifacts remaining were scattered bumps $\lambda/2$ in height due to secondary fringe detection (visible in figure 5(b)(ii)), mainly arising from the reduced signal to noise ratio in measuring the buried interface. The results would be "cleaner" on a bare chip without a passivation layer. Otherwise, the axial measurement uncertainty is limited by the scanning step, which is 40 nm in these examples. Slope artifacts are also visible on leading and trailing square step edges due to motion blur.

### *4.2. Measurements with the FSA algorithm*

Results of measuring the microfluxgate with the FSA algorithm for large and small image sizes for an axial step size of 76 nm are given in figure 6.

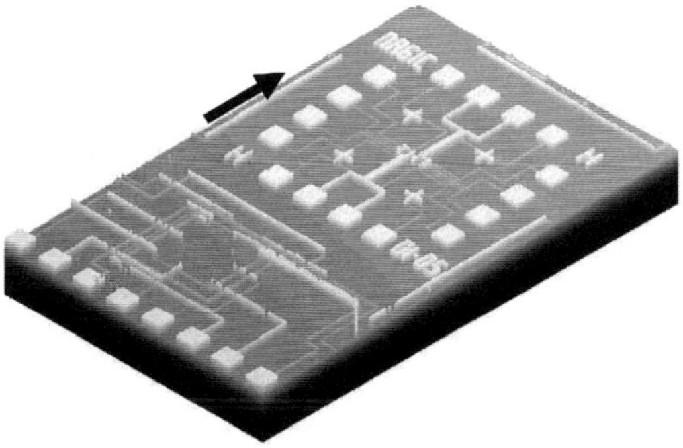

(i) $I_0$ t = 0 s, dx = 0 µm

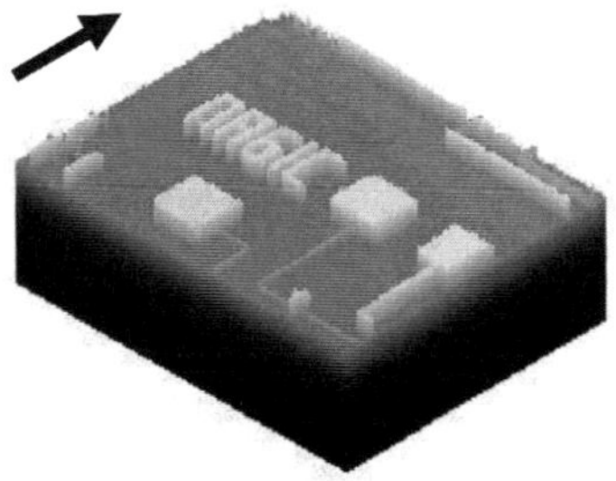

(i) $I_0$ t = 0 s, dx = 0 µm

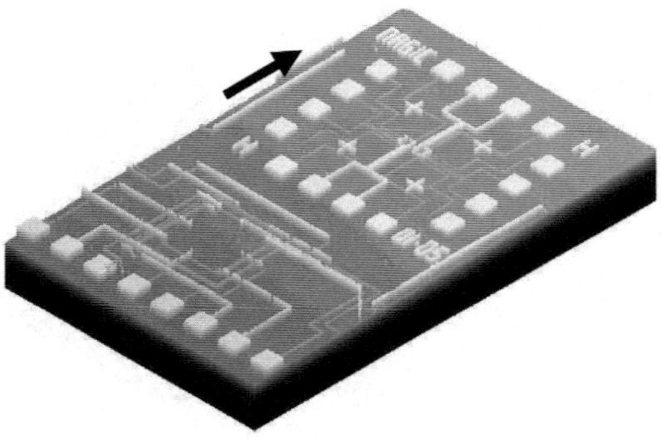

(ii) $I_{12}$ t = 4 s, dx = 80 µm
(a) Large image (600x1024 pixels), 76 nm step, lateral speed 20 µm/s, sampling rate 190 i/s and 3D rate 3 i/s

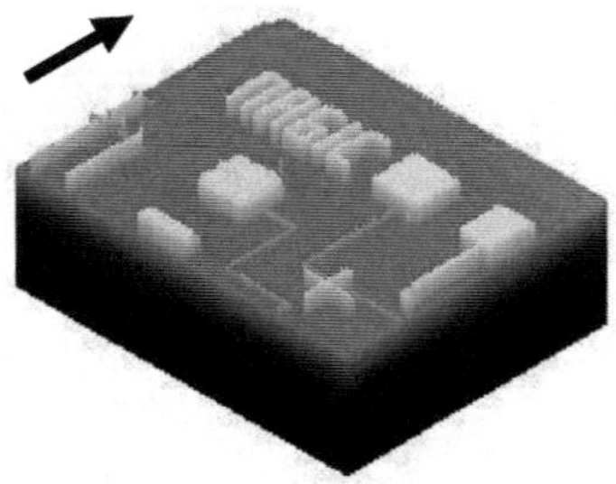

(ii) $I_6$ t = 0.43 s, dx = 86 µm
(b) Small image (320x256 pixels, 750x600 µm), 76 nm step, lateral speed 200 µm/s, sampling rate 924 i/s and 3D rate 20.52 i/s

**Figure 6.** Two sequences of measurements for different images sizes and sampling rates with the FSA algorithm of the microfluxgate chip being displaced sideways at different speeds (x5 Michelson objective, axial range of 5 µm)

For both image sizes, the x5 Michelson objective was used with a 5 µm scan. A fringe acquisition image rate of 190 i/s was used for the large image size (640x1024 pixels, 1.5x2.4 mm), giving a 3D image rate of 2.96 i/s. For the small image size (320x256 pixels, 750x600 µm), a fringe image acquisition rate of 924 i/s was used, giving a 3D image rate of 20.52 i/s. The measurements contained less axial noise than those made with the PFSM algorithm. This is to be expected as the FSA algorithm is more robust than the PFSM algorithm. The axial measurement uncertainty is again limited by the scanning step, which in this case is 76 nm. Slope artifacts are also visible on leading and trailing square step edges due to motion blur.

*4.3. Immersion head for in situ growth measurement*
Hydroxyapatite (HA) is the mineral part of bones, teeth and shells. Growth of thick layers (1-20 µm) of synthetic HA is important for medical implants and biosensors in the field of new biomaterials. The ISSP of the Bulgarian Academy of Sciences has developed a novel growth method (laser-liquid-solid interaction) in which substrates (steel, Si, glass, polymers…) are immersed in a simulated body fluid solution for layer growth. The effects of laser illumination, surface nano-structuring and the addition of diamond nanoparticles are being studied [1]. These layers are characterized using a series of

classical techniques (SEM, FTIR, AFM, XRD...). The surface roughness is measured at InESS using WLSI that has been optimized for these particular layers that are thick, rough and translucent (figure 7(b)). The aspect is very similar to that found with MEB and AFM, the layer consisting of micron sized sphere-like nodules.

In order to be able to measure the growth of such layers *in situ*, an immersion head is being specifically developed, shown in figure 7(a). A pair of immersion objectives is used in a Linnik configuration. The sample observation objective will be placed in a protective head containing distilled water and a window that will be immersed in the SBF solution. In this way, the observation window can be cleaned or replaced between growth experiments. The head is still under development. The combination of the real time system and immersion head will enable the study of the growth of HA layers *in situ*.

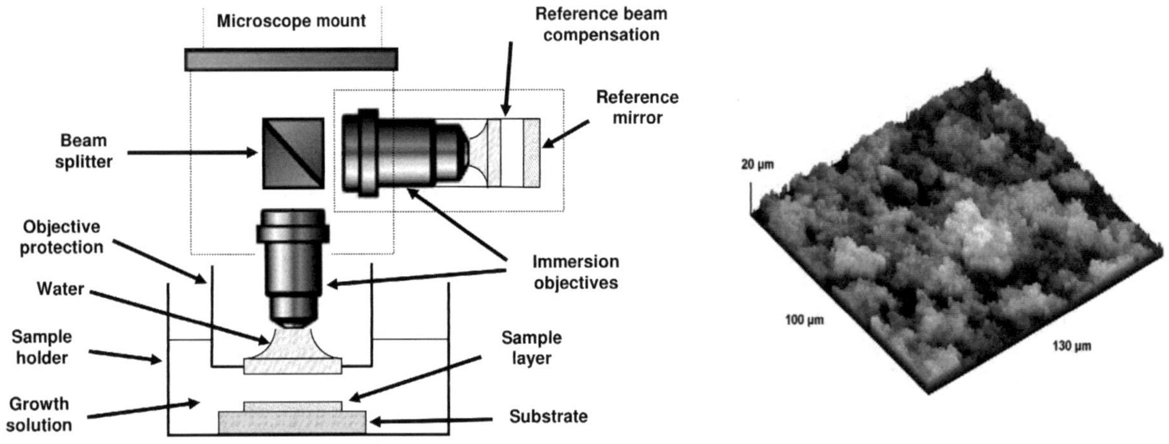


(a) (b)

**Figure 7.** Future adapted immersion Linnik interferometer head (a) for characterizing *in situ* layer growth in liquid conditions (b)

## 5. Conclusions

The latest results of a new system for measuring real time 3D surface relief for layer growth *in situ* using continuously scanned WLSI has been presented. The system consists of a high speed CMOS camera and cabled logic image processing system.

Two different algorithms for fringe processing have been implemented on the FPGA based board and characterized in terms of performance. The fastest algorithm is by simple detection of the peak intensity (PFSM), giving 3D image rates of 2.93 i/s to 152 i/s for image sizes of between 640x1024 pixels and 160x128 pixels and scan depths of 1 to 10 µm. For the same sampling step of 76 nm, an algorithm based on the detection of the fringe visibility (FSA) gives slightly slower 3D image rates of 1.48 i/s to 148.20 i/s for similar conditions. Tests performed on a microfluxgate chip being translated sideways between speeds of 20 µm/s and 200 µm/s demonstrated successful on-line measurements over a scan depth of 4-5 µm and 3D image rates of between 1.9 i/s and 20.52 i/s. The FSA algorithm proved to be more robust, with less artifacts than the PFSM algorithm.

Future work will consist of completing the immersion head and testing the combined system on *in situ* layer growth of HA layers. An improved real time camera and processor is also being developed to improve the performance rates still further.

## Acknowledgements

This work has been supported by the Oséo Conectus project "CAM4D" and the CNRS PICS project N° 4848. Thanks are also extended to Luc Hébrard for providing the microfluxgate chip.

## References


[1] Pecheva E, Montgomery P, Montaner D and Pramatarova L 2007 *Langmuir* **23** 3912

[2] Schick A and Breitmeier U 2004 *Proc. Int. Conf. on Optical Metrology in Production Engineering (Strasbourg, France, 27-30 April 2004) (Proc. SPIE vol 5457)* ed W Osten and M Takeda (Bellingham: SPIE) pp 115-25

[3] Horber J K H and Miles M J 2003 *Science* **302** 1002

[4] Humphris A D L, Hobbs J K and Miles M J 2003 *Appl. Phys. Lett.* **83** 6

[5] Dubois A and Boccara AC 1999 *Opt. Lett.* **24** 5 309

[6] Olszak A 2000 *Appl. Opt.* **39** 22 3906-13

[7] Charrière F, Kühn J, Colomb T, Montfort F, Cuche E, Emery Y, Weible K, Marquet P and Depeursinge C 2006 *Appl. Opt.* **45** 5 829

[8] Grigg D, Felkel E, Roth J, Colonna de Lega X, Deck L and de Groot P 2004 *Proc. Int. Conf. on MEMS, MOEMS and Micromachining (Strasbourg, France, 29-30 April 2004) (Proc. SPIE vol 5455)* ed H Urey and A ElFatatry (Bellingham: SPIE) pp 429-35

[9] Beer S and Seitz P 2005 *Proc. Int. Conf. on PhD Research in Microelectronics and Electronics (Lausanne, Switzerland, 2005) (PRIME 2005, vol 1-2)* (New York: IEEE) pp 335-38

[10] Montgomery P C, Montaner D, Manzardo O, Flury M and Herzig H P 2004 *Thin Sol.Films* **450** 79

[11] Dubois A, Vabre L, Boccara AC, Montgomery P C, Cunin B, Reibel Y and Draman C 2002 *Eur. Phys. J. - Appl. Phys.* **20** 169

[12] Montgomery P C, Anstotz F, Johnson G and Kiefer R 2008 *Mat. Sci.: Mat. in Elec.* **19** Suppl. 1 194

[13] Petitgrand S 2005 *Méthodes de microscopie interférométrique 3D statiques et dynamiques, pour la caractérisation de la technologie et du comportement des microsystèmes*, Ph.D. thesis *(Institut d'Electronique Fondamentale, Orsay, France, 29 Sept. 2005)* (Orsay: Université Paris-Sud XI)

[14] Montgomery P C and Fillard J P 1993 *Proc. Int. Conf. on Interferometry: techniques and analysis (San Diego, USA, 20-21 July 1992) (Proc. SPIE vol 1755)* ed G Brown (Bellingham: SPIE) pp 12-23

[15] Larkin K G 1996 *J. Opt. Soc. America* **13** 832-43,

[16] Malinowski G, Hehn M, Kammerer J B, Sajieddine M, Jouguelet E, Hébrard L, Alnot P, Braun F and Schuhl A 2005 *Eur. Phys. J. – Appl. Phys.* **30** 113

# Phase-sensitive reflectometer using a single-frequency laser diode and an Er-doped fibre amplifier

**T Eftimov[1], I Stefanov[2], A Popov[3] and S Vodenicharov[3]**

[1]Plovdiv University, "P.Hilendarski", 24 Tsar Asen St., Plovdiv 400, Bulgaria
[2]Sofia University, "St. Kliment Ohridski", 5 James Bouchier, Sofia 1000, Bulgaria
[3]Instiute of Metal Sciences, Bulgarian Academy of Sciences, Sofia 1000, Bulgaria

E-mail: teftimov@uni-plovdiv.bg

**Abstract.** We report on the observation of phase-sensitive Rayleigh backscattering from a single-mode fiber excited by light pulses obtained from a highly coherent single-frequency laser diode and an Erbium doped fiber amplifier (EDFA). The laser diode stability was tested using a highly imbalanced fiber optic Mach-Zehnder interferometer. The CW laser light was first modulated using fiber-optic EO modulator which formed 100ns to 500 ns pulses that correspond to 20 to 100 m of pulse length in the fiber. The backscattered light power is estimated to be about -51 dB lower than the launched power at the input and about $N$=500 averages are needed for a sensing length of $L_s$ = 40 km to be in the useful dynamic range.

## 1. Introduction

Fiber optic distributed sensing has emerged as a highly perspective technology in a number of applications especially those related to perimeter intrusion sensing [1,2] and pipe-line monitoring [3]. The alternative of this approach is to use a combination of a Mach-Zehnder and a Sagnac fiber interferometer [4]. For similar applications a phase-sensitive optical time domain reflectormeter ($\phi$-OTDR) is of particular interest since it allows the detection of weak acoustic pressure waves by a sensing cable buried underground [1] or attached to a pipeline. A phase-sensitive OTDR is in essence a fiber optic interferometer in reflectometric arrangement and requires high power, single-frequency lasers and low-noise avalanche photodiodes (APD) in combination with signal averaging equipment.

In [1] a single-frequency fiber laser is used in combination with and EDFA as a high coherence high power source to observe phase-sensitive response while a combination of a single-frequency laser diode and an EDFA is reported in [2].

In this paper we estimate the required signal-to-noise ratio (SNR) and the usable dynamic range (UDF) needed for the proper functioning of $\phi$-OTDR and report on the observation of phase-sensitive Rayleigh backscattering from a single mode fiber using single frequency laser diodes and Erbium doped fiber amplifier.

## 2. Optical time domain reflectometry

### *2.1. Rayleigh scattering in optical fibres*

Optical fibres are made of fused silica ($SiO_2$) and by adding different doping materials a desired index profile can be obtained so that the fibre acquires the desired transmission characteristics. Since the material is amorphous, its local density fluctuates which causes loss along the lightguiding structure caused by Rayleigh scattering. The scattering coefficient $\alpha_s$ is given as:

$$\alpha_s - \frac{(2\pi)^3}{3}.\frac{1}{\lambda^4}.n^8 p^2 kT_f \beta \quad (1)$$

where $p$ is the photoelastic coefficient of glass, $k$ Boltzmann's constant and $\beta$ is the isothermal compressibility of the material, while $T_f$ is a fictive temperature, representing the temperature at which the density fluctuations are "frozen" in the glass material.

*2.1.1. Rayleigh back-scattering in single mode optical fibres.* In optical fiber reflectometers a series of light pulses are injected into the fibre, which in the ideal case are rectangular. If at the initial moment $t = 0$ a pulse of width $dt$ is injected into the fibre its length along the fiber will be $dz = v_g.dt$

As an approximate estimation of the group velocity a value of $v_g \approx 2.10^8$ m/s can be assumed which means that a pulse of a duration $dt$ = 1 ns = $10^{-9}$ s will span over $dz$ = 0.2 m along the fiber. The elementary backscattered optical power $P_{bs}$, obtained at the fiber input is given as [5]

$$dP_{bs} = 0.5P_0.\alpha_s.S.dz \quad (2)$$

where $P_0$ is the peak optical, $\alpha_s$ [1/km] is the attenuation caused by Rayleigh scattering (see (1), $S$ is a scattering coefficient. The total optical power is obtained after integration over the physical length of the light pulse which is twice the one-way length.

$$P_{bs} = \frac{1}{2}P_0.\alpha_s.S.\int_0^D \exp(-\alpha z)dz \quad (3)$$

In the above equation $\alpha$ [1/km] is the attenuation in the fiber, which for single-mode fibers is practically the same as the scattering coefficient $\alpha \approx \alpha_s$ over the 1.3 μm to 1.55 μm range. The solution of (3) can be written as $P_{bs} = 0.5P_0.\alpha_s.S.[1 - \exp(-\alpha D)]/\alpha$. In the above specified spectral region $a << D$ which is of the order of meters or tens of meters, so $D << 1/\alpha$. Expansion into series of the exponent in (5) ($e^x \approx 1+x$) yields:

$$P_{bs} = 0.5S.\alpha_s.D.P_0 = A.P_0 \quad (4)$$

which means that the back-scattered signal is proportional to the peak power $P_0$,, the scattering coefficient $\alpha_s$ and the pulse length $D$ and the scattering coefficient $S$ given as:

$$S = (NA/n)^2/q \quad (5)$$

For single mode communication fibers $n$ = 1.465-1.468, $NA$ = 0.13, $q$ = 4.55, so $S = 1.76\times10^{-3}$. For $\lambda$ = 1550 nm, $\alpha \approx$ 0,26 dB/km, $\alpha_s$ = 0,2 dB/km. Taking into account that

$$\alpha_{s\,[1/km]} = 0{,}23\ \alpha_{s\,[dB/km]} \quad (6)$$

we obtain for D/2 = 0.02 km (20 m) $A = S.\alpha_s.D/2$ the value $A = 1.619\times10^{-6}$, i.e. $A \approx 58dB$

Since the resolution of a reflectometric sensor is defined by the pulse length $\delta l = D/2$ in the fiber the value of the coefficient $A$ will have different values as summarized in Table 1.

**Table 1**. Attenutation coefficient $A$ [dB] for different values of the pulse duration $\delta\tau$ [ns] and the corresponding spatial resolution of the sensor $\delta l$ [m].

| δτ [ns] | δl [m] | A | A [dB] |
|---|---|---|---|
| 200 | 40 | 3,24E-06 | -54,9 |
| 300 | 60 | 4,86E-06 | -53,1 |
| 400 | 80 | 6,48E-06 | -51,9 |
| 500 | 100 | 8,10E-06 | -50,9 |

Thus for a 100 m spatial resolution, and a 1 mW of input peak power, the peak power of the back-scattered Rayleigh light will be 8 nW (for a 500 ns pulse duration); for 10 mW - 80 nW, while for a 25 mW peak input power there will be correspondingly 200 nW at the receiver.

*2.1.2. Phase sensitive OTDR.* In the traditional OTDR scheme, which is used to detect losses and reflections along a fibre communication line which is as long as possible, it is desirable to avoid noise due to laser's high coherence. With a coherent laser, noise is generated because of the interference of two consecutive backscattered pulses (1) and (2) which arise at some small perturbation $\delta n$ of the fibre refractive index induced, for example by an external pressure wave.

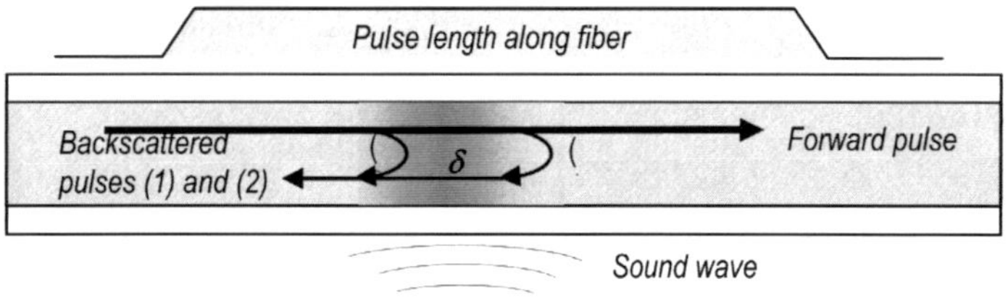


**Figure 1**. Illustration of the mechanism that generates coherent noise in the backscattered light.

The intensity of the superimposed waves with individual backscattered intensities $I_{bs.1}$ and $I_{bs.2}$ would be described by the well known interference equation

$$I_{bs} = I_{bs,1} + I_{bs,2} + 2\sqrt{I_{bs,1} I_{bs,2}} \cos\phi \tag{7}$$

where $\phi$ is the phase accumulated by wave (1) before it interferes with (2). So the intensity of the backscattered light will fluctuate in the case of a coherent source if the coherence length is $L_c > D$. To suppress this type of noise a lower coherence laser having a large number of longitudinal modes is used in conventional reflectometry.

In phase-sensitive OTDRs what is a noise in the conventional OTDR is a useful signal to detect small local modulations of the fibre refractive index caused by external fields such as sound pressure waves. Signals reflected at different consecutive moments will vary as shown in Figure 1 due to the phase changes $\phi(t)$ over time. Taking the phase difference $\Delta\phi(t)$ of two consecutive moments will produce a signal that will bear information on external influences:

$$\Delta\phi(t) = \phi(t) - \phi(t_0) \tag{8}$$

If at a given position the fibre has experienced an index perturbation $\delta n$ from an external cause, then a difference $\Delta I_{bs}$ of the backscattered intensities will be observed:

$$\Delta I_{bs}(t) = I_{bs}(t) - I_{bs}(t_0) \tag{9}$$

$$\Delta I_{bs}(t) = I_{bs,1}(t) - I_{bs,1}(t_0) + I_{bs,2}(t) - I_{bs,2}(t_0) + 2\left[\sqrt{I_{bs,1}(t) I_{bs,2}(t)} \cos\phi(t) - \sqrt{I_{bs,1}(t_0) I_{bs,2}(t_0)} \cos\phi(t_0)\right]$$

If the total backscattered intensities do not change considerably over time, i.e.

$I_{bs,1}(t) \approx I_{bs,1}(t_0) \approx I_{bs,1}$ and $I_{bs,2}(t) \approx I_{bs,2}(t_0) \approx I_{bs,2}$, then we obtain

$$\Delta I_{bs}(t) \approx I_{bs,1} + I_{bs,2} - 4\sqrt{I_{bs,1} I_{bs,2}} \sin\overline{\phi(t)}.\sin\left[\Delta\phi(t)/2\right] \tag{10}$$

with $\overline{\phi(t)} = \left[\phi(t) + \phi(t_0)\right]/2$.

Thus the differential signal is proportional to the phase difference between two sequential moments.

*2.2. Signal-to-noise ratio and Dynamic Range*

*2.2.1. Signal-to-Noise Ratio (SNR) and Dynamic Range (DR).* Since the received optical power is too low (Figure 2) and subject to fluctuations averaging over as large a number $N$ of pulses should be performed to increase the SNR defined as the ratio of the average optical power $\overline{P}$ to the noise level. Assuming the noise follows a Gaussian distribution of expectation μ and standard deviation $\sigma$,

$$f(x,\mu,\sigma)=\left[\sigma\sqrt{2\pi}\right]^{-1}\exp\left\{-\left[(x-\mu)/\sigma\right]^2/2\right\} \quad (11)$$

the SNR can be defined as

$$SNR=\overline{P}/\sigma \quad (12)$$

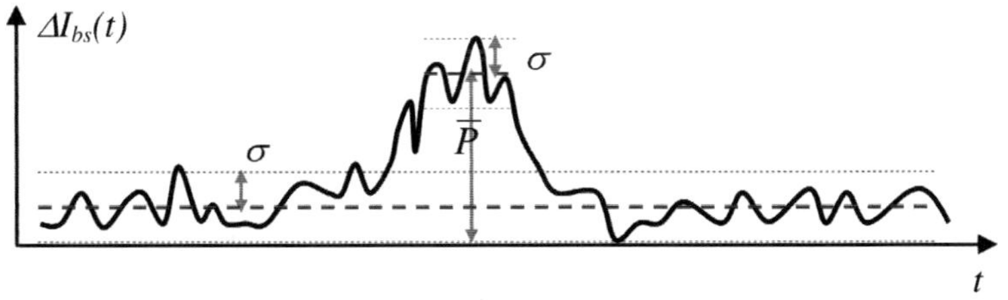


**Figure 2.** Illustration of a weak pulse buried in noise.

Averaging over $N$ pulses will yields a standard deviation of the sampled mean as:

$$\overline{\sigma}=\sigma/\sqrt{N} \quad (13)$$

SNR in the case of averaging over $N$ pulses will be

$$SNR_N=\overline{P}/\overline{\sigma}=\overline{P}\sqrt{N}/\sigma=\sqrt{N}.SNR \quad (14)$$

Since SNR increases, this leads to an increase of the dynamic range (DR) which is the ratio of the scattered signal to the noise level. If SNR is expressed in dB, then

$$DR_{dB}=SNR_{dB}=\frac{1}{2}.10\log SNR=5\log SNR \quad (15a)$$

From (17) the SNR in dB after averaging over N pulses we obtain:

$$SNR_{N,dB}=SNR_{dB}+5\log\sqrt{N}=SNR_{dB}+\Delta(DR)_{dB} \quad (15b)$$

The factor ½ in (15a) accounts for the fact that propagation is considered only in one direction – back to the input. For example averaging over $N$ =10 signals raises the SNR by 2.5 dB, that is 1.77 times, but averaging over $N$=100 signals increases SNR by 5 dB, i.e. 3.16 times.

*2.2.2. Usable dynamic range (UDR).* The usable dynamic range is the range within which measurements are done on signal that exceeds a definite level above noise. When peak-to-peak noise becomes close to the signal level obtained as the difference between two averaged signals of the phase-sensitive OTDR. This means SNR ≥ 1. Signals greater than this value are in the so called usable dynamic range (UDR). In analogy with conventional OTDRs, the UDR can be defined relative to a 0.2-dB losses as follows [5]:

$$0.2dB=5\log\frac{I_S+N}{I_S-N} \quad (16)$$

$I_S$ being the signal level, while $N$ is the noise level. The SNR is then obtained as:

$$SNR\approx 6{,}68dB \quad (17)$$

which means that the UDR is practically 7 dB above the level, below which are 98% of noises.

*2.2.3. Maximum number of averages $N_{max}$.* Since increasing the UDR needs averaging over N signals we need to evaluate what would be the maximum possible number $N_{max}$ of signal averages for a phase-sensitive OTDR.

Unlike conventional OTDRs in which millions of pulses can be averaged in the course of several minutes, phase-sensitive OTDR used as a distributed sensor has severe limitations coming from the integration time, sensing distance and the spatial resolution. To evaluate $N_{max}$ we must assess the averaging time $T_\Sigma$. If we set as $L_{crit}$ as the critical distance from the sensing cable to an intruder, approaching at a velocity $v$, the time period over which the critical distance would be covered is $L_{crit}/v$ which we set as the time $T_\Sigma$ available for integration. So we let

$$T_\Sigma = L_{crit} / v \tag{18}$$

Should we accept $L_{crit}$ as 10 m, which is of the order of the detectable signals [1], and the indruder's speed within the limits of $v_{min}$ = 3.6 km/h = 1m/s (calm walk) and $v_{max}$ = 72 km/h = 20m/s (car speed), the integration time would correspondingly vary from $T_{\Sigma\min}$ = 0.5 s to $T_{\Sigma\text{Max}}$ = 10 $s$. Reducing $L_{crit}$ to 2 m would then yield $T_{\Sigma\min}$ = 0.1 s to $T_{\Sigma\max}$ = 2 $s$. Thus for 2m ≤ $L_{crit}$ ≤ 10 m and 1m/s ≤ $v$ ≤ 20 m/s, the integration time would vary over two orders of magnitude and be 0.1s ≤ $T_\Sigma$ ≤ 10s.

Denoting by $L_{sens}$ the sensing length of the fiber and $v_g$ the group velocity of light along the fiber, the round-trip time $T$ for the signal is

$$T = 2L_s / v_g \tag{19}$$

And for $L_s$ = 1 km, then $T$ = 2 μs. So the maximum possible number of averages can be determined from the ratio of the averaging time $T_\Sigma$ to the round trip time which is the pulse repetition period with a duration of $\delta\tau < T$.

$$N_{\max} = T_\Sigma / T \tag{21}$$

For δτ = 500 ns with $\delta l$ = 100 m, we obtain the following possible averaging numbers and the corresponding addition to the dynamic range $\Delta(DR)_{dB}$. Figure 3 shows that if the signal level is close to the noise level (SNR = 1) then the maximum possible sensor length (no other restriction factors excluded) is about $L_s$ = 40 km so that it can be within the usable dynamic range. This length is satisfactory for distributed optical fiber intrusion sensor. Other limiting factors are input optical signal level, ADC response time, APD sensitivity. The graph in Figure 3 shows that we need at least 500 averages to be in the usable dynamic range

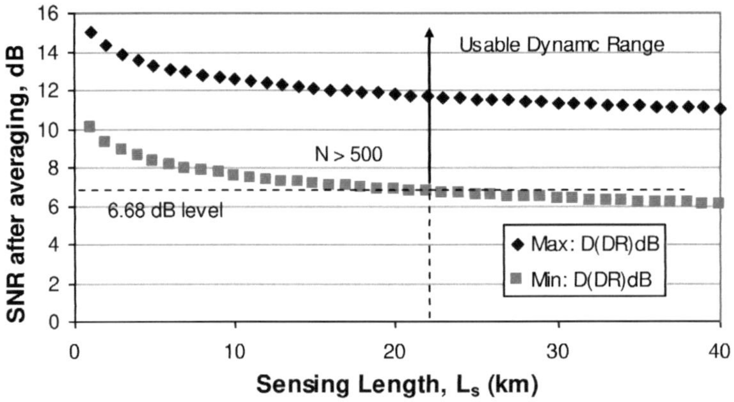


**Figure 3**. Dependence on the fiber length of the minimum and maximum SNR after averaging

## 3. Experimental scheme and results

### *3.1. Experimental scheme*

So far we have considered only the issue of SNR and the length restriction caused by the maximum possible number of averages. It follows that to obtain a satisfactory signal level of 100 nW or above

for a sensing length of $L_s \leq 40$ km light pulses of duration $\delta\tau = 500$ ns should be generated to obtain a resolution of $\delta l = 100$ m. We can have a SNR within the UDR if pulses of tens of nW can be detected. The input peak optical power in this case must be greater than 10 mW.
However, to ensure high coherence a highly stable single-frequency laser diode must be used. During operation the center wavelength of the laser must be stable within tens of kHz/s drifts. Since direct modulation causes chirp, external modulation must be applied. External electro-optical modulators typically introduce additional losses of up to 6 dB, which would reduce the input power level to unacceptably low levels. Therefore optical amplifiers should be used at the fiber input.

The basic experimental set up is shown in Figure 4. A single-frequency thermally stabilized laser diode with a $\Delta\nu = 20$ kHz bandwidth was used as highly coherent source. The output CW power at 1550 nm provided is 10 mW. Prior to tests the stability of the laser was tested using a highly imbalanced fiber Mach-Zehnder interferometer with a fiber length in the signal arm of 9.7 km. The results obtained showed that a stability was better than 24 kHz/s.

A 10 GHz in-line fiber optic modulator (Photline Technologies) was used as an external electro-optic modulator. (EOM). Since modulator performance is polarization-dependent, two in-line fiber polarizers were used before and after the EOM. An Er-doped fiber amplifier (EDFA) providing 14 dBm (25 mW) output optical power that was launched into port (1) of an optical circulator and through port (2) into the test fiber with $L_s = 3$ km. An avalanche photodiode model HCA-S-200M-IN (Femto) was used and responses were observed using a digital scope.

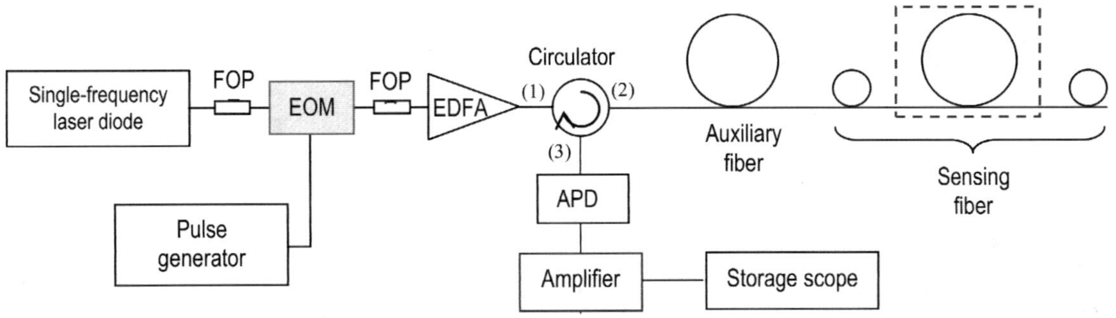


**Figure 4**. Experimental set-up.

*3.2. Obtained results*
We have varied the pulse duration from 100 ns to 500 ns and have observed the phase sensitive backscattered light on the oscilloscope. Figure 5 a-f) shows several responses for different sensitivities and number of averages. As we see on increasing the number of averages the noise level significantly reduces and at $N = 16$ (Figure 5e) we can see the back reflection from the end of fiber at a 15 μs time delay which corresponds to the 3 km fiber length. Even with this level of peak input optical power (25 mW), it was still insufficient to reduce noise below the level so as to enter in the usable dynamic range. As calculated in the first part to obtain a good performance responses, at least $N = 500$ signals must be averaged, for which purpose an ADC and a signal averaging scheme is under development.

## 4. Conclusions
We have estimated the signal averaging requirements for the normal functioning of a phase-sensitive OTDR and have considered intruder to fiber distances ranging between 2 m to 10 m and approaching

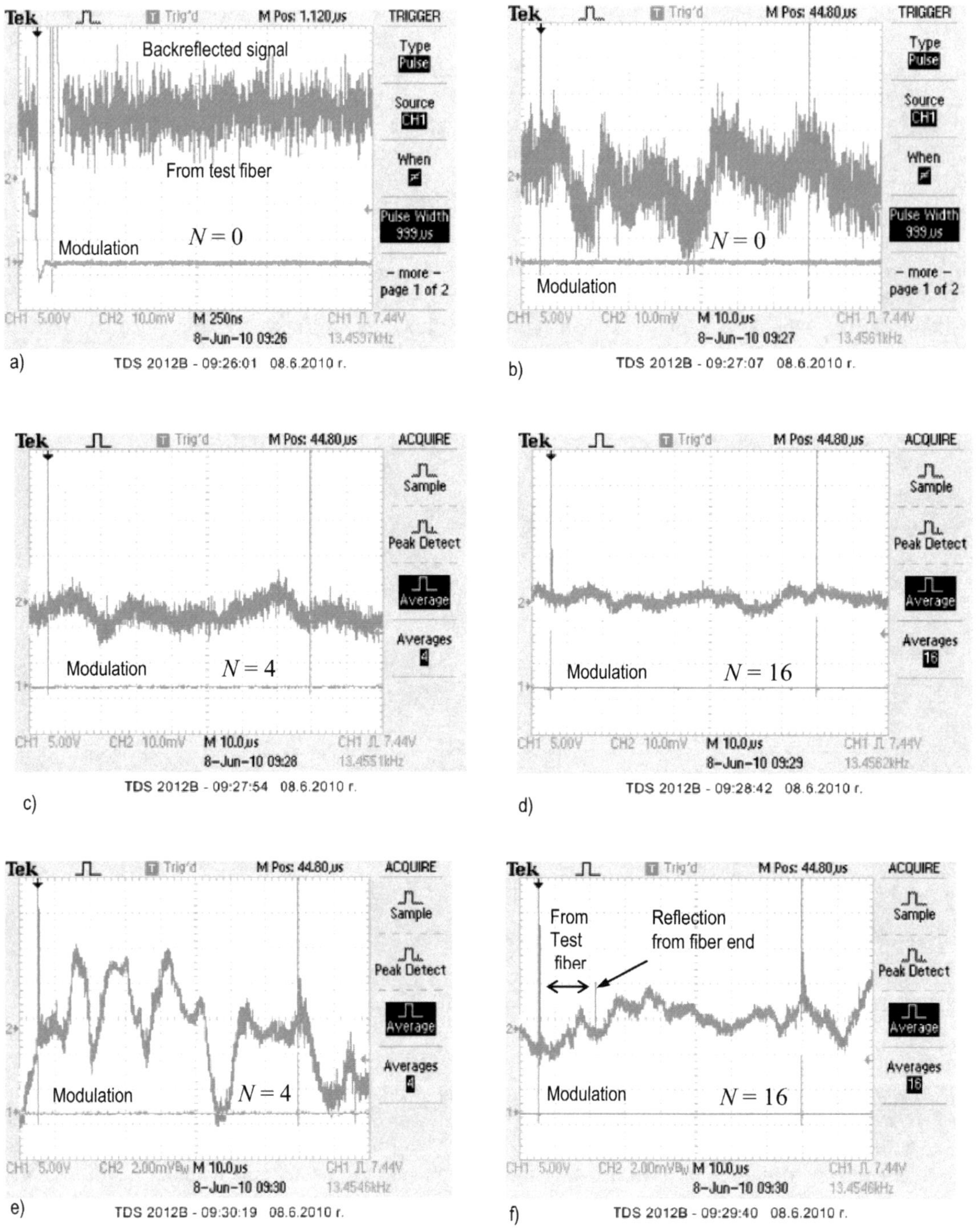


**Figure 5**. Observed responses: a) Pulses injected of duration 100 ns at 250 ns resolution and the backscattered light; b) same as a) but at 10 μs resolution and no averaging observed Rayleigh backscattered response with N = 16 averages; c) same as *b)* but with $N = 4$; d) same as c) but with $N = 16$; e) and f) same as c) and d) but with 2 mV/div vertical scale.

speed between 1 m/s and 20 m/s and have estimated the minimum number of averages as close to 500 so as to be in the usable dynamic range. We have also tested an optical scheme using a highly coherent single-frequency laser and an EDFA providing a 25 mW peak output power in combination with an APD detector with which we have observed the backscattered phase-sensitive response.

**References**

[1] Juarez J C, Maier E W, Choi K N and Taylor H F, 2005 *J. Lightwave Technol.* **23** 2081
[2] Choi K N, Juarez J C and Taylor H F 2003 *Proceedings of SPIE* **5090** 134
[3] Gorshkov B G, Paramonov V M, Kurkov A S, Kulakov A T and Zazirnyi M V 2006 *Quantum Electronics* **36** 963
[4] Zyczkowski M, Ciurapinski W, Kondra M, 2005 *Proceedings of SPIE* **5952** 59521E
[5] Anderson D, Johnson L and Bell F 2004 *Troubleshooting Optical-Fiber Networks* (London: Elsevier)

Journal of Physics: Conference Series **253** (2010) 012019 doi:10.1088/1742-6596/253/1/012019

# Vacancy-impurity nanoclusters in solid solutions $^3$He – $^4$He

**Dimitar I Pushkarov**
Institute of Solid State Physics Bulgarian Academy of Sciences, 1784 Sofia, Bulgaria
E-mail: Dimitar.Pushkarov@gmail.com

**Abstract.** Structural effects streaming from the quantum nature of the vacancies in solid helium are considered. As known, the point defects in quantum crystals are delocallized and turn into quasiparticles – defectons (vacancions, impuritons etc). An important feature is that the bottom of the quasiparticle energy band lies lower compared to the activation energy of a localized defect – as a rule by a half band width. However, the delocalization requires good periodicity and identity of the atoms around the defect. This may result both in repulsion between a vacancy and impurities, and in attraction between them. This leads to different effects – removal of impurities, new type of vacancy-impurity nanoclusters, creation of nuclei of new phases,vacancy induced local phase transition, change in the phase separation line in solid solutions. Experimental results are analyzed.

## 1. Introduction

The main property of the quantum crystals is the relatively large amplitude of zero-point vibrations of their atoms. This means that the wave functions of nearest atoms overlap and a nonzero probability of tunnel transitions appears. So, a new type of motion appears, and new carriers of this kind of motion arise. These are the new quasiparticles (defectons, impuritons, vacancions, interstitials, crowdions etc.). They appear due to a plain physical reason. The state of an ideal crystal with a localized defect cannot serve as a ground state because of the degeneracy with respect to the defect position (all lattice sites are equivalent). The bottom of the energy band of the delocalized defect lies lower than the activation energy of a localized defect. In most cases the lowering is by a half bandwidth, $\Delta/2$. In helium crystals with largest molar volumes the vacancion bandwidth $\Delta$ is of the order of several degrees and may even be comparable with the activation energy of an isolated vacancy. Therefore, the energy reduction is sufficient. In contrast, the bandwidth, e.g. of $^3$He impuritons in h.c.p. $^4$He is $10^4$ times less [1]. The situation described is closely dependent on the identity of lattice cites. This means not only that the geometrical structure must be well periodic, but also that the corresponding lattice sites are occupied by *identical particles in identical states.* For example, a $^3$He atom delocalizes in $^4$He matrix, but a $^4$He (or a vacancy) can delocalize in $^3$He matrix only if the spins of all $^3$He are parallel. Otherwise, the localized vacancy will try to orient the nearest spins. Andreev [4] has supposed that a vacancy should be surrounded by such magnetic region and turn into "ferrovacancy". All attempts to observe this structure have failed (we shall discuss the reason later on, see [10] as well). We shall not discuss these works here in detail because the magnetic effect mentioned has not been confirmed yet and concerns pure helium, not solid mixtures. The point is that the vacancy behavior is different in $^4$He and $^3$He crystals.

Journal of Physics: Conference Series **253** (2010) 012019 doi:10.1088/1742-6596/253/1/012019

Another kind of effects of vacancions on the crystal lattice structure in solid solutions were predicted by this author [7, 8, 9, 10]. It was shown that in the presence of impurities, a vacancy cleans the surrounding region. We called this *removal effect* or *rid effect* [7]. Another effect is the *vacancy induced local phase transition* (see below). More complicated effects were discussed in [10].

Another important phenomena is the phase separation of solid solutions $^3$He – $^4$He. The nature of this phenomena is still not well understood. Of special interest for us will be solutions with small concentration of $^4$He.

At first glance, a delocalized vacancy should have a ferromagnetic structure proposed by Andreev [4]. We show, however, that vacancies 'prefer' to attract $^4$He atoms and create vacancy-impurity clusters (VIC) instead orienting spins of the nearest host atoms, $^3$He.

The paper is organized as follows. In section 2 we give some information on quantum vacancies in helium crystals necessary for further understanding, in section 3 we consider effects in a rare solid solutions *far from the phase separation transition* (the removal effect) and an effect on impuriton diffusion, in section 4 a simple model for vacancy-impurity clusters in phase separated mixtures is considered, and a comparison with the experimental data is made, in section 5 a semi-quantitative analysis is given and another effect, *the local vacancy stimulated phase transition*, is considered. Results are discussed in Conclusion.

## 2. Vacancion spectrum in b.c.c. and h.c.p. $^4$He

The defecton bandwidth $\Delta$ is proportional to the tunnel probability amplitude $A$ (called sometimes exchange integral $J$; we prefer notation $A$ because of some uncertainty in defining $J$ in the literature). In simple lattices $\Delta = zA$ ($z$ being the number of the nearest neighbors in the lattice). The bandwidth for $^3$He impuritons in h.c.p. $^3$He as determined from the diffusion experimental data is $\Delta_{imp} \approx 10^{-4}\,\mathrm{K}$ [1, 11]. Unfortunately, there is still no reliable experimental data for vacancions. First theoretical estimations [2, 3, 12] gave $A \approx 1\,\mathrm{K}$ and this value has been used till now as a good approximation. The vacancion dispersion law for b.c.c. $^4$He has the form:

$$\varepsilon_b(\mathbf{k}) = \Phi + 4A_b(\cos\frac{k_x a}{2}\cos\frac{k_y a}{2}\cos\frac{k_z a}{2}) \tag{1}$$

where $\Phi$ is the activation energy of a localized vacancy, and k is the quasiwavevector. The bandwidth is therefore $\Delta_{bcc} = 8A_{bcc}$.

The dispersion relation for h.c.p. lattice is more complicated [13, 12]:

$$\varepsilon(\mathbf{k})_{\frac{ac}{opt}} = \varepsilon_0 + A\left\{7 - 4\cos\frac{\mathbf{k_1}.\mathbf{a_1}}{2}\cos\frac{\mathbf{k_2}.\mathbf{a_2}}{2}\cos\frac{\mathbf{k_3}.\mathbf{a_3}}{2}\right.$$
$$\left.\mp\cos\frac{\mathbf{k}.\mathbf{c}}{2}\left[1 + 8\cos\frac{\mathbf{k_1}.\mathbf{a_1}}{2}\cos\frac{\mathbf{k_2}.\mathbf{a_2}}{2}\cos\frac{\mathbf{k_3}.\mathbf{a_3}}{2}\right]^{1/2}\right\} \tag{2}$$

where

$$\mathbf{a_1} = a\,[1,0,0]\,, \quad \mathbf{a}_{2,3} = a\left[-\frac{1}{2}, \pm\frac{\sqrt{3}}{2}, 0\right], \quad = \mathbf{c} = a\left[0, 0, \frac{\sqrt{8}}{3}\right] \tag{3}$$

are the basic lattice vectors. This spectrum consists of two branches called conditionally *acoustic* and *optical.* The dispersion law (2) is written in a form which fixes the bottom of the acoustic band at $k = 0$, so that $\varepsilon(0) = \varepsilon_0$ is its lowest level. The bandwidth of the acoustic branch is $\Delta_{ac} = 7A$ while that of the optical one is $\Delta_{opt} = 5A$. The bottom of the optical branch lies higher by $3A$. Hence, the total width of the two partially overlapping branches is $\delta_h = 8A$. The form of the energy surfaces $\varepsilon = \varepsilon(\mathbf{k})$ at different fixed values of $k_z$ for h.c.p. structure are given on figures. 1, 2, 3, 4.

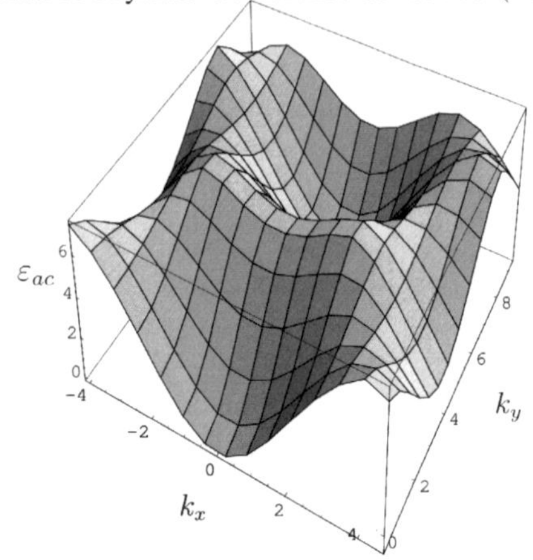


**Figure 1.** The acoustic branch of the vacancion dispersion law $\varepsilon(\mathbf{k})$ at constant $k_z = 0$ for h.c.p. lattice.

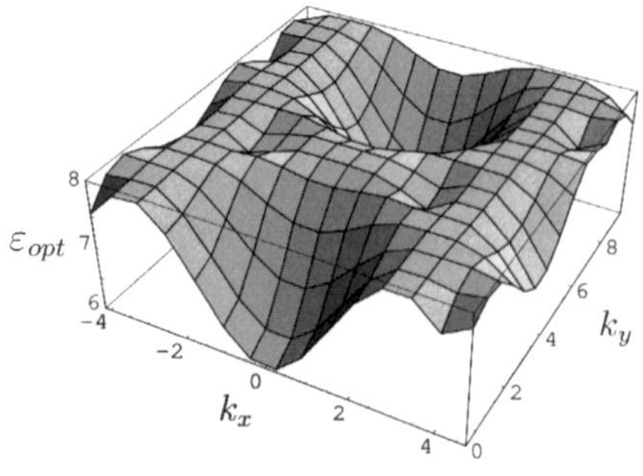


**Figure 2.** The optical branch of the vacancion dispersion law $\varepsilon(\mathbf{k})$ in (2) at constant $k_z = 0$ for h.c.p. lattice.

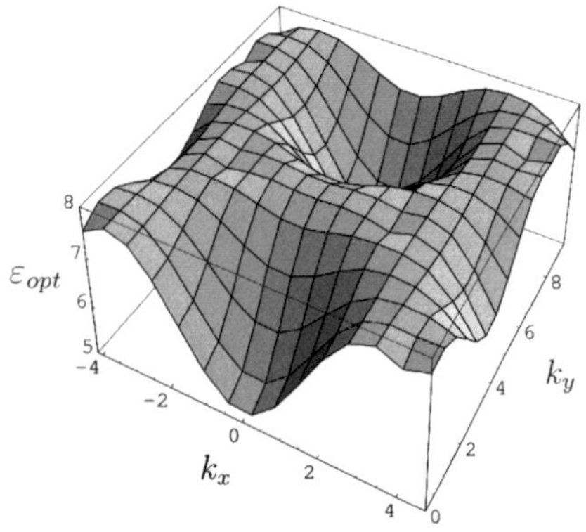


**Figure 3.** The optical branch of the vacancion dispersion law $\varepsilon(\mathbf{k})$ at constant $k_z = \pi/2$ for h.c.p. lattice.

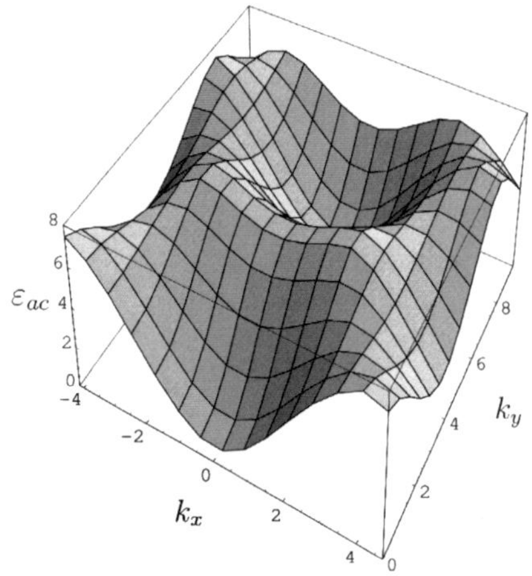


**Figure 4.** The acoustic branch of the vacancion dispersion law $\varepsilon(\mathbf{k})$ at constant $k_z = \pi/2$ for h.c.p. lattice.

Since we consider low temperatures typical of the phase separation quasiparticles from the acoustic branch can only participate. The corresponding effective mass is

$$m_{hcp} = \hbar^2/(2Aa^2) \tag{4}$$

while for the b.c.c. case one has

$$m_{bcc} = \hbar^2/(A_{bcc}a^2). \tag{5}$$

As to the activation energy $\varepsilon_0$ its value varies sufficiently (see e.g. [14]) since there are no direct measurements. For h.c.p. $^4$He of largest molar volumes a more or less reliable value is 6 K.

The tunneling amplitude as well as the vacancion bandwidth are larger in the b.c.c. phase. This may have effect on the b.c.c.–h.c.p. phase transition [7, 8, 9, 10]. Let us consider pure $^4$He near the b.c.c.–h.c.p. phase transition line. The free energy in the two phases has one and the same value. If a vacancy appears in h.c.p. phase the energy increases by $\varepsilon_0$. However, the energy of the same vacancy in b.c.c. phase is lower because of the larger bandwidth and hence, larger

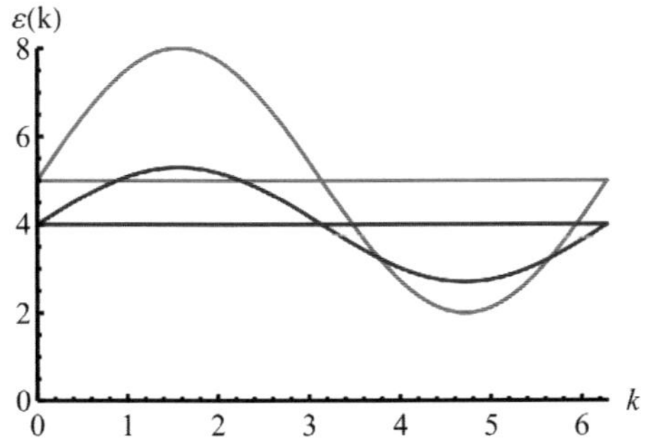


**Figure 5.** Even if the activation energy $\phi$ is larger, the bottom of the vacancion band is lower.

reduction of the activation energy (see figure 5). As a result, a cluster with b.c.c. structure can appear in the h.c.p. phase (local phase stratification) [7]. This reflects in a hysteresis in the phase transition and other properties when crossing the phase separation line [15]. We shall turn to this effect later on.

## 3. Quantum vacancy far from phase separation

In this section we consider a simple model of a quantum vacancy appropriate for $^4$He with a small concentration $x_3$ of $^3$He impurities. The presence of impurities breaks the lattice periodicity, so the vacancy should clean the surrounding region from them. This results in its localization in a region of radius $R$ (measured in interatomic distances $a_0 = (V_m/N_A)^{1/3}$, $V_m$ and $N_A$ being the molar volume and the Avogadro number, respectively). The lowest energy $E_0$ of a vacancion in such a region can be estimated by means of the Heisenberg uncertainty principle. This yields, having in mind relations (5):

$$E_0 = \pi^2\hbar^2/mR^2 = \pi^2 A/R^2. \tag{6}$$

The removal reduces the entropy by an amount of $4/3\pi R^3 S$ with $S = x\ln(e/x_3)$ being entropy per unit lattice site. Hence, the change of the free energy is

$$F = E - TS = \Phi - \Delta/2 + \pi^2 A/R^2 + \frac{4}{3}\pi R^3 TS. \tag{7}$$

This expression has minimum at

$$R = \left(\frac{\pi A}{2TS}\right)^{1/5}. \tag{8}$$

The volume of the cleaned region is hence

$$V_0 = \frac{4}{3}\pi\left(\frac{\pi A}{2TS}\right)^{3/5}. \tag{9}$$

This result is restricted by the requirement that the free energy be not lower than the bottom of the vacancion zone, $\varepsilon_0 = \Phi - \Delta/2$, as well as by the obvious condition $R > a$. The latter is stronger and reads $TS < A$. In addition, the energy of the first quantum level (6) must be less than $\Delta/2$. The minimal value of $F$ is

$$F_{min} = \Phi - \frac{\Delta}{2} + \frac{5}{2}TSV_0. \tag{10}$$

Hence, the variation of $F$ depends linearly on the number of the lattice sites, $V_0$, in the cleaned region. This number is limited by the condition $5TSV_0 < \Delta$.

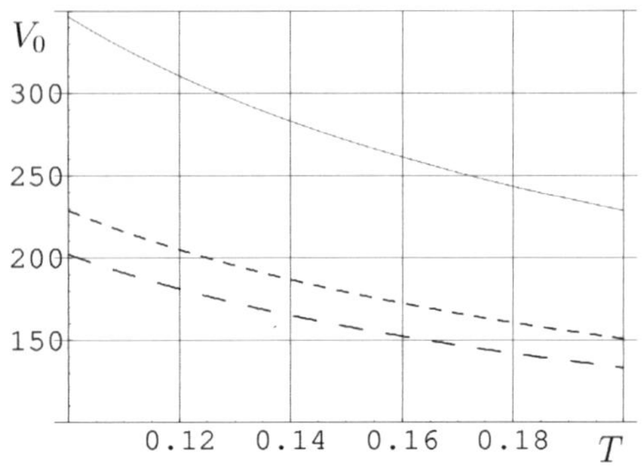


**Figure 6.** The number of atoms $V_0$ versus temperature at different concentrations. Solid line: $x' = 1\,\%$ in (21), dashed line: $x' = 2\,\%$ in (21), long-dashed line: $x' = 2\,\%$ in (9).

The above consideration does not take into account the change of the entropy outside the cleaned region due to the change of impurity concentration. A straightforward analysis [8, 10] yields

$$R = \left(\frac{\pi A}{2Tx_3'}\right)^{1/5}, \quad V = \frac{4}{3}\pi\left(\frac{\pi A}{2Tx_3'}\right)^{3/5} \tag{11}$$

where $x'$ is the impurity concentration in the solution after cleaning. The free energy $F_{min}$ becomes then

$$F_{min} = \Phi - \frac{\Delta}{2} + \frac{5}{2}Tx_3'V = \Phi - \frac{\Delta}{2} + \frac{5}{2}Tn_x \tag{12}$$

and one sees that the free energy depends on the number of impurities removed by one vacancy, $n_x = x_3'V$. This results also in a decreasing of the equilibrium vacancy concentration, $x_v$, compared to its value in an impurity free crystal, $\tilde{x}_v$:

$$\begin{aligned} x_v &= \exp\left\{-\frac{\Phi - \Delta/2}{T} - \frac{10\pi}{3}\left(\frac{\pi A}{2T}\right)^{3/5} x_3^{2/5}\right\} \\ &= \tilde{x}_v \exp\left\{-\frac{5}{2}n_x\right\}. \end{aligned} \tag{13}$$

In fact, the number of the removed impurities by a vacancy is $n_x \sim 2 \div 5$ and the change of the concentration $\delta x_3/x_3$ is of the order of several percents. However, this can lead to a decreasing of the equilibrium vacancy concentration by more than 2 orders of magnitude. For typical values $A = 1\,\mathrm{K}$, $T = 0.1\,\mathrm{K}$ and $x_3 = 1\%$, one finds from (9) $V = 346$ lattice sites. The dependence of the number of atoms in the cleaned region on temperature (at constant $x'$) corresponding to formulas (9) and (21) are shown in figure 6.

Let us note that the cleaned regions are not clusters. They consist of solvent atoms and do not have well expressed boundaries, surface energy etc. They have, however effect on the kinetic properties. The removal effect can be considered as an effective repulsion which prevents impurities and vacancies to be closer to each other. The impuriton energy band is very narrow ($\sim 10^{-4}\,\mathrm{K}$) and impuritons are scattered even on a small barriers. The corresponding cross-sections are very large [2, 1, 11]. A rough estimation can be made using $R$ as a measure of the scattering distance. Then, the cross-section is of the order of $\sigma \sim \pi R^2 \sim \pi(\pi A/2Tx_3)^{2/5} \sim 60\,a^2$ (at $T = 100\,\mathrm{mK}$). In fact, the cross-section can be smaller, because it depends on the way the wave function vanishes at distance $R$ and this asymptotic can be smooth. The diffusion coefficient, however, has a specific dependence on temperature and concentration:

$$D = D_0 \frac{T^{2/5}}{x_3^{3/5}}. \tag{14}$$

If such dependencies could be confirmed experimentally this will be an evidence for a new mechanism of impuriton-vacancion scattering.

Concluding this section let us emphasis once more that all the consideration was made for concentrations far from the saturation concentration $x_s$. If $x' > x_s$ then a new phase of impurities appears. The effect of vacancies will be analyzed in the next section.

## 4. Vacancy-impurity clusters

### *4.1. A simple model*

Let us consider a solid solution of $^4$He impurities with a concentration $x = n/N$ where $n$ is the number of impurities, and $N$ is the number of solvent ($^3$He) atoms. If $x < x_s$ the Gibbs potential $G$ may be written in the form [20]

$$G_1 = N\mu_0 + nT\ln\frac{n}{Ne} + n\psi \tag{15}$$

where $\mu_0$ is the chemical potential of the pure solvent, and $\psi = \psi(P,T)$. The chemical potentials $\mu_3$ and $\mu_4$ of the solvent atoms and impurities are derivatives of $G$ with respect to $N$ and $n$:

$$\mu_3 = \mu_0 - Tx, \quad \mu_4 = T\ln x + \psi. \tag{16}$$

If one moves $\delta n$ impurities from the solution to their "pure" condensed phase, the change of the Gibbs potential is

$$\Delta G = -\delta n(T\ln x + \psi) + \delta n\mu_4^0. \tag{17}$$

The chemical potential $\mu_4^0$ of the pure phase corresponds in equilibrium to a saturation concentration $x_s$ (otherwise this phase will dissolve). Hence, $\mu_4^0 = T\ln x_s + \psi$, and

$$\Delta G = -\delta nT\ln\frac{x}{x_s} = -NT\delta x\ln\frac{x}{x_s}, \quad \delta x = \delta n/N. \tag{18}$$

Therefore, the entropy difference per lattice site is

$$\Delta S = -\delta x\ln\frac{x}{x_s}. \tag{19}$$

Suppose that all $\delta n$ impurities are confined to $n_v = x_v N$ vacancies. Then $\delta n = Nx_vV$ where $V = 4/3\pi R^3$ is the vacancy cluster volume (the number of impurities around one single vacancy) and $\Delta S = -4/3\pi R^3 x_v \ln\frac{x}{x_s}$. With this expression for the entropy the free energy (c.f. (7)) takes the form

$$F = x_v\left(\Phi - \frac{\Delta}{2} + \pi^2\frac{A}{R^2}\right) + \frac{4}{3}\pi R^3 x_v T|\ln\frac{x}{x_s}| \tag{20}$$

and reaches its minimum at

$$R = \left(\frac{\pi A}{2T|\ln(x/x_s)|}\right)^{1/5}, \quad V = \frac{4}{3}\pi R^3. \tag{21}$$

Equation (21) is restricted by the conditions $T|\ln(x/x_s)| < A$ and $x \neq x_s$. When the concentration $x$ approaches the saturation concentration $x_s$ the entropy term vanishes and *the vacancy is fully delocalized inside the impurity phase.* This effect could be observed as a rapid increase of its mobility and growth of $^4$He domains. In terms of a phase transition from localized to delocalized motion, the domain size $R$ may be considered as an order parameter. The phase transition is continuous and may find analogy with the observed transitions in [16, 14] as well as to the revealed fast travelling gap nonphonon excitations.

### *4.2. Experimental Evidence*

Although the vacancy-impurity clusters were predicted by this author already in 1978 [7] they were dicovered experimentally by the Kharkov group only in 2001 [17] (see also [18, 19, 21]) when investigating phase separation in solid helium mixtures by precise pressure measurements. In their experiments Ganshin et al. [17] found an essential difference between the pressure behavior upon temperature cycling in phase separated $^4$He-$^3$He solid solutions in the two limiting cases: i) small concentration of $^4$He, and ii) small concentration of $^3$He. In the latter case the appearance of solid domains of $^3$He was accompanied with pressure lowering. Results were reproducible with temperature cycling. If the cycling starts from a solution with liquid $^3$He inclusions, then the pressure increases monotonically with the cycle number.

In contrast, the pressure amplitude at temperature cycling in the case of $^3$He with $^4$He inclusions decreases, and after several cycles stabilizes with an amplitude approximately 3 times less than the initial one. The initial pressure amplitude was restored only after heating the system to a temperature appreciably higher (by several tens mK) than the phase-separation temperature. The qualitative explanation was given in the frame of the above theory. The first cooling starts separating and some $^4$He rich domains appear. This results in increasing pressure. On subsequent heating these domains dissolve and the pressure rapidly decreases initiating creation of nonequilibrium vacancies. If a vacancy appears inside or on the boundary of the domain, a vacancy-impurity cluster arises. This cluster is more stable and survives (if temperature is not much higher the separation temperature, $T_s$) in contrast to the usual domains which dissolve. The dissolution process is stimulated in addition by the lowering of the concentration of $^4$He atoms in the solution due to the capturing atoms in VIC. After several cycles almost all impurities are sucked out. In order to restore the initial situation the clusters must dissolve at $T > T_s$.

## 5. Quantitative analysis

The simple model of VIC described in section II gives a correct order of magnitude for the number of atoms involved in a cluster. Equation (21) can be simplified for rough estimations. If the initial concentration is $x_0$, then the actual concentration is $x = x_0 - x_v - n_s/N \approx x_s - x_v V$ where $n_s$ is the number of atoms separated in the pure phase and the small concentration of solvent atoms, $^3$He in the pure phase $^4$He is neglected. The initial concentration drops as it should be in a case of a phase separated mixture in equilibrium. Then

$$|\ln(\frac{x}{x_s})| = |\ln\left(1 - \frac{x_v}{x_s}\frac{4}{3}\pi R^3\right)| \approx \frac{x_v}{x_s}\frac{4}{3}\pi R^3,$$

and (21) takes the form

$$R = \left(\frac{3A}{8T}\frac{x_s}{x_v}\right)^{1/8}. \tag{22}$$

For $T = 150\,\text{mK}$, $x_v = 10^{-5}$, $x_s = 5\times10^{-3}$ one has $R \approx 2.5$ and $V \approx 60$ $(x_v V/x_s = 0.12 \ll 1)$. Unfortunately this approximation requires $x_v V \ll x_s$ and is not applicable to the most interesting low temperature case where $x_s$ can be smaller than $x_v V$.

The main shortcoming in the derivation of (21) is that it does not take into account the surface energy. There are at least three problems with its introducing into consideration: i) VIC can appear both inside an existing domain, and as a separate inclusion, ii) the small size of the VIC makes its definition not clear enough, and iii) the structure of the VIC is not known. A quite sophisticated analysis was made in [21]. The authors supposed that the cluster has h.c.p. structure, and that the number of clusters equals the number of vacancies. They found for the cluster radius in the temperature region considered $R \approx (2,1 \div 4.4)\,a$. These values may have slightly changed because in the evaluation of the vacancion energy inside the cluster. The authors used the $E_0 = \pi^2\Delta/R^2$ instead of $\pi^2 A/R^2$ (see (6)). As a result, the first quantum level $E_0$ turns out to be out of the potential well (with depth $\Delta/2$), i.e. $\pi^2\Delta/R^2 > \Delta/2$ for all values of $R$ obtained and arbitrary $\Delta$. As for the surface energy and vacancy concentration fitted with $\Delta = 4\,\mathrm{K}$, the best fit has been obtained for a surface tension $\sigma = 1,43 \times 10^{-2}\,\mathrm{erg/cm^2} \approx 2 \times 10^{-2}\,\mathrm{K/Å^2}$ and $x_v = 8 \times 10^{-5}$.

All this consideration was made assuming that VICs have h.c.p. structure while the solvent is in b.c.c. phase. However, as shown in [12] the bandwidth for a vacancion in b.c.c. $^4$He is larger than that in the h.c.p. phase. In addition, the activation energy for a localized vacancy in b.c.c. lattice is about a half of the activation energy in the h.c.p. one.[24] Hence, a vacancy can gain more energy for delocalization if the cluster undergos a phase transition from h.c.p. to b.c.c. [7]. The cubic lattice of VIC is more favorable also because it coincides with the solvent structure with the corresponding reduction of the surface energy. An estimation in [17] shows a reduction more than two orders of magnitude: $\sigma < 10^{-4}\,\mathrm{erg/cm^2} \approx 10^{-3}\,\mathrm{K/particle}$. With neglecting surface energy and using the same procedure as in section II one obtains easily [7] for the number of particles in the VIC:

$$V = \frac{4}{3}\pi \left(\frac{\pi A}{2q}\right)^{3/5} \qquad (23)$$

where $q = T\delta S$ is the latent heat per particle, and $\delta S$ is the entropy change at the transition. Extrapolating the data from [22, 23] one finds $\delta S \approx 10^{-2}$, and $q \approx 10^{-3}\,\mathrm{K}$. This yields for the radius of VIC $R \approx 4,4\,a$ and for the total number of atoms in a cluster $V \approx 350$. Let us note that the local phase transition into b.c.c. phase will appear even if the free energy is lower in the h.c.p. phase because the bottom of the vacancy band is lower in b.c.c. [7]. In our opinion these were the clusters observed by the Kharkov group.

## 6. Conclusion

We considered in this work several effects arising from the quantum nature of vacancies in $^4$He – $^3$He solid solutions. For small concentration of $^3$He impurities in $^4$He quantum effects lead to an effective repulsion between vacancies and impurities which may express itself in cleaning the surrounding region from impurities, as well in significant disturbing the impurity diffusion. Near the phase separation line in $^4$He – $^3$He mixtures with a small fraction of $^4$He atoms vacancies form clusters. Unlike the intuitive feeling that a vacancy will delocalize creating a well ordered ferromagnetic region of $^3$He atoms with aligned spins, it prefers to create a cluster (VIC) built from impurities ($^4$He). The VIC formation undergoes a local phase transition to b.c.c. structure favorable due to the larger vacancion bandwidth and smaller surface tension. This makes the cluster even more stable. The evaluated cluster size is in good agreement with the experimental observations. On the phase separation line the cluster size tends to infinity, vacancies become fully delocalized and can move fast inside the $^4$He separated phase. The effect of vacancy assisted local phase stratification should be observable and can be useful for explaining hysteresis effects in phase transitions between different solid phases as well as at liquid-solid ones.

## Acknowledgments

The author acknowledges the financial support received from the Bulgarian National Science Fund, Contract F-1517.

## References

[1] Pushkarov D I 1974 *JETP Lett.* **19** 386
[2] Pushkarov D I 1970 *Zh.Eksp.Teor.Fiz.* **59** 1755

[3] Pushkarov D I 1972 PhD Thesis Moscow State University
[4] Andreev A F 1976 *JETP Lett.* **24** 564
[5] Krivoglaz M A 1974 *Sov.Phys.-Uspekhi* **6** 856
[6] Chaddah P 1980 *J.Low Temp.Phys.* **39** 131
[7] Pushkarov D I 1978 *Pisma v Zh.Eksp.Teor.Fiz.* **27** 359
[8] Pushkarov D I 1990 *Quasiparticle Theory of Defects in Solids* (Singapore: World Scientific Publ.)
[9] Pushkarov D I 1993 *Defectons in Crystals. Quasiparticle Approach to Defects in Crystals* (in Russian) (Moscow: Nauka)
[10] Pushkarov D I 2004 *Phys.Rev.* B **70**
[11] Pushkarov D I 2004 *Central Europ. J. Phys.* **2** 420
[12] Pushkarov D I 1975 *Sov.J.Low.Temp.Phys.* **1** 280
[13] Meierovich A E 1974 *Zh.Eksp.Teor.Fiz.* **67** 744 (1974)
[14] C.A. Burns and J. M. Goodkind 1994 *J.Low Temp. Phys.* **95** 695
[15] Greenberg A S and Armstrong G 1980 *Phys.Rev.* B **22** 4336
[16] Goodkind J M 2002 *Phys. Rev. Lett.* **89** 095301
[17] Ganshin A N, Grigor'ev V N, Maidanov V, Penzev A, Rudavskii E Ya, Rybalko A S and Syrnikov E V 2001 *JETP Letters* **73** 289
[18] Maidanov V, Ganshin A, Grigor'ev V N, Penzev A, Rudavskii E Ya, Rybalko A S and Syrnikov E V 2002 *J. Low Temp. Phys.* **126** 133
[19] A. Ganshin A, Grigor'ev V, Maidanov V, Penzev A, Rudavskii E, Rybalko A and Syrnikov Ye 2003 *Physica* B: *Condensed Matter* **329-333** 390
[20] Landau L D and Lifshitz E M 1976 *Statistical Physics* (Course of Theoretical Physics vol. 5) (Moscow: Nauka) (in Russian)
[21] Ganshin A, Grigor'ev V, Maidanov V, Penzev A, Rudavskii E, Rybalko A and Syrnikov Ye 2003 *Low Temp. Phys.* **29** 362
[22] Edwards D O and R.C. Pandorf R C 1966 *Phys.Rev.* **144** 143
[23] Edwards D O and S. Balibar S 1989 *Phys.Rev.*B **39** 4083
[24] Chaudhuri B, Pederiva F and Chester G.V. 1999 *Phys. Rev.* B **60** 3271

# On the quantum critical behaviour of a model of structural phase transitions with long-range interaction

**E S Pisanova**[1] **and N S Tonchev**[2]

[1]Department of Physics, University of Plovdiv, 24 Tzar Assen Street, Plovdiv 4000, Bulgaria

[2]Institute of Solid State Physics, Bulgarian Academy of Sciences, 72 Tzarigradsko chaussee Blvd., Sofia 1784, Bulgaria

E-mail: pisanova@uni-plovdiv.bg

**Abstract**. The pure quantum limit of an exactly solvable lattice model describing structural phase transitions in an anharmonic crystal with long-range interaction is considered. At the upper quantum critical dimension the free energy density at $T=0$ in the neighbourhood of the quantum critical point is exactly calculated in terms of the Lambert W-function. For the three real physical dimensions, the exact results, obtained here, and the asymptotic ones are compared. It is pointed out that the Lambert W-function turns out to be a very effective tool for an exact computation of non-universal characteristics in the upper critical dimensions, especially in a broader neighbourhood of the critical region.

## 1. Introduction

The aim of this study is to obtain an exact expression for the free energy density at $T=0$ of an exactly solvable model of structural phase transitions near the quantum critical point at the upper quantum critical dimension. The importance of this special point for broad classes of materials has rapidly increased in recent years (see [1] and refs. therein). The quantum critical point occurs at $T=0$ when the growth of the zero point quantum fluctuations associated with Heisenberg's uncertainty principle leads to a change in the physical state of the system. In this case the critical behaviour is controlled by the variation of a quantum parameter $\lambda$ (instead of the temperature $T$) which is a function of the external pressure or the amount of impurity doping. The model under consideration, in spite of its high idealization, retains many fundamental properties of the real systems related to the presence of both quantum and classical fluctuations, depending on the temperature $T$, the quantum parameter $\lambda$, the long-range interaction exponent $\sigma$ and the spatial dimensionality $d$. The model and its generalizations, in both bulk and finite-size cases, have been studied exactly in a variety of papers (see e.g. [2], Chapter 3 of [3] and references therein). For rigorous analysis of the specific properties of its quantum fluctuations one can see [4]. The finite-size corrections to the free energy for the pure quantum version ($T=0$) of the model have been considered in [5].

The present paper is a continuation of the application of the Lambert W-function [6] in studying the critical properties of some systems [7] in their upper critical dimensions. In [7] it has been shown that at the upper quantum critical dimension $d=d_{\mathrm{u}}^{\mathrm{q}}=3\sigma/2$ in the vicinity of the quantum critical point the self-consistent equation for the inverse susceptibility can be written in a form with leading logarithmic terms and the Lambert W-function can be successfully used to obtain exact solutions of this equation. Here, we will obtain an exact analytical expression for the free energy density of the model at $T=0$ when $d=d_{\mathrm{u}}^{\mathrm{q}}=3\sigma/2$ near its quantum critical point $\left(\lambda \to \lambda_{\mathrm{c}}^{+}\right)$ in terms of the Lambert W-function. Due to the fact that we consider systems with long-range interaction, i.e. $0<\sigma\leq 2$, the upper quantum critical dimension may coincide with one of the three real physical dimensions: for chains $(d=1)$ if $\sigma=2/3$, for planar systems $(d=2)$ if $\sigma=4/3$ as well as for 3-dimensional systems in the case of short-range interaction $(\sigma=2)$. For these physical dimensions the exact results, obtained here, and the asymptotic ones, obtained previously, for the free energy density near to the quantum critical point, are compared on the basis of the graphic presentations of the absolute and the relative errors as functions of the deviation from the quantum critical point. It is pointed out that the Lambert W-function turns out to be a very effective tool for an exact computation of non-universal characteristics in the upper critical dimensions, especially in a broader neighbourhood of the critical region.

## 2. The Lambert W-function

The Lambert W-function of the complex variable $z$ is defined as a function, satisfying the transcendental equation [6]

$$W(z)e^{W(z)}=z. \tag{2.1}$$

This equation has always an infinite number of solutions $W_k(z)$, where $k=0,\pm1,\pm2,$ etc. identifies the branch of this multivalued function. There is always a special interest in solutions that are purely real. When $z\equiv x$ is a real variable, then for $-1/\mathrm{e}\leq x<0$ there are two possible real values of $W(x)$: the branch $W_0(x)$ satisfying $-1\leq W_0(x)$ and the branch $W_{-1}(x)$ satisfying $W_{-1}(x)\leq -1$ as $W_{-1}\left(x\to 0^{-}\right)\to -\infty$. The basic mathematical properties of the Lambert W-function, some of its recent applications, as well as interesting historical remarks, can be seen in [6].

A variety of equations can be solved in the terms of this transcendental function, satisfying equation (2.1). Below, we have to solve an equation of the form

$$x\ln x+ax=b. \tag{2.2}$$

By the substitution $\ln x=t$, the above equation can be written in the form of equation (2.1), hence its solutions are

$$x=\exp\left(W\left(be^{a}\right)-a\right). \tag{2.3}$$

## 3. The model

The Hamiltonian reads

$$H=\frac{1}{2}\sum_{\mathbf{r}}\left(\frac{P_{\mathbf{r}}^2}{m}+AQ_{\mathbf{r}}^2\right)+\frac{1}{4}\sum_{\mathbf{r},\mathbf{r}'}\varphi(\mathbf{r}-\mathbf{r}')(Q_{\mathbf{r}}-Q_{\mathbf{r}'})^2+\frac{B}{4N}\left(\sum_{\mathbf{r}}Q_{\mathbf{r}}^2\right)^2. \tag{3.1}$$

Here $Q_{\mathbf{r}}$ and $P_{\mathbf{r}}$ are operators of displacement and momentum, respectively, of the particle of mass $m$ at site $\mathbf{r}$ of a $d$-dimensional hypercubic lattice. The parameter $A=\nu_0^2 m>0$ determines the frequency of the mode which is unstable in the harmonic approximation, and the parameter $B>0$ introduces into the model an anharmonic interaction which is inversely proportional to the particle number $N$. The harmonic force constants, $\varphi(\mathbf{r}-\mathbf{r}')$, which are assumed to decrease at large distances $r=|\mathbf{r}-\mathbf{r}'|$ as $r^{-d-\sigma}$, describe a short-range ($\sigma=2$) or a long-range ($0<\sigma<2$) interaction.

By using the Approximating Hamiltonian Method for the free energy density of the model (3.1) in the thermodynamic limit ( $N \to \infty$ ), one obtains [2, 3]

$$f(\Delta_0) = \max_{\Delta} f(\Delta), \tag{3.2}$$

where $f(\Delta)$ has the form

$$f(\Delta) = \frac{tS_d}{(2\pi)^d} \int_0^{x_D} x^{d-1} \ln\left[2\sinh\left(\frac{\lambda\sqrt{x^\sigma + \Delta}}{2t}\right)\right] dx - \frac{1}{4}(1+\Delta)^2. \tag{3.3}$$

In (3.3), $t = T/4E_0$ is the dimensionless temperature and $\lambda = \hbar\nu_0/4E_0$ is a parameter which switches on quantum fluctuations, $E_0 = A^2/4B$ is the barrier height of the double-well potential in equation (3.1), $x_D = 2\pi(d/S_d)^{1/d}$ is the radius of the effective sphere replacing the Brillouin zone and $S_d = 2\pi^{d/2}/\Gamma(d/2)$ is the surface of the $d$-dimensional unit sphere ($\Gamma(z)$ is the Euler gamma function). Note that $\lambda$ is large, hence the quantum effects are important, not only at large zero-point energy $\hbar\nu_0$ but also at small depth of the well $E_0$.

Since the function $f(\Delta)$ is differentiable and $f''(\Delta) < 0$ for all real $\Delta$, it has just one maximum. Thus, $\Delta = \Delta_0$ is the solution of the self-consistent equation $f'(\Delta) = 0$ which has the explicit form

$$1 + \Delta = \frac{\lambda}{2}\frac{S_d}{(2\pi)^d} \int_0^{x_D} \frac{x^{d-1}}{\sqrt{x^\sigma + \Delta}} \coth\left(\frac{\lambda}{2t}\sqrt{x^\sigma + \Delta}\right) dx. \tag{3.4}$$

The model (3.1) exhibits a phase transition of second kind at the critical temperature $t_c(\lambda)$. At the critical point the solution $\Delta_0$ of the equation (3.4) vanishes. In the disordered phase, $t > t_c(\lambda)$, $\Delta$ is finite and the susceptibility $\chi = \Delta^{-1}$.

## 4. The exact analytical expression

Setting $t = 0$ in the equation (3.3) we obtain

$$f(\Delta) = \frac{S_d}{(2\pi)^d}\frac{\lambda}{2} \int_0^{x_D} x^{d-1}\sqrt{x^\sigma + \Delta}\, dx - \frac{1}{4}(1+\Delta)^2. \tag{4.1}$$

At the upper quantum critical dimension $d = d_u^q = 3\sigma/2$ equation (4.1) may be presented in the form

$$f(\tilde{\Delta}) = x_D^\sigma \frac{\lambda}{\lambda_c}\left[\frac{2}{3}\frac{\tilde{\Delta}}{\sqrt{1+\tilde{\Delta}}} F\left(\frac{1}{2}, 1; \frac{5}{2}; \frac{1}{1+\tilde{\Delta}}\right) + \frac{2}{5}\frac{1}{\sqrt{1+\tilde{\Delta}}} F\left(\frac{1}{2}, 1; \frac{7}{2}; \frac{1}{1+\tilde{\Delta}}\right)\right] - \frac{1}{4}\left(1 + x_D^\sigma\tilde{\Delta}\right)^2, \tag{4.2}$$

where $\tilde{\Delta} \equiv \Delta/x_D^\sigma$ is the new renormalized variable and $F(\alpha, \beta; \gamma; z)$ is the hypergeometric function. The critical value $\lambda_c$ of the quantum parameter $\lambda$ is obtained from the condition that the solution $\Delta_0$ of equation (3.4) vanishes at the quantum critical point ($t = 0, \lambda = \lambda_c$), $\lambda_c = 2^{2+\sigma/2}\pi^{\sigma/4}\sqrt[3]{\Gamma(3\sigma/4+1)}/3$.

Sufficiently close to the quantum critical point ($\lambda \to \lambda_c^+$), i.e. when $\tilde{\Delta} \ll 1$, retaining the leading terms, including $\tilde{\Delta}^2$-term, after some algebra, we can write (4.2) in the form

$$f(\tilde{\Delta}) \approx \frac{x_D^\sigma}{2}\frac{\lambda}{\lambda_c}\left[1 + \tilde{\Delta} + \frac{1}{4}\tilde{\Delta}^2\ln\tilde{\Delta} + \frac{1}{8}(1 - 2\ln 4)\tilde{\Delta}^2\right] - \frac{1}{4}\left(1 + 2x_D^\sigma\tilde{\Delta} + x_D^{2\sigma}\tilde{\Delta}^2\right). \tag{4.3}$$

In this limit the self-consistent equation $f'(\Delta)=0$ has the form of equation (2.2),

$$\tilde{\Delta}\ln\tilde{\Delta}+a\tilde{\Delta}=b\,, \tag{4.4}$$

where

$$a=1-\ln 4-2x_{\mathrm{D}}^{\sigma} \quad \text{and} \quad b=-\delta\lambda\equiv-2\left(1-\frac{\lambda_{\mathrm{c}}}{\lambda}\right). \tag{4.5}$$

The exact solution $\tilde{\Delta}_0$ of equation (4.4) is [7]

$$\tilde{\Delta}_0=\exp\left[W_{-1}\left(-\delta\lambda\,\mathrm{e}^{a}\right)-a\right]. \tag{4.6}$$

Let us note that we choose the solution with the real branch $W_{-1}(x)$ among the solutions of (4.4) because at the critical point ($t=0,\ \lambda=\lambda_{\mathrm{c}}$) the renormalized inverse susceptibility $\tilde{\Delta}$ must vanish which takes place, due to the property $W_{-1}\left(x\to 0^{-}\right)\to-\infty$. The needed expression for the free energy density can be obtained after the substitution of the solution (4.6) into (4.3). Thus, for dimensions $d=d_{\mathrm{u}}^{\mathrm{q}}=3\sigma/2$, taking into account $\tilde{\Delta}<<1$ when $\delta\lambda\to 0^{+}$ and neglecting in the expansion (4.3) terms of order $O\left(\tilde{\Delta}^2\ln\tilde{\Delta}\right)$, we finally get

$$f=f_{\mathrm{c}}+\delta f \quad \text{with} \quad \delta f=\frac{1}{4}x_{\mathrm{D}}^{\sigma}\delta\lambda\exp\left[W_{-1}\left(-\delta\lambda\mathrm{e}^{a}\right)-a\right], \tag{4.7}$$

where $f_{\mathrm{c}}=x_{\mathrm{D}}^{\sigma}/2-1/4$ is the free energy density at $\lambda=\lambda_{\mathrm{c}}$, see figure 1.

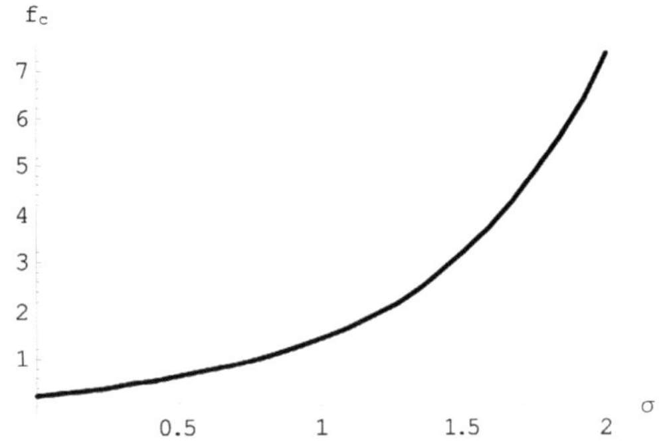


**Figure 1.** The dependence of the free energy density at the quantum critical point, $f_{\mathrm{c}}$, on the long-range interaction exponent $\sigma$.

We can get an asymptotic expression for $\tilde{\Delta}_0$ from (4.6), by using the series for the branch $W_{-1}(x)$ for small $x<0$ [6]. Thus, for the leading term in the asymptotic expression for $\tilde{\Delta}_0$ one obtains [7]

$$\tilde{\Delta}_0\approx\delta\lambda\left|\ln\delta\lambda\right|^{-1}. \tag{4.8}$$

By using the asymptotic solution (4.8) for the free energy density $f_{\mathrm{appr.}}$ we obtain

$$f_{\mathrm{appr.}}=f_{\mathrm{c}}+\delta f_{\mathrm{appr.}} \quad \text{with} \quad \delta f_{\mathrm{appr.}}=\frac{1}{4}x_{\mathrm{D}}^{\sigma}\left(\delta\lambda\right)^2\left|\ln\left(\delta\lambda\right)\right|^{-1}. \tag{4.9}$$

In the theory of phase transitions, it is a typical behaviour at the upper critical dimension and the logarithmic solution (4.8) of the self-consistent equation (4.4) is well known (see, e.g. [2, 3]). It was obtained simply by solving the self-consistent equation (4.4) (with $a=0$) iteratively. The dependences of the correction terms $\delta f$ and $\delta f_{\text{appr.}}$ as well as of the absolute error $\Delta f = |f - f_{\text{appr.}}|$ and the relative one $\Delta f / f$ on the parameter $\delta\lambda$, at different values of the long-range exponent $\sigma$, are graphically presented on figure 2.

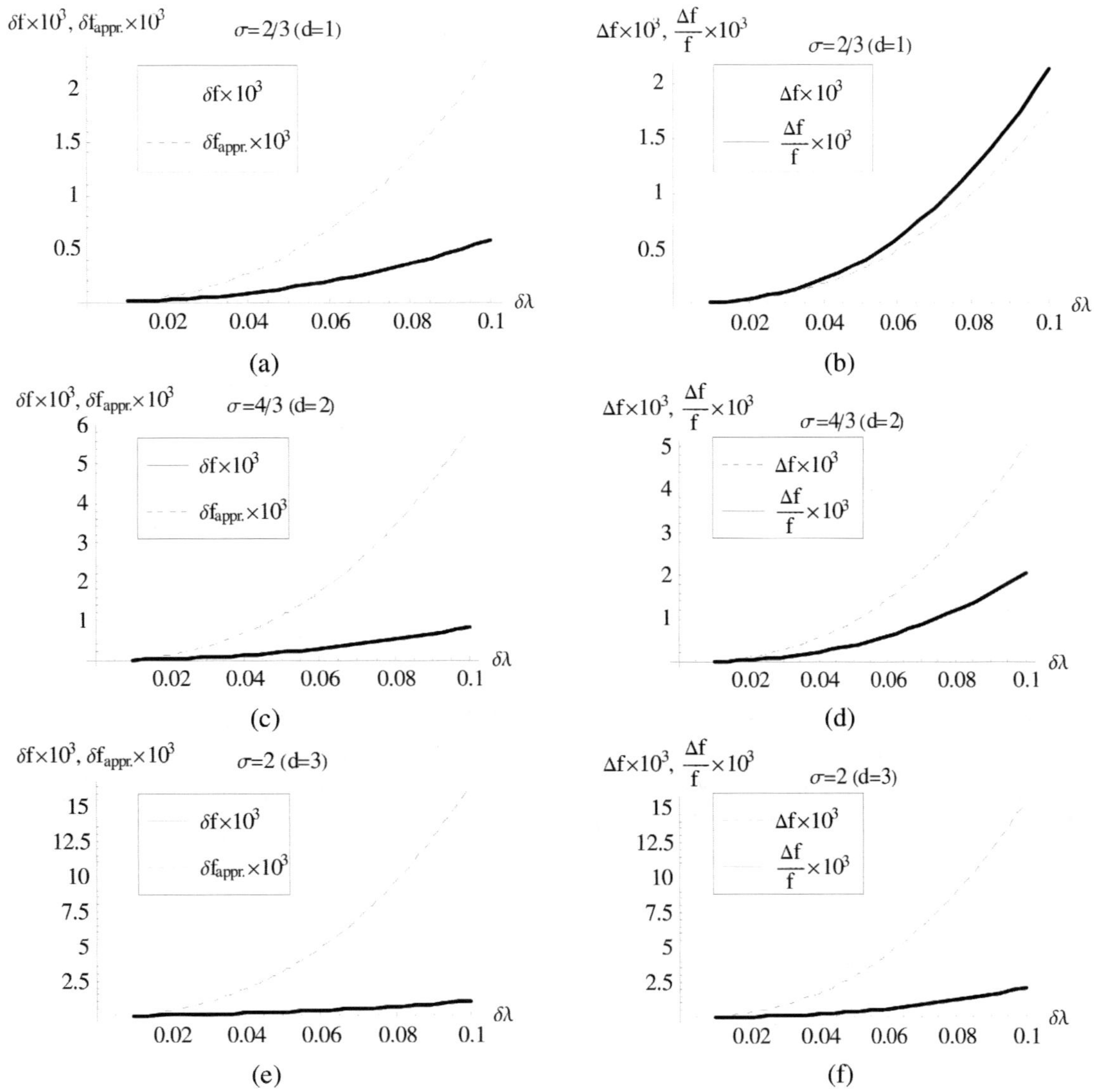


**Figure 2.** The dependence of the correction terms $\delta f$ and $\delta f_{\text{appr.}}$ as well as of the absolute error $\Delta f$ and the relative one $\Delta f / f$ on the parameter $\delta\lambda$ for the three physical dimensions.

For a more precise analysis of the behaviour of the relative error $\Delta f / f$ as a function of the long-range exponent $\sigma$, i.e. of the dimensionality $d$ of the system, some numerical data are presented on table 1.

**Table 1.** Some numerical data for $\tilde{\Delta}_0$, given by the equation (4.6), and the relative error.

| $\delta\lambda$ | $\sigma = 2/3$ $(d=1)$ | | $\sigma = 4/3$ $(d=2)$ | | $\sigma = 2$ $(d=3)$ | |
|---|---|---|---|---|---|---|
| | $\tilde{\Delta}_0$ | $(\Delta f/f)$ % | $\tilde{\Delta}_0$ | $(\Delta f/f)$ % | $\tilde{\Delta}_0$ | $(\Delta f/f)$ % |
| 0.02 | 0.0018 | 0.004 | 0.0011 | 0.004 | 0.0005 | 0.005 |
| 0.04 | 0.0039 | 0.022 | 0.0023 | 0.022 | 0.0011 | 0.023 |
| 0.06 | 0.0061 | 0.059 | 0.0036 | 0.059 | 0.0016 | 0.061 |
| 0.08 | 0.0085 | 0.121 | 0.0048 | 0.118 | 0.0022 | 0.122 |
| 0.10 | 0.0109 | 0.212 | 0.0061 | 0.205 | 0.0027 | 0.210 |
| 0.20 | 0.0238 | 1.306 | 0.0129 | 1.226 | 0.0056 | 1.227 |
| 0.40 | 0.0525 | 9.881 | 0.0270 | 8.972 | 0.0114 | 8.773 |
| 0.60 | 0.0839 | 41.312 | 0.0417 | 36.939 | 0.0172 | 35.771 |

## 5. Conclusions

At the upper quantum critical dimension $d = d_u^q = 3\sigma/2$ near to the quantum critical point ($\lambda \to \lambda_c^+$) the exact expression for the free energy density of the model under consideration is obtained in the Lambert W-function terms, equation (4.7). From figure 1 and figure 2 it follows that the free energy density of the system is an increasing function of the long-range exponent $\sigma$ and of the deviation from the quantum critical point, measured by the parameter $\delta\lambda$.

It is easily seen from figure 2 that for each value of the parameter $\sigma$ when the deviation from the quantum critical point $\delta\lambda$ increases both correction terms, $\delta f$ and $\delta f_{\text{appr.}}$, increase but $\delta f_{\text{appr.}}$ increases more quickly than $\delta f$. Thus, at each fixed $\delta\lambda$ when $\sigma$ increases, up to $\sigma = 2$, i.e. the dimension of the system increases up to $d = 3$, the absolute error $\Delta f$ increases. As one can see from table 1, sufficiently close to the quantum critical point ($\delta\lambda$ is small), when $\sigma$ increases, the relative error $\Delta f/f$ increases also. But, at a certain distance $\delta\lambda$ from the quantum critical point, the relative error has minimum for two-dimensional systems. Further, when $\delta\lambda$ increases the relative error $\Delta f/f$ as a function of the dimension of the system decreases (due to the form of the curve $f_c = f_c(\sigma)$, see figure 1), but its values are too large.

Finally, at each fixed dimension $d$ ($\sigma$ is fixed) of the system, both the absolute error $\Delta f$ and the relative one $\Delta f/f$, as functions of the deviation from the quantum critical point, $\delta\lambda$, increase.

**Acknowledgments**
This study was supported by Grant BYX 308 of the Bulgarian National Scientific Found.

**References**

[1] Coleman P 2010 Quantum Criticality and Novel Phases: A Panel Discussion *Preprint* cond-mat/1001.0185
[2] Plakida N M, Radosz A and Tonchev N S 1991 *Phase Transitions* **297** 179
[3] Brankov J G, Danchev D M and Tonchev N S 2000 *The Theory of Critical Phenomena in Finite-Size Systems – Scaling and Quantum Effects* (Singapore: World Scientific)
[4] Verbeure A and Zagrebnov V A 1995 *J. Stat. Phys.* **79** 377
[5] Chamati H 1994 *Physica* A **212** 357
[6] Corlles R M, Gonnet G H, Hare D E, Jeffrey D J and Knuth D E 1996 *Advances Computational Maths* **5** 329
[7] Pisanova E S, Tonchev N S and Kisov H T 2009 $7^{th}$ *Int. Conf. of the Balk. Phys. Union* (American Institute of Physics) ed A Angelopoulos and T Fildisis 255

# Vibrational vector solitons in biaxial crystals

**K T Stoychev[1,2], R S Kamburova[2] and M T Primatarowa[2]**

[1]Plovdiv University, 24 Tsar Assen Str., 4000 Plovdiv, Bulgaria
[2]Georgi Nadjakov Institute of Solid State Physics, Bulgarian Academy of Sciences, 1784 Sofia, Bulgaria

E-mail: stoychev@issp.bas.bg, prima@issp.bas.bg

**Abstract.** Vector solitons formed of pairs of nonlinear optical vibrations with orthogonal polarizations in biaxial crystals are investigated. They are described by a system of coupled nonlinear Schrödinger equations, which take account of linear birefringence, cross-phase-modulation and group-velocity-dispersion (GVD) mismatch. The equations are solved numerically for different initial conditions and parameters. The stability of the coupled solution are investigated. The GVD mismatch, which has been neglected in previous studies, modifies the properties of vector solitons and seems to stabilize the fast-mode component.

## 1. Introduction

Scalar solitons formed of anharmonic vibrations have been investigated in [1-3]. The dynamics of vibrational vector solitons (pairs of conjugated circularly-polarized pulses) propagating along the optical axis of gyrotropic crystals has been studied in [4, 5] on the basis of a system of coupled nonlinear Schrödinger (NLS) equations. The presence of circular birefringence yields a walk-off effect for the components and for weak nonlinear cross-interaction, a linearly polarized pulse splits into two conjugated circularly polarized solitons propagating with different velocities. For strong enough attractive cross-interaction the solitons modify their individual carrier wave-numbers (or central frequencies) and are trapped to a common group velocity. An important result is that group-velocity locking yields also equal GVD coefficients for the two modes. Exact coupled-soliton solutions exist in this case. The presence of inelastic cross-interaction, which allows energy transfer between the components, leads to shape oscillations and instability of the fast mode. Optical vector solitons have been extensively investigated in birefringent fibers [6-9]. The difference in the GVD coefficients for the two polarizations has usually been neglected and polarization instabilities of the fast mode have been obtained.

The present work is devoted to the study of the dynamics of vector solitons formed of nonlinear optical vibrations with orthogonal polarizations in biaxial crystals and in oblique directions in uniaxial crystals. The modes are characterized by different dispersion coefficients and phase velocities. Contrary to the gyrotropic and the optical cases however, trapping of the partial pulses and group-velocity locking yields appreciable difference in the GVD coefficients which has to be accounted for. The corresponding system of coupled NLS equations describing vector solitons in biaxial crystals is intrinsically nonintegrable. The equations are solved numerically for different soliton amplitudes and material parameters. The evolution and stability of the vector solitons in biaxial crystals show considerable differences compared to previously investigated cases.

Journal of Physics: Conference Series **253** (2010) 012021 doi:10.1088/1742-6596/253/1/012021

## 2. The coupled NLS equations

The Hamiltonian of the system of two anharmonic optical vibrations with orthogonal polarizations propagating along oblique directions in crystals can be cast in the following form:

$$\begin{aligned} H &= \omega_0 \sum_n (A_n^\dagger A_n + B_n^\dagger B_n) + \frac{1}{2}\sum_n [M_1(A_{n+1}^\dagger A_n + A_n^\dagger A_{n+1}) + M_2(B_{n+1}^\dagger B_n + B_n^\dagger B_{n+1})] \\ &+ \frac{g_1}{2}\sum_n (A_n^{\dagger 2} A_n^2 + B_n^{\dagger 2} B_n^2) + g_2 \sum_n A_n^\dagger A_n B_n^\dagger B_n + \frac{g_3}{2}\sum_n (A_n^{\dagger 2} B_n^2 + B_n^{\dagger 2} A_n^2), \end{aligned} \tag{1}$$

where $A_n^\dagger$, $A_n$ and $B_n^\dagger$, $B_n$ are Bose creation and annihilation operators of vibrational states at site $n$ for the two modes, $\omega_0$ is the harmonic intramolecular energy ($\hbar = 1$) and $M_1$ and $M_2$ are the transfer matrix elements between neighboring sites for the two modes. Note that the Hamiltonian (1) conserves the total number of quasiparticles in compliance with the assumption of narrow energy bands of optical phonons ($M_1$, $M_2 \ll \omega_0$). The quartic terms in (1) describe the anharmonic interactions for the two modes. The term $\sim g_1$ describes anharmonic interaction between excitations of the same type, while the terms $\sim g_2$ and $g_3$ describe anharmonic interaction between excitations of different types. The term $g_2$ conserves the number of particles of each type (elastic cross-intraction), while the term $\sim g_3$ conserves only the total number of particles and allows energy transfer between the modes (inelastic cross-interaction).

The equations of motion for the averaged vibrational amplitudes $\alpha_n \equiv \langle A_n \rangle$, $\beta_n \equiv \langle B_n \rangle$ are:

$$\begin{aligned} i\frac{\partial \alpha_n}{\partial t} &= \omega_0 \alpha_n + M_1(\alpha_{n+1} + \alpha_{n-1}) + (g_1|\alpha_n|^2 + g_2|\beta_n|^2)\alpha_n + g_3 \alpha_n^* \beta_n^2 \\ i\frac{\partial \beta_n}{\partial t} &= \omega_0 \beta_n + M_2(\beta_{n+1} + \beta_{n-1}) + (g_1|\beta_n|^2 + g_2|\alpha_n|^2)\beta_n + g_3 \beta_n^* \alpha_n^2 \end{aligned} \tag{2}$$

Seeking solutions in the form of Bloch waves with slowly-varying envelopes

$$\alpha_n(t) = \varphi_n(t) e^{i(k_1 n - \omega_1 t)}, \quad \beta_n(t) = \psi_n(t) e^{i(k_2 n - \omega_2 t)} \tag{3}$$

the following system of coupled NLS equations for the envelopes are derived in the continuum limit:

$$\begin{aligned} i\frac{\partial \varphi}{\partial t} &= (\varepsilon_1 - \omega_1)\varphi - 2i\frac{\partial b_1}{\partial k_1}\frac{\partial \varphi}{\partial x} + b_1 \frac{\partial^2 \varphi}{\partial x^2} + (g_1 \varphi^2 + g_2 \psi^2)\varphi + g_3 e^{2i(\Delta k x - \Delta \omega t)} \psi^2 \varphi \\ i\frac{\partial \psi}{\partial t} &= (\varepsilon_2 - \omega_2)\psi - 2i\frac{\partial b_2}{\partial k_2}\frac{\partial \psi}{\partial x} + b_2 \frac{\partial^2 \psi}{\partial x^2} + (g_1 \psi^2 + g_2 \varphi^2)\psi + g_3 e^{-2i(\Delta k x - \Delta \omega t)} \varphi^2 \psi \end{aligned} \tag{4}$$

where $k_i$ and $\omega_i$ are the wave numbers and the frequencies of the carrier waves (the lattice constant equals unity), $\varepsilon_i = \omega_0 + 2b_i$, $b_i = M_i \cos k_i$, $\Delta k = k_2 - k_1$, $\Delta\omega = \omega_2 - \omega_1$.

The imaginary parts of equation (4) govern the group velocities of the solitons. Neglecting small nonlinear corrections associated with $g_3$, the GVD coefficients $b_i$ corresponding to a common group velocity $v$ are derived:

$$b_i = M_i \sqrt{1 - v^2/4M_i^2}, \quad v = -2M_i \sin k_i, \quad i = 1, 2 \tag{5}$$

Note the difference in the GVD coefficients of the components of the vector soliton in the present case.

### 3. Numerical results

We carried out numerical simulations of equations (2) for different initial conditions and parameters. The input pulses were of the type:

$$\varphi_n(t) = \varphi_0 \mathrm{sech}\frac{n - vt}{L}, \quad \psi_n(t) = \psi_0 \mathrm{sech}\frac{n - vt}{L}, \tag{6}$$

where $L$ and $v$ are the width and the velocity of the solitons. For wide solitons ($L \geq 10$) the solutions of (2) converge on the solutions of the continuous equations (4).

Figure 1(a) shows a stable symmetric vector soliton corresponding to equal initial amplitudes and group velocities of the two components in the case of elastic cross-interaction only ($g_1 = g_2 \neq 0$, $g_3 = 0$). The presence of inelastic cross-interaction ($g_3 \neq 0$, figure 1(b)) yields positional and shape oscillations of $|\alpha|$ and $|\beta|$ with a spatial period of $\pi/2\Delta k \approx 78$ accompanied by emission of radiation. Over large timescales, the slow mode transfers part of its energy to the fast mode and an asymmetric vector soliton is formed.

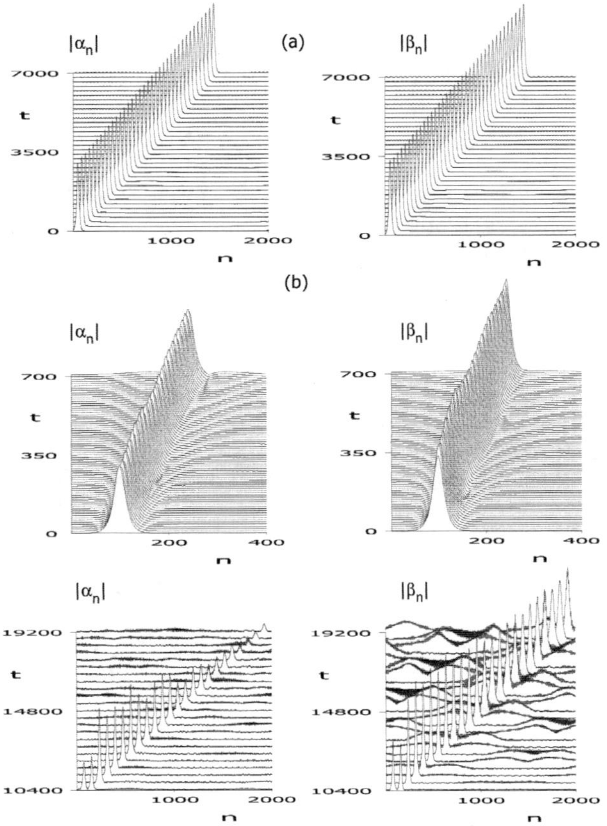


**Figure 1.** Evolution of a vector soliton with $L = 10$, $M_1 = -0.9$, $M_2 = -1.1$, $v = 0.2$, $k_1 = 0.111$, $k_2 = 0.091$, $g_1 = g_2 = -0.05$ and (a): $g_3 = 0$, (b): $g_3 = -0.4$. Time is in units $\omega_0^{-1}$.

Under the action of the exchange cross-interaction, the system supports also asymmetric vector solitons up to a certain difference between the amplitudes. Figure 2 shows such a case with 10 times smaller amplitude of one of the components. The evolution is symmetric with respect to the input amplitudes of the fast and slow modes and results in broadening of both partial pulses.

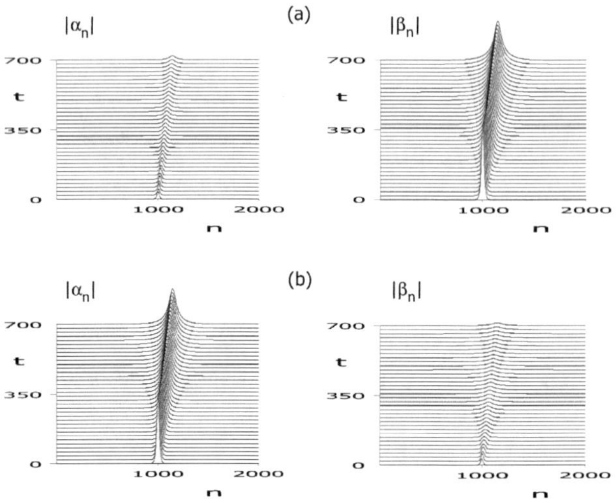


**Figure 2.** Evolution of a soliton pair with $g_3 = -0.4$ and (a): $\varphi_0 = 0.1\psi_0$, (b): $\psi_0 = 0.1\varphi_0$. All other parameters are the same as in figure 1.

The evolution of the asymmetric vector soliton depends strongly on the magnitudes of the input amplitudes as well as on the dispersion coefficients mismatch. For larger initial amplitudes, a remarkable difference is observed depending on whether the fast or slow mode amplitude is larger. When the fast mode amplitude is larger, the initial system remains practically unchanged [(figure 3(a)]. However, when the slow mode amplitude is larger, after a short time the slow mode transfers part of its energy to the fast mode which than splits into three - a weak soliton locked to the larger slow-mode to form an asymmetric vector soliton, and two scalar solitons with considerable amplitudes propagating in opposite directions [(figure 3(b)].

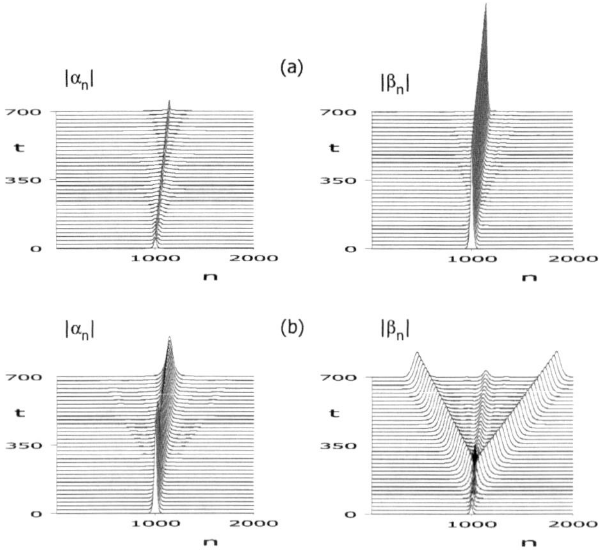


**Figure 3.** Evolution of a soliton pair with $g_3 = -0.4$ and (a): $\varphi_0 = 0.1\psi_0$, (b): $\psi_0 = 0.1\varphi_0$. The initial amplitudes are 40% larger than in figure 2.

The inelastic cross-interaction terms represent a nonlinear potential with a spatial period $\pi/\Delta k$. The effect of these terms depends strongly on the ratio of this period to the soliton width $L$. For $\pi/\Delta kL \ll 1$ the potential oscillates many times over the soliton width and its effect on the soliton dynamics is negligible. In the opposite case of slowly-varying potential however ($\pi/\Delta kL \gg 1$), effective energy transfer between the components of the vector soliton takes place. It is worth noting, that in all cases of the present study, the fast mode remained stable, while the slow mode showed some minor instabilities. This is a remarkable difference from the properties of vector solitons studied before and seems to be associated with the unequal GVD coefficients for the two modes.

The role of the GVD coefficients mismatch for the nonintegrability of the system (4) even in the near-Manakov case ($g_1 = g_2 \neq 0$, $g_3 = 0$) is illustrated in figure 4 in the process of scattering of two scalar solitons with orthogonal polarizations, launched from different places with different velocities. The scattering is inelastic, each scalar soliton splits off part of the other one to form two asymmetric vector solitons propagating with different velocities.

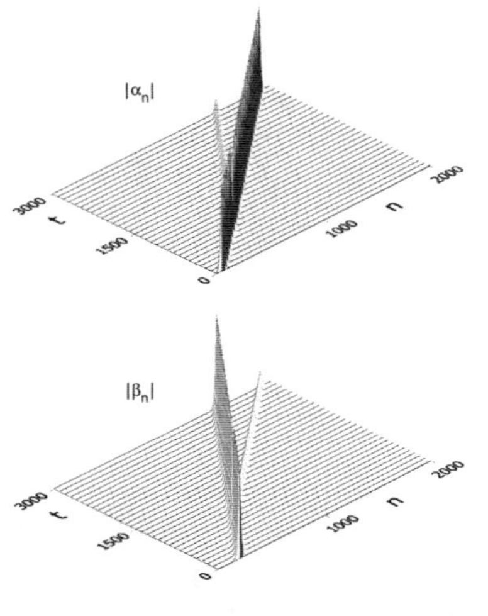


**Figure 4.** Scattering of scalar solitons with $k_1 = 0.35$, $k_2 = 0.2$, $g_1 = g_2 = -0.05$ and $g_3 = 0$.

## Acknowledgments

This work is supported in part by the National Science Fund of Bulgaria under Grants No. TK-X-1715 and No. DO 02-264.

## References

[1] Flitzanis N, Pnevmatikos St and Remoissenet M 1985 *J. Phys.* C **18** 4603
[2] Pnevmatikos St, Flitzanis N and Remoissenet M 1986 *Phys. Rev.* B **33** 2308
[3] Hochstrasser D, Mertens F G and Büttner H 1989 *Phys. Rev.* A **40** 2602
[4] Primatarowa M T and Stoychev K T 1994 *Phys. Rev.* B **49** 6634
[5] Stoychev K T, Primatarowa M T and Kamburova R S 2000 *J. Phys.: Condens. Matter* **12** 10429
[6] Blow K J, Doran N J and Wood D 1987 *Opt. Lett.* **12** 202
[7] Menyuk C R 1987 *Opt. Lett.* **12** 614
[8] Christodoulidis D N and Joseph R J 1988 *Opt. Lett.* **13** 53
[9] Malomed B A *Phys. Rev.* A 1991 **43** 410

Journal of Physics: Conference Series **253** (2010) 012022 doi:10.1088/1742-6596/253/1/012022

# Dark soliton dynamics in classical one-dimensional spin systems

**M T Primatarowa and R S Kamburova**

Georgi Nadjakov Institute of Solid State Physics, Bulgarian Academy of Sciences, 1784 Sofia, Bulgaria

E-mail: `prima@issp.bas.bg`

**Abstract.** Nonlinear effects in ferromagnetic Heisenberg chains of classical spins with both inter-site and on-site anisotropy have been studied. We obtained the condition for the appearance and stability of dark solitons in these systems. For narrow excitations the effect of the discreteness on the soliton properties and propagation is considered. The scattering of dark solitons from point defects in the discrete spin chain is investigated numerically and the role of the anisotropy is demonstrated.

## 1. Introduction

Considerable attention has been devoted to the study of solitary waves in solids. A large amount of theoretical and experimental research has been dedicated to one-dimensional magnetic systems [1]. Models corresponding to real quasi-one-dimensional magnets are broadly investigated as first they were considered as somewhat simpler than the really interesting three-dimensional systems but then turned out to be interesting in their own right. There was the availability of rear magnetic compounds which because of extremely small coupling between neighboring chains, could be considered as reasonable realizations of magnetically one-dimensional materials. Soliton solutions for classical sine-Gordon chains [2] and Heisenberg chains with various anisotropies [3-6] were obtained and analyzed. In recent years there is a renewed interest to the topic due to their application. Spin-wave dark solitons were predicted and experimentally generated [7,8]. Solitons in magnetic thin films [9] and in ferromagnets with biquadratic exchange [10] were investigated.

Bright and dark solitons as exact solutions of the nonlinear Schrödinger (NLS) equation have been studied intensively [3,4,11]. But, the models describing the physical effects in solids are mainly discrete and the question arises how the discreteness modify the properties of the nonlinear excitations [12,13]. Another interesting topic of research with practical importance is the interaction of solitons with impurities [14-17].

In the present paper we study the condition for the appearance of dark solitons in an anisotropic ferromagnetic chain. The discreteness effects for narrow excitations are considered and the scattering on point defects is investigated.

## 2. Hamiltonian of the system

We consider a ferromagnetic Heisenberg chain of $N$ spins with magnitude $S$ described in the nearest-neighbor approximation by the following Hamiltonian:

$$H = -\frac{J}{2}\sum_n (S_n^+ S_{n+1}^- + S_n^- S_{n+1}^+) - \tilde{J}\sum_n S_n^z S_{n+1}^z - A\sum_n (S_n^z)^2 - H_0 \sum_n S_n^z \,, \tag{1}$$

where the spin operators $S_n^\pm$, $S_n^z$ obey the commutation relations

$$[S_i^\pm, S_j^z] = \mp S_i^\pm \delta_{ij}, \quad [S_i^+, S_j^-] = 2S_i^z \delta_{ij} \,. \tag{2}$$

$J > 0$ and $\tilde{J} > 0$ are the exchange integrals and $A$ is the on-site anisotropy constant which can be positive (easy axis) or negative (easy plane). If $\tilde{J} \neq J$ the inter-site anisotropy is included. For $J = \tilde{J}$ and $A = 0$ the model is isotropic. $H_0$ is the external magnetic field applied along the $z$-axis.

The equations of motion for the operators $S_n^+$ yield ($\hbar = 1$):

$$i\frac{\partial S_n^+}{\partial t} = (H_0 - A)S_n^+ - JS_n^z(S_{n+1}^+ + S_{n-1}^+) + \tilde{J}(S_{n+1}^z + S_{n-1}^z)S_n^+ + 2AS_n^z S_n^+ \,. \tag{3}$$

In the quasiclassical approximation, where the components of the spin operators $S_n$ are complex amplitudes, $\alpha_n = S_n^+/S$, $\alpha_n^* = S_n^-/S$ and $S_n^z/S = \sqrt{1-|\alpha_n|^2}$, we have

$$\begin{aligned} i\frac{\partial \alpha_n}{\partial t} &= (H_0 - A)\alpha_n - JS(\alpha_{n+1} + \alpha_{n-1})\sqrt{1-|\alpha_n|^2} + \tilde{J}S\alpha_n\left(\sqrt{1-|\alpha_{n+1}|^2} + \sqrt{1-|\alpha_{n-1}|^2}\right) \\ &+ 2AS\alpha_n\sqrt{1-|\alpha_n|^2} \,. \end{aligned} \tag{4}$$

The set of differential equations (4) describes our system. In the continuum limit (wide excitations), it transforms into a perturbed NLS, where the perturbing terms are due not only to the discreteness but also to the complicated nonlinear interactions. For $\tilde{J} = A = 0$ and wide excitations, (4) turns to the discrete Ablowitz-Ladik equation which is also completely integrable and has soliton solutions. So, for narrow excitations it is necessary to investigate the whole set (4). Equations similar to (4) has been derived for exciton solitons in molecular crystals [18].

## 3. Dark soliton solutions

We shall look for solutions in the form of amplitude-modulated waves

$$\alpha_n(t) = \varphi_n(t)e^{i(kn - \omega t)} \,, \tag{5}$$

where $k$ and $\omega$ are the wave number and the frequency of the carrier wave (the lattice constant equals unity) and the envelope $\varphi_n(t)$ is a slowly varying function of the position and time. In the continuum limit, when the soliton width is much larger than the lattice spacing ($L \gg 1$), equation (4) transforms into the following NLS equation for the envelope:

$$i\left(\frac{\partial \varphi}{\partial \tau} + 2J\sin k\frac{\partial \varphi}{\partial x}\right) = (\varepsilon - \Omega)\varphi - J\cos k\frac{\partial^2 \varphi}{\partial x^2} + g|\varphi|^2\varphi \,, \tag{6}$$

where

$$\tau = tS \,, \quad \Omega = \omega/S \,, \quad \varepsilon = (H_0 - A)S^{-1} - 2g \,, \quad g = J\cos k - \tilde{J} - A \,. \tag{7}$$

Depending on the sign of the expression $gJ\cos k$ equation (6) possesses soliton solutions of different types. For negative values bright solitons exist while for positive values dark solitons are possible. It is interesting to point out that for the isotropic case ($J = \tilde{J}$, $A = 0$) equation (6) has bright-soliton solutions for $0 < k < \pi/2$, while for the strong anisotropic case ($\tilde{J} = A = 0$) dark-soliton solutions are possible for arbitrary $k$.

In what follows, we shall consider dark-soliton solutions which appear for

$$J\cos k(J\cos k - \tilde{J} - A) > 0 \tag{8}$$

and have nonvanishing boundary conditions, $|\varphi(x)|^2 \to \text{const}$ at $x \to \pm\infty$. Note that the condition (8) depends not only on the anisotropy constants but also on the wave number $k$. The dark soliton has the form

$$\varphi(x,\tau) = \varphi_0 \tanh\frac{x - v\tau}{L} + iB\,, \tag{9}$$

where $L$ and $v$ are its width and velocity. $B$ determines the value of the amplitude at the center, i.e the minimum intensity of the soliton $|\varphi(0)|^2 = B^2$, while the maximum intensity is $|\varphi(\pm\infty)|^2 = \varphi_0^2 + B^2$. As independent parameters of the soliton we can choose $L$, $k$ and $B$. Then $\varphi_0$, $\Omega$ and $v$ are given by the relations:

$$\varphi_0^2 = \frac{2J\cos k}{gL^2}\,, \quad \Omega = \varepsilon + g(\varphi_0^2 + B^2)\,, \quad v = 2J\sin k + B\sqrt{2gJ\cos k}\,. \tag{10}$$

The dark soliton can be considered as an excitation of the background (carrier) wave. It is important to note that its velocity depends not only on $k$ as in the case for bright solitons but also on $B$. For fixed $\varphi_0^2 + B^2$ and $k$ solitons moving faster have smaller amplitude.

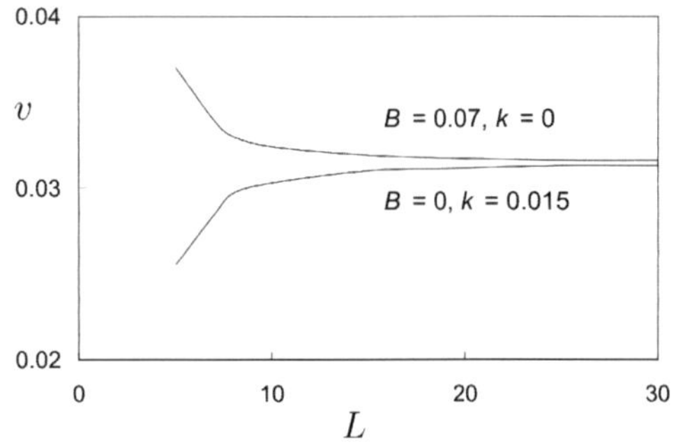


**Figure 1.** Final dark-soliton velocity as a function of the width $L$ for $v_0 = 0.0314$ and $J = \tilde{J} = 1$, $A = -0.1$.

## 4. Numerical results

Our numerical simulations are based on the general discrete equation (4). As initial function we put for $t = 0$ the form

$$\alpha_n(0) = \left(\frac{\sqrt{2J\cos k/g}}{L}\tanh\frac{n}{L} + iB\right)e^{ikn} \tag{11}$$

which is exact solution of the quasicontinuum equation, and observe the evolution for different anisotropic cases. For wide solitons ($L \gg 1$) this solution remain stable and its form and velocity are preserved. When the soliton width decreases the discreteness effects of equation (4) become important and the parameters of the dark soliton change.

The most significant and visible changes are in the velocity and they are shown on figure 1 for the case with on-site anisotropy ($J = \tilde{J}$, $A < 0$). The initial velocity is $v_0 = 0.0314$, which according to (10) results from the following two pairs of the parameters $k$ and $B$: $k = 0.015$, $B = 0$ and $k = 0$, $B = 0.07$, i.e. we investigate separately the influence of the two parameters on which the velocity depends. For these two cases the modification of the velocity with $L$ is obtained. The lower curve shows that for $k \neq 0$ the velocity decreases with the decrease of

$L$. In this case with the decrease of the width the form of the soliton changes, there are small radiations, the kinetic energy decreases and as a consequence the velocity becomes smaller. The effect is similar as for narrow bright solitons. More interesting is the behavior of the upper curve for $B \neq 0$. In this case a decrease of the width $L$ leads to an increase of the velocity $v$. This effect is due to the fact that the changes in the form of the soliton for smaller $L$ lead to a decrease of the amplitude and immediately to an increase of $B$, respectively $v$. This phenomenon is absent for bright solitons which have the amplitude and velocity as two independent parameters. We observed similar dependence $v(L)$ for the other anisotropy cases of the discrete equation.

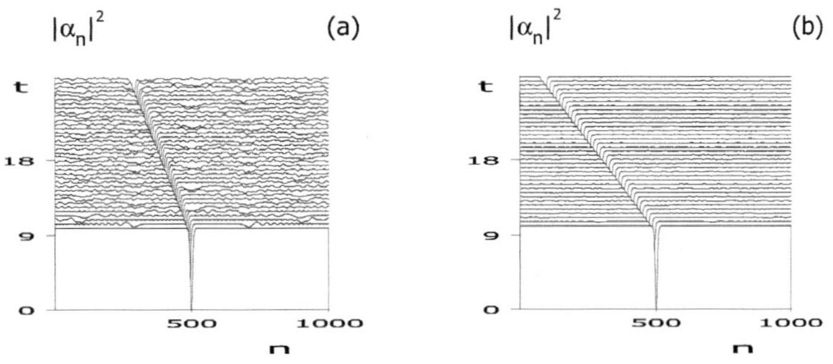


**Figure 2.** Time evolution of a narrow dark soliton ($L = 5$) with $J = \tilde{J} = 1$, $k = 0$, $B = -0.0222$ and (a): $A = -0.1$, (b): $A = -0.5$. The time is in units of $10^3/(JS)$.

The evolution of a dark soliton with $L = 5$ and $B = -0.0222$ for different on-site anisotropy values is presented in figure 2. Here the velocity can be estimated as $v_0 = B\sqrt{2A}$. For $A = -0.1$ $v_0 = -0.0099$, while the velocity evaluated from the data in figure 2(a) is $v = -0.0117$. For $A = -0.5$ $v_0 = -0.0222$ and $v = -0.0228$, respectively [figure 2(b)]. Obviously higher values of $|A|$ lead to higher values for $v$ and the parameters of the exact solution of the discrete equation (4) are closer to the initial solution originated from the continuum approximation.

## 5. Scattering on point defects

We studied the interaction of the dark solitons with impurities in the chain whose size is small compared to the soliton extent. Then the equation of motion becomes

$$i\frac{\partial \alpha_n}{\partial t} = (H_0 - A)\alpha_n - JS(\alpha_{n+1} + \alpha_{n-1})\sqrt{1-|\alpha_n|^2} + \tilde{J}S\alpha_n\left(\sqrt{1-|\alpha_{n+1}|^2} + \sqrt{1-|\alpha_{n-1}|^2}\right)$$
$$+ \; 2AS\alpha_n\sqrt{1-|\alpha_n|^2} + d\delta_{n,n_0}\alpha_n \,. \qquad (12)$$

The parameter $d$ describes the strength of the impurity at site $n_0$, which can be attractive ($d < 0$) or repulsive ($d > 0$). In our numerical simulations we investigate the interaction of a propagating dark soliton of the form (11) with a defect placed at the middle of the chain $n_0 = 500$. We have observed that for negative values of $d$ (attraction) the solitons pass through the defect for arbitrary initial velocity. More interesting is the interaction with repulsive impurities.

For $d > 0$ the soliton can be transmitted or reflected by the impurity. For a given initial velocity $v_0$ there is a critical value of the defect $d_c$ below which the soliton is transmitted and above which it is reflected. Figure 3 shows the scattering patterns for the case of inter-site anisotropy ($J = 1$, $\tilde{J} = A = 0$). The initial velocity $v_0 = 0.0314$ is determined by the wave number $k$ [figures 3(a),(b)] or by the amplitude $B$ [figures 3(c),(d)] and is conserved through the simulations. Figures 3(a) and (b) are for $d$-values around $d_c$ and we observe transmission and reflection, respectively. The results on figures 3(c) and (d) are for the same $d$ values but the soliton is only reflected. The difference is in the determination of the initial velocity and this is the reason for the different behavior. The determination of the velocity through $B$ leads to changes in the amplitude and as the energy balance for the process depends on the amplitude, the corresponding value of $d_c$ in this case is lower.

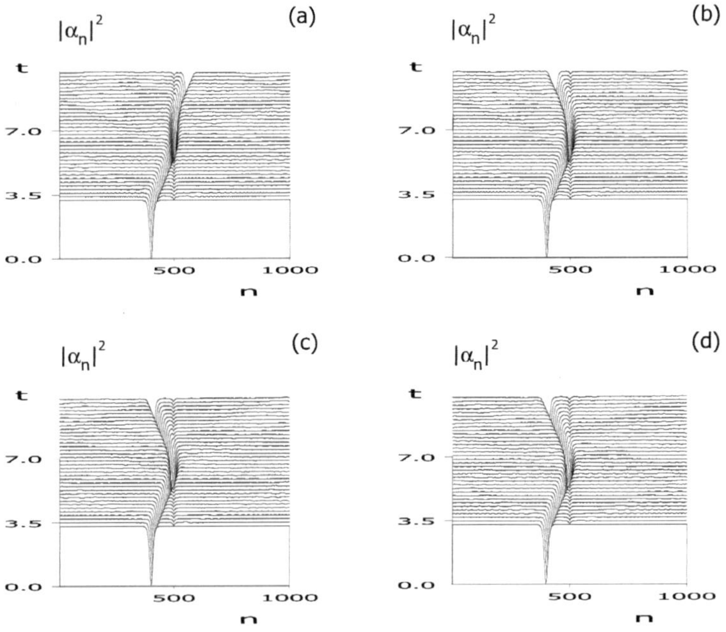


**Figure 3.** Scattering of a dark soliton with $L = 10$ and $J = 1$, $\tilde{J} = A = 0$ for $B = 0$, $k = 0.0157$ [(a), (b)] or $B = 0.0222$, $k = 0$ [(c),(d)]. $d = 0.0108$ [(a), (c)], $d = 0.0109$ [(b), (d)]. The time is in units of $10^3/(JS)$.

Similar behavior is presented on Figure 4 for the case of on-site anisotropy ($J = \tilde{J}$, $A < 0$). In this case the discreteness effects become important and the actual values for the initial velocity differs from $v_0 = 0.0314$. For figures 4(a),(b) it is $v = 0.0303$ and for figures 4(c),(d) it is $v = 0.0324$. So, the critical value $d_c$ in the second case is higher.

## 6. Conclusion

We have investigated the existence and stability of dark solitons in a discrete ferromagnetic chain with different anisotropies. The velocity of the dark solitons is characterized by the wave number and the amplitude at the center of the soliton. This dependence of the velocity leads to new effects for narrow dark solitons which do not appear for bright solitons. The interaction of solitons with impurities was studied as well. Depending on the initial velocity and the strength of the repulsive impurities the soliton can be transmitted or reflected. For attractive impurities only transmission is observed.

## Acknowledgments

This work is supported in part by the National Science Fundation of Bulgaria under Grant No. DO 02-264.

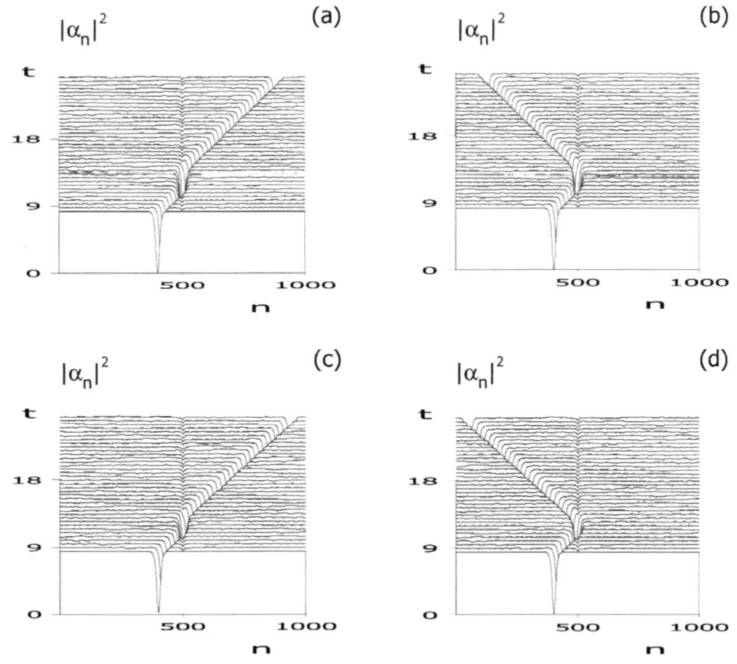


**Figure 4.** Scattering of a dark soliton with $L = 10$ and $J = \tilde{J} = 1$, $A = -0.1$ for (a): $B = 0$, $k = 0.0157$, $d = 0.00944$, (b): $B = 0$, $k = 0.0157$, $d = 0.00945$, (c): $B = 0.0702$, $k = 0$, $d = 0.0104$, (d): $B = 0.0702$, $k = 0$, $d = 0.0105$. The time is in units of $10^3/(JS)$.

## References


[1] Mikeska H-J and Steiner M 1991 *Adv. Phys.* **40** 191-356
[2] Tjon J and Wright J 1977 *Phys. Rev.* B **15** 3470-76
[3] Pushkarov D I and Pushkarov Kh I 1977 *Phys. Lett.* **61A** 339-40
[4] Huang G, Shi Z-P, Dai X and Tao R 1990 *J. Phys.: Condens. Matter* B **2** 8355-64
[5] Wallis R F, Mills D L and Boardman A D 1995 *Phys. Rev.* B **52** R3828-31
[6] Rakhmanova S and Mills D L 1998 *Phys. Rev.* B **58** 11458-64
[7] Slavin A N, Kivshar Yu S, Ostrovskaya E A and Benner H 1999 *Phys. Rev. Lett.* **82** 2583-6
[8] Bischof B, Slavin A N, Benner H and Kivshar Yu S 2005 *Phys. Rev.* B **71** 104424(7)
[9] Zaspel C E, Mantha J H, Rapoport Yu G and Grimalsky V V 2001 *Phys. Rev.* B **64** 064416(5)
[10] Ivanov B A, Galkin A Yu, Khymyn R S and Merkulov A Yu 2008 *Phys. Rev.* B **77** 064402(11)
[11] Kivshar Yu S and Luther-Davies B 1998 *Phys. Reports* **298** 81-197
[12] Kivshar Yu S, Królikowski W and Chubykalo O A 1994 *Phys. Rev.* E **50** 5020-32
[13] Kevrekidis P G, Kevrekidis I G, Bishop A R and Titi E S 2002 *Phys. Rev.* E **65** 046613(13)
[14] Forinash K, Peyrard M and Malomed B 1994 *Phys. Rev.* E **49** 3400-11
[15] Burtsev S, Kaup D J and Malomed B A 1995 *Phys. Rev.* E **52** 4474-81
[16] Sukhorukov A A, Kivshar Yu S, Bang O, Rasmussen J J and Christiansen P L 2001 *Phys. Rev.* E **63** 036601(18)
[17] Frantzeskakis D J, Theocharis G, Diakonos F K, Schmelcher P and Kivshar Yu S 2002 *Phys. Rev.* A **66** 053608(9)
[18] Primatarowa M T, Stoychev K T and Kamburova R S 1995 *Phys. Rev.* B **52** 15291-7

Journal of Physics: Conference Series **253** (2010) 012023 doi:10.1088/1742-6596/253/1/012023

# Multiplexed long period gratings with differential interrogation

**T Eftimov[1], W Bock[2], P Mikulic[2] and K Nikolova[3]**

[1] Faculty of Physics, Plovdiv University "P.Hilendarski", 24 Tsar Asen Str, Plovdiv 4000, Bulgaria

[2] Laboratoire de Recherche en Photonique, Université du Québec en Outaouais, P.O. Box 1250, Gatineau, Québec J8X 3X7, Canada

[3] University of Food Technology, 26 Maritza Blvd, Plovdiv 4002, Bulgaria

kr.nikolova@abv.bg

**Abstract**. In this work we present experimental results in which we simultaneously detect spectral changes of three long period fiber gratings (LPG) connected in series. The gratings can independently measure surrounding refractive index (SRI) changes $\Delta n$, temperature $T$ and strain $\varepsilon$ in any combination. We have successfully tested them with to simultaneous moisture/evaporation sensing based on SRI change measurements. Such arrangements can be used for simultaneous tracking of hazardous liquid ingress or spill in structural health monitoring applications.

## 1. Introduction

Long period fiber gratings (LPG) have generated intensive research since their discovery [1] and the most promising applications are related to their high sensitivity to changes in the surrounding refractive index [2]. This allows the development of unique refractometers with applications in chemical and biochemical sensing as well as fiber moisture/evaporation sensors for application in structure health monitoring [3].

We report here the results from experiments with multiplexed LPGs used to monitor wetting and evaporation. Such sensing set-ups can be used to monitor moisture/evaporation of liquids with refractive indices up to 1.41 such as fluorine refrigerant R-22 (1.26), fluorine refrigerant R-12 (1,29), water (1,333), methanol (1,33), propane (1,34), ether (1,35), ethanol (1,36), acetone (1,36), acetic acid (1,37), propyl alcohol (1,38), octane (1,40) and decane (1,41).

## 2. Long period gratings

A straight unperturbed LPG has a centre wavelength $\lambda_c$ defined as:

$$\lambda_c = \Delta n_{eff} \Lambda \qquad (1)$$

where $\Delta n_{eff}$ is the effective core-cladding mode refractive index difference and $\Lambda$ is the pitch of the grating. Strain, temperature and surrounding refractive index (SRI) changes $\delta l$ , $\delta T$ and $\delta n$ applied to the grating will induce a shift $\Delta\lambda$ in the centre wavelength accompanied by a decrease of the depth of the transmission notch. The centre wavelength changes are expressed as

$$\Delta\lambda = K_l\delta l + K_T\delta T + K_n\delta n \tag{2}$$

where $K_i$ $(i = l, T, n)$ are the corresponding grating sensitivities depending on the grating parameters.

## 3. Experimental set-up and results

We studied the spectral shifts of three tapered LPGs manufactured by the arc-fusion method [4]. The gratings were manufactured from an SMF-28 Corning optical fiber using a standard splicer (FITEL S182K). The LPGs were them spliced in series and kept straight.

The gratings were centered correspondingly at 1542 nm, 1581 nm 1626.6 nm and thus the separation between them is 38 nm and 45.6 nm non wetted.

The measurement experimental set-up is shown in figure. 1.

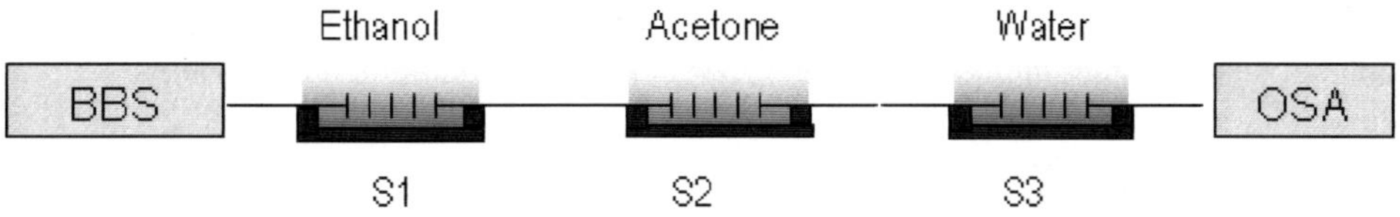


**Figure 1**. Experimental set-up using a broad-band source (BBS from 1250nm to 1650 nm) and an optical spectrum analyzer (OSA) in combination with three LPGs.

The sensing gratings were placed on supports and covered with a Kimwipe paper blanket which is a porous tissue to help keep the liquid used to wet the LPGs. When an LPG is wetted the SRI changed from that of air $n_a$ to $n_0$. If we denote the surfaces on the fiber grating exposed to air and to the liquid by $s_a$ and $s_0$, then the effective refractive index sensed by the grating is:

$$n = \frac{s_a.n_a + s_0.n_0}{s_a + s_0} \tag{3}$$

During evaporation, the surfaces $s_a$ and $s_0$ change and correspondingly the effective refractive index changes with time. At the initial moment when the LPGs are dry, the spectral distribution $P_0(\lambda)$ is taken as a reference. On wetting the gratings with different liquids such as water, ethanol and acetone, the corresponding spectral distribution $P(\lambda)$ would be measured. In the case of logarithmic units, the optical power $P$ is measured in *dBm*, defined as $I = 10\log[P/1\text{mW}]$ as shown in figure 2. We then calculate the differences for each wavelength of the spectrum

$$\Delta I_i = I(\lambda_i, n) - I(\lambda_i, n_0) = 10\log\frac{P(\lambda_i, n)}{P(\lambda_i, n_0)} \tag{4}$$

which are shown in figure 3. Next the differences are summed over two spectral intervals $\Delta\lambda_1$ and $\Delta\lambda_2$ as shown by the shado areas in figure 2.:

$$I_1 = \sum_{\Delta\lambda_1}\Delta I_i \qquad I_2 = \sum_{\Delta\lambda_2}\Delta I_i$$

Finally we calculate the differences between these

$$\Delta I_{12} = I_1 - I_2 = \sum_{\Delta\lambda_1} 10\log\frac{P(\lambda_i, n)}{P(\lambda_i, n_0)} - \sum_{\Delta\lambda_2} 10\log\frac{P(\lambda_i, n)}{P(\lambda_i, n_0)} \tag{5}$$

for each separate measurement at a given $n$. Thus measuring the spectral shape of $\Delta I_{12}$ it is possible to detect small centre wavelength changes of the LPG caused by external perturbations such as the SRI $n$.

The final result is shown in figure 4 representing the time dependence of $\Delta I_{12}$ which is proportional to the degree of wetting of the particular LPG. The spectrally integrated differential loss reaches a maximum when the grating is fully wetted and reduces to zero when completely dry. In our experiments we added drops of liquids and measured the wetting/evaporation responses in time. Since ethanol and acetone have lower specific heats of evaporation they vaporize faster and we have added two drops twice at different moments. The cross-channel noise was estimated to be less than 1.5 %.

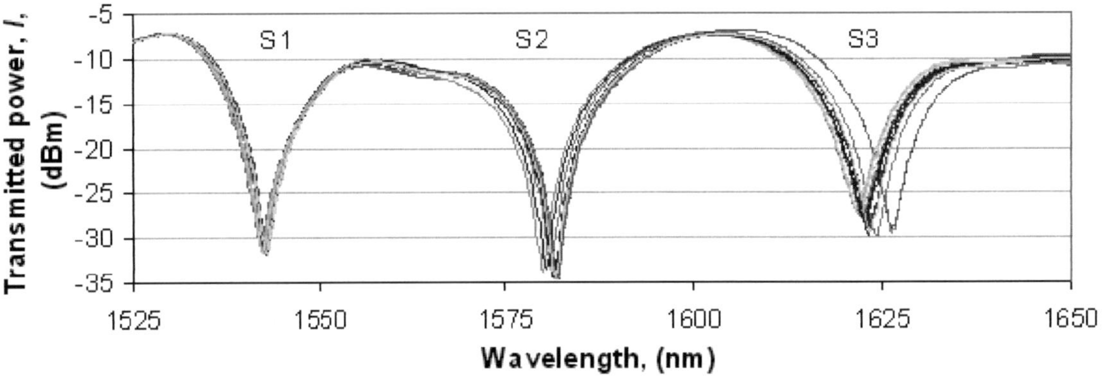


**Figure2**. Spectral dependencies of the losses for the three sensing LPGs S1, S2 and S3 at different moments of evaporation. The spectral shifts are clearly visible for all gratings.

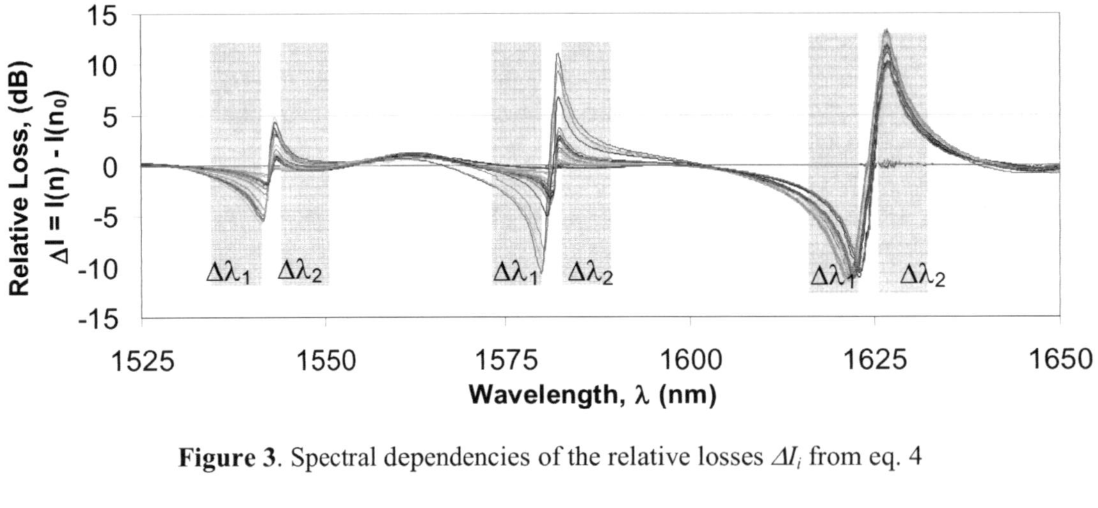


**Figure 3**. Spectral dependencies of the relative losses $\Delta I_i$ from eq. 4

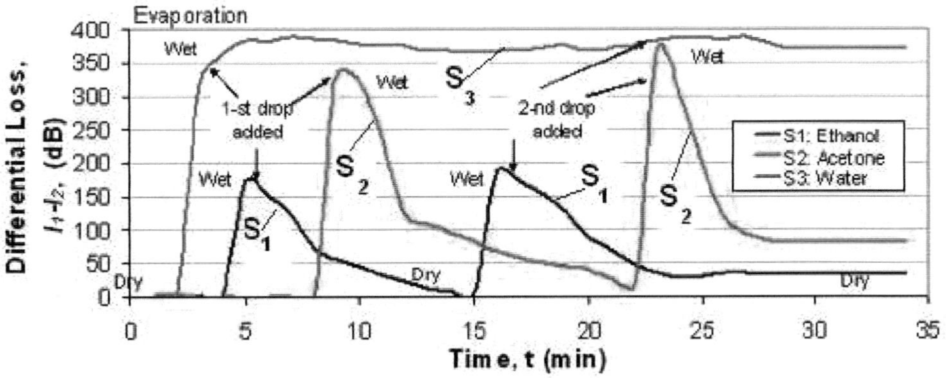


**Figure 4**. Integrated differential loss which is a measure of the center wavelength shifts of each grating during wetting/evaporation

## 4. Conclusion

We have observed the time responses of the evaporation process of ethanol, acetone and water from three spectrally multiplexed LPGs. The interrogation method used is based on differential loss measurement with a reference to the spectral distribution at the initial moment. The differential losses are summed over spectral intervals Dl1 and Dl2 around the center wavelength of each LPG which allows an easier to implement and cheaper detection system using either separate photodetectors or an InGaAs CCD array [5].

**Acknowledgements**

The authors gratefully acknowledge support for this work by the Natural Sciences and Engineering Research Council of Canada and by the Canada Research Chair Program. Support from the EES 303-2007 grant from the Bulgarian Ministry of Education and Science is also acknowledged.

## References

[1] Vengsarkar A M, Lemaire P J, Judkins J B, Bhatia V, Erdogan T, Sipe J E 1996 *Journal of Lightwave Technology* **14** 58-65

[2] James S W and Tatam R P 2003 *Measurement Science and Technology* **14** (5) 49-53

[3] Venugopalan T, Sun T and Grattan K T V 2008 *Sensors and Actuators A* **148** 57–62

[4] Humbert G and Malki A 2002 *J. Opt. A: Pure Appl. Opt.* **4** 194-198

[5] Baldziev P, Arnaudov R and Eftimov T A 2010 Study of the Performance of an InGaAs CCD Linear Photodiode Array for Fiber-Optic Grating Sensors, *ICEST- Conf. Proc., Ohrid, Macedonia*

Journal of Physics: Conference Series **253** (2010) 012024 doi:10.1088/1742-6596/253/1/012024

# Long period grating based refractometer with polarization-sensitive interrogation

**T Eftimov[1], W Bock[2], P Mikulic[2] and K Nikolova[3]**

[1] Faculty of Physics, Plovdiv University "P.Hilendarski", 24 Tsar Asen Str, Plovdiv 4000, Bulgaria

[2] Laboratoire de Recherche en Photonique, Université du Québec en Outaouais, P.O. Box 1250, Gatineau, Québec J8X 3X7, Canada

[3] University of Food Technology, 26 Maritza Blvd, Plovdiv 4002, Bulgaria

kr.nikolova@abv.bg

**Abstract.** We propose a new scheme for the interrogation of long-period fiber gratings (LPGs) which makes use of their polarization properties. Polarization-sensitive interrogation was applied to detect changes due to changes of the external refractive index by using three wavelengths on the International Telecommunication Union (ITU) grid. We show that the new approach can allow for a greater sensitivity and can be used in combination with spectral multiplexing schemes

## 1. Introduction

The polarization properties of long-period fibre gratings (LPGs) have caused considerable interest mainly because of the unwanted polarization-dependent losses (PDL) and their wavelength dependence [1]. The causes and magnitude of PDL have been studied for $CO_2$-written [2], arc-induced [3], and photonic-crystal-based LPGs [4,5] and several methods have been suggested for their reduction [6,7] or compensation [8]. Knowledge of PDL is essential in LPG-based components for communication as well as for sensor applications [9]. In the latter case, the existence of PDL can be made inessential only if a non-polarized broadband source is used. This condition, in turn, eliminates the possibilities to use single-wavelength sources such as standard communication lasers for LPG interrogation schemes.

The polarization properties of LPGs, however, are not restricted to the PDL. Recently [4,5] we have shown that linearly birefringent LPGs have a characteristic wavelength-dependent trajectory in the form of a circle on the Poincaré sphere when a highly polarized tunable laser is used. Consequently introducing external perturbations on an LPG the state of polarization at a given wavelength follows a unique trajectory [10].

In the present paper we show that when a linearly birefringent LPG is subjected to external perturbations such as torsion and change of the surrounding refractive index (SRI), the spectrally-dependent polarization trajectory on the Poincaré sphere, which lies approximately on a plane, changes orientation. The orientation of this plane can be determined by a single vector normal to it.

External perturbation causes the end of this vector to follow a unique trajectory in the three-dimensional Stokes space.

We show that by measuring the Stokes parameters of an LPG at three different wavelengths we can determine the orientation of the polarization trajectory for a given value of the external perturbation. This in fact represents a polarization-sensititive method for interrogating LPGs which is compatible with multiwavelength lasers [11] emitting on the standard International Telecommunication Union (ITU) grid.

## 2. Spectrally-dependent polarization evolution of an LPG

The experimental set-up used to measure the Stokes parameters $S_0$, $S_1$, $S_2$, $S_3$ and the degree of polarization (DOP) $P$, as well as to observe their spectral evolution on the Poincaré sphere, is shown in figure 1. A highy polarized laser (Agilent 8164A), tunable in the range from 1525 nm to 1584 nm, was used as a monochromatic source. A lightwave polarization analyzer (Agilent 8509C) was used to measure the total optical power $I$ in $dBm$, the Stokes parameters $S_{1,i}$, $S_{2,i}$, $S_{3,i}$ and $P_i$ at each wavelength $\lambda_i$. Simultaneously, each polarization state is represented in a three-dimensional Stokes space by a point with normalized coordinates $s_{1,i}$, $s_{2,i}$ and $s_{3,i}$ such that

$$s_{k,i} = S_{k,i} / P_i \qquad (k = 1,2\ 3) \tag{1}$$

$$P_i = \left(S_1^2 + S_2^2 + S_3^2\right)^{1/2} . S_0^{-1} \tag{2}$$

where $S_0$ is the total power, which if expressed in $dBm$, we note as $I$.

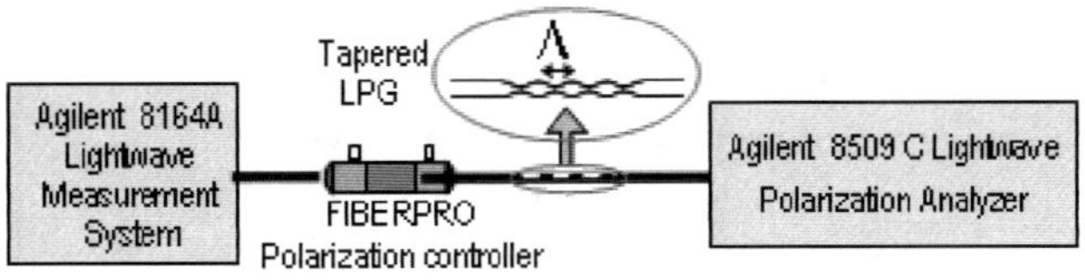


**Figure 1**. Experimental set-up used for measuring the wavelength dependence of the Stokes parameters and the DOP

The three-dimensional Stokes space is defined by $S_1$, $S_2$, $S_3$. When light is completely polarized $P = 1$, all points lie on a sphere, known as the Poincaré sphere. For partially polarized light ($0<P<1$), the state of polarization can be represented either by a point inside the Poincaré sphere or by the normalized Stokes parameters [1], in which case the state is projected to the Poincaré sphere as in our experiments. Depolarized light with $P = 0$ is represented by a point in the origin, or by infinite number of points evenly distributed on the Poincaré sphere.

Several tapered LPGs were fabricated from an SMF-28 Corning optical fiber using a standard splicer (FITEL S182K). In the experiment the SRI around the LPG was varied from that of air $n_a = 1$ and distilled water $n_a = 1.3323$ to that of 100% glycerine $n = 1.4714$. For each value of the SRI, the Stokes parameters $s_{k,i}$ ($k = 1, 2, 3$) were measured at wavelengths $\lambda_i$ ($i = 1, ...n$) and the polarization states were automatically plotted on the Poincaré sphere. Figure 2 shows the spectral evolution of the state of polarization of LPG-2 for two different ambient media, air and distilled water. We see that the spectral trajectory on the Poincaré sphere can be considered to lie on a plane, which we shall refer to as 'Stokes plane'. In reality, the trajectory lies on a slightly curved surface. However, for all practical purposes we assume that the error is negligible. Comparison between the two trajectories $T_1$ and $T_2$ reveals that $T_2$ is rotated with respect to $T_1$ through an angle $\Delta\Theta$, which we define as the angle between the vectors $\mathbf{N}_1$ and $\mathbf{N}_2$, normal to the plane of each trajectory as illustrated in figure 2c.

To describe the orientation of these vectors we need to find the equation for each plane [11], which in the Stokes space can be expressed as a linear equation with respect to the variables $s_1$, $s_2$, and $s_3$:

$$N_1 s_1 + N_2 s_2 + N_3 s_3 + D = 0 \tag{3a}$$

where $N_1$, $N_2$ and $N_3$ are the coordinates of the vector **N**, normal to the Stokes plane, namely:

$$\mathbf{N} = (N_1, N_2, N_3) \quad ,|\mathbf{N}| = \left(N_1^2 + N_2^2 + N_3^2\right)^{1/2} \tag{3b}$$

and $D$ is a constant. If we use the Hessian normal form, then we can introduce the orthonormal unit vector $\mathbf{n} = \mathbf{N}/|\mathbf{N}|$ [11]. To determine the coefficients $N_1$, $N_2$, $N_3$ and $D$ we need to know the Stokes parameters at three different wavelengths $\lambda_1$, $\lambda_2$ and $\lambda_3$, namely $S_{1,1}$, $S_{2,1}$ and $S_{3,1}$ for $\lambda_1$; $S_{1,2}$, $S_{2,2}$ and $S_{3,2}$ for $\lambda_2$ and $S_{1,3}$, $S_{2,3}$ and $S_{3,3}$ for $\lambda_3$. Then, using well-known relations [11], we can calculate the coefficients $N_1$, $N_2$ and $N_3$.

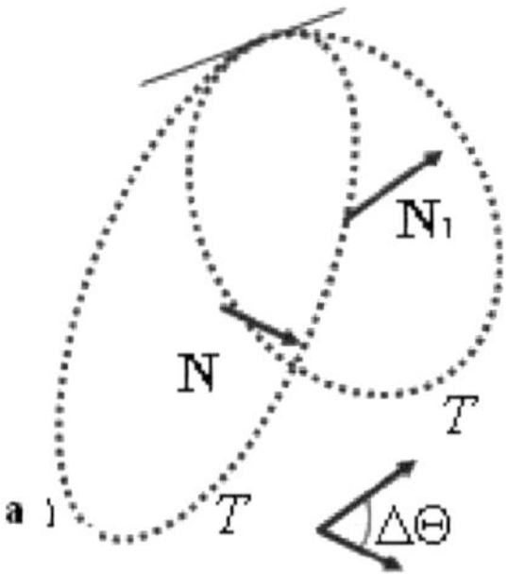


**Figure 2**. Spectral trajectories of the state of polarization at the output of LPG-2 at two different ambient refractive indices where each point represents a state of polarization at a separate wavelength: a) trajectories $T_1$ and $T_2$, the vectors $\mathbf{N}_1$ and $\mathbf{N}_2$ normal to them and the angle between them $\Delta\Theta$

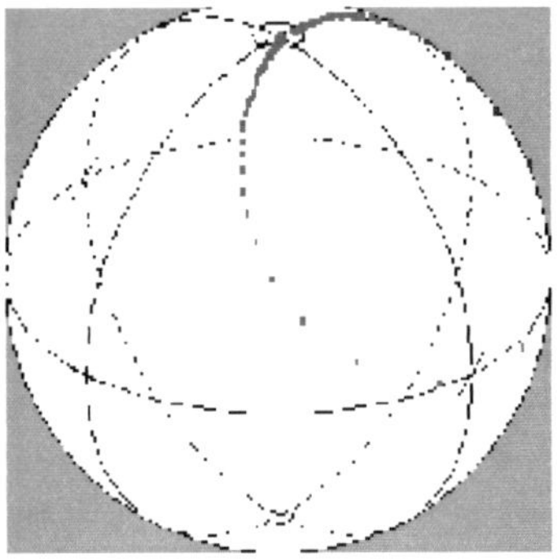

**Figure 2** b) $T_1$ - in air;

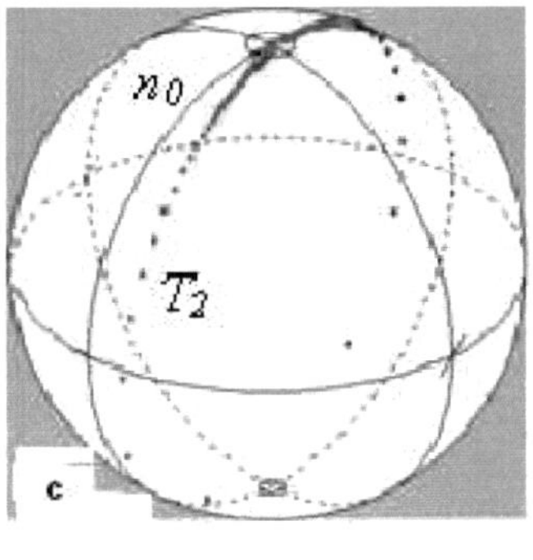


**Figure 2 c)** $T_2$ – in distilled water**;**

Figure 3 shows the rotation of the Stokes plane due to successive twists through different angles applied to LPG-3. If the external perturbation changes by equal increments then the ends of the vectors $\mathbf{N}_i$ change orientation and define a polarization trajectory (figure 3b). If we denote by $\Delta\Theta_i$ the angles between $\mathbf{N}_i$ and $\mathbf{N}_{i-1}$, then the total rotation angle accumulated is given by $\Theta = \sum \Delta\Theta_i$. For infinitesimal Stokes plane rotation angles $d\Theta$, summation will be replaced by an integral.

There are several ways of defining a line trajectory associated with the angular movement of the Stokes plane. One is to use the end of the normal **N**, another is to use the unit vector **n** to indicate a point on the trajectory and the elementary arcs will accordingly be $dS$ and $ds$ (Table 1) while $S$ and $s$

will be the lengths of the corresponding trajectories. A second way is to multiply the distance $d$ from the plane to the point of origin by **n** and define the vector $v = dn$ which uniquely describes the orientation of the Stokes plane by tracing a trajectory inside the Stokes phase space as shown in figure 3b). The length of the trajectory will then be $\delta$. A fourth way is to use the degree of polarization $P$ from equation (2), which was shown to vary depending on the phase detuning between the coupled polarization modes [4]. The elementary arc $dp$ and the total trajectory $P_{DOP}$ are also given in Table 1.

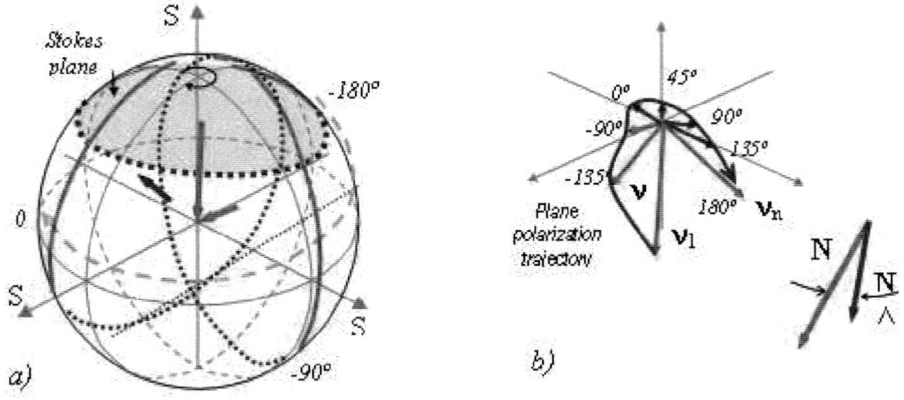


**Figure 3**. Rotation of the Stokes planes: a) spectral trajectories of the state of polarization at the output of LPG-3 for three different twist angles $\theta_1$ = -180°, $\theta_2$ = -90°, $\theta_3$ = 0° b) evolution of the normal vectors $N_i$

**Table 1**. Elementary arcs and lengths of trajectories for each of the four cases.

| | | | | |
|---|---|---|---|---|
| Elementary arc | $dS = \lvert\mathbf{N}\rvert\, d\Theta$ | $ds = \lvert\mathbf{n}\rvert\, d\Theta$ | $d\delta = d.\lvert\mathbf{n}\rvert\, d\Theta$ | $dp = P\mathbf{n}\, d\Theta$ |
| Length of trajectory | $S = \int \lvert\mathbf{N}\rvert\, d\Theta$ | $s = \int \lvert\mathbf{n}\rvert\, d\Theta$ | $\delta = \int d\lvert\mathbf{n}\rvert d\Theta$ | $P_{DOP} = \int P\lvert\mathbf{n}\rvert d\Theta$ |

Ultimately, because the orientations of the vectors **N** and **n** depend on the value $x$ of the external perturbation, then $S(x)$, $s(x)$, $\delta(x)$ and $P_{DOP}(x)$ are functions of $x$ as well. More importantly, each point on any of the plane polarization trajectories is uniquely related to a particular value $x$ of the measurand. Thus identifying the position on the plane polarization trajectory is equivalent to measuring the value of the particular external perturbation. In practice, a sensor based on this type of interrogation would be using the mapping of the trajectory vs. values of the external perturbation. In this case, to avoid ambiguity, it is important that the polarization trajectory does not cross itself. While this may happen to the trajectory defined by the normal unit vector **n**, it will not happen to the other trajectories.

## 3. Experimental results and analysis

We investigated two tapered LPGs made from standard SMF-28 Corning fiber by introducing periodic microtapering of pitch Λ and a total number of $k$. Fusion-induced LPGs have been known to exhibit linear birefringence and hence, polarization-dependent properties. The gratings studied were labeled as LPG-2 ($\lambda_c$ = 1544 nm) and LPG-3 (Λ = 770 μm, $k$ = 67, $\lambda_c$ = 1559 nm). The whole grating region was submerged in different concentrations of water solutions of glycerine to obtain a total of eleven different values for the SRI ranging from 1.3327 to 1.4653. These refractive indices were measured in the visible range ($\lambda$ = 589 nm) using a standard Abbe refractometer.

Using the set-up shown in figure 1 we measured the spectral evolution of the Stokes parameters for the different refractive indices. The wavelength of the tunable laser was changed by increments of $\delta\lambda$ = 0.1 nm. The results obtained are summarized in the plots in figure 4. Before taking the measurements, we set the state of polarization of the light close to right circular state (figure 2a,b) and figure 3a) at wavelengths away from the LPG resonance. This ensures that both polarizations are almost equally excited at the LPG input.

We found that changes of the SRI lead to changes in the spectral dependencies of the Stokes parameters, to wavelength shifts of their extrema (minima and maxima) and, consequently, to shifts of the PDL and DOP minima. We compared these polarization-dependent wavelength shifts with the standard centre wavelength shifts of the resonance wavelengths as shown in figure 4a,b. The results showed that, with the exception of the shifts of some extrema of Stokes parameters, the responses to changes in the SRI are practically identical. Fig. 4a shows that the wavelength shift of the maximum of the $S_3$ parameter is twice as large as the wavelength shifts of the central resonance wavelength and the minima of the DOP and PDL. Also, the sensitivity for SRIs from 1.38 and 1.45 is much higher than the standard response. In the 1.33 to 1.43 range, the averaged sensitivity for these center wavelength shifts is 0.087 r.i.u./nm which with a 10 pm spectral resolution would yield a sensitivity 8.7 x $10^{-4}$ r.i.u., where r.i.u. stands for "refractive index unit". Results for LPG-3 were similar.

To determine the orientation of the Stokes plane we took the values of the Stokes parameters at three different wavelengths, each of them coinciding with or very close to an ITU 100 GHz grid channel used in DWDM. These three wavelengths used to calculate the angular increments $\Delta\Theta_i$ are referred to as a "trio". For LPG-2 the central wavelength of the middle channel coincides with ITU channel #44. The channel combinations were as follows: trio #1 - ITU channels 43, 44, and 45; trio #2 - 42, 44, and 46, and trio #3 - 41, 44, and 47. If the spectral trajectories $T_i$ on the Poincaré sphere were ideal circles, all trios would yield identical responses. However, as was indicated earlier, the trajectories do not lie on a plane but on a slightly curved surface which results in different responses for each of the trios, as is clearly seen in figure 4b.

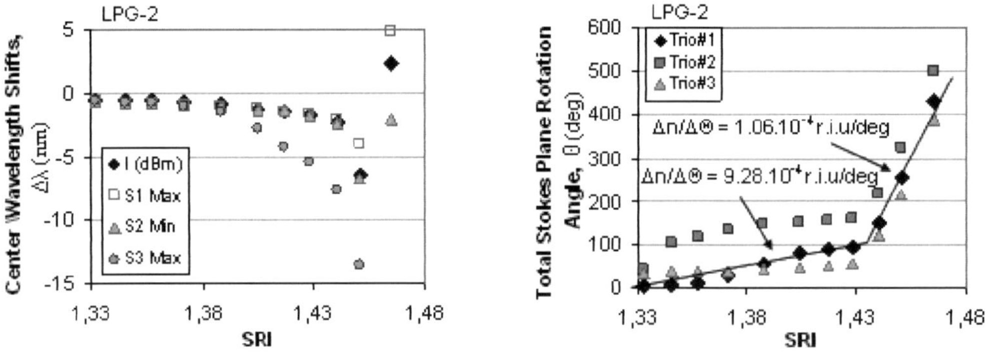


**Figure 4**. Responses of SRI changes in LPG-2 : a) wavelength shifts of the transmission minimum of the total power I and of chosen extrema (minimum or maximum) of each of the Stokes parameters; b) total Stokes plane rotation angle for three different trios of ITU channel wavelengths

The curves for the total Stokes plane rotation angle are quite similar in that they are essentially characterized by two regions of different sensitivity $\Delta n/\Delta\Theta$. The sensitivities for trio #1 are correspondingly $\Delta n/\Delta\Theta=9.28\times10^{-4}$ r.i.u./deg for the 1.33 to 1.43 range and $\Delta n/\Delta\Theta=1.06\times10^{-4}$ r.i.u./deg above 1.43. The orientation of the trajectories has been found to be stable and repeatable within about $\delta\Theta = 0.2$ deg, so the resolution of SRI measurements would be correspondingly $\Delta n = 1.86\times10^{-4}$ r.i.u. ($n = 1.33$ to 1.43) and $\Delta n = 2.12\times10^{-5}$ r.i.u. ($n > 1.43$) which is comparable to or better than the sensitivity obtained from the wavelength shifts.

## 4. Conclusions

We observed a rotation of the spectrally-dependent polarization trajectory of an LPG on the Poincaré sphere when the grating was subjected to changes in the SRI. Based on this finding, we have proposed a polarization-sensitive method for the interrogation of LPGs which makes use of their birefringence. The method uses three wavelengths on the ITU grid and offers a higher sensitivity than conventional methods based on centre wavelength shifts. It can also allow polarization-sensitive wavelength division multiplexing.

**Acknowledgments**

The authors gratefully acknowledge support for this work by the Natural Sciences and Engineering Research Council of Canada and by the Canada Research Chairs Program as well as the Ministry of Education in Bulgaria for support with research grant VVU 303/07.

## References

[1] Lee B H, Cheong J and Paek U C 2002 *Optics Lett.* **27** 1096-98
[2] Bachim B L and Gaylord T K 2003 *Appl. Opt.* **42** 6816-23
[3] Rego G M, Santos J L and Salgado H M 2006 *Opt. Comm.***262** 152-156
[4] Eftimov T A, Bock W J, Chen J and Mikulic P, "Muller-Stokes analysis of long-period gratings. Part I: Uniformly birefringent LPGs", *J. Lightwave Technol,* (accepted for publication)
[5] Eftimov T A, Bock W J, Chen J and Mikulic P, "Muller-Stokes analysis of long-period gratings. Part II: Uniformly birefringent LPGs", *J. Lightwave Technol,* (accepted for publication)
[6] Oh S T, Han W T, Paek U C and Chung Y 2003 *Opt. Express* **11** 3087-3090
[7] Oh S T, Han W T, Paek U C and Chung Y 2004 *Microwave and Opt. Techn. Lett.* **41** 188-190
[8] Kim Ch-S, Choi B, Nelson J, Li Q, Dashti P and Lee H 2005 *Opt. Letts.* **30** (1) 20-22
[9] Bhatia V 1999 *Opt. Express* **4** 457-466
[10] Eftimov T A, Bock W J, Mikulic P, Chen J and Nikolova K 2010 *Meas. Sci. Techn*, (accepted)
[11] Brochu G, LaRochelle S and Slavik Radan 2005 *J. Lightwav. Technol.***23** 44-53
[12] Harris J W and Stocker H 1998 *Handbook of Mathematical and Computational Science*, **New York:** Springer-Verlag

# UV light-induced thermoluminescence of Er + Li doped $ZrO_2$

**H Hristov**[1], **N Arhangelova**[1], **V Velev**[1], **I Penev**[2], **M Bello**[3], **G Moschini**[3] **and N Uzunov**[1, 3, 4]

[1] Faculty of Natural Sciences, Shumen University "K. Preslavsky", 115 Universitetska str., 9712 Shumen, Bulgaria

[2] Institute for Nuclear Research and Nuclear Energy, 72 Tzarigradsko chaussee blvd., 1784 Sofia, Bulgaria

[3] National Laboratories of Legnaro, INFN, Viale dell'Universita 2, 35020 Legnaro (PD), Italy

E-mail: Nikolay.Uzunov@lnl.infn.it

**Abstract**. Analysis of the thermoluminescence (TL) properties of $ZrO_2$ doped with Eu and Li has been conducted. Different quantities of lithium co-dopant have been added as $Li_2CO_3$ to a mixture of $ZrO_2$ with 1wt% Er. Pellets sintered at a temperature of $1200^0C$ have been prepared and the kinetic parameters of the phosphors have been studied after irradiation with UV light. It has been shown that the addition of 8 to 10 wt% of Li to the mixture of $ZrO_2$ with 1wt% Er yields a maximum intensity of the peaks at $65^0C$ and at $105^0C$. Spectral emission and spectral sensitivity of the phosphors have been studied. The analysis applied to TL glow curves, obtained from the UV irradiated phosphors and kept after the irradiation at different times in a dark storage, revealed that the peaks at $65^0C$ and $105^0C$ have relatively long fading. It is concluded that the phosphors thus obtained possess a good sensitivity to the UV emission and could be appropriate phosphors for detection and quantitative measurements of UV light.

## 1. Introduction

Thermally stimulated emission (thermoluminescence) of some crystals can be stimulated by irradiation with photons with lower than the X-rays' and gamma-rays' energies for instance ultra-violet (UV) emission. This case is interesting with respect to possible applications such as detection and quantitative measurements of the UV emission, holographic storage, etc.

It is known that an enhancement of the thermoluminescence (TL) emission of some phosphors can be achieved after co-doping with lithium [1, 2]. In this article we present the results from a study, conducted on the enhancement of the TL emission of $ZrO_2$ doped with Er and co-doped with Li, as a potential indicator, sensible to the UV light in the range of 290 nm – 400 nm.

## 2. Materials and methods

UV sensible phosphors have been produced from sintered pellets of $ZrO_2$ doped with 1 wt% of Er as $Er_2O_3$ according the procedure described in [3]. Different quantities of Li have been added to the

[4] To whom any correspondence should be addressed.

mixture as $Li_2CO_3$, thus obtaining samples with Li concentrations of 2.5 wt%; 5 wt%; 10 wt% and 15 wt%. The mixtures, pressed in pellets with a diameter of 8mm and thickness of 1 mm, have been sintered at $1200^0$C for 8 hours.

A setup for precise TL measurements (TL reader), developed in the Laboratory for Nuclear Physics and Radioecology (LNPR) at the University of Shumen, has been used to obtain the TL glow curves of the phosphors [4]. The system consists of a setup for sample heating, a glow curve measuring device on the basis of photomultiplier tube, and a setup for sample irradiation with higher-intensity ultraviolet (UV) light. The operation and control of the system has been realized on the basis of PIC16F876 microcontroller as well as on original forward/backward feedback method based on the so-called model predictive controllers in order to achieve higher temperature stability and repeatability of the analyses conducted.

The kinetic parameters of the prepared phosphors have been studied analyzing their TL glow curves (GC) after irradiation with a xenon short-arc lamp type XBO 75 W/2 from OSRAM. The crystals have been placed at a distance of 10 cm and the time of irradiation was in the interval 3 - 5 min. The alimentation of the XBO lamp was 75W maintained at lamp current of 5.4A, recommended by the manufacturer.

Kinetic parameters such as the activation energy and the kinetic order have been obtained applying glow curve decomposition analysis on each GC obtained. A single glow peak of a general order of kinetics was fitted using the equation [5]:

$$I(T) = I_m b^{\frac{b}{b-1}} \exp\left[\frac{E}{kT}\cdot\frac{T-T_m}{T_m}\right]\left[(b-1)(1-\Delta)\frac{T^2}{T_m^2}\exp\left[\frac{E}{kT}\cdot\frac{T-T_m}{T_m}\right]+Z_m\right]^{-\frac{b}{b-1}}, \quad (1)$$

where I(T) is the glow-peak intensity, $I_m$ is the maximum glow-curve intensity, E (eV) is the activation energy, k is the Boltzmann constant, T (K) is the temperature, $T_m$ (K) is the value of the temperature at the peak maximum, with $\Delta = 2kT / E$, $\Delta_m = 2kT_m / E$; $Z_m = 1+(b-1)/\Delta_m$ and the parameter b ($1 < b \leq 2$) is the so-called kinetic order.

When the glow peak is a composite peak, compounded by m overlapping glow peaks, the method consists of a minimization of the $\chi^2$ function:

$$\chi^2 = \sum_{i=1}^{n}\left(I_i - \sum_{j=1}^{m} I_j(T_j)\right)^2, \quad (2)$$

where $I_i$ are the measured GC intensity values, $I_j(Ti)$ are the values of each partial GC at the temperature $T_i$, n is the number of the experimental points and m is the number of the partial glow peaks compounding the observed peak.

The fading of the peaks in the TL glow curve (GC) has been obtained calculating the intensity of the corresponding partial TL emission (the net peak area) from the phosphors kept for a certain time after the irradiation in a dark storage at a temperature of $20^0$C. Glow curves obtained with the TL reader have been analyzed using a developed in the LNPR computer program implementing data smoothing using a Savitzky-Golay filtering and subsequently applying a minimization procedure exploiting conjugated gradients method for the peak decomposition technique described above.

## 3. Experimental results

A TL glow curve of $ZrO_2$, doped with $Er_2O_3$ and measured immediately after 3 minutes of irradiation with UV light from the XBO lamp at a distance of 10 cm, is shown in figure 1a). The GC, presented with a thick black line, has been obtained without annealing at a heating rate of $0.16^0$/s. Three peaks have been revealed by the decomposition analysis: peak 1 at 65 $^0$C, peak 2 at 105 $^0$C and peak 3 at 150 $^0$C, indicated in the figures of this article as 1, 2 and 3 respectively. The resulting fit is presented with white line in the same figure.

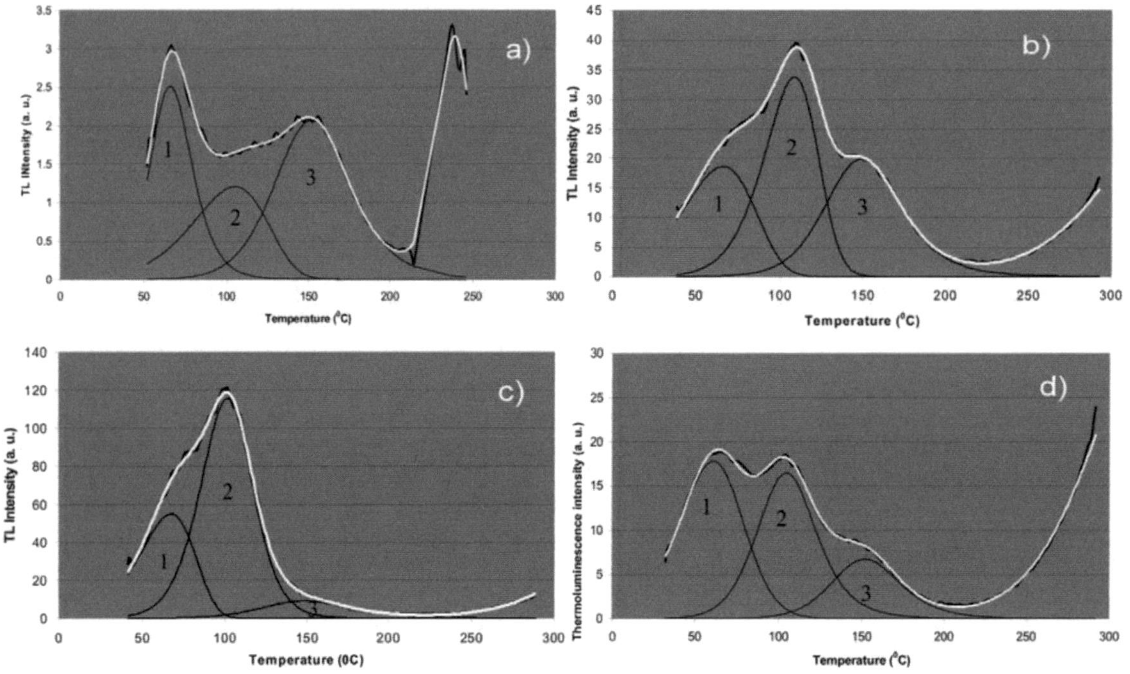


**Figure 1.** Glow curves of $ZrO_2$ doped with 1 wt% $Er_2O_3$ and co-doped with different quantities of Lithium: a) without $Li_2CO_3$; b) with 5 wt% $Li_2CO_3$; c) with 10 wt% $Li_2CO_3$; d) with 15% $Li_2CO_3$.

The TL glow curves obtained from $ZrO_2$ doped with 1 wt% of Er and co-doped with 5 wt%; 10 wt% and 15 wt% of Li, are shown in figure 1b) - 1d) respectively. The curves have been measured immediately after the irradiation with UV light under the same conditions (time of UV irradiation and distance to the XBO lamp). The glow curves, presented with thick black lines, have been acquired without annealing at a heating rate of 0.16$^0$/s. The resulting three peaks at 65 $^0$C, at 105 $^0$C as well as at 150 $^0$C have significantly higher TL yields as it can be seen from the figures.

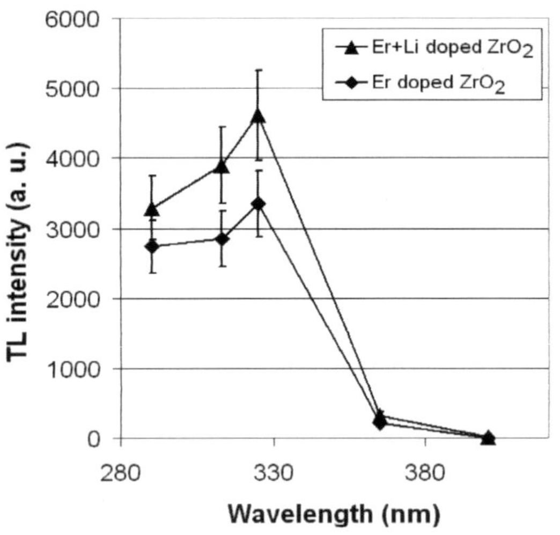


**Figure 2.** Spectral sensitivity of $ZrO_2$ doped with Er(1%) and Er(1%) +Li(10%).

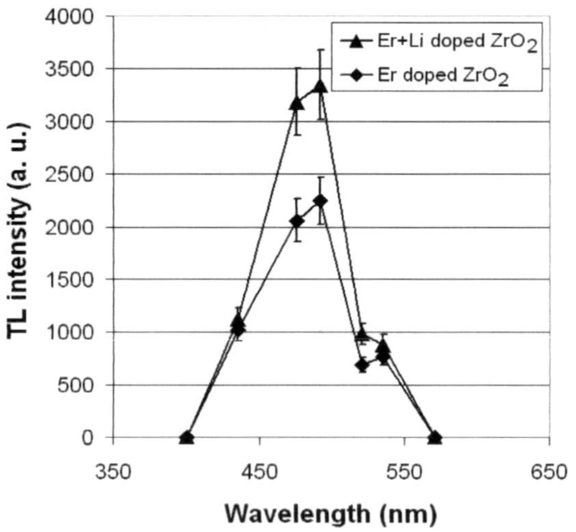


**Figure 3.** Spectral emission of $ZrO_2$ doped with Er(1%) and Er(1%) +Li(10%).

The dependence of the TL emission with respect to the wavelength of the irradiating light (the spectral sensitivity) has been studied measuring the intensity of the whole TL glow curve for $ZrO_2$ doped with Er and for $ZrO_2$ doped with Er+Li. Interference filters for different wavelength were placed consecutively between the XBO lamp and the phosphors during each irradiation. The measured total TL glow curves intensities versus the wavelength of the irradiating light are shown in figure 2. Thus the obtained spectral sensitivities of the crystals reveal a wavelength limit for the irradiating light around 380 nm. The later fact shows that the TL emission of the phosphors is caused only by the UV component of the emission of the XBO lamp.

Figure 3 shows the emitted total TL light intensity of Er doped $ZrO_2$ and Er+Li doped $ZrO_2$ with respect to the wavelength (the spectral emission). The TL emission has been measured during the crystal heating after 5 min irradiation of the crystal with XBO lamp. A set of interference filters has been placed in front of the photomultiplier tube of the TL reader during the acquisition and the total number of counts for a heating cycle ($30^0$C – $200^0$C) has been registered immediately after the irradiation. Corrections for the light attenuation in the filters and for the spectral response of the photomultiplier tube have been made to the measured total number of counts.

The analysis of the measured peak intensities from the glow curves of the sintered $ZrO_2$ doped with 1% Er form $Er_2O_3$ and the co-dopant Li from $Li_2CO_3$ in quantities of 2.5 wt%, 5 wt%, 10 wt% and 15 wt% shows that the quantities around 10wt% of Li co-dopant generate maximum intensities of the GC peaks. A plot of the peak intensities (integrated counts in the corresponding peak areas) for Peak 1 and Peak 2 from fig 1, as a function of the quantity of Li co-dopant, is shown in figure 4.

The calculated kinetic parameters for the peaks appearing in the TL glow curves of $ZrO_2$ doped with $Er_2O_3$ and $Li_2CO_3$ are shown in Table 1.

**Table 3.** Kinetic parameters of the peaks revealed in $ZrO_2$ doped with $Er_2O_3$ and $Li_2CO_3$

| Peak reference | Peak position ($^0$C) | Activation energy (eV) | Kinetic order |
|---|---|---|---|
| Peak 1 | 65 ± 2.0 | 0.52 ± 0.07 | 1.1 ± 0.15 |
| Peak 2 | 105 ± 1.8 | 0.96 ± 0.03 | 1.6 ± 0.29 |
| Peak 3 | 150 ± 1.3 | 1.03 ± 0.09 | 2.0 – 0.02 |

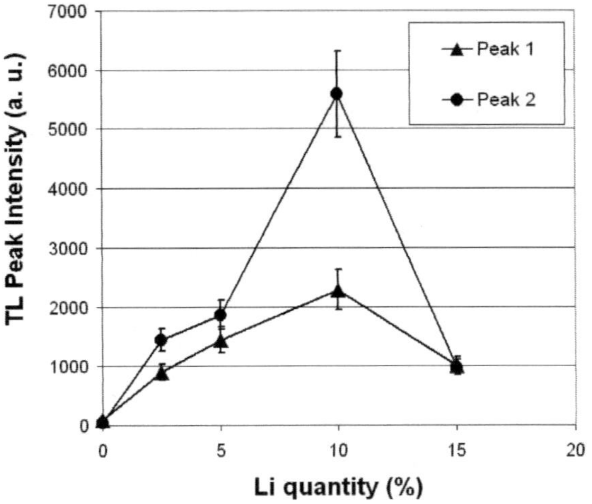


**Figure 4.** Measured net area of Peak 1 and Peak 2 of the TL glow curves from Er-doped $ZrO_2$ as a function of the quantity of Li co-dopant.

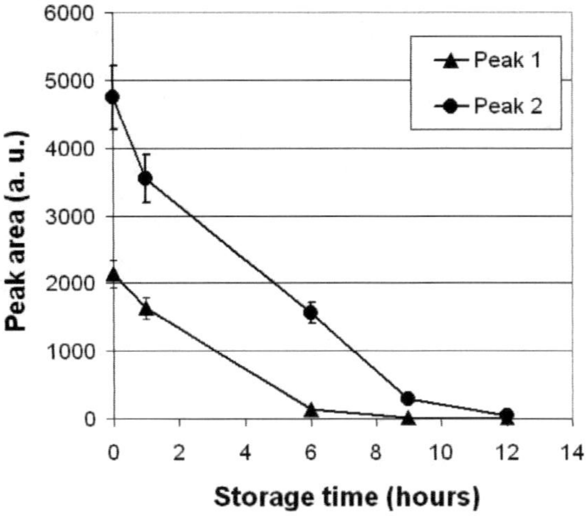


**Figure 5.** Fading of Peak 1 and Peak 2 of phosphors from sintered $ZrO_2$ doped with 1 wt% Er and 10 wt% Li.

Finally the fading measurements about the observed Peak 1 and Peak 2 in fig 1 have been performed plotting the peak areas from the TL glow curves versus the storage time of the phosphors from sintered $ZrO_2$ doped with 1 wt% Er and 10 wt% Li. (figure 5). To perform these measurements the phosphors have been irradiated each time with an equal amount of light (5 min) from the XBO lamp. After the irradiation the crystals have been kept for certain time periods in dark place at a temperature of $20^0$C. The maximum storage time interval was 12 hours. Form figure 5 is seen that peak 1 and peak 2 fade completely for about 12 hours.

## 4. Conclusion

The analysis of the TL glow curves of the sintered crystals shows that the presence of Li ions as a co-dopant enhances significantly the TL emission of $ZrO_2$ doped with $Er_2O_3$ with respect to the UV light. Quantities between 8 wt% and 10 wt% of $Li_2CO_3$ induce higher thermoluminescence emission of the peaks at 65 $^0$C and 105 $^0$C.

An interesting fact, analogical to the one, described in [2] is observed in the TL glow curves in figure 1: the increase of the TL emission of the two peaks is followed by the decrease of the higher-temperature TL peaks, where the deconvolution procedure is difficult to be conducted due to the higher thermal emission of the samples.

The measurements carried out for the fading of the peaks showed that the peak 1 has a fading of approximately 9 hours and the peak 2 has slightly longer fading – about 12 hours. This fact together with the higher sensitivity with respect to the UV light, shows on the possibility to use the sintered pellets of $ZrO_2$ doped with 1 wt% Er and co-doped with 8 to 10 wt% of Li as a potential phosphor for UV thermoluminescence measurements.

**Acknowledgements**

One of the authors (N Uzunov) would like to acknowledge the support received from the ICTP, Trieste, Italy and in particular the Program for Training and research in Italian Laboratories. Part of this work has been sponsored by Grants 5326 and 5323, RD – 07-1003/30.03.10 from the University of Shumen, Bulgaria.

**References**

[1] Hatayama Y, Fukumoto S and Ibuki S 1992 Photoluminescence on Li-codoped ZnS:Tm phosphor *Jpn. J. Appl. Phys.* **31** 3383 – 88

[2] Prokic M 2000 Effect of Lithium co-dopant on the thermoluminescence response of some phosphors *App. Rad. and Isotopes* **52** 97 – 103

[3] Hsieh W and Su C UV induced thermoluminescence in $ZrO_2$ doped by $Er_2O_3$, 1994 *J. of Physics D: Appl. Physics* **27** 1763 – 68

[4] Hristov H, Arhangelova N, Penev I and Uzunov N 2008 A setup for thermoluminescence analysis of crystals *Ann. of Konstantin Preslavski University B1* **16** 30 – 38

[5] Kittis G, Gomez-Ros J and Tuyn J 1998 Thermoluminescence glow-curve deconvolution functions for first, second and general orders of kinetics *J. Appl. Phys.* **31** 2636 – 41

# Birefringence in heat-mechanical modified freshly moulded polyester fibers

**V Velev[1], T Dimov[1], A Popov[2], Y Denev[2], H Hristov[1], T Angelov[3], K Markova[2], M Zagortcheva[2], N Arhangelova[1] and N Uzunov[1]**

[1]Department of Natural Sciences, Shumen University "Konstantin Preslavsky", 115 Universitetska Street, 9712 Shumen, Bulgaria
[2] University "Prof. Dr. Assen Zlatarov"-Burgas, 1 "Prof. Jakimov" Street, 8000 Burgas, Bulgaria
[3] Lukoil-Neftohim-Burgas AD, 8000 Burgas, Bulgaria

E-mail: v.velev@shu-bg.net

**Abstract**. The article submits new experimental data concerning to the role of combined thermo-mechanical treatments on the structural development of freshly moulded uncrystallized but crystallizable poly (ethylene terephthalate) (PET) fibers. The object of the present work is PET as a thermoplastic polymer with a large practical application. The report is devoted to the influence of the heat-mechanical modification temperature on the structure rearrangement in uniaxially orientated amorphous PET. The heat-mechanical modification of the investigated yarns and the optical measurements were realized by specialized gears constructed and built in the author's laboratories. The fibers heat-mechanical modification includes samples annealing at constant temperature above their glass transition temperature ($T_g$) without strain stress. The yarn annealing has been followed from well defined uniaxially strain-loading with values from 0 *MPa* up to 30 *MPa* during two minutes. The optical measurements were carried out by an optical system using a polarization microscope and a CCD camera. The obtained experimental data has been analyzed by Mocha-1.2 (Jandel Scientific) software. There are established dependences between the heat-mechanical modification mode and the structural rearrangements running in the studied PET samples.

## 1. Introduction

Poly (ethylene terephthalate) (PET) is a crystallizable thermoplastic polymer, which is widespread in many high-volume usages very often in orientated forms such as filaments and folios. The widely applications of PET are founded on its crystallization ability, relatively low crystallization velocity and comparatively high glass transition temperature. It is known that the end properties of the orientated polymeric systems are direct consequence of their super molecular structure [1-3]. The final structure of the polymeric filaments depends mainly by the melt spinning conditions and by the subsequent heat-mechanical modifications, too. The desirable super molecular structure and the needed final morphologies and properties of the polymer fibers can be obtained by controlling of the forming process parameters as well as the subsequent treatments [4-6]. Several experimental aspects of the deformation behavior and super molecular structure rearrangement in PET fibers caused by heat and mechanical treatments have been investigated in the recent years including participation of the authors of this article. Part of the experiments have been carried out in non-isothermal conditions [7,8] while in other of them the structural development of PET fibers are investigated at constant temperatures without application of stress [9] and with application of strain stress [10].

The objective of the ongoing experiments with polymer filaments in our laboratories is to establish links between the results collected at different experimental conditions and the basic parameters of the thermal deformation experiments and namely:

regime of the specimens heat treatment;
heating (cooling) rate in non-isothermal conditions;
temperature of the samples thermal treatment in isothermal conditions;
method of the mechanical treatment;
degree and speed of fibers withdrawal;
value of the applied tension stress.

Nevertheless the effects of the combined tension stress and thermal treatments on the structure development, relaxation process and phase transitions of the amorphous PET yarns remains unclear.

With purpose to clarify the role of the temperature and tension stress values on the structure development in uniaxially oriented PET in our labs were carried out thermal deformation experiments with amorphous PET filaments at constant temperature of 80 $^0C$ and under precisely defined strain stresses from 0 *MPa* to 30 *MPa* with increment step of 3 *MPa* [11]. The structural investigations of the heat mechanically modified specimens show very important results concerning the role of the temperature and the applied stress and in particular:

appearance of "resonance" filaments birefringence with a pronounced maximum at tension stress value of 6 *MPa*;

significant increase of more than 30 % of the fibers degree of crystallinity with the strain stress increasing from 3 *MPa* to 6 *MPa* [11].

Therefore it was very interesting for us to continue the investigations of the same PET specimens at enhanced temperature of 85 $^0C$. In the present work we submit new experimental data pertaining to the part of the simultaneous thermal and mechanical modification on the deformation behaviour and structure development of amorphous PET fibers. The experiment includes annealing of an as-spun PET yarns at temperature 85 $^0C$ above its glass transition temperature defined in our other work at 74 $^0C$ [9] while the fiber bundle is subjected to a well defined strain stress from 0 *MPa* to 30 *MPa* with step of growth 3 *MPa*.

## 2. Experimental

### *2.1. Materials*

Amorphous PET as-spun multifilament yarns prepared on the industrial spinning installation Furnet (France) have been selected as a precursor samples. The original filament features and formation conditions were as follows:

spinning speed 1150 *m/min*;
number of spinnerets in yarn 32;
single filament diameter 44 $\mu m$.
density $\rho = 1338\ kg/m^3$ ;
degree of crystallinity $\alpha = 1.7$ %;
birefringence $\Delta n = 0.008$;
coefficient of amorphous orientation $f_a = 0.029$.

### *2.2. Methods*

The density of the raw fibers was determined using a density gradient column filled with a tetra chloromethane and n-heptane mixture. The samples degree of crystallinity was calculated on the basis of the obtained filaments density data. The birefringence of the untreated fibers was measured using an Amplival POL D (Karl-Zeiss Jena) polarizing microscope, provided with Bereck compensators. The coefficient of amorphous orientation of the rough filaments was calculated using the obtained birefringence data and the Stein's equation [12].

The samples heat-mechanical modification was carried out using a specialized device designed and produced in our laboratory. The thermal deformation experiment includes rapidly sample heating from a room temperature up to temperature of $85^0C$ with subsequent annealing for ten minutes. After annealing time the studied yarn was loaded during two minutes with well-defined tension stress from 0 *MPa* to 30 *MPa* with increment step of 3 *MPa*. The heat mechanical filaments treatment was followed by a simultaneously taking down of the applied strain stress and yarn moving from the oven to room temperature.

The birefringence measurements of the heat mechanically treated PET fibers were performed using a specialized set-up developed and made in the author's laboratory. The gear involves polarizing interference microscope with a CCD camera attached [9]. The keystone of the experimental device is the system of a polarizer (P), analyzer (A) and birefringent filament (F) in between and it is the so-called "P-F-A" system. The transmitting directions of the polarizers P and A are mutually perpendicular (crossed polarizers). The fiber can be rotated round the optical microscope axis.

## 3. Results and discussion

With purpose to specify the occurred in the fibers structure developments the transmission of linearly polarized monochromatic light with wavelength $\lambda$ = 590 *nm* across the P-F-A system has been investigated. The obtained birefringence data of the heat mechanical treated PET samples depending on the applied to the yarn tension stress are present in figure 1.

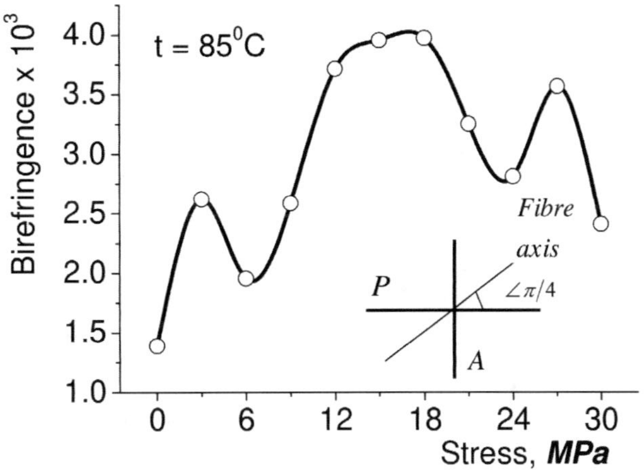


**Figure 1.** Birefringence of the heat mechanical treated PET filaments depending on the applied strain stress.

The experimental results shown on the figure 1 allow to see the role of the temperature of the heat mechanically treatments of the same PET filaments. In contrast to the birefringence received at temperature $80^0C$ where exist "resonance" dependence with pronounced peak at 6 *MPa* [11] here we can see a wide basic peak at stresses of 12 *MPa*, 15 *MPa* and 18 *MPa* with two attendant smaller peaks at stresses of 3 *MPa* and 27 *MPa* correspondingly.

The specimen's deformation behaviour during the thermal deformation experiments and more specifically the fibers diameter changes are shown in figure 2.

The alterations of the treated filaments length $(L-L_0)$ are present in the "stress-deformation" diagram in figure 3, where $L_0$ is the initial length of the fibers bundle. Here we detect the affect of the temperature on the PET filaments extension. The relatively few temperature increasing from $80^0C$ in our previous experiments [11] to $85^0C$ in the present heat mechanical fibers treatments leads to a tenfold yarns elongation growth (figure 3). As it can be seen from the figures 1, 2 and 3 the filaments' birefringence changes are in a good agreement with the samples' deformation behaviour.

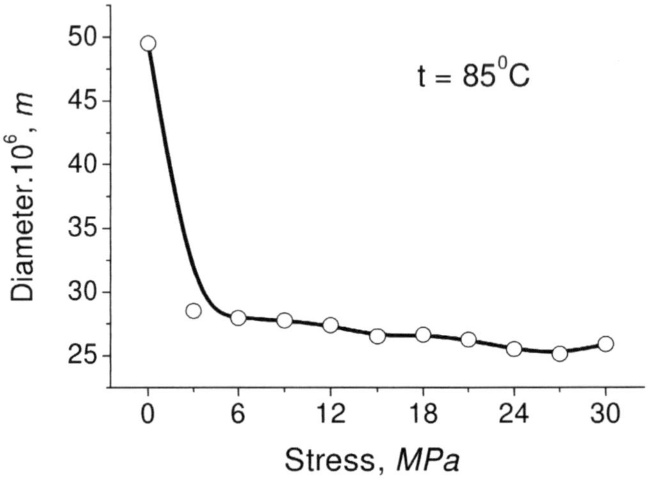


**Figure 2.** Fibers diameter.

The increasing of the amorphous orientation function and respectively the fibers birefringence at the applied strain stress from 0 *MPa* up to 3 *MPa* (figure 1) is in accordance with the collapse of the samples diameter (figure 2) and with the degree of withdrawal (figure 3).

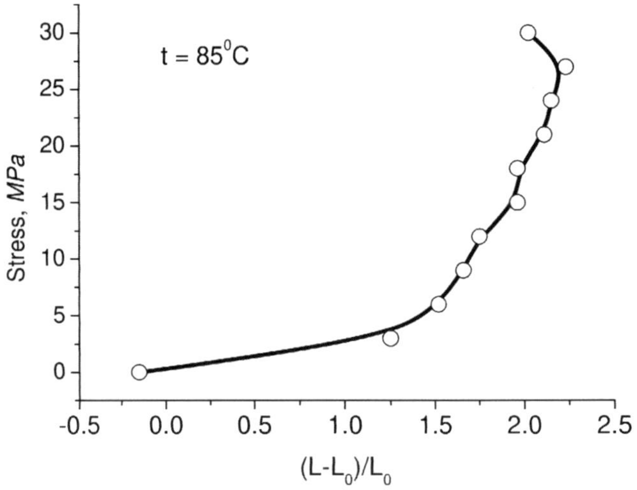


**Figure 3.** Tensile stress – extension diagram.

The specimens' birefringence reduction in the interval of the tension stress from 3 *MPa* to 6 *MPa* (figure 1) and the appropriate deformation behaviour (figures 2, 3) can be explained with the additional segments stretching and with the rupture of the pre-stressed segments, too. The strain stress values increasing from 9 *MPa* up to 27 *MPa* leads to the smooth filaments diameter (figure 2) reduction and to the gradually yarns length increase (figure 3). The possible reason for the observed

Journal of Physics: Conference Series **253** (2010) 012026 doi:10.1088/1742-6596/253/1/012026

deformation behaviour and the accompanying changes of the birefringence (figure 1) are the further fibrils withdrawal and slip accompanied from destructions and subsequent drop of the segments orientation. It should be mentioned the deformation behaviour of the studied filaments subjected to tension stress of 30 *MPa* (figures 2, 3) where in contrast to the bearing of the previous samples the increasing of the strain stress value leads to fibers diameter increasing and to yarns length decrease.

## 4. Conclusions

Correlations between the heat-mechanical modification parameters and the occurred structure rearrangement in the investigated uniaxially orientated amorphous PET have been established.

The temperature increase of the heat-mechanical treatment from 80 $^{0}C$ to 85 $^{0}C$ causes a tenfold PET filaments elongation increment.

A satisfactory explanation of the obtained results needs of additional structural investigations and they are coming.

The experiments realized at temperature of 85 $^{0}C$ are the next step of the planned structural studies in amorphous PET filaments in the temperature range from 80 $^{0}C$ to 95 $^{0}C$ caused by combined heat-mechanical treatments.

**Acknowledgement**

Grant 5323, RD – 07-1003/30.03.2010 from the University of Shumen, Bulgaria, supported the present study.

## References

[1] Dupaix R B and Boyce M C 2005 *Polymer* **46** 4827 – 37
[2] Betchev Ch 1994 *Crystallization of Polymers* (M. Dosiere, Ed.) *Kluwer Academic Publ.*
[3] Clauss B and Salem D R 1995 *Macromolecules* **28** 8328 – 33
[4] Shabana H M 2004 *Polymer Testing* **23** 291 – 97
[5] Ziabicki A and Jarecki L 2007 *J. Appl.Polym. Sci.* **105** 215 – 23
[6] Keum J and Song K H H 2005 *Polymer* **46** 939–45
[7] Velev V, Androsch R, Radusch H-J and Betchev Ch 2001 *Macromol. Mater. Eng.* **286** 752-55
[8] Betchev Ch, Velev V, Androsch R and Radusch H-J 2002 *Polymerwerkstoffe '2002, Merseburg, Germany* 195-98
[9] Dimov T, Velev V, Iliev I and D. Betchev Ch 2007 *Journal of Optoelectronics and Advanced Materials* **9** 2 275 – 77
[10] Velev V and Betchev Ch 2008 *Thermochimica Acta* **473** *50–53*
[11] Velev V, Dimov T, Popov A and Iliev I 2009, *Journal of Optoelectronics and Advanced Materials* **1** 3 563-65
[12] Stein R and Wilkes G 1975, Ward I M (ed.) *Applied Science* London 57

# Optical properties of PVA films with diamond and titania nanoparticles

**I Vlaeva[1], T Yovcheva[2], S Sainov[1], V Dragostinova[1] and S Stavrev[3]**

[1] Institute of Optical Materials and Technologies "Acad. J. Malinovski", Bulgarian Academy of Sciences, Acad. G. Bonchev Street, Bl. 101/109, 1113 Sofia, Bulgaria
[2] Department of Experimental Physics, University of Plovdiv "P. Hilendarski", 24 Tsar Assen Street, 4000 Plovdiv, Bulgaria
[3] "Nano" SS Ltd., 4700 Smolyan, Bulgaria

E-mail: vlaeva@optics.bas.bg

**Abstract**. Polymer nanocomposite materials have attracted increasing attention because of their unique optical properties. Of practical interest is the controllable increase in the refractive index (RI) values of the polymer nanocomposite materials by using nanoparticle inclusions with high refractive index values. This paper presents a study of the optical characteristics of polymer nanocomposites with diamond and titanium dioxide (titania) spherical nanoparticles with different diameters (2 nm and 15 nm) and very close refractive index values – 2.50 and 2.42, respectively. The investigated samples are with different concentration of the particles varying from 0 wt% to 3 wt.%. The refractive indices are measured by a laser refractometer at two wavelengths, 532 nm and 632.8 nm. The results obtained show that the RI increases with the increase in the concentration of the inclusions of titania and diamond in the PVA matrix.

## 1. Introduction

Organic-inorganic hybrid materials have attracted considerable attention in recent years due to their novel physical and chemical properties. By using titania inclusions the RI is increased and the new materials have an optical dispersion behavior between that of organic polymers and inorganic glasses [1]. The refractive index ($n$) increases linearly with the increase in the titania content (more then 0.1 wt.%) of the aminoalkoxysilane-capped pyromellitic dianhydride (PMDA) – titania hybrid optical thin films [2]. The same behavior is observed in the polyamide-nanocrystalline-titania hybrid materials with titania contents from 20 wt.% to 90 wt.% [3]. The linear RI variation is also measured in acrylic polymer-titania hybrid films with titania content (up to 5 wt.%) before and after hydrothermal treatment [4], in titania nanocrystals into (3-glycidyloxypropyl) trimethoxysilane organic silica sol (10-50 wt.% titania inclusions) [5] and in photosensitive polyimide-nanocrystalline titania hybrid (up to 30 wt.% titania inclusions) [6].

In [7, 8] the optical transmittance and RI of the titania/epoxy nanocomposites coatings were presented and the authors noted that the transmittance and RI of pure epoxy were 99 % and 1.48, respectively. The refractive index of pure titania nanoparticles – i. e. 100 wt.% titania content was measured to be 2.6, while the RI of hybrid nanocomposite films at 633 nm increased from 2.19 (10 wt.% titania content) to 2.38 (30 wt.% titania content). When adding amorphous-titania nanoparticles to the epoxy matrix, where titania is dispersed directly into the polymer matrix without

Journal of Physics: Conference Series **253** (2010) 012027 doi:10.1088/1742-6596/253/1/012027

the use of organic surfactant, the transmittance and RI of nano-titania/epoxy nanocomposite coating increase more than 90 % and become 1.607 (for 10 wt.% titania) and 1.668 (for 30 wt.% titania), respectively (see [8]).

The refractive index of acid-modified poly (bisophenol A carbonate) with titania nanocomposites at 486 nm, 589 nm and 656 nm is investigated and the experimental results are predicted based on Maxwell-Garnett and Lorentz-Lorenz theories [9]. The concentrations of titania in the cited references vary from 10 to 90 wt.%. At small concentrations (0-0.5 wt.%) the reverse relation is observed – a decrease in the refractive index at 514.5 nm [10].

As far as we know, diamond has not been used for polymer nanocomposites yet. The optical properties of nanocrystalline diamond films are investigated in [11] by a single-oscillator Sellmeier model that we also use in the present work.

In this paper we present optical investigations of polyvinyl alcohol thin films with small concentrations of both titania (up to 1 wt.%) and diamond (up to 3 wt.%) nanoparticles.

## 2. Experimental details

### *2.1. Sample preparation*

The films used in the experiments contained particles of titania with a diameter D = 15 nm and diamond nano-particles with a diameter of 2 nm. The nano-particles were imbedded in a matrix of polyvinyl alcohol (PVA), which is a solid colourless polymer:

$$\left[-CH_2 - CH(OH)_2 -\right]_n .$$

In the experiment, polyvinyl alcohol with a molecular mass 22 000 of the "Fluka" company was used, from which suspensions of 6 wt.% water solution of PVA with different quantities of titania and diamond nano-particles were prepared.

The titania and the diamond nano-particles are in various weight proportions relative to PVA. Their concentrations vary from 0 wt.% to 1 wt.% and from 0 wt.% to 3wt.%, respectively.

Glass substrates are used with a conductive coating of tin dioxide of the "Merck" company production. The glass substrates are washed and dipped in a 1 % – alcohol solution of polyvinyl pyrolidone (PVP). This contributes to the improvement of the film adhesion. On a leveled table at a temperature of 20°C certain quantities of the prepared suspensions are deposited which are then dried for 24 hours at room temperature. The films obtained are 10 μm thick.

### *2.2. Refractive index measurement*

The surface refractive indices *n* for all the samples were measured by the method of the disappearing diffraction pattern using two wavelengths laser refractometer. The principle set-up of the two wavelengths laser refractometer is shown in figure 1.

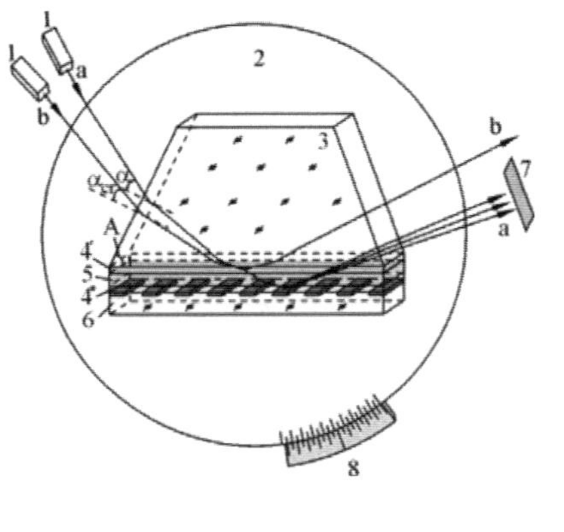

**Figure 1.** Experimental set-up: 1 - laser; 2 - rotary stage with 1 arcmin resolution; 3 - glass prism with RI = N; 4′, 4″ - liquid layer, 5 - sample; 6 - reflecting grating; 7 - screen; 8 - vernier.

Two lasers were used as light sources: He – Ne "Spectra Physics" (0.5 mW), and Laser pointer (0.1 mW) generating at 632.8 nm and 532 nm, respectively. A sample was placed between the glass prism and the reflecting grating. Benzyl alcohol was used as a contact liquid in order to minimize Fresnel losses. The thickness of the sub-surface layer where n was measured co-related to the evanescent wave penetration into the sample was estimated to be about 1 μm [12]. The value of the critical angle $\varphi_{cr}$ was measured in the air at a chosen wavelength. After that $n$ was calculated from the formula:

$$n = N \sin\left[ A \pm \arcsin \frac{\sin \varphi_{cr}}{N} \right]. \tag{1}$$

The signs (+) and (-) correspond to clockwise and counter-clockwise determination of $\varphi_{cr}$, respectively. In our case the sign is (-). In equation (1) $A = 65^{\circ}$ is the refraction angle of the prism, $N$ is the refractive index of the prism for the used wavelength – 1.7347 and 1.7480 for 632.8 nm and 532 nm, respectively.

The uncertainty in determination of the critical angle $\Delta\varphi_{cr}$ is the main source of error in evaluation of the refractive index. In our experiments "Micro Control" rotary stage with 1 arcmin resolution was used. The theoretical RI uncertainty is:

$$\Delta n = N \cos\left( A - \arcsin \frac{\sin \varphi_{cr}}{N} \right)\left( N^2 - \sin^2 \varphi_{cr} \right)^{-1/2} \cos \varphi_{cr} \Delta\varphi_{cr}. \tag{2}$$

In our experiments $\Delta\varphi_{cr} = \pm 2.10^{-3}$ (± 5 arcmin) and uncertainty is $\Delta n = \pm 1.10^{-3}$.

## 3. Results and discussion

In figures 2 and 3 the dispersion curves in the range 400 nm – 800 nm are plotted.

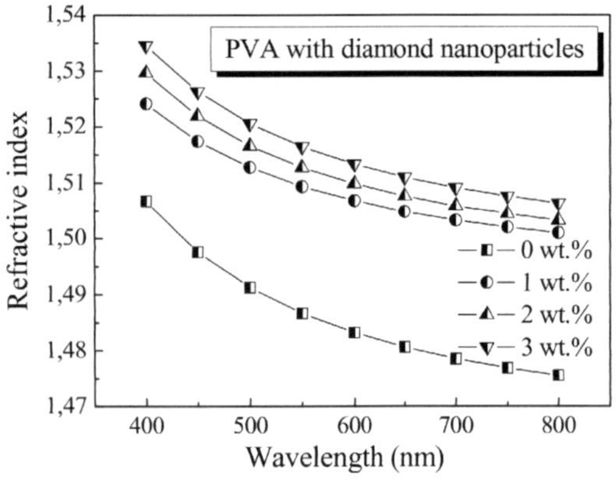


**Figure 2.** RI dispersion curves of the PVA matrix with diamond nanoparticles (D = 2 nm).

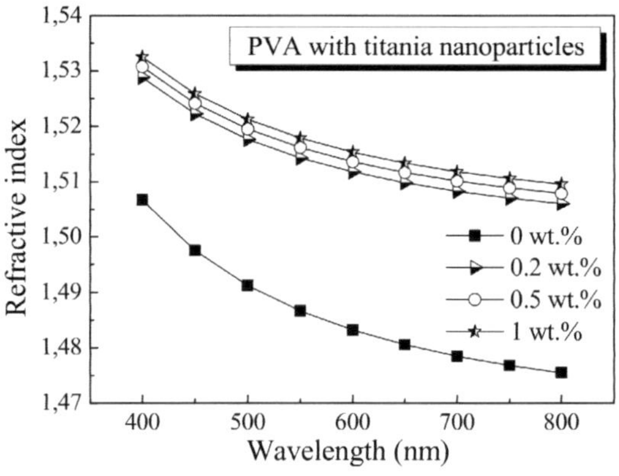


**Figure 3.** RI dispersion curves of the PVA matrix with titania nanoparticles (D = 15 nm).

The surface refractive index was measured at 532 nm and 632.8 nm and the dispersion relationships were presented by Sellmeier one-oscillator model:

$$n^2 - 1 = \frac{s}{1 - (\lambda_s / \lambda)^2}, \tag{3}$$

where $s$ and $\lambda_s$ are Sellmeier's coefficients.

In comparison to the samples without nanoparticles an increase of RI values was observed when nano-sized particles of diamond and titania were embedded into the PVA matrix. It was established that the percentage rise of the nanoparticles results in an increase of the RI values. This effect was stronger for the samples with 1 wt.% that contained titania nanoparticles, and weaker for those containing diamond ones. The relationships for the investigated samples as a function of the nanoparticles concentration in the PVA matrix are presented in figures 4 and 5.

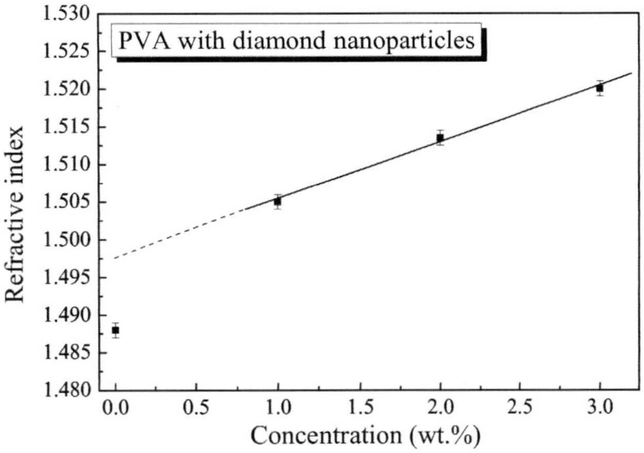


**Figure 4.** Dependence of the refractive index on the concentration for the PVA matrix with diamond nanoparticles at a wavelength of 532 nm.

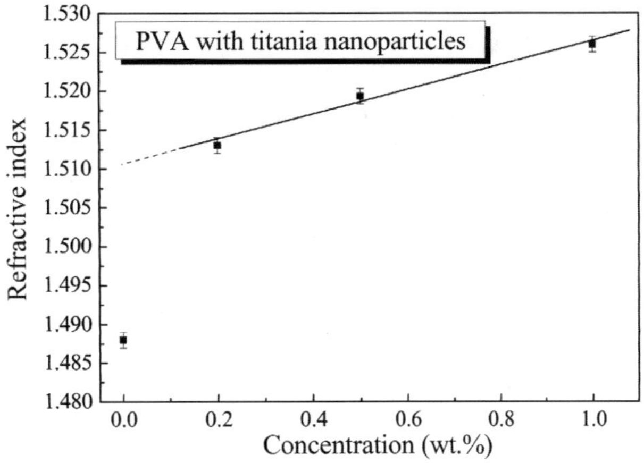


**Figure 5.** Dependence of the refractive index on the concentration for the PVA matrix with titania nanoparticles at a wavelength of 532 nm.

Our experimental results demonstrate that for concentrations of the diamond nanoparticles over 1 wt.% and of the titania nanoparticles over 0.2 wt.%, the concentration relationships are linear which is in good agreement with the reference data [2-6]. For lower nanoparticle concentrations, however, the concentration relationships do not have a clearly expressed linear nature, which, we assume, is due to the very low nanoparticle concentrations used in our investigations.

Journal of Physics: Conference Series **253** (2010) 012027 doi:10.1088/1742-6596/253/1/012027

Investigations on the transmission spectra in the range 500 nm – 800 nm with "Cary 5E" spectrophotometer for the 10 μm polymer films are carried out, the data being presented in figure 6. The experimental uncertainty is ± 0.5 %.

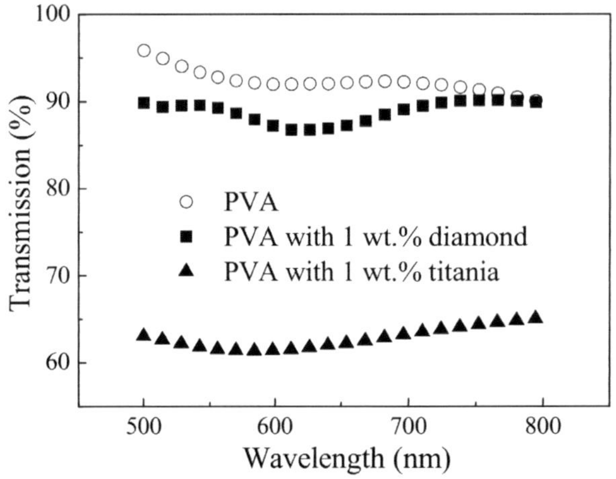


**Figure 6.** Transmission spectra for the PVA matrix with titania and diamond nanoparticles.

The experimental results show that the PVA film transmittance diminishes when nano-sized particles are introduced in the polymer matrix. The samples with diamond nanoparticles have higher transmittance value than those with titania. We suggest that this is due to scattering caused by the introduction of nanoparticles with different sizes in the polymer matrix. The lower transmittance of the samples containing titania can be attributed to the larger diamater of titania particles (15 nm), as compared to that of the diamond particles (2 nm).

## 4. Conclusion

Based on the experimental results obtained the following conclusions can be drawn:

- the nano-sized particles embedding in a polymer matrix causes an increase in the refracftive index of the polymer-nanoparticles composite material – from 1.488 to 1.505 at 532 nm for the samples that contain 1 wt.% diamond nanoparticles, and from 1.488 to 1.526 at 532 nm for the samples that contain 1 wt.% titania. The experimental uncertainty is ± 0.001;
- at very low concentrations of the nano-sized particles, the relationship of RI on the concentration at a wavelength of 532 nm has a non-linear nature, and at higher concentrations, the relationship is clearly linear;
- the samples that contain titania nanoparticles have lower transmittance as compared to those containing diamond particles. The reason for this is the different diameters of the investigated nanoparticles in the PVA matrix that causes increased scattering.

**Acknowledgements**

I. Vlaeva, S. Sainov, T. Yovcheva and V. Dragostinova are grateful for the financial support under Contract DO 02/155 by Ministry of Education and Science, Bulgaria

## References

[1] Wang B, Wilkes G L, Hedrick J C, Liptak S C and McGrath J E 1991 *Macromol.* **24** 3449-50
[2] Chang Ch-Ch and Chen W-C 2001 *Inc. J. Polym. Sci. Part* A: *Polym. Chem.* **39** 3419-27
[3] Su H-W and Chen W-Ch 2008 *J. Mater. Chem.* **18** 1139-45
[4] Su H-W and Chen W-Ch 2008 *Macromol. Chem. Phys.* **209** 1778-86

[5] Liu Y, Lü Ch, Li M, Zhang L and Yang B 2008 *Coll. Surf.* A: *Physicochem. Eng. Asprects* **328** 67-72
[6] Chang W-L, Su H-W and Chen W-Ch 2009 *Europ. Polym. J.* **45** 2749-59
[7] Chau J L H, Lin Y-M, Li A-K, Su W-F, Chang K-Sh, Hsu S L-Ch and Li T-L 2007 *Mat. Lett.* **61** 2908-10
[8] Chau J L H, Tung Ch-T, Lin Y-M and Li A-K 2008 *Mat. Lett.* **62** 3416-18
[9] Imai Y, Terahara A, Hakuta Y, Matsui K, Hayashi H and Ueno N 2009 *Europ. Polym. J.* **45** 630-38
[10] Jiménez Riobóo R J, De Andrés A, Kubacka A, Fernández-García M, Cerrada M L, Serrano C and Fernández-García M 2010 *Europ. Polym. J.* **46** 397-403
[11] Hu Z G, Prunici P, Hess P and Chen K H 2007 *J. Mater. Sci.: Mater. Electron.* **18** S37-41
[12] Yovcheva T, Sainov S and Mekishev G 2007 *J. Optoelectron. Adv. Mat.* **9** 2087-2090

# Temperature dependence of refractive characteristics of optical plastics

**S N Kasarova[1] , N G Sultanova[1] and I D Nikolov[2]**
[1] Dept. of Mathematics and Physics, University "Assen Zlatarov"- Burgas,
1 Prof. Yakimov Str., Burgas-8010, Bulgaria,
[2] Dept. of Optics and Spectroscopy, Faculty of Physics, Sofia University,
5 J. Bourchier Blvd., Sofia-1164, Bulgaria

E-mail: kasarova_st@yahoo.com

**Abstract**. In this study we investigate the variation of refractive characteristics of optical plastics with temperature in the range of 10 to 50 °C. We have measured the indices of several optical polymers as Optorez 1330, Zeonex E48R, Polymethyl methacrylate, S-low styrene and Bayer by means of the deviation angle method. The obtained refractive index data is presented at five standard spectral lines in the visible region. A decrease of refractive index values with increasing temperature is observed. Dependence on measuring wavelength is also noticed. We have calculated the temperature coefficients of refractive indices of the studied optical plastics. Charts of refractive index of some of the materials versus temperature are presented. Temperature dependence of dispersion curves based on Cauchy-Schott's approximation and Abbe numbers is also considered.

## 1. Introduction

Application of optical plastics (OPs) in the design of optical elements and devices requires the knowledge of their refractometric and dispersive characteristics. We have studied more than twenty types of plastic materials including principal and some new development polymers. Refractive index data has been obtained for both bulk and thin film samples [1-3]. Published refractive characteristics have been presented mainly at reference temperature of 20 °C which conforms to the European measuring standard [4] while the USA standard requires measuring temperature of 22 °C [1]. Usually, the most important operating temperature range for optical devices is between 10 and 40 °C though extreme conditions may sometimes exceed this interval.

Thermal properties of polymers are rather different compared to glasses. Service temperature limits for OPs (60 to 250 °C) are substantially less than for optical glasses (400 to 700 °C). The thermal expansion coefficient is about one order of magnitude greater than glass. For example, the linear thermal expansion coefficient of Acrylic has average value of $7\times10^{-5}$ $K^{-1}$ [5] while for the N-BK7 glass the corresponding value is $0.71\times10^{-5}$ $K^{-1}$ [4]. Another important parameter for optical materials is the variation of refractive index with temperature $dn/dt$. The change of this parameter depends on two counteracting effects: the change in density caused by positive coefficient of thermal expansion and the increase in polarizability with temperature. Evaluation of d$n$/d$t$ is possible by differentiation of the Lorentz-Lorenz equation with respect to temperature. The temperature gradient of refractive index d$n$/d$t$ is referred in literature as a thermo-optic coefficient [6] or temperature coefficient of refractive index [4]. The absolute values of d$n$/d$t$ of polymer materials are with orders of magnitude larger than those for optical glasses and exhibit usually negative deviations compared to the positive thermo-optic coefficients for most of glasses. For example, the change of refractive index with temperature for Acrylic materials is approximately $-8.5\times10^{-5}$ $K^{-1}$ [7] and for the N-BK7 glass with analogous refraction, the value is $1.6\times10^{-5}$ $K^{-1}$ [4]. Though the analysis of d*n*/d*t* is an essential issue for optical plastics' applications in photonic devices and optical fibre communication systems, the literature data on their thermo-optic properties of OPs is rather limited.

In this paper we present measured refractive indices of some plastic materials at different temperatures in the range between 10 and 50 °C. Five spectral lines of illumination in the visible spectral region were utilized. Temperature variation of the refractive characteristics of the studied OPs had been analysed.

## 2. Measuring the refractive indices of optical plastics

### 2.1. *Equipment and sample preparation*

Refractive index measurements were carried out using the Carl Zeiss Jena Pulfrich refractometer PR2 with an accuracy of $2.10^{-5}$. The illuminating system consists of three spectral lamps: a mercury source with a green e-line 546.07 nm and a blue g-line 435.83 nm, a helium source for the yellow d-line 587.56 nm and a hydrogen source for the blue F-line 486.13 nm and red C-line 656.27 nm [8]. We have used the V-type SF3 prism (VoF3-prism) which is suitable to the shape of the examined samples. The VoF3-prism is produced as a glass block assembled into a massive metal body with thermostatic housing which assures stable temperature during the measuring. Besides that, the prism cap is made of stainless steel and has a mirrored internal surface, so temperature fluctuations are cut down and the matching liquid cannot evaporate. Measuring temperature was maintained by a MLW Thermostat U4 with waterbath, made in Germany. Water from the reservoir circulated through the prism block and temperature was monitored by a thermometer with accuracy of 0.2°C.

Measuring with the VoF3 prism requires two mutually perpendicular surfaces of the examined solid body. The examined OPs samples were prepared as injection moulded plates with thickness varying from 1.5 mm to 2.5 mm and two adjacent surfaces were well polished to obtain a good refractometric data. We used a saturated aqueous solution of zinc chloride ($n_e$ =1.51), silicon oil ($n_D$ = 1.56) and a saturated water solution of $KHgJ_3$ ($n_e$ = 1.73) as a contacting liquid depending on the refraction of the examined material.

We have measured the transmittance of all examined plastic samples using the spectrophotometer SPECORD UV-VIS (VEB Carl Zeiss, Jena). The obtained results showed that these polymer materials are transparent in the considered spectral range, a normal dispersion is exhibited [1, 2] and the extinction coefficient can be neglected.

### 2.2. *Principle of measurement*

Determination of refractive index by means of the VoF3 prism of the Pulfrich refractometer is based on measurement of the deviation angle. The plastic sample is placed and covered into the prism (figure 1) and the desired temperature is adjusted. The collimated beam from the lighting system illuminates monochromatically the prism block and refracts by the examined material. The deviation angle $\gamma$ is determined and the refractive index $n_\lambda$ of the examined plastic is calculated as follows:

$$n_\lambda = \left(N_\lambda^2 - cos\,\gamma\left(N_\lambda^2 - cos^2\,\gamma\right)^{1/2}\right)^{1/2} \tag{1}$$

where $N_\lambda$ is the refractive index of the VoF3 prism. The value $N_\lambda$ of the SF3 glass is calculated by the data published in [8] at standard spectral wavelengths applying our program OptiColor [2] involving the Caushy–Schott's dispersion approximation. Each of the obtained results was corrected by involving the value of the temperature coefficient of the SF3 glass at the corresponding wavelength.

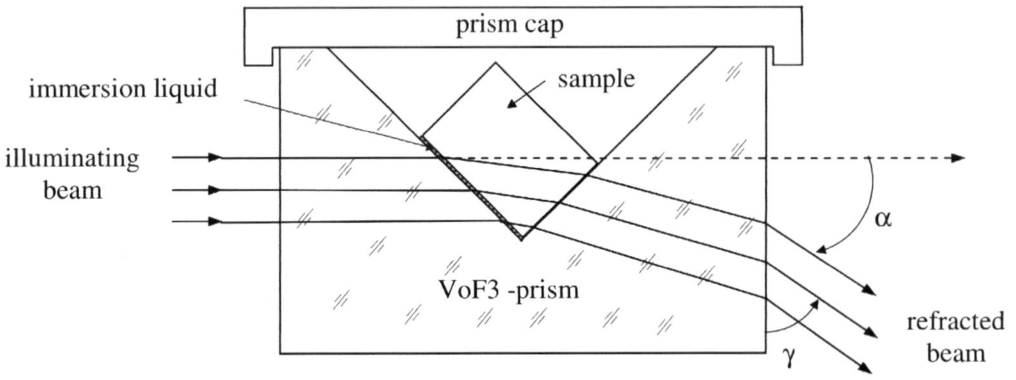


**Figure 1.** Deviation of the illuminating beam in the sample.

### *2.3. Measuring results*

We have obtained five different OPs, namely: polymethyle methacrylate (PMMA) and S-low styrene produced by Eastman Chemical Company, and some trade-marks as Zeonex E48R®, Optorez1330® and Bayer®. All of the samples were measured at five spectral lines at nine different temperatures in the interval between 10 and 50 °C which match the usual service temperatures of optical devices used in practice. The results for two of the samples are plotted in figure 2. A decrease of refractive indices with increasing temperature has been established for all of the samples. In figure 2-a) the results for the PMMA sample are presented. Graphs for all of the illuminating wavelengths show some nonlinearity especially at lower temperatures. Similar dependence has been obtained for the rest of the samples except for the Zeonex material which graph is almost linear – figure 2-b).

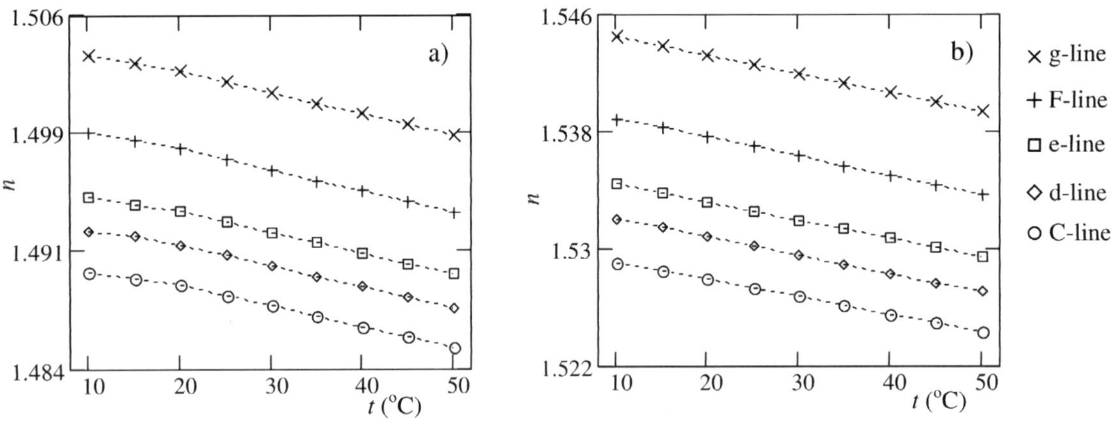


**Figure 2.** Temperature dependence of refractive index for a) PMMA; b) Zeonex E48R.

## 3. Discussion and conclusions

### *3.1. Variation of refractive index of optical plastics with temperature*

Refractive indices of polymers are affected by temperature. Furthermore, the change of refractive index as a function of temperature depends on the specific material. For a better comparison among studied plastics, these changes are evaluated as $\Delta n(t) = n(t) - n_{20}$, where $n_{20}$ is the measured refractive index at reference temperature of 20 °C which conforms to the European measuring standard. The obtained results for the e-line are plotted in figure 3. At lower temperatures (less than 20 °C) changes of refractive index are smaller in comparison to the values obtained at 40 °C or 50 °C. Greatest variation of refractive index is established for the S-low Styrene sample and lowest for the Bayer material. The dependence of $\Delta n(t)$ for the PMMA and Zeonex E48R is quite similar at temperatures higher than 20 °C, and at lower temperatures the PMMA curve coincides with the Optorez 1330 and Bayer graphs. Nonlinearity of the presented curves is noticed especially for the Bayer, Optorez 1330 and PMMA. Almost a linear graph is obtained again for the Zeonex E48R.

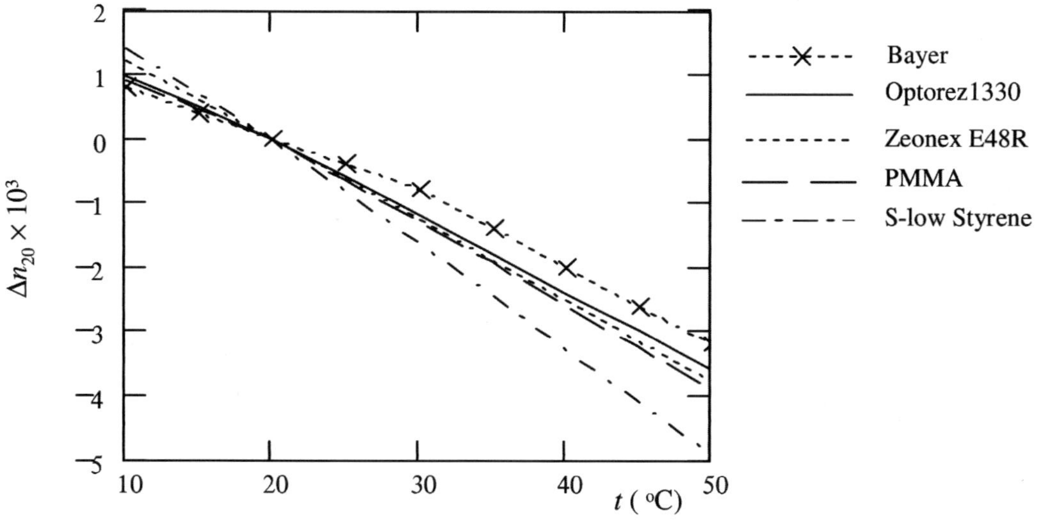


**Figure 3.** Changes of refractive index $\Delta n$ at e-line with temperature.

It is interesting to note that the magnitude of $\Delta n$ in the considered temperature interval is of similar order to the refractive index deviation $\Delta n_\lambda$ due to dispersion. For example, in case of the S-low Styrene material the $\Delta n$ value equals $\Delta n_\lambda$ in the spectral range between 560 and 700 nm. In case of the high refractive Bayer polymer which shows much greater dispersion than rest of the plastics, lowest change $\Delta n$ with temperature is established. Analysis of dispersion properties of more than twenty OPs has been published in our previous papers [2, 9].

*3.2. Temperature coefficients of refractive indices*

Linear approximation of the temperature dependence of refractive index is possible only in comparatively narrow temperature intervals. In figure 4 computed temperature coefficients d*n*/d*t* at e-line of the examined samples are presented. The obtained results for the Optorez 1330, Zeonex E48R, S-low Styrene and PMMA show that two temperature ranges in the considered interval can be distinguished. In case of the Bayer material the value of d*n*/d*t* is constant between 10 and 30 °C and drops down in the interval between 30 and 50 °C. The magnitudes of d*n*/d*t* for the examined samples are close and for the e-line they vary from $0.8\times10^{-4}$ for Bayer to $1.4\times10^{-4}$ for S-low Styrene. In case of the Zeonex plastic the d*n*/d*t* value is almost constant for the entire measuring temperature region.

The thermo-optic coefficient depends also on the wavelength. It decreases with increasing wavelength but the change of magnitude of d*n*/d*t* is considerably smaller than the deviation of the refractive index due to the wavelength variation d*n*/d$\lambda$. Our results show that the variation of the thermo-optic coefficient with wavelength in the visible spectrum has absolute values between $4\times10^{-6}$ and $2\times10^{-5}$. In table 1 some of the obtained thermo-optic coefficients at two wavelengths are presented.

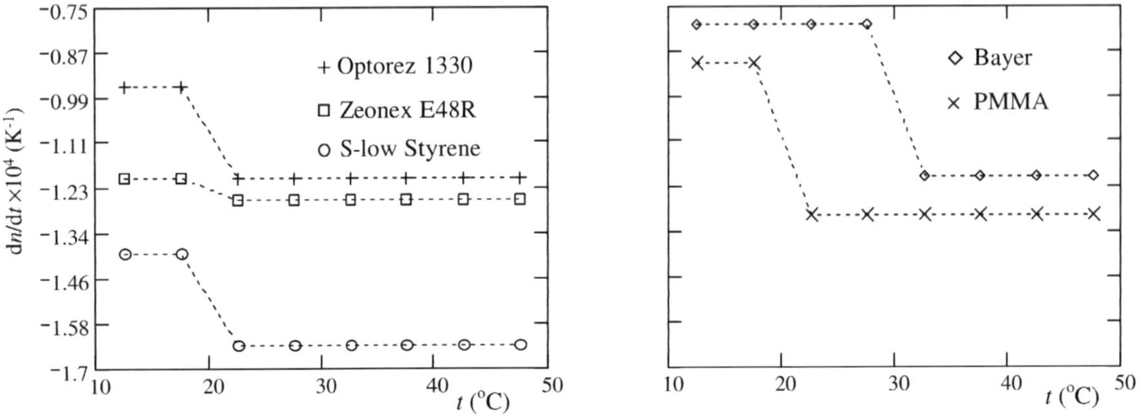


**Figure 4.** Change of the thermo-optic coefficient of optical plastics with temperature.

**Table 1.** Thermo-optic coefficient of OPs at two wavelengths.

| Optical plastics | d*n*/d*t* ($\times10^{-4}$ K$^{-1}$) | | | |
|---|---|---|---|---|
| | g-line (10 ÷ 20 °C) | d-line (10 ÷ 20 °C) | g-line (20 ÷ 50 °C) | d-line (20 ÷ 50 °C) |
| PMMA | -1.00 | -0.90 | -1.32 | -1.30 |
| Optorez 1330 | -1.00 | -0.96 | -1.22 | -1.20 |
| S-low Styrene | -1.50 | -1.40 | -1.70 | -1.64 |

*3.3. Temperature influence on dispersion*

The variation of refractive index values with temperature alters the dispersion characteristics of OPs. We have used the Cauchy-Schott approximation usable in case of normal dispersion and our program OptiColor for evaluation of the results. Dispersion curves of the Optorez 1330 material in the VIS range obtained at four of the measuring temperatures are presented in figure 5.

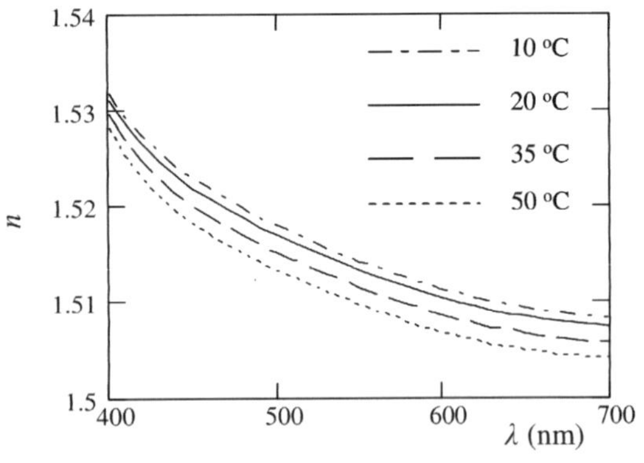


**Figure 5.** Dispersion curves of Optorez 1330 at different temperatures.

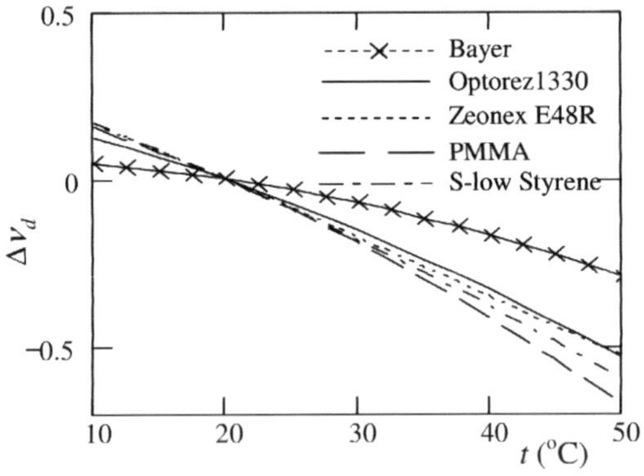


**Figure 6.** Change of the Abbe numbers $\Delta\nu_d$ as a function of temperature.

A change of the dispersive behaviour of the materials is noticed. As it can be seen a greater dispersion is observed at higher temperatures in the VIS region. Such dependence is observed for all of the studied materials. Usually the Abbe number is used to characterize the dispersive properties of optical materials. We have calculated $\nu_d$ at all measuring temperatures. A decrease of the $\nu_d$ values with increasing temperature is observed for all investigated OPs. For a better comparison among the results changes of the Abbe numbers are calculated in respect to the value at 20 °C: $\Delta\nu_d(t) = \nu_d(t) - \nu_{d_{20}}$. Figure 6 illustrates variations of the Abbe numbers for the examined OPs. Deviation of $\nu_d$ in respect to $\nu_{d_{20}}$ also reveals nonlinearity. Lowest variation of the Abbe number, respectively dispersion, with the temperature is observed again for the Bayer material and greatest for the PMMA polymer.

### *3.4. Conclusions*

Temperature dependence of refractive characteristics of five plastic materials has been investigated in the interval between 20 °C and 50 °C. Refractive indices of the polymer samples have been measured at 9 measuring temperatures at five spectral lines using the deviation angle method. A decrease of refractive indices with increasing temperature is established for all of the OPs. Nonlinearity of the obtained $n(t)$ curves at all measuring wavelengths is noticed (figure 2). Only the Zeonex material exhibits a linear dependence of $n(t)$ in the entire temperature interval. A better comparison of refractive index changes with temperature is achieved by calculation of $\Delta n(t)$ in respect to the indices at reference temperature of 20 °C (figure 3). The S-low Styrene shows greatest deviation of $\Delta n(t)$ and smallest value is obtained for the Bayer material. Changes of refractive index with temperature could be very substantial. For the S-low Styrene sample the value of $\Delta n(t)$ is almost equal to the change of refractive index due to dispersion beyond 560 nm in the VIS region.

Thermo-optic coefficients of the examined materials have been calculated. Only the Zeonex material has almost a constant value in the entire measuring interval. For the rest of materials more complex behavior of the thermo-optic coefficients is noticed. Their values could be estimated for much narrower temperature intervals (figure 4). A polynomial interpolation of $dn/dt$ dependence is possible when larger measuring temperature intervals are demanded.

Influence of temperature on dispersion properties of OPs is studied. The shape of dispersion curves at different temperatures varies substantially (figure 5). Abbe numbers values decrease with increasing temperature (figure 6). Greatest deviation of $\Delta\nu_d$ in respect to the value at 20 °C is observed for the PMMA plastic.

Our results show that temperature dependence of refractive characteristics of OPs can not be neglected in the design and construction of optical elements and devices when required service temperatures vary significantly. This dependence is specific for each optical polymer. Among studied materials the Bayer plastic shows best temperature stability of its refractive and dispersive properties. The Zeonex material could be also recommended for its almost constant value of the thermo-optic coefficient.

## References

[1] Sultanova N, Nikolov I and Ivanov C 2003 *Opt. Quant. Electron.* **35** 21-34
[2] Kasarova S, Sultanova N, Ivanov C and Nikolov I 2007 *Opt. Mater.* **29** 1481-90
[3] Kasarova S, Sultanova N, Petrova T, Dragostinova V and Nikolov I 2009 *J. Optoelectron. Adv. Mater.* **11** 1440-3
[4] SCHOTT AG, http://www.us.schott.com/
[5] Pfeffer M 2010 *Handbook of plastic optics* ed S Baumer (New York: Wiley–VCH) p 23
[6] Gorachand G 1998 *Handbook of optical constants of solids* ed E D Palik (London–Academic Press) p 118
[7] U. S. Precision Lens, Inc. 1973 *The handbook of plastic optics* (Ohio, Cincinnati) p 13
[8] Carl Zeiss JENA Jena D-77830 1976 *The manual of Pulfrich-Refraktometer PR2* (Germany)
[9] Sultanova N, Kasarova S and Nikolov I 2009 *Acta Phys. Pol. A* **116** 585-7

Journal of Physics: Conference Series **253** (2010) 012029 doi:10.1088/1742-6596/253/1/012029

# Factors determining optical absorption of variously doped single crystals of $Bi_{12}SiO_{20}$

**B Kostova**[1] **and L Konstantinov**

Institute of Mineralogy and Crystallography "Acad. I. Kostov", Bulgarian Academy of Sciences, 1113 Sofia, 107 Acad. G. Bonchev str. Bulgaria

E-mail: bilyana_k@abv.bg

**Abstract**. $Bi_{12}SiO_{20}$ are widely used in optical devices, spatial-time light modulators and as holographic media depending on their properties in a wide spectral range. This work deals with the optical absorption coefficient of pure and doped $Bi_{12}SiO_{20}$ (BSO) crystals measured in the energy region 1.5 – 3.2 eV and generalise our already published and new experimental data on crystals doped with Al, P, Cr, Mn, Fe, Co, Ni, Cu, Se, Ru. It is shown that such doping affects differently the optical absorption in both the near-fundamental-edge shoulder (3.2 – 2.2 eV) and on impurity levels in the band gap (2.2 – 1.5 eV). The main factors determining the absorption are concentration and segregation coefficient, electro-negativity of the doping ion and its difference from that of oxygen (DEN), the oxidation states, possible coordination and corresponding ionic radius. The role of some of these factors to the spectral dependence of the optical absorption of BSO is discussed from the view point of a suitable doping of BSO to obtain and modify the optical transparency.

## 1. Introduction

Sillenite-type $Bi_{12}SiO_{20}$ (BSO) pure and doped single crystals are widely used in optical devices, spatial-time light modulators and as holographic media [1,2]. BSO are cubic crystals, I23 space group symmetry, built up of $Si\text{-}O_4$ tetrahedrons and deformed Bi-On polyhedrons [3], defined by some authors as Bi-On (n = 7) - octahedrons [4] or as Bi-O-pseudo-octahedrons [5]. The chemical bonds Bi-O and Si-O in BSO are covalent [3]. In pure BSO the defects occur in the $Si\text{-}O_4$ tetrahedron, creating a spectral feature known as absorption shoulder, which result most probably due to energy levels from silicon vacancies [6] or holes from substitution of $Si^{4+}$ by $Bi^{3+}$ ions [7,8]. After reported literature data most of doping ions in BSO are placed at metal position in $Si\text{-}O_4$-tetrahedron [9]. For Ru, Cu and Mn - ions there are data for occupation of both metal position in tetrahedron (replacing Si) and in pseudo-octahedron (replacing Bi) [5,9,10].

Many studies on the optical absorption coefficient of pure BSO and doped with Al, P, Cr, Mn, Fe, Co, Ni, Cu, Se and Ru crystals in the UV-VIS spectral region have been reported so far and the effect of doping were discussed mainly in respect to: (i) absorption edge shifts [4,9-19], (ii) influence on the absorption shoulder region in 2.2-3.2 eV region [4,10-13,17,18] and (iii) possible oxidation states of doping ions, position of allowed electron transitions in the 1.5 – 2.2 eV region and the dependence of corresponding absorption coefficient on dopant concentration [5,10,16-24].

[1]Corresponding author: bilyana_k@abv.bg

Nevertheless, until now there are no systematic complex investigations on doping effects in both the near-fundamental-edge absorption shoulder (3.2 – 2.2 eV) and on impurity levels in the band gap (2.2 – 1.5 eV). This work deal with the optical absorption coefficient of pure and doped BSO crystals measured in the energy region 1.5 – 3.2 eV, generalizing both our already published [25,26] and new experimental data on crystals doped with Al, P, Cr, Mn, Fe, Co, Ni, Cu, Se, Ru. We attempt to present the effect of doping as well as to specify some of main parameters controlling the observed effects, namely concentration and segregation coefficient, electrical-negativity of the doping ion and its difference from that of oxygen (DEN), the oxidation states, possible coordination and corresponding ionic radius. The role of some of these factors to the spectral dependence of the optical absorption of BSO is discussed from the view point of a suitable doping of BSO to obtain and modify the optical transparency of doped crystals. .

## 2. Materials and methods

We investigated pure and doped BSO crystals with Al, P, Cr, Mn, Fe, Co, Ni, Cu, Se and Ru. All the crystals were grown from stoichiometric melts $Bi_2O_3$:$SiO_2$ = 6:1 using the Czochralski method under conditions described in detail elsewhere [12,14]. High purity $Bi_2O_3$, $SiO_2$, $Al_2O_3$, $P_2O_5$, $Cr_2O_3$, $MnO_2$, $Fe_2O_3$, $Co_3O_4$, $Ni_2O_3$, CuO, $SeO_2$ and $RuO_2$ were used for synthesis and doping. The concentration of doping was determined by flame (Zeeman 3030) and electrical-thermal atomic (Varian 240) absorption spectrometry as well as by inductively coupled plasma atomic emission spectrometry (Jobin Yvon, ULTIMA 2).

## 3. Results and discussion

The optical absorption were measured at room temperature in the energy interval 1.5 – 3.2 eV on bi-side polished parallel plates using a Varian's UV-VIS spectrophotometer Carry 100. Prior to measurement, the crystals were de-excited either by illumination for 20 min with a He-Ne laser Calr Zeiss Jena, HNA 188-S (the surface power of about 90mW/cm$^2$) or by annealing at 500°C in air for 20 min [10]. In order to attain better spectral resolution and accuracy we express some times the data on optical absorption coefficient in terms of its first and second derivatives with respect to photon energy. Three are the main spectral changes provoked in optical absorption of BSO by doping: (i) shift of the fundamental absorption edge, (ii) change of the shape position of the near-edge shoulder and (iii) appearance of impurity states and corresponding energy levels in the band-gap. In what follows we present representative dependencies illustrating the character of these three spectral changes.

There are numerous results published on the shift of the fundamental absorption edge [10-12,14,19] which coincide in principle with our experimental data, indicating that its position depends mainly on the type and concentration of doping ions. In sillenite-type $Bi_{12}TiO^{20}$ crystals doping with ions of lower electrical-negativity leads to an edge-shift towards longer wavelengths [18]. As electrical-negativity describes the ability of an atom to attract electrons/electron density towards itself and determines the type and strength of chemical bonds in a compound, we calculated the differences between electrical-negativity of the doping ions and that of oxygen (DEN) (Table 1) [29]. When DEN is between 1.7 and 2.0, the doping ion forms ionic bond with oxygen (this holds for Mn, Al and Cr), whereas for DEN between 0.5 and 1.7, the bond is polar covalent (Fe, Co, Cu, Ni, P, Ru, Se as well as for Si-O and Bi-O bonds). Respectively, the absorption edge upon doping with the first three elements shifts to higher, while with the other seven to lower energy. This behavior can be clearly seen from the spectra of the first derivative of absorption coefficient with respect to energy (Fig.1), where peak positions indicate the edge and their change describes the variation in its slope upon doping. Figure 1 have only illustrative character showing three typical cases of doping, namely very large low-energy edge-shift (Cu, reducing the edge position of pure BSO from 3.12 eV down to about 2.6 eV), no or very small edge-shift (Ru) and significant high-energy edge-shift (Mn, rising the edge position up to about 3.2 eV). These effects are of importance and have to be taken into account when model and modify the region of transparency of doped BSO crystals.

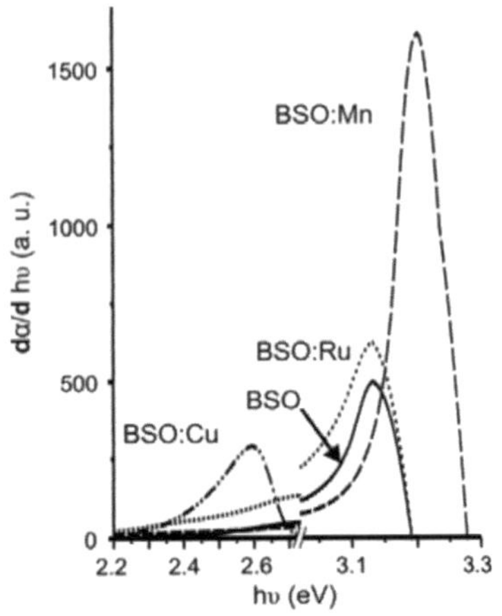


**Figure 1.** Energy dependence of first derivative of absorption coefficient for BSO, BSO:Mn, BSO:Ru and BSO:Cu in the absorption edge vicinity.

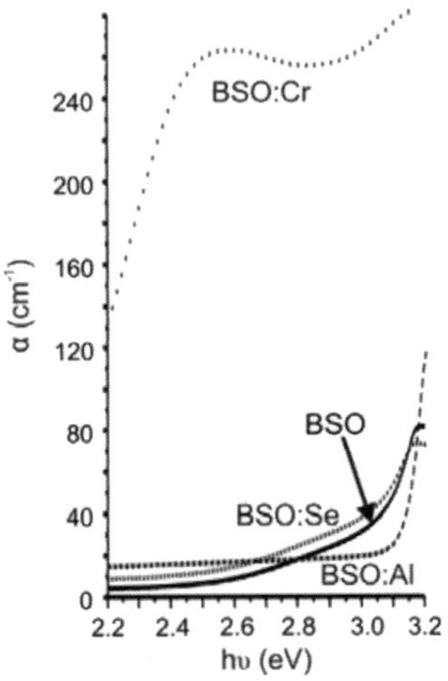


**Figure 2.** Energy dependence of absorption coefficient for BSO, BSO:Cr, BSO:Al and BSO:Se in the region of fundamental edge shoulder.

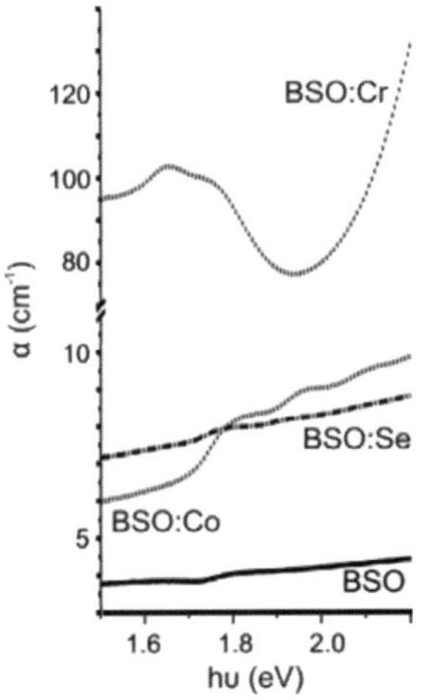


**Figure 3.** Energy dependence of absorption coefficient for BSO, BSO:Cr, BSO:Co and BSO:Se in the sub-band-gap optical range.

**Table 1.** Doping influence and parameters of dopants and major elements (Bi and Si) for BSO crystals.

| crystal | Doping Influence | | | Dopant Parameters | | | | | |
|---|---|---|---|---|---|---|---|---|---|
| | edge | shoulder | band gap | $\chi^a$ | DEN | oxidation state [b] | electronic configuration | coordination type[a] | i. r.[a] [pm] |
| BSO:Mn | strong | weak | — | 1.55 | 1.89 | $2^+$ | [Ar]$3d^5$ | Tetrahedral/Octahedral | 80/81 |
| | | | | | | $3^+$ | [Ar]$3d^4$ | Octahedral | 72 |
| | | | | | | $4^+$ | [Ar]$3d^3$ | Tetrahedral/Octahedral | 53/67 |
| | | | | | | $5^+$ | [Ar]$3d^2$ | Tetrahedral | 39.5 |
| BSO:Al | strong | strong | — | 1.61 | 1.83 | $3^+$ | [Ne] | Tetrahedral/Octahedral | 53/67.5 |
| BSO:Cr | weak | strong | + | 1.66 | 1.78 | $3^+$ | [Ar]$3d^3$ | Octahedral | 75.5 |
| | | | | | | $4^+$ | [Ar]$3d^2$ | Tetrahedral | 55 |
| | | | | | | $5^+$ | [Ar]$3d^1$ | Tetrahedral | 48.5 |
| | | | | | | $6^+$ | [Ar] | Tetrahedral | 40 |
| BSO:Fe | weak | strong | | 1.83 | 1.61 | $3^+$ | [Ar]$3d^5$ | Tetrahedral | 63 |
| BSO:Co | strong | weak | + | 1.88 | 1.56 | $2^+$ | [Ar]$3d^7$ | Tetrahedral | 72 |
| BSO:Cu | strong | strong | — | 1.90 | 1.54 | $1^+$ | [Ar] $3d^{10}$ | Tetrahedral/Octahedral | 74/91 |
| | | | | | | $2^+$ | [Ar] $3d^9$ | | 71/87 |
| BSO:Ni | weak | weak | — | 1.91 | 1.53 | $2^+$ | [Ar]$3d^8$ | Tetrahedral | 69 |
| BSO:P | strong | strong | — | 2.19 | 1.25 | $5^+$ | [Ne] | Tetrahedral | 31 |
| BSO:Ru | weak | weak | — | 2.20 | 1.24 | $3^+$ | [Kr]$4d^5$ | Octahedral | 82 |
| | | | | | | $4^+$ | [Kr]$4d^4$ | | 76 |
| | | | | | | $5^+$ | [Kr]$4d^3$ | | 70.5 |
| BSO:Se | weak | weak | — | 2.55 | 0.89 | $6^+$ | [Ar] $3d^{10}$ | Tetrahedral | 42 |
| element | | | | Bi and Si parameters | | | | | |
| Si | | | | 1.90 | 1.54 | $4^+$ | [Ne] | Tetrahedral | 41.0 |
| Bi | | | | 2.09 | 1.42 | $3^+$ | [Xe]$4f^{14}5d^{10}6p^1$ | Octahedral | 117 |
| | | | | | | $5^+$ | [Xe]$4f^{14}5d^{10}$ | | 90.0 |

[a] literature data [12-14,28,29]
[b] possible oxidation states of dopants, proven by other with different experimental methods [5,9,10,13,14,17,18,21-24]

The edge of fundamental absorption is the high-energy end of the so-called optical absorption shoulder which for doped BSO crystals covers the spectral range 3.2-2.2 eV where the absorption takes place on energy levels from both intrinsic and more or less shallow electron states as result of doping. Figure 2 shows the spectra of absorption coefficient of pure BSO and crystals doped with Cr, Se, and Al. Again, the spectra illustrate the general observed peculiarities of the effect of doping. The doping with Se does not change seriously the absorption of pure BSO, while that with Al modifies the shapes of absorption shoulder making it almost stepwise due to compensation of intrinsic states (most probably structural defects) of the pure material. Totally different is the spectrum of BSO:Cr revealing the presence of a very intense wide band in the spectral range 2.5 – 3.1 eV, which is typical for this ion only. Such a strong and unusual effect of doping of BSO with Cr can be expected due to the suitable oxidation states in tetrahedral coordination with electronic configuration allowing crystal-field electron transitions. Table 1 presents schematically the significance of doping for the three spectral regions, the electrical-negativity X, DEN, oxidation state, electronic configuration, coordination type and ion radius for the ten doping elements as well as for Bi and Si. Thus, among the four possible oxidation states of chromium, most appropriate for forming optical absorption levels would be $4^+$ and $5^+$ with tetrahedral coordination.

The doping of BSO with elements, involving oxygen in ionic bonds, shifts as a rule the absorption edge of doped crystal to higher energies, while for these forming polar covalent bonds the edge shifts to lower energies. On the other hand, elements that affect strongly the absorption shoulder (Mn, Fe, Cu, P and Al in our case) probably substitute $Bi^{3+}$ in tetrahedral position and in oxidation states forbidden for electron transitions. Contrary, doping elements not affecting this spectral range most probably substitute $Si^{4+}$ in tetrahedral position as their oxidation state and corresponding electronic configuration allow crystal-field transitions. For instance, ions $Al^{3+}$ and $P^{5+}$ are with noble-gas electronic configurations,$Cu^{1+}$ with $d^{10}$-configuration (filled 3d-electronic shell), $Fe^{3+}$ and $Mn^{2+}$ with $d^5$-configuration (half-filled shell), and for all these elements the electron transitions are forbidden or with lower probability and low intensity.

In the spectral range 2.2 – 1.5 eV, optical absorption takes place on deep impurity levels formed upon doping, which in some cases, particularly at high values of segregation coefficient for a given doping ions assuring high density of states and concentration of doping, may lead to very intense absorption. Figure 3 shows as typical illustrations the spectra of optical absorption coefficient of BSO, BSO:Cr, BSO:Co, and BSO:Se in this range.

The spectra of pure BSO and of BSO:Se (continuous lines) seem quite alike in shape, being structureless. This holds also for other investigated doped samples except for BSO:Cr and BSO:Co (doted lines). The latter two exhibit intense spectral bands in the optical absorption coefficient at about 1.56, 1.65, 1.74 eV for BSO:Cr and at 1.78, 1.95 and 2.10 eV for BSO:Co. These peak positions are determined either via decomposition of the spectrum using Gauss functions [10] or, to enhance the spectral resolution, by calculating its second derivative, both methods giving results in a good agreement with published data [11-14,23,26]. In addition, careful inspection of the spectra of BSO and BSO:Se reveal the presence of very weak absorption peaks near 1.51, 1.70, 1.88 and 2.18 eV which could be due to undesirable doping of the crystals with Fe and Mn [9.26,27]. Practically, except Cr and Co, doping with all other elements under study do not affect the optical transparency of BSO.

## 4. Conclusions

On the basis of experimentally measured optical absorption spectra and of literature data for ionic electrical-negativity, oxidation states, electronic configuration, coordination type and ionic radii, we were able to draw some general conclusions and prognosis about the effect of doping on the optical absorption of variously doped crystal of BSO :

(i) The doping of BSO with elements, involving oxygen in ionic bonds, shifts the absorption edge of doped crystal to higher energies, while for such forming polar covalent bonds the edge shifts to lower energies.

(ii) Elements that upon doping of BSO affect strongly the absorption shoulder probably substitute $Bi^{3+}$ in tetrahedral position and in oxidation states forbidden for electron transitions. Contrary, doping elements not affecting this spectral range most probably substitute $Si^{4+}$ in tetrahedral position as their oxidation state and corresponding electronic configuration allow crystal-field transitions.

(iii) Only doping of BSO with Cr and Co forms absorbing states in the sub-band-gap spectral range.

**Acknowledgments**

Authors gratefully acknowledge the financial support of the Bulgarian National Scientific Research Fund by contract TK-X-1715/2007.

**References**

[1] Gunter P and Huignard J 2006 *Photorefractive materials and their applications.* Part 1, New York, Springer Science+Buisness Media

[2] Gunter P and Huignard J 2007 *Photorefractive materials and their applications.* Part 2 and 3, New York, Springer Science+Buisness Media

[3] Mihailova B, Gospodinov M and Konstantinov L 1999 *J. Phys. Chem. Solids* **60** 1821

[4] Marinova V, Veleva M and Petrova D 2001 *JAP* **89** 2686

[5] Ramaz F, Rakitina L, Gospodinov M and Briat B 2005 *Opt. Mat.* **27** 1547

[6] Burkov V, Kargin Y, Kizel V, Sitinkova V and Skorikov A 1983 *JETP Let.* **38** 390

[7] Reyher H, Hewllwing U and Thiemann O 1993 *Phys. Rev.* B **47** 5638-564

[8] Ahmad I, Marinova V and Goovaerts E 2009 *Phys. Rev.* B **79** 033107

[9] Kargin Y, Burkov V, Maryin A and Egorysheva A 2005 Alphabet, Moscow 180 (in Russian)

[10] Marquet H, Tapiero M, Merle J, Zielinger J and Launay J 1998 *Opt. Mater.* **11** 53

[11] Gospodinov M, Doshkova D, Sampil A and Aneva Z 1993 *Comp. Rend Acad. Bulg. Sci.* **46** 45

[12] Gospodinov M and Doshkova D 1994 *Cryst. Res. Technol.* **29** 603

[13] Gospodinov M and Doshkova D 1994 *Mat. Res. Bull.* **29** 681

[14] Petrova D, Gospodinov M and Sveshtarov P 1995 *Mat. Res. Bull.* **30** 1201

[15] Petrova D and Gospodinov M 1996 *Cer. Intern.* **22** 45

[16] Nechitailov A, Krasinkova M, Mokrushina E, Petrov A, Kartenko N and Prokofiev V 2000 *JNORG* **36** 820

[17] Egorisheva A, Burkov V, Kargin Y, Vasiliev A, Volkov V and Skorikov V 2001 *JNORG* **37** 817

[18] Egorisheva A 2005 *JNORG* **50** 407

[19] Marinova V, Lin S, Hsieh M, Liu R and Hsu K 2009 *Photorefractive Matt, Effects, and Devices PR'09* P1-39

[20] Egorisheva A, Volkov V, Coya C and Zaldo C 1998 *Phys. Stat. Sol.* B **207** 283

[21] Potera P and Pircuch A 2007 *Physica* B **387** 392

[22] Kovacs L, Capelli R and Gospodinov M 2008 *Vib. Spec.* **46** 69

[23] Zmija J and Malchowski M 2008 *Arh. Mat. Sci. Eng.* **33** 101

[24] Ahmad I, Marinova V and Goovaerts E 2009 *Photorefractive Matt, Effects, and Devices PR'09* **P1**-38

[25] Kostova B, Milenov T, Rafailov P, Egorysheva A, Yankova L, Veleva M, Dobreva S and Skorikov V 2009 *AIP Conf. Proc.* **1203** 188

[26] Kostova B, Gospodinov M and Konstantinov L 2009 *AIP Conf. Proc.* **1203** 193

[27] Leonite L, Caraman M, Delibas M and Rusu G 2001 *Mat. Res. Bull.* **36** 1629

[28] Skorikov V, Milenov T, Egorisheva A, Rafailov P, Dudkina T, Veleva M, Vasiliev A and Gospodinov M 2003 *Phys. Stat. Sol.* B **235** 3292

[29] http://www.webelements.com

# Structural, optical and electrical properties of V doped ZnO thin films deposited by r.f. magnetron sputtering

**K Lovchinov[1], H Nichev[1], O Angelov[1], M Sendova-Vassileva[1], V Mikli[2] and D Dimova-Malinovska[1]**

[1]Central Laboratory of Solar Energy and New Energy Sources, Bulgarian Academy of Sciences, 72 Tzarigradsko Chaussee, 1784 Sofia, Bulgaria
[2]Centre for Materials Research, Tallinn University of Technology, Ehitajate tee 5, 19086 Tallinn, Estonia

E-mail: lov4@abv.bg

**Abstract**. Structural, optical and electrical properties of V doped ZnO thin films deposited by r.f. magnetron co-sputtering on glass substrates at different temperature, $T_s$, between $150^0$C and $500^0$C are studied. The EDS analyses indicate that the average vanadium content in the films is in the range of 0.86-0.89 at. %. XRD spectra demonstrate preferential (002) crystallographic orientation with *c*-axis perpendicular to the substrate surface and grains sizes of the films about 21-29 nm. The band gap energy, $E_g$, values are in the range of 3.44-3.47 eV. The deposited V doped ZnO films have low resistivity - (2-8).$10^{-3}$ Ω cm. Raman spectra show vibrational phonons modes typical for ZnO. Comparison with the structural, optical and electrical properties of thin films ZnO and ZnO:Al is given.

## 1. Introduction

The systematic study of the optical, structural, electrical and other properties of thin ZnO films in dependence on the deposition technology is topical nowadays in the development of new electronic thin film devices [1], magnetic memories [2], transparent conductive oxides (TCO) with applications in thin film solar cells [3]. ZnO doped with transition elements provide variety of applications in gas sensors [4], spintronics [5], diluted magnetic semiconductors (DMS) [6] etc. The most frequently used ZnO growth techniques are MOCVD, pulsed laser deposition (PLD), magnetron sputtering and molecular beam epitaxy (MBE). R.f. magnetron sputtering is an attractive technique for deposition of undoped and doped ZnO thin films with different concentration of doping metals [7, 8].

In this work, a study of the influence of V and of the substrate temperature, $T_s$, on the optical, structural and electrical properties of V doped ZnO thin films deposited by magnetron r.f. sputtering in Ar atmosphere is reported. The results are compared with data for undoped and Al doped ZnO films.

## 2. Experimental

Thin ZnO films doped with V (ZnO:V) were prepared by r.f. magnetron co-sputtering of ZnO target with pieces of vanadium plate in the maximum erosion zone of its surface in Ar atmosphere at a pressure of 0.5 Pa and r.f. power of 180W. The films were deposited on glass substrates at different temperatures, $T_s$, between $150^0$C and $500^0$C. The thickness of the layers was 600 nm.

The film structure was studied by XRD and Raman spectrometry. XRD spectra were obtained using a Brucker D8 Advance spectrometer with Cu $K_\alpha$ radiation: λ Cu $K_{\alpha 1}$ =1.540560 Å and λ Cu $K_{\alpha 2}$ = 1.544426 Å (intensity half of that of λ Cu $K_{\alpha 1}$). The instrumental broadening in 2θ geometry was $0.04^0$. Raman spectra were recorded with a Horiba Jobin Yvon LabRam HR800 spectrometer using the 600 l/mm grating and a HeNe laser for excitation. Raman spectra were measured in back scattering geometry with resolution 1 $cm^{-1}$. Transmittance and reflectance spectra were measured by a spectrophotometer Shimadzu UV.3600 in the range 300-2600 nm. The vanadium content in the films was determined by Energy Dispersive X-ray Analysis (EDAX) using Link AN10000 system analysis. SEM pictures were obtained by Jeol JSM-840A with $LaBa_6$ cathode. The resistivity of the films was measured by the four point probe method using VEECO instrument.

## 3. Results and discussion

The concentration of vanadium in the obtained ZnO:V thin films varies little with the substrate temperature, $T_s$, (table 1). The presented values are averaged after measurements of the concentration at four different points on the surface of the samples. In table 1 the obtained electrical and structural parameters of the films under investigation are presented, as well. For comparison the data of undoped and Al-doped ZnO films are given too.

**Table 1.** The values of the optical band gap, $E_g$, Urbach tail, $E_0$, position of the (002) peak in the XRD spectra, θ, the FWHM of 2θ, Δ2θ, the average grains sizes, D, the concentration of the dopants (Al or V), c, the values of the stress, σ, and the resistivity, ρ, of the films deposited at different $T_s$.

| Sample | $T_s$, [$^0$C] | $E_g$, [eV] | $E_0$, [meV] | 2θ, [deg.] | Δ2θ, [deg.] | D, [nm] | c, [at.%] | σ, [GPa] | ρ, [Ω.cm] |
|---|---|---|---|---|---|---|---|---|---|
| ZnO* | 150 | 3.33 | 63 | 34.30 | 0.52 | 16.0 | | -0.81 | 31 |
| ZnO* | 275 | 3.30 | 59 | 34.40 | 0.47 | 17.7 | | -0.18 | 20 |
| ZnO* | 500 | 3.27 | 60 | 34.40 | 0.33 | 25.2 | | -0.18 | 4300 |
| ZnO:Al* | 150 | 3.36 | 100 | 34.20 | 0.62 | 13.6 | 1 | -1.52 | 18 |
| ZnO:Al* | 275 | 3.41 | 120 | 34.15 | 0.61 | 13.4 | 2 | -1.79 | 5.4 |
| ZnO:Al* | 500 | 3.33 | 91 | 34.27 | 0.62 | 13.4 | 1 | -0.98 | 6.2 |
| ZnO:V | 150 | 3.44 | 91 | 34.18 | 0.39 | 21.3 | 0.86 | -1.62 | 0.007 |
| ZnO:V | 275 | 3.47 | 98 | 34.14 | 0.33 | 25.2 | 0.88 | -1.89 | 0.002 |
| ZnO:V | 350 | 3.47 | 86 | 34.19 | 0.35 | 23.7 | 0.89 | -1.54 | 0.004 |
| ZnO:V | 500 | 3.44 | 83 | 34.19 | 0.29 | 28.6 | 0.87 | -1.54 | 0.008 |

*Data from reference [8]

XRD spectra of the films are presented in figure 1. The diffraction patterns show that the deposited ZnO:V films are polycrystalline with reflection corresponding to the (002) plane of the wurtzite structure with c-axis perpendicular to the substrate surface. The peak position of the ZnO:V films is shifted to the lower 2θ, compared to the undoped ZnO films, and are closed to that for Al doped films. This shift to lower 2θ is evidence for the presence of the tensile stress in the films, probably due to the presence of the doping atoms. The (002) peck position shifts slightly to the lower value of 2θ for the ZnO:V sample deposited at $T_s = 275^0$C. The stress in ZnO:V films are higher than in undoped and Al doped ZnO films. With increasing $T_s$, the Full Wide at Half Maximum (FWHM) of the (002) peak, Δ2θ, and the value of the stress, σ, decreases and the average grain sizes increase, which demonstrates an improvement of the structure with $T_s$, as in the case of undoped and Al doped ZnO films [8].

The values of the resisitivity of V doped ZnO films are lower than in the case of undoped and Al-doped ZnO deposited at the same conditions and are in the range of $2 . 10^{-3} - 8 . 10^{-3}$ Ω.cm. These

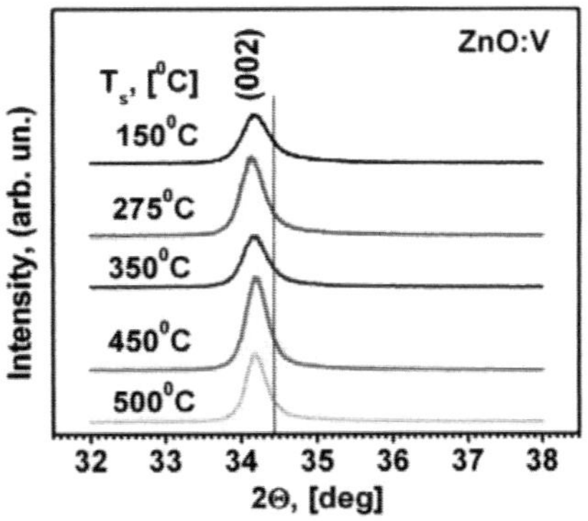


**Figure 1.** XRD spectra of thin films ZnO:V at different $T_s$.The black line indicates the position of (002) peak in crystalline ZnO [9].

values are in agreement with results reported in [10]. The average grain size is about 21 – 29 nm and increases with $T_s$.

Transmittance spectra of the ZnO:V thin films are presented in figure 2. The spectra are corrected for the transmittance of the glass substrate. All spectra demonstrate transmittance value higher than 85% in the range 500 – 1000 nm, independently of $T_s$. In the IR region beyond 1000 nm where the absorption due to the free carrier concentration (plasma resonance) takes a place, the transmission decreases. An absorption peak at about 850 nm appears for the ZnO:V deposited at $T_s > 275^0$ C more clearly pronounced for the sample with lower resistivity. The peak is attributed to the d-d absorption of $V^{2+}$ ion in tetragonal crystal field [11].

The absorption coefficient, α, was calculated as [12]:

$$\alpha(\lambda) = 1/d \,.\, \ln\,[(1 - R(\lambda))^2 / T(\lambda)]$$

where d is the film thickness, T – transmittance and R – the reflectance.

The spectral dependence of α exhibits two regions: a power law one at high photon energies and an exponential one at lower energies. The formula for direct allowed transitions can be used to obtain the optical gap, $E_g$, [13]:

$$\alpha(h\nu) = A\,[(h\nu - E_g)^{1/2} / h\nu]$$

In the lower energy range ($h\nu < E_g$), where α varies exponentially with photon energy, the spectral dependence of the absorption edge follows the Urbach formula [14]:

$$\alpha(h\nu) = \alpha_0 \exp\,[(h\nu - E_1) / E_0]$$

where $\alpha_0$ is the Urbach absorption at the edge ($E_1$), and $E_0$ is the Urbach energy. The Urbach region ($h\nu < E_g$) is due to the perturbation of the parabolic density of the states at the band edge. The increasing of the structural disorder results in an increase in Urbach energy.
The spectral dependences of α for sample deposited at different $T_s$ are shown in figure 3. The values of the optical band gap of V doped ZnO films are in the range of 3.44 – 3.47 eV and are typical for ZnO. However they are higher than in the case of undoped and Al-doped ZnO films. The optical energy gap increases with $T_s$ increasing until 275 °C, above which it decreases. It has to be noted that this sample demonstrates the lowest value of the resistivity (see table 1). The observed widening of $E_g$ in ZnO:V films as compared to undoped ZnO films, could be due to an increase in the donor concentration, related to shallow V donors, as it is in the case of Al doped ZnO films [14]. The Urbach energy decreases with $T_s$ increasing due to the improved structure in accordance with the XRD data. The values of $E_o$ in V doped ZnO are lower than in the Al doped ZnO, however higher than in undoped

Journal of Physics: Conference Series **253** (2010) 012030 doi:10.1088/1742-6596/253/1/012030

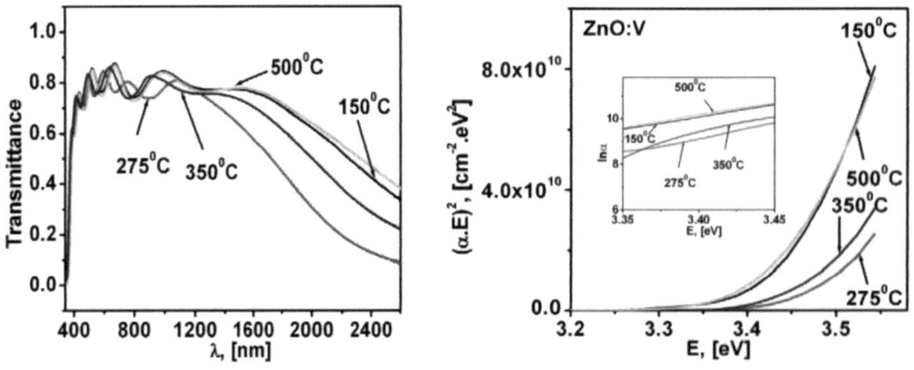


(a) (b)

**Figure 2.** Transmittance spectra (a) and $(\alpha.E)^2$ versus energy, E, (b) for ZnO:V films deposited at different $T_s$.The insert in (b) shows the plot of lnα vs hν.

ZnO. Possibly, the presence of V results in lower deterioration of the structural order in ZnO compared to Al.

Raman spectra of ZnO:V films are presented in figure 3. The glass luminescence background is substrated before drawing of plots. In the Raman spectra typical for ZnO bands are observed: 276 ($B_2$), 380 ($A_1$), 439 ($E_2$), 580 (LO). The band at 276 ($B_2$) is related to the build-in electric field in the depletion region in the grains of the films [15]. Additional bands at 509, 607 and 630 $cm^{-1}$ are present as well. These bands cannot be attributed to bulk phonon modes and are probably due to localized ones. The band at 509 $cm^{-1}$ could be assigned to the phonon mode highly localized near the grain boundaries as it is in the case of Al doped ZnO films [15]. This mode was observed when the crystallite size was smaller than 30 nm as in the present case. The 607 and 630 $cm^{-1}$ bands probably have similar origin to the $I_1$ and $I_2$ modes in [16] related to dopant complexes or host defects.

The surface morphology of the films was studies by Scanning Electron Microscopy (SEM) and Atomic Force Microscope (AFM). The AFM pictures (figure 4) and SEM micrographs (figure 5) show that the films have column structures. The surface roughness decreases and the homogeneity improve with increasing of the $T_s$.

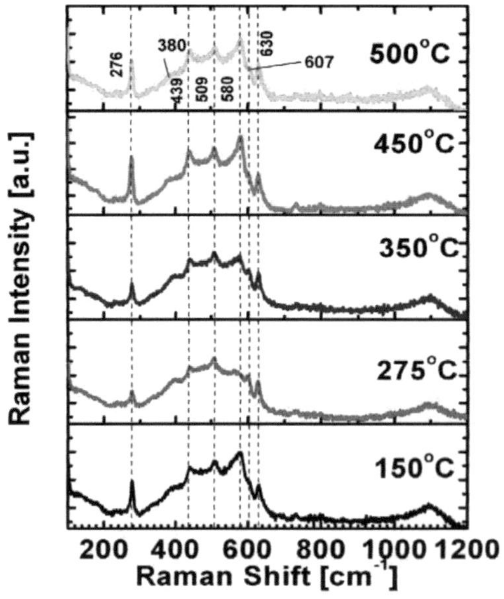


**Figure 3.** Raman spectra of ZnO:V films deposited at different $T_s$.

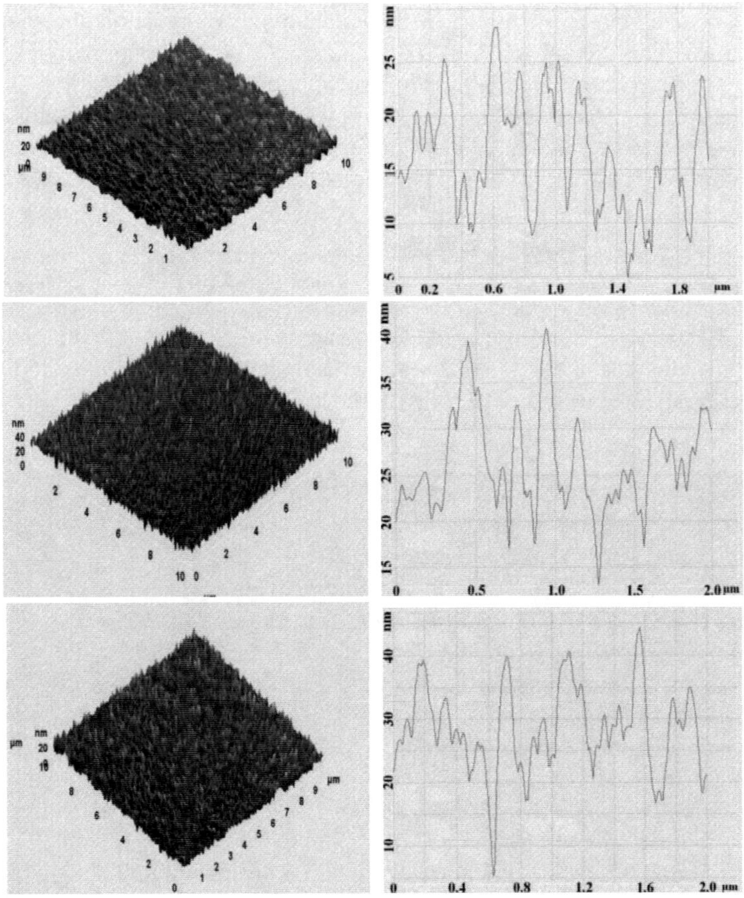

**Figure 4.** AFM picture of the ZnO:V deposited at $T_s$ 150$^0$ C (a), 350$^0$ C (b) and 500$^0$ C (c).

## 4. Conclusion

The study of the structural properties of V doped ZnO thin films deposited by magnetron co-sputtering at different substrate temperatures shows wurtzite ZnO structure with c-axis perpendicular to the substrate surface. XRD analyses demonstrate that the crystalline structure improves with $T_s$

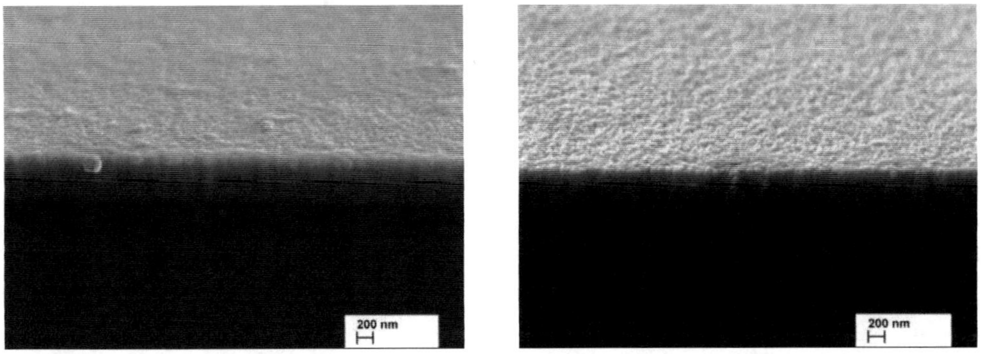


**Figure 5.** SEM micrographs of the ZnO:V films deposited at $T_s$ 150$^o$ C and 500$^o$C.

Journal of Physics: Conference Series **253** (2010) 012030 doi:10.1088/1742-6596/253/1/012030

increasing and the grain size increases from 21 to 29 nm. The V concentration does not change with $T_s$ and is about 0.86-0.89%. The thickness of the films is about 600 nm. The optical band gap ranges from 3.42 to 3.47 eV. The Urbach energy is higher than in undoped ZnO films, however lower than in the ZnO:Al films. The Raman spectra contain bands typical for ZnO, however bands due to the grain structure of the films are observed as well. The V doped ZnO films have transparency about 85% and low resistivity in the order of $(2\text{-}8).10^{-3}$ Ω cm. This demonstrate a potential for application of ZnO:V films as transparent conductive oxide in thin films solar cells and different optoelectronic devices.

**Acknowledgement**
This work was supported by the National Science Fund of Bulgaria -project DO02-207/2008), FP7 – NMP-2009-SMALL-3 (project NanoPV) and the Human Resources Development Programme (contract BG 051PO001/ 07/3.3-02/58/ 17.06.2008.The Raman equipment is used in the framework of project Integrated Research Centres at Universities № DO02-167/2008.

**References**

[1] Li Y, Gong J, McCune M, He G and Deng Y 2010 *Synthetic Metals* **160** 499
[2] Seshadri R 2005 *Current Opinion in Solid State and Materials Science* **9** 1
[3] Minami T 2005 *Semicond. Sci. Technol.* **20** S35
[4] Koshizaki N and Oyama T 2000 *Sensors and Actuators B* **66** 119
[5] Saeki H, Tabata H and Kawai T 2001 *Solid State Communications* **120** 439
[6] Fukumura T, Jin Z, Ohtomo A, Koinuma H and Kawasaki M 2000 *Appl. Phys. Lett.* **75** 3366
[7] Yin Z, Chen N, Yang F, Song S, Chai C, Zhong J, Qian H and Ibrahim K 2005 *Solid State Communications* **135** 430
[8] Dimova-Malinovska D, Angelov O, Nichev H, Kamenova M and Pivin J C 2007 *J.Optoelectr. Adv. Mat.* **9** 2512
[9] Hickenell F S 1975 *J. Vac. Sci. Technol.* **12** 879
[10] Miyata T, Suzuki S, Ishii M and Minami T 2002 *Thin Solid Films* **411** 76
[11] Saeki H, Tabata H and Kawai T 2001 *Solid State Commun.* **120** 439
[12] Pankove J 1971 *Optical Processes in Semiconductors* (New Jersey: Prentice-Hall Inc)
[13] Dragoman D and Dragoman M 2002 *Optical Characterization of Solids* (Heidelberg: Springer-Verlag)
[14] Dimova-Malinovska D, Nichev, Angelov O, Grigorov V and Kamenova M 2007 *Superlattices and Microstructures* **42** 123
[15] Tzolov M, Tzenov N, Dimova-Malinovska D, Kalitzova M, Pizzuto C, Vitali G, Zollo G and Ivanov I 2000 *Thin Solid Films* **379** 28
[16] Ke X, Zou C, Li M, Liu C, Guo L, Fu D 2010 *Jap. J. Appl. Phys.* **49** 033001

# Influence of the processing conditions on the structural properties of ZnO layers obtained by PECVD

**S Kitova and G Danev**

Institute of Optical Materials and Technology "Acad. J .Malinowski", Bulgarian Academy of Sciences, Acad. G. Bonchev str., bl.109, 1113 Sofia, Bulgaria

E-mail: skitova@clf.bas.bg

**Abstract**. The plasma enhanced chemical vapor deposition (PECVD) is a powerful and flexible instrument for depositing thin layers, nanocomposites or nanostructures. In this work ZnO layers have been grown by metal-organic PECVD (RF - 13.56 MHz) on Si wafers. Zn acetylacetonate has been used as a precursor and oxygen as oxidant. The influence of the oxygen content in gas mixture, the total pressure, substrate temperature and ZnO seed layer on the structural properties of the layers deposited on Si wafers has been studied. ZnO layer properties were investigated by X-ray diffraction (XRD) and scanning electron microscopy (SEM). XRD data have shown that all layers are crystalline with hexagonal wurtzite structure. The crystallites are preferentially oriented along c-axis direction perpendicular to the substrate surfaces. The results obtained indicate that by controlling the oxygen content in gas mixture, the total pressure and substrate temperature during the film growth one can control the formation of c-axis phase and the crystallite grain size. Nanorods with good alignment, vertically orientated to the substrate surface can be observed in the layers deposited at low content of $O_2$ in plasma at substrate temperature of 400 °C. Due to their structural characteristics these layers are potential materials for preparing chemical- and biosensors where inherently large surface to volume ratio of structured materials are important prerequisite for enhanced sensitivity.

## 1. Introduction

So far, various deposition techniques such as thermal evaporation, pulsed laser deposition, sputtering, sol-gel deposition, template assisted and solution processes, molecular beam epitaxy, metal-organic chemical vapor deposition (MOCVD) have been used to grow high crystalline ZnO layers on a practical substrate such as sapphire and Si wafer (for reviews see [1, 2] and references cited there in). Among these, MOCVD can offer high growth efficiency, large area uniformity and the suitability for growing components containing volatile elements [3].

Plasma enhanced MOCVD (PECVD) technology is one of the most attractive for a synthesis of ZnO layers at low and moderate temperatures of substrates and minimal impact upon them. Diethylzinc and dimethylzinc are usually applied as precursors in PECVD process [4, 5]. Nevertheless, these precursors are less desirable because of their high toxicity and flammability and high price. Zinc acetylacetonate ($Zn(acac)_2$) should be a good candidate for use in low temperature fabrication of crystalline ZnO layers because of its non-toxicity and inflammability, low vaporization temperatures (250 °C), low thermal decomposition temperatures (150-500 °C), and high vapor pressure [4]. However, $Zn(acac)_2$ is rarely used as precursor in PECVD.

It is well known that structures of synthesized ZnO are dependent on processing conditions such as working pressure, substrate temperatures, deposition power and gas mixture. Several authors have already synthesized different crystalline nanostructured ZnO by adjusting substrate temperatures, reaction gases, and starting material during CVD [1, 5-8]. ]. A selective growth of high quality ZnO nanorods on Si substrates at 800-900 °C was demonstrated using a pattern thin film of ZnO as seed layer [9].

In this paper, the effect of the oxygen partial pressure, the total pressure, the substrate temperature and ZnO seed layer on the structural properties of ZnO layers deposited on Si wafers has been studied. The layers are grown by PECVD, using $Zn(acac)_2$ and $O_2$ as reactants.

## 2. Experimental

ZnO layers were grown in PECVD equipment (GENUS 8720). The gas plasma was excited by a 13.56 MHz RF generator. The RF power was 2000 W. The reaction chamber was evacuated to 8 mTorr before the process. Oxygen gas and $Zn(acac)_2$ ($Zn(C_5H_7O_2)_2.xH_2O$ - Sigma-Aldrich) were used as precursors. A system attached to the one of the chambers windows for dosed injection of the precursor to the plasma reactor was developed. The Zn source was sublimated by heating up to 140 °C and carried by nitrogen gas at constant flow rate to the growth chamber. The stable precursor temperature resulted in a stable precursor flow rate, even if we could not directly measure it. $O_2$ gas was introduced separately into the reaction chamber in order to prevent any pre-reaction of the two precursors. Additional Ar gas was injected through a separately connected gas line in order to vary total pressure. All gas flows were controlled by mass flow meters controllers. By varying the gas flow rate the chamber pressure was kept at different ranges from 70 to 900 mTorr. The substrates were heated in the range 400-600 °C. The deposition time was between 60 - 90 min. The thickness of the layers of about 500 μm was measured by a profilometer type Talystep.

The substrates used to deposit the layers were p-type Si wafers with (100) orientation. Preliminarily cleaning of the substrate surface was very important to grow ZnO layers with a good crystal quality and to enhance ZnO nucleation. For the clean surface, the Si wafers were dipped in a 10% dilute HF acid solution and then rinsed in deionized water and isopropyl alcohol.

The crystal structure of the layers was characterized by XRD diffraction (XRD), using a Philips (PW 1710) apparatus with Cu-Kα radiation separated by a graphite focusing monochromator. The 2θ range used in XRD measurements was from 20° to 60° in step of 0.05° $s^{-1}$ and fixed counting time of 5 s per step. A smoothing process of 5-point Fast Fourier transform followed by an appropriate baseline correction was applied to all diffractograms. The intensity and full width at half maximum (FWHM) of a (hkl) plane were determined by profile fitting procedure, using a Voight or Gaussian peak shape. Scherrer's equation was used for estimating the average crystallite size. The value of λ was taken as 0.15406 nm.

The surface morphology of the layers was examined by a scanning electron microscope (SEM Philips 515).

## 3. Results and discussion

Our previous study have showed that the layers obtained at the highest RF power of 2000 W, $O_2$ gas ambient pressure of about 300 mTorr and low substrate temperature (~ 40 °C) are amorphous, with high optical transmission in the visible region and $E_g$ of 3.36 eV, which is very close to that of bulk ZnO material [10]. Figure 1 presents X-ray diffraction patterns of the layers grown on Si wafers at different partial oxygen pressure $P_{O_2}$, constant total pressure $P_{tot}$ of 280 mTorr and substrate temperate $T_s$ of 400 °C. The XRD data demonstrate that the all ZnO layers are polycrystalline. Three peaks at 31.7°, 34.3°, and 36.2° are observed. These peaks can be indexed to $(10\bar{1}0)$, (0002) and $(10\bar{1}1)$ planes of the hexagonal wurtzite structure of ZnO crystal [11]. The results show that the crystallites are preferentially oriented along the c-axis direction perpendicular to the substrate surface, as the (0002) reflection is greatly enhanced relatively to the usual $(10\bar{1}1)$ maximum reflection of ZnO.

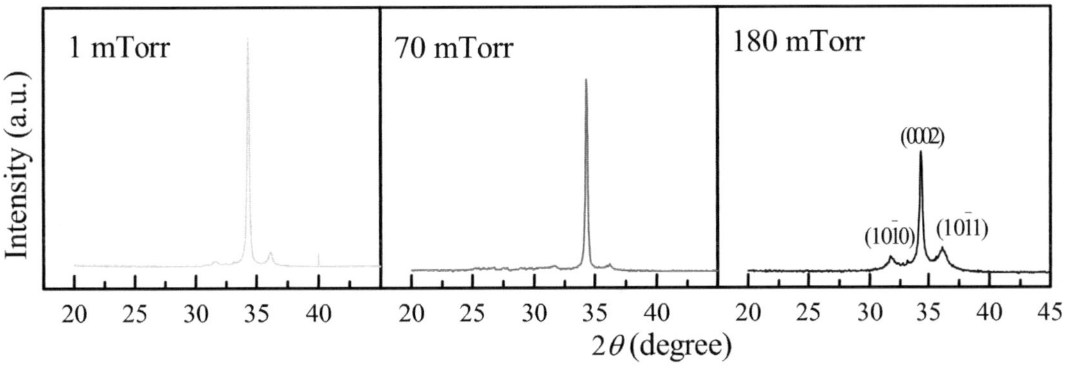


**Figure 1.** X-ray diffraction patterns of the layers grown on Si wafers at indicated partial oxygen pressure $P_{O_2}$ and constant total pressure $P_{tot}$ of 280 mTorr and substrate temperature $T_s$ of 400 ºC.

In order to quantify the preferred orientation (PO), often called texture, in the crystalline films, the texture coefficient (Tc) was defined as [12]:

$$Tc(hkl) = \frac{I(hkl)/I_o(hkl)}{N^{-1}\sum_n I(hkl)/I_o(hkl)} \quad (1)$$

where $I(hkl)$ is the measured intensities of a (hkl) plane converted into the corresponding intensity of infinitely thick sample. $I_o(hkl)$ is the intensity of a random oriented ZnO powder standard measured at the same conditions, $N$ is the reflection number and $n$ is the number of the diffraction peaks. A sample with randomly oriented crystallites presents Tc(hkl) = 1, while the larger value, the larger abundance of crystallites oriented at the (hkl) direction.

Figure 2 (a-c) presents the influence of the partial oxygen pressure, the total pressure and the substrate temperate on Tc, FWHM and grain size calculated from XRD patterns of the highly oriented (0002) peaks. The high value of Tc of about 4.5, the very narrow full width of 0.22-0.27 and grain size of about 35 nm are the indications of predominant c-axis phase with better crystal quality formed on pure Si wafers at oxygen partial pressure up to 150 mTorr. As it is seen in figure 2a, the further increase in $P_{O_2}$ in plasma gases leads to decreasing Tc value to 2.2 and to broadening FWHM value to 0.97. Correspondingly, the average crystallite size, which is inversely proportional to FWHM steeply, decreases to the smallest value of 9 nm. These results denote less c-axis preferred phase with smaller size grown at $P_{O_2}$ higher than 150 mTorr. As a rule with increasing the oxygen partial pressure, the crystalline quality degrades from single oriented crystal to polycrystalline. We have found that there was no film to grow on Si wafers with further increasing the oxygen pressure. The data presented in figure 2a indicate that in the presence of ZnO seed films on Si wafer Tc and FWHM are less dependent on partial $P_{O_2}$. Figure 2b shows that the increasing the total pressure up to 280 mTorr favors the c-axis phase formation with better crystal quality since Tc has high value and FWHM decreases from 0.5 to 0.22. Further rising $P_{tot}$ degrades the c-axes PO and leads finally to more random orientation with lower crystal quality. Figure 2c shows that the rising the substrate temperature degrades the c-axes PO and leads finally to more random orientation. The growth of the grain along substrate surface is enhanced and thus grain size of c-phase decreases.

It should be noted that the results illustrated in figure 2 are consistent with the data obtained by SEM observations. Figure 3 shows top- and cross sectional- view SEM images of the layers grown on Si wafers at different partial oxygen pressure and at constant $P_{tot}$ and $T_s$. The images show a well developed grain structures of layers deposited at low $P_{O_2}$ in plasma and denser and smoother surface of layers deposited at high $P_{O_2}$ in plasma. All layers exhibit columnar structure. Nanorods with good alignment, vertically orientated to the substrate surface can be seen in layers deposited at low content of $O_2$ in plasma gases.

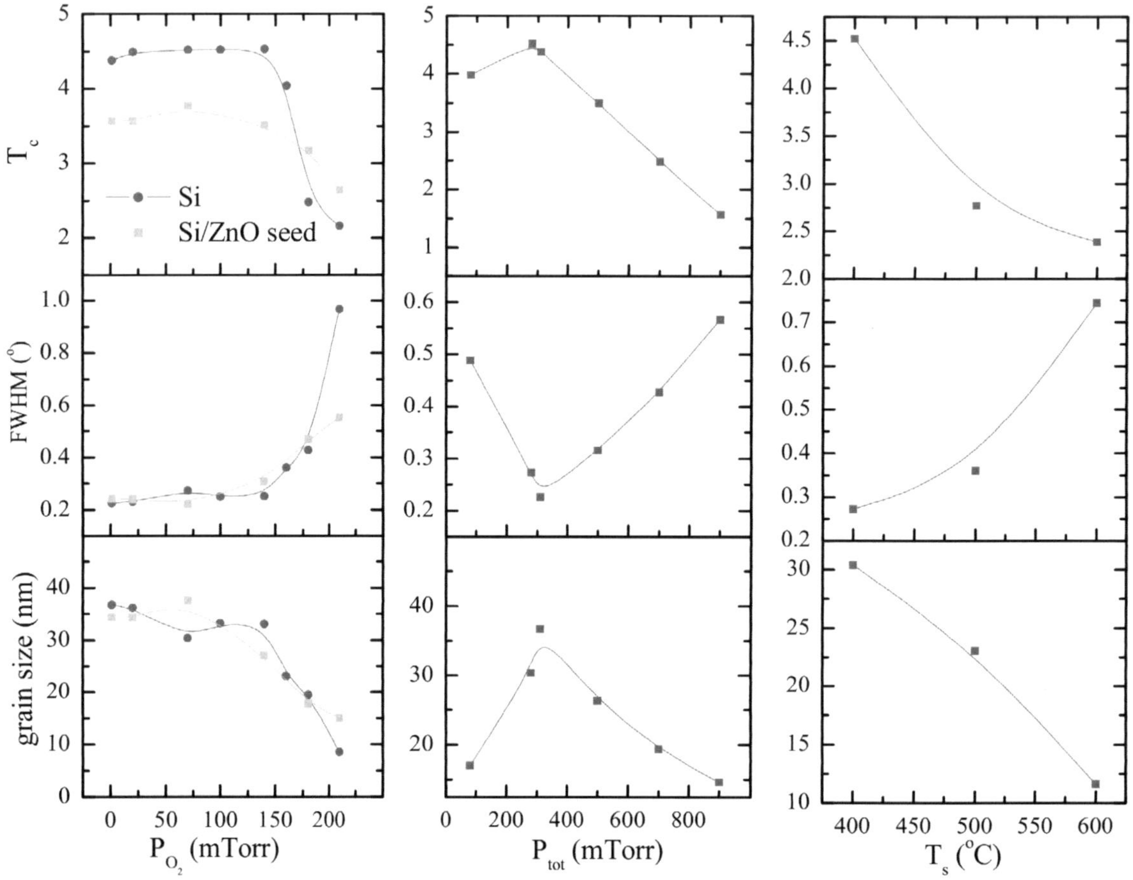


**Figure 2.** Tc, FWHM and grain size calculated from XRD patterns of the highly oriented (0002) peaks in dependence of:

(a) the partial oxygen pressure $P_{O_2}$ - for the layers grown on Si wafers without and with ZnO seed layer at constant $P_{tot}$ of 280 mTorr and $T_s$ of 400 °C.
(b) the total pressure $P_{tot}$ - for the layers grown on Si wafers at constant $P_{O_2}$ of 20 mTorr and $T_s$ of 400 °C.
(c) the substrate temperature $T_s$ - for the layers grown on Si wafers at constant $P_{O_2}$ of 70 mTorr and $P_{tot}$ of 280 mTorr.

Most probably, the observed preferred orientation (PO) of deposited crystalline layers and their related structure are connected with so called preferential nucleation and preferential grain growth. The preferential nucleation is a process in which initial nuclei with a specific crystallographic direction preferentially nucleate on the substrates. For the case of ZnO, c-axis PO nucleation has been considered to arise because of the lowest surface free energy of the plane among hexagonal crystallographic planes [8]. Therefore, it could be proposed that the initial nuclei with predominant c-axis PO are formed at the beginning of the layer growth and their PO have determined the PO and the columnar structure of the layers obtained. The results obtained indicate that the $O_2$ content in the plasma influences the nucleation process and the growth of ZnO. In plasma with high $O_2$ concentration ZnO has a high nucleation density and a rapid growth, which leads to formation of dense films with smaller grain size. On the other hand, at low $O_2$ concentration the activation energy of nucleation is higher which makes the nucleation of ZnO more difficult. Under this oxygen deficient condition, ZnO will grow preferentially on the pre-existing ZnO nuclei that have been formed during the beginning of the layer growth. In addition due to the faster vertical growth rate along c-axis compared to the lateral direction rod shaped ZnO crystal are developed.

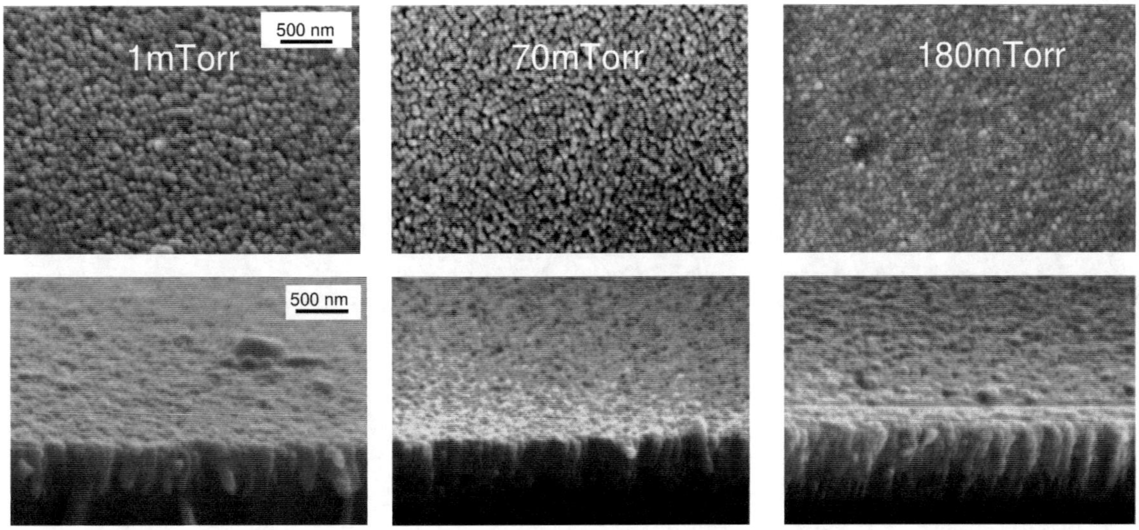


**Figure 3.** Top- and cross sectional- view SEM images of the layers grown on Si wafers at indicated partial oxygen pressure $P_{O_2}$ and at constant $P_{tot}$ of 280 mTorr and $T_s$ of 400 °C.

## 4. Conclusion

The crystalline ZnO layers have been obtained by PECVD at $T_s$ within the range of 400 to 600 °C on Si substrates without and with ZnO seed layer. The XRD data indicate that all layers exhibit the hexagonal wurtzite structure. The crystallites are preferentially oriented along c-axis direction perpendicular to the substrate surfaces. The result obtained show that by controlling the oxygen content in gas mixture during the film growth as well as the total pressure and substrate temperature one can control the formation of c-axis phase and grain size. It has been found that ZnO layers deposited at low $O_2$ content in the plasma gases and at $P_{tot}$ of 200 -300 mTorr and $T_s$ of 400 °C have well developed grain structures with predominant c-axis phase and better crystal quality than that of the samples obtained in plasma with high oxygen concentration and high $P_{tot}$. However, in the presence of ZnO seed films on Si wafer the formation of c-axis phase and its quality are less dependent on partial $P_{O_2}$. Nanorods with good alignment, vertically orientated to the substrate surface can be observed in the layers deposited at low content of $O_2$ in plasma at substrate temperature of 400 °C. Due to their structural characteristics, these layers are potential materials for preparing chemical- and biosensors where inherently large surface to volume ratio of structured materials are important prerequisite for enhanced sensitivity.

**Acknowledgements**
The authors thank Prof. G. Schmahl, Dr. J. Thieme and Dr. B. Niemann from Roentgenphysik-Uni of Gottingen-Germany for the gift of the PECVD installation, and Sc.D. I. Tomov for carrying out the XRD analysis, and Dr. K. Starbova for the SEM observations.

This study was supported by National Science Fund Grant # DO02-207.

**References**

[1] Mende L and Driscoll L 2007 *Materialstoday* **10** 40
[2] Lu J, Chang P and Fan Zh 2006 *Mater. Sci. Eng.* **R52** 49
[3] Ma Y, Du G, Yin J, Yang T and Zhang Y 2005 *Semicond. Sci. Technol.* **20** 1198
[4] Karpina V, Lazorenko V, Lashkarev C, Dobrowolski V, Kopylova L, Baturin V, Pustovoytov S, Karpenko A, Erimin S and Lytvyn P 2004 *Cryst. Res. Technol.* **39** 980
[5] Ozgur U, Alivov Y, Liu C, Teke A, Reshchikov M, Dogan S, Avrutin V, Cho S, and Morkoc H 2005 *J. Appl. Phys.* **98** 041301
[6] Iwanaga H, Shibata N, Nittono O and Kasuga M 1978 *J.Crys.Growth* **45** 228
[7] Liu X, Wu X, Cao H and Chang R 2004 *J. Appl. Phys.* **95** 3141
[8] Kajikawa Y 2006 *J. Crystal Growth* **289** 387
[9] Conley J, Stecker L and Ono Y 2006 *J. Electronic Mater.* **35** 795
[10] Minchev M, Kitova S and Danev G 2009 *JOAM* **11** 1312
[11] Powder Diffraction File 89-1397 for hexagonal Zinc oxide (2001 JCPDS-International center for Diffraction data)
[12] Barret C and Massalski T 1980 *Structure of Metals* (Oxford: Pergamon Press)

Journal of Physics: Conference Series **253** (2010) 012032 doi:10.1088/1742-6596/253/1/012032

# Ellipsometric characterization of AlN films synthesized by Pulsed-Laser-Deposition

**A Szekeres[1,*], E Vlaikova[1], A Cziraki[2], P Petrik[3], G Socol[4], C Ristoscu[4] and I N Mihailescu[4]**

[1]Institute of Solid State Physics, Bulgarian Academy of Sciences, 72, Tzarigradsko Chaussee blvd., 1784 Sofia, Bulgaria

[2]Eotvos Lorand University, 1 Pazmany Peter setany, 1117 Budapest, Hungary

[3]Research Institute for Technical Physics and Materials Science, Hungarian Academy of Sciences, Konkoly Thege Miklós út 29-33, H-1121 Budapest, Hungary

[4]National Institute for Lasers, Plasma, and Radiation Physics, PO Box MG-54, RO-77125, Magurele, Ilfov, Romania

E-mail: szekeres@issp.bas.bg

**Abstract.** AlN films were synthesized on Si(100) by pulsed laser deposition at 800°C and different incident laser fluences in vacuum and in nitrogen at 0.1 Pa. Two KrF* ($\lambda$= 248 nm) excimer laser sources were used generating pulses of 7.4 and 25 ns duration, respectively. The incident laser intensity on target was kept constant in all experiments within (3-4)x$10^8$ W/cm$^2$. The films were studied by ellipsometry (SE) in the spectral range $\lambda$=190÷900 nm and the optical parameters, such as refractive index, high frequency dielectric constant and single oscillator energies were estimated. The analysis of the SE results revealed that the films, deposited with short laser pulses and low laser fluence (3.7 J/cm$^2$) were characterized with refractive index values higher than 2 and an optical band gap energy value of ~ 5 eV, suggesting that their structure was polycrystalline with cubic crystallites. With longer laser pulses and large fluence the refractive index values decreased and the values of the energetic parameters increased suggesting that the films, deposited in vacuum were polycrystalline with hexagonal phase, while those, deposited in nitrogen at 0.1 Pa were amorphous.

## 1. Introduction

Aluminium nitride thin films are promising for various technological applications, such as electroacustic and optical devices, protective coatings and insulating layers. It has been established that preferential orientation of polycrystalline AlN films affects and often improves the properties of practical use such as piezoelectricity [1] or hardness [2,3].

AlN generally crystallizes in either hexagonal wurtzite (h-AlN) structure or cubic zinc-blende (c-AlN) polytype, which is reported theoretically as metastable phase [4]. On the other hand, because of its higher crystallographic symmetry, cubic AlN is expected to have several advantages for device applications, such as lower carrier scattering and higher doping efficiency. Since the growth of cubic AlN is quite difficult due to its metastable nature, the quality of cubic nitrides has been inferior to that of hexagonal ones. Theoretical investigations of the thermal conductivity of AlN have shown that the

* To whom any correspondence should be addressed.

energy difference between two kinds of phase structure of AlN is so small that both hexagonal and cubic phases are able to coexist in a realistic material [5]. Still, relevant data on the formation of each of the crystalline phases of AlN have been scant. Because of that, the properties of AlN have been extensively studied in the last decade, both theoretically and experimentally. The main accent has been placed on development of new deposition techniques and optimization of the growth conditions of AlN films. One of the most efficient and versatile techniques is pulsed laser deposition (PLD) possessing many advantages like the ability to grow thin films with good crystallinity and stoichiometry at low temperatures. There is still no straightforward theoretical or experimental model of the processes during deposition and the resulting film properties. Hence, the characterization of film growth and the mechanisms governing the film synthesis is an important task in all application areas of AlN films.

Our research deals with pulsed laser deposited AlN films, with a focus on their structure and optical and electrical properties. The recent investigations revealed that the structure of the growing AlN films strongly depends on the laser fluence. Thus, at low fluences (2.7-3.7 $J/cm^2$) the films were polycrystalline with 30-60 nm sized predominantly cubic crystallites [6-8]. When increasing the laser fluence to 10 $J/cm^2$, the prevalent crystalline phase turned from cubic to hexagonal [9]. The nitrogen pressure in the gas ambient during deposition played a decisive role in the crystallization process.

Here we present results of spectroscopic ellipsometric (SE) studies of the effect of deposition conditions on the growth of PLD AlN films and on their optical properties. The dependence of the films structure on the laser fluence and nitrogen content in the vacuum chamber during deposition is discussed.

## 2. Experimental details

AlN films were synthesized by pulsed laser deposition on crystalline Si substrates. Prior to deposition, the chamber was evacuated to $10^{-4}$ Pa and the substrates were heated up to 800°C, a temperature at which the native $SiO_2$ covering the wafers is known to decompose. The target consisting of polycrystalline AlN with 99% purity was rotated with 0.3 Hz to avoid piercing and to ensure a clean surface during multi-pulse laser irradiation. Two pulsed KrF* excimer laser sources ($\lambda$ = 248 nm) were used for the ablation of the targets. The first one, model M-1701, generated pulses of 7.4 ns, while the second, a COMPEXPro 205, delivered pulses of 25 ns duration. 20,000 subsequent pulses were applied for the synthesis of each structure. The films were synthesized on substrates kept at a temperature of 800°C during deposition. Two different deposition regimes were applied: (i) laser pulse duration of 7.4 ns, repetition rate of 2 Hz, target-substrate distance of 4 cm, incident laser fluence of 3.7 $J/cm^2$, and (ii) laser pulse duration of 25 ns, repetition rate of 3 Hz, targetsubstrate distance of 5 cm, incident laser fluence of 10 $J/cm^2$. Accordingly, the laser intensity on the target was in all cases similar, within the narrow range of $(3 - 4) \times 10^8$ $W/cm^2$. This way we could independently study the effects of laser pulse duration and incident laser fluence.

The AlN films synthesis was carried out in vacuum ($5 \times 10^{-4}$ or $10^{-4}$ Pa) or in nitrogen at dynamic pressures of 0.1 Pa.

Depending on deposition conditions, the film thickness in the first set was in the range of 600-800 nm, while for the second set of experimental conditions the film thickness was in the range of 400-600 nm. There was a tendency for the thickness to decrease with nitrogen addition.

The ellipsometric measurements were carried out either on a Rudolph 436 ellipsometer covering the wavelength range from 300 to 820 nm or on a Woollam ellipsometer in the wavelength region of 190-900 nm. The angle of light incidence was $50^o$. Based on our [10] and other observations [11] that the PLD AlN films are transparent in the spectral range under consideration, the determination of the optical constants were performed using the Cauchy's equation in the wide spectral range of 400-900 nm. Below 400 nm some light scattering due to surface roughness can arise and, therefore the optical constants were inferred by solving the inverse problem of ellipsometry in this spectral region.

## 3. Results and discussion

The spectral dependences of the refractive index *n* of the studied AlN films are presented in figure 1. In a wide wavelength region of visible light all the *n* values are within 2-2.2, being characteristic for polycrystalline AlN material [12], as the larger the n value the higher the order of crystallinity.

The shape of the curve for the films, deposited at 3.7J/cm$^2$ laser fluence differs from that for films, deposited at more than twice higher laser fluence, suggesting different film structure. For low laser fluence, the refractive index values are very close, which is inherent for similar film structure. Contrary to that, films deposited at high laser fluence (10 J/cm$^2$) have different refractive index values being evidence that film structures should differ from each other. The index data of AlN film, deposited at high laser fluence and in small amount of nitrogen at a dynamic pressure of 0.1 Pa, shows n values below 2 in almost the whole spectral region studied. Such spectral dependence is characteristic for amorphous AlN material [12].

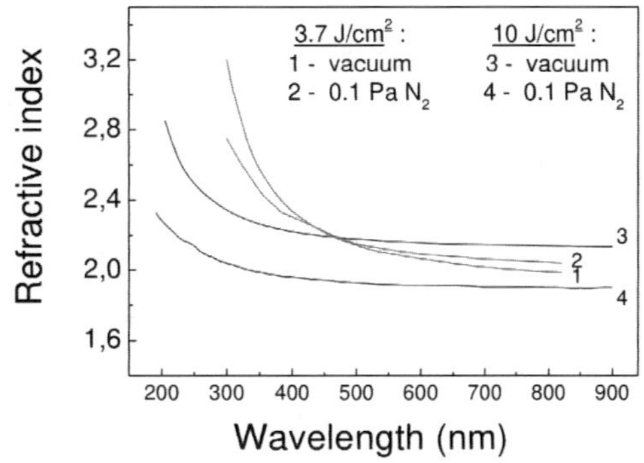


**Figure 1.** Dispersion curves of the refractive index of AlN films deposited at different laser fluence and in different gas ambient.

The extinction coefficients for all the films remained below the value 0.02 even when approaching the near ultraviolet region, showing that the PLD AlN films are transparent in the studied spectral range. This makes it impossible to calculate the optical band gap energy $E_{og}$ values. However, we could estimate how much the optical bandgap should be and what the influence of deposition conditions on its value is. For this purpose we analysed the refractive index dispersion curves using the concept of the single oscillator [13] and applying two approaches.

The first approach considers a single oscillator at wavelength $\lambda_0$. It includes the high frequency region, where the dielectric constant $\varepsilon_l$ can be calculated using the simple classical dispersion equation $(n_0^2-1)/(n^2-1)=1-(\lambda_0/\lambda)^2$, where $n_0$ is the refractive index of an empty lattice at infinite wavelength and it gives the high frequency dielectric constant $\varepsilon_\infty = n_0^2$. These values were deduced from the plots of $1/(n^2-1)$ versus $1/\lambda^2$, presented in figure 2, from the linear constants of the resulting straightline equation $y = a-bx$ ($a=1/(n_0^2-1)$ and $b=\lambda_0^2/(n_0^2-1)$). The wavelength $\lambda_0$ is related to the energy at which direct interband electron transitions take place ($\lambda_0=hc/E_0$). The calculated quantities are summarized in Table 1.

In the second approach the refractive index data were analyzed below the interband absorption edge using the single-effective-oscillator equation $n^2-1=E_0E_d/(E^2_0-E^2)$ [13]. $E_0$ is the oscillator energy and it represents the average band gap being a distance between the "centers of gravity" of the valence and conduction bands. $E_d$ is the dispersion energy of a single oscillator, it is a measure of the intensity of the inter-band optical transition and does not depend significantly on the band gap. These energetic parameters were determined from the intercept and slope resulting from the extrapolation of the curves in figure 3, which represents the plots of $1/(n^2-1)$ versus $E^2$. The values of $E_0$ and $E_d$ are also given in Table 1. The values of $n_0$ and $E_0$, obtained by the two approaches are in good coincidence.

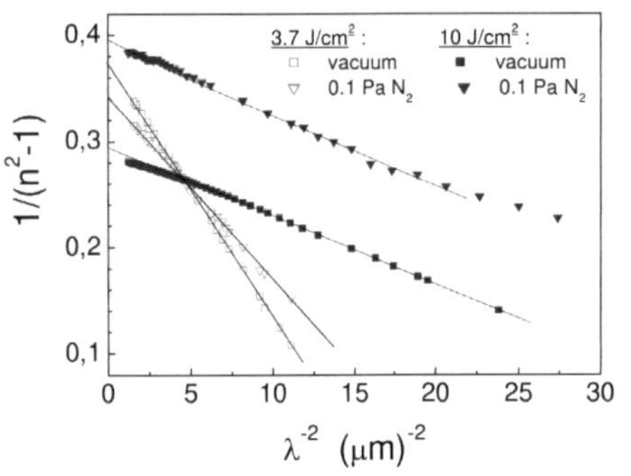


**Figure 2.** Plots of $1/(n^2-1)$ against $\lambda^{-2}$ for AlN films deposited at different laser fluences and in different gas ambient.

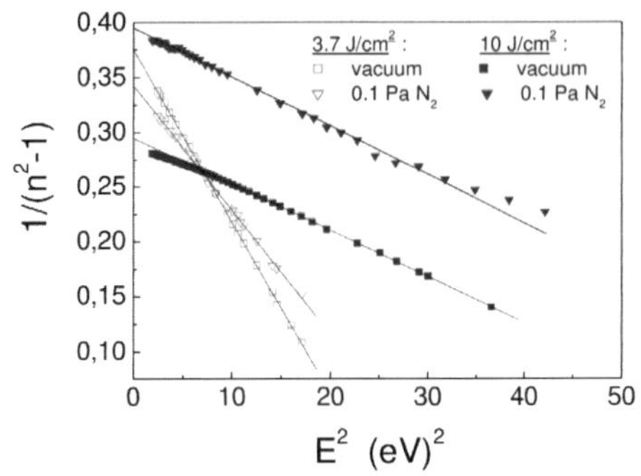


**Figure 3.** Plots of $1/(n^2-1)$ against $E^2$ for AlN films deposited at different laser fluences and in different gas ambient.

Table 1. Optical constants and structure of PLD AlN films in dependence on deposition conditions.

| | | | | | |
|---|---|---|---|---|---|
| Deposition condition (at constant laser intensity (3-4)x$10^8$ W/cm$^2$ and number of pulses 20 000) | Pulse duration | 7.4 ns | | 25 ns | |
| | Laser fluence | 3.7 J/cm$^2$ | | 10 J/cm$^2$ | |
| | Repetition frequency | 2 Hz | | 3 Hz | |
| | Gas pressure | Vacuum (5x$10^{-4}$ Pa) | Nitrogen (0.1 Pa) | Vacuum ($10^{-4}$ Pa) | Nitrogen (0.1 Pa) |
| Optical constants | $(n_0^2-1)/(n^2-1)=1-(\lambda_0/\lambda)^2$ | | | | |
| | $n_0$ | 1.91 | 1.98 | 2.1 | 1.88 |
| | $\varepsilon_\infty = n_0^2$ | 3.65 | 3.92 | 4.41 | 3.53 |
| | $\lambda_0$ | 253.4 nm | 240 nm | 148 nm | 135 |
| | $E_0=hc/\lambda_0$ | 4.89 eV | 5.17 eV | 8.38 eV | 9.18 |
| | $n^2-1=E_0E_d/(E^2_0-E^2)$ | | | | |
| | $n_0$ | 1.91 | 1.98 | 2.1 | 1.88 |
| | $E_0$ | 4.88 eV | 5.16 eV | 8.37 eV | 9.2 eV |
| | $E_d$ | 13.01 eV | 15.06 eV | 26.9 eV | 23.17 |
| Film structure | | Cubic AlN phase | | Hexagonal AlN phase | Amorphous AlN phase |
| | | Polycrystalline AlN | | | |

The oscillator energy $E_0$ is empirically related to the lowest direct band gap energy $E_{og}$ as $E_0 \approx 1.5E_t$ [13], where $E_t$ is an optical threshold energy level at which electron transitions begin and cause absorption to appear in the film. The optical band gap energy $E_{og}$ is expected to be higher than $E_t$, as the relation between $E_t$, $E_0$ and $E_{og}$ is $E_t < E_0 < E_{og}$ [13].

According to the limitation $E_0 < E_{og}$, for the AlN films, deposited at lower laser fluence, the real bandgap energy value is higher than 4.8 eV. For cubic AlN structure, bandgap values of 5.1 eV [14] and 4.5-6 eV [15] for direct electron transitions and, $E_{og}$ ~3.5 eV for indirect electron transitions [16] are reported. Based on this and on the obtained refractive index values, a conclusion can be made about the film structure, namely that PLD AlN films, deposited at these conditions are polycrystalline with cubic phase crystallites. This conclusion was recently confirmed by the XRD studies on these films [6,9].

In the case of PLD deposition with longer laser pulses and high fluence (Table 1), the gas ambient obviously played more decisive role in the crystallization process. From the refractive index data analysis one can conclude that the AlN films, deposited in vacuum are polycrystalline, while the films grown in nitrogen ambient at 0.1 Pa have an amorphous structure. Since the refractive index of cubic AlN is somewhat larger than that of hexagonal modifications when approaching the absorption edge [12, 14] this suggests that in the PLD AlN films deposited in vacuum the crystallites are grown in hexagonal phase. This can be an explanation for the larger values of the energetic parameters registered for these films. Our XRD studies on PLD AlN films, deposited in vacuum at similar conditions and with high laser fluence of 8.6 J/cm$^2$, have revealed that films had a prevalent hexagonal polycrystalline structure [9].

## 4. Conclusions

The analysis of the SE results revealed that by varying the pulsed laser deposition conditions the optical parameters of the PLD AlN films changed to parameters characteristic for either cubic or hexagonal polycrystalline AlN. The PLD AlN films, deposited with short laser pulses and low fluence, are polycrystalline with cubic phase crystallites. In the case of PLD deposition with longer laser pulses and high fluence the PLD AlN films structure has a stronger dependence on the amount of nitrogen during the synthesis, as AlN films deposited in nitrogen at a dynamic pressure of 0.1 Pa grow in amorphous phase.

## Acknowledgements

The authors acknowledge the support of this work under the 2007-2009 Collaboration Agreement between the Bulgarian, Hungarian and Romanian Academies of Sciences.

## References

[1] Liufu D and Kao K C 1998 *J. Vac. Sci. Technol.* A **16** 2360
[2] Takikawa H, Kimura K, Miyano R, Sakakibara T, Bendavid A, Martin P J, Matsumuro A and Tsutsumi K 2001 *Thin Solid Films* **386** 276
[3] Szekeres A, Bakalova S, Cziraki A, Grigorescu S, Socol G, Axente E, Ristoscu C and Mihailescu I N 2006 *Proceedings of Intern. Conf. „New trends in development of nano-structured thin films, hard and superhard coatings"*, Eds. R. Kakanakov and L. Kolaklieva 41
[4] Paisley M and Davis R 1993 *J. Cryst. Growth* **127** 136
[5] Kitagawa H, Shibutani Y and Ogata S 1995 *Model. Simul. Mater. Sci. Eng.* **3** 521
[6] Bakalova S, Szekeres A, Cziraki A, Lungu C P, Grigorescu S, Socol G, Axente E and Mihailescu I N 2007 *Appl. Surf. Sci.* **253** (19) 8215
[7] Bakalova S, Szekeres A, Cziraki A, Grigorescu S, Socol G, Mihailescu I N 2007 *Nanoscience &Nanotechnology* **7**, eds. E. Balabanova, I. Dragieva Heron Press Sofia 119
[8] Bakalova S, Szekeres A, Cziraki A, Huhn G Havancsak, K Grigorescu S, Socol G, Axente E, Mihailescu I N and Gavrila R 2009 *J. Optoelectron.&Adv.Mater.* **10** 1479
[9] Szekeres A, Bakalova S, Grigorescu S, Cziraki A, Socol G Ristoscu and C Mihailescu I N 2009 *Appl.Surf. Sci.* **255** 5271
[10] Bakalova S, Szekeres A, Grigorescu S, Axente E, Socol G and Mihailescu I N 2006 *Appl. Phys.* A **85** 99

[11] Narayan J, Wu H and Vispute R D 1996 *J.Electronic Mater.* **25** 143
[12] Hadis Morkoc: Properties of Nitride Semiconductors, http://www.engineering.vcu.edu/fac/morkoc/learning/mahajan_short.pdf
[13] Wemple S H and DiDomenico M 1971 *Phys. Rev.* B **3** 1338
[14] Suzuki T, Yaguchi H, Okumira H, Ishida Y and Yoshida S 2000 *Jpn J. Appl. Phys.* **39** L497
[15] Lawniczak-Jablonska K, Suski T, Gorczyca I, Christensen N E, Attenkofer K E, Perera R C C, Gullikson E M, Underwood J H, Ederer D L and Liliental Weber Z 2000 *Phys. Rev.* B **61** 16 623
[16] Yim W M 1973 *J. Appl. Phys.* **44** 292

16 ISCMP: Progress in Solid State and Molecular Electronics, Ionics and Photonics IOP Publishing
Journal of Physics: Conference Series **253** (2010) 012033 doi:10.1088/1742-6596/253/1/012033

# Single-oscillator description of spectral dependence of the refractive index of new chalcogenide Ge-Sb-(S,Te) glasses

**V Pamukchieva, A Szekeres and K Todorova**[1]
Institute of Solid State Physics, Bulgarian Academy of Sciences,
Tzarigradsko Chaussee 72, 1784 Sofia, Bulgaria

E-mail: kalina26@abv.bg

**Abstract.** New quaternary telluride glasses based on Ge-S system with addition of Sb and partial substitution of S by Te have been synthesized and their optical properties in dependence on composition have been studied by spectroscopic ellipsometry in 400-820 nm spectral range. The dispersion behavior of the refractive index of these glasses with composition of $Ge_xSb_{40-x}S_{50}Te_{10}$ and $Ge_xSb_{40-x}S_{55}Te_5$ (x= 10, 20, and 27) was examined applying the single-oscillator theory and using the Wempel-DiDomenico's model. From the data analysis the optical constants ($n$, $k$, $\varepsilon$), optical bandgap energy $E_{og}$ and film thickness, as well as the oscillator energies $E_0$ and $E_d$ have been determined. The observed increase of $E_0$ and decrease of $E_d$ with increasing the Ge content or decreasing the Te content are explained by the change of amount and strength of heteropolar chemical bonds and by the less covalent character of chemical bonding in these materials. Correlation between the oscillator energy $E_0$ and optical band gap energy $E_{og}$ is established.

## 1. Introduction

Telluride glasses from quaternary systems are potential candidates for integrated optics due to their high refractive index and photosensitivity as well as their transparency in the infrared spectral region. This explains the growing interest in these semiconducting materials for manufacturing filters, anti-reflection coatings and, in general, a wide range of optical devices [1-3]. In recent years the number of papers, dealing with such telluride glasses has increased which reflects the growing interest in these materials [4-11]. Still, there is no comprehensive study of the influence of Te substitution for S on the optical properties of the glasses belonging to the quaternary Ge-Sb-S-Te system. Knowledge of optical constants of these materials is quite essential to examine the material's potential for opto-electronic applications.

Recently, we have synthesized new quaternary telluride glasses based on Ge-S system with addition of Sb and partial substitution of S by Te and we have started studying the compositional dependences of their physicochemical and optical properties [12, 13]. In this paper we examine the dispersion behavior of the refractive index of these glasses with composition of $Ge_xSb_{40-x}S_{50}Te_{10}$ and $Ge_xSb_{40-x}S_{55}Te_5$ (x= 10, 20, and 27), the optical constants being calculated from the spectroscopic ellipsometric (SE) measurements. It should be emphasized, that all considered optical parameters are

[1] To whom any correspondence should be addressed.

characteristic for an amorphous glassy material as derived from the neutron diffraction measurements performed on these glasses [13].

## 2. Experimental

The bulk glasses with four compositions, namely $Ge_xSb_{40-x}S_{50}Te_{10}$ (x=10 and 27) and $Ge_xSb_{40-x}S_{55}Te_5$ (x=20 and 27), were synthesized from 5N purity elements by the conventional melt-quenching method. The synthesis was performed in a rotary furnace, as the glass components of a proper composition were placed in a quartz ampoule which was evacuated down to $10^{-3}$ Pa. The ampoules were loaded in the furnace and were heated up to 950°C. They were kept at this temperature for 24 h rotating the furnace for homogeneous melting and after ending the process the ampoules were pull out and were quenched in ice water. The samples were cut into thick slices and were polished on one side for optical measurements.

The ellipsometric measurements were performed in the spectral region of 400 - 820 nm (1.5 - 3.1 eV) on a manual Rudolf Research ellipsometer with PCSA configuration and at an angle of light incidence of 50°. The accuracy of the polarizer, analyzer and incidence angles was within ±0.01°. The systematic errors in the experimental ellipsometric angles were minimized through averaging the four zones measurements. From the SE data analysis, the complex refractive index (ñ=n-jk, where n is the refractive index and k is the extinction coefficient) of the glasses was evaluated.

## 3. Results and discussion

The refractive index *n* and extinction coefficient *k* of the synthesized glasses with composition of $Ge_xSb_{40}\text{-}xS_{50}Te_{10}$ (x=10 and 27) and $Ge_xSb_{40-x}S_{55}Te_5$ (x=20 and 27) are presented in figure 1(a) and 1(b), respectively. With changing the glass composition by increasing the Ge content or by decreasing the Te content a tendency is observed for these optical parameters to increase.

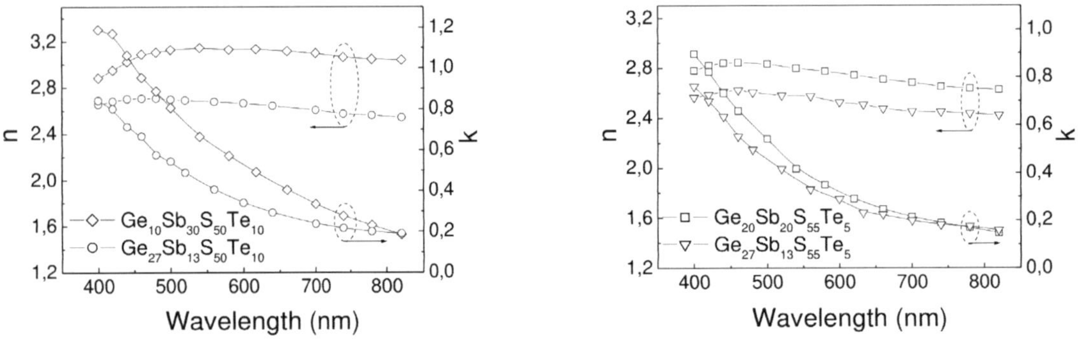


(a) (b)

**Figure 1.** Dispersion curves of the refractive index *n* and extinction coefficient *k* of the studied glasses with compositions $Ge_xSb_{40-x}S_{50}Te_{10}$ (a) and $Ge_xSb_{40-x}S_{55}Te_5$ (b).

From the obtained *n* and *k* data (figure 1), the complex dielectric function $\tilde{\varepsilon}$ ($\tilde{\varepsilon}= \varepsilon_1 -i\varepsilon_2$,, where $\varepsilon_1=n^2-k^2$ and $\varepsilon_2=2nk$) can be determined. In the high frequency range $\varepsilon_1$ can be expressed by a simple classical dispersion equation $(n_0^2-1)/(n^2-1)=1-(\lambda_0/\lambda)^2$ [14]. The index $n_0$ is the refractive index of an empty lattice at zero photon energy (at infinite wavelength) and is called static refractive index. It determines the high-frequency component of the relative permittivity $\varepsilon_\infty$, in other words the infinite wavelength dielectric constant $\varepsilon_\infty= n_0^2$ for a given material. The quantity $\lambda_0$ is the average oscillator wavelength and it is related to the energy at which direct interband electron transitions take place. By plotting $1/(n^2-1)$ versus $1/\lambda^2$ (Figure 2) these quantities were determined, as the intercept of the approximated line with the vertical axis at zero $1/\lambda^2$ gave $n_0$ and the slope of this line gave $\lambda_0$. Their values are presented in Table 1. There was a clear tendency for the $n_0$ values, respectively for the $\varepsilon_\infty$ values, to decrease with increasing the Ge content or decreasing the Te content in the studied glasses.

Further, the refractive index dependences were analyzed in the spectral region below the interband absorption edge by the single-oscillator theory using the Wempel-DiDomenico's (WDD) model [15]. The refractive index is connected with the energetic parameters of a single oscillator by the relationship $n^2\text{-}1 = (E_d E_0) / (E_0^2 - E^2)$, where $E$ is the photon energy; $E_0$ is the oscillator energy related to an average band gap and it corresponds to the distance between the "centres of gravity" of the valence and the conduction bands; and $E_d$ is the dispersion energy, a measure of the interband transitions strength. The values of $E_0$ and $E_d$ parameters were calculated by plotting $1/(n^2\text{-}1)$ versus $E^2$ (Figure 3) and by linear approximation of the low-energy part of the curves. The results are also given in Table 1. On the other hand, $E_0$ is connected with $\lambda_0$ through the relation $E_0=hc/\lambda_0$, where $h$ and $c$ are Planck's constant and velocity of light in vacuum respectively. The $E_0$ value that was calculated from $hc/\lambda_0$ and the one deduced from the $1/(n^2\text{-}1)$ versus $E^2$ plots deviated from each other less than 1 % on average.

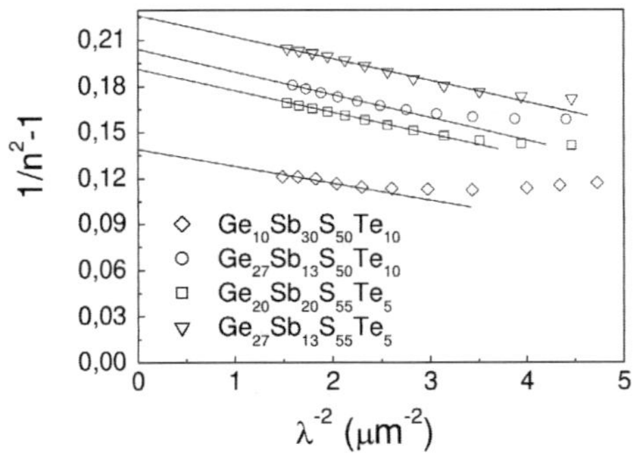


**Figure 2.** Plots of $1/(n^2\text{-}1)$ versus $\lambda^{-2}$ for different compositions of the studied glasses.

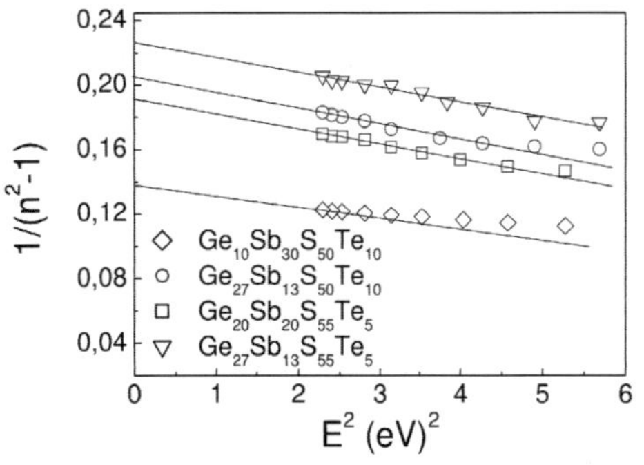


**Figure 3.** Plots of $1/(n^2\text{-}1)$ versus $E^2$ for different compositions of the studied glasses.

**Table.** 1 Parameters derived from the refractive index dispersion data for the studied glasses.

| Glass composition | $n_0$ | $\varepsilon_\infty$ | $\lambda_0$ (nm) | $E_0(\lambda_0)$ (eV) | $E_0$ (eV) | $E_d$ (eV) | $E_{og}$ eV |
|---|---|---|---|---|---|---|---|
| $Ge_{10}Sb_{30}S_{50}Te_{10}$ | 2.86 | 8.2 | 279.04 | 4.44 | 4.44 | 31.6 | 1.30 |
| $Ge_{27}Sb_{13}S_{50}Te_{10}$ | 2.43 | 5.9 | 270.55 | 4.58 | 4.65 | 22.8 | 1.42 |
| $Ge_{20}Sb_{20}S_{55}Te_{5}$ | 2.50 | 6.3 | 271.96 | 4.56 | 4.56 | 23.9 | 1.47 |
| $Ge_{27}Sb_{13}S_{55}Te_{5}$ | 2.33 | 5.4 | 247.67 | 5.0 | 4.97 | 22.0 | 1.49 |

The observed compositional dependences of the refractive index can be explained by the different polarizability of the glass constituent elements, as the higher the polarizibility the larger the index $n_0$ value [16]. It is known that when the covalent atomic radius is larger the electron clouds of atoms are easier to be polarized. Since Ge and S possess smaller covalent radius than Sb and Te atoms [12], the polarizability of Ge and S is lower compared to Sb and Te atoms. This explains why the $n_0$ and $\varepsilon_\infty$ values decrease by increasing the Ge content or decreasing the Te content.

The dependences of the extinction coefficient $k$ (Figure 1) were also considered. The absorption coefficient was determined from the $k$ data ($\alpha=4\pi k/\lambda$) and described by the relation $(\alpha E)=B(E\text{-}E_{og})^m$ [17,18], where the parameter $B$ depends on the electron transition probability. The power $m$ is an integer number characterizing the transition process and, for chalcogenide glasses with indirect type electron transitions, it is equal to 2. By plotting the $(\alpha E)^{1/2}$ versus $E$, the energy band gap $E_{og}$ was determined, as the interception of the extrapolated linear part of the curve with the photon energy axis provides $E_{og}$ with an accuracy of $\pm$ 0.05 eV. The corresponding dependences are presented in Figure 4 and 5, while the optical bandgap values are given in Table 1. As it is seen, the same tendency as the one for $E_0$ is observed, namely that the $E_{og}$ value increases with increasing the Ge content or decreasing the Te content.

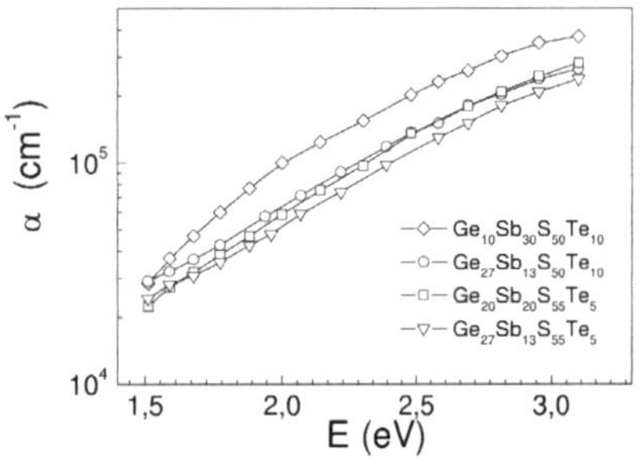


**Figure 4.** Absorption coefficient $\alpha$ versus photon energy $E$ for the studied glasses.

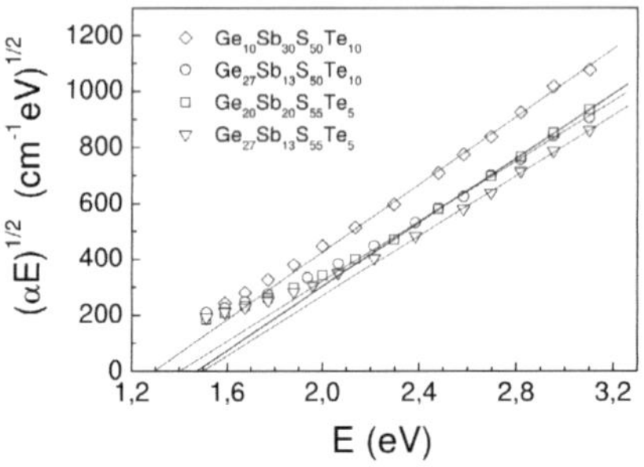


**Figure 5.** Plots $(\alpha E)^{1/2}$ versus photon energy $E$ for the studied glasses.

The behavior of the optical band gap $E_{og}$ with varying the glass composition (Table 1) has been explained by the change of the amount and strength of heteropolar chemical bonds in these materials [13]. For better understanding, we calculated the percent of each possible chemical bond in the glassy composition taking into account the chemical bond ordering, i.e. that the strongest bonds are created first followed by those with decreasing bond strength. The results are presented in Table 2, where from the left column to the right the strongest Ge-S bonds are given followed by the weaker bonds up to the weakest Ge-Ge bonds. According to the chemical bond ordering, the formation of chemical bonds proceeds in strict sequence toward decreasing bond energy until all the valences of the atoms are saturated [19]. The bond energy and covalent character of the possible heteropolar chemical bonds were taken from [12], where the calculation procedure is discussed in detail [12]. As is seen from Table 2, with increasing the Ge or decreasing Te content the number of stronger chemical bonds (such as Ge-S and Te-S) increases making the chemical bonding less covalent.

Table 2 Amount of possible chemical bonds for the corresponding glassy compositions, given in percentage.

| Glassy composition | Amount of possible chemical bonds (%) | | | | | |
|---|---|---|---|---|---|---|
| | Ge-S | Te-S | Sb-S | Sb-Te | Sb-Sb | Ge-Ge |
| $Ge_{10}Sb_{30}S_{50}Te_{10}$ | 20.0 | 20.0 | 26.8 | - | 33.2 | - |
| $Ge_{27}Sb_{13}S_{50}Te_{10}$ | 57.7 | - | - | 23.1 | 14.5 | 4.6 |
| $Ge_{20}Sb_{20}S_{55}Te_{5}$ | 44.4 | 11.1 | 14.9 | - | 29.6 | - |
| $Ge_{27}Sb_{13}S_{55}Te_{5}$ | 65.8 | 2.4 | | 6.5 | 25.2 | - |

In the case of constant S and Te content, the increase of the optical band gap $E_{og}$ can be attributed to an increase of the amount of stronger chemical bonds at the expense of the amount of weaker ones. This leads to narrowing of the conduction band tail, while the valence band-tail remains the same, as it is defined mainly by the amount of chalcogenide (S and Te) atoms [1]. When the Te content increases from 5 to 10 at.%, the increase of the number of the much weaker Ge-Te and Sb-Te bonds (Table 2) results in a reduction of the average stabilization energy of chemical bonds [12] and, therefore, in a decrease of the bandgap energy (Table 1).

Since the oscillator energy $E_0$ represents the average band gap energy, its value follows the same compositional dependence as the one for the optical band gap $E_{og}$. Thus, in respect to $E_{og}$, the oscillator energy increases by increasing the number of stronger Ge–S bonds at the expense of the amount of weaker Sb–S bonds, as well as by increasing the number of stronger Ge–S and Sb–S bonds at the expense of weaker Ge–Te and Sb–Te bonds (Tables 1 and 2).

Quite the contrary is observed for the dispersion energy $E_d$, as its value decreases with increasing the Ge content or decreasing the Te content (Table 1). This behavior can be also explained by the change of the number and type of chemical bonds and their bonding energy. In these glasses the formed bonds are predominantly covalent since the constituent elements have close values of electronegativity [20]. $E_d$ is connected with bond ionicity, as it can be also expressed by an empirical formula $E_d = \beta N_c Z_a N_e$ (eV) [15], where $\beta$ is a parameter which reflects the character of chemical bonding, $N_c$ is the coordination of the cation nearest neighbour to the anion, $Z_a$ is the formal chemical valency of the anion, and $N_e$ is the total number of valence electrons (cores excluded) per anion. For ionic chemical bonds the $\beta$ parameter is equal to $\beta_i = 0.26 \pm 0.03$ eV, while for covalent bonds $\beta_c = 0.37 \pm 0.04$ eV [15]. We can conclude that an increase of the number of stronger chemical bonds leads to weakening of the covalent bond character and, as a consequence, to a decrease of $\beta$ value. As the dispersion energy is proportional to the $\beta$ parameter ($E_d = \beta N_c Z_a N_e$), the $E_d$ value decreases accordingly. On the other hand, by increasing the Ge or Te content, the compactness of the glassy structure decreases [12], which causes a decrease of the effective cation coordination number $N_c$ which also contributes to the decrease of $E_d$ value.

## 4. Conclusion

The dispersion behavior of the refractive index of new amorphous quaternary telluride glassy materials with composition of $Ge_xSb_{40-x}S_{50}Te_{10}$ and $Ge_xSb_{40-x}S_{55}Te_5$ (x= 10, 20, and 27) has been studied by spectroscopic ellipsometry in the spectral range of 400-820 nm. By applying the single-oscillator theory, from the refractive index dispersion data analysis the values of the oscillator energy $E_0$ and oscillator strength $E_d$ are obtained. The observed changes in these parameters are explained in terms of chemical bonds formation and change of the bonding energy and bond covalency.

## References

[1] Tauc J 1974 *Amorphoua and Liquid Semiconductors* (Plenum Press, London/NY) Chap. 4
[2] Savage J A 1985 *Infrared Optical Materials and their antireflection coatings* (Adam Hilger, Bristol) ISBN 0-85274-790-X
[3] Kosek F, Cimpl Z, Tulka J and Chlebny J 1987 *J. Non-Cryst. Solids* **90** 401
[4] Pamukchieva V and Szekeres A 2008 *Opt. Mater.* **30** 1088
[5] Shiryaev V S, Adam J-L, Zhang X H, Boussard-Plerdel C, Lucas J and Churbanov M F 2004 *J. Non-Cryst. Solids* **336** 113
[6] Michel K, Bureau B, Pouvreau C, Sangleboeuf J C, Boussard-Plédel C, Jouan T, Rouxel T, Adam J-L, Staubmann K, Steinner H, Baumann T, Katzir A, Bayona J and Konz W 2003 *J. Non-Cryst. Solids* **326&327** 434
[7] Keirsse J, Boussard-Plédel C, Loreal O, Sire O, Bureau B, Turlin B, Leroyer P and Lucas J 2003 *J. Non-Cryst. Solids* **326&327** 430
[8] Le Coq D, Boussard-Plédel C, Fonteneau G, Pain T, Bureau B and Adam J L 2003 *Mater. Res. Bull.* **38** 174
[9] Dongol M, Hafiz M M, Abou-Zied M and Elhady A F 2001 *Appl. Surf. Sci.* **185** 1
[10] Pamukchieva V, Szekeres A and Todorova K 2004 *Proceeding of SPIE* **5581** 608
[11] Pamukchieva V and Szekeres A 2005 *J. Optoelectron.&Adv. Mater.* **7** 1277
[12] Pamukchieva V, Szekeres A, Todorova K, Fabian M, Svab E, Revay Zs and Szentmiklosi L 2009 *J. Non-Cryst. Solids* **355** 2485
[13] Pamukchieva V, Szekeres A, Todorova K, Svab E and Fabian M 2009 *Opt. Mater.* **32** 45
[14] Zemel J N, Jensen J D and Schoolar R B 1965 *Phys. Rev.* **140** A330

[15] Wemple S H and DiDomenico Jr M 1971 *Phys. Rev.* B **3** 1338
[16] Dimitrove V and Sakka S 1996 *J. Appl. Phys.* **79** 1736
[17] Tauc J, Grigorovici R and Vancu A 1966 *Phys. Status Solidi* **15** 627
[18] Mott N F and Davis E A 1970 *Philos. Mag.* **22** 903
[19] Jozef B, Stanford O, Mahadevan S, Gridhar A and Singh A K 1985 *J. Non-Cryst. Solids* **74** 75
[20] http://www.knovel.com/web/portal/periodic_table

# Electrical characterization of MOS structures with self-organized three-layer gate dielectric containing Si nanocrystals

**N Nedev[1], D Nesheva[2], M Curiel[1], E Manolov[2], I Petrov[3], B Valdez[1] and I Bineva[2]**
[1] Institute of Engineering, Autonomous University of Baja California, Benito Juarez Blvd., s/n, C.P. 21280, Mexicali, Baja California, Mexico
[2] Institute of Solid State Physics, Bulgarian Academy of Sciences, 72 Tzarigradsko Chaussee Blvd., 1784 Sofia, Bulgaria
[3] Center for Microanalysis of Materials, Frederick Seitz Materials Research Laboratory, University of Illinois, Urbana, Illinois 61801

E-mail: nicola@iing.mxl.uabc.mx

**Abstract.** $SiO_x$ (x = 1.3) films with thicknesses of 50 and 100 nm deposited on c-Si by thermal evaporation of SiO in vacuum were subjected to $N_2$ or two steps $N_2$, $90\%N_2+10\%O_2$ annealing at 1000 °C for 60 min. The time of the second ($N_2+10\%O_2$) annealing step was varied in order to vary the depth to which the $SiO_x$ films were intentionally oxidized. The I-V and XTEM results obtained confirm that the suggested two steps annealing leads to a formation of three-layer $SiO_2$/$SiO_2$-Si nanocrystals/$SiO_2$ gate dielectric, which contains Si nanocrystals. The nanocrystals are with a diameter of a ~ 4-5 nm and are embedded in a stoichiometric $SiO_2$ matrix. The I-V measurement show that the current through the gate dielectric is limited by a $SiO_2$ region, close to the top surface, formed during the second annealing step. MOS structures subjected to a two step annealing show larger retention times when charged with electrons/holes in comparison with the control or annealed only in $N_2$ structures.

## 1. Introduction

Dielectric layers containing silicon nanocrystals (Si NCs) have attracted much attention in the recent years because they are promising for application in opto- and microelectronic devices. One possible application is in Si-based light sources [1,2]. The fabrication process of such a source would be compatible with the existing manufacturing infrastructure for silicon integrated circuits. Dielectric layers containing nanocrystals are also promising for usage in "third generation" solar cells [3]. Another possible application of Si nanocrystals is as charge storing nodes in non-volatile memory devices [4]. All these applications require a good knowledge of the structural and electrical properties of the films containing nanocrystals. Normally in non-volatile memory structures based on silicon nanocrystals the gate dielectric is composed of at least three different films: high quality tunnel silicon oxide ($SiO_2$), another silicon oxide film, which contains nanocrystals ($SiO_2$-Si NCs) and finally a third (control) $SiO_2$ layer, which insulates the nanocrystals from the control gate. In most cases each of these layers is obtained in a different deposition system. Another approach is to vary the deposition conditions and/or gas ratio in order to vary the film composition. For example, $SiO_2$/$SiO_2$-Si NCs superlattice structures have been fabricated by thermal evaporation of SiO powder either under high vacuum or in low pressure of oxygen [5-7].

Journal of Physics: Conference Series **253** (2010) 012034 doi:10.1088/1742-6596/253/1/012034

Our previous results [8-11] obtained by Transmission Electron Microscopy (TEM), infra red (IR) spectroscopy and X-ray Photoelectron Spectroscopy (XPS) have shown that furnace annealing of $SiO_x$ ($x = 1.15$ or 1.3, $d_{SiOx} \sim 15$ nm) films at 1000 °C in an inert gas ambient for 60 min causes phase separation and growth of Si nanocrystals in an amorphous matrix by a self-assembling process. The annealed samples exhibit nanocrystals with a diameter of ~ 4-5 nm ($x = 1.3$) or ~ 5-6 nm ($x = 1.15$) positioned almost in a single plane closer to the c-Si wafer than to the top surface and separated from the c-Si wafer/dielectric interface by an amorphous region with thickness ≥ 3 nm [10,11]. It has been assumed that the Si nanocrystals grown in the region close to the top surface (~ 5-6 nm) get transformed into $SiO_2$ due to native oxide formation after exposure to air. XPS has shown [10] that the spectra of the Si 2p core level of $SiO_x$ films with $x = 1.3$ changes significantly after high temperature annealing. In the annealed films the amount of excess Si atoms bonded to O in the suboxide network is strongly reduced and only clearly separated peaks corresponding to stoichiometric $SiO_2$ and to pure Si phase were observed. Combining the results obtained by TEM, IR and XPS it was concluded that the high temperature annealing leads to formation of Si nanocrystals in a stoichiometric $SiO_2$ matrix.

The obtained results motivated us to study the influence of an intentional oxidation step during the annealing process in order to fabricate a self-organized three-layer gate dielectric ($SiO_2$/$SiO_2$-Si NCs/$SiO_2$) structure in a controllable way and thus to avoid the deposition of the third $SiO_2$ control layer.

## 2. Experimental details

Metal-Oxide-Semiconductor (MOS) structures were fabricated by thermal evaporation of SiO at a vacuum of $1\times10^{-3}$ Pa on top of n-type crystalline silicon ($\rho = 4$-$6$ Ω cm). The deposited $SiO_x$ films were with composition of $x = 1.3$ and thicknesses of 50 and 100 nm. For each thickness four sets of samples were fabricated. The samples of the first set were annealed at 250 °C for 30 min in Ar only for stabilization (control samples), while those of the second set were subjected to an additional annealing at 1000 °C in pure $N_2$ atmosphere for 60 min. The 1000 °C annealing of sets 3 and 4 was carried out in two steps keeping the total annealing time equal to 60 min. The first step was an annealing in $N_2$ and the second step was annealing in 90% $N_2$ + 10% $O_2$ atmosphere. The $N_2$, 90%$N_2$+10%$O_2$ annealing times of the thinner films were 55, 5 min (set 3) and 50, 10 min (set 4), respectively, while those of the thicker films were 50, 10 min (set 3) and 45, 15 min (set 4). Thus by varying the time of the second ($N_2$+10%$O_2$) annealing step the depth to which the $SiO_x$ films were intentionally oxidized was varied. Aluminium metallization was carried out through a mask and MOS capacitors with area of ~ $2 \times 10^{-3}$ $cm^2$ were formed. The structures were characterized electrically by current-voltage (I-V) and capacitance–voltage (C-V) measurements using Agilent B1500A Semiconductor Device Analyzer controlling Agilent E4980A Precision LCR Meter. The polarity of the applied voltage used below concerns the top Al electrode.

## 3. Results and discussion

Figures 1 (a) and (b) show cross-sectional micrographs (XTEM) obtained by lattice-resolution transmission electron microscopy (TEM) of a control and annealed (1000 °C in $N_2$ for 60 min) $SiO_x$ films with $x = 1.3$ and thicknesses of ~ 15 nm. The control layer (figure 1 (a)) exhibits an amorphous structure and a nearly atomically flat c-Si wafer/dielectric interface, while the annealed one (figure 1 (b)) exhibits nanocrystals with a diameter of a ~ 4-5 nm positioned almost in a single plane and a rougher interface, 3-4 monolayers [10,11].

Figures 2 (a) and (b) show I-V characteristics of sets 2, 3 and 4 of samples with 50 nm and 100 nm thick $SiO_x$ film, respectively. For both thicknesses the current flow through the dielectric starts at lower voltages for $N_2$ annealed samples (sets 2). For the samples subjected to a 90%$N_2$+10%$O_2$ annealing step (sets 3 and 4) the curves are shifted to higher voltages, respectively higher electric fields, which indicates formation of $SiO_2$ in the region close to the top surface. The I-V shift is better pronounced for the samples with longer oxidation times (see the inset in figure 2 (a), which shows similar I-V characteristics in a linear scale). The difference between the currents through the samples from sets 3 and 4 at a given positive or negative voltage is more than one order of magnitude for both thicknesses, which can be explained assuming larger depth to which the NCs are oxidized at larger oxidation times.

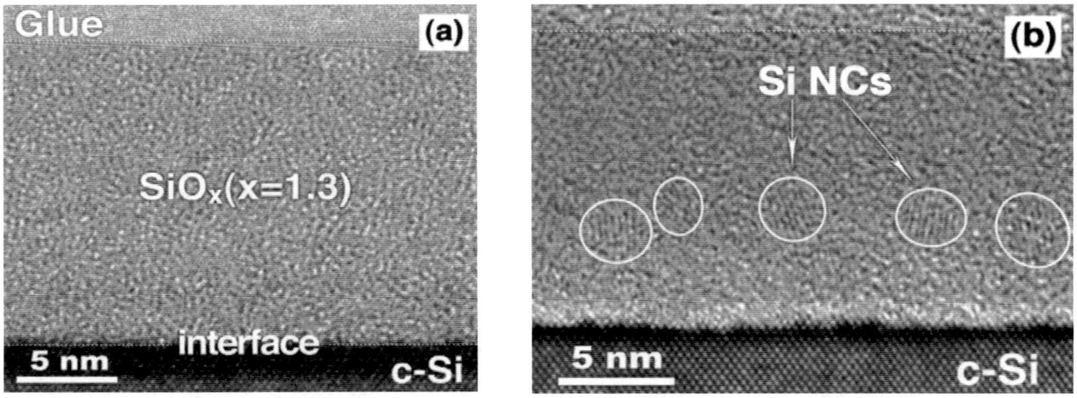


**Figure 1**. XTEM micrographs of a control $SiO_x$ film showing an amorphous structure and abrupt interface (a) and an annealed film with nanocrystals in a dielectric matrix and rougher interface, ~ 0.8 nm or 3-4 monolayers (b).

The difference between the I-V curves of sets 2 and 3 is not so well pronounced, which is consistent with our previous observation [10] that the Si nanocrystals grown close to the top surface get transformed in $SiO_2$ to a thickness of ~ 5-6 nm, due to native oxidation after exposure to air. Curves 3 and 2 in figure 2 (b) intersect each other at a positive voltage of ~ 20 V (current of ~ $4\times10^{-9}$ A) because of the smaller curve slope of the $N_2$ annealed sample.

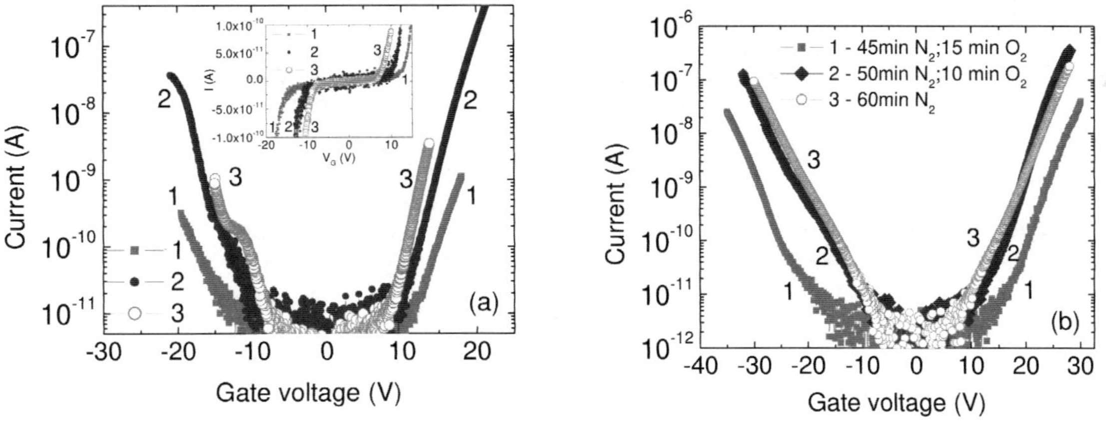


**Figure 2**. I-V characteristics of MOS structures with: (a) a 50 nm $SiO_x$ film annealed in $N_2$ (curves 3), $N_2$+5 min $O_2$ (curves 2) and $N_2$+10 min $O_2$ (curves 1) and (b) a 100 nm film annealed in $N_2$ (curves 3), $N_2$+10 min$O_2$ (curves 2) and $N_2$+15 min $O_2$ (curves 1). The inset in figure 1 (a) shows the onset of current injection in a linear scale.

For the samples from sets 3 and 4 (both thicknesses) the current measured at a given positive voltage is much larger than the corresponding current measured at the same negative voltage. For example for set 3 sample with 100 nm thick $SiO_x$ the current at ±25 V is $I_{(25V)} = 7.1\times10^{-8}$ A and $I_{(-25V)} =$

$-3.5\times10^{-9}$ A, while for sample from set 4 the values are $I_{(25V)} = 2.8\times10^{-9}$ A and $I_{(-25V)} = -8.9\times10^{-11}$ A. It can be assumed that the main component of the current flowing through the gate dielectric at negative gate voltage is due to electrons, because they have to overcome a much lower barrier at the Al/$SiO_2$ interface (~ 3 eV) than the holes at the c-Si wafer/dielectric interface (~ 5 eV). Then, the observed large difference between the currents can be explained keeping in mind the increased roughness of the annealed c-Si/$SiO_x$ interface and assuming that the injection of electrons at positive voltage in the top $SiO_2$ (formed after 90%$N_2$+10%$O_2$ annealing) is from nanocrystals. The increased roughness and NC injection lead to a local enhancement of the electric field [4,12] and thus to a larger current at a given positive applied voltage.

In figure 3 I-V characteristics, measured in positive direction, of control MOS structures and structures annealed in $N_2$ for 60 min (set 2) are shown for both thicknesses. For the 50 nm structures the current through the control sample starts at a lower voltage compared to the $N_2$ annealed one, while for the 100 nm samples the conduction starts at approximately the same voltage. For both thicknesses the control samples have shown worse insulating properties compared to the annealed ones, lower voltage and lower current density breakdowns.

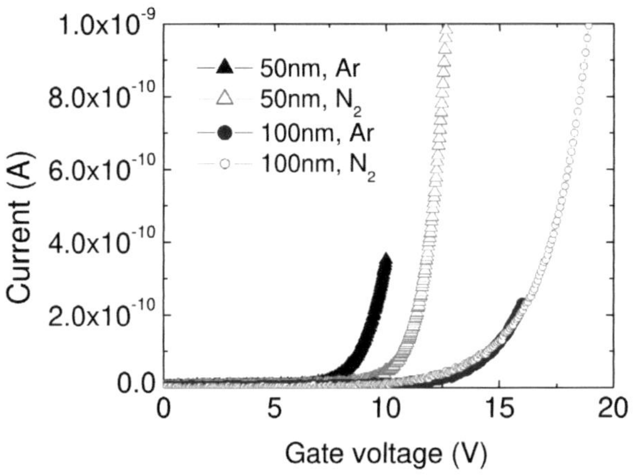


**Figure 3.** I-V characteristics of control MOS structures (Ar) and structures annealed in $N_2$ for 60 min.

In figure 4 the I-V characteristics of samples with the same duration of the 90%$N_2$+10%$O_2$ annealing step and having different thicknesses of the $SiO_x$ layer are shown. As it is seen in spite of the large difference between both dielectric thicknesses, 100 nm and 50 nm, the structures start to conduct approximately at one and the same voltage for both polarities. This observation implies formation of a stoichiometric $SiO_2$ region, close to the top surface, which limits the current through the gate dielectric in a similar way.

Thus the I-V and XTEM results obtained confirm that the suggested two step $N_2$, 90%$N_2$+10%$O_2$ high temperature annealing leads to formation of three-layer $SiO_2$/$SiO_2$-SiNCs/$SiO_2$ gate dielectric containing Si nanocrystals.

Figures 5 (a) and (b) show high frequency C-V curves of a control and annealed (50 min $N_2$, 10 min 90%$N_2$+10%$O_2$) samples having 50 nm thick $SiO_x$ layer. Curves 1 and 2 in figure 5 (a) were measured when sweeping the gate voltage from 0 to -4 V and then in the reverse direction. A hysteresis of ~ 0.6 V is seen, which is due to charging and discharging of traps in the $SiO_x$ layer. After measuring the initial curve, pulses with amplitudes of +10 V, +15 V, -10 V, -15 V, -20 V and duration of 1 s were applied and after each pulse the position of the C-V curve was monitored. The application of the first pulse (+10 V) makes the slope of the C-V curve steeper and shifts it slightly in the positive

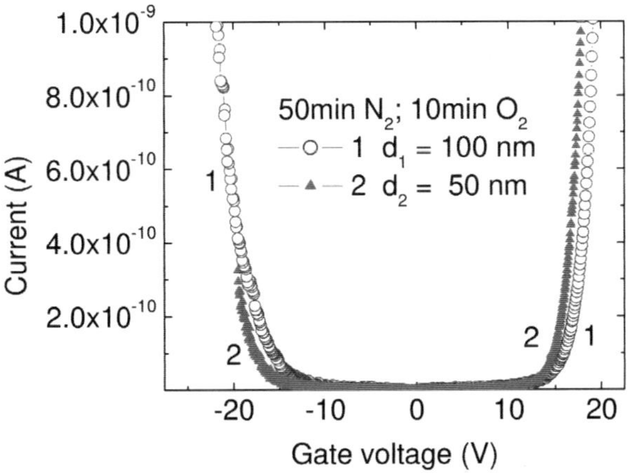


**Figure 4.** I-V characteristics of MOS structures with 10 min 90%$N_2$ + 10%$O_2$ annealing time and thicknesses of the $SiO_x$ film 100 nm (curve 1) and 50 nm (curve 2).

direction. The observed change in the curve shape implies charging of c-Si/$SiO_x$ interface states and some of the $SiO_x$ bulk traps. The application of the next pulses leads to similar changes; the only difference is the direction in which the C-V curve shifts (positive pulses shift it to higher positive voltages, while negative pulses shift it in the opposite direction). However, even after the highest pulse applied (-20 V) the C-V curve remains to the right of its initial position (curves 1, 2). Another important feature of the control MOS structure is that it loses the bulk trapped charge quite fast (after several minutes).

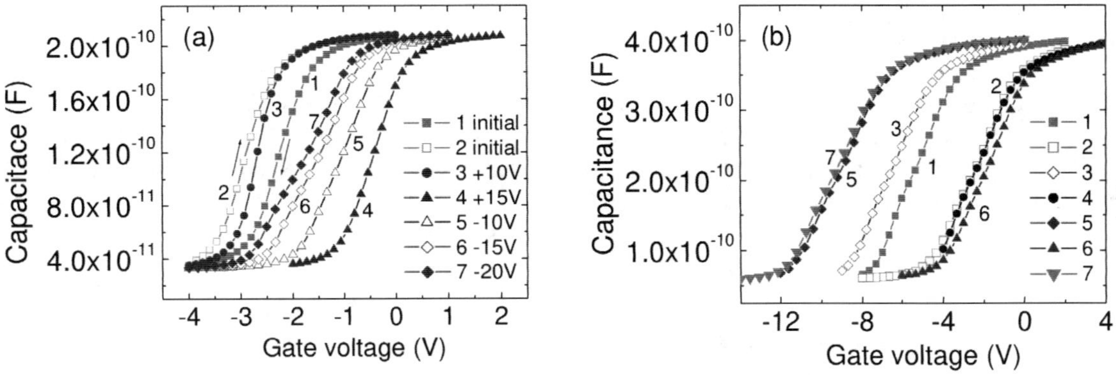


**Figure 5.** High frequency (1 MHz) C-V curves of: (a) a control and (b) annealed (50 min $N_2$, 10 min 90%$N_2$+10%$O_2$) samples with 50 nm thickness of the $SiO_x$ layer.

Figure 5 (b) shows the initial C-V dependence (curve 1) and the curves obtained after charging the structure with +10 V, 1 s (curve 2) or -10V, 1 s (curve 3) pulses. Parallel shifts, corresponding to charging of the nanocrystals with electrons/holes are seen after the application of positive/negative pulse. Before applying each of these pulses the structure was brought to its initial state by applying appropriate bias. The larger shift caused by positive pulse is in agreement with the I-V results shown in figure 2. After increasing the charging pulse amplitudes to ±12 and ±15 V curves 4, 5 and 6, 7 were measured. Besides the parallel shifts on both sides of the initial curve the oxygen annealed MOS structure shows a larger retention time in comparison with the control one; one hour after charging the structure (positively or negatively) the C-V curve remains in the same position.

## 4. Conclusions

The I-V and XTEM results obtained confirm that the suggested two step $N_2$, 90%$N_2$+10%$O_2$ annealing of $SiO_x$ films (x = 1.3) at 1000 °C for 60 min leads to a formation of three-layer $SiO_2/SiO_2$-SiNCs/$SiO_2$ gate dielectric, which contains Si nanocrystals. The nanocrystals are with a diameter of a ~ 4-5 nm and are embedded in a stoichiometric $SiO_2$ matrix. The I-V measurement show that the current through the gate dielectric is limited by the $SiO_2$ region, close to the top surface, formed during the second (90%$N_2$+10%$O_2$) annealing step. The MOS structures subjected to a two step annealing can be charged with electrons/holes by applying positive/negative voltage pulses at the control gate and show larger retention times in comparison with the control structures or structures annealed only in $N_2$.

## Acknowledgements

The authors gratefully acknowledge the financial support of the Bulgarian Ministry of Education, Youth and Science under grant NT-4-1, Autonomous University of Baja California and Consejo Nacional de Ciencia y Tecnologia CONACyT, Mexico. The measurements were carried out in part in the Frederick Seitz Materials Research Laboratory Central Facilities, University of Illinois, which are partially supported by the U.S. Department of Energy under grants DE-FG02-07ER46453 and DE-FG02-07ER46471.

## References

[1] Pavesi L, Dal Negro L, Mazzoleni C, Franzò G and Priolo F 2000 *Nature* **408** 440
[2] Ossicini S, Pavesi L and Priolo F 2003 *Light Emitting Silicon for Microphotonics, Springer tracts in modern physics*, Vol. 194 (Berlin: Springer)
[3] Conibeer G et al 2006 *Thin Solid Films* **511-512** 654
[4] Tiwari S, Rana F, Hanafi H, Hartstein A, Crabbé E and Chan K 1996 *Appl. Phys. Lett.* **68** 1377
[5] Zacharias M, Heitmann J, Scholz R, Kahler U, M. Schmidt M and Blasing J, 2002 *Appl. Phys. Lett.* **80** 661
[6] Jivanescu M, Stesmans A and Zacharias M 2008 *J. Appl. Phys.* **104** 103518
[7] Hiller D, Jivanescu M, Stesmans A and Zacharias M 2010 *J. Appl. Phys.* **107** 084309
[8] Nedev N, Nesheva D, Manolov E, Brüggemann R, Meier S, Kirilov K and Levi Z 2007 *J. Optoelectron. Adv. Mater.* **9** 182
[9] Nesheva D, Nedev N, Levi Z, Brüggemann R, Manolov E, Kirilov K and Meier S 2008 *Semicond. Sci. Technol.* **23** 045015
[10] Curiel M, Nedev N, Nesheva D, Soares J, Haasch R, Sardela M, Valdez B, Manolov E, Bineva I and Petrov I 2010 *Materials Science & Engineering B* **174** 132
[11] Curiel M, Petrov I, Nedev N, Nesheva D, Sardela M, Murata Y, Valdez B, Manolov E and Bineva I 2010 *Materials Science Forum* **644** 101
[12] DiMaria D J, DeMeyer K M and Dong D W 1980 *IEEE Electron Device Lett.* **EDL-1** 179

# Composition and structure of $Zn_xCd_{1-x}Se$ single layers prepared by thermal evaporation of ZnSe and CdSe

**D Nesheva[1], Z Aneva[1], M J Šćepanović[2], I Bineva[1], Z Levi[1], Z V Popović[2] and B Pejova[3]**
[1]Institute of Solid State Physics, Bulgarian Academy of Sciences, 72 Tzarigradsko Chaussee Blvd., 1784 Sofia, Bulgaria
[2]Center for Solid State Physics and New Materials, Institute of Physics, Pregrevica 118, Belgrade 11080, Serbia
[3]Institute of Chemistry, Faculty of Natural Sciences and Mathematics, Sts. Cyril and Methodius University, POB 162, 1000 Skopje, Macedonia

E-mail: nesheva@issp.bas.bg

**Abstract**. Single layers of $Zn_xCd_{1-x}Se$ with five different compositions and thickness of 400 nm have been prepared by thermal vacuum evaporation, through alloying of ultra thin ZnSe and CdSe films with equivalent thickness of 0.12, 0.25 or 0.37 nm. The deposition was carried out on rotating substrates kept at room temperature. The layer composition was varied by alloying ZnSe and CdSe films with different equivalent thicknesses. The film composition $x$ = 0.39, 0.52, 0.59, 0.69 and 0.8 has been determined by Energy-Dispersive Spectroscopy and confirmed with Raman scattering data. The microstructure of $Zn_xCd_{1-x}Se$ has been investigated by Atomic Force Microscopy and Raman scattering measurement. The Atomic Force Microscopy results have revealed that the layers are nanocrystalline and the grain size is ≤ 20 nm. The Raman scattering data have shown four replicas of the longitudinal optical phonons, thus confirming the conclusion for the layer crystallinity. The obtained results have shown that the applied deposition technique makes possible preparation of ternary nanocrystalline $Zn_xCd_{1-x}Se$ layers with desired compositions..

## 1. Introduction

Ternary alloys of ZnCdSe are an important class of semiconductors which are applicable in fabrication of optoelectronic devices operating in the entire visible spectrum, as well as in fabrication of solar cells, thin film transistors, etc. [1]. The routes for preparation of ternary II-VI semiconductor layers are based mainly on application of molecular beam epitaxy [2,3] and electrodeposition [4,5] techniques.
Ternary alloys of II-VI semiconductors (ZnCdSe, ZnSSe) included in zinc-selenide based quantum structures have demonstrated considerable promise as short-wavelength light sources, fast switching devices, etc. [6-8]. The interest in inclusion of such alloys in low-dimensional structures is high [3,9-12] since they offer possibilities to change their lattice constant and optical properties by changing the composition and this gives them advantages over binary compounds (ZnSe, CdS). For preparation of multilayers (MLs) with good stability and periodicity or production of nanocrystals in the quantum structures with narrow size distribution, molecular beam epitaxy (MBE) or metal-organic chemical vapour deposition (MOCVD) are normally applied. However, both techniques are quite expensive and, in addition, MOCVD normally uses dangerous gases. Besides, high quality crystalline ZnSe or

GaAs substrates are required for the epitaxial growth. This has motivated a considerable research activity in the field of preparation of low-dimensional structures by nonepitaxial film growth and application of less expensive vapour deposition techniques [13 and References therein] and electrodeposition [4].

In several previous papers [13-15] we have reported on preparation o f amorphous/nanocrystalline multilayers with good artificial periodicity and smooth interfaces nanostructured thin films by thermal evaporation in vacuum. In this study, a technique for preparation of microcrystalline $Zn_xCd_{1-x}Se$ single layers with controlled composition is developed, based on thermal vacuum evaporation. It can be successfully applied for production of $ZnSe/Zn_xCd_{1-x}Se$ MLs. Energy-Dispersive Spectroscopy (EDS), Atomic Force Microscopy (AFM) and Raman scattering measurements are carried out to determine the composition and to explore the film microstructure.

## 2.Experimental

### *2.1 Sample preparation*

Thermal evaporation of ZnSe and CdSe powders (Merk, Suprapure) from two independent, simultaneously heated tantalum crucibles was applied for the preparation of $Zn_xCd_{1-x}Se$ layers with five different compositions and thickness of 400 nm. The crucibles were disposed in the bottom of two cylindrical screens (not intentionally heated) whose top is close to the substrates. Thus, the deposition of each material was carried out in a quasi-closed volume.

The layers of $Zn_xCd_{1-x}Se$ were deposited on crystalline c-Si substrates (Wacker, p-type, 11 Ω/cm) kept at room temperature. The native oxide on the c-Si substrates was not removed. During the film deposition the substrates were rotated at a rate $V_r = 20$ turns/min and during each turn pass they spent over each crucible 1/12 of the turn time. The time $\Delta t$ between the consecutive deposition of CdSe and ZnSe is 5/12 of the turn time. The substrate rotation was started after fixing the desired deposition rates for both CdSe and ZnSe. The rates were controlled by two independent preliminary calibrated quartz microbalance systems MIKI FFV. The layer composition was varied by changing the ratio of the deposition rates of both materials; five ratios $V_d^{CdSe}$(nm/s)/$V_d^{ZnSe}$(nm/s) were used (0.5/1.5, 1.0/1.5, 1.5/1.5, 1.5/1.0, 1.5/0.5). Thus, five kinds of $Zn_xCd_{1-x}Se$ single layers were prepared by alloying of the consecutively deposited small portions of CdSe and ZnSe with equivalent layer thicknesses $d_e = 0.12$ nm (at 0.5 nm/s), 0.25 nm (at 1.0 nm/s), 0.37 nm (at 15nm/s).

It has been shown [16] that CdSe single layers prepared at the same deposition conditions (i.e. application of cylindrical screens, same distance between the crucible and the substrates, similar deposition rates and rates of substrate rotation) are stoichiometric. Based on this and keeping in mind the ratio of the deposition rates measured, the following compositions were expected: $x_e \approx 0.25$ ($V_d^{CdSe}/V_d^{ZnSe}$ = 15 nm s$^{-1}$/5 nm s$^{-1}$), $x_e \approx 0.4$ (15/10), $x_e \approx 0.5$ (15/15), $x_e \approx 0.6$ (10/15), $x_e \approx 0.75$ (5/15).

### *2.2 Sample characterization*

The layer thickness was determined by optical transmission measurements [17] which showed that the deviation of the film thickness from the expected one did not overcome 5%. The sample composition was determined by EDS measurements using a JEOL Scanning Electron Microscope (JSM-6390) operating at an accelerating voltage of 25.00 kV. The samples were uncoated; the element used for optimization was Iron. At least two different points on the sample surface were explored.

The surface morphology of the layers was investigated by AFM measurements that were carried out using a Multimode V (Veeco, Santa Barbara, CA). Imaging was performed in tapping mode and height, amplitude, and phase images were recorded. The scan rate was in the range 1-2 Hz, the images resolution was 512 lines per scan direction. At least three different points on the sample surface were explored. Silicon cantilevers with a super sharp spike tip (TESP-SS, Veecoprobes),d a nominal resonance frequency of 320 kHz and a typical force constant of 42 N/m were used. The tip radius for these probes is in the range 2-5 nm. Image processing was performed by means of the Nanoscope 7.30 programme. All the images were flattened and no further processing was carried out.

Raman scattering measurements of $Zn_xCd_{1-x}Se$ single layers were performed in the backscattering geometry, by Jobin-Yvon T64000 triple spectrometer system, equipped with a confocal microscope and a nitrogen-cooled CCD detector. All spectra were collected by using 514.5 nm (2.41 eV) line of mixed $Ar^+/Kr^+$ ion laser at room temperature in the air. A low output laser power (≤5 mW) was used to avoid local heating and subsequent $Zn_xCd_{1-x}Se$ layer decomposition.

## 3. Results and Discussion

Table 1 presents the average Zn, Cd and Se atomic percentages determined from the EDS data with a maximum error of 5%. It is seen from the table that some films have a slight Se-enrichment; the maximum Se/metal ratio obtained for these films is 1.27. It is smaller than the Se/Zn value (1.42) reported in Ref 18 for 260 nm thick ZnSe layers prepared by thermal vacuum evaporation at room substrate temperature. The lower Se-enrichment of our $Zn_xCd_{1-x}Se$ layers can be related to the fact that both ZnSe and CdSe were deposited in a quasi-closed volume.

**Table 1.** Average EDS data for the chemical composition of the prepared $Zn_xCd_{1-x}Se$ thin films.

| | $V_d^{CdSe}/V_d^{ZnSe}$ | Zn (at.%) | Cd (at.%) | Se (at.%) | $x_e$ | $x_m$ | $x_m$-$x_e$ |
|---|---|---|---|---|---|---|---|
| 1 | 3 | 19.6 | 30.5 | 49.9 | 0.25 | 0.39 | 0.14 |
| 2 | 1.5 | 23 | 20.9 | 56.1 | 0.4 | 0.52 | 0.12 |
| 3 | 1.0 | 25.8 | 18 | 56.2 | 0.5 | 0.59 | 0.09 |
| 4 | 0.667 | 32.8 | 15 | 52.2 | 0.6 | 0.69 | 0.09 |
| 5 | 0.333 | 34.4 | 8.4 | 57.2 | 0.75 | 0.80 | 0.05 |

According to the model applied to ternary $A_xB_{1-x}C$ alloys, in particular to $Zn_xCd_{1-x}Se$ [2,9], two sublattices exist in such alloys: one consisting solely of Se atoms and the other one consisting of Zn and Cd atoms in a random arrangement. The experimental $x_m$-values given in Table 1 have been obtained assuming that in the metal sublattice no Se atoms are included and, in accordance with the $Zn_xCd_{1-x}Se$ formula, the sum of the metal atom percentages has been considered as 100 % of metal atoms. Thus obtained $x_m$-values indicate that layers of five different compositions were successfully produced. One can see from figure 1 that the experimental points lie on a line $x_m = A + Bx_e$ ($A = 0.19 \pm 0.02$, $B = 0.82 \pm 0.04$) which indicate that a linear relation exists between the expected and experimental compositions. Using this relation, it is possible to produce films with desired compositions by means of the above described deposition approach. Further studies will define the range of $x$-values for which this linear relation is preserved.

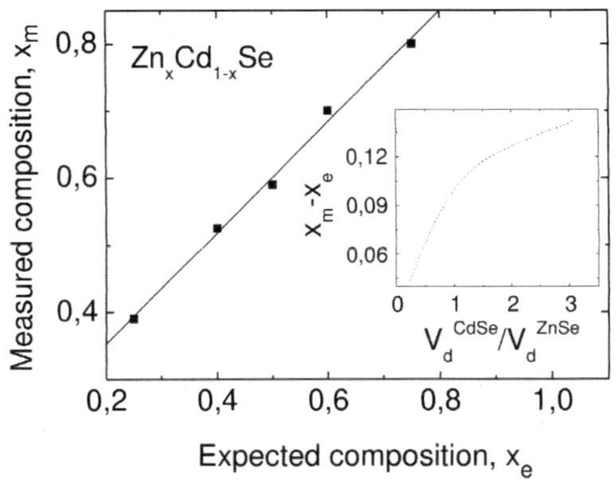


**Figure 1.** Relation between the expected $x_e$ and EDS determined $x_m$ composition of $Zn_xCd_{1-x}Se$ thin films, prepared by thermal evaporation of ZnSe and CdSe in vacuum. The solid line represents a linear fit of the experimental points. Inset: Deviation of the measured layer composition from the expected one versus the ratio of the deposition rates of CdSe and ZnSe. The line is guide to the eye.

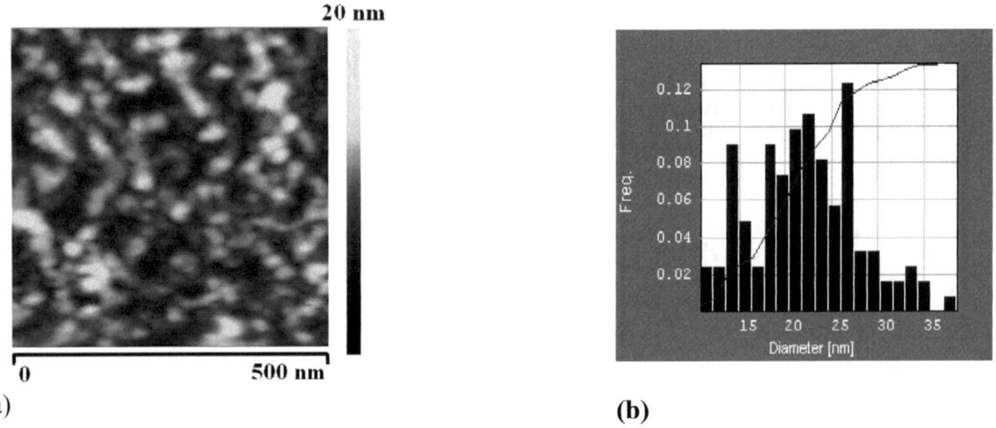


**Figure 2.** Two-dimensional AFM surface image (a) and size distribution (b) of a 400 nm thick $Zn_xCd_{1-x}Se$ single layer with $x = 0.59$.

The observation that the *B*-value is < 1 means that there exists a systematic deviation of the experimental composition from that expected; for all compositions the Cd inclusion is less than that expected on the basis of the ratio of the deposition rates. One can see from Table 1 and the inset in figure 1 that the Cd-deficiency increases with increasing ZnSe deposition rate and decreasing the ratio of the deposition rates. An enhanced redesorption of the Cd deposit has been reported [19] during overgrowth of CdSe layers by ZnSe which has been related to Cd segregation and replacement of Cd by Zn in the topmost surface. A similar process may be responsible for the observed deviation of the experimental composition from that expected. Besides, this process causes [19] a strong intermixing of the ZnSe and CdSe layers, which is in favour of the preparation technique described here, since the process will facilitate fabrication of layers with a quite homogeneous composition at the nanoscale.

AFM surface image and grain size distribution of a 400 nm thick $Zn_xCd_{1-x}Se$ single layer with '$x = 0.59$' are depicted in figure 2 (a) and (b), respectively. The surface images of the layers with the other compositions are very similar. They reveal that the layers are nanocrystalline with an apparent grain size of ~ 23 nm. Normally the apparent AFM size depends on the tip size; the smaller the tip size, the closer the apparent grain size to the real one. In our experiments the tip apex size was 8 nm and hence one can assume that the average grain size is around or less than 20 nm. The AFM experiment has also revealed a tendency for the $Zn_xCd_{1-x}Se$ layer hardness to increase with increasing Zn content.

Micro-Raman spectra of five layers with different compositions, which were taken under excitation with the 514.5 nm laser line, are shown in figure 3. The longitudinal optical (LO) phonons demonstrate a single phonon behaviour, expected for $Zn_xCd_{1-x}Se$ [2,3]. A series of 4 LO Raman peaks (at ~ 240, ~ 480, ~ 720, ~ 960 $cm^{-1}$) is observed in the spectra of all samples. This series originates from so-called LO-phonon replicas, well known in polar crystalline semiconductors [18]. The appearance of four LO peaks in all spectra is an indication that the films are crystalline rather than amorphous and this observation is in agreement with the AFM results. The FWHM of the 1LO bands is quite large (~ 17 $cm^{-1}$) which could be due to both imperfect film crystallinity and some variations in the film composition at the nanoscale. According to the structural model of $Zn_xCd_{1-x}Se$ [2,9], Zn and Cd atoms in the metal sublattice are randomly arranged and no short-range order is imposed. However, experimental studies of ZnSe/ZnCdSe/ZnSe structures, produced by submonolayer insertion of Cd into ZnSe matrices, have shown that the resulting distribution of Cd along the growth plane is strongly inhomogeneous in many cases. Formation of nanosized Cd-rich regions has been observed in $Zn_xCd_{1-x}Se$ grown by different MBE systems [20-23]. It is considered that the formation of those Cd-enriched islands is caused by a strong mismatch of the lattice constants of ZnSe and CdSe, resulting in an increase of the diffusion length of Cd [20,23]. Hence, one can think that the relatively large FWHM

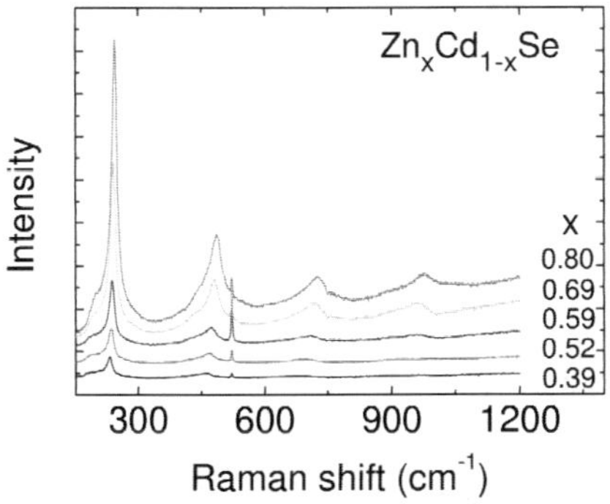


**Figure 3.** Micro-Raman spectra of five $Zn_xCd_{1-x}Se$ layers measured under excitation with the 514.5 nm line of an $Ar^+/Kr^+$ laser. All spectra correspond to the same scale but offset for clarity.

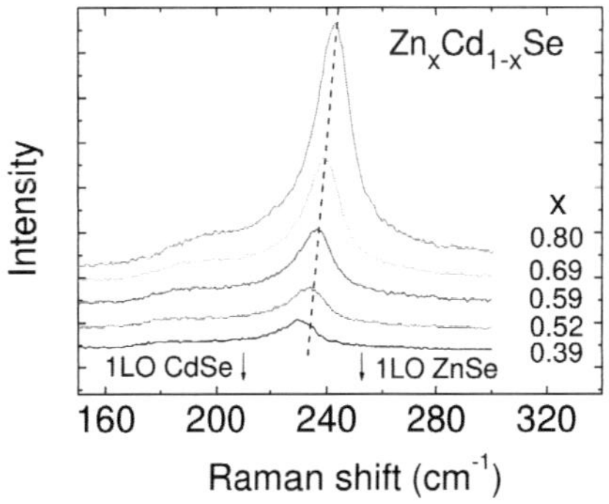


**Figure 4.** Shift of the 1LO in the Raman spectra of five $Zn_xCd_{1-x}Se$ layers measured under excitation with the 514.5 nm line of an $Ar^+/Kr^+$ laser. All spectra correspond to the same scale but offset for clarity.

of the Raman band is due to existence of Cd-rich regions in our layers.

The 1LO Raman bands in the Raman scattering spectra of all samples are disposed between the the 1LO bands of stoichiometric ZnSe (252 cm$^{-1}$) and CdSe (210 cm$^{-1}$) (Figure 4). As expected, a gradual red-shift is seen with increasing Cd content, which confirms that $Zn_xCd_{1-x}Se$ thin films of various compositions have been successfully prepared by the applied deposition technique. Figure 5 compares the observed 1LO position changes with experimental results reported for $Zn_xCd_{1-x}Se$ films produced by MBE [2] and a theoretical curve [9]. One can see that the 1LO positions we determined are close to both experimental and theoretical curves. This coincidence gives compositions which are close to those determined by EDS and indicates that the internal stress in these films is not excessively high.

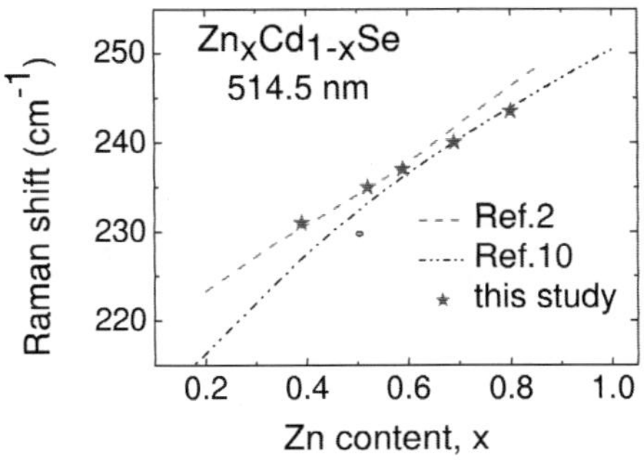


**Figure 5.** Comparison of the composition-induced shift of the 1LO band obtained in this study (stars) with experimental [2] and theoretical [9] data reported by other authors for $Zn_xCd_{1-x}Se$ layers.

## 4. Conclusions

Single layers from $Zn_xCd_{1-x}Se$ with five different compositions have been prepared by parallel thermal evaporation of ZnSe and CdSe in vacuum and substrate rotation. Alternating ZnSe and CdSe (island type) films have been deposited on the substrates at room temperature having equivalent thicknesses of 0.12, 0.25, 0.37 nm. Both materials alloy and thus form ternary $Zn_xCd_{1-x}Se$ layer. The EDS investigations performed have given values of $x = 0.39$, 0.52, 0.59, 0.69 and 0.8, while values of $x = 0.25$, 0.40, 0.50, 0.60 and 0.75, respectively, were expected from the deposition conditions applied. The observed discrepancy has been related to Cd redesorption during deposition of ZnSe on CdSe surface.

The microstructure of the $Zn_xCd_{1-x}Se$ layers prepared has been explored by AFM and Raman scattering measurements. The AFM results have shown that the layers are nanocrystalline with a grain size $\leq 20$ nm. The Raman scattering results have confirmed the film composition obtained from the EDS data thus proving that the film composition can be successfully varied with varying the deposition conditions. They indicate that the internal stress in the $Zn_xCd_{1-x}Se$ layers in not very high. The results obtained have proven that this deposition technique allows preparation of ternary $Zn_xCd_{1-x}Se$ layers with controlled compositions.

**Acknowledgments**
This work has been partially supported by the Bulgarian Ministry of Education and Science under grant BM – 1, the Serbian Ministry of Science and Technological Development under the Project No. 141047 and Bilateral project between BAS and SASA.

**References**

[1] Chandramohan R and Mahalingam T 1999 *Proc. SPIE* **3896** 107
[2] Alonso R G, Suh E-K, Ramdas A K, Samarth N, Luo H and Furdyna J K 1989 *Phys. Rev.* B **40** 720
[3] Melnik N N, Sadofyev Yu G, Zavaritskaya T N and Vodop'yanov K L 2000 *Nanotechnology* **11** 252
[4] Swizer J A 2001 Electrodeposition of Superlattices and Multilayers *Electrochemistry of nanomaterials* ed G Hodes (Weinheim: Wiley-VCH) chapter 3 p 67
[5] Sutrave D S, Shahane G S, Patil V B and Deshmukh L P 2000 *Turk. J. Phys.* **24** 63
[6] Hines M A and Guyot-Sionnest P 1998 *J. Phys. Chem.* B **102** 3655
[7] Suemune I, Tawara T, Saitoh T and Uesugi K 1997 *Appl. Phys. Lett.* **71** 3886
[8] Harris Liao M C, Chang Y H, Chen Y F, Hsu J W, Lin J M and Chou W C 1997 *Appl. Phys. Lett.* **10** 2256
[9] Avendano-Lopez J, Castillo-Alvarado F L, Escamilla-Esquivel A, Contreras-Puente C, Ortiz-Lopez J and Zelaya-Anhel O 1996 *Solid State Commun.* **100** 33
[10] Balocchi A, Curran A, Graham T C M, Bradford C, Prior K A and Warburton R J 2005 *Appl. Phys. Lett.* **86** 011915
[11] Gupta L, Rath S and Abbi S C 2005 *Mat. Sci. Eng. B* **119** 171
[12] Sweiti A, Medina F, Martinez L and Lopez-Rivera A 2008 *Semicond. Sci. Technol.* **23** 035019
[13] Nesheva D 2004 Nanoparticle Layers in Multilayers *Encyclopedia of Nanoscience and Nanotechnology* ed H S Nalwa (California: American Scientific Publishers) **7** p 105
[14] Ionov R and Nesheva D 1992 *Thin Solid Films* **213** 230
[15] Nesheva D 2001 *Handbook of Surfaces and Interfaces of Materials* ed H.S. Nalwa (San Diego: Academic Press) **3** chapter 6 p 239
[16] Nesheva D, Arsova D and Ionov R 1993 *J. Mater. Sci.* **28** 2183
[17] Nesheva D, Levi Z, Aneva Z, Miloushev I and Pejova B 2009 Optical properties of $Zn_xCd_{1-x}Se$ single layers prepared by thermal vacuum evaporation *AIP Conf. Proc.* **1203** 358
[18] De CK and Misra N K 1997 *Indian J. Phys.* A **71** 535
[19] Passow T, Heinke H, Schmidt T, Falta J, Stockmann A, Selke H, Ryder P L, Leonardi K and Hommel D 2001 *Phys. Rev.* B **64** 193311
[20] Permogorov S, Klochikhin A and Reznitsky A 2002 *J. Luminescence* **100** 243
[21] Schikora D, Schwedhelm, As D J, Lischka K, Litvinov D, Rosenauer A, Gerthsen D, Strassburg M, Hoffmann A and Bimberg D 2000 *Appl. Phys. Lett.* **76** 418
[22] Perano N, Rosenauer A, Gerthsen D, Sorokin S V, Sedova I V and Ivanov S V 2000 *Phys. Rev.* B **61** 16015
[23] Valakh M Ya, Vuychik N V, Strelchuk V V, Sorokin S V, Shubina T V, Ivanov S V and Kop'ev P S 2003 *Semiconductors* **37** 699; Translated from 2003 *Fizika i Tekhnika Poluprovodnikov* **37** 724

Journal of Physics: Conference Series **253** (2010) 012036 doi:10.1088/1742-6596/253/1/012036

# Study of the charge transport mechanism in pulsed laser deposited AlN:Cr films

**I P Minkov[1,2,4], S Simeonov[1], A Szekeres[1], C Ristoscu[3], G Socol[3], S Grigorescu[3], I N Mihailescu[3]**

[1]Georgi Nadjakov Institute of Solid State Physics, Bulgarian Academy of Sciences, 72 Tsarigradsko Shausse, 1784 Sofia Bulgaria

[2]Technical University – Sofia, 8 Kliment Ochridski blvd, 1000 Sofia, Bulgaria

[3]National Institute for Lasers, Plasma and Radiation Physics, PO Box MG-54, Ro-77125, Bucharest – Magurele, Romania

E-mail: ipmincov@tu-sofia.bg

**Abstract.** AlN thin films doped with Cr (AlN:Cr) were prepared on p-Si(100) substrates by pulsed laser deposition. Al–AlN:Cr–Si MIS structures were formed and their current–voltage and 1 MHz admittance characteristics were measured and analyzed. The current through the MIS structure is proportional to the square of the applied voltage, which is evidence that the current is limited by space charge at traps in the AlN:Cr films. These traps are responsible also for the observed negative differential resistance and hysteresis of the current–voltage characteristics. The *C–V* and *G–V* characteristics measured at 1 MHz and the impedance measurements in 500 Hz–500 kHz frequency range without applying bias voltage are also influenced by the traps in the AlN:Cr films.

## 1. Introduction

The possibility for applying diluted magnetic semiconductors (DMS) in spintronics devices and structures has led to preparation and investigation of different DMS thin films. DMS is formed when metal ions are replaced by magnetic ions of transition metals such as Co, Cr or Mn. Ferromagnetism with Curie temperatures as high as 900°C is observed in Cr-doped AlN (denoted further as AlN:Cr) films [1]. For preparation of these films, the incorporation of Cr ions is accomplished usually by Cr co-evaporation during AlN MBE deposition [2] or by Cr ion implantation into already deposited AlN film [3], or by reactive dc co-sputtering of Al and Cr in a nitrogen gas ambient [4]. The pulsed laser deposition (PLD) technique, due to its capability of reproducing the target stoichiometry in the deposited film, is very suitable for preparation of AlN:Cr films. Prospective AlN:Cr devices and structures formed on Si substrates will be compatible with devices and integrated circuits in contemporary Si planar technology. Because of this it is interesting to prepare AlN:Cr–Si structures and to investigate their properties. For spintronics applications, such as resonant tunneling diodes and

[4] Corresponding author

tunneling magnetoresistance structures, the knowledge of charge transport mechanism in AlN:Cr films on Si substrates is of great importance.

In this paper we present results of the study of the charge transport mechanism in the AlN:Cr films prepared on p-Si(100) substrates using the pulsed laser deposition technique. For this purpose, MIS structures with PLD AlN:Cr films were formed and their DC current-voltage (*I–V*) and admittance (capacitance and conductance) characteristics were measured. Through detailed analysis of these electrical characteristics, the charge transport mechanism was established.

## 2. Experimental details

Thin AlN films doped with Cr were synthesized on single crystalline Si(100) p-type substrates by pulsed laser deposition [5]. The PLD targets were prepared by homogeneously mixing AlN and Cr (10 at.%) powders followed by pressing and sintering. The films were grown using a KrF* ($\lambda$ = 248 nm) excimer laser generating pulses of $\tau_{FWHM} \leq 25$ ns at a repetition rate of 10 Hz. During ablation, the target was rotated and translated along two orthogonal axes to avoid punctures and improve the morphology of the deposited films. The incident laser beam was incident at 45°, while ablation plasma was evolving normally to the target surface. The separation distance target–substrate was set to 5 cm. After loading the substrate and target, the deposition chamber was pumped down to a residual pressure of ~$10^{-4}$ Pa. Before deposition, the substrate was heated up to 800°C for 1 hour for the purpose of decomposing native $SiO_2$ on the Si surface. The substrate temperature was kept at 800°C during deposition, providing conditions for crystalline growth of AlN. The laser fluence on the target was 10 J/cm$^2$. For the deposition of one film 15 000 subsequent laser pulses were applied yielding a ~300 nm thick films. The AlN:Cr film synthesis was undertaken in vacuum at a residual gas pressure of $10^{-3}$ Pa.

In order to measure electrical characteristics, MIS capacitors with Al–AlN:Cr–Si structures were formed by vacuum evaporation of Al dots with 0.5 mm diameter on film surface and of a continuous Al film on Si-wafer backside. Room temperature 1 MHz capacitance–voltage (*C–V*) and conductance–voltage (*G–V*) measurements were carried out by a digital LCR meter E7-12, while the admittance measurements were carried out by a Tesla BM-507 impedance meter in the frequency range of 500 Hz – 500 kHz.

## 3. Results and discussion

The DC *I–V* characteristics of these AlN:Cr MIS structures were measured for electrical fields up to $10^5$ V/cm. The measurement sequence began from 0 V toward negative or positive voltages applied to the contact on the AlN:Cr film surface followed by a voltage reversal toward zero applied bias voltage. The measured *I–V* curves are given in figure 1. In both forward and reverse bias directions the current is non-ohmic and a hysteresis appears. During the initial voltage passage from zero to minimal negative or maximal positive voltage the current density is larger than that obtained by going back to zero voltage.

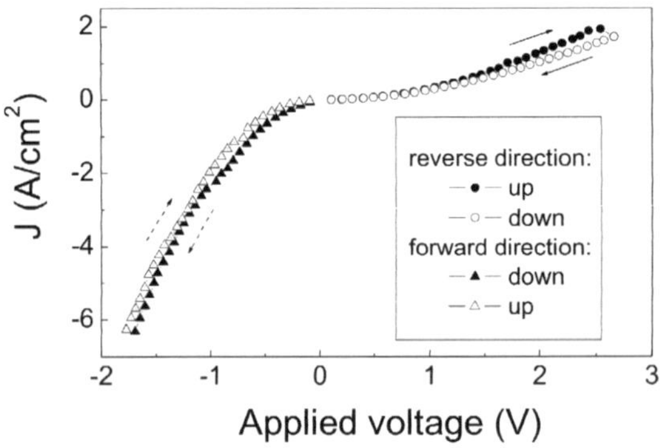


**Figure 1.** Current density vs. applied voltage characteristics of AlN:Cr MIS structure.

For all the four recording directions, when the current density is re-plotted as ln*J* versus ln*V* plots the slope of the curves approaches 2, (as an example see figure 2). Such current–voltage dependence is evidence for trap space-charge limited current [6]. In that case, the charge trapped in the insulator is proportional to the applied voltage and the effective velocity of released charge from traps under the applied electric field is also proportional to the voltage. In these circumstances the current is proportional to $V^2$, as we observed.

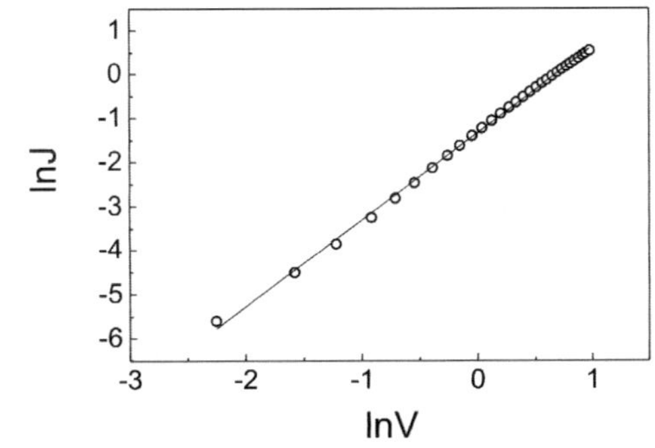


**Figure 2.** ln*J* vs. ln*V* plot of current density in reverse down direction on applied voltage of AlN:Cr MIS structure.

The presence of hysteresis in the *I–V* curves is another evidence of trap-connected conductivity in the investigated AlN:Cr films. Starting the *I–V* measurements (so called “up” direction for positive voltages and “down” for negative voltages in figure 1), some of the charges are captured at traps which have time constants in order of tens of minutes, being commensurable to the time for a *I–V* measurement performance. This trapped charge opposes the external electric field of the applied voltage during the voltage passage toward zero. Because of this, in the second voltage passage (so called “down” direction for positive voltages and “up” for negative voltages in figure 1) the current at a given voltage is lower. The density of this trapped charge, $N_t$, can be estimated from the expression $\Delta V = qN_t d/\varepsilon_f$, where $\Delta V$ is the voltage difference of up and down curves for a given current density in the *I–V* characteristics (figure 1), *d* is the film thickness and $\varepsilon_f$ is the film dielectric constant. With $\Delta V = 0.1$ V at $J = -2.43$ A/cm$^2$ and $\varepsilon_f = 31\varepsilon_0$, obtained from the *C–V* measurement, and $d = 300$ nm the density of trapped charge in the AlN:Cr film is estimated as $1.9\times10^{17}$ cm$^{-3}$.

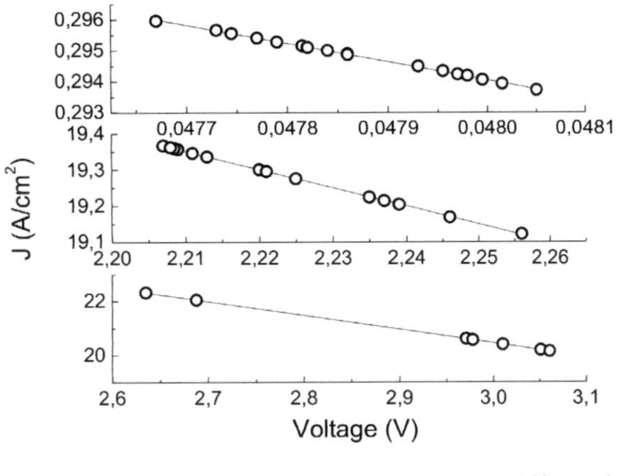


**Figure 3.** *I–V* characteristics with negative differential resistance of AlN:Cr MIS structure.

In figure 3 the change of the current density in the forward direction with the voltage applied to the AlN:Cr MIS structure is shown. These dependences were obtained by registering the current alteration through the structure during 15–20 minutes, as the total voltage applied to this structure and the current measuring serial resistor was kept constant. The current density decreases when the electric field increases. This decrease is less than 1% at electric field around $1.6\times10^3$ V/cm and it is approximately 10% at electric field around $10^5$ V/cm. In [7] for electric fields above $2\times10^6$ V/cm the negative differential mobility in AlN films has been calculated and it has been attributed to polar optical phonon scattering of electrons at these high electric fields. However, in our AlN:Cr–Si structures the negative differential resistance, observed at electric fields ranging from $1.6\times10^3$ to $10^5$ V/cm differs from that discussed in [7]. The presence of traps in the AlN:Cr films with very long capture and emission time-constants, in the order of minutes, can be the reason for the observed negative differential resistance. After applying the voltage to the AlN:Cr–Si structure, additional electrons are captured at these traps and this additional charge opposes the external electric field leading to the observed negative differential resistance. Decrease of the electric field $\Delta E$ can be expressed as $\Delta E = qN_{t,a}d/\varepsilon_f$, where $N_{t,a}$ is the density of additional charges at traps with long time constants. Therefore, at 15% decrease of $10^5$ V/cm electric field, and with $\varepsilon_f = 31\varepsilon_0$ and $d = 300$ nm the estimated $N_{t,a}$ density is $8.6\times10^{15}$ cm$^{-3}$. Both densities of trapped charges $N_t$ and $N_{t,a}$ are reasonable. The difference between these values arise from the fact that different traps are responsible for the hysteresis in the *I–V* characteristics and for the negative differential resistance of the studied AlN:Cr–Si structures.

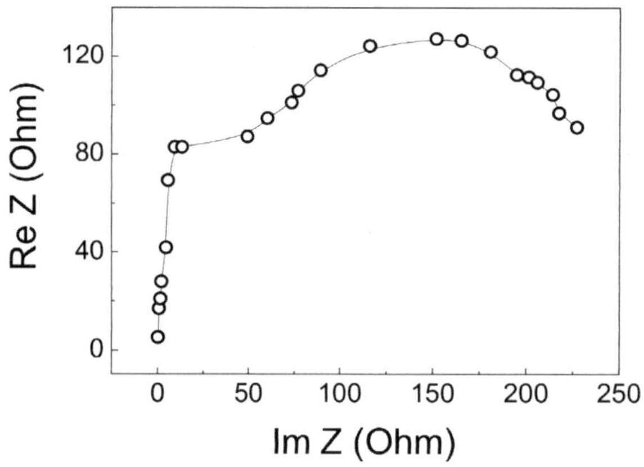


**Figure 4.** *Re Z* vs. *Im Z* plot of the impedance measurement of AlN:Cr MIS structure.

The impedance of the AlN:Cr MIS structure was measured at test signal voltages in the 500 Hz – 500 kHz frequency range without applying bias voltage. The values of $Re\ Z = |Z|\cos\varphi$ and $Im\ Z = |Z|\sin\varphi$ are plotted in figure 4. As it is seen, the maximal values of $Re\ Z$ = 124–128 Ohm correspond to values of $Im\ Z$ = 115–152 Ohm, measured at 1 and 2 kHz. From this follows that $\omega\tau = 1$, where $\tau$ is the response time of traps related with impedance data and, it is equal to $\tau = 80$ μs at 2 kHz.

The *C–V* and *G–V* characteristics of the AlN:Cr–Si MIS structures, measured at 1 MHz are presented in figures 5 and 6, respectively. The sweep direction is given with an arrow, starting from negative to positive voltages ("up") and contrariwise ("down"). One may notice that up and down *C–V* curves in the sequence from –20 V to 20 V and the reversal to –20 V almost coincide with each other. At applied voltages below –7.2 V and above 7 V both kinds of characteristics saturate, which one may expect for a MIS structure. The observed saturation in accumulation and deep inversion regions is due to the complete filling of traps in the AlN:Cr film at these bias voltages.

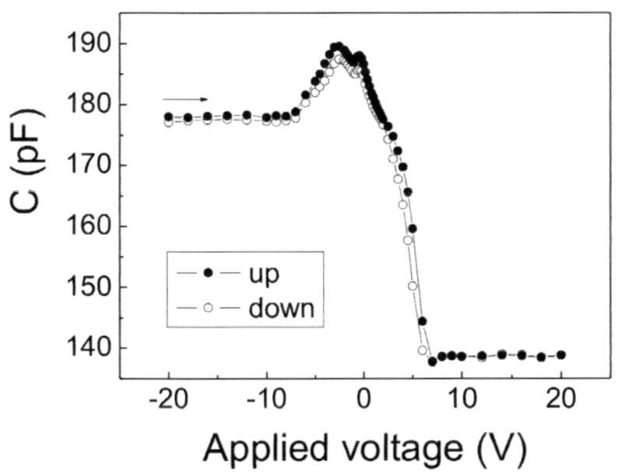


**Figure 5.** *C–V* characteristics of AlN:Cr MIS structure.

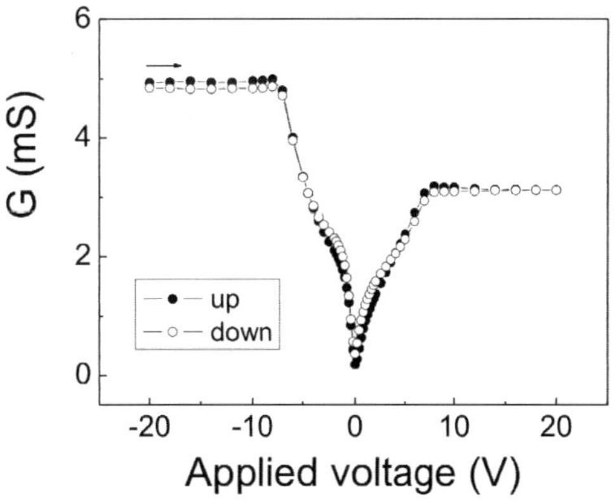


**Figure 6.** *G–V* characteristics of AlN:Cr MIS structure.

In the voltage region between –7.2 V and 1.3 V a bump appears in the *C–V* curves. Similar capacitance feature is observed in many semiconductor structures and it is connected with the back-to-back Schottky model of depletion regions on both sides of the grain boundary [8]. But the sharp transition to the accumulation capacitance at –7.2 V together with the sharp decrease of the capacitance above 1.3 V indicate that this capacitance bump is connected with traps, which are able to respond to the bias voltage at 1 MHz. From the shape of the *G–V* curves (figure 6) one can conclude that the traps, responding to 1 MHz test voltages, are filled with charge carriers to different extent in the voltage region between –7.2 V and 7 V and, they become completely filled in the regions, where the capacitance and the conductance are constant. There are not empty traps below the charge carrier quasi-Fermi levels in these AlN:Cr films.

The permittivity $\varepsilon_f$ of the AlN:Cr film was calculated from the capacitance value in accumulation and it was equal to $31\varepsilon_0$. This value is higher than the value $\varepsilon_f \approx 10\varepsilon_0$, estimated from the optical measurements of AlN films in [9]. In [10] $\varepsilon_f$ has reached even $400\varepsilon_0$ and this high value has been explained by defects related to nitrogen vacancies in AlN films. Therefore, it is reasonable that due to the high density of traps in the AlN:Cr films the trapped charges contribute to additional capacitance in accumulation regime.

## 4. Conclusions

It has been established that the current through the AlN:Cr–Si MIS structures in both forward and reverse directions is proportional to square of applied voltage and it is governed by the space charge at filled traps in the AlN:Cr films. The density of these traps with long emission time-constants is estimated from the hysteresis of the *I–V* curves and it is $1.9\times10^{17}$ cm$^{-3}$. The density of traps with long capture time-constants, which are responsible for the observed negative differential resistance, is $8.6\times10^{15}$ cm$^{-3}$. Both capacitance and conductance values saturate in accumulation and deep inversion regime revealing that the traps are completely filled below the quasi-Fermi level in these AlN:Cr films.

## Acknowledgements

The authors acknowledge with thanks the support of this work under the 2007–2009 Collaboration Agreement between the Bulgarian Academy of Sciences and Romanian Academy.

## References

[1] Liu H X, Wu S Y, Singh R K, Lin G, Smith D J, Newman N, Dilley N R, Montes L and Simmonds M B 2004 *Appl. Phys. Lett.* **85** 4076

[2] Frazier R M, Thaler G T, Leifer J Y, Hite J K, Gila B P, Abernathy C R, Pearton S J 2005 *Appl.*

*Phys. Lett.* **86** 052101

[3] Pearton C J, Abernathy C R, Fan Ren 2006 "*Gallium nitride processing for electronics, sensors and spintronics*" (Berlin: Springer) p. 291

[4] Kumar D, Antifakos J, Blamire M G, Barber Z H 2004 *Appl. Phys. Lett.* **84** 5004

[5] Szekeres A, Bakalova S, Grigorescu S, Cziraki A, Socol G, Ristoscu C and Mihailescu I N 2009 *Appl. Surf. Sci.* **255** 52271

[6] Lampert M A, Mark P 1970 *"Current injection in solids"* (Academic Press: New York)

[7] Bulutay C, Ridley B K, Zakleniuk N A 2003 *Phys. Rev. B* **68** 115205

[8] Bardhadi A, Amziel H, Muller J C, Siffert P 1989 *Springer Proc. Physics, Polycryst.Semicond.* **35** 158

[9] Mohammad S N, Morkoc H 1996 *Progress Quantum Electr.* **20** 361

[10] Bi Z X, Zheng Y D, Zhang R, Gu S L, Xiu X Q, Zhou L L, Shen B, Chen D J, Shi Y 2004 *J. Mater. Sci.:Materials in Electronics* **15** 317

# Electroreflectance spectroscopy study of hydrogen plasma immersion ion implanted silicon with ultrathin oxide film

**S Alexandrova[1], A Szekeres[2], R Yu Holiney[3] and L Matveeva[3]**

[1]Technical University-Sofia, 8 Kl. Ohridski Blvd., Sofia 1797, Bulgaria
[2]Institute of Solid State Physics, 72 Tzarigradsko Chaussee, Sofia 1784, Bulgaria
[3]V. Lashkarev Institute of Semiconductor Physics, 45 Prosp. Nauki, Kiev 03028, Ukraine

E-mail: salex@tu-sofia.bg

**Abstract.** Results are presented from a study of (100)Si substrates, hydrogenated by plasma immersion ion implantation with $H^+$ ions with energy of 2 keV and fluence varying from $10^{13}$ to $10^{15}$ $cm^{-2}$. The implanted Si substrates were oxidized at considerably low Si oxidation temperatures of 700 and 750°C. The formed structures are characterized by electroreflectance (ER) spectroscopy. The implanted Si surface was accessed through the ultrathin oxide layer. From the ER spectra the characteristic energetic bands of direct electron transitions in Si and the corresponding broadening parameters are elaborated. The shifts of the direct energy gap are related to the process induced stress in implanted Si that does not exceed $10^9$ $N/m^2$. The phenomenological broadening parameters $\Gamma$ are below 130 meV and $H^+$ implanted Si with fluence of $10^{13}$ $cm^{-2}$ has a value of 100 meV. The slight increase of $\Gamma$ as compared to the unimplanted Si is due to incomplete anneal of ionization induced defects. The oxidation rates of the $H^+$ implanted Si increases to values being characteristic for Si wet oxidation.

## 1. Introduction

The behavior of hydrogen in semiconductors has been extensively studied for the last decades. The efforts of the researchers were provoked not only by a scientific interest, but were motivated by the possibility to improve device performance. Hydrogen is a highly reactive species in semiconductors and can have a profound effect on the electrical and optical properties of a device. It has been found that dangling bonds on the surface and interface, defects in the bulk of the material, as well as at grain boundaries can successfully be passivated by formation of electrically neutral complexes with H species. Therefore hydrogenation can be used as a highly effective technological tool to enhance the performance of electronic and optical devices. Over last two decades, a number of hydrogenation methods have been empirically optimized for device passivation. These include forming gas anneal, ion implantation and dry plasma processing. Despite its success, the fundamental understanding of H passivation is still incomplete, especially when a production scheme of a hydrogenation process for maximum passivation is required.

In an experiment where controllable hydrogenation combined with low defect concentrations has to be investigated, H can be introduced in the surface region of the sample using low energy $H^+$ ion implantation. Ion implantation provides accurate control into Si depth by the choice of acceleration energy and ion fluence. In the present study Si wafers were subjected to hydrogenation of by plasma ion immersion (PIII) technique, which ensures the requirements stated above. On the implanted surface an ultrathin silicon oxide was grown by thermal oxidation at temperatures between 700 and 800°C. The oxidized Si surface was investigated by electroreflectance spectroscopy (ERS), the implanted Si surface being accessed through the oxide layer. This optical technique, as highly sensitive to interface properties [1,2], is very well suited for characterization of the implanted Si surface. From the analysis of the electroreflectance (ER) spectra conclusions can be drawn about the properties of the interfacial region including defect concentrations and stress resulting from the oxidation process. The results enable us to show that hydrogenation through PIII can be effective tool for growing high quality oxide with lowered structural strains at considerably low Si oxidation temperatures.

The oxide growth temperatures are high enough to serve as annealing of implantation-induced defects to a level allowing the record of ERS signals. However, these temperatures lie well below the temperatures needed for relaxation of macroscopic compressive strain in the bulk of the $SiO_2$ film (990°C) and interface relaxation and bond rearrangement process (~ 900°C) [3]. The oxidation temperatures are lower than that used in the present Si-based industry, so that conclusions can be drown for the possibility to use such low temperatures to obtain structures with smaller defect densities.

## 2. Experimental details

### *2.1 Sample preparation*

The studied samples were Si/oxide structures, prepared onto (100)p-type Si substrates with a resistivity of 5-8 Ohm.cm. The Si wafers were subjected to $H^+$ ion implantation with energy of 2 keV in a PII installation unit. The hydrogen fluence varied from $10^{13}$ to $10^{15}$ $cm^{-2}$. Further, following standard RCA cleaning, the implanted Si substrates were oxidized at temperatures between 700 and 800°C in dry oxygen atmosphere for 30 min. No further temperature treatments were applied.

### *2.2. Measurements*

The method used to characterize the samples was the electroreflection spectroscopy. Analysis of the ER spectra in a specific region in the *k*-space of the Brillouin zone makes it possible to determine the transition energy $E_g$ the phenomenological Lorentz broadening parameter $\Gamma$, which allows for dissipation processes in the electron transition, and the energy relaxation time of the light excited charge carriers. The energy of transition and the value of the phenomenological broadening parameter are used to make conclusion about the implanted Si surface region and as indicators of Si/oxide interface perfection.

The ERS consists in recording the modulation of the reflectivity of the sample that appears when an external modulating electric field is applied to the sample. The measured quantity is the relative reflectivity modulation, $\Delta R/R$. The ER spectra were measured through the oxide layer in the photon energy region of 3.0 - 3.8 eV, over the spectral range of the Si direct bandgap energy. The ER response of the structures was measured by a standard technique using 0.1 N KCl solution in an electrolytic cell to apply modulating voltage of 1 V. DMR-4 monochromator was used with automatic recording of the spectra with a 0.003-eV resolution. The sensitivity of the device in measuring $\Delta R/R$ was one part in a million, and the accuracy of measuring the signal strength was 2%. The measurements were performed at room temperature in the low-field mode condition of $\hbar\Omega < \Gamma/3$, where $\hbar\Omega$ is the electro-optical energy and $\Gamma$ is the phenomenological parameter of broadening [4,5]. The values of the direct transition $E_g$ and $\Gamma$, were calculated from the analysis of the ER spectra line shape using the Aspnes three-point technique [4,6].

The thickness of the oxides was determined by ellipsometry at $\lambda = 632.8$ nm with an accuracy of ±0.2 nm. The oxide thickness at 750°C was below 9 nm, varying depending on hydrogen ion fluence.

By applying SRIM code (Stopping and Range of Ions in Matter) simulations [7] the distributions of the implanted ions, the ionization effects and vacancy formation were modelled.

## 3. Results and discussion

Figure 1 presents the ER spectra from hydrogen implanted Si surface oxidized at 700 and 750°C. According to electron band structure of silicon, the spectra correspond to the direct transition, which occurs between the valence and the conduction band at the center of the Brillouin zone. In this range the Si known to exhibit two neighbouring direct electron transitions around the center of Brillouin zone: $E_0^{'}(\Gamma_{25}^{'} - \Gamma_{15})$ at 3.3 eV and $E_0(\Gamma_{25}^{'} - \Gamma_2^{'})$ at 3.37 eV. In view of broadening, the signals from the two transitions overlap and usually are not resolved at room temperature [8,9] and are viewed as a single transition at $E_g$ = 3.4 eV. As reported, silicon surface in unstressed conditions is characterized by 3.37 eV of the direct transition [10,11]. The record of the near 3.4 eV structure in the ER signal in our spectra is evidence that the ER signal originates from the Si substrate and the observed shifts and shape changes with ion fluence are connected to the condition at the interface region.

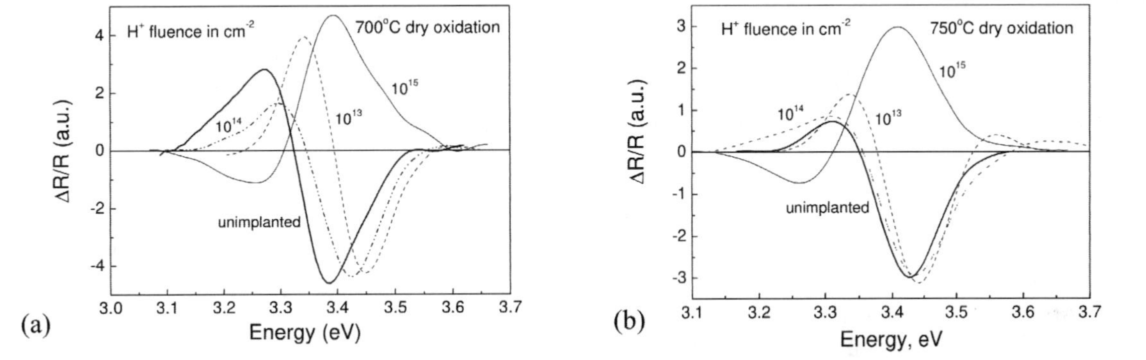


**Figure 1.** Electroreflectance spectra of the oxidized Si surface for different $H^+$ implantation fluences. Oxidation was performed at 700°C (a) and at 750°C (b).

The transition energies and phenomenological broadening parameters for Si wafer with different implantation fluences are given in figure 2.

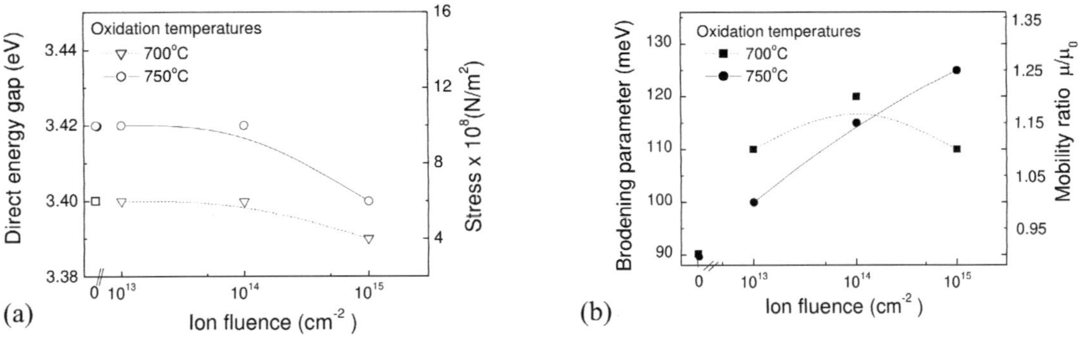


**Figure 2.** Transition energies (a) and phenomenological broadening parameters (b) for Si wafer with different implantation fluences. The values for unimplanted Si are given at zero fluence. The lines are given to show the tendency behaviour.

The transition energies are shifted relative to the unstressed position and are indicative for the stress revealed in the Si during oxidation. On the left axis of figure 2a the energy shifts are transferred to the stress scale. For that purpose, the correlation of the energy shift with the stress level is needed. Photoreflectance study of Si oxidized at 850°C [12] showed a correlation coefficient 2.09x10$^7$ N/m$^2$/meV. Applying Raman spectroscopy for light-emitting anisotropically etched Si yielded a value of 1x10$^7$ N/m$^2$/meV [13]. Here we used the data from the photoreflectance because the resemblance to the experimental technique used in the present study. Moreover, in that case the Si was also subjected to a high temperature, although higher (850°C) than in out experiments (≥700°C). It can be seen that the stress in implanted Si does not exceed 10$^9$ N/m$^2$.

The broadening parameter parameter Γ is inversely proportional to the relaxation time and mobility of the electrons μ ~ e h /m*Γ [14], where m* is the reduced effective mass and h and e have their usual meaning of the Plank constant and elementary charge, and therefore characterizes the scattering process at the Si interface region. In figure 2b the mobility is given at the right axis, normalized to the unstressed mobility $\mu_o$.

The referent Si without implantation oxidized at both 700 and 750°C show a shift of $E_g$, although very small, to 3.37 eV for unstressed Si. This indicates a low level of tensile stress at the Si surface, compressive stress at the oxide, respectively, which is a result of the oxidation. It is known that the relaxation of oxide stress is observed at temperatures about 900°C and above. For comparison, mechanical stress of the order of 10$^9$ N/m$^2$ is typical for oxides grown thermally at 850°C as obtained from spectroscopic ellipsometry analysis [15]. The broadening parameter Γ of 95 meV is relatively small. Being inversely proportional to the carrier relaxation time, this value characterizes the scattering process at the unimplanted Si interface region and its small value is an indication of a high quality region.

The spectra for the $H^+$ implanted Si with fluences of 10$^{13}$ and 10$^{14}$ cm$^{-2}$ reveal a shape and polarity typical for oxidized Si surface [10]. The implantation at these fluences causes no measurable changes of $E_g$, while changes are observed for the broadening parameter Γ (Figure 2). This is reasonable since implantation with the light $H^+$ ions is not expected to cause severe damage in the Si lattice. However, the ionization effects can be substantial. Simulation using SRIM code shows the development of ionization events inside the Si substrate due to $H^+$ PIII as evident from Figure 3a.

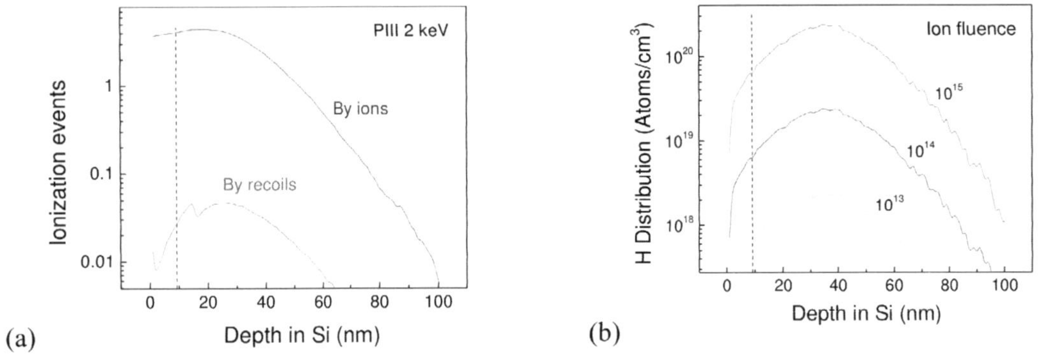


**Figure 3.** Hydrogen ion distributions into Si substrate after implantation (a) and ionization defects due the implantation process (b).

Ionization leads to defect generation in the Si space charge region and at the interface. These defects have contribution to scattering processes and result in larger Γ. As a consequence, the mobility decreases (figure 2b). The annealing process during oxidation is not sufficient to anneal out completely the defects. A high hydrogen concentration in the Si subsurface region is evident from SRIM results in figure 3b, which however, seems not to result in complete hydrogen saturation of the dangling bonds at the interface.

For ion fluence $10^{15}$ $cm^{-2}$ the observed red shift of $E_g$ approaching 3.39, which is very close to 3.37 eV, can be attributed to relaxation of the substrate stress due to annealing effects. The decrease of the broadening Γ also indicates development of the interface region with higher structural perfection and less scattering centers. Hydrogen RF plasma has been shown to cause stress relaxation in our earlier study using spectral ellipsometry [16]. Also, passivation of the surface with hydrogen has lead to strain relaxation upon treatment of a porous Si surface in hydrogen plasma [17].

The inversion of the polarity of the ER spectra, found only for $10^{15}$ $cm^{-2}$ ion fluence, for both oxidation temperatures should be noted. The opposite ER signal is usually attributed to formation of a thin defective Si layer near the surface [10,11]. However, even if such a layer is present, the defect concentration cannot be high, since otherwise it would not be possible to record ER signal. The $E_g$ and Γ also indicate high quality interface as discussed above. Most probably, at this fluence the ER shape has to be considered as a unipolar signal typical for intrinsic Si surface. This could be due to formation of hydrogen complexes with the substrate doping atoms in the space charge region. Similar effect has been found in RF hydrogen plasma treated oxides [18].

An interesting point is the influence of the oxidation temperature. It can be seen from figure 2 that the transition energies for all fluences are higher for the higher oxidation temperature. At higher oxidation temperature larger Si region is consummated during oxidation moving the interface deeper into the implanted region. From figure 3 it can be inferred that this region contains higher net concentrations of hydrogen ions and ionized defects causing the bigger $E_g$ shift.

The oxidation of the implanted region of the Si is expected to exhibit increased oxidation rates. Due to the implantation defects the diffusion of the oxidizing species is facilitated. The presence of H-atoms in the Si surface layer can promote the surface reaction rate of the oxidation process. An illustration of the increased rate is shown in figure 4. For the oxides, grown on implanted Si at a certain oxidation temperature, the experimental points refer to the different ion implantation fluences.

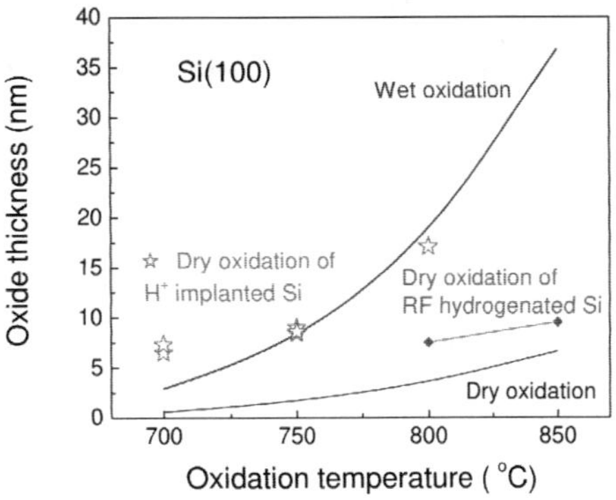


**Figure 4.** Comparative data for Si oxidation rates. The oxidation time is 30 min.

In the figure the data (open asterisks) for oxidation of implanted Si is compared with oxidation rates of Si according to Deal-Grove model (full lines) [19] in wet and dry oxygen atmosphere. The substantial increase of the oxidation rate of the $H^+$ PII implanted Si up to values characteristic for wet oxidation is evident, the increase being even more pronounced at 700°C. No strong dependence of the oxidation rates has been found on the implantation fluence as evident from the close experimental points for a given oxidation temperature. Increase of oxidation rates was also found in our earlier experiments on oxidation of Si hydrogenated in RF plasma [19], as is demonstrated in figure 4 (diamonds).

## 4. Conclusions

The direct energy gap $E_g$ of Si wafer hydrogenated through hydrogen plasma immersion ion implantation and oxidized in dry $O_2$ at 700° and 750°C was examined by means of electroreflectance spectroscopy. The Si surface is under small tensile stress up to $H^+$ fluence of $10^{14}$ cm$^{-2}$, while at $10^{15}$ cm$^{-2}$ stress relief is observed. The increased broadening parameter is due to incomplete anneal of ion induces interface defects.

Hydrogenation through plasma immersion ion implantation allows for a substantial increase of the oxidation rate at temperatures as low as 700°C up to the values typical for wet oxidation. The values of the direct energy gap and the broadening parameter indicate the possibility to obtain small structural strains and low concentration of defects in the interface region being of importance for the Si technology.

The results show that hydrogenation through PIII shows a potential for growing high quality oxide with lowered structural strains at considerably low Si oxidation temperatures suitable for practical use.

**Acknowledgments**
This work was supported in part by a Research grant from the Technical University Sofia.

## References

[1] Kapil Dev, Jung M Y L, Gunawan R, Braatz R D and Seebauer E G 2003 *Phys. Rev.* B **68** 195311
[2] Vlasenko A I, Gentsar P A and L.A. Demchyna 2005 *Ukr. J. Phys.* **50** 9 997
[3] Lucovsky G and Phillips J C 2004 *Appl. Phys.* A **78** 453
[4] Aspnes D E and Rowe J E 1972 *Phys. Rev.* B **5** 4022
[5] Misiewicz J, Sitapek P and Sek G 2000 *Opto-Electron. Rev.* **8** 1
[6] Aspnes D E 1972 *Phys. Rev. Lett.* **28** 913
[7] http://www.srim.org/
[8] Venger E F, Gorbach T Ya, Matveeva L A and Svechnikov S V 1999 *J. Exper. Theor. Phys.* **89** 948
[9] Grower J W and Handler P 1974 *Phys. Rev.* B **9** 2600
[10] Matveeva L A, Venger E F and Holiney R Yu 1999 *Semicond. Phys. Quantum Electron. & Optoelectron.* **2** 10
[11] Gorbach T Ya, Holiney R Yu, Matiyuk I M, Matveeva L A, Svechnikov S V and Venger E F 1998 *Semicond. Phys. Quantum Electron. & Optoelectron.* **1** 66
[12] Gorbach T Ya, Rudko G Yu, Smertenko P S, Svechnikov S V and Valakh M Ya 1994 *Applied Physics A: Materials Sci. & Processing* **58** 183
[13] Fitch J T, Bjorhmann C H, Lucovsky G, Pollak F H and Yin X 1989 *J. Vac. Sci. Technol.* B **7** 775
[14] Begishev A P, Galiev G B, Kapaev VV and Mokerov V G 1982 *Phys. Tech. Polup.* **16** 426
[15] Szekeres A 1998 in *Fundamental Aspects of Ultrathin Dielectrics on Si-based Devices* Eds Garfuncel E., Gusev E and Vul' A (Kluwer Academic Publishers) NATO Science Series 3 High Technology vol 47 p 65
[16] Szekeres A, Paneva A, Alexandrova S, Lisovskyy I and Litovchenko V 2003 *Vacuum* **69** 355
[17] Venger E F, Holiney R Yu, Matveeva L A and Vasin A V 2003 *Semiconductors* **37** 103
[18] Szekeres A 1991 *Radiat. Eff. and Defects in Solids* **116** 145
[19] Deal B E and Grove A S 1965 *J. Appl. Phys.* **36** 3770
[20] Szekeres A, Alexandrova S, Lytvyn P and Kompitsas M 2005 *J. Phys: Conf. Ser* **10** 246

# High-*k* HfTaO stacks in response to microwave irradiation

**E Atanassova and D Spassov**

Institute of Solid State Physics, Bulgarian Academy of Sciences, Tzarigradsko Chaussee 72, Sofia 1784, Bulgaria

E-mail: d_spassov@abv.bg

**Abstract.** The effect of microwave treatment (6-15 s) at room temperature on the leakage current and mechanisms of conductivity in mixed HfTaO (10 nm) stacks has been studied by temperature dependent (20-100°C) current-voltage characteristics. It was established that the short 6 s irradiation affects the electrically active centers in the stack, provokes modification of the dominant conduction mechanism and improves the temperature stability of capacitors manifesting as low level of current at high temperatures (current decrease up to 4 orders of magnitude at 100°C after the treatment is detected). The traps involved in the conduction process in pre- and post-irradiated capacitors are identified.

## 1. Introduction

The doped high-*k* materials and the composite films of two or more high-*k* films were recognized as the next step in the way of extending the potential of pure high-*k* dielectrics [1]. A doping or mixing-induced improvement of these insulators, allowing further scaling of equivalent oxide thickness, was observed and the mechanism of improvement was considered to be oxygen vacancy compensation by the ions of dopant. The focus on $Ta_2O_5$ (the most attractive high-*k* alternative for nanoscale dynamic random access memories) is currently shifted on $Ta_2O_5$ doped with different elements or mixed with other high-*k* films [2,3]. Recently, [4] we have reported that mixing of $Ta_2O_5$ with $HfO_2$ modifies the traps resulting in a change of the dominant conduction mechanism in comparison to pure $Ta_2O_5$ stack. We have also shown that seconds of microwave irradiation of pure $Ta_2O_5$ [5] at room temperature could be successfully used as an alternative to high temperature annealing steps, and this is a way to avoid some high-temperature processes obstacles as partial crystallization of the films. The big advantage of this treatment is the strongly reduced thermal budget (room temperature, extremely short exposure times). The results imply that the specific manifestation of the effect of microwave treatment is a function of the initial parameters of high-*k* stack, and hence we are motivated here to study the impact of this irradiation on the trap parameters and conduction mechanism in the composite HfTaO films.

## 2. Experimental procedure

Chemically cleaned p-type (100) 15 Ωcm Si wafers were used as substrates. The HfTaO mixed layers were obtained by consecutive deposition of 5 nm $HfO_2$ film on the top of 5 nm $Ta_2O_5$ film. The $HfO_2$ and $Ta_2O_5$ films were deposited by rf sputtering of Hf and Ta targets, respectively, in Ar +10% $O_2$ atmosphere. Ta and Hf targets with purity of 99.99% were used. Optimized conditions [6] guaranteeing $Ta_2O_5$ with high dielectric constant (up to 37) and low leakage currents ($< 10^{-9}$ A/cm$^2$ at

1 MV/cm) were used for the fabrication of the composite films. The wafer temperature during the deposition of both films was 200°C; the working gas pressure was 0.33 Pa and rf power density was 3.6 W/cm$^2$. Details on the deposition conditions are presented in [4]. Post-deposition annealing in $N_2$ at 400 °C for 30 min was performed, in order to mix the two layers. The total film thickness $d$ and the refractive index $n$ were measured by ellipsometry, ($d$ = 10 nm, $n$ = 2.1). MIS capacitors with Al gate (area of $1\times10^{-4}$ cm$^2$) were used as test structures. A post-metal annealing was performed in a forming gas at 450 °C for 1 h. The capacitors were exposed to microwave irradiation in a magnetron (2.45 GHz, power density of 1.5 W/cm$^2$) at irradiation time of $t_i$ = 6-15 s. The temperature of the samples during the exposure was close to the room one. The heating due to the irradiation itself does not exceed ~ 50 °C, and consequently does not lead to some annealing effect. This heating is considered as a feature of the irradiation effect. The irradiation affects neither $d$ nor the refractive index of the films. According to X-ray diffraction analysis pre- and post- irradiated films are amorphous. Temperature dependent (20-100°C) current-voltage (*J-V*) curves were recorded before and after irradiation with a ramp rate of 0.1 V/s using Keithley 236 source meter unit in a dark chamber.

**3. Results and discussion**

Figure 1 compares the *J-V* curves at room temperature before and after microwave treatment. There is no measurable effect of 6 s exposure on the curves. J reduces with increasing $t_i$ from 6 to 10 s for the whole applied field range and further the current does not longer change with increasing $t_i$, i.e. ~ 10 s irradiation is effective enough for the current reduction to the constant value below $10^{-7}$ A/cm$^2$. This confirms our previous results [5] indicating the presence of optimal exposure time to improve the current; the exact value of this optimal time depends on the composition of high-*k* film and d, but in all cases $t_i$ is of the order of seconds. The behavior of the curves for 10-15 s treated stacks (the current is nearly constant up to the appearance of breakdown events) does not enable to invoke any of the commonly observed in high-*k* layers conduction mechanisms. With increasing the temperature the shape of the curves is preserved, only the start of the breakdown events moves to lower applied voltage reaching about -1 V at T ~ 90-100° C. With this in mind we will focus on the *J-V* characteristics of the initial and 6 s irradiated capacitors (figure 2). The asymmetry of the curves observed between forward (negative voltage) and reverse bias (positive voltage) is mainly due to the difference in the barrier heights at Al and Si interfaces, and the saturation of J at reverse biases is assigned to the limited amount of available minority carriers in the space charge region of the substrate. The curves at positive applied voltage are actually an electrical manifestation of the properties of the interfacial layer at Si. The differences in the shape of the right part of the curves in figure 2 suggest a measurable effect of the microwave treatment on the interface layer at Si, which manifests itself as low level of current at high temperatures. The current as low as $10^{-7}$ A/cm$^2$ is

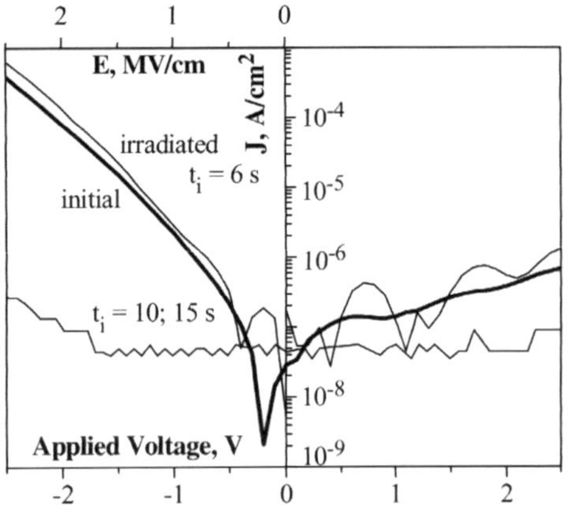


**Figure 1.** Current density as a function of voltage across the stacks before and after microwave treatment. The upper X-axis shows the corresponding values of applied field, E.

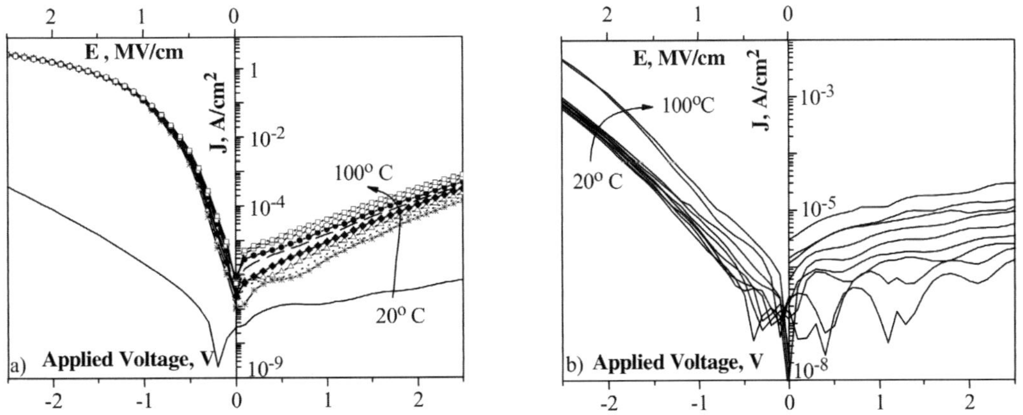


**Figure 2.** Temperature dependent J-V curves for HfTaO stacks before a) and after b) microwave exposure, $t_i = 6$ s.

detected in the very low voltage region around zero bias; this current and the observed current fluctuations are attributed to a transient conductivity which has weak field dependence and can be detected only when the steady state conduction is negligible [7]. So the treatment provides lower $J$ at high temperatures for reversely biased electrode and changes the shape of curves: the saturation level of the current at 100°C is two orders of magnitude lower than the corresponding level of non-irradiated samples. Improved temperature stability after irradiation is also observed at forward biases. The current strongly increases (~ 6 orders of magnitude) in the non-irradiated capacitors as $T$ changes from 20 to ~ 40°C; further the current rise with increasing $T$ to 100°C is negligible. The increase of $J$ is due to temperature stimulated detrapping of electrons and as is seen a thermal activation takes place even at 40°C. The shape of the curves measured at different $T$ is significantly changed after the treatment, indicating radiation-induced modification of the mechanism of conductivity. The change of $T$ from 20 to 100°C increases by an order of magnitude the current of irradiated capacitors but finally $J$ at 100°C is more than 100× smaller than $J$ of the initial samples. Two regions in the curves of irradiated stack can be identified: a gradual increase of the current with the temperature in the low voltage region; when the applied voltage becomes more negative the leakage current is not affected by the temperature in the interval of 20-80°C, and further at T ~ 90-100°C it increases almost 10×. The irradiation lowers the current at high (~80-100°C) temperatures ~ 3 orders of magnitude at -0.5 V applied voltage and keeps a constant current at fields of ~ 2 MV/cm in the temperature interval 20-80°C. The improved temperature stability is generally assigned to the irradiation-induced annealing of both the damaged interface at Al gate (as a result of reaction between Al and the mixed film, quite possible scenario for layer containing $Ta_2O_5$) [7] and the oxygen vacancies in the bulk of high-$k$ layer. The conduction in $Ta_2O_5$ is usually interpreted with Poole-Frenkel (PF) effect, Schottky emission and trap-assisted-tunneling (TAT). To verify the presence of any of these mechanisms, the curves at forward biases for all temperatures have been plotted in PF ($\ln J/E$ vs. $E^{1/2}$) and Schottky ($\ln J$ vs. $E^{1/2}$) scales, where $E$ is the applied field. The electric field in the layer is assumed to be homogeneous and is determined by dividing the applied voltage on the stack by the thickness. $E$ is described in an assumption of absence of trapped charge at the border between high-k layer and interfacial layer at the substrate. PF emission is expressed by the equation:

$$J = C_t E\exp[-q(\phi_t-(qE/\pi\varepsilon_0 r k_r)^{1/2})/kT],$$

where $q$ is the charge of electron; $C_t$ is a trap density related constant; $\phi_t$ is the trap barrier height; $\varepsilon_0$ is the free space permittivity; $k_r$ is the dynamic dielectric constant; $r$ is the compensation factor; $k$ is Boltzmann constant and $T$ is the absolute temperature. The current governed by Schottky process is:

$$J = AT^2\exp[(q^3E/4\pi\varepsilon_0 k_r)^{1/2}/kT],$$

$A = C_{RD}\exp(-\phi_b/kT)$, ($C_{RD}$ is Richardson constant and $\phi_b$ is the Schottky barrier height). Plotting ln($J/E$) vs. $E^{1/2}$ (PF) and ln$J$ vs. $E^{1/2}$ (Schottky emission) should lead to a straight line, from the slope of which the value of the dynamic dielectric constant, $k_r$, ($k_r = n^2$) is derived. Since $n \sim 2.1$ before as well as after irradiation, $k_r$ should be ~ 4.4 In the case of modified PF conduction, the effect of compensating traps on the curves is presented by the compensation factor, r, ($1 \leq r \leq 2$). As far as the precise modelling of the PF and Schottky conduction is beyond the aim of this work, the standard formulas were used having in mind that the charging effects can modify slightly the values of $r$ and $k_r$. The conduction mechanisms was also investigated by plotting the data in TAT process scale, $J \sim \exp(V)$. This equation describes two-step tunneling of injected electrons via trap. The mechanisms of conductivity of the initial films were discussed in [4]. At room temperature, the TAT process dominates at low applied fields, and at $E > 1$ MV/cm the experimental results are fitted very well by both PF emission with compensation (r = 1.9) and TAT, and it is not possible to give preference to one of them. With the increase of $T$ the contribution of PF mechanism ($r$ = 1.6) to the conductivity increases and at high temperatures it fully governs the current at E of 0.7-1.3 MV/cm; at lower fields TAT dominates. The results for irradiated capacitors are as follows: the modified PF process controls the leakage current at fields of ~1-2.8 MV/cm of all the curves of irradiated capacitors, but contrary to the fresh stack, with temperature-dependent compensation factor: $r$ = 1.8 at 20°C and further it decreases to 1.45 with increasing T to 100°C (figure 3a). The irradiation does not influence the mechanism of conductivity at E ≥ 1 MV/cm – it is PF emission with compensation. As a result of the treatment, however, the proportion between the density of compensation and donor centers

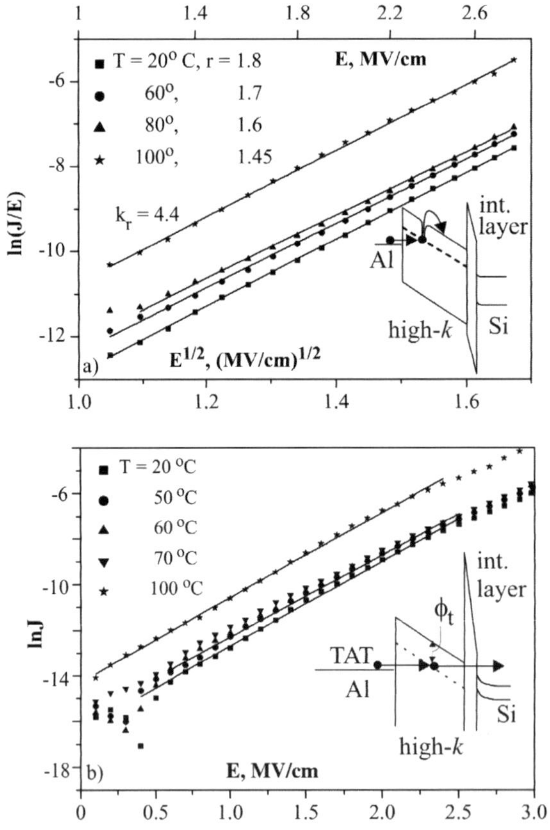


**Figure 3.** PF plots (a) and TAT plots (b) of J-V curves at different T of post-irradiated stacks, $t_i$ = 6 s.

participating in PF process changes with the temperature. The modification of the mechanism towards weak compensation with increasing $T$ means a decrease of the part of acceptors available for compensation, i.e. the microwave treatment affects the electrically active defects in the composite films, and the high temperature facilitates the manifestation of this irradiation-induced trap modification. The slopes of the Schottky plot (not shown) of irradiated capacitors for all $T$ are not consistent with Schottky effect since the values of $k_r$ do not agree with $n$. TAT, however, describes also very well all curves in the interval from 20 to 100°C, (figure 3b). So, two mechanisms of conductivity can be indentified for irradiated samples: TAT at a wide field range (0.4-2.5 MV/cm) and PF with compensation at fields of 1-2.8 MV/cm. In the range of $E \sim$ 1-2.5 MV/cm the current fits well PF as well as TAT mechanisms and it is no possible to detect transition from one to another mechanism, suggesting a regime PF+TAT exists. The energy location $\phi_t$ of traps for PF emission and activation energies $E_a$ for TAT were obtained from the Arrhenius plots of $J$ (figure 4) at different applied voltages across the stack, $V_{stack}$. $\phi_t$ of traps accounting for PF emission is ~ 0.32 eV for both the initial and the irradiated films, implying that the same traps are involved in the conduction. The trap energy shows, however, field dependence and the value of 0.31 eV is obtained at $V_{stack}$ = -1.5 V after irradiation. This correlates with irradiation provoked modification of the transport mechanism at -1.5 V from the conductivity implying contribution of different mechanisms before irradiation to PF effect with compensation after microwave exposure. At $V_{stack}$ = -0.9 V the PF conductivity with compensation for non-irradiated samples turns into TAT for the irradiated ones. TAT operates for fresh and irradiated samples at E up to 0.7 ~ 1 MV/cm, with $E_a$ which is field dependent. Deep traps, $E_a$ = 0.32; 0.33 eV define the conductivity before and after irradiation at very low applied voltages (-0.2 V). The conductivity is via the traps with $E_a$ = 0.13 eV and with 0.32 eV in the initial and irradiated stacks, respectively at $V_{stack}$ = -0.5 V, i.e. irradiation tends to create deeper traps. In some cases the trap energy level obtained for PF emission (~0.3 eV) is comparable to the activation energy of TAT. This means that the same bulk traps which govern the current through PF effect at high fields above 1 MV/cm participate also in the lower field conductivity but through TAT, as is illustrated in

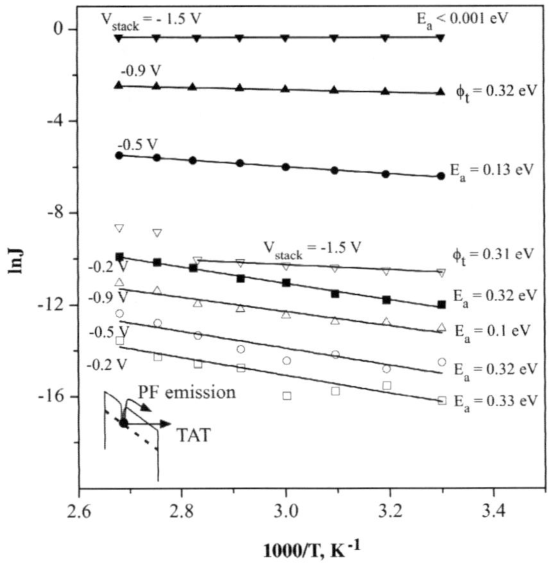


**Figure 4.** Arrhenius plots for initial and irradiated capacitors. Solid symbols represent data before and open symbols after microwave irradiation. The energy depth of electron traps and the activation energies from experimental fits are given.

the inset of figure 4. The activation energy of 0.32; 0.33 eV is very close to the energy of the localized Ta-d state acting as an electron trap at ~ 0.3 eV above the Si conduction band edge of Ta-aluminates [8]. The assignment of the activation energy of ~ 0.3 eV to trapping into Ta d-states, however, works only if one assumes that the energies of Ta-introduced localized d-states are independent of the specific dielectric composition ($HfO_2$-$Ta_2O_5$ mixed layer in our case instead of Ta-aluminates in [8]).

The traps with $\phi_t \sim 0.3$ eV could be also Si/O vacancy complex single donor [9]. The irradiation initiates a change of the conduction at fields about and above 1 MV/cm ($V_{stack}$ = -0.9 and -1.5 V): it modifies the PF mechanism into TAT with $E_a = 0.1$ eV in the former case ($E \sim$ 1MV/cm) and the fairly temperature insensitive current into the PF effect controlled current in the last case ($E > 1$ MV/cm). The constant current region ($T \sim$ 35-100°C) means field ionization of electrons (process is essentially independent of the temperature) into the conduction band presumably from the same centers as for Poole-Fenkel conduction (field-enhanced thermal excitation). Therefore, the 6 s irradiation suppresses PF emission at low fields and stimulates the tunneling process; most likely the same traps participate in both processes. There is not evidence, however, for irradiation-induced traps annealing leading to current reduction (figure 1) unlikely the case of longer treatment (10-15 s). Most likely 6 s treatment provokes changes of the local environment of defects manifesting as electron traps and this results in a change in their energy position. Since the current reflects mainly the changes in the cathode electric field it seems that the traps responsible for conductivity are localized in the dielectric near the Al-gate interface. Previously [7,10] we have reported that these defects can seriously compromise leakage and reliability characteristics of $Ta_2O_5$ and Ti-doped $Ta_2O_5$ capacitors. Generally, the microstructural nature of the defects is weak and strained Ta-, Si- and Hf-related bonds; the presence of gate-induced defects as a result of reaction between Al and high-*k* layer [6] is also possible. The complementary information here is that these defects are affected by microwave treatment, and the strongly reduced leakage after 15 s exposure could be addressed to their annealing.

## 4. Conclusion

The short term (~ 6 s) microwave treatment at room temperature changes the dominant conduction mechanism(s) in HfTaO based capacitors: PF conduction turns into TAT at fields < 1 MV/cm and modifies towards weaker compensation at higher fields. The radiation-modified conductivity is accompanied by better stability of the irradiated capacitors at elevated temperatures. Therefore, a short term microwave irradiation may potentially serve as an engineering solution to current stability at high temperatures in these high-*k* stacks. Although, the exact mechanism of microwave treatment effect on the electrically active centers is not addressed in this work it can be assigned to some kind of reconstruction of the defects in both the bulk of high-*k* layer and the interfacial region during the exposure. Longer treatment (10-15 s) strongly reduces the current down to 3-4 orders of magnitude indicating that by using an appropriate exposure time the leakage current in the stacks could be kept low enough.

**Acknowledgments**

The work is supported under projects DTK02/50 (Bulgarian National Science Foundation) and BK09/08 (Institute of Solid State Physics).

## References

[1] Intern. Techn. Roadmap for Semicond. (ITRS), http://publick.itrs.net
[2] Lu J, Kuo Y, Tewg J Y, 2006 *J. Electrochem. Soc.* **153** G410
[3] Atanassova E, Spassov D, Paskaleva A, Georgieva M, 2008 *Thin Solid Films* **516** 8684
[4] Atanassova E, Georgieva M, Spassov D, Paskaleva A, 2010 *Microelectron Eng.* **87** 668
[5] Atanassova E, Mitin V F, Konakova R V, Spassov D, Shinkarenko V V, 2008 *Semicond. Sci. Technol.* **23** 035004
[6] Atanassova E, 1999 *Microelectron. Reliab.* **39** 1185
[7] Spassov D, Atanassova E, Virovska D, 2006 *Appl. Phys. A* **82** 55
[8] Johnson R S, Hinkle C, Hong J G, Lukovsky G, 2002 *Solid St. Electronics* **46** 1799
[9] Paskaleva A, Spassov D, Atanassova E, 2007 *J. Phys. D: Appl. Phys.* **40** 6709
[10] Paskaleva, Atanassova E, Novkovski N, 2009 *J. Phys. D: Appl. Phys.* **42** 025105

Journal of Physics: Conference Series **253** (2010) 012039 doi:10.1088/1742-6596/253/1/012039

# Laser - induced modification of hydrogenated amorphous carbon films

**Savcho Tinchev**[1], **Petranka Nikolova**[1], **Evgenia Valcheva**[2], **Yuliana Dyulgerska**[1] **and Elitsa Petrova**[1]

[1] Institute of Electronics, Bulgarian Academy of Sciences, 72 Tzarigradsko Chaussee Blvd., 1784 Sofia, Bulgaria

[2] Faculty of Physics, Sofia University, 5, blvd. J.Bourchier, Sofia-1164, Bulgaria

Corresponding author's e-mail address: stinchev@ie.bas.bg

**Abstract**. Laser-induced modification of thin films of amorphous carbon (a-C:H) was investigated. The films have been deposited by DC PECVD on silicon substrates using different bias voltages: 1 kV, 1.5 kV, 2 kV, and 2.5 kV. Raman spectra of the films measured in a micro-Raman system at 633 nm with different laser power were recorded. It was observed an increased D-peak for laser power density of $10^{10}$ W/m$^2$ indicating modification of the film by the laser, which does not happened for smaller laser power. This modification is stronger in the films produced at higher bias voltages, which have stronger absorption at 633 nm. This observation suggests that laser modification of a-C:H films is due to thermal effects.

## 1. Introduction

Hydrogenated amorphous carbon films (a-C:H) possess interesting physical and chemical properties [1-3]. Therefore there is a lot of interest of their investigation and applications. The most common fabrication process for a-C:H films is plasma-enhanced chemical vapour deposition (PECVD). The macroscopic properties of the resulting a-C:H films depend on the deposition energy. Changing only the bias voltage in a DC PECVD system it is possible to change the microstructure and thus the optical properties of the fabricated films. In this way we succeeded in fabrication of solar absorber matched to the solar spectrum [4].

During our work on solar absorber we noticed that these films are easily modified by the laser irradiation during the Raman spectra measurements. Actually such effect was observed earlier [5]. In this paper we report laser-induced modification behavior of a-C:H films fabricated at different bias voltages in a DC PECVD system.

## 2. Experimental

Hydrogenated amorphous carbon films (a-C:H) were deposited in a DC plasma CVD reactor. Benzene diluted with argon was used in this process. The films were deposited on silicon substrates placed upon the water-cooled cathode. The metal vacuum chamber acted as anode. Benzene vapors were obtained by boiling of benzene at constant temperature of 25°C. Before the film deposition the substrates were cleaned in acetone, ethanol and deionized water in ultrasonic bath followed by a 10 min Ar ion sputtering at 1 kV. Four series of samples were deposited at different bias voltages: 1 kV, 1.5 kV, 2 kV, and 2.5 kV.

Raman spectra of the samples were measured with a LabRAM HR spectrometer at 633 nm in the near backscattering geometry. The diameter of the laser spot on the sample is about 1 μm and the spectral resolution is 1 $cm^{-1}$. The laser power was 9 mW and set of filters were used in order to reduce the laser power on the sample surface.

## 3. Results

Figures 1÷4 show Raman spectra of the films deposited at different bias voltages. For each voltage two Raman spectra are given – one at full laser power of 9 mW and another one taken with 10% transmission filter. All films measured with this filter (i.e. at laser power 0.9 mW) showed no modification. This was concluded from measurements with filters with higher attenuation, which give practically the same Raman spectra as the spectra taken with 10% transmission filter. All samples – figures 1 ÷ 4 (deposited at 1 kV, 1.5 kV, 2 kV, and 2.5 kV) showed increased D-peak at the full laser power. This is an indication for increased graphitic clusters dimensions [6] caused by the laser power modification.

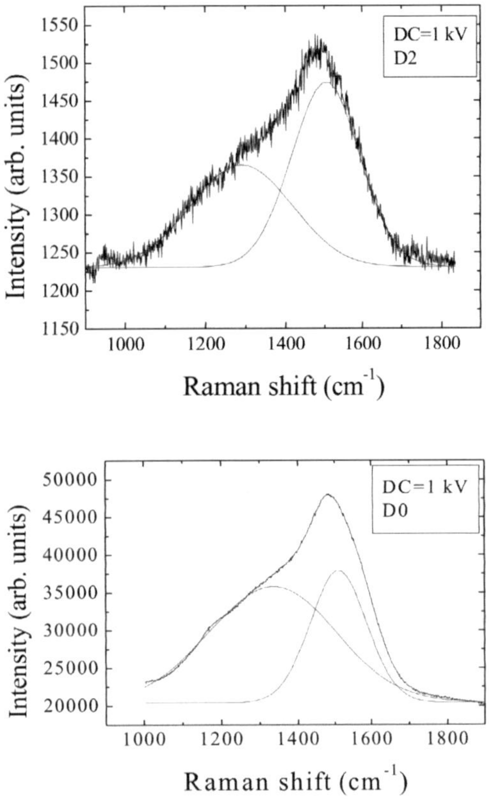


**Figure 1.** Raman spectra of a-C:H films deposited at 1 kV bias voltage and measured without filter (D0) and with 10% transmission filter (D2).

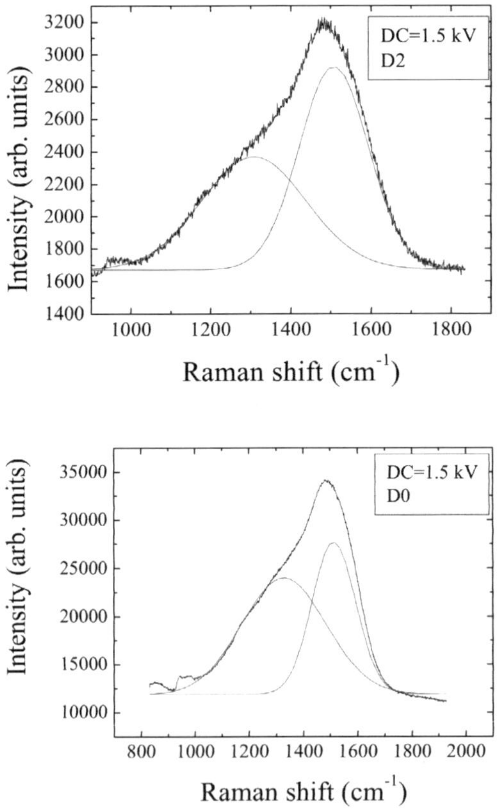


**Figure 2.** Raman spectra of a-C:H films deposited at 1.5 kV bias voltage and measured without filter (D0) and with 10% transmission filter (D2).

Journal of Physics: Conference Series **253** (2010) 012039 doi:10.1088/1742-6596/253/1/012039

We were not able to determine the threshold laser power, where the modification starts for each film. However, it could be noticed that the films deposited at higher bias voltages were easier to modify. This is understandable because they have much higher absorption at this laser wavelength. Nevertheless an estimation using the laser spot diameter of ~ 1 μm gives a laser power density of $10^{10}$ W/m$^2$, in good agreement with other published data [5].

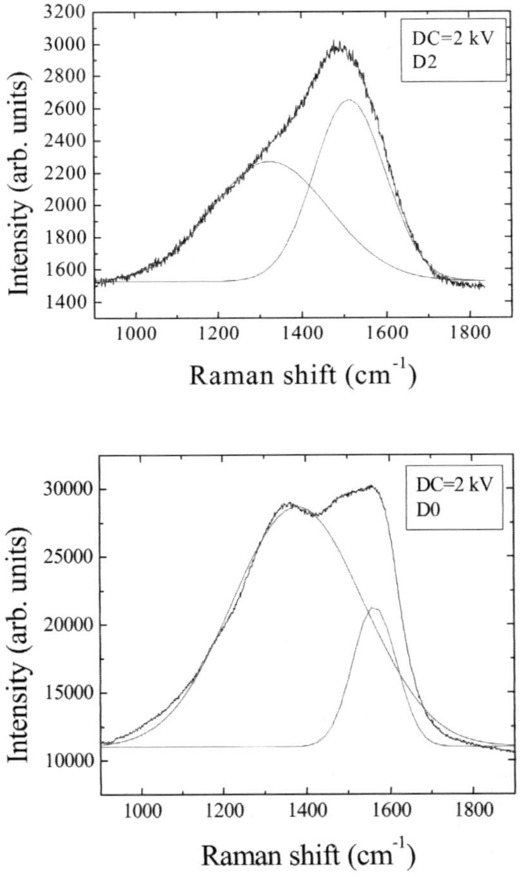


**Figure 3.** Raman spectra of a-C:H films deposited at 2 kV bias voltage and measured without filter (D0) and with 10% transmission filter (D2).

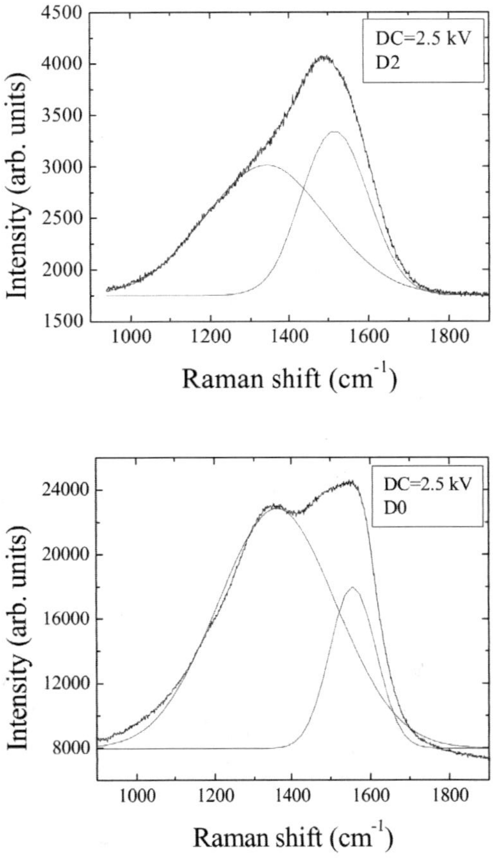


**Figure 4.** Raman spectra of a-C:H films deposited at 2.5 kV bias voltage and measured without filter (D0) and with 10% transmission filter (D2).

## 4. Discussion

The parameters of the measured Raman peaks (position, width and their intensity ratio) were used for the characterization of the films. The interpretation and evaluation of the Raman spectra was first based on the decomposition of the curve into two components. Later decomposition into four peaks was also used. The Raman spectra of all our films measured at filtered (reduced) laser power showed a peak at 1510 cm$^{-1}$ typical for a hydrogenated amorphous carbon (a-C:H), which however, is different from the pure G-peak observed in graphite. We observed also the well-known peak at about 1360 cm$^{-1}$ referred to as the D-peak associated with in-plane vibrations and resulted from structural imperfections of the material. In all films the I(D)/I(G) ratio was about 1 or smaller. After Raman measurement at full laser power a significant increase in I(D)/I(G) ratio was observed for all samples (figure 5). The increase in I(D)/I(G) is an evidence for the development of graphitization in the films. It was also found a shift in a G-peak position from 1510 to 1573 $cm^{-1}$, an effect, which can be explained with the films becoming more graphite-like.

These effects are probably due to the high absorption in the films inducing a very high local temperature, which cannot be dissipated quickly enough by the substrate, leading to graphitization and even to material ablation. The power dissipated by the substrate can be estimated by the simple thermal conductivity equation $Q = kA(T_2\text{-}T_1)/d$, where $k$ is the thermal conductivity of the silicon substrate, and $A$ is the area of the laser spot. The temperature $T_1$ = 300 K is the room temperature and as $T_2 \approx$ 4000 K sublimation temperature of graphite is used. Because the thermal conductivity of the silicon is temperature dependent a value of 98.9 W/(m.K) at 400 K was used. Thermal flux through the carbon was neglected since the thermal conductivity of amorphous carbon is small (about 0.2–0.3 W/m.K) [7]. Inserting these values an estimated power of about 1 mW was found which is in good agreement with the experimental data. Obviously even some mW of the laser power absorbed in the film cannot be dissipated by the substrate leading to temperature increase, graphitization and even to material ablation. We consider this laser power as threshold power above which the amorphous carbon will be modified. Of course the exposition time and the film thickness are also important, but we did not investigate their influences in details.

When no filter was used, we did not notice a reduced intensity of the spectrum of the films deposited at 1 kV, 1.5 kV and 2 kV, which could be possible explanation of laser ablation and reduction in the volume of scattering material in the irradiated area. Thus, it appears that this will happened at higher laser intensities. However, in the 2.5 kV sample one can notice such a reduction, i.e. laser ablation already occur.

Comparisons of the Raman spectra with those of thermally annealed samples [8, 9] suggest that the observed modification is primarily due to local temperature effects. In the annealing experiments as the temperature increased, the Raman spectra of annealed DLC films show a narrowing of the G – peak bandwidth and a G – peak position shift toward the graphitic peak at 1590 $cm^{-1}$. Furthermore, increasing the annealing temperature increases the D-peak intensity and the intensity ratio I(D)/I(G). Similar effects could be observed in the samples modified by the Raman laser. In general, we can conclude that the increase of the bias voltages of our samples produces similar structural changes as the increase of annealing temperature. This can be explained by the increased absorption of these films at the laser wavelength. Although our results show decrease of D – peak and G – peak intensities with the bias voltages, Raman spectra of our films obtained with 1 – 1.5 kV and 2 – 2.5 kV are similar to those of in-situ annealed with atomic hydrogen DLC films at about 400°C and 700°C, respectively [9, 10].

As mentioned already decomposition into four peaks was also used. This was presented (figure 6 and figure 7) for films deposited at 2 kV and 2.5 kV. The main peaks are D-peak at about 1535 $cm^{-1}$ - disorder peak, dependent on physical size of the $sp^2$ aromatic domains and G - disorder peak (random-network clusters). Additional peaks: D' at 1220÷1300 $cm^{-1}$ and $G_2$ at about 1585 $cm^{-1}$ [8, 9] could be identified [8, 9]. D'-disorder peak was also observed on the Raman spectrum of heavily ion-implanted highly oriented pyrolytic graphite [11]. $G_2$ peak is ordered graphite [12].

## 5. Conclusions

Laser-induced modification of amorphous carbon (a-C:H) films during the Raman spectra measurements was investigated. Our results confirmed earlier observations that a film modification can occur if proper care is not taken when using Raman spectroscopy and as a result "real" Raman spectra of materials cannot be obtained. The threshold power density above which the amorphous carbon is modified was estimated to be about $10^{10}$ $W/m^2$. This modification is stronger in the films produced at higher bias voltages, which have stronger absorption at laser wavelength 633 nm. Our observation suggests that laser-induced modification of a-C:H films is due to thermal effects.

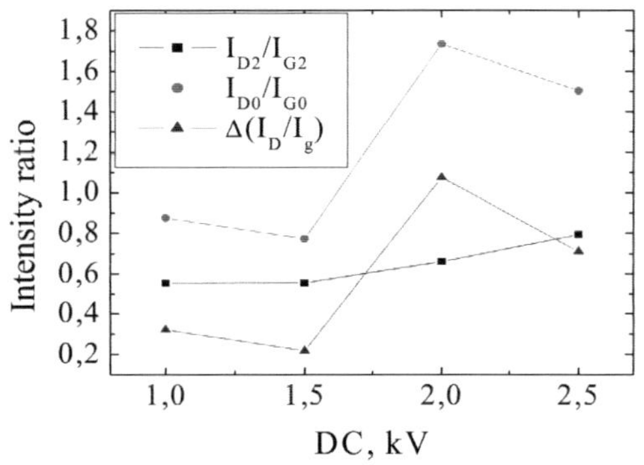


**Figure 5.** Dependence of the ratio of D and G peak intensities on bias voltage before (D2) and after (D0) laser annealing

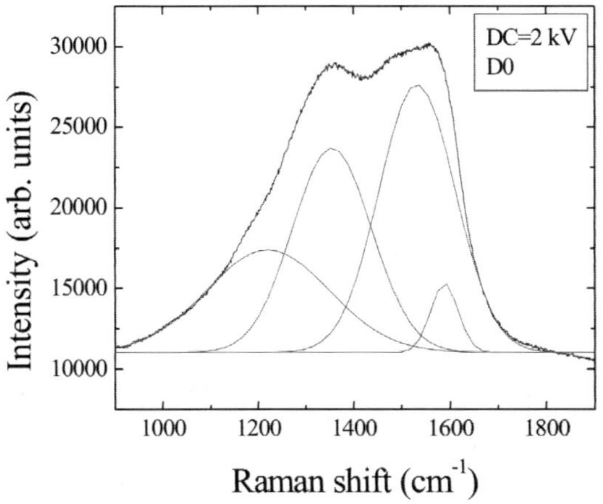


**Figure 6.** Raman spectra of a-C:H films deposited at 2 kV bias voltage and measured without filter (D0) fitted with four peaks.

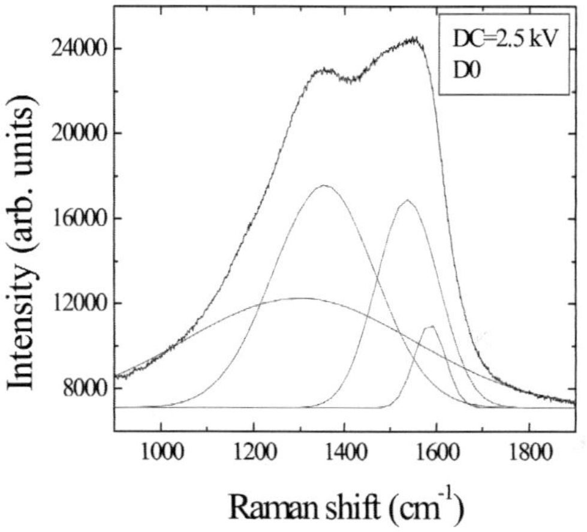


**Figure 7.** Raman spectra of a-C:H films deposited at 2.5 kV bias voltage and measured without filter (D0) fitted with four peaks.

**References**

[1] Robertson J 2002 *Materials Sci. and Engineening* R **37** 129

[2] Grill A 1999 *Diamond and Related Materials* **8** 428

[3] Buijnsters J G, Camero M, Vázquez L, Agulló-Rueda F, Gómez-Aleixandre C and Albella J M 2007 *Vacuum* **81** 1412

[4] Tinchev S, Nikolova P, Dyulgerska J, Danev G and Babeva Tz 2005 *Sol. Energy Mater. Sol. Cells* **86** 421

[5] Lamberton R, Morley S, Maguire P and McLaughlin J 1998 *Thin Solid Films* **333** 114

[6] Ferrari A and Robertson J 2000 *Physical Review* B **61** 14095

[7] Shamsa M, Liu W L, Balandin A A, Casiraghi C, Milne W I and Ferrari A C 2006 *Applied*

*Physics Letters* **89** 161921
[8] Oral B, Huert R, Muller U and Ernst K H 1995 *Diamond & Related Materials* **4** 482
[9] Yang P, Chen J Y, Leng Y X, Sun H, Huang N and Chu P K 2004 *Surface & Coating Technology* **186** 125
[10] Cheng C-L, Chia C-T, Chiu C-C, Wu C-C and Lin I-N 2001 *Diamond & Related Materials* **10** 970
[11] Elman B S, Shayegan M, Dresselhaus M S, Mazurek H and Dresselhaus G 1982 *Phys. Rev.* B **25** 4142
[12] Huong P V 1991 *Diamond & Related Materials* **1** 33

# Characterization of reactive sputtered $TiO_2$ thin films for gas sensor applications

**Stefan Boyadzhiev**[1, 2], **Velichka Georgieva**[1] **and Milka Rassovska**[2]

[1] Institute of Solid State Physics, Bulgarian Academy of Sciences, 72 Tzarigradsko Chaussee Blvd, 1784 Sofia, Bulgaria
[2] Department of Microelectronics, Technical University of Sofia, Kl. Ochridski Blvd. 8, 1756 Sofia, Bulgaria

Email: stefan_b@mail.bg

**Abstract**. The technology for preparing and characterisation of titanium dioxide ($TiO_2$) thin films with properties appropriate for usage as gas sensors are discussed. For preparing the samples the methods of reactive radio frequency (RF) and direct current (DC) magnetron sputtering were used. The composition and microstructure of the films were studied by XPS, EPMA, XRD and Raman spectroscopy and the surface of the films was observed by high resolution SEM. Thorough profile analyses on the structure changes were performed by XPS. Interactions with the substrate and changes between the different modifications of the crystal structure also were investigated. For measuring the thickness and to identify the refractive indices of the films laser ellipsometry was used. The research was focused on the sensing behaviour of the sputtered $TiO_2$ thin films. Films of various thickness were deposited on quartz resonators and the quartz crystal microbalance (QCM) method was used. This enables highly sensitive gas sensor capable of detecting changes in the molecular range to be constructed. Prototype QCM sensors with $TiO_2$ sensitive films made in our laboratory, showed good sensitivity to ammonia at room temperature, and are currently being tested for sensitivity to other gasses.

## 1. Introduction

Titanium dioxide attracts attention due to its remarkable chemical, electrical and optical properties such as large energy gap, high refractive index and dielectric constant, excellent chemical stability, nontoxicity, good mechanical hardness and high transparency in the visible and near infra-red range. [1,2] Additionally $TiO_2$ exhibits excellent photocatalytic and superhydrophilic properties [3-5] and is a promising material for humidity, chemical and gas sensors, as it shows good adsorption of nitric oxides, ammonia, CO, $H_2S$ and many organic compounds including alcohols, hydrocarbonic and aromatic gases and many other [6-11].

Various techniques are employed for deposition of $TiO_2$ thin films, such as electron beam evaporation [1], various forms of sputtering [1-2,7,9,11], sol–gel [4], chemical vapor deposition (CVD) [5], pulsed laser deposition (PLD) [12] and others [1,8]. Wide variations were found in the films' properties, even among films produced under nominally the same deposition conditions [1]. Among these techniques, reactive sputtering has gained significant importance because it enables control of the structure, composition and properties of $TiO_2$ films by adjusting the deposition conditions. This results in high quality thin films with thickness uniformity over large areas and well controlled stoichiometry. Moreover, sputtering methods are widely used in industrial processes because high quality films can be obtained at low substrate temperatures. These advantages, as well as the possibility of building QCM sensors with $TiO_2$ thin films with only one additional operation encouraged us to use this method in the present research. The technological parameters were tuned to obtain films with optimal properties at different substrates – silicon and quartz and PZT resonators with gold and silver electrodes. The control silicon samples were used for XPS, XRD, EPMA, Raman and ellipsometry analysis. The resonators were used for preparing sensors based on the QCM method. QCM is a well-established tool for monitoring the adsorption of nano-amounts. This means that QCM sensors are capable of measuring mass changes as small as a fraction of a monolayer or a single layer of atoms. The high sensitivity and the real-time monitoring of mass changes on the sensor crystal make QCM a very attractive technique for gas sensors. Compared with others sensors, the advantages of QCM are: simple technological implementation, capability of operating at room temperature, good sensitivity and chemical process reversibility [13].

## 2. Experimental

For both RF and DC sputtering depositions titanium targets (purity 99.99%) in the presence of oxygen as reactive gas were used. The films were deposited on unheated substrates. The influence of technological conditions during deposition, such as the oxygen partial pressure and deposition time, on the films' structure and properties was studied. $TiO_2$ films of various thickness (30 nm – 120nm) were deposited on resonators and silicon wafers. The deposition time was between 30 and 120 min.

The composition of the films as well as thorough profile analyses on the structure changes were performed by Ulvac-Phi "Scanning ESCA Microprobe Quantum2000" XPS system. The composition of the films was studied also by Shimadzu "Electron Probe Microanalyzer EPMA-1610". The surface of the films was observed by a high resolution FE-SEM (Field Emission SEM Hitachi S-4800). The structural properties of the films were characterized using Philips "X-Pert-MRD" XRD system and Raman spectroscopy study performed by "SPEX 1403" Raman double spectrometer. The thickness and refractive indices of the films were measured by multiangle four zone null ellipsometry. More detailed information about the RF sputtering deposition technology and conditions as well as the Raman spectroscopy and laser ellipsometry apparatus can be found in our previous publication [14].

To determine films' sensing properties first adsorption of ammonia was tested using a special laboratory constructed measurement system. Most of the test sensor devices were based on 8-mm polished AT-cut quartz plates with gold electrodes (diameter 4 mm and thickness 120 nm) evaporated on both sides. QCM resonance frequency around 14 MHz was thus obtained. The piezoelements thus prepared were covered with $TiO_2$ thin films on both sides. Equivalent dynamic parameters of the QCM, including the static capacitance, $C_0$, and the equivalent dynamic resistance, Rq, were measured by a Selective Level Meter. Other parameters, as the dynamic capacitance, Cq, the dynamic inductance, Lq, and the quality factor, Q, were calculated [15]. The sorption properties of $TiO_2$ were defined from the frequency-time characteristics (FTC) when saturation over aqueous ammonia solutions for each concentration investigated was achieved. To eliminate the influence of $H_2O$ molecules on the $NH_3$ sorption, each sample was first saturated with water vapor preceding the measurement over the ammonia solution. This frequency was taken as an initial value for subsequent measurements. Besides being highly sensitive to mass changes, the QCM frequency is also sensitive to temperature variations. To eliminate temperature influence on the resonant QCM frequency, f, the piezoelements were prepared on thermostable AT-cut quartz plates. The temperature was maintained at 25 °C ± 0.5 °C during the experiments. The QCM resonant frequency shift, Δf, was measured and the sorbed mass was calculated by Sauerbery's equation, which gives a linear relationship between the Δf and the mass adsorbed on the electrode surface [16].

$$\Delta f = \frac{-2.26 \times 10^6 \times f^2 \times \Delta m}{S_\perp} \qquad (1)$$

According to the Sauerbery's equation developed for an AT-cut quartz crystal (1), for each value measured of Δf; the $NH_3$ mass sorbed (Δm) can be calculated when one knows the initial frequency f in MHz, the measured Δf in Hz and the area of the electrodes (S⊥) in $cm^2$. For our experiments S⊥ = 0.1256 $cm^2$. The experimental system and the methodology of the measurement are described in detail in [17].

## 3. Results and discussion

The XRD investigations showed that both RF and DC sputtered $TiO_2$ films are predominantly amorphous. These results were also supported by the Raman spectroscopy study. But the XPS profile analyses showed that the structure changes between the surface and the bulk. In both RF and DC sputtered film profiles we observed close to the substrate very thin but significant metallic titanium phase followed by growth of close to stoichiometric amorphous titanium dioxide. The films are predominantly amorphous, with weak lines showing crystalline structure (both $TiO_2$ and TiO) throughout the film, almost without any significant changes in composition. However on the surface we observed clear crystalline structure. The results are shown in figures 1 and 2. Both profiles were made at sputtering energy 2 kV with 1 min sputtering intervals. The sputtering conditions and film thickness are similar. We observed slightly better expressed crystalline phase in the bulk for the RF sputtered films but there is no significant difference.

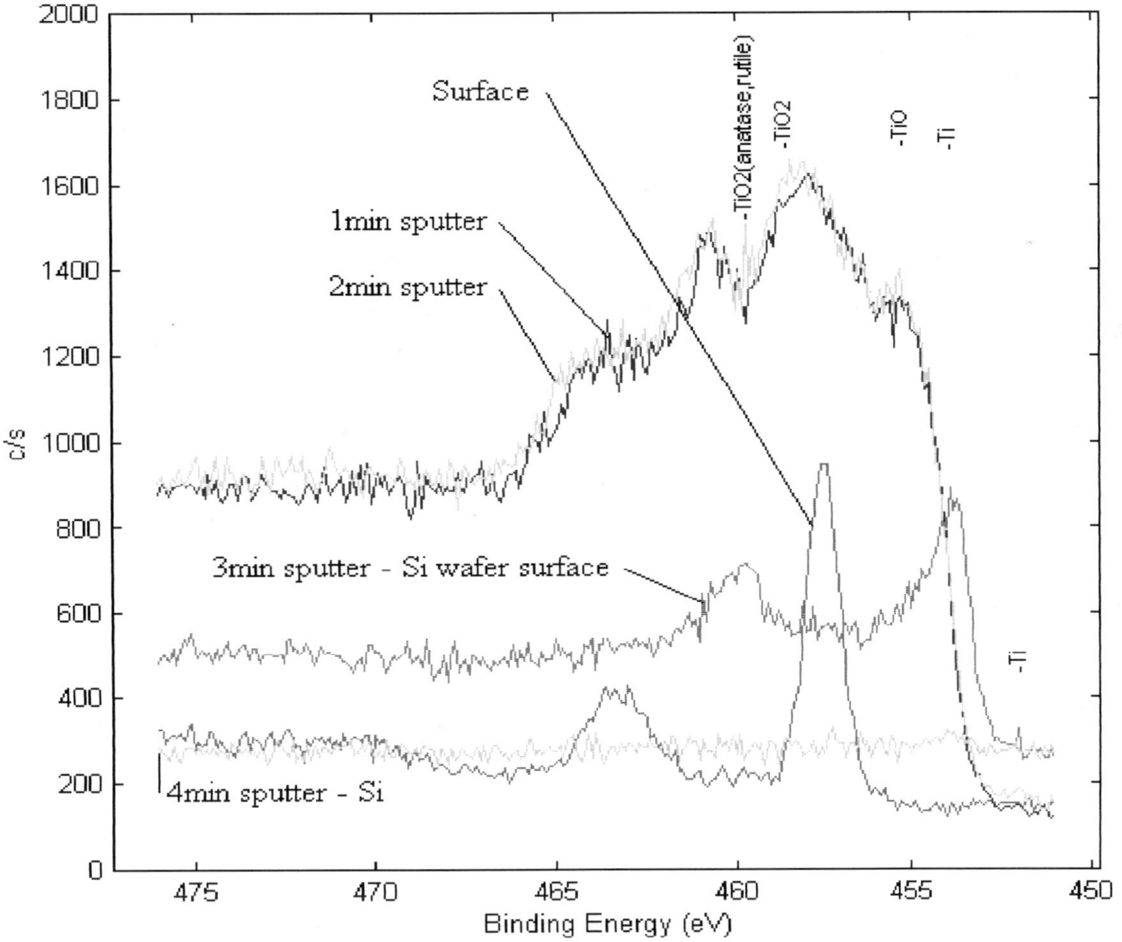


**Figure 1.** XPS profile of 50nm thick RF sputtered $TiO_2$ thin film.

Journal of Physics: Conference Series **253** (2010) 012040 doi:10.1088/1742-6596/253/1/012040

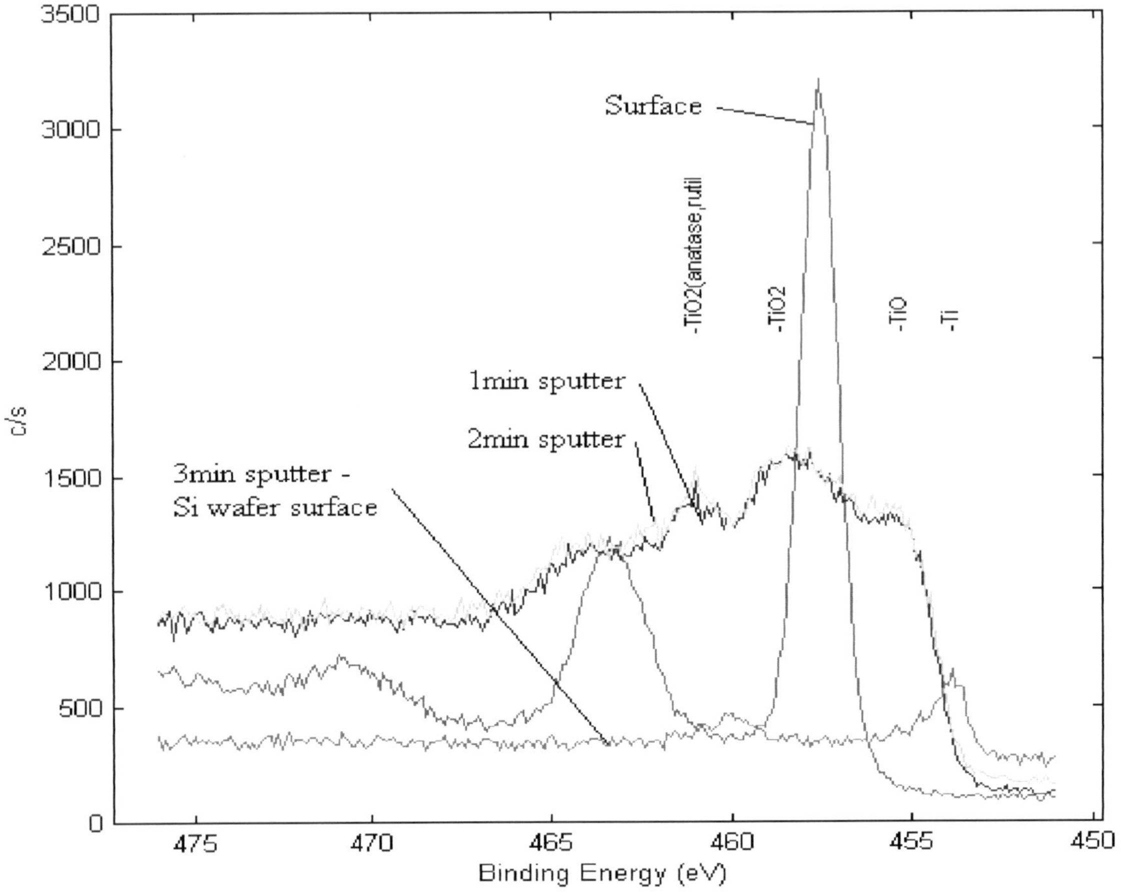


**Figure 2.** XPS profile of 50nm thick DC magnetron sputtered $TiO_2$ thin film.

The composition study results for the same films are presented in table 1.

**Table 1.** XPS composition study of RF and DC magnetron sputtered $TiO_2$ thin films

| Sputtering | RF sputtered film | | DC magnetron sputtered film | |
|---|---|---|---|---|
| | Ti [at. %] | O [at. %] | Ti [at. %] | O [at. %] |
| 0 (surface) | 24.2 | 75.8 | 34.0 | 66.0 |
| 1 | 35.8 | 64.2 | 35.8 | 64.2 |
| 2 | 36.0 | 64.0 | 35.4 | 64.6 |
| 3 | 29.6 | 70.4 | 33.2 | 66.8 |

It was not possible to corroborate the measured by XPS composition with EPMA because the films are too thin and there is an inaccuracy in the oxygen concentration caused by additionally detected oxygen from the $SiO_2$ film of the substrate surface. The refractive indices of the films are around 2,32 (at wavelength 632.8 nm) and also support a composition close to stoichiometric $TiO_2$. To investigate the surface, which is considered to be the most important region for the sorption properties of the films, we applied an ultra-high resolution FE-SEM. The images are shown in figure 3. The films are homogeneous and uniform. They are amorphous or with very small nano-sized grains of a few nanometers. The object shown in the images was only used to enable improved focusing of the microscope.

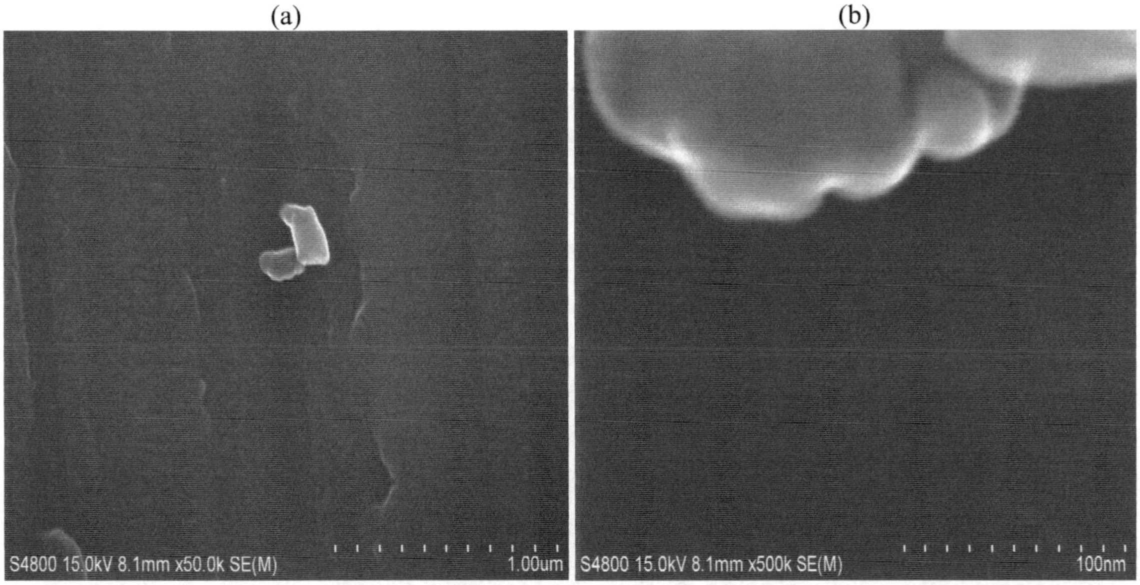


**Figure 3.** FE-SEM images of DC magnetron sputtered $TiO_2$ thin film magnified 50 000 (a) and 500 000 (b) times.

The present research was focused on the sensing behavior of the sputtered $TiO_2$ thin films. Prototype QCM sensors with other transition metal oxides sensitive films were also made by our team [18] and showed good sensitivity to ammonia. Similar results for good sensitivity to $NH_3$ of the $TiO_2$ thin films were observed. Figure 4 shows the sorption properties of a QCM coated on both sides with 50 nm $TiO_2$ thin film when exposed to water vapour.

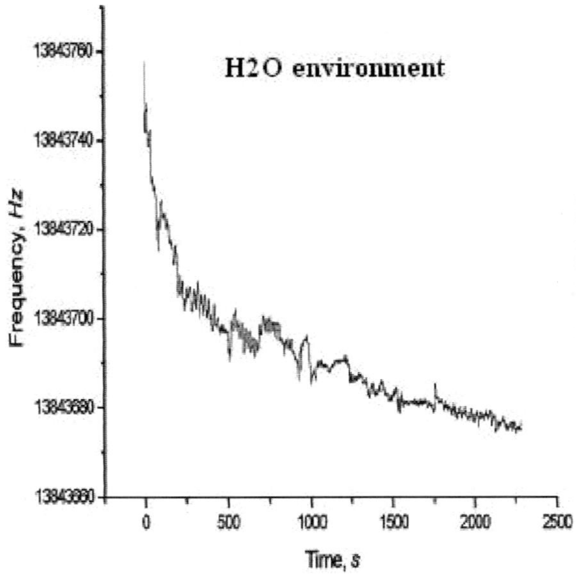


**Figure 4.** FTC of QCM with RF sputtered $TiO_2$ thin film during exposure to water vapour.

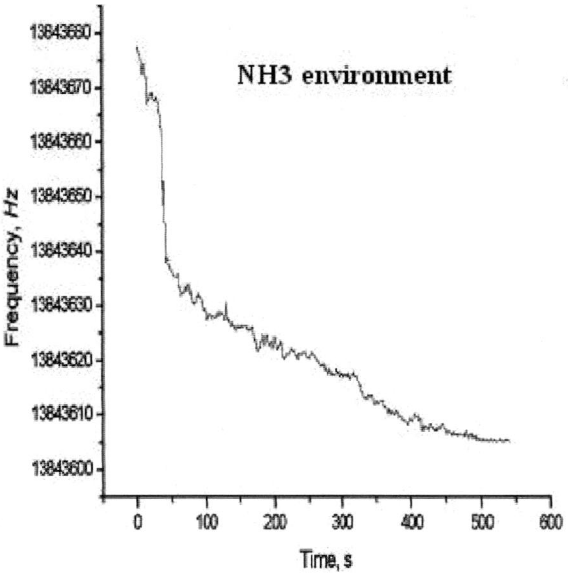


**Figure 5.** FTC of QCM with RF sputtered $TiO_2$ thin film over 1000 ppm aqueous $NH_3$ solution.

The behavior of the curve indicates that during the first 30 s the structure is abruptly loaded, which results in a rapid frequency decrease. During the following 150 s the curve falls more smoothly and after 200 s a dynamic equilibrium is gradually established between the number of sorbed and desorbed water molecules, which corresponds to a state of saturation. Once being saturated over $H_2O$, the sample was placed above aqueous solutions of $NH_3$ with 1000 ppm concentration. Again in the first 35 s there is fast sorption leading to rapid frequency changes, followed by smooth curve during the next 300 - 350 s. After 400 s the system reaches saturation. The results are presented in figure 5. Once the experiments were completed, the samples were placed in a neutral environment, which resulted in recovering the initial frequency of the samples, thus proving the reversible nature of the sorption process.

## 4. Conclusions

The properties of RF and DC magnetron sputtered titanium oxide thin films were studied. The films were predominantly amorphous and nanocrystalline with grain size of a few nanometers. Non-uniformity between the surface and sub-surface layers of the films was observed at the XPS profile study. The films consist of stoichiometric $TiO_2$ on the surface and amorphous-like, including many phases like TiO, $TiO_{2-x}$, in the bulk.

The reactive sputtering method is found to be suitable for deposition of $TiO_2$ thin films for sensor applications. Prototype QCM sensors with $TiO_2$ sensitive films were made and showed good sensitivity to ammonia. These films even in as-deposited state and without heating the substrates show good sensing properties and additional thermal treatment is not necessary, making manufacturing of QCM gas sensor simple and cost-effective, as it is fully compatible with the technology for producing the initial resonator. The future development of these sensors as $NH_3$ detection devices is therefore possible.

**Acknowledgements**
This work was partially supported by the Bulgarian National Scientific Fund - Grant NT3/2006.

## References

[1] Bennett J M, Pelletier E, Albrand G, Borgogno J P, Lazarides B, Carniglia C K, Schmell R A, Allen T H, Tuttle-Hart T, Guenther K H and Saxer A 1989 *Appl. Opt.* **28** 3303
[2] Karunagaran B, Kim K, Mangalaraj D, Yi J and Velumani S 2005 *Sol. Energy Mater. Sol. Cells* **88** 199
[3] Fujishima A and Honda K 1972 *Nature* **238** 37
[4] Yu J, Zhao X and Zhao Q 2000 *Thin Solid Films* **379** 7
[5] Brevet A, Fabreguette F, Imhoff L, Marco de Lucas M C, Heintz O, Saviot L, Sacilotti M and Bourgeois S 2002 *Surf. Coat. Technol.* **151-152** 36
[6] Madou M J and Morrison S R 1989 *Chemical Sensing with Solid State Devices* (Boston: Academic Press)
[7] Tang H, Prasad K, Sanjines R and Lévy F 1995 *Sens. Actuators* B **26** 71
[8] Zhang Y, Fu W, Yang H, Qi Q, Zeng Y, Zhang T, Ge R and Zou G 2008 *Appl. Surf. Sci.* **254** 5545
[9] Guidi V, Carotta M C, Ferroni M, Martinelli G, Paglialonga L, Comini E and Sberveglieri G 1999 *Sens. Actuators* B **57** 197
[10] Idriss H and Seebauer E G 2000 *J. Mol. Catal.* A: Chem. **152** 201
[11] Chow L L W, Yuen M M F, Chan P C H and Cheung A T 2001 *Sens. Actuators* B **76** 310
[12] Syarif D G, Miyashita A, Yamaki T, Sumita T, Choi Y and Itoh H 2002 *Appl. Surf. Sci.* **193** 287
[13] Cunningham A 1998 *Introduction to Bioanalytical Sensors* (New York: Wiley)
[14] Boiadjiev S I, Dobrikov G H and Rassovska M M 2007 *Thin Solid Films* **515** 8465
[15] Manolov S and Tihchev H 1982 *Generators* (Sofia: Tehnika) 128 (in Bulgarian)
[16] Sauerbery G 1959 *Z. Phys.* **155** 206
[17] Lazarova V, Spassov L, Andreev S, Manolov E and Popova L 1996 *Vacuum* **47** 1423
[18] Boyadjiev S and Rassovska M 2007 *Proceedings of the 16-th Scientific and Applied Science Conference Electronics ET'2007* vol 4 p 121

Journal of Physics: Conference Series **253** (2010) 012041 doi:10.1088/1742-6596/253/1/012041

# Multilayer organic based structures with enhanced hole transport

**D Mladenova[1, 2], V Sinigersky[4], D Budurova[4], T Dobreva[3], D Karashanova[1], D Dimov[1] and I Zhivkov[1*]**

[1] Institute of Optical Materials and Technologies "Acad. J. Malinowski", Bulgarian Academy of Sciences, Acad. G. Bonchev Str. bl. 101/109, 1113 Sofia, Bulgaria
[2] Department of Microelectronics, Technical University of Sofia, Kl. Ochridski Blvd. 8, 1756 Sofia, Bulgaria
[3] Institute of Mechanics, Bulgarian Academy of Sciences, Acad. G. Bonchev Str. bl. 4, 1113 Sofia, Bulgaria
[4] Institute of Polymers, Bulgarian Academy of Sciences, Acad. G. Bonchev Str. bl. 103A, 1113 Sofia, Bulgaria

Email: izhiv@clf.bas.bg

**Abstract**. Multilayer Organic Based Devices (OBDs) were constructed by subsequent casting of organic films (from polymers, soluble in the same organic solvent). The problem with dissolution of the underlying layer was avoided by using electrophoretic deposition technique. Optimized conditions for electrophoretic deposition (EPD) of thin films with homogeneous and smooth surfaces, as confirmed by SEM, were found. The EPD, carried out at constant current, requires continuous increase of the voltage between the electrodes. In this way the decreased deposition rate caused by the decreased concentration of the material in the suspension and the increased thickness of the film deposited is compensated. The SEM images and the current voltage characteristics recorded, show that the hole transport polyvinylcarbazole (PVK) underlayer survive the treatment with the suspension used for the electrophoretic deposition of the active poly[2-methoxy-5-(3,7-dimethyloctyloxy)-1,4-phenylene vinylene] electroluminescent layer. The PVK hole transport layer increases the device current, as confirmed by the current-voltage measurements. The results obtained demonstrate the possibility of OBDs preparation for electroluminescent and photovoltaic applications.

## 1. Introduction

Organic Based Devices (OBDs) [1, 2] like solar cells, thin film transistors and light emitting diodes often suffer from poor charge carrier injection due to contact barrier formed between the metal (or ITO) electrode and the active semiconductor (conjugated polymer). This problem is usually solved by the deposition of an intermediate layer having a work function between the electrode work function and the energy of the corresponding transport band. Thus the contact barrier is overcome on intermediate steps - so called 'ladder' effect [3]. The preparation of multilayer structures by vacuum

* To whom any correspondence should be addressed

Journal of Physics: Conference Series **253** (2010) 012041 doi:10.1088/1742-6596/253/1/012041

deposition techniques is relatively easy. Unfortunately, this method is not applicable to most of the conjugated polymers, due to their high molecular weight. Polymer layers are usually prepared by 'wet' deposition techniques like spin and deep coating, ink-jet printing, etc. The preparation of multilayer structures is almost impossible when the polymeric materials used are dissolved in the same solvent. In this paper we present a possible solution of this problem by using a combination of spin coating for casting of the polyvinylcarbazole (PVK) hole transport layer and a subsequent electrophoretic deposition of poly[2-methoxy-5-(3,7-dimethyloctyloxy)- 1,4-phenylene vinylene] (MDMO-PPV) on the ITO|PVK structure.

## 2. Electrophoretic deposition technique

The phenomenon of electrophoresis has been known since the beginning of the $19^{th}$ century and it has found applications during the last 40 years mainly in the traditional ceramic technology. A DC electric field is applied across a colloidal suspension of charged particles depositing a film of particles on the oppositely charged electrode. Electrophoretic deposition (EPD) is essentially a two-step process: electrophoretic transport and deposition. In the first step charged particles in a liquid are forced to move toward an electrode by applying an electric field across the solution. These charged particles may be generated by precipitating of the material mixing solvent and non-solvent.

In the second step the particles, deposited on the electrode surface, form a homogeneous film via coagulation.

A post-EPD processing step is usually applied, involving a suitable heat-treatment (firing or sintering) in order to densify the deposits and to decrease the porosity. The drying of the film, carried out in non-solvent vapors, is expected to yield polymer films with novel microscopic structures. In this case the stage of film formation is separated from the solidification process [4].[5]:

EPD requires stable suspension of charged particles in a way to respond to the applied electric field. This stability of the suspension can be realized in several ways [6]:

(i) Selective adsorption of ions onto the solid particle from the liquid;

(ii) Dissociation of ions from the solid phase into the liquid;

(iii) Adsorption of dipoles at the particle surface;

(iv) Electron transfer between the solid phase and the liquid phase due to differences in the work functions.

Examples for electrophoretic thin film deposition are reported in the literature, related to metals, polymers, carbides, oxides, borides and glasses [7].

Several successful attempts for depositing polymer materials like polymer photonic crystals [8] and polymeric donor-acceptor composites [9] have been reported. Polymer EPD can facilitate the formation of nanostructure in thin polymer films [10]. Additionally, EPD provides substantial control over the film thickness, marked enhancement of the deposition rate, and improved film composition homogeneity, compared to other wet casting-nanostructured thin film deposition methods [11]. Electrophoretic deposition is one of the most widely used coating methods capable of patterning.

Although the mechanism of the electrophoretic deposition is not clearly understand, EPD is already successfully used. However, a better understanding would reduce experimental work required to determine the optimal parameters of EPD.

Recently, considerable efforts are involved in the production of large area low-cost organic solar panels and cheap organic light emitting diodes. The EPD techniques suggest genuine solutions of most of the problems already mentioned. It could be expected that separating the stage of film formation from the solidification step in a mixture of solvent and non-solvent will allow the construction of a multilayer structure from films generally dissolved in the same solvent.

## 3. Experimental details

Toluene solution of MDMO-PPV (Aldrich) was heated at 70 °C for 20 min. After cooling down to RT the solution was filtered and non-solvent (acetonitrile) was added. The suspension obtained in this way was used for EPD immediately after preparation. After 15-20 min, due to coagulation of the precipitated MDMO-PPV particles, the suspension becomes unusable. Both the solution concentration and the toluene/acetonitrile volume ratio were varied in order to optimize film quality. The EPD current was kept constant during the deposition of one sample by increasing the voltage applied between the electrodes. The current was varied from sample to sample in the range of 0.1 – 1 mA. The time for deposition of a single sample was 1 – 3 min.

The film thicknesses were measured by Taylor Hobson Talystep, Model 223-7.

SEM images were taken from Philips 515 microscope.

Current-voltage (*I-V*) characteristics were measured in dark, in vacuum of 2.66 Pa with Keithley 617 electrometer. The preparation of samples for electrical measurements is presented on figure 1. On ITO glass (figure 1a) two strips each 3 mm wide, were formed by etching in acid solution (figure 1b). The sample was cleaned in deionized water, isopropyl alcohol and chloroform, and then dried in a stream of nitrogen. PVK was dissolved in toluene (70 g $l^{-1}$), stirred (magnetic stirrer) at 70 °C for 1 h. PVK films of approximately 500 nm thickness were casted by spin-coating at 500 rpm for 30 s. It was found that the quality of the deposited films strongly depends on the preparation conditions, e. g. rotation speed solution concentration and substrate temperature.

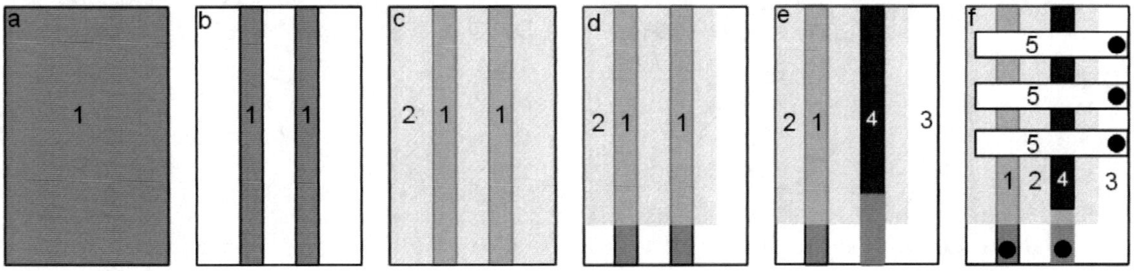


**Figure 1**. Consecutive stages (a ÷ f) of preparation of samples for electrical measurements: 1– commercially purchased ITO film deposited on glass substrate, 2 – spin-coated PVK on the ITO electrodes, 3 – cleared area on the glass substrate removing the PVK film by scratching, 4 – electrophoretically deposited MDMO-PPV film, 5 – vacuum deposited aluminium top electrodes; black circles present silver paste spots for connecting copper wires.

The films deposited cover both ITO strips on the sample uniformly (figure 1c). Then the PVK film was removed from the area outside the active region in a way to prepare ground for a contact formation (figure 1d). One of the ITO strips was connected to the voltage source and EPD of the MDMO-PPV film (thickness of 500 nm) was deposited selectively on it (figure 1e). Top aluminum electrode strips were deposited in vacuum of $1.33\times10^{-3}$ Pa trough a mask in a way to cross perpendicularly both ITO electrodes (figure 1f). Finally Al and ITO electrodes were connected with copper wires by silver paste. In this way two types of samples were prepared on the same substrate – ITO|PVK|Al and ITO|PVK|MDMO-PPV|Al. For comparison ITO|MDMO-PPV|Al samples were prepared on separate substrates following the procedures presented on figure 1 a, b, e, f. The electrode area is about 9 $mm^2$.

## 4. Results

### *4.1. Film morphology and structure*

SEM study was carried out to estimate the influence of the suspension parameters on the film structure and morphology. 0.066 g $l^{-1}$ MDMO-PPV solution in toluene with acetonitrile (volume ratio 1:4) was used as starting material (suspension), further denoted as Suspension I. In this way the experimental conditions were kept close to the literature data [12] for a similar material (MEH-PPV). SEM morphology image of 500 nm thick MDMO-PPV film, deposited from Suspension I for 5 s at current of 1 mA, is presented in figure 2. The presence of cavities of about 100 - 300 nm diameter and relatively wide smooth areas in between can be clearly seen. Cavities could result from the fast speed of the deposition, where the relatively big particles from the solution have no time to spill in a uniform film on the substrate surface.

**Figure 2.** SEM image of 500 nm EPD film (sample tilted at 40°), deposited from Suspension I for 5 s at current of 1 mA.

**Figure 3.** SEM image of 500 nm EPD film (sample tilted at 40°), deposited from Suspension II for 150 s at current of 0.1 mA.

The general trend found shows that the film structure becomes smoother with increasing the amount of the solvent in the suspension. This result confirms the literature data, obtained for a similar material [13]. The best deposition conditions found are achieved at toluene/acetonitrile volume ratio 1:1. No film improvement was attained at further increase of the toluene concentration. This is why this ratio was kept constant but the MDMO-PPV concentration was decreased. The optimized conditions found, involve 0.033 g $l^{-1}$ MDMO-PPV solution in toluene mixed with acetonitrile (volume ratio 1:1), denoted further as Suspension II. Further decrease of MDMO-PPV concentration leads to strong increase of the suspension resistance even during the first few seconds of the deposition, resulting in unsuccessful film deposition process.

Uniform films were deposited from Suspension II for a time of 150 s at constant current of 0.1 mA. SEM image of 500 nm EPD film deposited from Suspension II for 150 s at current of 0.1 mA is presented on figure 3.

The films obtained in this way exhibit smooth surfaces without pinholes, suitable for electrical measurements.

### *4.2. Electrical characteristics of the suspension*

During the EPD process the concentration of the material in the suspension decreases, the thickness of the film deposited on the electrode increases, leading to an increase of the electrophoretic cell resistance. The voltage changes, as a function of the time elapsed (at a constant current of 1 mA) during the deposition of MDMO-PPV from Suspension I, are presented on figure 4. At the beginning of the deposition the voltage increases almost linearly. This suggests a single process of decreasing the MDMO-PPV concentration during material deposition on the electrode.

The second nonlinear part of the curve could result from a simultaneous action of two processes - solution dilution and increased film thickness on the electrode. The second process could cause redistribution of the electric field between the increased film thickness and the suspension. This effect agrees with the theoretical considerations [6] for a deposition of wide-gap or insulator materials.

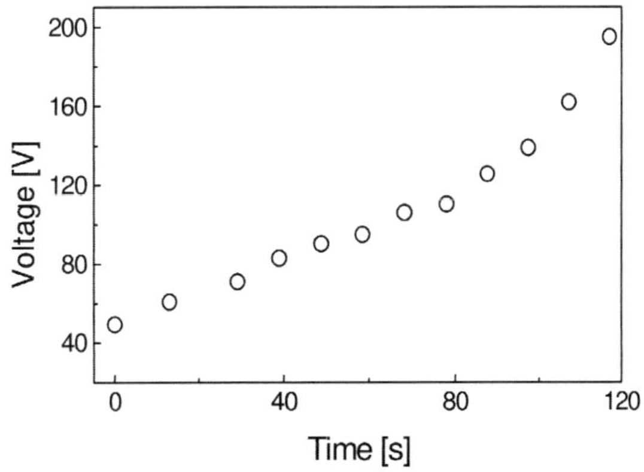


**Figure 4.** Dependence of the voltage applied on the time elapsed for the electrophoretic deposition of MDMO-PPV film from Suspension I; deposition current – 1 mA

In agreement with the literature data [13] the film deposition takes place at the anode, as the particles are negatively charged during the preparation of the toluene/acetonitrile suspension.

*4.3. Current voltage measurements*

Current-voltage measurements (figure 5) on ITO|PVK|Al structure show that the toluene/acetonitrile (Suspension II) mixture used for MDMO-PPV EPD (see figure 1e) does not destroy the preliminarily deposited PVK film (figure 1c). This result shows that a combination of spin coating and electrophoretic deposition methods could be used for the preparation of multilayer structures, comprising thin films cast from materials, soluble in the same solvent.

The shape of the curve in forward direction shows a typical diode behavior [14]. The symmetry of the *I-V* curve in both directions on the voltage scale should not be expected as the electrodes used have different work functions (ITO – 4.8 eV, Al – 4.17 eV). One possible reason for the symmetry observed could be due to the influence of high density dangling bonds, which could induce a broad distribution of localized states in the band gap [14]. Therefore the *I-V* characteristic could be controlled by hopping of charge carriers on the localized states, rather than direct injection from the electrodes into the corresponding transport band.

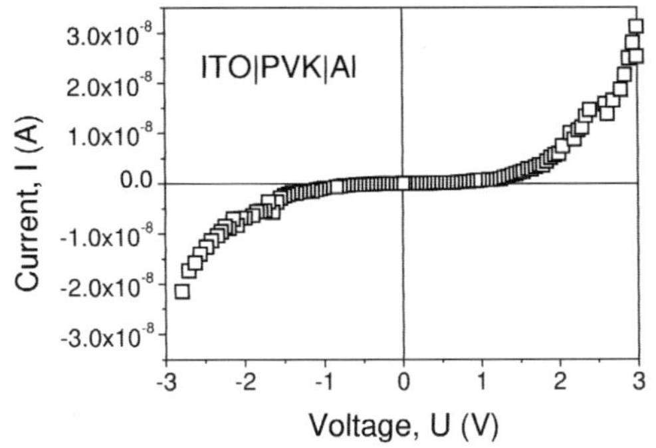


**Figure 5.** I-V characteristic of ITO|PVK|Al structure measured in dark in vacuum of 2.66 Pa.

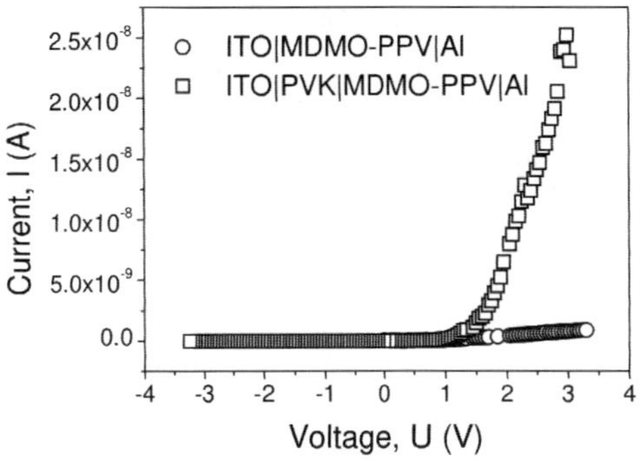


**Figure 6.** I-V characteristic of ITO|MDMO-PPV|Al (circles) and ITO|PVK|MDMO-PPV|Al (squares) samples measured in dark in vacuum of 2.66 Pa.

*I-V* characteristics, measured from ITO|MDMO-PPV|Al and ITO|PVK|MDMO-PPV|Al, are compared on figure 6. Both curves show typical diode behavior.

The introduced PVK film between the ITO electrode and the active MDMO-PPV semiconductor does not reduce the contact barrier but leads to an increase of the current. PVK has lower value of its HOMO orbital (-5.7 eV) comparing with that of MDMO-PPV (-5.2 eV) and should not reduce the contact barrier in ITO|PVK|MDMO-PPV|Al structure. Oppositely the higher conductivity of the PVK film, facilitate the hole transport from the ITO electrode to the MDMO-PPV semiconductor acting as a hole transport layer. This result proves that multilayer structures, consisting of films, deposited from materials, soluble in the same solvent, can be successfully used to increase the current in OBDs, like electroluminescent devices and solar cells.

*I-V* characteristics measured in the reverse direction in both cases looked similar. This points out that the electrical current flown trough the sample in both cases is most probably controlled by the high Al|MDMO-PPV contact barrier.

## 5. Conclusions

Optimized conditions for electrophoretic deposition of thin films with homogeneous and smooth surface (as evidenced by SEM) were found using 0.033 g $l^{-1}$ MDMO-PPV solution in toluene mixed with acetonitrile in volume ratio 1:1.

The EPD deposition was carried out at constant current, which required continuous increase of the voltage between the electrodes to compensate the decreased concentration of the material in the suspension and the increased thickness of the film deposited.

A combination of spin coating and electrophoretic deposition methods was used for the preparation of multilayer structures with thin films, deposited from materials, soluble in the same solvent.

The results obtained demonstrate the possibility for preparation of multilayer structures with improved electrical parameters in OBDs, like electroluminescent diodes and solar cells.

## Acknowledgements

The authors would like to thank the National Science Fund of the Ministry of Education and Science, Bulgaria (grants DO 02-254 and BU-TH-205/2006) for the financial support of this study.

## References

[1] Garnier F 1999 *Acc. Chem. Res.* **32** 209
[2] Peisert H, Knupfer M, Schwieger T, Fuentes G G, Olligs D, Fink J and Schmidt Th 2003 *J. Appl. Phys.* **93** 9683
[3] Lee S, Wang Y, Hou X and Tang C 1999 *Appl. Phys. Lett.* **74** 670
[4] Landfester K, Montenegro R, Scherf U, Gqnter R, Asawapirom C, Patil S, Neher D and Kietzke T 2002 *Adv. Mater.* **14** 651
[5] Boccaccini A and Zhitomirsky I 2002 *Curr. Opin. Solid State Mater. Sci.* **6** 251
[6] Sarkar P and Nicholson P 1996 *J. Am. Ceram. Soc.* **79** 1987-2001
[7] Trung Vu Q 2006 *PhD Thesis*
[8] Dziomkina N V, Hempenius M A and Vancso G Julius 2005 *Adv. Mater.* **17** 31
[9] Tada K and Onoda M 2005 *Current Applied Physics* **5** 5–8
[10] Tada K and Onoda M 2006 *Thin Solid Films* **499** 19
[11] Maenosono S, Okubo T and Yamaguchi Y 2003 *J. Nano Res.* **5** 5
[12] Tada K and Onoda M 2004 *Adv. Func. Mater.* **14** 139
[13] Tada K and Onoda M 2008 *J. Phys. D: Appl. Phys.* **41** 032001
[14] Djobo S Ouro, Bernede J, Napo K and Guellil Y 2002 *Materials Chemistry and Physics* **77** 476–483

# Study of thin-film silicon solar cells at irradiances above ten thousand suns

**Yeping Lu, Amin Abdolvand and Steve Reynolds**

School of Engineering, Physics and Mathematics, University of Dundee, Nethergate, Dundee DD1 4HN, United Kingdom

E-mail: s.z.reynolds@dundee.ac.uk

**Abstract**. We used a tightly focused Gaussian beam of a HeNe laser to study accelerated light-induced degradation (Staebler-Wronski effect) and high photocarrier generation rates in amorphous and microcrystalline silicon thin-film solar cells, at up to13 MW/m$^2$ irradiance. For the experiments, the spot radius was varied from a minimum of 8.6 μm in the focus to around 1 mm away from the focus. According to COMSOL® simulations, even at these high power densities heat diffusion into a glass substrate aided by spreading conduction via the Ag back-contact restricts the temperature rise to less than 14 K. Short-circuit current can be measured directly over a range of irradiances, and the J-V characteristic may be estimated by taking into account shunting by the inactive part of the cell.

## 1. Introduction

Degradation in performance of solar modules over their operational lifetime is an important factor shaping the commercial future for large-area photovoltaics (PV). In amorphous silicon solar cells, light-induced mid-gap defects generated over the first few months of operation may reduce PV conversion efficiency by up to 20%. This is termed the Staebler-Wronski Effect (SWE) [1]. These defects are metastable silicon dangling bonds, which may be largely eliminated by annealing in the dark for a few hours at > 160°C. The two most widely-quoted models of SWE are the weak bond-dangling bond conversion model of Stutzmann, Jackson and Tsai [2] and the hydrogen collision model of Branz [3], although neither is entirely satisfactory, and after more than 30 years this important topic remains the subject of intensive research.

Typical light-soaking laboratory test conditions comprise 1000 hours exposure to AM1.5 radiation at 50 °C, in keeping with those met in service. However, there are reasons for wishing to study solar cell properties at elevated irradiance. Degradation is more rapid, opening up the possibility of faster initial screening of materials and devices prior to formal longer-term test. It is also useful to test theories of PV operation or degradation over a range of generation rates. However, the temperature rise accompanying higher irradiances usually requires the substrate to be cooled to control it. In the case of amorphous silicon, studies on 1 cm$^2$ cells at up to 140 suns [4] have been conducted under simultaneous forced cooling with nitrogen gas pre-cooled by liquid nitrogen. Another similar study [5] employed an air-cooled Al heat sink.

We believe it is possible to gain useful information at much higher irradiances, without the need for special cooling measures, simply by using a tightly-focused beam of laser light incident on a small area of the cell. Consider the case of a semi-infinite slab (the substrate) of thermal conductivity *k* coated with a thin optically-absorbing film (such as amorphous silicon). If a circular beam of radius $a$ and uniform irradiance $H$ is fully absorbed by the film, giving a disc-shaped heat source, the steady-state temperature rise at the centre of the disc (the hottest point) is [6]:

$$\Delta T = Ha / k \tag{1}$$

Taking $H$ = 10 kWm$^{-2}$ (10 suns), $k$ = 1 Wm$^{-1}$K$^{-1}$ (typical for glass and silicon films) and $a$ = 1 cm yields $\Delta T$ = 100 °C. The need for cooling even at quite modest irradiances is apparent. However, if the disc is 10μm in radius, $\Delta T$ is only 0.1 K. We demonstrate by means of a COMSOL® [7] numerical thermal model that equation (1) is a good approximation for a film subjected to a beam whose radius is less than the thickness of the glass substrate. The model is extended to a solar cell structure by including an Ag film back-contact, which greatly reduces the peak temperature by spreading heat into the substrate. While the short-circuit current may be measured directly, the complete *I-V* characteristic requires any current flowing through the dark part of the cell to be accurately subtracted. To test this novel approach we used focused light from a HeNe laser to investigate both amorphous (aSi:H) and microcrystalline (μcSi:H) *pin* solar cells, which behave quite differently in respect of both PV properties and degradation behavior.

## 2. Experimental

### *2.1. Solar cell deposition*

Amorphous silicon cells were deposited in *pin* sequence by RF PECVD: Substrate temperature of 180 °C, power of 3 W, pressure of 5 mbar and silane:hydrogen process gas flow rate of 6.25:60 sccm. Microcrystalline silicon cells were deposited in *pin* sequence by hot-wire CVD: Substrate temperature of 220 °C, wire temperature of 1650 °C, pressure of 3 Pa, silane concentration 6% in hydrogen. In both cases the *i*-layer thickness was nominally 1 μm. Raman spectroscopy determined that the crystalline volume fraction of the microcrystalline absorber layer in the cell was 40%.

### *2.2. Optical system*

To produce a spot of light of the requisite size, the beam from a Uniphase 1101 1.5 mW HeNe laser was firstly expanded using a reverse telescope, and then focussed by a converging lens to a Gaussian beam of waist radius $w_0$ = 8.6 μm (at the 1/e$^2$ points) and Rayleigh range $z_R$ = 0.37 mm. The solar cell was mounted on a translation stage, enabling it to be positioned orthogonal to the beam and moved along the beam axis $z$ to vary the spot size [8]:

$$w(z) = w_0(1 + (z / z_R)^2)^{1/2} \tag{2}$$

$z$ = 0 is the position of the beam waist. The peak irradiance varies as $H(z) = 2P_0 / \pi w^2(z)$ [8], where $P_0$ is the beam power. A micrometer screw adjustment enabled the solar cell to be positioned accurately along the $z$-axis and for the irradiance to be tuned from 13,000 kWm$^{-2}$ at the beam waist to 1 kWm$^{-2}$ at z = 42 mm. Solar cell current falls to a minimum value in the vicinity of the calculated position of the beam waist ($z$ = 0) and increases symmetrically on either side. The current minimum was taken to signify that the cell was aligned precisely at $z$ = 0. The device physics has not yet been investigated, but bearing in mind the very high generation rates (> 10$^{25}$ cm$^{-3}$s$^{-1}$) at the focus this behaviour may be associated with reduced carrier collection due to a high density of trapped space-charge [9]. Since spot size is normally much smaller than the cell contact area, new un-degraded regions can readily be found by moving the beam elsewhere on the cell.

### *2.3. COMSOL® simulations*

The COMSOL® 3.3 heat transfer module (2-d axisymmetric geometry) was used for steady-state and transient simulations [7]. As temperature changes here are generally quite small, no temperature-dependence of material constants was included. The basic model substrate consists of a 10 mm radius glass disc 1 mm thick, coated on one side with a 1 μm silicon film. Heat was generated uniformly within the 'illuminated' part of the film to facilitate direct comparison with equation (1). It is possible within COMSOL® to model a generation depth profile, and also to include a Gaussian radial distribution, but this would significantly increase run times and would not change the key outcomes of this work. Two additional scenarios were investigated; a 1 μm film of Ag was added on top of the silicon film to give a structure resembling a solar cell with back contact, and the glass slide was exchanged for a sapphire one, with thermal conductivity some 30 times higher than glass. The mesh was refined to provide sufficient spatial resolution close to the generation volume.

## 3. Results and discussion

### *3.1. Thermal modelling*

Figure 1(a) shows the simulated temperature profile produced by a 10 μm radius beam of irradiance $10^6$ $Wm^{-2}$ (1000 suns equivalent), fully absorbed by a 1 μm thick silicon film deposited on a 1 mm thick glass disc of radius 10 mm. Only one half-plane is shown, since the system is axially symmetric. For simplicity, both film and substrate have the same thermal conductivity of 1 $Wm^{-1}K^{-1}$. As this system is finite, boundary conditions must be imposed at all free surfaces. This has been done in two different ways, to investigate the sensitivity of the model. These are as follows: (i) Free film surface perfectly insulated, free substrate surfaces maintained at $T_0$ (set at 273.15 K here); (ii) All free surfaces subject to heat loss $h\Delta T$ per unit area to the surroundings, where $h$ = 4 $Wm^{-2}K^{-1}$, typical for free convection in air. Case (ii) represents a minimum (worst-case) thermal coupling to the surroundings under normal laboratory test conditions. It is found that the maximum temperature rise above $T_0$ is 9.5 K in case (i) and 9.6 K in case (ii); both are in excellent agreement with the value of 10.0 K obtained from equation (1). In case (ii) the surfaces of the substrate are slightly above ambient, as required to maintain convective heat flow into the surroundings.

Having confirmed agreement with the analytic solution, a solar cell structure with a 1 μm Ag 'back-contact' on top of the silicon film was then modelled. The result under the same conditions as case (ii) above is shown in figure 1(b). It is seen that the maximum temperature rise in the silicon film is dramatically reduced, by a factor of 10, to 1 K. The temperature distribution is considerably

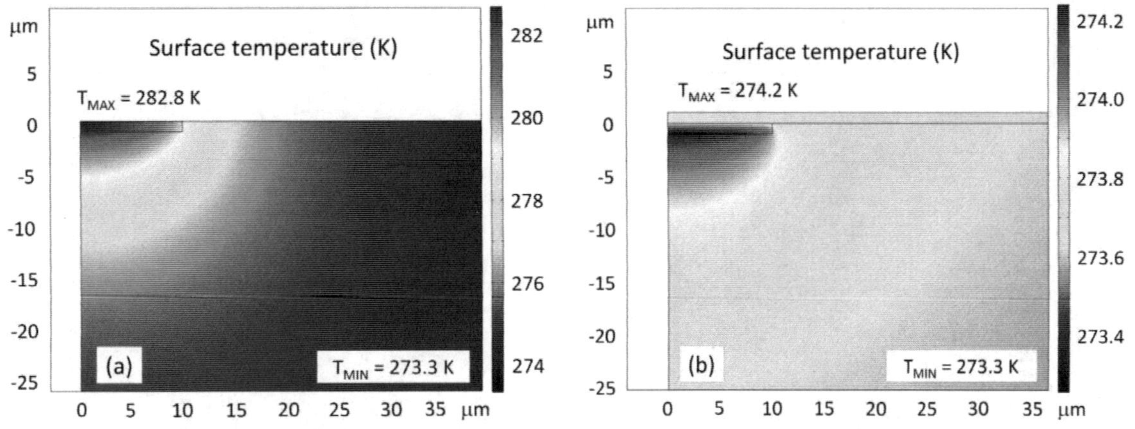


**Figure 1.** (a) COMSOL® thermal simulation of 1 μm aSi:H film on glass substrate. Laser spot radius 10 μm, irradiance $10^6$ $W/m^2$; (b) as (a), but with 1 μm Ag film on top of aSi:H film as in solar cell.

modified in comparison to figure 1(a); the region of highest temperature now lies at the generation volume - substrate interface, heat is conducted along the Ag film to more distant regions of the substrate, increasing its temperature. This 'spreading' effect afforded by the Ag film is clearly beneficial in reducing the temperature of the generation volume. The TCO top-contact behaves similarly, but as its thermal conductivity is around one tenth that of Ag the effect is less pronounced.

We have also simulated a sapphire substrate in place of glass. This has a similarly dramatic effect on the maximum temperature rise, and a combination of Ag film and sapphire substrate gives a reduction in $\Delta T$ of a factor of around 20. The thermal time-constant with regard to achieving steady-state conditions in the generation volume is less than one millisecond in all cases. Substrate geometry is found to have little bearing on the temperature profile in the generation volume, provided the spot size is less than the minimum dimension of the substrate.

The solar cells studied here were deposited on borosilicate glass substrates ($k$ = 0.9 $Wm^{-1}K^{-1}$ ), and the Ag layer is nominally 0.7 μm thick. These parameters, and a value of $k$ = 1.3 $Wm^{-1}K^{-1}$ for the silicon layer, were used to simulate the maximum temperature rise likely to be encountered in this work, *i.e.* with the beam most tightly focussed ($a$ = 8.6 μm). This was found to be some 14 K above ambient. It follows that in this work thermal effects should have negligible impact in comparison to optically-driven phenomena such as SWE and the influence of space-charge [9] on carrier transport.

### 3.2. *Effects of light-soaking*

The results of light-soaking on amorphous and microcrystalline silicon solar cells up to a maximum irradiance of 13 $MWm^{-2}$ are shown in figure 2. This was carried out under short-circuit conditions, and the currents shown are those measured under full irradiance. Each trace is normalised to the initial current prior to degradation. As expected, in the case of amorphous silicon higher irradiance results in more rapid degradation. At maximum irradiance, short-circuit current falls by 10% in just one second. The results at lower irradiance are broadly in keeping with Yang *et al* [4], who monitored reduction in efficiency with time at up to 140 suns. However it is not possible to make a direct comparison with their data, as we have not yet measured fill-factor, which is a better indication of degradation, and open-circuit voltage (see section 3.3 below), and so cannot calculate efficiency.

The quality of the data at higher irradiance is somewhat poorer, with oscillations and less well-defined fluctuations being apparent. As these greatly exceed the variations in laser amplitude after warm-up, the most likely explanation is mode instability, which causes changes in the shape of the irradiance curve, particularly for a tightly-focussed beam. The microcrystalline silicon cell is far more resilient to degradation than the amorphous cell. At maximum irradiance, it takes approximately one day for the current to degrade by 10%. Previous reports [10] indicate that exposure for 1000 hours (40 days) is normally needed to obtain a similar reduction under AM1.5.

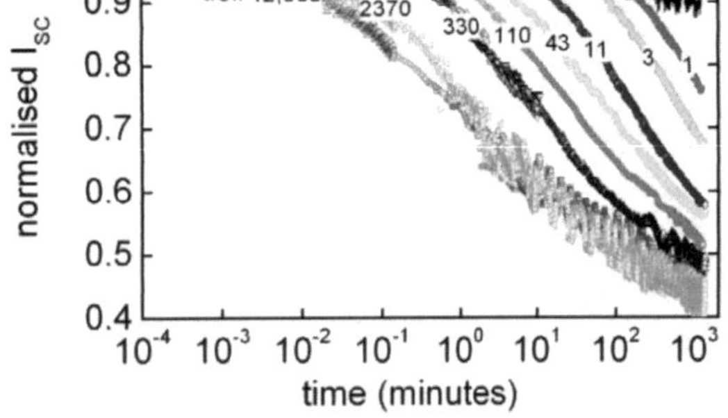


**Figure 2.** Degradation of short-circuit current *vs.* time for a range of irradiance values (in $kWm^{-2}$).

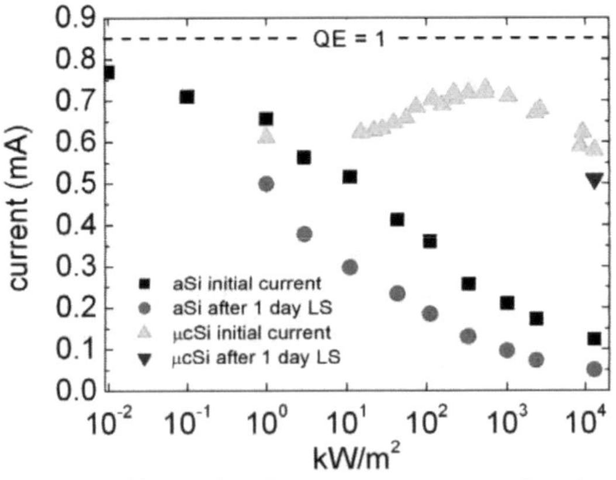


**Figure 3.** Short-circuit current measured prior to, and following, degradation, *vs.* irradiance.

Figure 3 summarises the effect of increasing irradiance on solar cell current prior to degradation, and after one day exposure. The aSi:H cell is strongly affected by increased volume generation rate (up to $10^{25}$ $cm^{-3}$ $s^{-1}$) by a mixture of the above and additional recombination via light-induced defects. To separate these effects it is necessary to measure the degraded spot at a lower irradiance, which we are unable to do at present. The μcSi:H cell shows quite different behaviour. The current increases with increasing irradiance, reaching a maximum at 420 $kWm^{-2}$, 15% higher than at 1 $kWm^{-2}$. The reason for this is presently unclear. The laser output was calibrated against a standard Si detector, and the current corresponding to an external quantum efficiency of 1 is 0.84 mA ±10%.

Figure 4 shows the photocurrent decay in an aSi:H cell under high irradiance at short times, and the effect on the current of interrupting the beam. The initial decay is very rapid, and it is likely that the current reaches even higher values at shorter times. It then decays roughly according to a power law. When the beam is interrupted and then re-applied, the decay recommences from the same current level, and follows continuously in time – there is no spike. Thus the decay is not due to a reversible cycle such as heating/cooling or charging/discharging of traps. The behaviour is in keeping with SWE, but we have not yet determined if the current is restored to its original value by annealing. Until then, the possibility of crystallisation, hydrogen effusion, persistent photoconductivity or other processes cannot be completely ruled out. Viera *et al* [11] have studied laser-induced crystallisation of aSi:H using Raman spectroscopy, and report no effect on the structure at $10^4$ $kWm^{-2}$, which is close to the maximum irradiance used here. In their case, the irradiance reached $90\times10^4$ $kWm^{-2}$ before crystallisation occurred. However, from a thermal point of view these results cannot be properly compared without knowledge of spot size, as is clear from equation (1).

### *3.3. I-V characteristic*

Figure 5 shows the *I-V* curve for a degraded aSi:H cell, in the dark, and under a 13 $MWm^{-2}$ spot. Also shown is the result of subtracting the dark trace to leave the *I-V* curve of the active volume. This is fairly successful until the forward current becomes large and device plus measurement noise exceeds the difference between the two traces. It seems unlikely that the entire fourth quadrant could be recovered in this way, but we are continuing to investigate this possibility.

## 4. Conclusions

COMSOL® thermal models of a thin-film silicon solar cell illuminated by a focused laser spot indicate that very high local irradiance can be delivered without incurring excessive temperature rise. The maximum temperature rise in the active volume of a typical solar cell under an 8.6 μm radius Gaussian

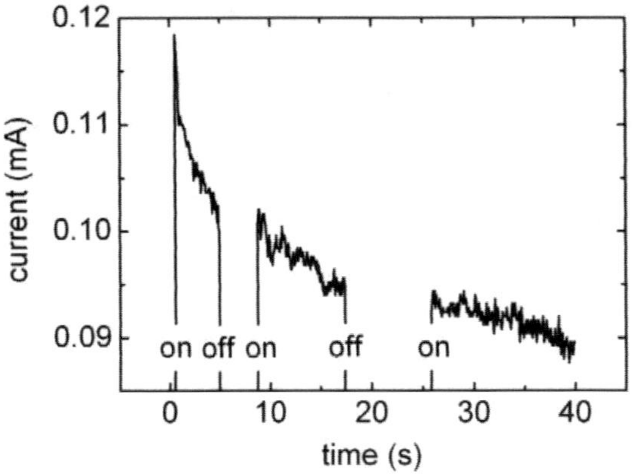


**Figure 4.** Effect of interruption of laser beam on current, at an irradiance of 13 $MW/m^2$.

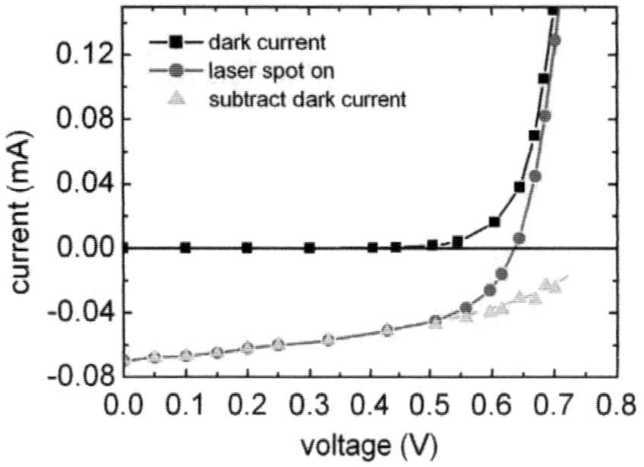


**Figure 5.** *I-V* characteristic of degraded aSi:H cell with and without laser beam, showing subtraction of dark current.

spot focused down from a 1.5 mW HeNe laser (peak irradiance 13 $MWm^{-2}$) is predicted to be 14 K. The Ag back contact has a beneficial effect in spreading the heat by conduction. This is in contrast to standard uniformly-illuminated 1 $cm^2$ test structures, which require forced cooling to attain much more modest irradiances. By moving the cell along the beam axis, this set-up provides adjustable irradiance.

The Staebler-Wronski effect in amorphous silicon solar cells has been investigated as a function of irradiance, by monitoring short-circuit current *vs.* time. Degradation is shown to be more rapid with increasing irradiance, in keeping with previous studies although these do not extend to such high irradiance values. There is evidence of fluctuations in beam profile at smaller spot size, which may be caused by laser mode instability. A microcrystalline silicon cell was also investigated, and found to be far more resistant to degradation. This approach could be used for initial rapid screening of promising low-degradation materials prior to longer-term tests at one sun.

This method was also used to study the effect of very high generation rates (over $10^{25}$ $cm^{-3}s^{-1}$) on solar cell operation. In the case of thick amorphous silicon cells, increased irradiance markedly reduces the quantum efficiency, even at 10 $kWm^{-2}$ (10 suns equivalent). Microcrystalline silicon cells however, show a 15% *increase* in quantum efficiency from the 1 sun value, at around 400 suns equivalent irradiance. It is possible to measure part of the *I-V* curve of the active region of the cell by subtracting the dark *I-V* curve, but unfortunately this is limited in its success by noise considerations.

Future work is planned, including more stable laser and optics to improve beam quality, faster and more accurate data-logging, and a temperature-controlled stage that will allow samples to be thermally annealed in situ, essential when working on a small spot on a much larger cell.

**Acknowledgments**
The authors are grateful to colleagues at IEF-5 Photovoltaik, Forschungszentrum Jülich, for provision of solar cell samples.

**References**
[1] Fritzsche H 2001 *Annu. Rev. Mater. Res.* **31** 47
[2] Stutzmann M, Jackson W and Tsai C C 1985 *Phys. Rev. B* **32** 23
[3] Branz H 1998 *Solid State Commun.* **105** 387
[4] Yang L, Chen Y and Catalano A 1991 *Appl. Phys. Lett.* **59** 840
[5] Isomura M, Hata N and Wagner S 1992 *Jpn. J. Appl. Phys.* **31** 3500
[6] Wilson J and Hawkes J F B 1987 *Lasers: Principles and Applications* (Hemel Hempstead, UK: Prentice Hall) p 171
[7] COMSOL® Heat Transfer module: http://www.comsol.com/products/heat-transfer/ (accessed 13 August 2010)
[8] Siegman A E 1986 *Lasers* (Mill Valley, CA: University Science Books) p 675
[9] Abel C D, Paes H R and Bauer G H 1991 *MRS Symp. Proc. Ser.* **219** 851
[10] Meillaud F, Shah A, Vallat-Sauvain E, Niquille X, Dubey M and Ballif C 2005 *Proc. 20th EU-PVSEC (Barcelona, Spain, 6-10 June 2005)* p 1509
[11] Viera G, Huet S and Boufendi L 2001 *J. Appl. Phys.* **90** 4175

Journal of Physics: Conference Series **253** (2010) 012043 doi:10.1088/1742-6596/253/1/012043

# Characterisation of multi-layer InAs/InP quantum wires by surface photovoltage and photoluminescence spectroscopies

**Ts Ivanov[1], V Donchev[1], T Angelova[2], A Cros[2], A Cantarero[2], N Shtinkov[3], K Borissov[1], D Fuster[4], Y González[4] and L González[4]**

[1] Faculty of Physics, Sofia University, 5, blvd. J. Bourchier, Sofia-1164, BULGARIA
[2] Institut de Ciència dels Materials, Univ. de València, P.O. Box 22085E-46071, València, SPAIN
[3] Department of Physics, University of Ottawa, Ottawa ON, CANADA K1N 6N5
[4] Instituto de Microélectronica de Madrid (CNM- CSIC), Isaac Newton 8, Tres Cantos, 28760, Madrid, SPAIN

Corresponding author: tsvetanivanov@lycos.com

**Abstract**. The optical properties of multi-layer InAs/InP quantum wires (QWRs) with two different spacer thicknesses have been investigated by means of room temperature surface photovoltage (SPV) and photoluminescence (PL) spectroscopies, combined with empirical tight binding electronic structure calculations and structural data. The SPV and PL spectra reveal several features, which energy positions are in good agreement. They have been ascribed to excitonic transitions, which take place in the QWR families with heights differing by an integer number of monolayers. Comparing the experimental results with the theoretical ones, we have estimated the QWR family heights and the average atomic concentration of phosphorus in the QWRs. From the simultaneous analysis of the SPV amplitude and phase spectra, based on our vector model for SPV signal representation, a deeper understanding of the SPV results and of the mechanisms of carrier separation in the sample is obtained.

## 1. Introduction

The interest in multi-layers of self-assembled InAs/InP quantum wires (QWRs) grown in the Stranski-Krastanov mode is motivated by their promising applications in light emitting devices in the 1.3 - 1.55 μm wavelength range used in optical fibre telecommunications [1]. The optical properties of these nanostructures have been studied by means of low-temperature photoluminescence (PL) spectroscopy, combined with electronic structure calculations in frames of the envelope function approximation. The PL spectra have revealed several wide bands, which have been generally attributed to emission of QWRs belonging to different families with heights differing by an integer number of monolayers (MLs) [2-5]. There are relatively few experimental investigations at room temperature (RT), which will be the operating temperature of a final device, based on the discussed multi-layers [6-8]. Due to its high sensitivity even at RT, the surface photovoltage (SPV) spectroscopy is an alternative to avoid the difficulties related to optical absorption measurements in nanostructures [9-11].

In this work we apply SPV spectroscopy to study multi-layer InAs/InP QWRs at RT in combination with PL measurements, structural data from previous studies [3], and comparison with empirical tight binding (ETB) electronic structure calculations.

## 2. Experimental and calculation details

Two samples were grown on (001) semi-insulating (SI) InP:Fe substrates by solid-source molecular beam epitaxy. They contain a 150 nm InP buffer layer and a stack of seven InAs QWR layers, separated by InP spacers of thickness $d$ = 10 and 20 nm in the two samples. The growth temperature was 515°C and 380°C for the QWRs and the InP spacers, respectively (figure 1). High resolution transmission electron microscopy (TEM) [3] has shown that the QWRs are oriented along [1-10] and their average height $H$ and lateral width are around 3.2 nm and 10.3 nm, respectively. The residual doping in the buffer and the QWR structure is n-type, of the order of $1\times10^{17}$ cm$^{-3}$.

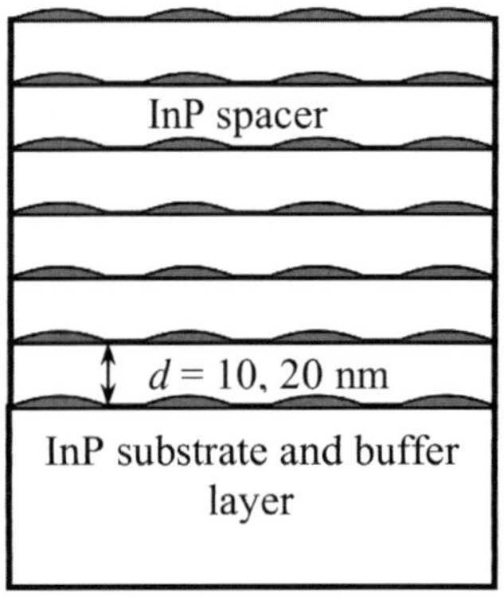


**Figure 1.** Scheme of the samples.

Details about the set-up used for SPV spectral measurements in metal-insulator-semiconductor operation mode and the measurement procedure can be found elsewhere [10, 12]. Here the semitransparent probe electrode was a $SnO_2$ film deposited on a quartz glass. It was adjusted to gently press the sample to a grounded copper platform. All measurements were performed with normal incident light chopped at 94Hz, keeping a constant photon flux density ($\Phi \approx 1.5\times10^{14}$ cm$^{-2}$s$^{-1}$ $\pm$ 2%) for all wavelengths, scanned from high towards low ones. The PL spectra were measured at RT using an Ar+ ion laser (514.5 nm line) a 0.19 TRIAX single monochromator (600 grooves/mm grating) and a $N_2$-cooled InGaAs detector connected to a lock-in amplifier.

The procedure for the ETB electronic structure calculations is described in detail in [13]. The intermixing that inevitably exists in such structures was taken into account by considering a nonzero phosphorus concentration [P] in the QWR, because this method is consistent with experimental observations for the InAs/InP system [13]. Since InAs/InP QWRs and dots are relatively flat, the effect of the lateral confinement on the transition energies is small [13] and was therefore neglected in our calculations. A temperature correction of 65 meV is subtracted from the calculated (at 0K) energies to account for the InAs band gap decrease between 0 and 300K [14].

## 3. Results and discussion

The SPV amplitude spectra of the samples with $d$ = 10 and 20 nm, respectively, are shown on figures 2(a) and 2(b). In the energy range 0.75–0.9 eV they reveal a few steps or peaks superimposed on a smooth background. The PL spectra, given on the same figures, represent broad bands, which can be decomposed to a few Gaussians. The energy positions (in eV) of the SPV (PL) features for the sample with $d$ = 10 nm are: 0.84 (0.83); 0.81 (0.80); 0.77 (0.77) and (0.72), while for the sample with $d$ = 20 nm they are: 0.85 (0.84); 0.81 (0.81) and (0.78); the lowest-energy peaks are absent in the SPV spectra. The good agreement between the SPV and PL features implies that they originate from electronic transitions between the same electronic states in the samples. We have checked that these features are absent in the PL spectrum of bulk InP. Thus, in accordance with our previous works on PL spectroscopy of InAs/InP QWRs [2-4, 8], we can also ascribe the SPV features to excitonic transitions arising in QWR families with heights differing by an integer number of MLs. The small difference of ~ 10 meV between the SPV and PL peak positions represents the Stokes shift for the QWR exciton.

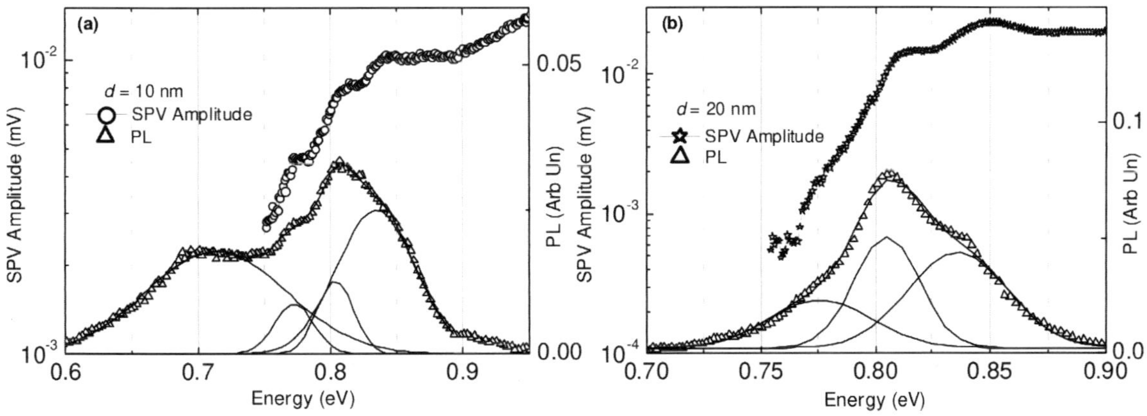


**Figure 2.** SPV amplitude and PL spectra of the samples with $d = 10$ (a) and 20 nm (b). The PL band is decomposed to Gaussians.

In order to determine the QWR family heights we have compared the experimental results with numerical calculations of the electronic structure performed by the ETB method. The calculated transition energies for different QWR heights as a function of [P] are shown with solid lines in figure 3. The horizontal dashed lines in the same figure represent the experimental peak energies for both samples. From the multiple intersections of the two line sets we have selected the ones giving the best agreement. The procedure is described and verified in [13]; it requires that i) H increase by 1 ML between peaks and ii) [P] remain nearly constant for the different QWR families. Hence, the results suggest that the peaks around 0.85, 0.81 and 0.77 eV originate form QWR families with heights 7, 8 and 9 MLs, respectively.

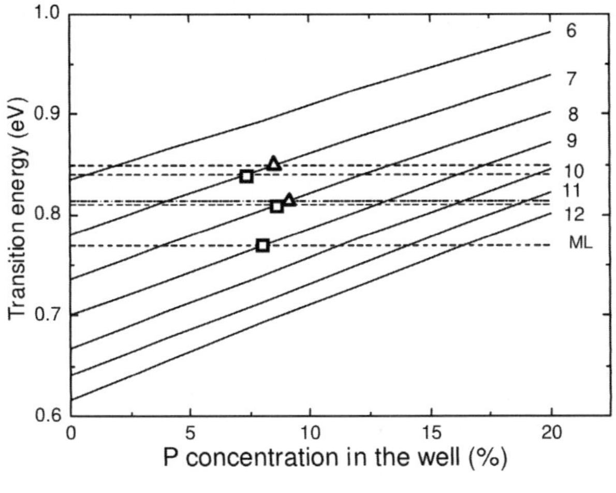


**Figure 3.** Transition energies of a single InAsP/InP QWR as a function of the P content for QWRs heights from 6 to 12 MLs. The dashed lines show the SPV peak values and their relevant intersections with the calculated curves are illustrated by squares ($d$ = 10 nm) and triangles ($d$ = 20 nm).

The PL peak with the lowest energy (0.72 eV) in the sample with $d$ = 10 nm is much broader and could account for a number of families (with heights 10-12 ML). Alternatively, this emission could originate from larger QWRs formed by the overlapping of two or more small wires. Such wires have been observed by atomic-scale cross-sectional scanning tunneling microscopy [15].

The intersections of the curves in figure 3 show also that a constant P concentration in the QWRs of the order of 8 % can effectively represent the effects of the InAs/InP intermixing on the optical transition energies. The error in [P] can be estimated to ± 4% and it results from the uncertainty in the

InAs/InP valence band offset (0.3-0.4 eV), which gives an error of ±(2-3)% [13], the experimental error in the transition energies (±5 meV) giving ±0.6% and the effect of the strain relaxation during the QWR formation (< 1% [13]) which is not taken into account in the calculations.

Similar samples with InP spacer layers, grown at higher temperatures (515°C), have been investigated in our previous study [8]. The comparison with the present results shows that the current QWRs are higher, but have a lower equivalent P concentration. This height difference is supported by the TEM results [3]. It is explained by the fact that at higher growth temperature of the InP spacers an enhanced P/As exchange takes place, which favours the formation of lower InAs wires with smaller cross section aspect ratio, as pointed out in [3]. The lower equivalent P concentration obtained for the present QWRs corresponds with their less efficient P/As exchange and larger cross section aspect ratio.

The slight difference in peak positions between the samples with $d$ = 20 nm and $d$ = 10 nm is probably due to the increase of the electron coupling between adjacent layers with decreasing $d$, as discussed in [4].

In order to obtain a deeper understanding of the SPV results and to study the mechanisms of carrier separation necessary for the SPV signal formation, we analyze in parallel the SPV amplitude and phase spectra shown on figure 4. For that purpose we have used our vector model [11], in which the SPV signal is represented by a radial vector, with magnitude equal to the SPV amplitude and angle with respect to the X-axis equal to the SPV phase. The occurrence of more than one SPV process is then represented by summation of the corresponding SPV vectors.

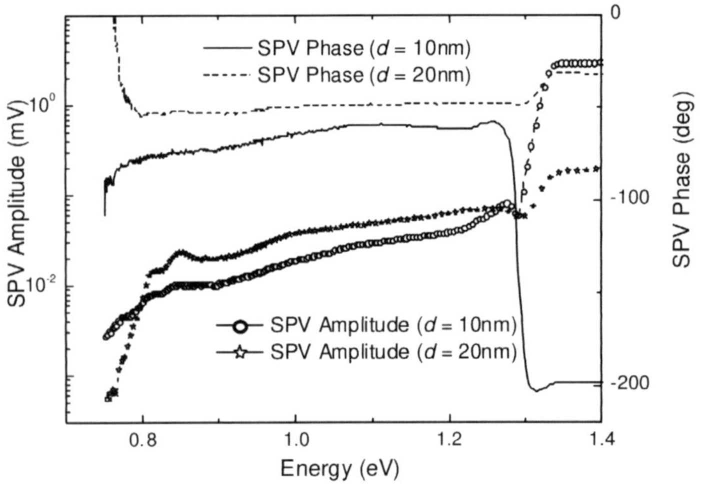


**Figure 4.** SPV amplitude and phase spectra of both samples in the whole spectral range under study.

In the range 0.75 – 0.9 eV, where the QWR spectral features are observed in the PL and SPV amplitude spectra, the SPV phase is in IV$^{-th}$ quadrant (see figure 4). This indicates an upward energy band bending in the region of the QWR structure with respect to the bulk [10, 11]. In fact an upward band bending at the sample surface is expected because of the n-type residual doping of the multi-layer structure and buffer layer. The QWR structure is then situated in the space charge region (SCR), which arises at the sample surface with a built-in electric field [see figure 5(b)]. The SPV signal from the QWRs is due to photocarriers, which thermally escape from the InAs potential wells into the InP barriers, where they are separated by the SCR built-in electric field. Since the carrier escape is easier for higher energy states, the SPV peaks at higher energies are stronger. Thus, the SPV spectra extend the information obtained by PL, where the lowest energy transition is predominantly observed. For energies from 0.9 to 1.27 eV the SPV phase remains nearly constant, while the amplitude smoothly increases, which suggests that in this range the optical absorption is associated with excited states in the multilayer structure.

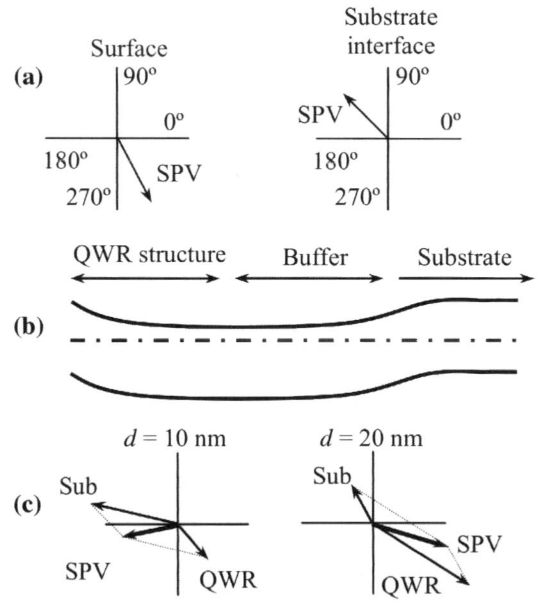


**Figure 5.** SPV vectors for both interfaces (a); alignment of the energy bands (b); summation of the SPV vectors of the substrate (Sub) and of the multilayer structure (QWR) for both samples (c)

At higher energies the SPV amplitude spectrum for both samples reveals a minimum around 1.29 eV followed by a strong step in the range 1.29 - 1.34 eV. In the same energy range a jump occurs in the SPV phase spectrum. The step is associated with band to band absorption in InP (bandgap energy 1.34 eV). The amplitude minimum and the phase jump at 1.29 eV can be explained taking into account the nearly opposite phases of the SPV signals from the multilayer structure and from the bulk InP. The first one includes the signals from the InP spacer layers and from excited states in the InAs QWRs. The phase in this case is in $IV^{-th}$ quadrant due to the upward energy band bending, as explained above. The second one includes the signal from the interface buffer/substrate, where the band bending is downward with respect to the bulk, because of the different layouts of the Fermi levels in the two materials [n-type InP buffer and SI InP substrate, see figure 5(b)]. Therefore, the SPV signal associated with the band to band absorption in the bulk InP is represented by a vector in $II^{-nd}$ quadrant (vector Sub) [10, 11], contrary to the SPV vector of the multilayer structure (vector QWR).

The superposition of the SPV signals from the multilayer structure and from the bulk InP is represented in figure 5(c) by summation of two SPV vectors nearly opposite in orientation (vector QWR in $IV^{-th}$ and vector Sub in $II^{-nd}$ quadrant). It is easy to understand that this superposition leads to a minimum in the SPV amplitude and a SPV phase jump, as observed around 1.29 eV. The different strength and direction of the SPV phase jump for the two samples can be explained by different orientation and magnitude of these sets of vectors. In the case of the sample with $d = 10$ nm, where the multilayer structure is thinner, the bulk InP vector is relatively strong as compared to the multilayer vector and forms an obtuse angle with it, as shown in figure 5(c) left. The growth of vector Sub leads to clockwise rotation of the overall SPV vector across $III^{-rd}$ quadrant. The bulk InP vector of the sample with $d = 20$ nm, where the multilayer structure is thicker, is relatively weak as compared to the multilayer vector and forms an obtuse angle with it, as shown in figure 5(c) right. Therefore the growth of vector Sub in this case leads only to a small anticlockwise rotation of the overall SPV vector and an initial SPV amplitude minimum at 1.29 eV. The following increase in the SPV amplitude is due to the increased contribution of the InP spacer layers [the QWR vector in figure 5(c) right].

## 4. Conclusion

We present a comparative study of the optical absorption and emission of multi-layer InAs/InP QWRs, performed by RT SPV and PL spectroscopies in combination with ETB calculations and structural data obtained in previous studies. The energies of the numerous features in the SPV spectra are in good agreement with the ones observed in PL. They are interpreted as due to excitonic transitions, which arise in QWR families with different heights. The values of the later (7, 8 and 9 MLs) are estimated from the comparison of the experimental and the calculated transition energies. Using this comparison we have found also that the degree of P intermixing in the wires is equivalent to a constant P concentration of the order of 8%.

The dependence of the QWR height on the growth temperature of the InP spacer layers is discussed and explained by the corresponding temperature dependent behaviour of the P/As exchange processes.

Due to the residual n-type doping of the QWR structure and buffer layer, the energy bands are bent upward (downward) at the sample surface (at the interface buffer/substrate), which results in non-trivial features in the SPV spectrum, which have been successfully explained with the help of our vector model for SPV signal representation.

Finally, it should be noted that this study demonstrates the abilities of the SPV spectroscopy for characterization of complex nanostructures at RT.

**Acknowledgements**
This work was supported by the Sofia University research fund (contract 149/2010), Ministry of Education and Science of Spain (MAT2009-10350, FEDER) and the Alexander von Humboldt Foundation. The authors would like to thank assoc. prof. Germanova for the critical reading and fruitful discussions.

**References**

[1] Schwertberger R, Gold D, Reithmaier J P and Forchel A 2003 *J. Crystal Growth* **251** 248–252
[2] Alen B, Martínez-Pastor J, Garcia-Cristobal A, González L and Garcia J M 2001 *Appl. Phys. Lett.* **78** 4025
[3] Fuster D, González M U, González L, González Y, Ben T, Ponce A, Molina S I and Martínez-Pastor J 2004 *Appl. Phys. Lett.* **85** 1424
[4] Fuster D, Martínez-Pastor J, González L and González Y 2006 *J. Phys.* D*: Appl. Phys* **39** 4940
[5] Sidor Y, Partoens B, Peeters F M, Ben T, Ponce A, Sales D L, Molina S I, Fuster D, González L and González Y 2007 *Phys. Rev.* B **75** 125120
[6] Angelova T, Cros A, Cantarero A, Fuster D, González Y and González L 2008 *J. Appl. Phys.* **104** 033523
[7] Mazur Y I, Dorogan V G, Bierwagen O, Tarasov G G, Jr E A D, Noda S, Zhuchenko Z Y, Manasreh M O, Masselink W T and Salamo G J 2009 *Nanotechnology* **20** 065401
[8] Donchev V, Ivanov T, Angelova T, Cros A, Cantarero A, Shtinkov N, Borisov K, Fuster D, González Y and González L 2010 *J. Phys.: Conf. Ser.* **210** 012041
[9] Kronik L and Shapira Y 1999 *Surf. Sci. Rep.* **37** 1
[10] Donchev V, Kirilov K, Ivanov T and Germanova K 2006 *Mat. Sci. & Eng.* B **129** 186
[11] Ivanov T, Donchev V, Germanova K and Kirilov K 2009 *J.Phys.D: App.Phys.* **42** 135302
[12] Ivanov T, Donchev V, Wang Y, Djie H S and Ooi B S 2007 *J. Appl. Phys.* **101** 114309
[13] Dion C, Desjardins P, Shtinkov N, Robertson M D and Schiettekatte F 2008 *Phys. Rev.* B **77** 075338
[14] Fang Z M, Ma K Y, Jaw D H, Cohen R M and Stringfellow G B 1990 *J. Appl. Phys.* **67** 7034
[15] Ulloa J M, Koenraad P M, Fuster D, González L and González Y 2008 *Nanothecnology* **19** 445601

Journal of Physics: Conference Series **253** (2010) 012044 doi:10.1088/1742-6596/253/1/012044

# Solar cell technology on the base of III-V heterostructures

**P Vitanov[1], M Milanova[2], E Goranova[1], Ch Dikov[1], Pl Ivanov[1] and V Bakardjieva[1]**

[1]Central Laboratory of Solar Energy and New Energy Sources, Bulgarian Academy of Sciences,72, Tzarigradsko Chaussee Boulevard, 1784 Sofia, Bulgaria,
[2]Central Laboratory of Applied Physics, 59, St.Petersrburg Boulevard, 4000 Plovdiv, Bulgaria

E-mail: vitanov@phys.bas.bg

**Abstract.** Low-cost single-junction GaAs solar cells on the base of p-AlGaAs/p-GaAs/ n-GaAs heterostructures grown by low-temperature liquid-phase epitaxy have been fabricated. For analyzing the factors contributing to solar cell efficiency the program products PC1D have been used. The optimal doping concentrations and thicknesses of the layers which correspond to the maximum values of the solar cell output parameters have been determined. The multilayer solar cell structure has been optimized on the base of this simulation. Low cost technological processes and masks, necessary for fabrication of the cells on the base of the grown heterostructures, have been developed, too. The front and back metallization of the cell is Ni/Al system deposited by magnetron sputtering. A double layer $Al_2O_3$ /$ZrO_2$ antireflection coating is deposited on the front side of a solar cell structure by the spin coating technique. The conversion efficiencies of 19 % at one sun AM1.5 conditions have been measured.

## 1.Introduction

As shown previously [1-3] high-efficiency solar cells with enhanced spectral response can be developed on $Al_xGa_{1-x}As$/GaAs heterostructures. Conversion efficiencies up to 25% at (AM1.5) one sun illumination and 27-28 % at concentrated light were obtained, as well. Recently, interest in the GaAs single junction is raised again. The new concepts is the use of films of GaAs or AlGaAs grown in thick, multilayer epitaxial assemblies, then separated from each other and distributed on foreign substrates by printing [4].

Melt-grown epitaxial films, prepared by low temperature Liquid Phase Epitaxy (LPE), are successfully used [5, 6] to fabricate high-quality multilayer AlGaAs/GaAs heterostructures containing layers as thin as 2-20 nm combined with several microns thick ones and having a smooth surface and flat interfaces.

In this paper, we investigate thin film multilayer heterostructures for solar cell application, grown by low temperature LPE. The aim of the investigations is to optimize the layer thickness, doping and composition, in order to ensure high efficiency solar cell preparation.

## 2. Heterostructure preparation

### *2.1. Epitaxy*

Journal of Physics: Conference Series **253** (2010) 012044 doi:10.1088/1742-6596/253/1/012044

The AlGaAs/GaAs multilayer heterostructures are grown in a horizontal LPE reactor using multiple bin "piston" boat on n-type (100) oriented 15×20 mm GaAs substrate. In order to reduce the growth rate for preparing thin layers we use Ga-melt only 500 µm thick and the growth temperature is lowered down to 480°C where the As-fraction in the melt becomes too low. In the used piston boat the substrate surface after the first wetting stays always covered by the melt. The latter ensures wetting of Al-containing melts at low growth temperature in the range 500-600°C which is a very difficult problem during the AlGaAs heterostructures preparing. The source materials used in the melt are pure Ga (6N) and Al and undoped boat-grown polycrystalline GaAs. The dopants are: Te for n-type GaAs and Ge or Mg for p-type GaAs and AlGaAs layers. The growth is carried out in a purified $H_2$ gas flow after pre-heating of the charged boat at 800°C for 1h in order to decrease the contaminations of the residual impurities in the melt. The initial epitaxy temperature was 670°C and the p-n junction was formed at 520°C.The thin AlGaAs window layer was deposited at 480°C.

*2.2.Samples structure*

The investigated heterostructures for high efficiency solar cell application contain 5 epitaxial films melt-grown on n-GaAs substrate: a AlGaAs buffer layer, a n- type GaAs layer as a base, a $p^+$ type GaAs emitter, an ultra- thin AlGaAs "window" layer , and a thin heavily doped $p^{++}$- type GaAs contact layer. The latter forms a stable low-resistance contact to the p-side and protects the window layer during the production processes. As back surface field, the introduction of potential barrier at the back of the base improves the collection efficiency of the minority carriers generated in the photoactive region. This barrier is made by growing a n- type $Al_xGa_{1-x}As$ buffer layer on a GaAs substrate. The surface recombination on the front side of the solar cell can be minimized by depositing an ultra-thin film (window layer) of wide band gap material and transparent to short- wavelength light. In our case that is $Al_xGa_{1-x}As$ with 80-85 % AlAs content.

For analyzing the factors, contributing to solar cell efficiency the solar cell simulator PC1D was used [7]. The effect of layers' thickness on the window and emitter layers as well as the doping of the window, the base and the emitter layers on the device efficiency was assessed. The multilayer structure of the solar cell was optimized on the base on these results. The scheme of optimized structure grown in this work is shown in the figure 1.

| | | |
|---|---|---|
| 0.2- 0.3 µm | $p^+$ - GaAs | CAP |
| 0.02-0.05 µm | p - $Al_{0.8}Ga_{0.2}As$ | WINDOW |
| 1.0 -1.5 µm | p - GaAs | EMITTER |
| 3.0 - 4.0 µm | n - GaAs | BASE |
| 2.0 - 3.0 µm | $n^+$ - $Al_{0.3}Ga_{0.7}As$ | BSF LAYER |
| 400 µm | $n^+$ - GaAs | SUBSTRATE |

**Figure 1.** Schematic diagram of the optimized multilayer AlGaAs/ GaAs structure for solar cell application.

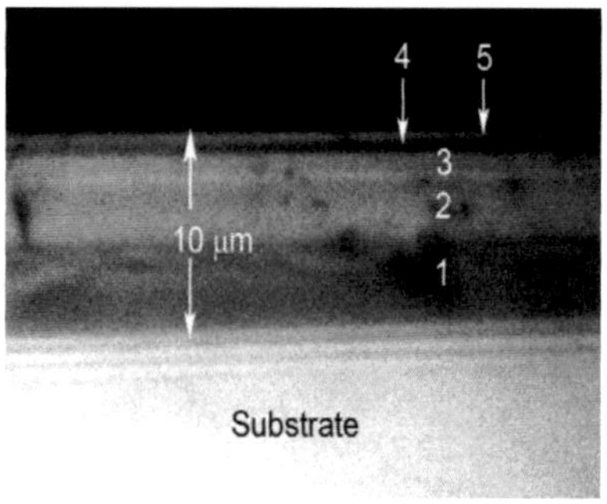


**Figure 2.** Cross-sectional view of the heterostructure: Substrate; 1 - BSF layer, 4µm; 2-Base, 3.1µm; 3 - Emitter, 1.5µm; 4 - Window layer; 5 - Capping layer , 0.2 µm

## 3. Heterostructure characterization

The layer thickness is measured on (110) cleaved cross-sections using SEM and optical microscopes. Figure 2 shows an optical photomicrograph of the heterostructure. The latter shows layers in a compositional contrast since their colours are different due to the different band gap energy. The first and the fourth wide-gap layers seem darker. They are nearly equal in colour, actually violet in the used colour gamma. The nanoscale window layer 4 is seen just like an interface between the area of the layers 3 and 4. The measured thicknesses of the layers are near the technologically predicted ones: 4 μm for the BSF layer 1, 3.1 μm for the base 2, 1.5 μm for the emitter 3 and 0.2 μm for the capping layer 5. The window layer 4 is estimated to be less than 0.1 μm in the limits of the optical measurements. Figure 2 shows also that the interfaces are flat and of a good quality without macroscopic defects.

The molar fraction of AlAs in the ternary AlGaAs solid solution measured on the same cross-section by EDAX technique in BSF regions x = 0.23.

The thicknesses and composition of the "window" layer are estimated by a combination of Raman scattering with anodic oxidation method [8]. The principle of this method is a precise step of the anodization structures and subsequent detection of the Raman spectra. Analysis of the Raman intensity and Raman shift on the thickness of the anodization material make it possible to estimate the thickness and composition values of the layer. Anodic oxidation of the heterostructures was performed in tartaric acid-glycol-water electrolytes. The consumption material rate was found to be 1nm/V. The Raman measurements were performed at room temperature in the backscattering configuration. The sample was excited by the 514.5 nm line of an $Ar^+$ ion laser. In order to determine the thickness and composition of the window layer the solar cell structures without heavily doped $p^+$ GaAs layer on the surface were grown.

Figure 3 presents the Raman spectra of the one as grown structure (curve 1) and after anodic oxidation at anodic voltage 30 and 40 V (curves 2 and 3, respectively). Since in back scattering geometry along [100] the transverse optical phonon (TO) frequency is forbidden for the zinc - blend compounds only the longitudinal optical (LO) phonon peak shifts appeared.

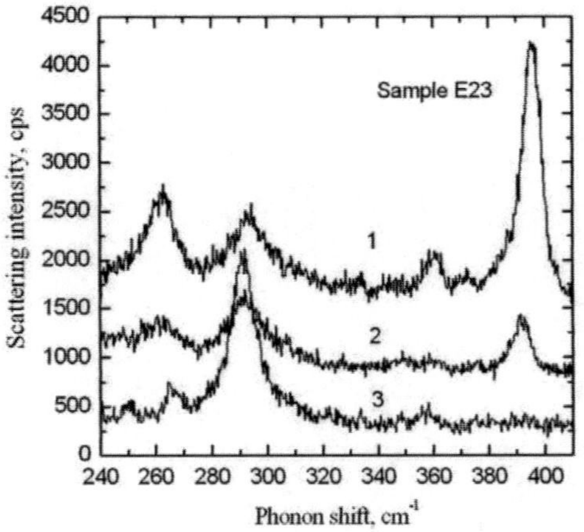


**Figure 3.** Raman spectra at 300 K from grown structure without heavily doped $p^+$ GaAs contact layer: 1 - as grown structure; 2 - after oxidation at 30 V; 3 - after oxidation at 40 V

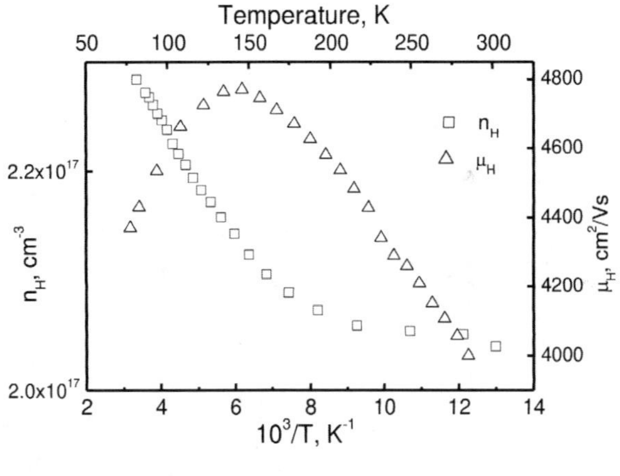


**Figure 4.** Temperature dependence of carrier concentration and Hall mobility of epitaxial n-type GaAs layer on a semiinsulating substrate.

Raman spectra of AlGaAs give two mode optical phonon frequencies of GaAs like LO1 phonon and AlAs like LO2 phonon. The peak shift at 262 $cm^{-1}$ of LO1 phonon and 396 $cm^{-1}$ of LO2 phonon were measured of the as grown structure. Using the dependence of the frequencies of the LO phonons in mixed $Al_xGa_{1-x}$ As crystals as a function of the composition x, established by the authors of ref. [8, 9] we determined the AlAs content in the window layer to be about 80%. The LO phonon peak shift seen at 292 $cm^{-1}$ is coming from the GaAs emmiter. Analysis of Raman spectra of the structure after precisely etching of the window layer at each stage of oxidation allowed us to evaluate the window thickness of about 40 nm. After step oxidation of the structure at 10, 20 and 30 V the intensity of the optical phonons of the window layer drops in accordance with the decrease of the window thickness. As seen in the figure after oxidation of 30nm of the structure (curve 2) a weak LO1 and LO2 peak shift from very thin graded composition AlGaAs window were measured. Further oxidation of the structures at 40 V (curve 3) leads to the fully etched of the window layer and the Raman spectra gives only LO phonon peak shift of the GaAs emitter layer with high intensity.

The separate control layers of heterostructures grown on semi insulating GaAs substrates are characterized by Hall-effect measurements. Carrier concentration $n_h$ and Hall mobility $\mu_h$ in the range of 80 - 300 K are measured. The temperature rates of $n_h$ and $\mu_h$ for the n-type base layer of the structures are shown in figure 4. Reasonable values and behaviour are obtained which correspond to good quality epitaxial layers.

These results have been applied at AlGaAs/GaAs solar cell fabrication.

## 4. Solar cell processing and results

### *4.1 Solar cell design*

A set of 4 masks for a chip with a multiplication step of 2.75 mm is developed. Solar cells (2.5 x 2.5) $mm^2$ in size with a circle illumination area of 3.14 $mm^2$ are prepared. With mask 1 the contact $p^+$ GaAs thin surface layer is etched off selectively until the window is open in the regions outside the contact grid. The front metal grid is defined by lift off photolithography with mask 2. The width of the metal lines is 8 μm, while the space (illuminated window) between the lines is 92 μm. The front and back metallization of the cell is Ni/Al system deposited by magnetron sputtering. A typical specific resistance of $5.10^{-5}$ Ohm.cm was achieved for n-type contact and higher than $10^{-4}$ Ohm.cm for p-type contact.

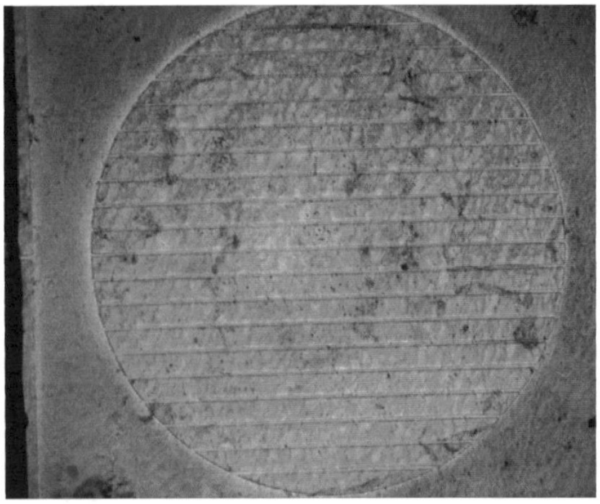

**Figure 5.** A photograph of a multilayer AlGaAs/GaAs based solar cell.

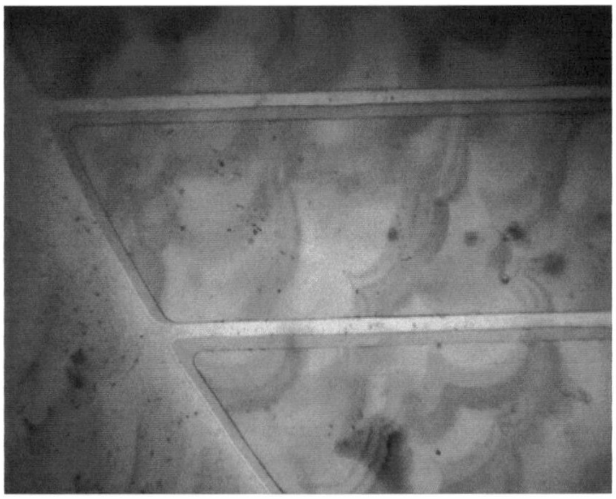

**Figure 6**. A detail view of the metal grid front side.

The cell has a double antireflection coating (ARC), $Al_2O_3$ (45 nm) on a top of $ZrO_2$ (49 nm) with the refractive index equal to 1.48 and 1.9 respectively. A double layer ARC is deposited on the front side of a AlGaAs/GaAs solar cell by the spin coating technique - a cheap, non expensive method with fairly reproducible results. The optimal thickness of the two layers was determined by minimization of the average weighted reflection [10]. A theoretical value of 2.17% is reached for top layer ($Al_2O_3$) thickness of 45 nm and bottom layer ($ZrO_2$) thickness of 49 nm. The double-layer ARC on the contact pads is removed with mask 3 and the coating remains only on the solar cell active area - the central circle with diameter 2 mm. Finally, a mesa etching is applied using mask 4 in order to isolate each solar cell on the wafer.

A solar cell structure without $p^+$GaAs contact layer on the surface and with two region emitter: the main region with optimal parameters determinated from the simulation process and an additional heavily doped emitter layer grown on the top of the main region in order to ensure good ohmic contact [2] is realized, too. This technological version is preferred because thus the total number of masks is reduced to three.

*4.2 Results*

An automatic measuring system under standard test conditions AM1.5 is used to assess the solar cell basic parameters: short-circuit current $J_{sc}$, open-circuit $V_{oc}$, fill factor FF and conversion efficiency η. Figures 7a and 7b show typical I-V curves of the two types of cells measured under standard test conditions.

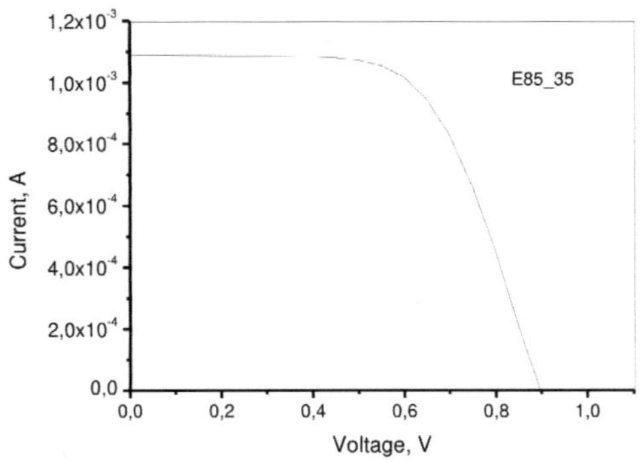


**Figure 7a.** Experimental I-V curve of typical AlGaAs/GaAs based solar cell with cap layer.

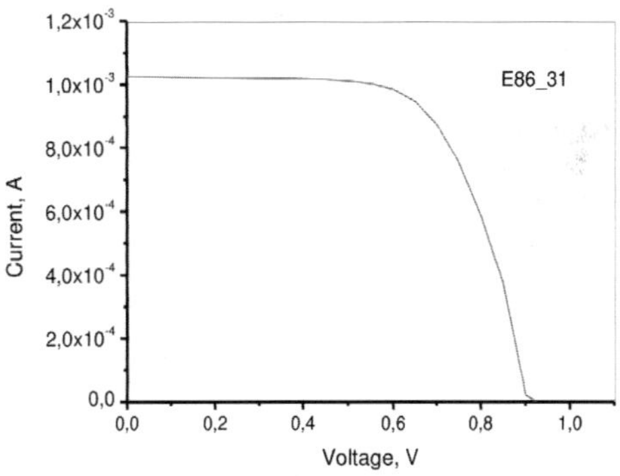


**Figure 7b**. Experimental I-V curve of typical AlGaAs/GaAs based solar cell without cap layer.

The substrates used for epitaxial growth are of dimensions 15×20 mm and the 22 chips of the solar cells are formed on them. Typical values of conversion efficiency η of 18.7 % ($J_{sc}$=29.87 mA, $V_{oc}$= 0.92V, FF=0.68) were obtained for the most chips for sample E86 which is without cap layer. The best measured values are 19.5% which is a very good result for one-junction solar cells. The deviation of 1.5 % for all 22 cells is due to the possible defects of the technological realization which is made at laboratory conditions and not at "clean room" conditions. Comparable conversion efficiencies exhibit the samples with $p^+$ GaAs cap on the surface.

## 5. Conclusions

Multilayer heterostructures have been grown by low temperature Liquid Phase Epitaxy on a heavily doped $n^+$GaAs substrate. The solar cell simulator PC1D 4.5 is used to optimize the epitaxial layer thickness and doping concentrations. Electron concentration and Hall mobility are investigated on test structures grown on semiinsulating substrates. Results show that epitaxial layers of very good crystalline properties are prepared by LPE. The technological processes and the masks, necessary for the fabrication of devices based on AlGaAs/GaAs heterostructures have been developed and solar cells with the best conversion efficiency of 19.5% under AM1.5 one sun conditions fabricated.

**Acknowledgements**

This work has been financed by the Ministry of Education and Science of Bulgaria under the contract D-01-369/2006.

**References**

[1] Tobin S P, Vernon S M, Bajgaret C, Wojtczuk S, Melloch M R, Keshavarzi A, Stellwag T B, Venkatensan S, Lundstrom M and Emery K A 1990 *IEEE Trans.Electr. Dev.* **37** 469
[2] Algora C, Oritz E, Rey-Stolle I, Diaz V, Pena R, Andreev V M, Khvostikov V P and Rumyantsev V D, 2001 *IEEE Trans. Elec. Dev.* **48** 840
[3] Andreev V M, Khvostikov V P, Larionov R, Rumyantsev V D, Paleeva E V and M. Shvartz M Z 1999 *Fiz. Tekh. Poluprovodn.* **33** 1070
[4] Yoon J, Jo S, Chun I S, Jung I, Kim H, Meitl M, Menard E, Li X, Coleman J J, Paik U, Rogers J A, 2010 *Nature* **465** 329
[5] Andreev V M, Kazantsev A B, Khvostiokov V P, Paleeva E V, Rumyantsev V D, Sorokina S V, 1996 *Mat. Chem. and Physics* **45** 130
[6] Milanova M and Khvostikov V 2000 *J. Cryst. Growth* **219** 193
[7] Ivanov P, Vitanov P, Goranova E, Milanova M, Kakanakov R and Alexieva T 2009 *J. Optoel. Adv. Mater.* **1** 238
[8] Andreev V M, Mintairov A M, Nahimovich V D, Rumyantsev V D, Khvostikov V P, Yakimov A Yu 1989 *Procc I$^{st}$. Sov. Conf. Solid-State Electronics (Leningrad, USSR ,1989* vol B) pp 102-103
[9] Andreev V M, Larionov V R, Mintairov A M, Prutskih T A, Rumyantsev V D, Smekalin K E and Khvostikov V P 1990 *Tech. Phys. Let. (Russ.)* **16** 7
[10] Alexieva Z I, Nenova Z S, Bakardjieva V S, Milanova M M and Dikov Hr 2010 *J. Phys.: Conference Series* **223** 01204

Journal of Physics: Conference Series **253** (2010) 012045 doi:10.1088/1742-6596/253/1/012045

# Study of melt-grown GaAsN and InGaAsN epitaxial layers

**P Vitanov[1], M Milanova[2], B Arnaudov[3], S Evtimova[3], E Goranova[1], G Koleva[2], V Bakardjieva[1] and G Popov[2]**

[1]Central Laboratory of Solar Energy and New Energy Sources, 72 Tzarigradsko Chaussee Blvd, 1784 Sofia,Bulgaria,
[2]Central Laboratory of Applied Physics,59 St.Petersrburg Blvd. 4000 Plovdiv,Bulgaria
[3]Faculty of Physics, Sofia University, 5 J.Baucher Blvd, 1164 Sofia, Bulgaria

E-mail: vitanov@phys.bas.bg

**Abstract.** This paper presents investigation of electronic transport properties of GaAsN and InGaAsN epitaxial layers with low nitrogen content, the so called dilute nitrides, grown by liquid-phase epitaxy (LPE). The layers up to 2 microns thick have been grown from Ga- and In-rich melt at different initial epitaxy temperatures in the range 660-620 °C. Polycrystalline GaN has been used as a source for nitrogen. As grown, unintentionally doped GaAsN and InGaNAs are n-type with free carrier concentrations one order of magnitude higher than those for the reference nitrogen free undoped GaAs and InGaAs layers. Lattice matched to GaAs substrate InGaAsN layers exhibit Hall mobility values higher than 2000 $V/cm^2.s$

## 1. Introduction

The quaternary solid solution InGaAsN proposed by Kondow [1] for the realization of long wavelength lasers based on GaAs has attracted considerable attention during the last few years. Numerous laboratories have demonstrated that the optical energy-gap of GaInAs decreases substantially when small amounts of N are incorporated into the solid. More importantly, InGaNAs alloys can be grown lattice-matched to GaAs because In and N have opposing strain effects on the lattice (N adds tensile strain, In - compressive). Using this quaternary system, it should also be possible to engineer a strain-free graded band gap base layer.

The incorporation of small quantity of nitrogen into GaAs also deteriorates the crystalline and optoelectronic properties of the dilute nitride materials, including reduction of the photoluminescence intensity and life-time, decreasing of electron mobility and increasing in the background carrier concentration. Even for the low N content the concentration of N interstitial is the major defect formed in nitrides-arsenides due to the large size mismatch between the As and N atom. The existence of high density of N interstitials and Ga vacancies in the as-grown structures will significantly affect device performance, since they would introduce band-gap levels affecting the carrier collection and recombination in transport process. GaAsN- based solid solutions and heterostructures are primarily grown by metaloorganic vapor-phase epitaxy (MOVPE) [2, 3] and molecular-beam epitaxy (MBE) [4, 5], but the material quality has been inferior to that of GaAs. Recently chemical-beam epitaxy (CBE) has been developed in order to improve the quality of the grown layer [6], but to date it has remained a challenge to grow dilute nitride materials for photovoltaic (PV) application. We propose low-temperature LPE [7] as a new growth method for these dilute nitrides. Moreover, it is a challenge to demonstrate thick, lattice-matched and of high electrical and structural qualitiy InGaAsN layers needed for development of multijunction solar cells. In this work, our focus is specifically on LPE growth conditions of thick InGaAsN films on GaAs (001) substrates with a polycrystalline GaN as N source.

## 2. Experiment

A series of GaAsN and InGaAsN epilayers have been grown on semi-insulating (100) GaAs substrates. The crystal growth is performed in a horizontal quartz tube using a graphite boat designed for 10×15 $mm^2$ substrate. Pure 99,9999 Ga and 99, 9999 In were used as solvents, and polycrystalline GaN and GaAs were used as sources for N and As, respectively. Epitaxial InGaAsN with In- content in the layer lower than 0.1 at% which acts as isovalent dopant were grown from Ga+(5-10%) In- melt. In order to obtain ternary and quaternary InGaAs and InGaAsN alloys with a few percent In content the growth was done from 80%In +20%Ga melt. Prior to growth, substrates were etched in $5H_2SO_4$: $H_2O_2$: $H_2O$ for 4 minutes and rinsed in deionized water. The charged boat was heated for 1 hour at 800 °C in order to reduce the contamination and to homogenize the melt. The layers with thicknesses of 1-2 μm were grown in the temperature interval of 10°C from initial epitaxy temperatures varying in the range 660-620 °C at a cooling rate of 0.6 °C/min.

The values of composition parameters x and y of the epitaxial $In_xGa_{1-x}N_yAs_{1-y}$ layers were determined by X-ray microanalisys and X-ray diffraction (XRD) methods.

The Hall electron mobility and free carrier concentration were measured in the temperature range 80-300K by conventional Van der Pauw method on layers grown on 5x5 $mm^2$ semi-insulating GaAs substrates.

## 3. Results and Discussion

Epitaxial GaAsN and InGaAsN layers grown from Ga and Ga+(5-10%)In -melt with 1.5 at.% GaN content from 660 °C initial epitaxy temperature are with thicknesses of 1 μm and N content in the layers varies from 0.1 at. % for GaAsN to about 0.2 at. % for InGaAsN layers. The presence of 5-10% In in the melt enhances the incorporation of N in the grown layers.

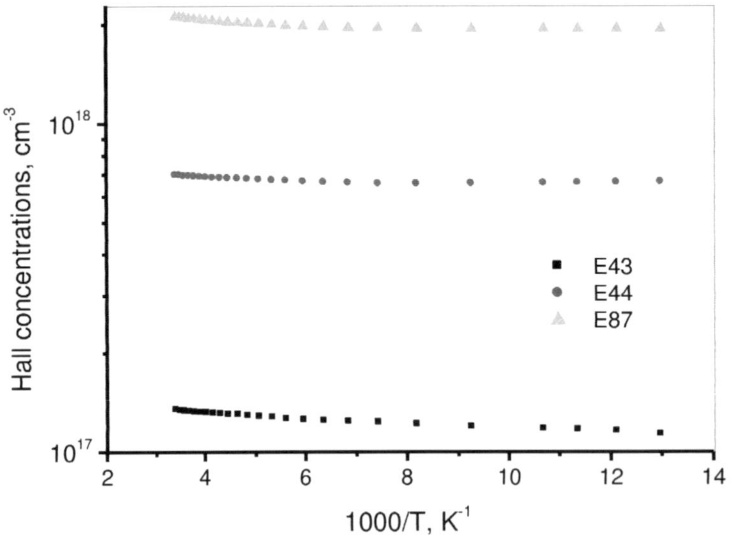


**Figure 1.** Free carrier concentration as a function of inverse temperature for as grown GaAs, GaAsN and InGaAsN layers grown from Ga-rich solution

Figure 1 shows the temperature dependence of the Hall-concentration $n_H$ on reciprocal temperature for as grown GaAsN (sample E44) and InGaAsN (sample E87) layers in comparison with undoped GaAs (sample E43). It is seen that all samples are of n-type and for layers containing nitrogen electron concentration increases about one order of magnitude. This could be explained by the assumption that nitrogen behaves mainly as an isoelectronic donor, which arises from the local heterojunction scheme GaAs-GaN according to Belliache et al [8]. The N-related point defects and Ga- vacancies have a dominant role on electrical properties and increase the background concentrations.

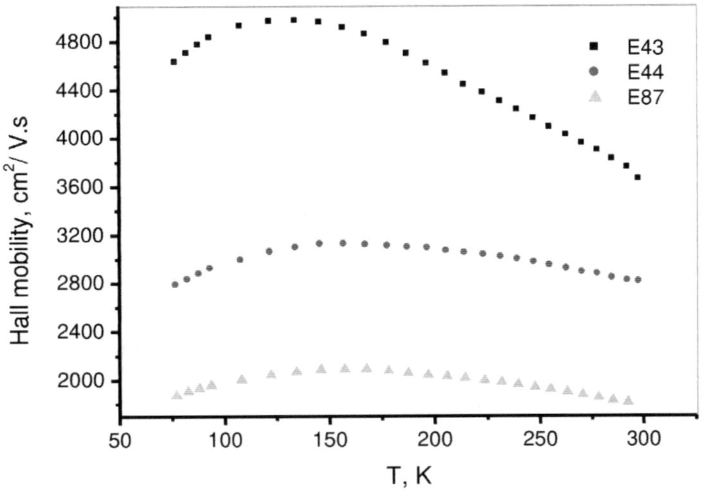


**Figure 2.** Temperature dependence of Hall electron mobility for GaAs (squares), GaAsN (circles), InGaAsN (triangles)

Figure 2 presents the temperature dependencies of the Hall-mobility for the same samples E43, E44 and E87. The mobility of the samples doped with nitrogen (E44 and E87) is considerably lower due to space charge scattering contributions induced by N-related defects added to well-known scattering mechanisms such as phonon and ionized impurity scattering. It is seen a well expressed low-temperature mobility decrease which could be explained by the temperature dependence of the GaAs conduction band edge energy, which is closer to the N defect levels at lower temperatures, increasing the scattering cross-section.

Another series of lattice matched InGaAsN and metamorphic InGaAs and InGaAsN layers were grown from In-rich melt. An intentionally undoped InGaAs layer and nitrogen doped one were grown under the same growth condition in the range 650-640 ˚C. X-ray microanalisys data show that In content in nitrogen free InGaAs layer is 3.75 at. % in the solid, while for N-doped layer it sharply decreases to 1.3 at.%. This result indicates that GaN in the melt suppresses the incorporation of In in Ga sub-lattice during epitaxial growth. These layers were lattice mismatched with rough surface as shown in figure 3a which makes it difficult to remove the melt from the surface after epitaxy.

Lattice matched InGaAsN with In content of 3.47 at. % were grown at lower epitaxy temperatures in the range 620-610 ˚C. They exhibit smoother growth surfaces due to the lack of a highly dislocated region (figure 3b).

The log of free carrier concentration $n_H$ vs. reciprocal temperature for intententionally undoped metamorphic InGaAs (sample E93) and lattice matched InGaAsN (sample E97) are plotted in the figure 4. For for N-containing films, two distinct temperature regimes with different temperature dependence of $n_H$ are observed. Hall-concentration decreases as the temperature decreases down to

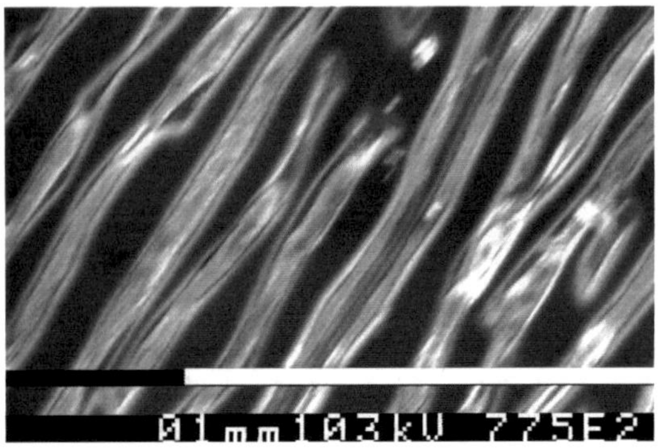


**Figure 3a.** The micrographs of the surface after epitaxy of undoped InGaAs (E93).

**Figure 3b**. The micrographs of the surface after epitaxy of lattice matched InGaAsN (E97).

about 200 K, indicating the presence of thermally activated deep donor levels within the dilute nitride bandgap. The saturation of $n_H$ at low temperature (T < 200K) is attributed to fully ionized shallow donors. This temperature dependence of the $n_H$ is similar to that observed in n-type AlGaAs and attributed to the co-existence of a shallow donors and a deep DX-center donors [8,9]. In our case this behavior could be explained by the presence of two donor levels in the InGaAsN bandgap, one being a shallow N isoelectronic donor and the second a thermally activated deeper donor, presumably N-related deep-level defects typically associated with different N-N pair and N-cluster states [10,11].

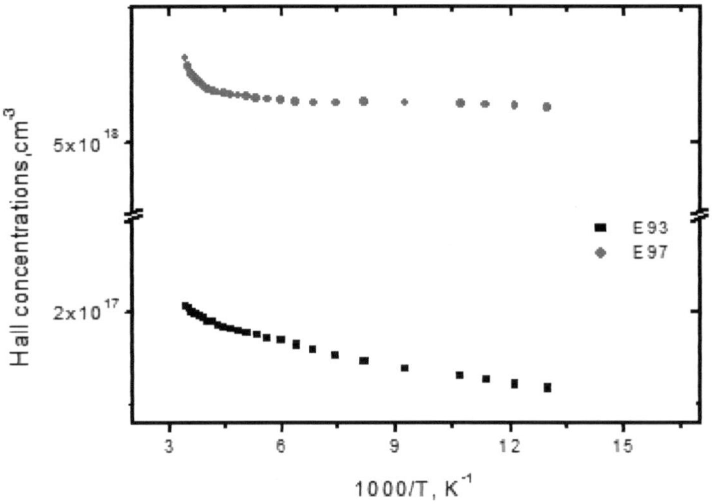


**Figure 4.** Temperature dependence of free carrier concentrations for undoped InGaAs (E93) and lattice matched InGaAsN (E97) samples

In the case of InGaAs layer, $n_H$ decreases almost linearity in the explored temperature range, 80 to 300K, typical for slightly degenerate III-V semiconductors. The deviation from linearity is due to the background N- doping of this layer grown at the same graphite boat as N-doped layers.

The temperature dependence of Hall mobility for undoped InGaAs E93 and InGaAsN E97 layers grown from In-solution is similar to those for the layers grown from Ga-rich solution as it is shown in figures 5a and 5b.

The mobility of the metamorphic InGaAs structure (E93) is low, down to 500 $cm^2/V.s$, since it possibly contains threading dislocations of high density and the latter causes relatively poor material quality. High values over 2000 $cm^2/V.s$ for Hall mobility exhibits the lattice matched to GaAs

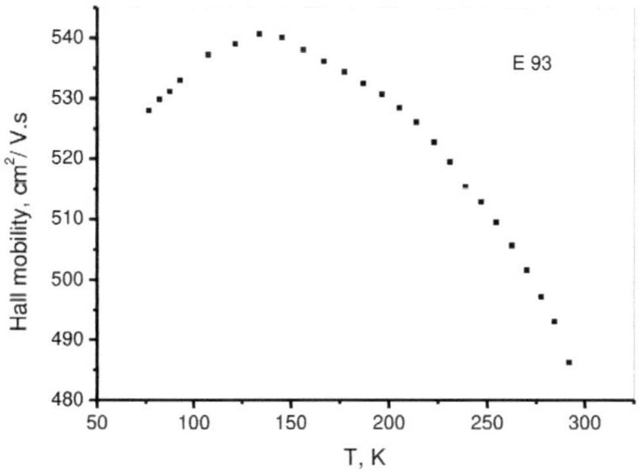


**Figure 5a.** Hall mobility as a function of temperature for undoped InGaAs (E93) sample

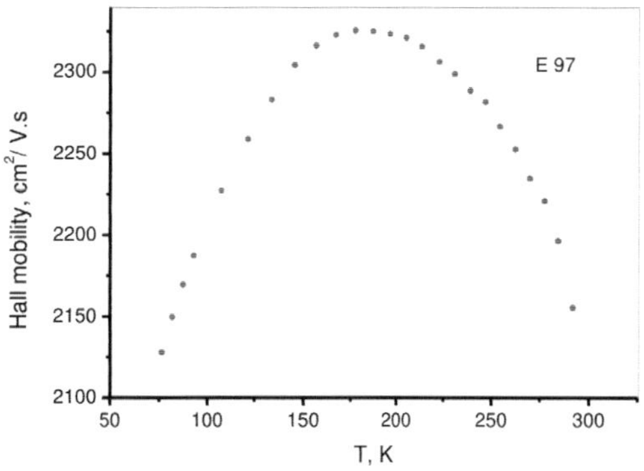


**Figure 5b.** Hall mobility as a function of temperature for lattice matched InGaAsN (E97) sample

substrate InGaAsN sample E97. These values are about the theoretical limit predicted by Fahy and O'Reilly [12] and the highest reported for lattice matched thick InGaAsN layers.

## 4. Conclusions

Dilute nitride GaAsN and InGaAsN epitaxial layers have been grown by LPE using polycrystalline GaN as a source for nitrogen. The layers grown from Ga-rich melt are with 0.1-0.2 at. % N content in the solid. Hall measurements reveal sharply increase of free carrier concentrations about one order of magnitude and decrease of Hall mobility for nitrogen doped samples in comparison with undoped GaAs.

Metamorphic and lattice matched InGaAsN layers with 3-4 at. % In content in the solid have been prepared from In-rich melt. Temperature dependent electronic transport measurements show a thermally activated increase in free carrier concentration at measurement temperatures higher than 200 K, suggesting the presence of a carrier trapping level below the GaAsN conduction band edge. The mobility for metamorphic layers is low (about 500 $cm^2/V.s$) possibly due to high density threading dislocations. Lattice matched InGaAsN with In content of 3,47 at % have been grown in temperature range 620-610 ˚C from 80%In+20%Ga melt. The measured Hall mobility values over 2000 $cm^2/V.s$ are the highest reported to date for thick InGaAsN layers lattice matched to GaAs substrate.

## References

[1] Kondow M, Uomi K, Niwa A, Kitatani T, Watahiki S and Yazawa Y 1996 *Jpn. J. Appl. Phys.* **35** 1273
[2] Kurtz S, Allerman A A, Jones E D, Gee J M, Banas J J and Hammons B E 1999 *Appl. Phys. Lett.* **74** 729
[3] Kurtz S R, Allerman A A , Seager C H, Sieg R M, and Jones E D 2000 *Appl. Phys. Lett.* **77** 400
[4] Ptak A J, Friedman D J, Kurtz S, and Keihl J *Proc. 31st IEEE PVSC( Lake Buena Vista, Florida, January 3-7 2005)* **31** p p 603-606
[5] Liua T, Chandrilb S, Ptakc A J, Korakakisa D and Myers T H 2007 *J. Cryst. Growth* **304** 402
[6] Yamaguchi M, Nishimura K, Sasaki T, Suzuki H, Arafune K, Kojima N, Ohsita Y, Okada Y, Yamamoto A, Takamoto T and Araki K 2008 *Solar Energy* **82** 173
[7] Milanova M and Khvostikov V 2000 *J. Cryst. Growth* **219** 193
[8] Watanabe M, Morizuka K, Mashita M, Ashizawa Y, and Zohta Y1984 Jap. J Appl. Phys 23 L103

[9] Schubert E F and Ploog K, 1984 *Phys. Rev.* B **30** 7021
[10] Zhang Y, Mascarenhas A 2000 *Phys. Rev.* B **61** 15562
[11] Francoeur S, Nikishin S A, Jin C, Y. Qiu Y and Temkin H 1999 *Appl. Phys. Letters* 751538
[12] Fahy S, O'Reilly E P 2004 *Physica* **E 21** 881

# Quartz roughness affect on $WO_3$ coated QCM

**V Georgieva[1], Z Raicheva[1], A Grechnikov[2], V Gadjanova[1], M Atanassov[1], J Lazarov[1] and E Manolov[1]**

[1]Georgi Nadjakov Institute of Solid State Physics, Bulgarian Academy of Sciences, 72 Tzarigradsko chaussee blvd., 1784 Sofia, Bulgaria

[2]Vernadsky Institute of Geochemistry and Analytical Chemistry of RAS, Kosygin str., 19, 119991 Moscow, Russia

E-mail: lazarova@issp.bas.bg

**Abstract**. The influence of the initial quartz roughness on the $WO_3$-QCM (Quartz Crystal Microbalance) parameters and sorption properties was investigated. Experiments were carried out using AT-cut quartz crystal with different surface roughness obtained after treatment with SiC abrasive having grain size of 3 µm, 7 µm, 14 µm and 20 µm respectively. The QCMs were covered with thin $WO_3$ films. The equivalent dynamic parameters of the $WO_3$- QCM were measured and a correlation with the surface roughness was determined. The equivalent dynamic resistance of the as–created QCM increases from 15.42 Ω to 117 Ω and the quality factor decreases as a result of the roughness changes. The surface morphology was observed by a Scanning Electron Microscopy (SEM), which showed higher increase of the $WO_3$ surface roughness compared to those of the initial quartz surface. The cross-section profiles of the investigated structures show different overlay of Au-electrodes. The sorption properties of $WO_3$-QCM system to $NH_3$ were studied in the range from 10 ppm to 1000 ppm. The response and recovering times were determined. The experimental results show an increase of the sorption ability with the increase of the initial quartz roughness. By increasing the roughness the mass-loading is improved.

## 1. Introduction

The adsorption properties of the gas sensing layers depend on the surface morphology. If they are nanostructured of significant importance is the particles' size, packing of nanoparticles etc. The adsorption properties of sensitive films formed by gold nanoparticles with different content are investigated using mass-sensitive (QCM) sensors [1]. It was established that the change of the surface concentration of nanoparticles leads to change of the sensors' sensitivity to different alcohols. Another way for enhancing the adsorption is formation of self-assembled monolayer (SAM) on the surface of gold QCM electrode. Such investigation was carried out by Kim and coworkers [2], who found that a circular and strip type structures of SAM show extremely high adsorption to the octadecanethiol solutions in hexane and alcohol, compared to the smooth surfaces. Mérian et al. have investigated the ammonia absorption properties of polyaniline [3]. The results show that the oxidative doping is very efficient to obtain a highly adsorbing and sensitive layer. The adsorption properties of several $IrO_2$ layers with different morphologies, such as nanoblade, layered-column, incomplete-nanotube and square-nanorod have been studied [4]. $IrO_2$ crystals with nanoblade and layered-column morphologies

Journal of Physics: Conference Series **253** (2010) 012046 doi:10.1088/1742-6596/253/1/012046

show higher sensitivities than those with incomplete-nanotube and square-nanorod morphologies. By replace of the conventional gold electrodes with porous Hieda et al. have found that the effective adsorbing surface area and hence sensitivity was enhanced by a factor of 40 without worsen of the quality factor or the stability of the resonant frequency [5]. $WO_3$ nanoflake functionalized QCM sensors were fabricated for detection of trace sarin stimulant and dimethyl methylphosphonate [6]. The sensors have better reproducibility and stability as well as much shorter response and recovery time than those functionalized by conventional flat $WO_3$ surface. Faster response has also been observed for ZnO functionalized QCM sensors for detection of ammonia [7]. In the recent years, one-dimensional ZnO nanostructures, such as nanowire, nanorods, nanobelts, and nanotetrapod, have been attracted much attention since the dimension or the surface-to-volume ratio has great influence on the performance [8]. In this work the influence of the initial quartz surface roughness, obtained by different grain size abrasive on the $WO_3$ roughness, $WO_3$-QCM equivalent parameters and sorption properties to $NH_3$ is studied.

## 2. Experimental

For the purpose of the present work, quartz plates with different surfaces were processed by using SiC abrasive with grain size of 3 μm, 7 μm, 14 μm and 20 μm. Samples of polished quartz plates were fabricated too. QCMs were formed on AT-cut quartz plates with diameter of 8 mm. Au-electrodes with 4 mm diameters and thickness of 60 nm were evaporated on both sides. Thus, QCMs with resonance frequency of 16 MHz were fabricated. QCM's quality was evaluated by measuring the main equivalent dynamic parameter - dynamic resistance ($R_q$), by a Selective Level Meter, the other parameters such as quality factor (Q), were calculated [9]. The created structures were coated with a $WO_3$ layer deposited by RF sputtering at 13.56 MHz from W target in ($Ar + O_2$) mixture containing 50% $O_2$. The process was carried out at a total pressure ($P_{Ar} + P_{O2}$) of $4.10^{-2}$ mbar and deposition rate of about 19.5 nm/min. The refractive index and thickness of the $WO_3$ films were determined by ellipsometric measurements on control silicon wafer, placed next to the resonator structure. A Rudolph 435 ellipsometer λ of 632.8 nm was used. The surface morphology of the $WO_3$ films deposited on the resonators was determined using a SEM-Philips 515. The samples were covered additionally with C and Au thin layers in order to increase the resolution and because of avoiding the charge. The cross-section profiles of all samples were studied too. The sorption abilities of the $WO_3$ films were determined by measuring the resonance frequency shift of QCM at different $NH_3$ concentrations. The frequency-time characteristics (FTCs) for all samples were measured. The sorption properties of the $WO_3$-QCMs to $NH_3$ are measured in laboratory set up. The experimental unit and the methodology of measurements are described in details in [10]. The main stages of the measurement consisted of:

- purging of the structures with dry air;
- creating a certain $NH_3$ concentration;
- reaching saturation of the frequency values;
- purging of the structures with dry air.

The $NH_3$ concentration in the test chamber was controlled by mass flow controllers (MFCs) for $NH_3$ and diluting gas flows. The experiments were carried out at 26 °C with an accuracy of ± 0.2 °C. The QCM frequency was registered by frequency counter Hameg 8123 connected to the QCM and a computer for data recording. The relative error for frequency measurement was ± $5.25.10^{-7}$. The sorption ability of the system $WO_3$-QCMs was evaluated on the basis of the measured FTCs. Correlation between the total frequency shift (ΔF) and the sorbed mass (Δm) for AT-cut quartz is given by Sauerbrey equation [11], thus the sorbed mass was calculated.

## 3. Results and discussion

The measured $R_q$ and calculated Q-factor for the samples prepared on different quartz surfaces showed that with the increase of the abrasive grain size the $R_q$ value grows up while the Q-factor decreases. In the investigated interval for 3 μm grain size Rq = 15.42 Ω and Q-factor = $27.10^3$, whereas for 20 μm sample the values reach $R_q$ = 117 Ω and Q-factor = $4.10^3$ respectively. With the increase of the abrasive size the QCM parameters worsen. Refractive index of 1.94 and thickness of 227 nm were determined for the $WO_3$ films by ellipsometric measurements.

*3.1. Sorption properties of $WO_3$-QCM structure*

Sorption in one stage was observed for all measured FTCs in the $NH_3$ range of 10 ppm - 50 ppm for QCM realized on surfaces treated with SiC abrasive having grain size of 3 μm and 7 μm. For QCM formed on wafers treated with 14 μm and 20 μm abrasives one stage sorption process was established only at 10 ppm $NH_3$, which is probably due to the lower number of active areas on the rougher surface. Figures 1 and 2 illustrate typical FTCs in the case of one and two stage sorption process, measured at two $NH_3$ concentrations-50 ppm and 500 ppm, when treatment with 3 μm abrasive was conducted. The influence of the initial quartz surface on the sorption properties of $WO_3$-QCM was determined by comparison between FTCs measured at 500 ppm $NH_3$ for all investigated samples. Changes in the frequency shift and frequency recovering for the first 120 s were determined concerning the sorption and desorption processes. The sorption and desorption rates were calculated. The comparison between figure 2 and figure 3 shows decrease in the sorption rate from 0.68 Hz/s to 0.59 Hz/s. The same trend was observed for the desorption process - three times decrease (from 0.45 Hz/s to 0.16 Hz/s) for samples treated with 3 μm and 14 μm. It was found that samples treated with 3 μm and 7 μm give very similar FTC. The one difference is that samples treated with 7 μm possess higher mass-loading ability. The recovery time increases dramatically as compared with samples treated with 14 μm, so such structures are not suitable for sensor application. The results show that the increased roughness embarrasses the sorption process and considerably increases the recovery time of the initial frequency of $WO_3$-QCM structure.

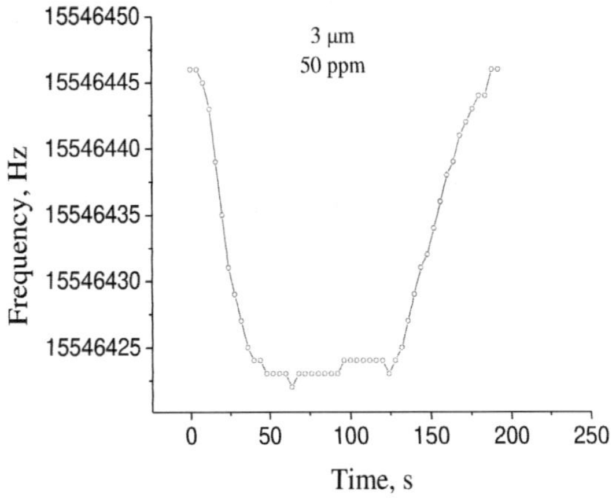


**Figure 1.** FTC of $WO_3$-QCM at 50 ppm $NH_3$ - quartz plate treated with 3 μm abrasive.

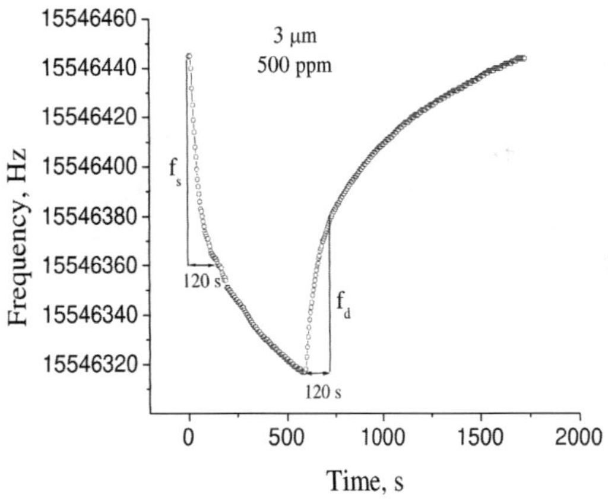


**Figure 2.** FTC of $WO_3$-QCM at 500 ppm $NH_3$ - quartz plate treated with 3 μm abrasive. $f_s$ and $f_d$ are frequency shifts for the first 120s ammonia sorption and desorption respectively.

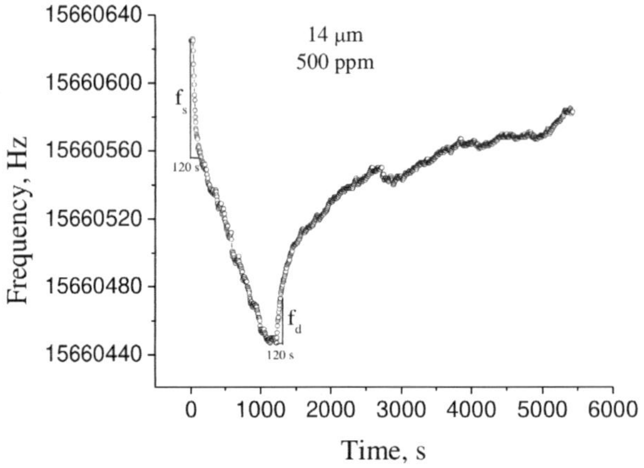


**Figure 3.** FTC of $WO_3$-QCM at 500 ppm $NH_3$-quartz plate treated with 14 μm abrasive.

The effect of the surface morphology on the maximal frequency shift for $WO_3$-QCM is shown in figure 4. The measurements were carried out in $NH_3$ range of 10 ppm to 1000 ppm. The maximum frequency shift for the samples treated with 14 μm abrasive is higher in comparison with those treated with 7 μm in the whole concentration range. With increasing of $NH_3$ concentration the difference in the frequency shift increases considerably and reaches three times higher values at 1000 ppm. The sorbed mass Δm calculated in accordance with the measured frequency shift and the Sauerbrey equation is shown in figure 5. The sorbed mass increases from 15.31 ng for samples treated with 7 μm abrasive to 42.37 ng for those treated with 14 μm at $NH_3$ concentration of 1000 ppm. This means that the rough surface increases the mass loading ability of a $WO_3$ - QCM structure.

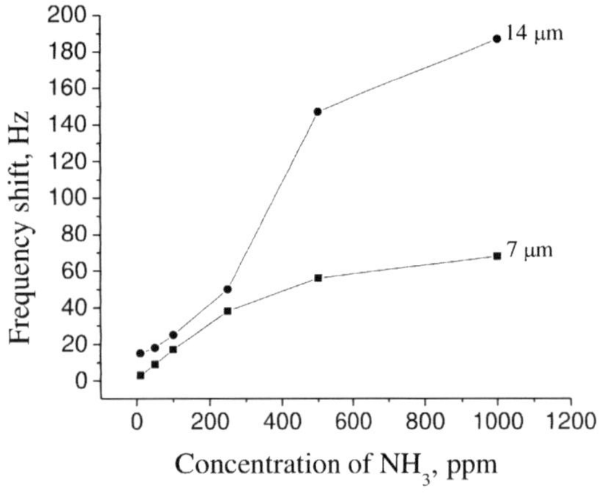


**Figure 4.** Dependence of the $WO_3$-QCM frequency shift towards $NH_3$ concentrations.

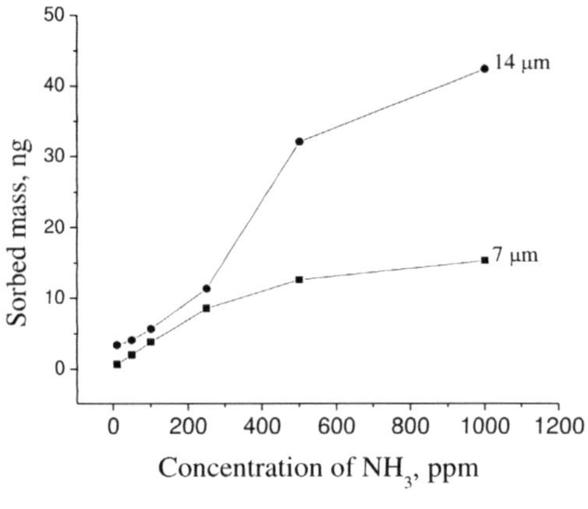


**Figure 5.** Mass-loading of the $WO_3$-QCM vs. $NH_3$ concentrations.

The measured sorption properties of the $WO_3$ - QCM structures can be partly explained with results obtained from SEM images for surface morphology and the cross-section profiles of the samples. Figures 6a and 6b illustrate the difference in $WO_3$ layer morphology deposited on the quartz plate and on Au electrode. Comparison between quartz - $WO_3$ and quartz - Au - $WO_3$ systems treated with 7 μm abrasive was made. It is obvious that the roughness of the $WO_3$ layer above Au is considerably higher, because it exactly follows the Au surface. The gradually increase of the $WO_3$ layer roughness of the system quartz - Au - $WO_3$ as a function of the different abrasives is shown in figure 7 a-d. For this purpose all samples were observed at a magnification of $5.10^3$. The surface changes from smooth for the polished samples to clusters. The predominantly grains size when 3 μm abrasive is used was 100 nm, at 7 μm this value increases to 250 nm and reaches size higher than 350 nm for 20 μm abrasive, i.e. the influence of the initial state of the surface is clearly shown.

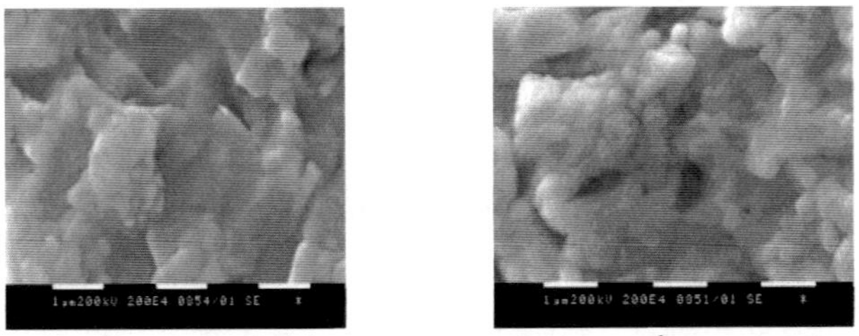

a b

**Figure 6.** Comparison of $WO_3$ surface roughness (7 μm ) at magnification $2x10^4$: a) deposited on quartz plate, b) deposited on Au electrode

**Figure 7.** SEM image of $WO_3$ thin film deposited on a) quartz plate - polished - Au; b) quartz plate treated with 3μm abrasive - Au; c) quartz plate treated with 7 μm abrasive - Au; d) quartz plate treated with 20μm abrasive - Au.

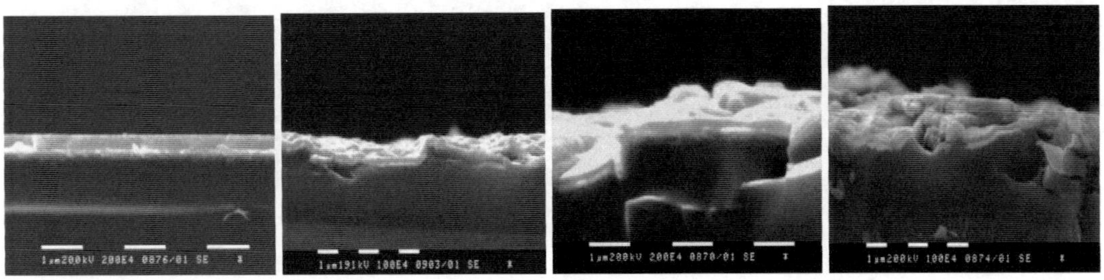

**Figure 8.** Cross section profiles image of the quartz - Au - $WO_3$ structures: a) polished quartz - Au - $WO_3$; b) quartz (3μm) - Au - $WO_3$; c) quartz (7μm) - Au - $WO_3$;
c) quartz (20μm) - Au - $WO_3$.

Deterioration of the equivalent parameters of the investigated $WO_3$ - QCM structures, as well as the difficulties in the sorption and desorption process are largely determined by the way of the Au electrode covering. Figures 8 (a)-(d) show quartz - Au - $WO_3$ profiles of the investigated samples. It can be seen that at the polished surface the $WO_3$ layer is smooth and follows the relief of the Au electrode (figure 8a). At the samples treated with 3 μm abrasive the $WO_3$ layer also follows the relief of the Au electrode, but it was observed presence of pinholes, which are filled. At the treated with 7 μm abrasive samples (figure 8c) the layer is continuous, but with rather rough surface. The quality of coating with 20 μm (figure 8d) is strongly worsened. The continuity of the mass sensitive layer is impaired with hollows. The distortion of the layer continuity causes high acoustic losses and strongly worsens the dynamical parameters of the structures. Discontinuity of the layers is a function of the quartz plate roughness - by its increasing the sorbing surface is enlarged, which results in higher sorbed mass. In the same time the sorption and especially the desorption processes are embarrassed because of the reduced layer's homogeneity. Depending on the application of the $WO_3$-QCM structures a compromise decision is necessary to be made for saving of the dynamical parameters and increasing of the structures' sorption ability by increasing of their sorbing surface.

## 4. Conclusion

In this study the influence of the initial quartz roughness on the $WO_3$-QCM parameters and sorption properties is presented. It is shown that both of them were strongly affected. By increasing the roughness, the mass-loading of the $WO_3$-QCM can be improved. On the base of the obtained results depending on the application a suitable condition for $WO_3$-QCM preparation can be chosen.

## Acknowledgements

This work was supported by Bulgarian Ministry of Science and Education under grant NT3/2006 and bilateral project (2009 - 2011) ISSP - BAS and GEOKHI - RAS.

## References

[1] Burlachenko J, Savchenko A, Shirshov Y, Kukla O, Kruglenko I and Belyaev O 2009 *Procedia Chemistry* **1** 301
[2] Kim D, Noh J, Hara M and H. Lee 2001 *Bull. Korean Chem. Soc.* **22** 276
[3] Mérian T, Debarnot D, Rouessac V and Poncin-Epaillarda F 2010 *Talanta* **81** 602
[4] Ohira S, Wanigasekara E, Rudkevich D and Dasgupta P 2009 *Talanta* **77** 1814
[5] Zhao Y, He J, Yang M, Gao S, Zuo G, Yan C and Cheng Z 2009 *Analytica Chimica Acta* **654** 120
[6] Wang X, Zhang J and Zhu Z 2006 *Appl. Surf. Sci.* **252** 2404
[7] Banerjee D, Lao J, Wang D, Huang J, Ren Z, Steeves D, Kimball B and Sennett M *Appl. Phys. Lett.* **83** 2061
[8] Sung W, Sung J and Lee S 2005 *Phys. Rev.* E **71** 031805
[9] Manolov S and Tihchev H 1982 *Generators* (Sofia Tehnika)
[10] Georgieva V, Stefanov P, Spassov L, Raicheva Z, Atanasov M and Tincheva T 2009 *J. Optoel. and Adv. Mat.* **11** 1363
[11] Sauerbrey G 1959 *Zeitschrift für Physik* **155** 206

Journal of Physics: Conference Series **253** (2010) 012047 doi:10.1088/1742-6596/253/1/012047

# Interface trap generation in MOS structures by high-energy electron irradiation

**E Halova[1], S Alexandrova[1], S Kaschieva[2] and S N Dmitriev[3]**

[1]Technical University-Sofia, 8 Kl. Ohridski Blvd., Sofia 1797, Bulgaria
[2]Institute of Solid State Physics, 72 Tzarigradsko Chaussee, Sofia 1784, Bulgaria
[3]Joint Institute of Nuclear Research, Flerov Laboratory of Nuclear Reactions, Dubna, Moscow region 141980, Russia

E-mail: ehalova@tu-sofia.bg

**Abstract.** The changes of defect characteristics of MOS structures induced by irradiation with 23 MeV electrons are studied. The samples were irradiated for 30 or 120 s. The distributions of radiation induced interface traps, located at the Si/$SiO_2$ interface obtained from capacitance-voltage (C-V) characteristics and decomposition of multi-frequency conductance-voltage (G-V) characteristics, are dominated by distinct defects, with different localized energy levels in the Si bandgap. The interface traps are related with $P_b$ centers, with concentrations depending on irradiation time. The $P_{b1}$ concentration shows an order of magnitude increase as compared to the reference unirradiated sample, while $P_{b0}$ are found only after electron irradiation. The C-V characteristics indicate an increase of the positive oxide charge in $SiO_2$ due to generation of E' defects. Defects in the Si bulk space charge region are also generated, which can be attributed to boron (V/B) and oxygen (V/O) vacancies complexes.

## 1. Introduction

In MOS devices the influence of ionising radiation on the electrical properties is of a special concern particularly with respect to reliability issues. In space or nuclear environment, the oxide layer of MOS structures may be subjected to high doses of ionising radiation, which is known to introduce traps in the oxide layers and at the interfaces. The generation of these traps subsequently may cause reliability problems and result in the degradation of the device electrical properties. The device parameters are affected by factors, such as type of ionizing radiation, time scale, dose, dose rate, device design, temperature etc. The electron beams in the present study have some special characteristics distinct from ions, neutrons and gamma irradiation.

The influence of radiation on the MOS structures has been extensively investigated through the years. Previous investigations on defect generation by irradiation with MeV electrons were concerned mostly to the low energy range [1, 2]. However, in order to achieve control over the radiation effects, it is important to gain detailed quantitative and qualitative information on the generated interface defect sites, when the MOS structures are exposed to high energy electron irradiation.

Our earlier experiments in the energy range of 11 to 23 MeV have shown creation of a spectrum of interface states in the silicon band gap. Some estimations of the energy depth in the Si band gap and capture cross section have been made [3] by means of TSC method. In the present contribution we describe the response of MOS structures to high-energy electron irradiation by assuming that it is the

Journal of Physics: Conference Series **253** (2010) 012047 doi:10.1088/1742-6596/253/1/012047

result of three effects: (i) generation of interface traps, (ii) generation of oxide traps, and (iii) generation of bulk Si traps.

## 2. Experimental details

### *2.1 Sample preparation*

The samples in this study were MOS capacitors prepared onto (100)p-type Si substrates with a resistivity of 44 Ohm. cm. Following standard RCA cleaning of silicon wafers, oxide layers with thickness of 50 nm, were grown at 1050 °C in dry oxygen ambient. After finishing the oxidation the $Si/SiO_2$ samples were cooled down in $N_2$ + $O_2$ ambient at a rate of 0.5 °C/s. MOS capacitors were formed by deposition of aluminium contacts with thickness 150 nm onto the oxide surface by thermal vacuum evaporation using a mechanical mask. To form an ohmic contact to the wafers, their back side was coated with a continuous thin Al layer.

The MOS structures were irradiated with 23 MeV electrons from the gate side parallel to the sample normal. The electron irradiation was carried out in vacuum at Microtron MT-25, in Flerov Laboratory of Nuclear Reactions of Joint Institute of Nuclear Research, (FLNR, JINR) Dubna, Russia. The distance between the Microtron window and the samples was 150 mm. The samples were bombarded with a flux of electrons, for different durations for 30 or 120 s, i.e. different irradiation doses. The average anode current of the electron beam at the samples was about 9 μA. At this energy, the electrons penetrate throughout the whole capacitor. No bias was applied to the devices during irradiation. Part of the MOS structures was left unirradiated to be used as reference samples for control measurements and comparative purposes.

### *2.2. Measurements*

The electrical properties of the MOS samples were examined by means of capacitance–voltage (C/V) and conductance-voltage (G/V) measurements performed at different frequencies from 500 Hz to 300 kHz at the room temperature. The measurement unit was a Precision Component Analyzer (WAYNE KERR 6425). These characteristics were analyzed in order to determine the distributions of radiation induced interface traps, located at the $Si/SiO_2$ interface, and to evaluate induced oxide charge in $SiO_2$ layer and the bulk traps in the Si space charge region after the high energy electron irradiation. The densities of the interface traps $D_{it}$ were inferred from standard HF C-V analysis comparing the measured and the ideal curves. The bulk traps were approached from analysis of the G-V curves.

The thickness of the oxides was determined by ellipsometry method, using a high precision Rudolf Research ellipsometer in the wavelength range of 280-820 nm, at an incidence angle of 70° with an accuracy of ±0.2 nm.

## 3. Results and discussion

Radiation effects in MOS devices occur in the relatively thin noncrystalline dielectric films and at the dielectric film/silicon interfaces. Accumulation of charges can be traced by analyzing the behavior of the electrical characteristics of the MOS structures. In figure 1 (a) and (b) the high frequency C-V and G-V curves, measured at 300 kHz, are displayed for the two irradiation durations. The C-V curve for the reference unirradiated oxide is also presented. The shape of the C-V curves and comparison with the reference oxide in figure 1 (a) shows typical for increased density of interface traps. The 120 s electron irradiation results in two overlapping peaks in the G-V curves in figure 1 (b). For the 30 s this is not obvious and the multiple peaks are visible only after decomposition as shown below.

The shifts of the C-V curve for irradiated oxides to more negative voltages indicate higher density of oxide traps. The fixed charge in the oxide of MOS structure was evaluated from the flat-band shifts between the experimentally measured C-V curves and theoretically calculated ideal C-V curves for the case of oxide without any interface and oxide charges. The oxide defect densities are given in table 1.It can be seen that the after irradiation the oxide charge increases over the initial fixed charge in the oxide of the reference sample. The increased positive oxide charge is usually attributed to the E' center, i.e. an oxygen-vacancy where each Si is back-bonded to three oxygen atoms. Note that the value for the smaller irradiation time is higher. However, this has to be attributed to the higher interface trap density in the lower part of the Si bandgap as will be seen below. The charge captured in these traps appears causes additional shift of the C-V curve, which results in higher density of the fixed oxide charge.

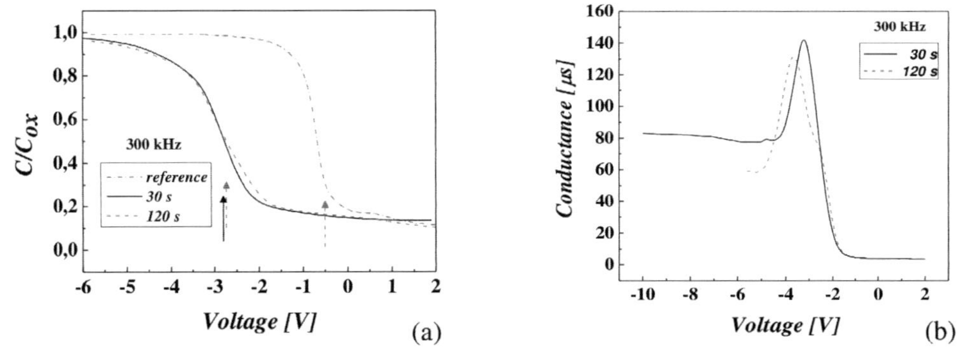


**Figure 1.** C-V (a) and G-V (b) characteristics of 23 MeV electron irradiated MOS structures for 30 and 120 s. The flatband voltages are marked with arrows.

**Table 1.** Fixed oxide charge $Q_{fb}$ and concentrations of $D_{it}$ at the $P_b$ defect centers.

| MOS structures | $Q_{fb}$ ($cm^{-2}$) | $P_{b0}$ ($eV^{-1}cm^{-2}$) | $P_{b1}$ ($eV^{-1}cm^{-2}$) |
|---|---|---|---|
| Reference | $1.5x10^{11}$ | - | $1.8x10^{11}$ |
| Irradiated 30 s | $9.0x10^{11}$ | $4.5x10^{12}$ | $1.8x10^{12}$ |
| Irradiated 120 s | $8.0x10^{11}$ | $6.8x10^{12}$ | $2.8x10^{12}$ |

The interface traps density distributions $D_{it}$ obtained from the 300 kHz C-V characteristics are shown in figure 2. Increase in interface defect concentrations with irradiation dose increasing is evident.

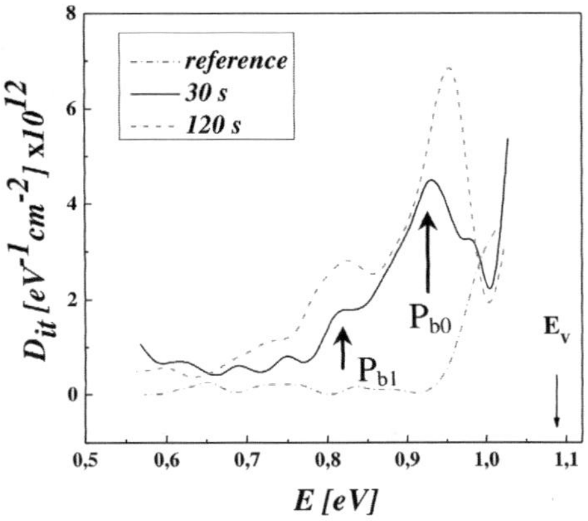


**Figure 2.** Interface trap profiles in the Si bandgap of irradiated MOS structures for 30 and 120 s. The profile for the reference sample is also given.

The steep rise of interface trap concentration near valence band edge for 30 s irradiation is seen as a higher density of the positive oxide charge in table 1. The interface trap distributions exhibit peaks in the lower part of the Si bandgap on side of the valence band. These peaks are attributed to that localized defect interface states. It can be seen that defects with close energy positions are observed for both irradiation times. This suggestion is confirmed by the analysis of the G-V characteristics. The shape of these curves in figure 1 (b) indicates the presence of localized states. This is especially evident for the MOS capacitor irradiated for 120 s. The localized states can be better seen after decomposition of the G-V curves as given in figure 3 (a) and (b).

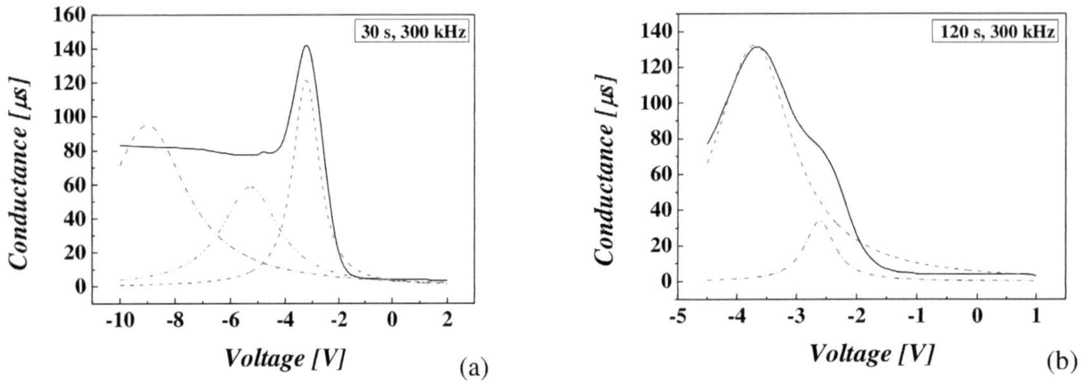


**Figure 3.** Decomposition of the G-V characteristics of irradiated by 23 MeV electron irradiation for 300 kHz for 30 s (a) and for 120 s (b) p-type MOS structures.

Usually radiation-induced interface states in oxides grown on silicon (defined as electronic levels located spatially at the oxide/Si interface and energetically within the band gap of the Si) are correlated with ESR active defects referred to as $P_b$ centers [4-6]. Their concentrations are given in table 1. It is well established that on (100)Si, a twofold $P_b$ centre, i.e. $P_{b0}$ and $P_{b1}$, are present [4]. Energetically they appear at the positions shown in figure 2, and are most probably donor type as discussed in ref. [7]. The $P_{b0}$ is a dangling Si bond with three Si back bonds (• Si ≡ $Si_3$), while the structure of the $P_{b1}$ center has remained controversial for a long time, and was first ascribed to dangling Si bonds with two Si and one O back bonds [4]. Recent studies [8] have confirmed the suggestion for the $P_{b1}$ to be a Si dangling bond in a strained Si-Si dimer configuration [9]. Irradiation is known to cause greater increases in the $P_{b0}$ than the $P_{b1}$ [10]. In our case the $P_{b1}$ center is better seen only after the irradiation for both durations, but is more pronounced at 120 s. The $P_{b1}$ concentration is about an order of magnitude greater as compared to the reference unirradiated sample, while $P_{b0}$ are found only after electron irradiation. The concentrations of both $P_{b0}$ than the $P_{b1}$ show 1.5-fold increase for 120 s as compared to 30 s irradiation.

Contribution of bulk traps [11], i.e. traps that reside in the space charge region in the Si substrate is detectable only by the G-V technique. The bulk traps manifest their presence as peaks in G/ω versus ω curves, with positions independent of the gate voltage, as shown in figure 4 (a) and (b). The capture cross section for 30 s irradiation is $1.2 \times 10^{-14}$ $cm^2$. Defects in Si substrate with similar parameters have been found in earlier TSC and DLTS experiments and attributed to di-vacancies and Si atoms recoiled from the $Si/SiO_2$ interfaces that migrated into the Si lattice [12, 13]. As it has been already shown the main defects created by high-energy electron irradiation in MOS structures are related to the main impurities in the Si substrates [12]. In our case (p-Si), boron is the main impurity of the Si substrates, and consequently, the defects are related to boron-vacancy and oxygen-vacancy defect complexes.

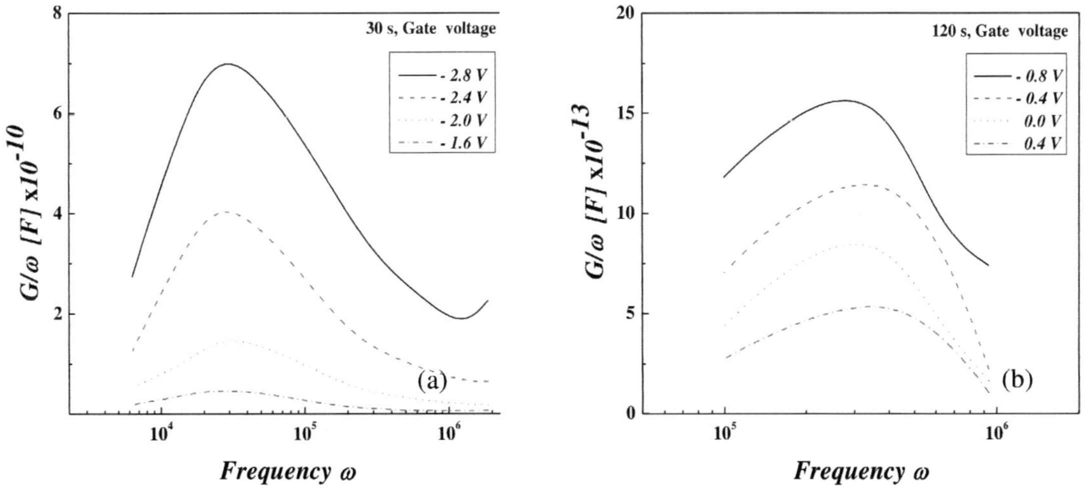


**Figure 4.** Frequency dispersion of the conductance of MOS structures irradiated for 30 s (a) and for 120 s (b).

## 4. Conclusions

Generation of three different types of defects was observed in MOS structures subjected to high energy electron irradiation of 23 MeV as revealed by the analysis of the C-V and G-V characteristics. The interface traps that are located at the interface exhibit distinct energetic levels in Si bandgap. The defects were identified as donor type $P_{b0}$ and $P_{b1}$ centers, with concentrations depending on irradiation dose. The $P_{b0}$ center was found only in irradiated samples. Positive oxide charge build-up was established upon irradiation due generation of E' centers. The higher density of this charge for the lower dose is attributed to the higher interface density near the valence band edge seen in the interface trap distribution. Bulk defect are also generated in the Si substrate which are ascribed to boron-vacancy and oxygen-vacancy defect complexes.

## References

[1] Klimenkov M, Matz V, Nepijko S and Lehmann M 2001 *Nucl. Instr. and Methods* B **179** 209
[2] Kachurin J, Ruault M, Gutakovsky A and Kaitaso O 1999 *Nucl. Instr. and Methods* B **147** 356
[3] Kaschieva S 1994 *Nucl. Instr. and Methods* B **93** 274
[4] Stathis J H and Dori L 1991 *Appl. Phys. Lett.* **58** 641
[5] Albohn J, Füssel W, Sinh N D, Kliefoth K and Fuhs W 2000 *J. Appl. Phys.* **88** 842
[6] Friedrich F, Boehme C and Lips K 2005 *J. Appl. Phys.* **97** 056101
[7] Lenahan P, Mishima T, Jumper J, Fogarty T and Wilkins R 2001 *IEEE Trans. Nuclear Science*, **48** 2131
[8] Kato K, Yamasaki T and Uda T 2006 *Phys. Rev.* B **73** 073302
[9] Stesmans A, Nouwen B and Afanas'ev V V 1998 *Phys. Rev.* B **58** 15801
[10] Kim Y Y and Lenahan P M 1988 *J. Appl. Phys.* **64** 3551
[11] Brunson K M, Sands D, Thomas C B and Reehal H S 1987 *J. Appl. Phys.* **62** 185
[12] Kaschieva S, Todorova Zh and Dmitriev S N 2004 *Vacuum* **76** 307
[13] Stefanov K G, Kaschieva S and Karpusov D 1998 *Vacuum* **51** 235

Journal of Physics: Conference Series **253** (2010) 012048 doi:10.1088/1742-6596/253/1/012048

# Effect of inserting of thin Rubrene layer on performance of Organic Light-Emitting Diodes based on $Zn(BTz)_2$

**R L Tomova, P K Petrova and R T Stoycheva-Topalova**

Institute of optical materials and technologies "Acad. J. Malinowski", Bulgarian Academy of Sciences, "Acad. G. Bonchev"str. bl. 109, 1113 Sofia, Bulgaria

E-mail: reni@clf.bas.bg

**Abstract**. Organic light-emitting diodes (OLEDs) with improved performances are fabricated using a thin (1 nm) yellow-emitting layer of 5,6,11,12-tetraphenylnaphthacene (Rubrene) inserted at different position in green emitting electroluminescent (EL) layer of bis-(2-(2-hydroxyphenyl) benzothiazole)zinc ($Zn(BTz)_2$) in configuration: ITO/PVK:TPD/ $Zn(BTz)_2$ (***x*** nm)/ Rubrene (1 nm)/ $Zn(BTz)_2$ (75-***x*** nm)/Al, where PVK:TPD is a hole transporting layer of N, N'-bis(3-methylphenyl)-N, N'-diphenylbenzidine (TPD) incorporated in poly(N-vinylcarbazole) (PVK) matrix and Al is a cathode. EL spectra predominantly influenced by Rubrene emission when the doping layer is close to (PVK:TPD)/ $Zn(BTz)_2$ ($x \rightarrow$ 0-15 nm) and to $Zn(BTz)_2$/Al ($x \rightarrow$ 70-75 nm) interfaces and shift toward emission of $Zn(BTz)_2$ increasing the distance of Rubrene from both interfaces ($x \rightarrow 35$ nm). The same dependence of the EL efficiency on the position of the doping Rubrene layer in the OLED structure was found.

## 1. Introduction

Organic light-emitting devices (OLEDs) have become particularly attractive because of their promising application, which offers advantages of low driving voltage, low power consumption, high contrast, wide viewing angle, low cost, and fast response. They are now emerging as the leading candidate among the many technologies under development for next-generation flat panel displays and solid state lightings. Today, various emission colors, including the three primaries of blue, green and red, can be obtained at high luminance levels, so these devices can be used in flat panel full color displays.

The doping technique has been widely used for fabricating OLEDs in order to improve device characteristics or to change the emission color. Doping the light emitting layer with appropriate fluorescent or phosphorescent dopants can tune the emission color or enhance the electroluminescence (EL) efficiency and the stability, compared with the undoped devices [1, 2]. The 5,6,11,12-tetraphenylnaphthacene (Rubrene) is one of the most useful red dopants extensively studied because of its higher photoluminescence quantum yield, which could enhance the efficiency and improve the stability of OLEDs [1-9]

While the doping via co-evaporation of host material and dopant is controlled very difficult the process is unacceptable in large-scale manufacturing of devices for commercialization. In 2005, Wang reported that OLEDs with better performance could be obtained by the deposition of an ultra thin dopant layer between hole-transporting and electroluminescent layer [3]. This idea was first used by

Tang in 1989 [1] and later by Li [4], Hao [5] and Beierlein [6] for the investigation of recombination and emission zones. OLEDs with ultra thin Rubrene layer also have been fabricated by Shi [7]. The enhancement of Rubrene's emission could be ascribed as an organic quantum well structure's trap effection [7].

In our previous paper we investigated the novel Zn complexes - acetylacetonate zinc 2-(2-hydroxyphenyl)benzothiazole (AcacZnBTz), zinc bis(2-(2-hydroxynaphtyl)benzothiazole) ($Zn(NBTz)_2$), zinc bis(2-(2-hydroxyphenyl) benzothiazole) ($Zn(BTz)_2$) as electroluminescent and electron transporting layers for OLED [10]. We have established that the novel Zn complexes are very promising for the development of OLEDs. In this paper we investigate the influence of the thin film of Rubrene inserted into green emitting electroluminescent layer of $Zn(BTz)_2$ on the chromaticity tuning and the basic characteristics of OLEDs.

## 2. Experimental

We investigated conventional structure: ITO/HTL/EL/M, where ITO was a transparent anode of $In_2O_3$:$SnO_2$, HTL – a hole-transporting layer of N, N'-bis(3-methylphenyl)-N, N'-diphenylbenzidine (TPD) incorporated in poly(N-vinylcarbazole) (PVK) matrix, EL - an electroluminescent layer of $Zn(BTz)_2$, and M - a metallic Al cathode. A thin layer of Rubrene (figure 1) was inserted into the $Zn(BTz)_2$ at different distances $\boldsymbol{x}$ from the (PVK:TPD)/$Zn(BTz)_2$ interface.

Devices with area 1cm$^2$ were prepared on commercial polyethylene terephtalate (PET) substrates coated with ITO (40 Ω/sq). The HTL of PVK:TPD film was formed by spin-coating 0.75 % solutions of PVK in dichloroethane containing 10 w% (relatively to PVK) of TPD at 2000 rpm. Layers of $Zn(BTz)_2$ (75 nm), Rubrene (1 nm) and Al cathode (120 nm) were prepared by thermal evaporation in vacuum better than $10^{-4}$ Pa at rates 0.2 - 0.5 nm/s. The rates of the depositions were controlled with a quartz crystal thickness monitor. In the investigated structure PET substrate, PVK, TPD and Rubrene were purchased from Aldrich, the electroluminescent compound $Zn(BTz)_2$ was synthesized in the Department of Applied Organic Chemistry at Sofia University by T. Deligeorgiev [11].

All measurements were performed at room temperature and under ambient atmosphere, without any encapsulation. The photoluminescence (PL) spectra were obtained by Spectrofluorimeter Perkin Elmer MPF 44, and the electroluminescence (EL) spectra by Ocean Optics HR2000+ spectrometer. The current-voltage (I-V) curves were measured by programmable with Labview power supply. The luminescence (L) was measured in continuous DC mode and the light output was detected using a calibrated Hamamatsu silicon photodiode S2281-01. The electroluminescent efficiency ($\eta_L$) was calculated by equation (1) and used for quantifying the properties of the OLEDs.

$$\eta_L = \frac{L}{I} \qquad (1)$$

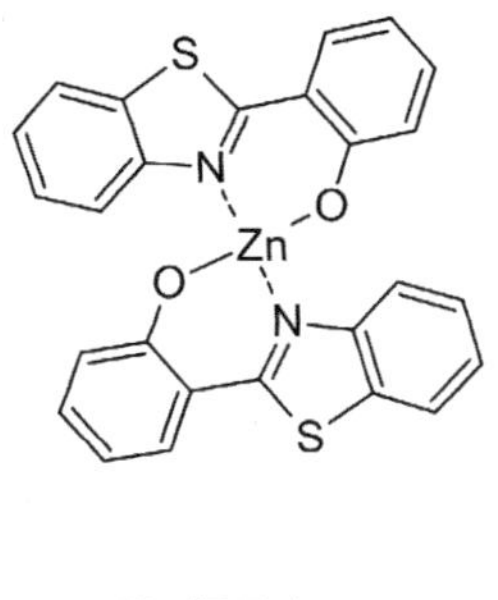

$Zn(BTz)_2$

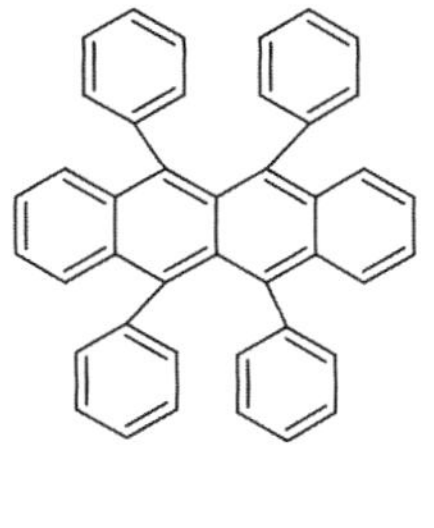

Rubrene

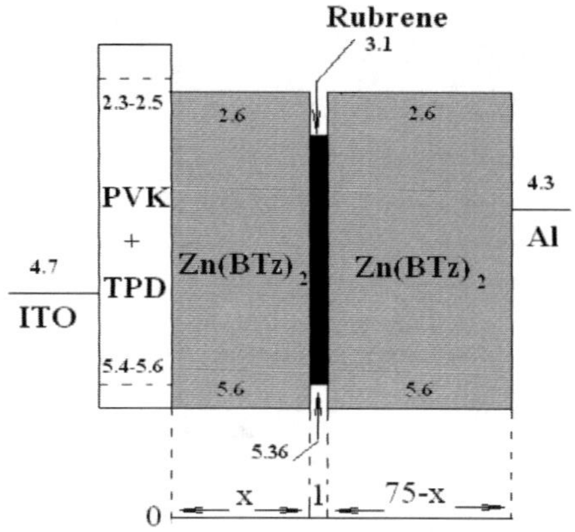


**Figure 1.** Chemical structures of $Zn(BTz)_2$ and Rubrene and energy diagram of investigated OLED.

## 3. Results and discussion

In our preliminarily investigation we studied the effect of the thickness $d$ ($0.5 \leq d \leq 15$ nm) of the inserted Rubrene layer on the performance of devices. The optimal electroluminescent efficiency was obtained for $d = 1$ nm. Here the ultrathin Rubrene layer may be thought as an intermediate case between the monolayer and a "localized doped" one (the size of a rubrene molecule is ~1.7 x 1.4 x 0.7 nm) in host of $Zn(BTz)_2$. That's why we started to investigate the influence of the distance $\boldsymbol{x}$ of 1 nm Rubrene layer from (PVK:TPD)/$Zn(BTz)_2$ interface in the structure ITO/(PVK:TPD)/$Zn(BTz)_2$(**x**)/Rubrene(1)/$Zn(BTz)_2$(75-**x**)/Al on the OLEDs characteristics.

Figure 2a presents the current density and the luminescence intensity versus voltage (figure 2b), and the electroluminescent efficiency versus luminescence (L) (figure 2 c) characteristics of the investigated devices. For comparison as references, curves for the structure: $Zn(BTz)_2$(75)/Al and Rubrene(64)/$Zn(BTz)_2$(10)/Al are added. We did not succeed to realize OLED with Rubrene only due to its recrystallization during the evaporation of Al cathode.

It is seen that inserting of Rubrene increases the current and the luminescence of the devices for each $\boldsymbol{x}$ except for $\boldsymbol{x}$=75 nm, where the layer is placed to cathode. At the same time the increase of the distance from $\boldsymbol{x}$(0) to $\boldsymbol{x}$(10) from the (PVK:TPD)/$Zn(BTz)_2$ interface decreases the electroluminescent efficiency, however for $\boldsymbol{x}$(10) to $\boldsymbol{x}$(70) increases $\eta_L$. Devices with $\boldsymbol{x}$(15) and $\boldsymbol{x}$(70) nm exhibit higher electroluminescent efficiency than the reference $Zn(BTz)_2$ in the whole interval of luminescence (figure 2 c). The best characteristics had the devices with $\boldsymbol{x}$(15) and $\boldsymbol{x}$(70) - at 15 V showed luminescence 375 and 1100 cd/m$^2$, respectively, corresponding to electroluminescent of 5.0 and 5.5 cd/m$^2$. The intensity of the luminescence was 1.5 and 4 times higher and the electroluminescent efficiency $\eta_L$ 1.75 and 2 times higher compared with the reference devices.

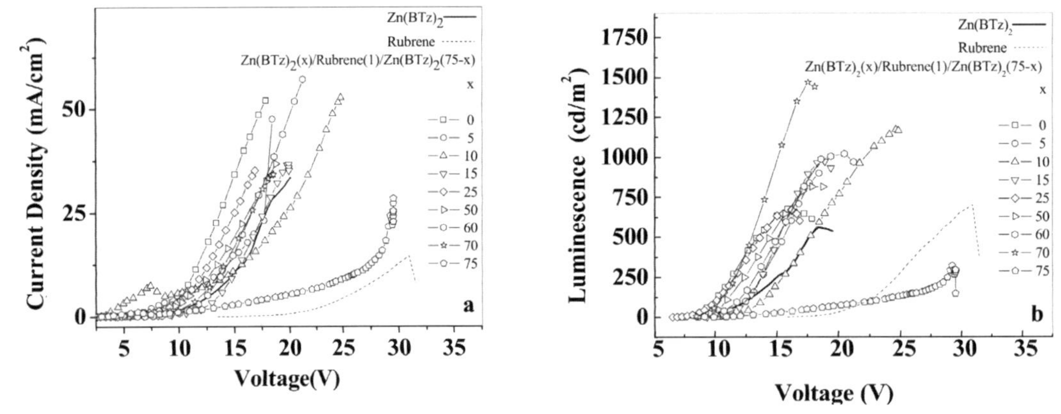


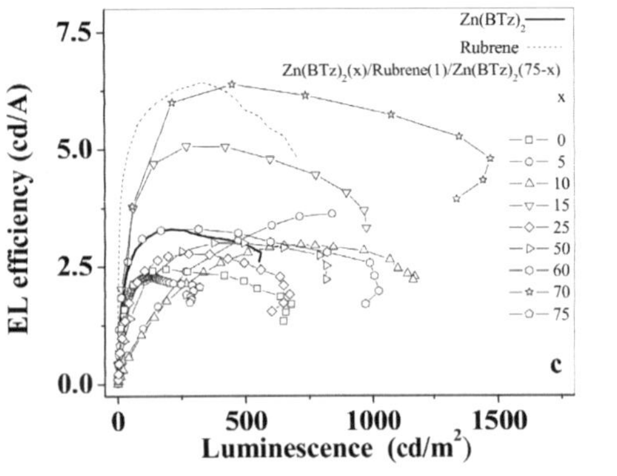


**Figure 2.** Current density - voltage (I-V) (a), luminescence – voltage (L-V) (b), and electroluminescent efficiency versus luminescence (c) characteristics of devices shown in set.

Figure 3 shows absorption and photoluminescence spectra (PL) of evaporated film (100 nm) of $Zn(BTz)_2$, absorption of evaporated (100 nm) film and PL Rubrene dissolved in $C_2H_4Cl_2$, and electroluminescence spectra (EL) of the reference device: (PVK:TPD)/$Zn(BTz)_2$(75)/Al; (PVK:TPD)/Rubrene(64)/$Zn(BTz)_2$(10)/Al at 20V. The PL peak wavelength of $Zn(BTz)_2$ is at 486 nm, but EL is red shifted to 531 nm. The data obtained for PL of $Zn(BTz)_2$ are quite close to the results reported by Wu (480 nm) [8] and Qureshi (485 nm) [9]. However while Wu showed that PL and EL spectra are basically similar, Qureshi found that the EL spectra are function of the thickness of $Zn(BTz)_2$. He reported for 95 nm thick film EL at wavelength of 528 nm, which agrees with our results. The PL and EL peaks for Rubrene are at 566 nm. The absorption spectrum of Rubrene shows bands at 462, 492 and 528 nm (curve 4) (the same values are reported by Zeng [12]) and overlaps the EL of $Zn(BTz)_2$.

The normalized EL spectra of the devices (PVK:TPD)/$Zn(BTz)_2$($\boldsymbol{x}$)/Rubrene(1)/$Zn(BTz)_2$(75-$\boldsymbol{x}$)/Al at 20 V, where $x$ nm ($0 \le \boldsymbol{x} \le 75$) is the distance between Rubrene layer and (PVK:TPD)/$Zn(BTz)_2$ interface are shown in figure 4. As expected, the emission is due to both Rubrene and $Zn(BTz)_2$. When Rubrene layer is close to the (PVK:TPD)/$Zn(BTz)_2$ ($\boldsymbol{x}\to 0$) or $Zn(BTz)_2$/Al ($\boldsymbol{x}\to 75$) interfaces the spectrum exhibits an emission peak at 566 nm and shoulder at 530 nm. This indicates that predominantly Rubrene emission is present - recombination of carriers, i.e., excitons formation and/or excitons diffusion, took place mainly at the Rubrene layer. With increasing the distance between Rubrene layer and both interfaces ($\boldsymbol{x}\to 35$ nm) the full-widths at half-maxima (FWHM) of EL spectra are widened, the relative intensity of the Rubrene emission is lower and the shapes of spectra are gradually shifted toward emission of pure $Zn(BTz)_2$. The 25 nm red shift of EL spectra of the doped devices was observed in comparison with the spectrum of pure $Zn(BTz)_2$.

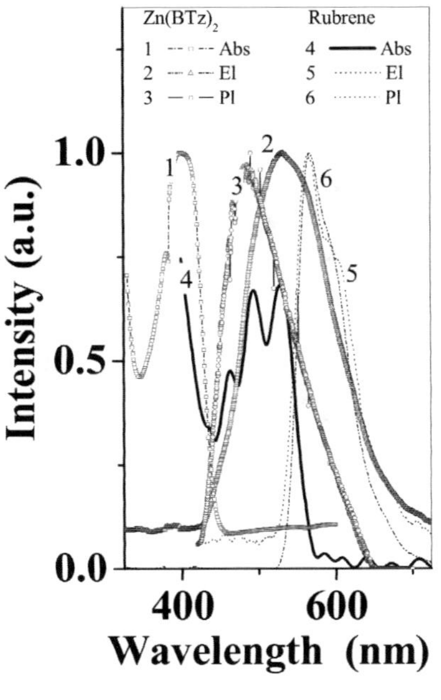


**Figure 3.** Absorption and PL of evaporated $Zn(BTz)_2$; Absorption of evaporated Rubrene and PL of $C_2H_4Cl_2$ solution; EL at 20 V for: (PVK:TPD)/$Zn(BTz)_2$(75)/Al; (PVK:TPD)/Rubrene(64)/$Zn(BTz)_2$(10)/Al.

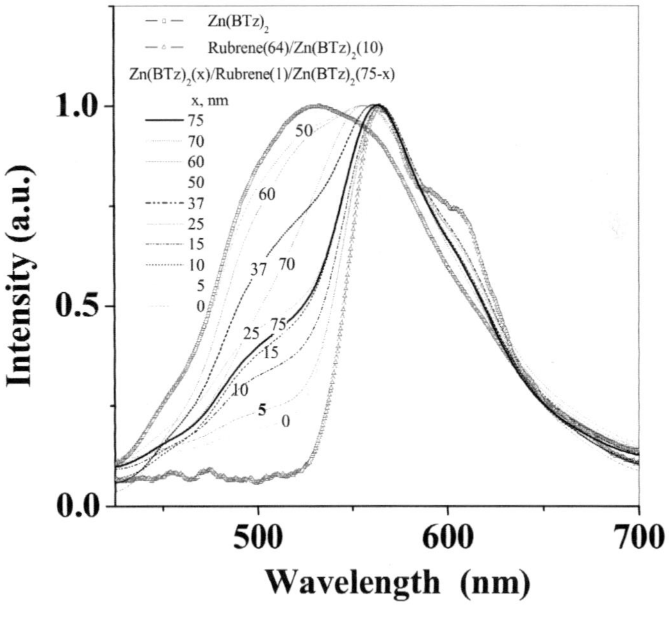


**Figure 4.** El spectra for structures (PVK:TPD)/$Zn(BTz)_2$($x$)/Rubrene(1)/$Zn(BTz)_2$(75-$x$)/Al at 20 V. $\boldsymbol{x}$ nm ($\boldsymbol{x}$ = 0–75) is the distance between Rubrene layer and (PVK:TPD)/$Zn(BTz)_2$ interface.

Obviously the resulting EL spectra are the superposition of $Zn(BTz)_2$ and Rubrene emission and can be deconvoluted as shown in figure 5. The relative Rubrene electroluminescence intensity obtained from EL spectra presented in figure 4 are shown in figure 6. For comparison, in the same figure are presented the results for $\eta_L$ at 200 cd/m$^2$ obtained from figure 2(c). It is seen that the trend of both curves is quite similar.

It is believed that EL of dye-doped OLEDs can be described via two approaches. One is so-called direct carrier trapping (DCT) process where the dye molecule directly captures charge carrier transported from the host emissive material, then the light emission of the dopant takes place when opposite carriers encounter each other on the dye molecule and recombine thereafter. The other approach is energy transfer (ET) process, i.e., the dopant molecule does not capture carriers but accepts energy released from host excitons through Forster ET and reaches its exciting state. The important requirements for the DCT-dopant is that the energies of the highest occupied molecular orbital (HOMO) and lowest unoccupied molecular orbital (LUMO) are located within those of the host material, while for the ET-dopant is required that the absorption spectrum of the dopant overlaps the PL or EL spectrum of the host.

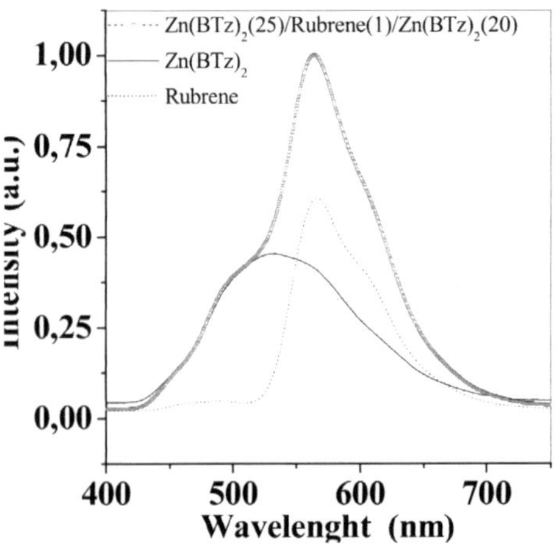


**Figure 5.** Deconvolution of EL spectrum of $Zn(BTz)_2(25)/Rubrene(1)/Zn(BTz)_2(50)$, measured at 20 V DC using pure $Zn(BTz)_2$ and Rubrene spectra.

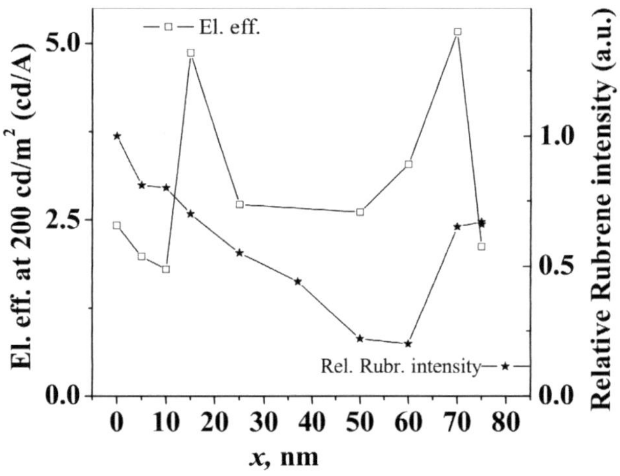


**Figure 6.** Electroluminescent efficiency at 200 cd/m$^2$ and relative Rubrene electroluminescence intensity at 20 V vs. different distances $x$ from the (PVK:TPD/$Zn(BTz)_2$ interface.

In our case both processes are possible, because the energy gap of $Zn(BTz)_2$ (3.00 eV) is broader than that of Rubrene (2.26 eV) (see figure 1) and the absorption spectrum of Rubrene overlaps the EL spectrum of $Zn(BTz)_2$ (figure 3). On the basis of the results we suppose that the DCT mechanism is more probable, because ET process cannot explain the increase of the Rubrene electroluminescence intensity for $x$ >50 nm.

In $Zn(BTz)_2$ reference device the recombination zone is near to (PVK:TPD)/$Zn(BTz)_2$ interface, because electrons are accumulated there ($LUMO_{Zn(BTz)2}$ is higher 0.3 eV then $LUMO_{PVK:TPD}$). When Rubrene layer is inserted in $Zn(BTz)_2$, electrons start to accumulate preferably at Rubrene/$Zn(BTz)_2$ interface ($LUMO_{Rubrene}$ is higher 0.5 eV then $LUMO_{Zn(BTz)_2}$). When $x$ becomes large, the Rubrene captures electrons from the cathode because the mobility of holes is 2 orders of magnitude smaller than that of electrons. The numbers of electrons and holes accumulated at the Rubrene site and PVK:TPD/$Zn(BTz)_2$ interface, respectively, rapidly increase since they have no opposite charges to recombine with. While concentration of accumulated carriers are large enough, two forces drive electron and hole moving towards each other: Coulombic attraction between electrons and holes and the concentration gradient of carriers. Accordingly, excitons of $Zn(BTz)_2$ are formed if electrons and holes meet each other in the bulk of $Zn(BTz)_2$ layer during their transport. Meanwhile, the holes do not recombine with electrons in $Zn(BTz)_2$ and are transported towards the cathode until they reach the dopant area. Here they recombine with accumulated electrons and, consequently, the light emission from the dopant is also observed in EL spectrum.

Consequently, we can conclude that the position of inserted Rubrene layer has a significant impact on carriers recombination and/or excitons diffusion area in $Zn(BTz)_2$ host layer. It could be stressed that a good correlation was established between the dependences of luminescent efficiency and the relative intensity of Rubrene peak on the position of inserted layer. Two times higher efficiency $\boldsymbol{\eta_L}$ was found for the devices with Rubrene layer inserted at $\boldsymbol{x}$(15) and $\boldsymbol{x}$(70) from (PVK:TPD)/$Zn(BTz)_2$ interface where the relative electroluminescence intensity of Rubrene peak is also higher.

## 4. Conclusions

OLEDs with improved performances on the basis of only 1 nm film of Rubrene inserted at different position in configuration: ITO/PVK:TPD/$Zn(BTz)_2$($\boldsymbol{x}$ nm)/Rubrene(1 nm)/$Zn(BTz)_2$(75- $\boldsymbol{x}$ nm)/Al are developed. Luminescence 1.5x and 4x higher and efficiency 1.75 and 2x higher to the reference structure are established for the devices with Rubrene doping layer at $\boldsymbol{x}$(15) and $\boldsymbol{x}$(70) nm.

EL spectra of the devices determines predominantly of Rubrene emission when the doping layer is close to (PVK:TPD)/$Zn(BTz)_2$ ($\boldsymbol{x}\to$ 0-15 nm) and to $Zn(BTz)_2$/Al ($\boldsymbol{x}\to$ 70-75 nm) interfaces and shifts toward emission of pure $Zn(BTz)_2$ increasing the distance of Rubrene to both interfaces ($\boldsymbol{x}\to$35 nm). The good correlation between efficiency and relative electroluminescence intensity of Rubrene peak is established.

We showed that by adjusting the both thickness and position of an ultrathin layer of rubrene a very accurate control of the emitted color can be obtained. The opportunity of inserting a thin layer in electroluminescent layer opens the door towards more complex structures using 3 or more different emitters at various wavelengths, with adjustable positions and thicknesses.

## References

[1] Tang C W, VanSlyke S A and Chen C H 1989 *J. Appl.Phys.* **65** 3610
[2] Baldo M A, O'Brien D F, Shoustikov Y You, Sibley A S, Thompson M E and Forrest S R 1998 *Nature* **395** 151
[3] Wang Y, Teng F, Xu Z, How Y, Yang S and Xu X 2005 *Materials Chemistry and Physics* **92** 291
[4] Li W, Yu J, Wang T, Jiang Y and Wei B 2008 *Materials Science and Engineering B* **149** 77
[5] Hao J, Deng Z and Yang S 2006 *Displays* **27** 108
[6] Beierlein T A, Ruhstaller B, Gundlach D J, Riel H, Karg S, Rost C and Rieß W 2003 *Synthetic Metals* **138** 213
[7] Shi Y, Deng Z, Xu D, Chen Z and Lip X 2007 *Displays* **28** 97
[8] Wu X, Hua Y, Wang Z, Zheng J, Feng X and Sun Y 2005 *Chinese Phys. Lett.* **22** 1797
[9] Qureshi M, Manohara S, Singh S and Mahapatra Y 2005 *Solid State Comm.* **133** 305
[10] Tomova R, Petrova P, Kaloianova S, Deligeorgiev T, Stoycheva-Topalova R and Buroff A 2008 Organic Light Emitting Devices Based on Novel Zn Complexes *A new light for a bright future: Proc. 14th International Workshop on Inorganic and Organic Electroluminescence & Int. Conference on the Science and Technology of Emissive Displays and Lighting, (Grand Hotel Duca D'Este, Bagni di Tivoli, Rome, Italy, 9-12 September 2008)* ed. G.Baldachini, R.Montereali and M.Vincenti pp 69-72
[11] Deligeorgiev T, Luminescent organic complexes available from UNIDYES GROUP http://www.chem.uni-sofia.bg/UNIDYES/oled.html
[12] Zeng X, Zhang D, Duan L, Wang L, Dong G and Qiu Y 2007 *Applied Surface Science* **253** 6047

Journal of Physics: Conference Series **253** (2010) 012049 doi:10.1088/1742-6596/253/1/012049

# Physical deposition of thin polyimide layers by applying an argon plasma assisted process

**D Dimov[1], A Georgiev[2], E Spassova[1], J Assa[1], P Dineff[3] and G Danev[1]**
[1]Acad. J .Malinowski Central Laboratory of Photoprocesses,
Bulgarian Academy of Sciences, Acad. G. Bonchev St., Bl.109, 1113 Sofia, Bulgaria
[2]University of Chemical Technology and Metallurgy, Department of Organic Chemistry, 8 Kl. Ohridski Blvd., 1756 Sofia, Bulgaria
[3]Technical University, "Kl. Ohridski" blv.,# 8, 1797 Sofia, Bulgaria

E-mail: dean@clf.bas.bg

**Abstract.** A novel method for physical deposition of thin polyimide layers by applying an argon plasma assisted process has been developed. The influence of the plasma on the combined molecular flux of the two thermally evaporated precursors – 4,4'- oxydianiline and pyromellitic dianhydride was investigated. The process parameters were changed in the limlts 0,4 – 2 A for the anode current and 80 – 170 V for the anode voltage. Their influence was studied using FTIR spectroscopy and electron microscopy techniques. It was proposed that the plasma flux crossing the molecular flows of the polyimide precursors enhances the imidization process by partly activating the precursor molecules in the gas phase.

## 1. Introduction

Polyimide (PI) is known to be a high performance aromatic polymer with outstanding mechanical and insulating properties, thermal and radiation stability [1]. This material is used in diversified areas such as space missions, nuclear, electronic and automotive industries [2]. For the last several decades, studies on the effects of high-energy radiation on polymer have become an important area of basic as well as applied research [3]. Several methods are available for the deposition of PI coatings on various substrates: from solution; vapour deposition polymerization (VDP) [4], ionized cluster beam deposition (ICBD) [5] and glow discharge induced VDP (GDVDP).The sputtering technique cannot be extended to the deposition of PI layers [6], because the ion bombardment of the target (a Kapton™ disk) is inapplicable for the ether and imide groups of the PI molecule. The chemical composition of the deposited film is different from the original PI. In the GDVDP method the evaporation of the monomers is not obtained by heating, but by the standard radio frequency magnetron sputtering equipment. In this paper we report a novel method for the physical deposition of thin PI layers by applying an argon (Ar) plasma assisted process in the deposition of the monomer vapours (obtained by heating) on a substrate taking place in a vacuum chamber. The aim of the paper is to describe the influence of the process parameters on the PI layer properties.

## 2. Experimental

The 500 nm PI layers were formed by vacuum co-deposition of the precursors 4,4'-oxydianiline (ODA) and pyromellitic dianhydride (PMDA) from two independently heated vessel sources (Figure 1). The deposition rate was 0.2 – 0.4 nm/sec and base pressure $4\times10^{-2}$ Pa. The layers were grown using an Ar plasma assisted process: the molecular fluxes of the evaporated precursors cross the Ar plasma. A part of the ionized Ar molecules and electrons collide with the precursor molecules and act on the growing deposited layers. In order to trace the influence of the assisted process we have studied PI layers, obtained at: a/ anode current (Ia) = 0,6 A and change of the anode potential difference (Ua) from 80 to 170 V; б/ Ua of 100 V and a shift in the Ia from 0,4 to 2 A. The obtained layers were transformed into PI by a thermal treatment 1h at 180°C followed by a 15 min at 300°C in air. The layers were characterized by Fourier Transform Infrared Spectroscopy (FTIR) and by Transmission electron microscopy (TEM), JEM 2100. The FTIR spectra (PI deposited on KBr substrate) were recorded by a Bruker Tensor 27spectrometer in the range of 4400 - 450 $cm^{-1}$ with a resolution of 2 $cm^{-1}$.

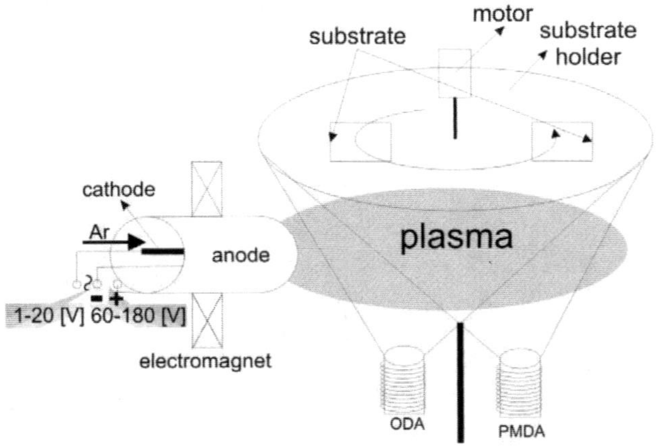


**Figure 1.** A detailed scheme of the system for plasma assisted deposition (PAD).

## 3. Results and discussion

The FTIR spectroscopy investigation results are shown in figures 2 and 3. The follow-up of the changes in the FTIR spectra shows that with the increase of Ua from 80 to 120 V (figure 2a, 2b, 2c) a gradual change of the band characterizing the imidization process (1376 $cm^{-1}$) is observed. Its intensity grows with the change of $U_a$ and at $U_a$ = 120 V (figure 2c) it is most intensive and most narrow. The same is also true in the band at 1241 $cm^{-1}$, characterizing the molecular vibrations of the PI. At $U_a$ = 170 V (figure 2d) the quality of the produced film is abruptly impaired. This is also corroborated by the non-characteristic bands for the PI in the long area of the stretching C-C vibrations at 1500 $cm^{-1}$ for the aromatic ring, the range of the molecular vibrations as well as by the band typical of the degree of imidization (δ) at 1376 $cm^{-1}$. A conclusion can be drawn that at $I_a$= 0,6 A and $U_a$ = 120 V the obtained PI is of the best quality. Upon applying an plasma assisted deposition (PAD) process at $I_a$=0.4A (figure 3a) one could presume that the imidization process is facilitate as early as in the precursor deposition process. The band at 1378 $cm^{-1}$ is shifted by 6 $cm^{-1}$ to the lower frequencies as opposed to the PI obtained without PAD and δ is respectively greater. With the increase of $I_a$ to 0.60A δ grows evenly and reaches maximum value (figure 3b). At $I_a$=1.0A (figure 3c) a shoulder at 1339 $cm^{-1}$ on the imide band (1377 $cm^{-1}$) emerges. This could possibly mark the onset of destructive processes in the molecular fluxes of the evaporated PI precursors. The obtained PI should be containing low-molecular fragments which would impair its physical and chemical properties. At $I_a$=1.25 A (figure 3d), the shoulder at 1339 $cm^{-1}$ on the imide band (1377 $cm^{-1}$) is clearly expressed, the band is slightly distorted which testified to a decrease of the effect of imidization and the formation of by-products as a consequence from the destruction of the precursors.

The entire PI spectrum is of low quality. With the increase of $I_a$ above 1,25 A the massive deformation and splitting of the imide band and that characterizing the stretching vibrations of the aromatic ring (1500 $cm^{-1}$) was established. It supports our idea about the chemical alteration of the initial precursors under the influence of PAD leading to the PI of an impaired qualities and unclear composition.

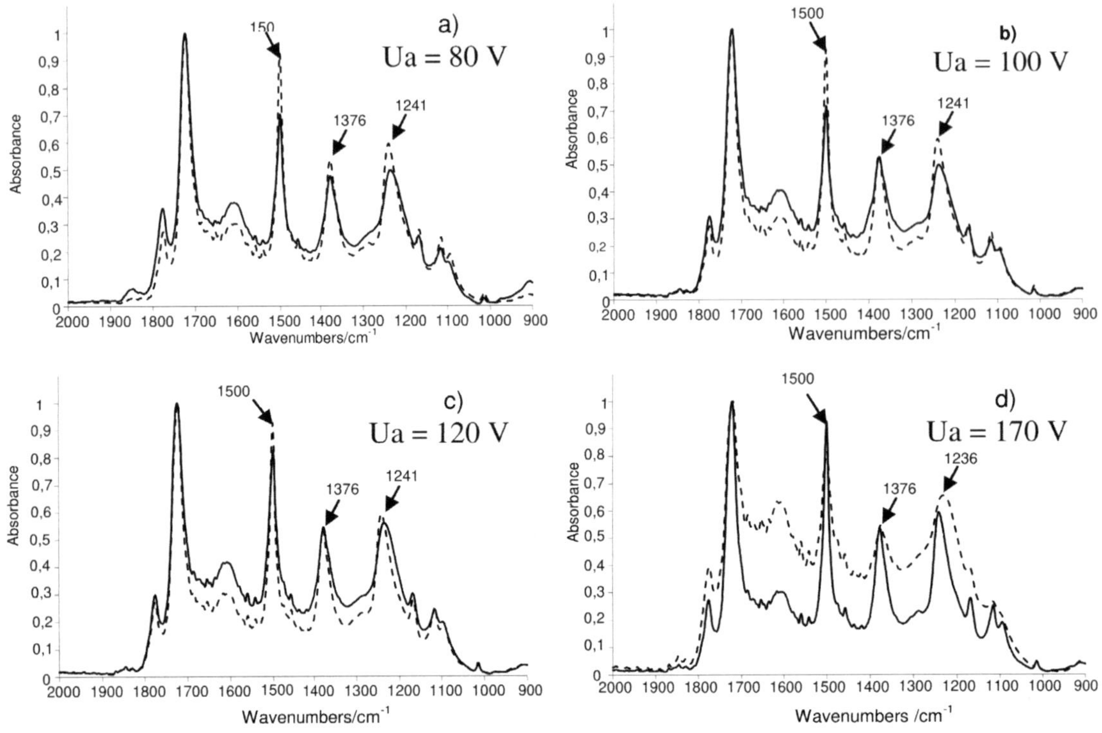


**Figure 2.** PI layers with PAD and $I_a$ = 0,6 A: full linear - thermally treated; dashed lines - thermally untreated.

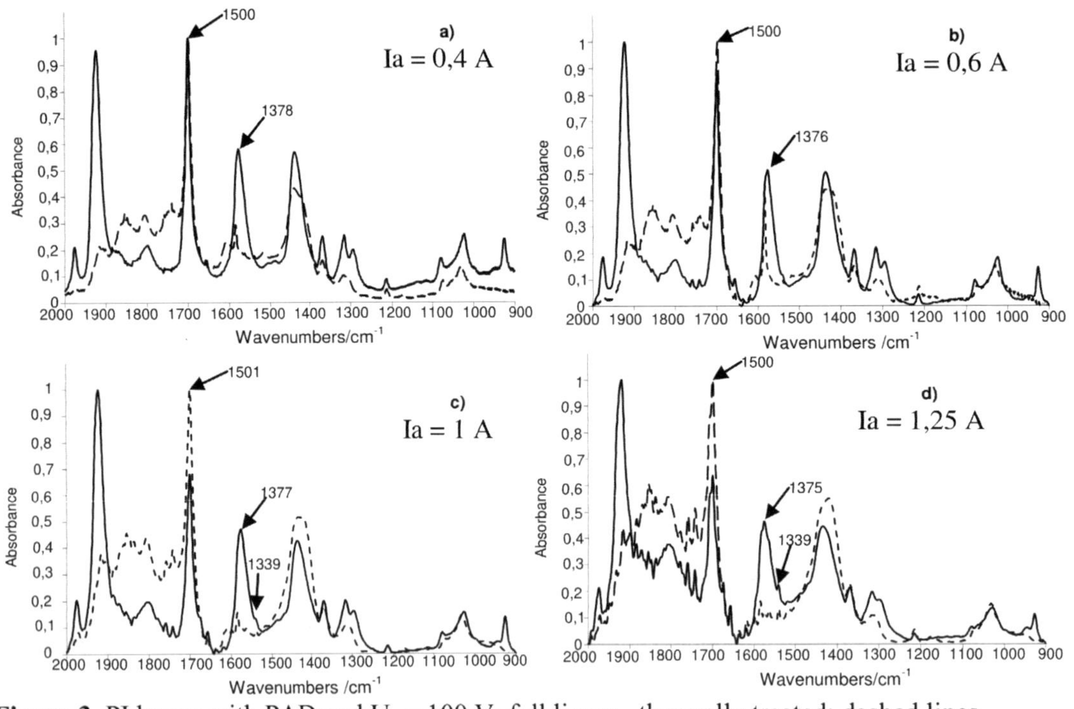


**Figure 3**. PI layers with PAD and $U_a$ = 100 V: full linear - thermally treated; dashed lines - thermally untreated.

The obtained results also show the effect of the $I_a$ on the properties of the produced layers is much stronger than that of the $U_a$. The PI layers obtained at $I_a$ =0,4 to 0,6 A and $U_a$ =100 to 120 V are of best quality wich is also confirmed by the TEM micrographs (figure 4). Upon applying PAD the structure of the produced films is of finer granularity and is smoother.

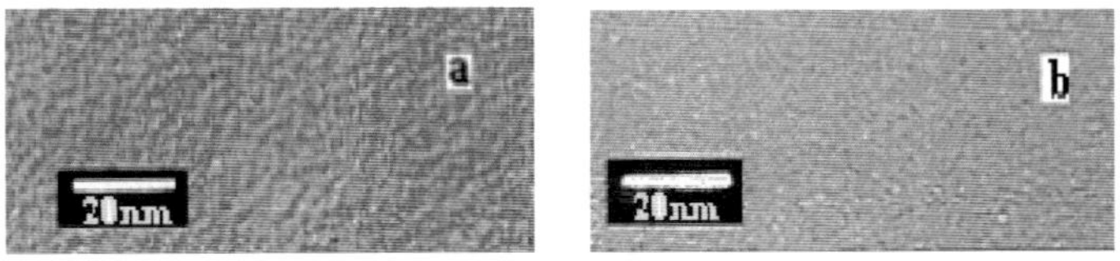


**Figure 4.** TEM micrographs of 50 nm VDP PI layers: a/ without PAD; b/ with PAD: $I_a$ = 0,6; $U_a$ = 100V

## 4. Conclusions

The results from the research carried out show that the effect of $I_a$ on the properties of the obtained layers is much stronger than that of $U_a$. The PI layers obtained at $I_a$ =0,4 to 0,6 A and $U_a$ =100 to 120 V are of best quality and unchanged chemical composition. The observed effects on the properties of the deposited films are explained with the increased energy of the precursor molecules resulting from the ion-molecular collisions. It can be assumed that the plasma flux crossing the molecular flows of the PI precursors enhances the imidization process by partly activating the precursor molecules in the gas phase. The presented new method, developed for the physical deposition of thin PI (PMDA/ODA) layers can probably be successfully used for the deposition of other polymeric films.

## Acknowledgements

The financial support of the National Fund of the Ministry of Education and Science, Bulgaria – contracts: ВУ- ТН- 205 /06 and DO 02-254/ 18.12.2008 are gratefully acknowledged.

## References

[1] Narendra L Mathakari, Vasant N Bhoraskar, Sanjay D Dhole 2010 *Material Science and Engineering* B **168** 122-126
[2] Svorcik V, Rybka V, Hnatowicz V, Novotna M and Vognar M 1997 *J.Appl. Polym. Sci.* **64** 2529
[3] Fintzou A T, Badeka A V, Kontominas M G and Ringanakos K A 2006 *Radiat. Phys.Chem.***75** 87
[4] Lamb R N,Bahter J, Grunze M, Kong C W, Unert W N 1988 *Lagmuir* **4** 249
[5] Kim K W, Hong C E, Choi S C, Cho S J, Whang C N, Shim T E, Lee D E 1994 *J. Vac. Sci. Technol.* A **12** 3180
[6] Kitoh M, Honda Y 1995 *Thin Solid Films* **271** 92
[7] Spassova E 2003 Vacuum deposited polyimide thin films *Vacuum* **70** 551-61

Journal of Physics: Conference Series **253** (2010) 012050 doi:10.1088/1742-6596/253/1/012050

# Realization of cantilever arrays for parallel proximity imaging

**Y Sarov[1], Tz Ivanov[1], A Frank[1], J-P Zöllner[1], N Nikolov[2] and I W Rangelow[1]**
[1] Department of Micro- and Nano-electronic Systems, IMNE, FEI, Ilmenau University of Technology, PF 100565, 98684 Ilmenau, Germany
[2] Microsystems Ltd., 9010 P.O. Box 147 Varna, Bulgaria

E-mail: ysarov@yahoo.com, yanko.sarov@tu-ilmenau.de

**Abstract**. This paper reports on the fabrication and characterisation of self-actuating, and self-sensing cantilever arrays for large-scale parallel surface scanning. Each cantilever is integrated with a sharp silicon tip, a thermal-driven bimorph actuator, and a piezoresistive deflection sensor. Thus, the tip to the sample distance can be controlled individually for each cantilever. A radius of the tips below 10 nm is obtained, which enables nanometre in-plane surface imaging by Angstrom resolution in vertical direction. The fabricated cantilever probe arrays are also applicable for large-area manipulation, sub-10 nm metrology, bottom-up synthesis, high-speed gas analysis, for different bio-applications like recognition of DNA, RNA, or various biomarkers of a single disease, etc.

## 1. Introduction

During the last decade the surface imaging by atomic force microscopy (AFM) has become a routine surface analysis method providing Angstrom resolution [1]. The AFM systems are based on micro-cantilever probes with sharp tips brought to the vicinity of the sample surface as the interaction between them is localized in a nano-sized region. Unfortunately, the applications of the cantilever systems for large-area surface proximity scanning are limited by the low scanning speed [2]. Complete scanning of larger samples can be realized by a small fast cantilevers hold by a small light stage [3-4] or by introducing of large 2-dimensional (2-D) array of parallel working, independently addressed cantilevers [5-6]. The latter approach offers a new quantitative way for large-area simultaneous scanning or analysis, manipulation and bottom-up manufacturing. It is clear, that the classical AFM actuation and read-out are slightly applicable for such an array. The piezoelectric actuation of the array is in contradiction with the independent addressing of the cantilevers. In addition, the standard read-out based on a laser-reflection is slightly feasible for very large arrays.

The development and the application of large parallel 2-D cantilever arrays for large area surface proximity scanning is the purpose of the European Project "PRONANO- Technology for the Production of Massively Parallel Intelligent Cantilever Probe Platforms for Nanoscale Analysis and Synthesis" Recently, the fabrication of such NEMS-chip (Nano Electro Mechanical System), incorporating 4x32=128 fully addressable proximal probes has been reported [7-8] as a part of the PRONANO activities. Later, new generation of 8x64 cantilever array with electrical contacting from the back side of the chip and flip-chip packaging has been demonstrated [9]. PRONANO appeared as an alternative to the famous IBM's project Millipede, headed by P. Vettiger and G. Binnig (the latter is a Nobel price winner for the discovery of the scanning tunnelling microscope), the aim of which is the development of a 2-dimentional arrays for ultra-dense record of information [5, 10-11].

Main challenges at PRONANO are the development of new technology for realization of large, 2-dimentional, individually addressed, self-actuated, piezoresistive nanosystems for proximity scanning, the application of new solutions for their packaging, development of specialized dedicated driving and measurements electronics, real-time transfer and processing of information (~ TBs/sec) and integration of all components in a complete system for nanopositioning and large-area surface proximity scanning. In this work some of our recent results on fabrication of large 1-D and 2-D arrays of cantilevers with integrated sharp AFM tips, thermally driven actuators and piezoresistive deflection sensors are presented. The development and the fabrication of the AFM arrays is considered in connection with the physical principles and the requirements of the AFM scanning, self-actuation of the probes, piezoresistive sensing and of the imaging by large 2-dimentional arrays.

## 2. Fabrication of the cantilever arrays

The fabrication of the piezoresistive AFM probes is illustrated in figure 1. The source material are <100> phosphorous doped Si wafers with small total thickness variation < 1µm (1). First, the wafers are oxidized and the AFM tips are formed (2) by the under-etch technique [12] using and a double photoresist + silicon dioxide surface protection and an anisotropic plasma reactive ion etching (RIE) or by oxide mask and an anisotropic chemical etching in potassium hydroxide (KOH). It should be noted, that the formation of the AFM tips in the beginning requires their further protection during rest of the fabrication by a mask of thick photo-resist.

The wafers are oxidized again. With the next fabrication steps the electrical structures on the chip are formed by three consequent implantations with resist mask: a) highly doped phosphorous doped areas for electrical shielding and contact to the substrate (3), b) highly doped boron implantation for low-resistive on-chip connections and contacts areas (4) and c) boron implantation of the piezoresistor deflection sensors (5). Then, an appropriate annealing activates the implanted elements. As a next step, the wafers are passivated by plasma enhanced chemical vapor deposited (PECVD) layer of SiNx (6). Contacts to the implanted structures are lithographically defined and chemically (wet) etched by buffered hydrofluoric acid (HF) (7). In the same time the passivation is removed near the tips. Again, lithographically patterned end structured aluminum is employed to form planar wiring interconnections, creating contacts to the piezoresistors and forming metallic bimorph micro-heaters for a thermal actuation of the cantilevers (8). The cantilever membrane is realized by bulk micromachining of Si, obtained by anisotropic KOH etch step. To form the cantilever beam and to cut the single sensor chip, a RIE step is employed with a thick resist mask. Finally, the resist mask is removed using microwave plasma striping and the cantilever arrays are separated mechanically (9).

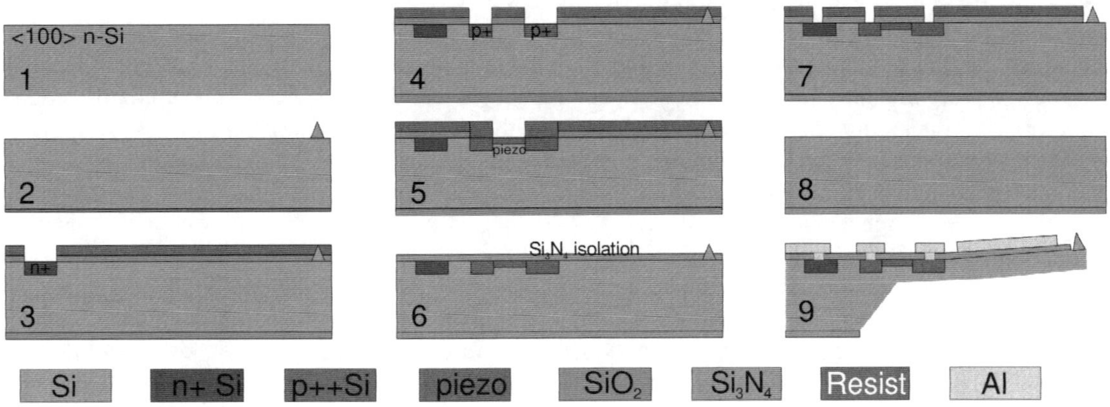


**Figure 1.** Fabrication sequence

## 3. Results and discussion

Scanning electron microscopy (SEM) pictures of the fabricated 1-D (1x32) and a 2-D (2x32) arrays are shown is figure 2 a) and b), respectively. The cantilevers are 240 μm long, 90 μm bright and 3-5 μm thick. The probes transversal pitch is 150 μm. The largest 2-D array realized within PRONANO has 8x64 cantilevers, following the same basic design as the chips presented in figure 2.

**Figure 2.** Cantilever arrays for parallel scanning and analysis a) 1x32 array, b) 2x32 array.

A fundamental requirement for cantilever arrays is independent actuation and deflection readout of each cantilever, which is needed for parallel proximity scanning of large areas by all cantilevers. This can be realized by integration of each cantilever with its own actuator and deflection sensor. Such integration provides an elegant and compact solution for high speed non-contact AFM as an alternative to the commercial cumbersome AFMs with optically detected devices with piezotube z actuator [13-14]. Piezoresistors, applied as stress gauges are widely used as cantilever deflection sensors. Especially important by the cantilever design is the piezoresistors to be situated at the cantilever areas of highest stress by bending (figure 3). This place is at the base of the cantilever and can be precisely determined by simulating the cantilever behavior by finite elements methods (e.g. the computer program ANSIS) [14]. In addition, the independent driving of each cantilever requires fabrication of microactuators, compatible with CMOS processing. The microactuators described in this article are based on the bimorph effect and are realized by the thin film technology. They are heaters, microstructured from an Al film in a meander-shape (figure 3). Again, the ANSIS simulations show the best situation for them at the front side of the cantilever [14]. The actuators are driven electrically, as the electrical to thermal conversion there causes bending of the cantilever, since the aluminum has significantly higher thermal expansion coefficient then the silicon. In standard feedback operation, the integration of the thermal actuator with the cantilever is used for both exiting the cantilever at resonance as well as applying the feedback actuation signal directly to the cantilever in order to control the position of the AFM tip with respect to the surface. This causes the resonance frequency to be much increased compared to the normally used piezoelectric-actuator stage in commercial systems. The high resonance frequency of the cantilever, feedback loop with smaller time constants and bigger gain determine the surface scan speed by which the tip can accurately follow the topography features. In our setup the cantilever is thermal driven at its resonance frequency $f_{res}$ by an AC current with frequency $f_{res}/2$, while z axis actuation in the z axis is provided by applying a DC current. The magnitude of the latter determines the deflection of the cantilever, controlled by the proportional integral differential (PID) feedback loop. A fundamental requirement for cantilever arrays is independent actuation and deflection readout of each cantilever, which is needed for parallel surface scanning of all cantilevers. The phase signal shift of the driving voltage and voltage drop response of the piezoresistive sensor is monotonic, ensuring a stable and reproducible control over the full z range of probe sample interaction.

The measured surface, obtained by proximity scanning is a conjunction between the real surface topography and the shape of the AFM tip. Hence, a sharper tip is required for better image. Figures 3 and 4 show cantilever arrays, which tips are microstructures by wet (chemical) and by dry (plasma) etching, respectively. The obtained tip's radius in the first case is below 100 nm, while by the second processing tips sharper then 10 nm are realized. On the other hand the chemical etching is easier to perform and results in a better prediction of the tip's shape since the KOH etching is defined by the <111> crystal planes of the silicon.

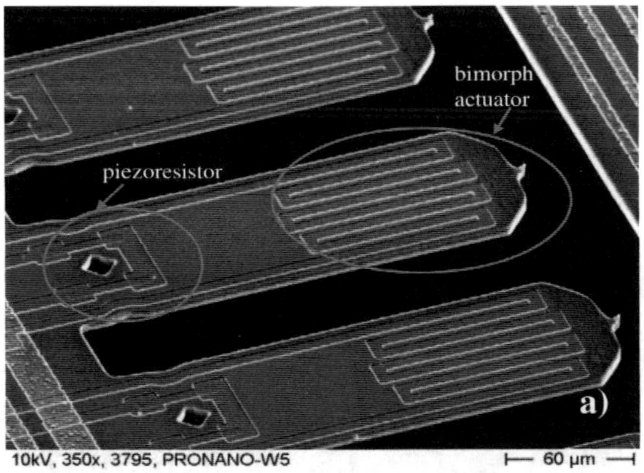


**Figure 3.** Array of thermal actuated, piezo-resistive cantilevers with integrated tips.

**Figure 4.** Arrays with integrated sharp AFM tips obtained by plasma under-etching.

The surface by an array of scanning probes requires that every cantilever can independently approach and follow the topography of the sample. For 1-D array (a row of cantilevers) the chip can be tilted for free access of the tips to the surface. For the case of 2-D array this is obviously not possible. To have the ability to scan surfaces with "hills" and "valley" the cantilevers of such 2-D array have to be off-plane of the wafer, i.e. to be pre-deflected (bent) towards the sample surface [8]. Such out-of-plane deflection has been recently achieved by passivation of the cantilevers with a film of LPCVD Si3N4 [10]. Such a layer of silicon nitride possesses string tensile stress (~750MPa) which causes the released beams to bend. The homogeneity of the bending is excellent, as can be seen from the SEM picture in figure 5.

The non-contact AFM operation the cantilevers is driven by DC and AC current. When a large 2D array is brought in proximity to the surface and scanned over it, all cantilevers have to bend up and down in order to follow the surface topography. That can be realized by pre-heating (quasi-static deflection) of beams before scanning by applying of certain DC power to the heater. In such a case by individual control of the actuator DC current each cantilever can be either heated additionally in order to approach the surface or cooled down in order to retract from it. That is realized using a topography feedback from piezoresistive sensors trough the PID controllers. The "DC" time constant is effectively the maximum imaging bandwidth of the cantilever defined as the maximal speed with witch the cantilever can follow the surface topology. Practically this is the time needed for the cantilever and its holder to reach a thermal equilibrium with the ambient. This time is usually defined as the inverse of the frequency at which the peak-to-peak resistance decreases to 3 dB of the maximum. The "DC" time constant for 240 µm long cantilever with uniformly distributed Al-heater is estimated to be below 0.5 ms, which corresponds roughly to a bandwidth of 2 kHz. The "AC" time constant corresponds to the frequency and mode used in the non-contact mode. Usually choosing first eigenmode the AC frequency is about several tens kHz and going up to about 1 MHz for the third eigenmode.

The technique of the strong off-plane bent cantilevers allows a simple way to fabricate AFM arrays. The tips can be just situated in the plane of the chip (figure 6) and formed during the last fabrication step (9) from figure 1- release of the cantilevers. In such a case the processing step (3) is skipped and all fabrication is made much simpler using standard a thin photo-resist, since there are no tips to protect. After the release of the cantilever it bends away of the wafer plane and the upper end of the tip forms a 3-dimentional AFM apex. Its radius is determined mainly by the narrower line, which can be created lithographically. In our case, although additionally sharpen by a short isotropic plasma etching before releasing of the cantilever, the tips curvature is between 200 and 250 nm. Despite, when no superior X-Y scanning resolution is required, such arrays are sometimes preferable due to the significantly simplified and faster fabrication. Moreover, shear-force surface scanning is possible when the cantilevers are strongly off-plane bend and the tips are directed orthogonal to the plane of the chip and to the surface of the sample [15]. In this configuration the oscillation of the tip is parallel to the surface and some parasitic effects like the air duping or sticking to the sample due to the surface dew are avoided.

**Figure 5.** Arrays of of-plane bent cantilevers.

**Figure 6.** Array with in-plane tips and strongly off-plane bent cantilevers.

Figure 7 shows a typical resonance curve of a thermally driven cantilever. The detection is from the piezoresistive sensors. The main eigene modes up to the third resonance are clearly resolved. The corresponding resonance frequencies $f_{res}$, the peak's width $\Delta f_{res}$ and Q-factor of the resonances are presented in Table 1. All measurements are performed by room conditions. The clear readability of the resonance spectrum determines such cantilevers applicable for parallel proximity imaging.

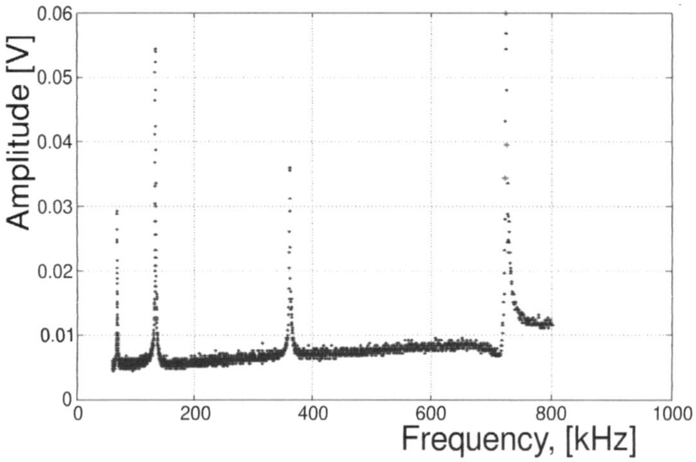


**Figure 7.** Resonance spectrum of a thermal actuated piezoresistive cantilever.

**Table 1.** Resonance frequencies $f_{res}$, width $\Delta f_{res}$ and Q-factor of the peaks for figure 7.

| $f_{res}$ [kHz] | $\Delta f_{res}$ [Hz] | Q |
|---|---|---|
| 67.72 | 428.92 | 157.9 |
| 135.40 | 775.44 | 174.6 |
| 362.28 | 1170.42 | 309.5 |
| 724.163 | 2231.07 | 324.8 |

## 4. Conclusion

This work describes the physical principles of the parallel surface proximity scanning. The requirements of the large-area fast AFM imaging by 2-dimentional cantilever arrays are shown and are related with the basics of the classical AFM scanning, self-actuation of the probes, piezoresistive deflection sensing. To realize that task, large 1-D and 2-D arrays of cantilevers with integrated sharp AFM tips, thermally driven actuators and piezoresistive deflection sensors are designed and successfully fabricated. The biggest microstructured AFM array up to the moment has 8x64 cantilevers. The successful realization of parallel fast AFM systems is crucial for development of future commercial nanotechnology products with an unprecedented throughput, for highly effective tools for analysis, manipulation and for synthesis (both top-down and bottom-up). Besides, an array of controllable proximal probes can also be used for the determination of large numbers of analytes such as various biomarkers of a single disease, for high-speed analysis, and the detection of explosive, combustible, and toxic components. Multi-parameter analysis of liquid or gas can also be done (e.g. simultaneous determination of temperature, pressure, humidity, concentration) if the cantilevers of an array are differently functionalized. The techniques presented in this paper are a part of a technology platform that can be used for the next generation of highly sensitive controllable proximal probes.

**Acknowledgements**
This work is supported by the EU Commission under the project IP 515739 - PRONANO.

**References**

[1] Eaton P and West P 2010 *Atomic force microscopy*, (New York: Oxford Universy Press)
[2] Crampton N, Yokokawa M, Dryden D, Edwardson M, Rao D, Takeyasu K, Yoshimura and Henderson R 2007 *Proc. Natl. Acad. Sci. USA* **104** 12755
[3] Schitter G 2009 *Technisches Messen* **76** 266
[4] Leang L and Fleming A 2008, *Proc. American Control Conference* (*Seattle, Washington, USA; 11-13 June 2008*) 3188
[5] Vettiger P, Cross G, Despont M, Drechsler U, Dürig U, Gotsmann B, Häberle W, Lantz M, Rothuizen H, Stutz R and Binnig G 2002 *IEEE Trans. Nanotechnol.* **1** 39
[6] Nama H, Kima Y, Lee C, Jin W, Jang S, Choa I, Bua J, Choi W and Choi S 2007 *Sens. Actuat. A*, **134** 329
[7] Rangelow I W, Ivanov Tz, Ivanova K, Volland B, Grabiec P, Sarov Y, Persaud A, Gotszalk T, Zawierucha P, Zielony M, Dontzov D, Schmidt B, Zier M, Nikolov N, Kostic I, Engl W, Sulzbach T, Mielczarski J, Kolb S, Latimier D, Pedreau R, Djakov V, Huq S, Edinger K, Fortagne O, Almansa A and Blom H 2007 *MicroElectron. Eng.* **84** 1260
[8] Sarov Y, Ivanov Tz, Frank A and Rangelow I W 2008 *Appl. Phys. A* **92** 525
[9] Sarov Y, Frank A, Ivanov Tz, Zöllner J, Ivanova K, Volland B, Rangelow I W, Brogan A, Wilson R, Zawierucha P, Zielony M, Gotszalk T, Nikolov N, Zier M, Schmidt B and Kostic I 2009 *J. Vac. Sci. Technol. B*, **27** 3132
[10] Despont M, Brugger J, Drechsler U, Dürig U, Häberle W, Lutwyche M, Rothuizen H, Stutz R,

Widmer R, Binnig G, Rohrer H and Vettiger P 2000 *Sens. Actuat. A.* **80** 100

[11] King W, Kenny T, Goodson K, Despont G, Durig U, Rothuizen H, Binnig G K and Vettiger P 2001 *Appl. Phys. Let.* **78** 1300

[12] Kim D, Lym S, Jung M, Jeon H and Choi S 1999 *Microelectr. Eng.* **46** 423

[13] Pedrak R, Ivanov Tz, Ivanova K, Gotszalk T, Abedinov N, Rangelow I W, Edinger K, Tomerov E, Schenkel T and Hudek P 2003 *J. Vac. Sci. Technol. B,* **21**, 3102

[14] Ivanov Tz 2004 *Piezoresistive cantilevers with an integrated bimorph actuator*; (Ph.D. thesis), IMA, Faculty of Natural Science, University of Kassel)

[15] Woszczyna M, Gotszalk T, Zawierucha P, Zielony M, Ivanov Tz, Ivanova K, Sarov Y, Nikolov N, Mielczarski J, Mielczarska E and Rangelow I W 2009 *MicroElectron. Eng.* **86** 1212

# Near-IR micro-fluidic sensing by total internal reflective diffraction

**Y Sarov[1,2], V Sarova[1], I Capek[3] and I Rangelow[1]**

[1] Department of Micro- and Nano-electronic Systems, IMNE, FEI, Ilmenau University of Technology, PF 100565, 98684 Ilmenau, Germany
[2] Institute of Optical Material and Technologies, Acad G. Bonchev Str., Bl. 109, 1113 Sofia, Bulgaria
[3] Polymer Institute, Institute of Measurement Science, Slovak Academy of Sciences, Dubravska cesta 9, Bratislava, Slovakia

E-mail: ysarov@yahoo.com, yanko.sarov@tu-ilmenau.de

**Abstract**. This work studies how the sensitivity of a recently reported novel integrated micro-fluidic system [1] can be improved optimizing the conditions of measurements. The considered lab-on-a-chip system is based on a diffraction grating, operating under total internal reflection and illuminated by near-IR light (1.55 μm). The interaction between the evanescent field of the totally reflected light with dipoles of the analyte results in phase and amplitude changes in the reflected diffraction orders. In this work the dependence of the sensing signal on the measured diffraction order, polarization of the light and angle of incidence are investigated. The conditions giving highest sensitivity are determined. The feasibility of the integrated micro-fluidic system for analysis of nano-dispersions is demonstrated by samples of oil-based dispersion of composite $CuFe_2O_4$ nanoparticles. A resolution toward the $CuFe_2O_4$ concentration of 0.1 wt.% is obtained. The possibilities for further increase of the sensitivity are discussed.

## 1. Introduction

The fast development in the modern field of micro-chemical and microbiological instrumentation is motivated mainly by two reasons. The first stimulus is related with the broad applications of the micro-fluidic on-chip systems in the biotechnology and bio-analysis, process control, environmental and medical sciences [1]. Moreover, the micro-reactors and micro-sensors benefit of the times of magnitude quicker processing and lower consumption of samples [1-2]. Second, the micro-fluidic transporting part can be easily and cheap integrated into the microchip applying the standard, parallel microstructuring and microfabrication [2]. A complete fabrication of integrated lab-on-a chip microsystems gives a requirement that the sensor part to be also capable of being integrated on the chip and its fabrication to be compatible with the standard micro-fabrication technologies. An optical analysis of micro-fluidic samples is especially interesting in the near-infrared (NIR) range as a result of the absorption peculiarities of the bio-substances due to the organic absorption bands of the infrared (IR) region.

Very recently, a novel integrated micro-optical system for micro-fluidic analysis has been reported [3-4]. Its principle of operation is based on the diffraction from a grating under Total Internal

Reflection (TIR) [5-7]. Later, the integrated sensor system has been modified and applied for near-IR micro-fluidic sensing and analysis [8].

This work presents the further optical investigation of the integrated IR micro-optical sensor and its application for micro- and nano-fluidic diagnostics and analysis. The sensor is calibrated with standard water solutions of sucrose, which absorption coefficient changes linear with the concentration at that spectral region. The dependence of the sensitivity of the detector on the angle of incidence and the polarization of the light is studied for the both $0^{th}$ and the $1^{st}$ diffraction orders. The possibilities for increasing of the sensitivity are discussed. The potential applications of integrated IR system for micro-fluidic investigation and concentration analysis are demonstrated in case of oil-based dispersions of composite $CuFe_2O_4$ nanoparticles with size below 10nm.

## 2. Principle of sensing and optical setup

Figure 1 presents the optical setup and explains the principle of operation of the integrated IR micro-fluidic sensor. The NIR light source is a diode laser with λ=1550nm (1). Its light is coupled with a visibly beam from another diode laser (2) by a beam-splitter (4), which allows easy optical adjustment. The direction of the light polarization is chosen by the rotator of the polarization (3). The incident IR beam is directed to the prism facet in such a way, that the beam inside the prism (5) fulfills the TIR condition for the interface between the micro-prism (5) and the fluid in the micro-channel (7). On this interface a metal diffraction grating (6) is deposited.

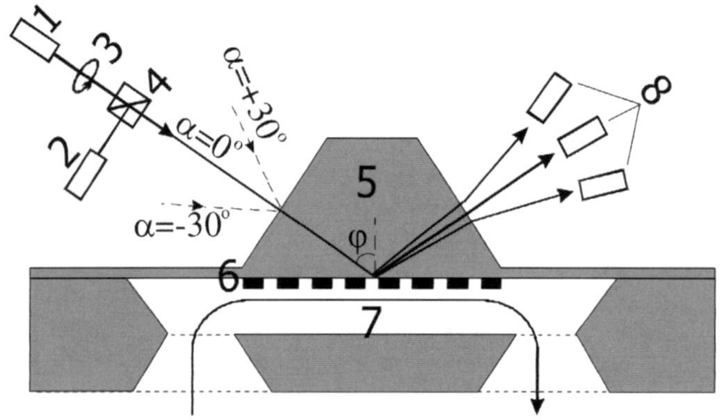


**Figure 1.** Near IR micro-fluidic sensor chip employing total internal reflection of light: 1) NIL diode laser, 2) visible diode laser, 3) rotator of the polarization, 4) beam-splitter, 5) silicon micro-prism, 6) diffraction grating, 7) microfluidic channel, 8) photodetectors.

The fluidic sensing is realized with the help of the evanescent electromagnetic field, which penetrates in the investigating medium from the spacing between the lines of the grid. The conversion of the harmonic waves into the evanescent TIR field at the two media interface, caused by the dielectric discontinuity of the media, is influenced by the optical constants of the micro-fluid. When the micro-channel is filled with absorbing fluid having a complex refractive $\hat{n}_2 = n_2(1+i\kappa)$, an effect of frustrated TIR occurs, i.e. the light reflects particularly due to the absorption of the evanescent TIR field. The reflection amplitude of the TIR $r$ is given by:

$$r_t^{s,p} = 1 - 2\pi n_2 \kappa . d^{s,p} / \lambda \tag{1}$$

where $n_2$ and $\kappa$ are the indices of refraction and absorption, respectively, λ is the wavelength and the relation $\kappa << 1$ is supposed. The superscripts $^s$ and $^P$ corresponds to s and to p polarization of the light, when transferal to the plane of incidence oscillates the electric or the magnetic vector, respectively. The quantity $d$ is the penetration depth of the evanescent filed [9], calculated by:

$$d_s = \frac{\lambda n \cos(\varphi)}{\pi . n_1 \sqrt{\sin^2 \varphi - n^2 (1 - n^2)}}, \qquad d_p = d_s \frac{2\sin^2 \varphi - n^2}{(1+n^2)\sin^2 \varphi - n^2} \tag{2}$$

Journal of Physics: Conference Series **253** (2010) 012051 doi:10.1088/1742-6596/253/1/012051

where $n=n_2/n_1$ is the relative refractive index, $n_1$ is the refractive index of the denser medium (silicon) and $\varphi$ is the angle of incidence at the base of the prism. The phase of the totally reflected light is determined by:

$$tg(\Phi^s) = -\frac{2\sqrt{\sin^2\varphi - n^2}\cos\varphi}{n^2+1-2\sin^2\varphi}, \quad tg(\Phi^p) = \frac{2n^2\sqrt{\sin^2\varphi - n^2}\cos\varphi}{n^4\cos^2\varphi + n^2 - \sin^2\varphi} \tag{3}$$

The (1-3) show that the amplitude of the TIR light is influenced by the absorption of the fluid, while the TIR phase depends on its refractive index. At the same time the reflection from the lines of the grating (figure 1) creates a constant reference signal. The TIR light from the lines spacing interfere with the light, reflected from the metallic lines. As a result a diffraction pattern (8) is crated. The Diffraction Efficiency (DE) of the $0^{th}$ ($\eta_0$)□ and the $1^{st}$ (□$\eta_1$) diffraction orders is given by [7]:

$$\eta_0 = \frac{1}{4}(r^2 + r_m{}^2 + 2rr_m\cos(\Phi - \Phi_m)) \quad \eta_{\pm1} = \frac{1}{\pi^2}(r^2 + r_m{}^2 - 2rr_m\cos(\Phi - \Phi_m)) \tag{4}$$

where the $r_m$ and $\Phi_m$ are the amplitude and the phase of the metallic reflection. Any change in the fluidic optical constants results in change in the values of $\eta_0$□ and □$\eta_1$. Thus, the fluidic sensing is realized by measuring of the energy diffracted in the 0th and the 1st order (figure 1). For that purpose fixed photodetectors (8) are used, since the angles of diffraction are constant [4-5].

## 3. Experimental

The micro-sensors are realized by fitting of two <100>silicon wafers, as reported in [3]. The microprisms are created in one of the wafers. First, it is oxidized, the $SiO_2$ is etched by hydrofluoric acid (HF) from the reverse side of the wafer and Cr diffraction gratings are deposited there by lift-off technique. Then, the $SiO_2$ layer front side of the wafer is patterned by photolithography and chemical oxide etching. The resist is stripped and the microprisms are micromachined by wet anisotropic potassium hydroxide (KOH) etching using the patterned $SiO_2$ as a mask. This process reveals the (111) crystal planes of the silicon, which are etched 100 times slower than the <100> direction. The resulting prisms have an angle of 54.7° (the angle between the (100) and (111) planes in a cubic crystal cell). The microchannels are structured by realized in the second wafer. It is oxidized, and the $SiO_2$ is patterned from the both sides of the wafer by photolithography and consequent $SiO_2$ chemical etching. Then, again bulk micromachining by KOH etching is performed until meeting of the both etch fronts. Finally, the two wafers are adjusted, bonded together and the sensor chips are separated mechanically. A picture of the micro-fluidic chip placed in its fluidic-supply holder is shown in figure 2. A smaller top-left situated box shows a top- and back-side view of the chip itself.

Oil soluble $CuFe_2O_4$ composite nanoparticles have been successfully synthesized. A suitable surface functionalization and choice of solvent are crucial to achieving sufficient repulsive interactions to prevent aggregation so as to obtain a stable colloidal solution of magnetic nanoparticles, a ferrofluid. In the magnetic nanoparticles, the carboxylate ends of long-chain carboxylic acids such as oleic acid bind well to the particles so that the aliphatic chains extending out from their surface of the particle make them effectively hydrophobic and dispersible in nonpolar solvents such as hexane. $Fe(acac)_3$ (1 mmol), $Cu(acac)_2$ (1 mmol), 1,2-hexadecanediol (10 mmol), oleic acid (6 mmol), oleylamine (6 mmol), and benzyl ether (20 mL) were mixed and magnetically stirred under a flow of argon for 30 min. The mixture was heated to 200 °C for 30 min and then, in argon atmosphere, heated to reflux (265 °C) for another 30 min. The black-brown mixture was cooled to room temperature in several hours by removing the heat source. Then, ethanol (40 mL) was added to the mixture, and a black material was precipitated and separated via centrifugation (ca. 5000 rpm, 10 min). The black product was dissolved in hexane in the presence of oleic acid (~ 0.05 mL) and oleylamine (~ 0.05 mL). Centrifugation (6000 rpm, 10 min) was applied to remove any undispersed residue. The product, $CuFe_2O_4$ nanoparticles, was then precipitated with ethanol, centrifuged (6000 rpm, 10 min) to remove the solvent, and redispersed into hexane. All precipitation and centrifugation runs were performed under an argon atmosphere.

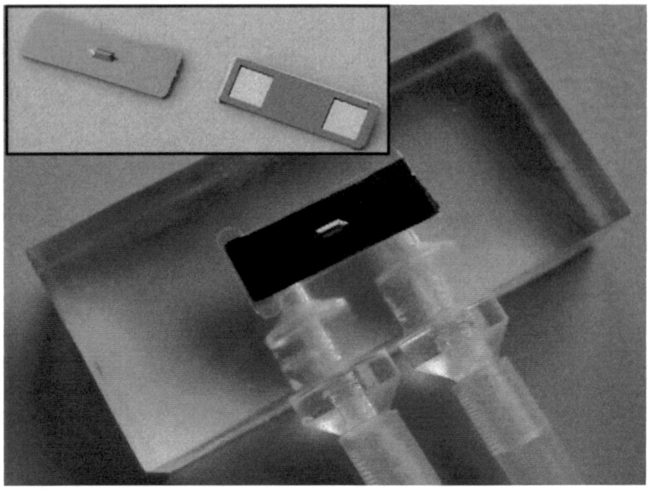

**Figure 2.** Micro-fluidic sensor: single chips and the micro-sensor placed in a fluidic holder.

## 4. Results and discussion

The dependence of $\eta_0$ and $\eta_1$ (the $0^{th}$ and the $1^{st}$ order diffraction efficiency) on the concentration $c$ of aqueous solutions of sucrose is shown in figure 3. Figures 3a) and 3b) correspond to s and p polarization of the IR laser beam. The signs ■, ● and ▲ are related to the $0^{th}$ order and note a deviation from normal incidence to the front prism's facet of α=-30°, 0° and 30°, correspondingly. The hollow signs □, ○ and Δ denote the $1^{st}$ order, respectively. The lines represent a linear fits of the experimental data, as expected before [3]. In all cases the measurements agree with the linear fits within the experimental error.

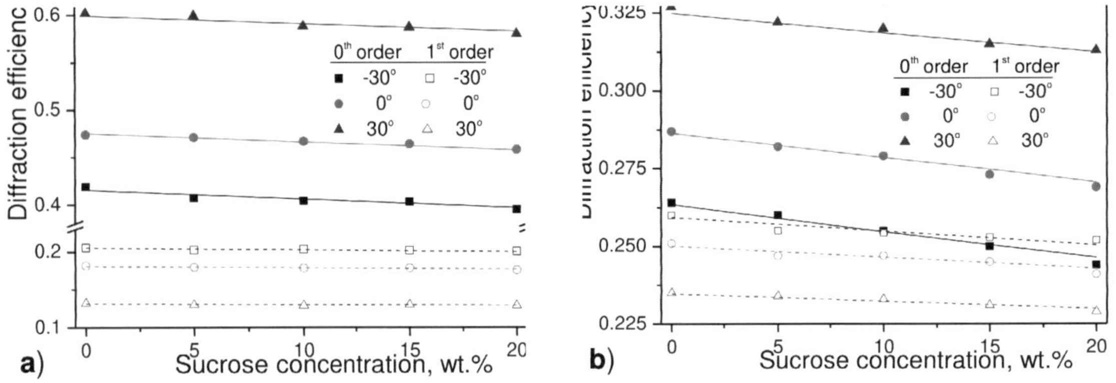


**Figure 3.** Diffraction efficiency of the $0^{th}$ (filled signs) and $1^{st}$ (hollow signs) diffraction orders with a variation of the analyte's concentration at different angles of incidence (square for -30°, circle for 0° and triangle for 30°) a) case of s-polarized light, b) p-polarized light.

It visible from figure 3, that the polarization of the light has a critical influence on the characteristics of the sensor. Figure 4 shows that the polarization behaviour of the diffraction orders is described by curves with a maximum and a minimum at ca 90° polarization angle. At TIR the reflected beam undergoes both amplitude and phase changes. The phase dependence is especially significant, as the difference between the phases for s and p polarization is close to π [9]. Hence, the maximum DE for a given polarization corresponds to a minimum for the orthogonal one. Moreover, the DE change with the polarization is about 2 times higher for the $0^{th}$ order, because both the $+1^{st}$ and $-1^{st}$ order contribute to the energy transfer. The polarization characteristic of the integrated micro-sensor is compared with a $\cos^2$ function, which is typical for the amplitude diffraction gratings. The good agreement between them confirms the validity of a theoretical description of such grating's system by an amplitude transmission model, as was made in the previous works [3-4, 8].

Based on the above investigations, it could be concluded, that the presented IR integrated system can be applied for concentration analysis of the micro-fluidic mixtures. The resolution of analyte concentration $\Delta c$ is given by $\Delta c=|d\eta/dc|^{-1}.\Delta\eta$. In all cases the absolute experimental uncertainty in the diffraction efficiency is $\Delta\eta < 0.001$, which is determined by multiple measurements and statistic processing of the data. Also, the experimental data coincide with the theory within such error. Table 1 summarizes the DE sensitivity $d\eta/dc$ from the experiments with sucrose solutions for all combinations between light polarization, observed diffraction order and deviation angle α. The concentration resolution is also calculated and presented. It is clearly visible, that the sensitivity of the micro-fluidic sensing system can be increased several times by properly choosing the conditions of the measurements. In such a case the concentration of sucrose can be determined with a precision, better than 1wt.%. The measurements are exacter by p-polarized light in the $0^{th}$ diffraction order and smaller possible value of the deviation angle α. The latter corresponds to incidence at the prism base at angle, closer, but still exceeding the critical angle. Then the penetration depth of the evanescent filed is higher, i.e. effective absorption path is longer and thus the sensitivity is higher. The measurements range of concentration practically does not depend on the angle of incidence because the total internal condition is fulfilled for any angle of incidence due to the higher Si refractive index. On the other hand, a value of α, smaller then -45° to -60° is not desirable due to the increased reflection losses at sloped incidence (even at normal incidence the reflection from the silicon is about 50%).

**Table 1.** DE sensitivity and resolution for sucrose concentration.

| | *s* polarization | | | | | | *p* polarization | | | | | |
|---|---|---|---|---|---|---|---|---|---|---|---|---|
| | $0^{th}$ order | | | $1^{st}$ order | | | $0^{th}$ order | | | $1^{st}$ order | | |
| | α=-30° | α=0° | α=30° | α=-30° | α=0° | α=30° | α=-30° | α=0° | α=30° | α=-30° | α=0° | α=30° |
| $d\eta/dc$ x$10^2$ | -8.5 | -7.3 | -6.4 | -2.2 | -1.7 | -1.4 | -11.2 | -9.0 | -7.9 | -4.8 | -3.7 | -2.5 |
| Δc, wt.% | 1.2 | 1.4 | 1.6 | 4.5 | 5.9 | 7.1 | 0.9 | 1.1 | 1.3 | 2.1 | 2.7 | 4 |

The next experiment with composite $CuFe_2O_4$ nanoparticles in hexane (figure 5) proves the feasibility of the micro-fluidic system for analysis of nano-dispersions. Such magnetic nanoparticles bear two attractive features, fluorescence and superparamagnetism. The size of the $CuFe_2O_4$ particles here is below 10 nm. The measurements are led in the most-beneficial configuration: $0^{th}$ order, p polarization, $\alpha$= -30°. The influence of the continuous medium could be excluded, because hexane has no absorption peculiarities at the considered wavelength [11]. A clear effect of decrease of the DE with the concentration of the nanoparticles can be reported. The data are compared with a linear function, as the angular coefficient defines the concentration sensitivity $|d\eta/dc|$=0.02. The experimental uncertainty of the considered results $\Delta\eta=2_x10^{-3}$ is determined by the statistical processing of the data of multiple-repeated measurements and determination of the deviation between the data points and the fit line. The corresponding resolution toward the concentration of the $CuFe_2O_4$ nanoparticles is $\Delta c=|d\eta/dc|^{-1}.\Delta\eta=0.1$ *wt. %.*

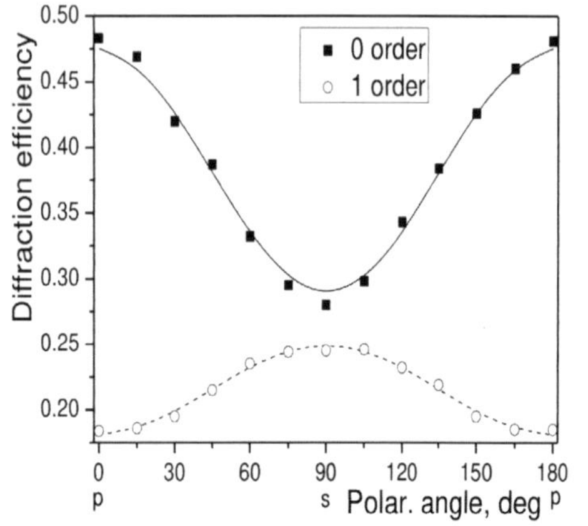


**Figure 4.** Diffraction efficiency of the $0^{th}$ and the $1^{st}$ orders at different polarization angle of the incident light.

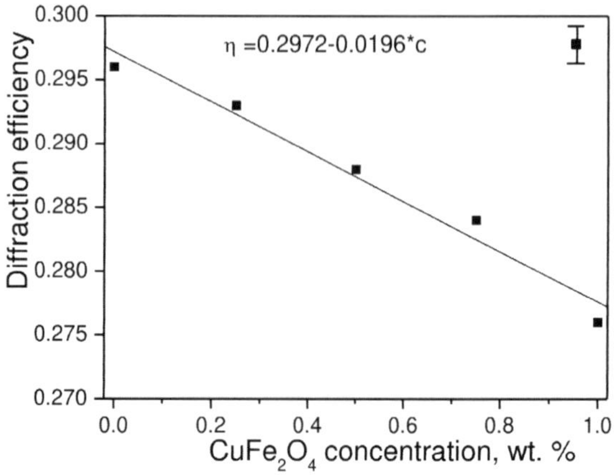


**Figure 5**. Diffraction efficiency of the $0^{th}$ order at p polarization and external angle $\alpha$= -30° from the concentration of $CuFe_2O_4$ nanoparticles.

A slight saturation effect (deviation from the linear behavior) in the DE can be observed. It could be explained with the increased influence of the multiple scattering at higher particles concentration. Other reason could be the specific photo-properties of such nanoparticles. $MnFe_2O_4$ and $Fe_3O_4$ nanoparticles are known to absorb light and transform to photoexcited once [10]. The photoexcitation of feromagnetic nanoparticles led to the demagnetization of ferromagnetic nanoparticles. The photoexcited $CuFe_2O_4$ nanoparticles may be deactivated by the nonradiative process. Thus, absorption of the incident light and light scattering by metal nanoparticles are expected to decrease additionally the reflected light intensity or DE. Furthermore the self-quenching of excited states of nanoparticles is known to increase with the higher concentration (or the number) of metal nanoparticles. We should exclude the participation of organic shell (composed of oleic acid and oleyl amin) on the excitation and deactivation processes (absorption of light, light-scattering and quenching of excited states). Anyway, a certain conclusion cannot be made due to the relative high error of the experiment.

Finally, we have proven that the analysis of the experiments leads to more precise measurements. On the other hand, the measurements can be made simultaneously in the both 0th and the 1st order and/or both with s and p polarization. Such investigation will additionally increase the precision of the measurements and will eliminate the influence of casual mistakes.

## 5. Conclusions

This work presents for a first time an investigation of the sensitivity of a novel integrated IR laser system for micro and nano-fluidic investigation on the measurement conditions. Increased sensitivity is found out for the most-beneficial configuration of p-polarized light, $0^{-th}$ diffraction order and minimal possible angle of incidence. Main advantages of the micro-fluidic detector are the simplified fabrication technique, the fixed positions of the optical elements (there are no moving parts) and the operation with a monochromatic NIR source (no spectral analyzer is required). Moreover, it is sensitive to transparent and absorbing fluids and is applicable for strongly absorbing, highly concentrated and nano-disperse media. The last application is proven for concentration analysis of oil-based dispersion of composite $CuFe_2O_4$ nanoparticles. In this case, a resolution below 0.1 wt. % toward the $CuFe_2O_4$ nanoparticles is demonstrated. Other applications of the integrated micro-system are chemical, medical and biological diagnostic of liquids and dispersions of drugs, serums, proteins, peptides, DNA, RNA, viruses, etc.

**Acknowledgement**
This work is supported by AvH Foundation, the VEGA Projects No. 2/0037/10 and 2/0160/10 and ASFEU project 26240220011 OPVaV-2008/4.2/01-SORO.

**References**
[1] Li P 2006 *Microfluidic lab-on-a-chip for chemical and biological analysis and discovery* (New York: Taylor & Francis, CRC Press)
[2] Zhang L, Dang F and Baba Y 2003 *J. Pharm. Biomed. Anal.* **30** 1645
[3] Sarov Y, Ivanova K, Ivanov Tz, Volland B and Rangelow I W 2006 *MicroElectron. Eng.* **83** 1294
[4] Sarov Y, Ivanov Tz, Ivanova K, Sarova V, Capek I and Rangelow I W 2006 *Appl. Phys. A* **84** 191
[5] Sainov S and Tonchev D 1989 *Opt. Las. Eng* **10** 17
[6] Sainov S 1994 *Sensors and Actuators A* **45** 1
[7] Sarov Y and Sainov S 2002 *J. Opt. A: Pure Appl. Opt.* **4** 382
[8] Sarov Y, Sarova V, Ivanov Tz, Ivanova K, Capek I and Rangelow I W 2008 *Chem.Eng.J.* **135S** 2847
[9] Harrick N 1967 *Internal refcelction spectroscopy* (New York: Interscience, Wiley)
[10] Vestal C and Zhang Z 2003 *J. Am. Chem. Soc.* **125** 9828

# X-ray photoelectron study of high-energy $He^+$ implanted a-SiC:H thin films

**T Tsvetkova[1], P Sellin[2], D Dimova-Malinovska[3], O Angelov[3], V Krastev[1], P Stefanov[4] and I Avramova[4]**

[1] Georgi Nadjakov Institute of Solid State Physics,
Bulgarian Academy of Sciences, 72 Tzarigradsko Chaussee, 1784 Sofia, Bulgaria
[2] University of Surrey, Guildford, Surrey GU2 7XH, United Kingdom
[3] Central Laboratory for Solar Energy,
Bulgarian Academy of Sciences, 72 Tzarigradsko Chaussee,1784 Sofia, Bulgaria
[4] Institute of General and Inorganic Chemistry, Bulgarian Academy of Sciences,
Acad.G.Bonchev Str., 1113 Sofia, Bulgaria

E-mail: tania_tsvetkova@yahoo.co.uk

**Abstract.** The implantation with high-energy $He^+$ ion beam results in modification of the optical band-gap and absorption coefficient of a-$Si_{1-x}C_x$:H films. X-ray photoelectron spectroscopy was used to characterize a-$Si_{1-x}C_x$:H thin films deposited by r.f. magnetron sputtering and implanted with high-energy $He^+$ ions. The non-implanted films show the expected variety of chemical bonds: silicon-silicon and silicon-carbon bonds, carbon being three- and fourfold coordinated. The X-ray and Raman spectra show that high-energy $He^+$ ion implantation leads to the introduction of additional disorder in the films. The X-ray photoelectron spectra of the implanted films show that, in addition to the already mentioned changes, the high-energy $He^+$ ion bombardment results in an increase of the threefold coordinated as compared to the fourfold coordinated carbon bonds, i.e. increased graphitization of the carbon content in the films. These structural modifications, due to the high-energy $He^+$ ion implantation, are the reasons for the observed changes in the optical properties of the films.

## 1. Introduction

Hydrogenated amorphous silicon-carbide (a-$Si_{1-x}C_x$:H) alloy films have been studied for many years as structural materials for high-temperature electronics and for various optoelectronic elements and devices operating in aggressive environment (acid vapors, radiation, open space, etc.) [1,2]. It is particularly attractive for various optoelectronic applications due to its comparatively wide band gap which may be tailored in the range 1.8-3.0 eV [2].

The studied material can be prepared in thin film form on various substrates by chemical vapour deposition [3], or by radio-frequency (RF) reactive magnetron sputtering method [4]. The mechanical, optical and electrical properties of the material can be varied controllably by changing the relative composition ratio of the constituent elements (x).

In addition, control of the material properties can also be achieved using ion implantation [5,6]. Recently, using the emerging technology of ion micro-beam implantation [7], promising results have been obtained for possible applications of these materials in high-density optical data storage [8,9].

This approach has been further developed by implementing ion doping with focused ion beam (FIB) systems that use $Ga^+$ or other chemically active species as implants [10-13]. Another approach was also used applying focused high-energy (MeV range) ion implantation with $H^+$ and $He^+$ ion beams for the formation of optical contrast of suitable size to be used for optical data storage in a-$Si_{1-x}C_x$:H films [14]. In order to study the underlying high-energy ion beam induced structural modification of the implanted material, additionally some broad-beam high-energy $H^+$ and $He^+$ ion implantation has been performed. The implanted films have been studied using optical transmission, reflection and photo-deflection spectroscopy (PDS) in the ultraviolet (UV) and visible range of the spectrum [15,16].

In the present paper X-ray photoelectron spectroscopy (XPS) was used to characterize a-$Si_{1-x}C_x$:H thin films deposited by r.f. magnetron sputtering and implanted with high-energy $He^+$ ions. Additionally, some optical transmission measurements in the infra-red (IR) and Raman spectroscopy have been used to characterize the ion beam induced structural modification of the implanted material.

## 2. Experimental

Thin (d ~ 1μm) a-$Si_{1-x}C_x$:H films (x=0.18) were deposited onto polished silicon substrates by RF (13.56 Hz) reactive magnetron sputtering. A composite target, composed of mono-crystalline (100) silicon wafer with chips of pure graphite placed on it, was sputtered in an Ar-20%$H_2$ gas mixture. Typical deposition conditions are RF power 150 W (power density 1.91 W/cm$^2$), total gas pressure 1P, substrate temperature 275°C, and graphite-to-silicon target ratio 0.025. The film thickness was determined by Talystep to be 1 μm. Rutherford backscattering spectrometry was used to determine the carbon content, x, of the films.

High energy ion implantation of $He^+$ was performed at energies 0.2, 0.5 and 1.0 MeV at room temperature and ion beam intensity 2-5 μA.cm$^{-2}$. Various ion doses were implanted in the range $1x10^{16}$ – $1x10^{17}$ cm$^{-2}$. Water cooling system for the sample holder was used to maintain room temperature during the implantation process.

The XPS studies were carried out on an ESCALAB MK II (VG Scientific) electron spectrometer with base pressure in the preparation and analysis chambers of $2\times10^{-8}$ Pa and $1\times10^{-8}$ Pa, respectively. The photoelectrons were excited using X-ray source Mg Kα (hν=1256.6 eV). The surface sensitivity was estimated as ~ 10 monolayers. A strip of gold, deposited on the surface of the sample, was used as a binding energy (BE) reference. The BE of Au $4f^{7/2}$ was assumed to be 83.8 eV. This allowed a determination of the binding energies of the C 1s, O 1s, Si 2p and Si 2s peaks with an accuracy of ±0.1 eV. The depth profiling was carried out by sputter etching at 90° using a defocused $Ar^+$ ion beam of 3 keV energy and a current density of 16 μA.cm$^{-2}$.

The Raman spectra were taken at room temperature using a SPEX 1403 double beam spectrometer. The instrument was equipped with a photomultiplier, working in a photon counting mode and a spectral resolution of 4 cm$^{-1}$.The 488nm excitation line of an $Ar^+$ laser was used for the measurements.

## 3. Results and discussion

Figure 1a shows the XPS Si2p spectra of unimplanted a-$Si_{1-x}C_x$:H thin film. Si2p peak recorded from the surface is centred at 101.7 eV. After 30 min sputtering it is positioned at 100.7 eV and in depth of the sample after 60 min and 120 min the peak shits back to 101.0 eV and 101.5 eV. Cubic silicon carbide crystal SiC characterised by the fourfold coordinated carbon bonds [17] shows similar values of binding energy (100.3 eV). Some small amount of bonds of the type Si-O, met in non-stoichiometric silicon oxide [18], are characterized by Si2p position at 103.2 eV which is confirmed by O1s peak (not shown) at 533.8 eV [19]. There is some contribution of Si-Si bonds characteristic of elemental silicon Si $2p^{3/2}$- 99.85 eV [20]. C1s peaks of the same sample are shown on Figure 1b. After sputtering in depth of the layer C1s peak is positioned at 283.4 eV characteristic of n-type beta-SiC [21] in which carbon is fourfold coordinated. In Figure 1c VB spectra of unimplanted a-$Si_{1-x}C_x$:H thin film are shown. In depth of the sample there is a broad peak at around 10 eV which is probably due to dangling bonds in SiC. The non-implanted films show the expected variety of chemical bonds: silicon-silicon and silicon-carbon bonds, carbon being three- and fourfold coordinated.

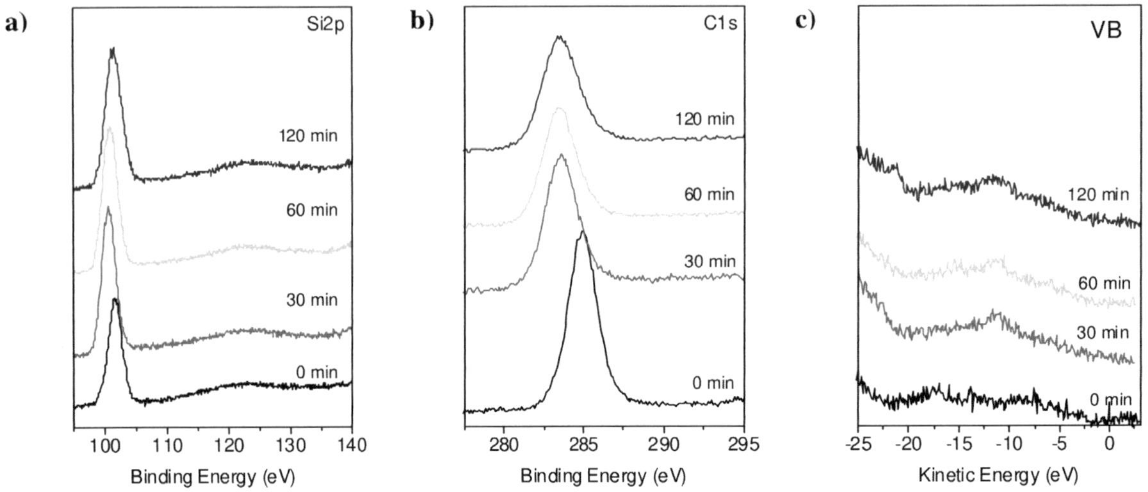


**Figure 1.** X-ray photoelectron spectra of (a) Si2p and (b) C1s electrons, and (c) Valence band (VB) of an unimplanted a-$Si_{1-x}C_x$:H sample.

Figure 2a shows Si2p spectra of a-$Si_{1-x}C_x$:H thin film implanted with 200 keV $He^+$ ions. The Si2p peak is broadened and shifted to higher energies compared to the peak recorded from the unimplanted a-$Si_{1-x}C_x$:H thin film. Figure 2b shows C1s peaks of the same sample. The peak is broadened and shifted to higher energies compared to the peak recorded from unimplanted sample. Figure 2c shows the VB spectra. A broad peak at around 7 eV is probably due to dangling bonds.

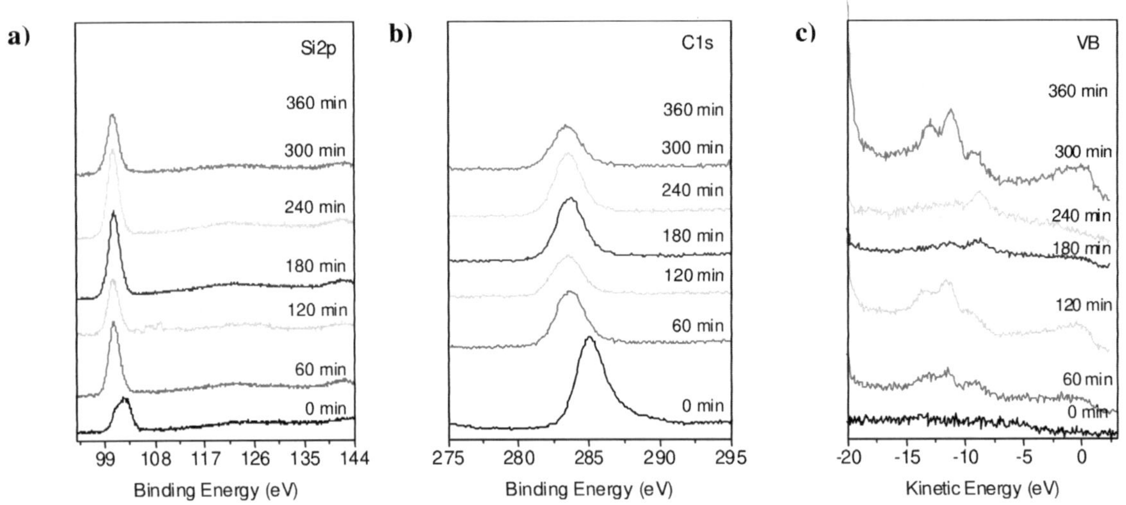


**Figure 2.** X-ray photoelectron spectra of (a) Si2p and (b) C1s electrons, and (c) Valence band (VB) of $He^+$ unimplanted a-$Si_{1-x}C_x$:H sample at ion beam energy of E = 200 keV and a dose D = $1\times10^{17}$ $cm^{-2}$.

Figure 3a shows Si2p spectra of a-$Si_{1-x}C_x$:H thin film implanted with 500 keV $He^+$ ions. Figure 3b shows C1s peaks of the same sample. The peak is broadened and shifted to higher energies around 284.0 eV due to threefold coordinated Si-C bonds characteristic of highly oriented pyrollytic graphite [22]. Figure 3c shows the VB spectra. A broad peak at around 7 eV probably due to dangling bonds appears only in depth of the sample.

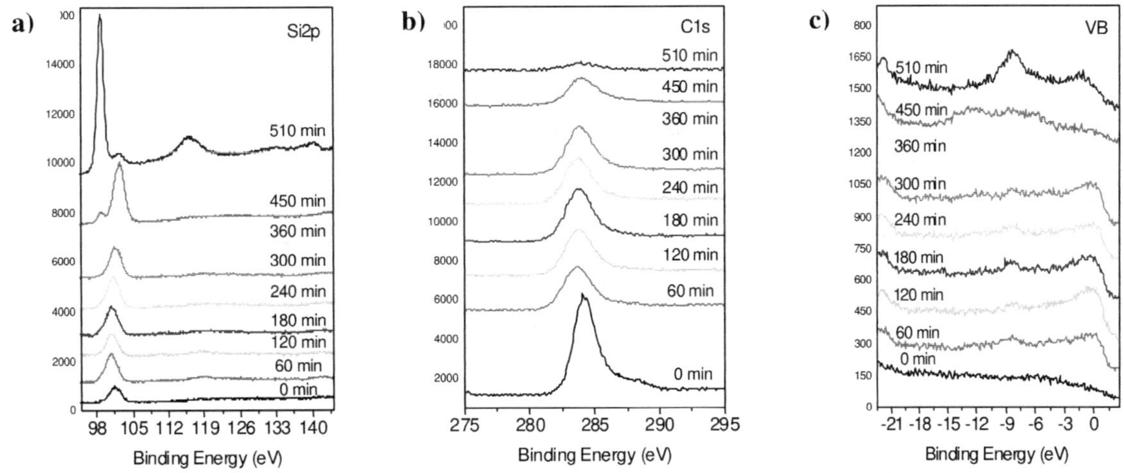


**Figure 3.** X-ray photoelectron spectra of (a) Si2p and (b) C1s electrons, and (c) Valence band (VB) of $He^+$ unimplanted a-$Si_{1-x}C_x$:H sample at ion beam energy of E = 500 keV and a dose D = $1\times10^{17}$ $cm^{-2}$.

Figure 4a shows Si2p spectra of a-$Si_{1-x}C_x$:H thin film implanted with 1 MeV $He^+$ ions. Figure 4b shows C1s peaks of the same sample. On the surface there is a peak centered at 284.9 eV which is similar to the peak position characteristic of highly oriented pyrollytic graphite [22] and in depth at 283.3 eV characteristic of n-type beta-SiC [21] in which carbon is fourfold coordinated. Figure 4c shows the VB spectra. A broad peak at around 7 eV probably due to dangling bonds is clearly visible from the surface of the sample.

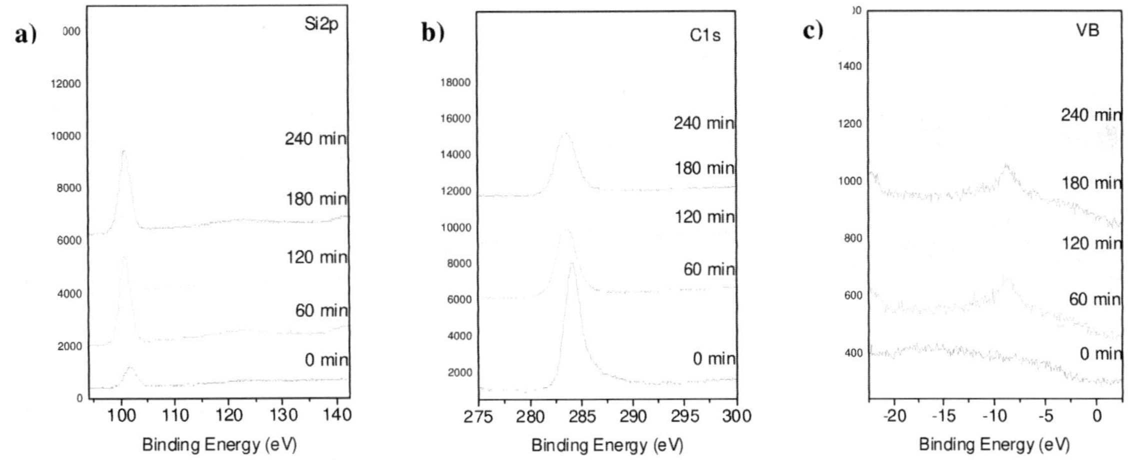


**Figure 4.** X-ray photoelectron spectra of (a) Si2p and (b) C1s electrons, and (c) Valence band (VB) of $He^+$ unimplanted a-$Si_{1-x}C_x$:H sample at ion beam energy of E = 1 MeV and a dose D = $1\times10^{17}$ $cm^{-2}$.

The XPS spectra show that high energy $He^+$ ion implantation leads to introduction of additional disorder in the films. The XPS spectra of the implanted films show that in addition to the already mentioned changes the high-energy $He^+$ ion bombardment results in an increase of the threefold coordinated as compared to the fourfold coordinated carbon bonds, i.e. increased graphitization of the carbon content of the films. The structural modifications due to the high-energy $He^+$ ion implantation, as revealed by the XPS measurement, essentially confirm the optical modification effects [15].

Journal of Physics: Conference Series **253** (2010) 012052 doi:10.1088/1742-6596/253/1/012052

As can be seen from the Full Width at Half Maximum (FWHM) of Si and C peaks, the damage level for both sublattices C and Si differs after implantation with 200 keV. According to the shape of peaks, the sublattice of C shows a maximum damage level located at a depth of 120 min sputtering beneath the surface. The damage in the sample implanted with 500 keV is still visible in both sublattices. However, the C1s peak is flatter and has a larger width, as compared to the case for 200 keV implantation, which makes the evaluation of the aligned spectrum rather difficult. After implantation with 1 MeV, the peaks of Si2p and C1s are narrow and positioned at energies characteristic of SiC. This could possibly be explained by some kind of annealing during implantation.

The results of the Raman measurements (Figures. 5 and 6) could also be considered as justification of the above reasoning. As it is seen, the Full Width at Half Maximum (FWHM) of the Si-Si TO-like peak at about 470 $cm^{-1}$ increases with the dose and energy (figure 5 a and figure 6 a), which is better expressed at the lowest dose and energy. With the further dose and energy increase this change is more subtle. The changes in FWHM imply that there is an effect of increased disorder with the dose in the Raman spectra for one and the same energy – an additional amorphisation of the films takes place during the implantation. In the higher energy region of the Raman spectra (figure 5 b and figure 6 b), the wide band corresponding to the C-C bonds is present. For the unimplanted material, the $sp^3$ (diamond-like) carbon bonding is predominant. After the ion implantation its partial transformation into $sp^2$ (graphitic) bonding is observed, particularly well expressed for the highest implantation energy and dose (figure 6 b).

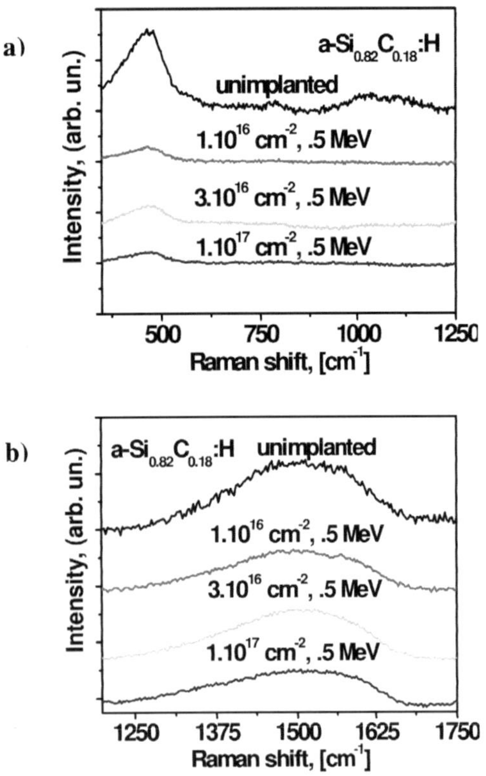


**Figure 5.** Raman spectra of unimplanted and $He^+$ implanted a-$Si_{0.82}C_{0.18}$:H films at ion beam energy E = 0.5 MeV and different doses in the range $1x10^{16}$ – $1x10^{17}$ $cm^{-2}$ , in the regions 350÷1250 $cm^{-1}$ (a) and 1250÷1750 $cm^{-1}$ (b).

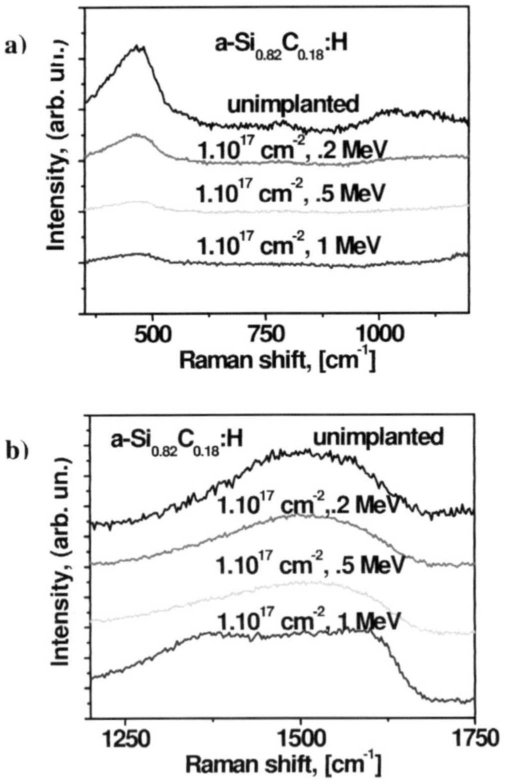


**Figure 6.** Raman spectra of unimplanted and $He^+$ implanted a-$Si_{0.82}C_{0.18}$:H films with ion dose D = $1x10^{17}$ $cm^{-2}$ and different energies in the range 0.2 – 1.0 MeV, in the regions 350÷1250 $cm^{-1}$ (a) and 1250÷1750 $cm^{-1}$ (b).

## 4. Conclusion

The XPS and Raman spectra show that high energy $He^+$ ion implantation leads to introduction of additional disorder in the films. The XPS spectra of the implanted films show that in addition to the already mentioned changes the high-energy $He^+$ ion bombardment results in an increase of the threefold coordinated as compared to the fourfold coordinated carbon bonds, i.e. increased graphitization of the carbon content of the films. Additionally, also Si dangling bonds are revealed by the valence band study. The structural modifications due to the high-energy $He^+$ ion implantation are the reasons for the observed changes in the optical properties of the films, which could potentially be of interest in relation to the recently developed high-energy ion micro-beam technologies.

## Acknowledgements

The support of the EPSRC, UK, is gratefully acknowledged. The authors wish also to thank their colleagues from the Ion Beam Centre at the University of Surrey, UK, for their experimental support with the ion beam processing of silicon carbide samples.

## References

[1] Kanicki J 1991 *Amorphous and Micro-crystalline Semiconductor Devices* (Boston: Artech House)
[2] Powell J A, Matus L 1989 *Amorphous and Crystalline Silicon Carbide*, ed G L Harris and C Y W Yang, (Berlin: Springer)
[3] Heinrich J, Hemeltjen S and Marx G 2000 *Microchim. Acta* **133** 209
[4] Bullot J and Schmidt M P 1987 *phys. stat. sol. (b)* **143** 345
[5] Hirvonen J K 1984 *Ion Implantation and Ion Beam Processing of Materials* (Amsterdam: North Holland)
[6] Ziegler J F 1988 *Ion Implantation* (New York: Academic Press)
[7] Bohringer K, Jonsten K and Kalbitzer S 1988 *Nucl. Instrum. & Meth. B* **30** 289
[8] Ruttensberger B, Krotz G, Muller G, Derst G and Kalbitzer S 1991 *J. Non-Cryst. Solids* **137-138** 635
[9] Muller G 1993 *Nucl. Instrum. & Methods B* **80-81** 957
[10] Tsvetkova T 1996 *Beam Processing of Advanced Materials*, ed J Singh, S Copley and J Mazumder (New York: Metals Park, ASM International) p 207
[11] Tsvetkova T, Tzenov N, Tzolov M, Dimova-Malinovska D, Adriaenssens G J and Pattyn H 2001 *Vacuum* **63** 749
[12] Bischoff L, Teichert J, Kitova S and Tsvetkova T 2003 *Vacuum* **69** 73
[13] Tsvetkova T, Angelov O, Sendova-Vassileva M, Dimova-Malinovska D, Bischoff L, Adriaenssens G J, Grudzinski W and Zuk J 2003 *Vacuum* **70** 467
[14] Tsvetkova T, Takahashi S, Sellin P, Gomez-Morilla I, Angelov O, Dimova-Malinovska D and Zuk J 2010 *Acta Physica Polonica* in press
[15] Tsvetkova T, Sellin P, Carius R, Angelov O and Dimova-Malinovska D 2007 *Journal of Optoelectronics and Advanced Materials* **9** 375
[16] Tsvetkova T, Sellin P, Carius R, Angelov O, Dimova-Malinovska D and Juk J 2009 *Nucl. Instrum. & Meth. B* **267** 1583
[17] Tabata A, Fujii S, Suzuoki Y, Mizutani T and Ieda M 1990 *J. Phys. D* **23** 316
[18] Aarnik W A M, Weishaupt A and van Silfout A 1990 *Appl. Surf. Sci.* **45** 37
[19] Montero I, Galan L, de la Cal E, Albella J M and Pivin J C 1990 *Thin Solid Films* **193** 325
[20] Magnusson K O, Wiklund S, Dudde R and Reihl B 1991 *Phys. Rev. B* **44** 5657
[21] Mizokawa Y, Geib K M and Wilmsen C W 1986 *J. Vac. Sci. Technol. A* **4** 1696
[22] Palchan I, Crespin M, Estrade-Szwarckopf H and Rousseau B 1989 *Chem.Phys.Lett.* **157** 321

# Study of $As_2Se_3$ and $As_2Se_2Te$ glass structure by neutron- and X-ray diffraction methods

**M Fábián[1], E Sváb[1], V Pamukchieva[2], A Szekeres[2], S Vogel[3] and U Ruett[4]**
[1]Research Institute for Solid State Physics and Optics, H-1525 Budapest, P.O.B. 49, Hungary
[2]Institute of Solid State Physics, Bulgarian Academy of Sciences, Tzarigradsko Chaussee 72, 1784 Sofia, Bulgaria
[3] Los Alamos Natl. Lab., Lujan Neutron Scattering Ctr, Los Alamos, NM 87545, USA
[4]Deutsches Elektronen-Synchrotron (DESY), Notkestr. 85, D-22603 Hamburg, Germany

E-mail: vesela@issp.bas.bg

**Abstract.** Neutron and high-energy X-ray diffraction measurements have been performed on $As_{40}Se_{60}$ and $As_{40}Se_{50}Te_{10}$ glasses. Both the traditional Fourier transformation technique and the reverse Monte Carlo (RMC) simulation of the experimental data have been applied to model the 3-dimensional atomic configurations. From the analysis of the partial atomic correlation functions and structure factors the first and second neighbour distances, coordination numbers and bond-angle distributions were calculated. It is established that substitution of Se by Te does not change the basic glassy network structure. For the As-Se bonding is revealed that the first neighbour distance is at 2.42 Å, the average coordination numbers are $CN_{AsSe}=2.6\pm0.1$ and $CN_{SeAs}=1.8\pm0.1$ atoms and the three-atom-bond angle for <Se-As-Se>, <As-Se-As> configurations is 95°, the same value within limit of error, as obtained from RMC calculations.

## 1. Introduction

Chalcogenide glasses are perspective materials for practical application in optoelectronics as high-speed optical elements, for applications such as data processing devices, electronic switches, and optical memories. In general, they are transparent from the visible or near-infrared regions. Among these glasses the $As_2Se_3$ and Te–As–Se compositions are perspective materials for manufacturing optical fibers because of their low phonon energy, good transparency, low optical losses and good thermal and chemical stability. The physical–chemical properties and exploitation possibilities of chalcogenide glasses are usually treated in dependence on their atomic structure [1–5]. The structure of chalcogenide glasses has been the subject of some debate for many years. The absence of translational symmetry in As-Se and Te–As–Se glasses suggests that particles are randomly arranged around each atom. Since there is no universal approach for revealing the glass structure, one can obtain a consistent picture of the glass structure only by combining several investigation techniques.

In this work we focus on the atomic-scale structural characterization of $As_{40}Se_{60}$ and $As_{40}Se_{50}Te_{10}$ chalcogenide glasses using neutron- and X-ray diffraction methods. Structure modeling by RMC simulation was also applied.

## 2. Experimental Details

### *2.1. Sample preparation*

The glassy samples $As_{40}Se_{60}$ and $As_{40}Se_{50}Te_{10}$ were synthesized by the conventional melt-quenching method from 5N purity elements. The syntheses were performed in evacuated quartz ampoules ($10^{-3}$ Pa) using a rotary furnace. The specimens were heated up to 950°C, kept at this temperature for 24 h meanwhile the furnace was rotated for homogeneous melting. The ampoules were quenched in air.

### *2.2. Neutron and X-ray diffraction measurements*

Neutron diffraction measurements were carried out by combining the data measured in the 2-axis 'PSD' diffractometer ($\lambda_0$=1.068 Å) [1] at the 10 MW Budapest research reactor and by the time-of-flight 'HIPPO' instrument [2] at the LANSCE pulsed neutron source. For both experiments, the powder specimens of about 3-4 g were filled in cylindrical vanadium sample holder of 8 mm and 10 mm diameter, respectively. The neutron structure factors, $S(Q)$s were revealed in a broad momentum transfer range, $Q$=0.45-30 Å$^{-1}$, which made possible high $r$-space resolution analysis ($\Delta r=2\pi/Q_{max}\approx 0.2$ Å).

The high-energy X-ray diffraction measurements were performed at the BW5 experimental station [3] at Desy/Hamburg using 109.5 keV radiation energy ($\lambda_0$=0.113 Å). The powdered samples were filled into quartz capillary of 2 mm in diameter (wall thickness of ~0.02 mm). The amplitudes of the small intensity physical oscillations at the relatively high Q-values were obtained with good statistic of about 1%, and the X-ray $S(Q)$s were revealed in the $Q$=0.5-19 Å$^{-1}$ momentum transfer range.

## 3. Results and discussion

Figure 1 shows the neutron $S(Q)$ and the total reduced atomic correlation functions, $G(r)$, calculated by Fourier transformation:

$$G(r)=\frac{2}{\pi}\int_{o}^{Q_{max}} Q[S(Q)-1]M(Q)\sin Qr\,dQ \quad , \qquad (1)$$

where $Q_{max}$ = 30 Å$^{-1}$ is the integration limit, and $M(Q)$ is a modification function to reduce the artificial oscillations caused by the finite $Q_{max}$ truncation (see [4] for more details). The overall character of the $S(Q)s$ is very similar for the studied glasses suggesting a strong similarity of the basic network structure. A small prepeak appears at $Q$ = 1.3 Å$^{-1}$, which can be interpreted as the existence of intermediate range ordering. The first sharp peak appears at $r_1$=2.42±0.01 Å in $G(r)$ for both the $As_{40}Se_{60}$ and $As_{40}Se_{50}Te_{10}$ compositions, which corresponds to the As-Se atom pairs. A small but significant peak appears for the ternary glass at ≈2.7 Å, which corresponds to the Te-As and/or Te-Se atomic pair distances. The second coordination sphere shows up a strong peak centered at $r_2$=3.7±0.05 Å. The three-atom-bond angle between As-Se-As and Se-As-Se atoms is estimated from the equation $\Theta=2\sin^{-1}(r_2/2r_1)$ [5, 6], which gives a value of $\Theta_{<As\text{-}Se\text{-}As>,\ <Se\text{-}As\text{-}Se>} = 99\pm3^{o}$ .

For the data treatment reverse Monte Carlo (RMC) simulation [7] of the diffraction experimental data was applied to get structural information about possible atomic configurations. As a RMC starting model a disordered atomic configuration was built up with a simulation box containing 10 000 atoms. In the RMC simulation procedure for the $As_{40}Se_{60}$ and $As_{40}Se_{50}Te_{10}$ glasses the atom densities of 0.03 and 0.035 atoms/cm$^3$ and half-box lengths of 34.668 and 32.931 Å, respectively, were taken. Two types of constraints were used, namely the minimum interatomic distances between atom pairs (cut-off distances) and connectivity constraints.

In order to check the quality of the experimental data, the neutron and X-ray diffraction datasets were modeled, both separately and simultaneously by the RMC method.

The constituent elements (As, Se, Te) possess rather similar neutron ($b_{As}$=6.58 fm, $b_{Se}$=7.97 fm, $b_{Te}$=5.8 fm [8]) and X-ray scattering amplitudes and interatomic distances. Consequently, the atomic correlation functions overlap, which makes difficult to separate the atomic positions. Using reasonable cut-off distances and connectivity constraints in the RMC simulation procedure, we obtained very good agreement between the experimental and the calculated S(Q) curves, as it is illustrated in figure 2. The cut-off distances were taken from the total G(r) functions (figure 1) as follows: 2.2 Å for As-Se, 2.45 Å for As-Te and for Se-Te, 3.1 Å for As-As, for Se-Se and for Te-Te.

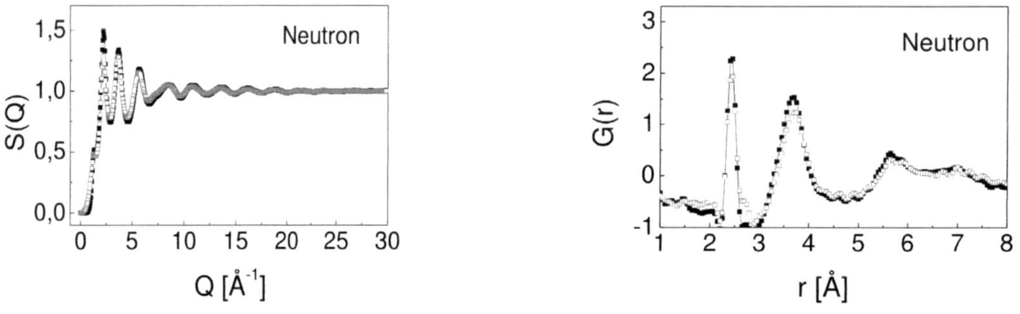


(a) (b)

**Figure 1.** Experimental neutron total structure factor (a) and atomic reduced correlation function (b) for $As_{40}Se_{60}$ (full square) and $As_{40}Se_{50}Te_{10}$ (open square) glasses

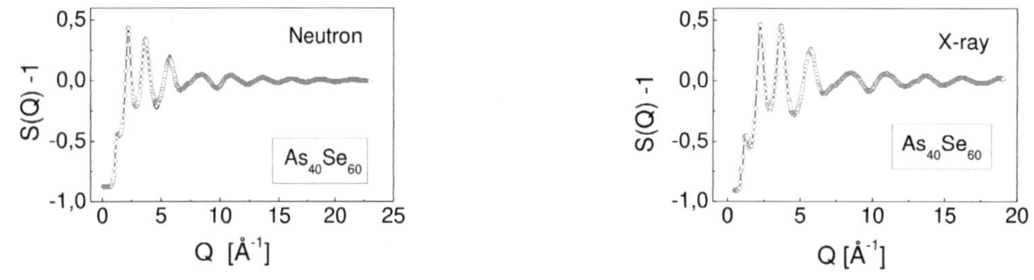


(a) (b)

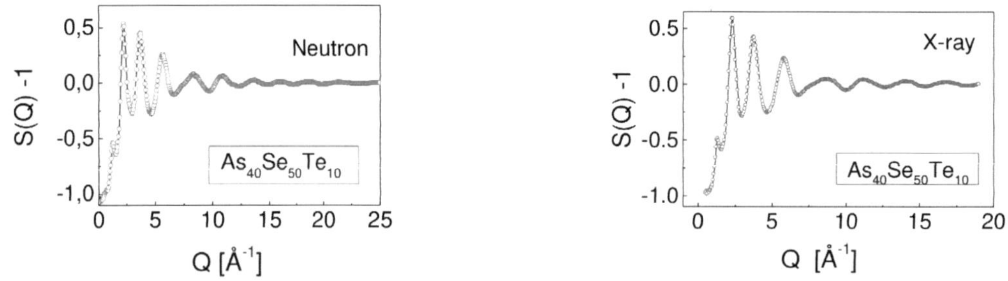


(c) (d)

**Figure 2.** Neutron (a and c) and X-ray (b and d) RMC models (solid line) and experimental structure factors (open circles) for $As_{40}Se_{60}$ and $As_{40}Se_{50}Te_{10}$ glasses.

In the first approach of the RMC modelling we consider only the possible heteropolar bonds, since homopolar bonds are probably absent. We have applied connectivity constraints for As atoms to be surrounded by three Se neighbours, and for Se atoms to be surrounded by two Se neighbours, as reported in many papers [5, 6, 9-15].

From the RMC modelling we revealed the partial correlation functions ($g(r)$), the coordination number ($CN$) and bond angle ($\Theta$) distributions with a fairly good reproducibility for the studied glasses. Figures 3-5 compare the results obtained for As-Se, As-As and Se-Se network former atom pairs. In general, the curves are almost the same for the two compositions. This suggests that substituting Te for Se atoms does not change the basic network structure within the limit of experimental error and modelling procedure. The As-Se correlation (figure 3a) shows a rather sharp first neighbour distance at 2.42±0.02 Å. The corresponding average coordination numbers are $CN_{AsSe}$=2.6±0.1 atoms and $CN_{SeAs}$=1.8±0.1 atoms (figure 4). These values are somewhat less than the expected (constrained) 3 and 2 atomic neighbours, which may be a consequence of a slight distortion from the ideal $AsSe_3$ trigonal pyramid units reported in the literature. From the RMC calculations for the three-atom-bond angle $\Theta_{<As\text{-}Se\text{-}As\,,\,<Se\text{-}As\text{-}Se>}$=95±5° is obtained (figure 5), which is in accordance with the value calculated from the G(r)'s distances, as discussed above. These results are in good agreement with reference [5] and are somewhat less than that reported in references [12, 14, 15]. The error of the actual structural parameters is calculated from the data obtained from different RMC runs.

The observed 2$^{nd}$ neighbour distance at 3.7 Å and 3$^{rd}$ neighbour distance at 5.6 Å for the As-Se bonds are an evidence for medium range correlation in the glassy structure. For the As-As bonds the characteristic distance is at 3.7 Å, while for Se-Se at somewhat shorter distance of 3.65 Å (figures 3b and 3c, respectively).

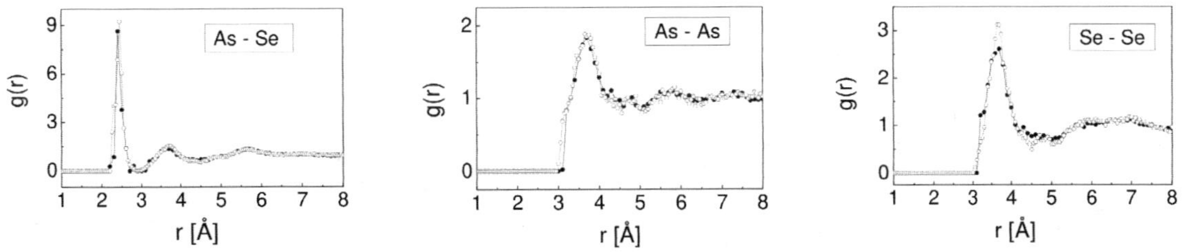


(a) (b) (c)

**Figure 3.** Partial atomic correlation functions obtained from the RMC modelling of As-Se (a), As-As (b) and Se-Se (c) atom pairs for $As_{40}Se_{60}$ (full circles) and $As_{40}Se_{50}Te_{10}$ (open circles) glasses.

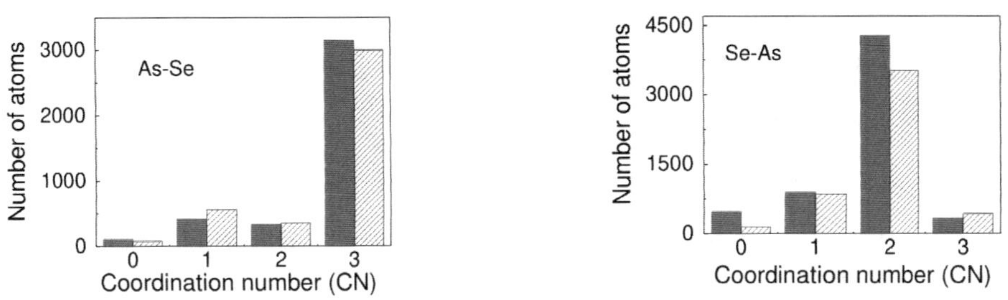


(a) (b)

**Figure 4**. Coordination number distribution for $As_{40}Se_{60}$ (filled) and $As_{40}Se_{50}Te_{10}$ (crossed) glasses from RMC modelling of As-Se (a) and Se-As (b) atom pairs.

The partial atomic correlation functions of the possible Te-containing chemical bonds revealed from the RMC modelling are presented in figure 6. Relatively low intensity first neighbour peak appears at 2.50 Å in the As-Te and Se-Te pair correlation functions, indicating a random substitution of Se atoms by Te atoms. It should be noted, that both the position and peak intensity depend on the constraints applied in the RMC modelling. For unconstrained case, the $g_{AsTe}(r)$ and $g_{SeTe}(r)$ fully overlap with $g_{AsSe}(r)$, while in the constrained RMC calculation $g_{AsTe}(r)$ is shifted to higher distances. The intense peak at 3.5-3.6 Å for all three As-Te, Se-Te and Te-Te partial atomic pair correlations is an evidence of second neighbour distance of these bonds.

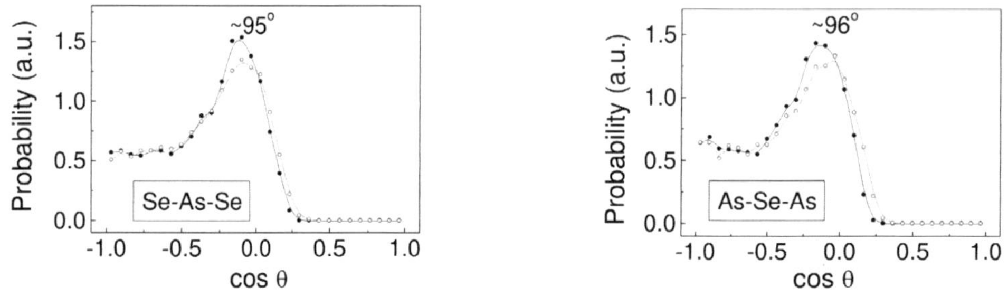


(a) (b)

**Figure 5**. Three-atom-bond angle distribution from RMC modelling for $As_{40}Se_{60}$ (full circles) and $As_{40}Se_{50}Te_{10}$ (open circles) glasses: a) Se-As-Se b) As-Se-As

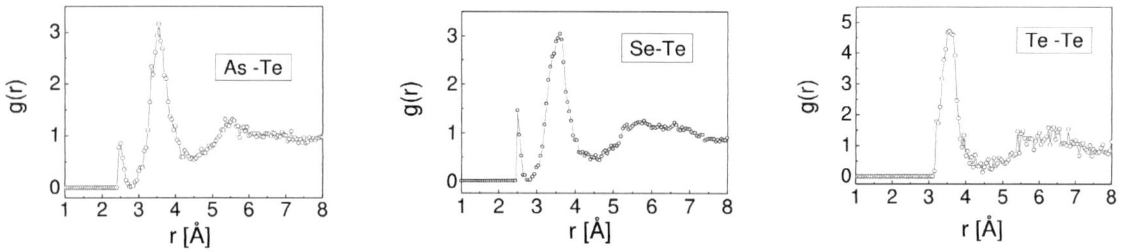


(a) (b) (c)

**Figure 6.** Partial atomic correlation functions, obtained from the RMC modelling of As-Te (a), Se-Te (b) and Te-Te (c) atom pairs for $As_{40}Se_{50}Te_{10}$ composition.

## 4. Conclusions

We have performed neutron and X-ray diffraction measurements and reverse Monte Carlo modelling on $As_{40}Se_{60}$ and $As_{40}Se_{50}Te_{10}$ glasses. A 3-dimensional structure model has been devised which is consistent with the experimental data. The partial atomic correlation functions, coordination number distributions and three-atom-bond angle distributions have been obtained. Our main findings are the following:

- addition of Te to the binary glass does not change the basic glassy network structure;
- for As-Se first neighbour distance 2.42 Å is revealed;
- the average coordination numbers of As and Se atoms are established, according to which As atoms are surrounded by 2.6±0.1 Se atoms, while Se atoms - by 1.8±0.1 As atoms;
- RMC calculation reveals an angle of 95° for the average three-atom-bond angle of <Se-As-Se> and <As-Se-As> .

**Acknowledgements**

The authors are indebted to Drs L. Kőszegi, Gy. Mészáros and P. Jóvári (Research Institute for Solid State Physics and Optics) for their helpful contribution in this work. This research project has been supported by the EC under the FP7 Grant Agreement N226507-NMI3 and n°226716. This study has benefited from the use of HIPPO at the LANSCE, by US-DOE Contract W-7405-ENG-36.

**References**

[1] Strite S and Morkoc H 1992 *J. Vac. Sci. Technol.* B **10** 1237
[1] váb E, Mészáros Gy and Deák F 1996 *Mater. Sci. Forum* **228** 247; http://www.bnc.hu/
[2] Wenk H R, Lutterotti L and Vogel S 2003 *Nucl. Instr. Methods* A **515** 575
[3] Fábián M, Sváb E, Proffen Th and Veress E 2008 *J. Non-Cryst. Solids* **354** 3299
[4] Poulsen H, Neuefeind J, Neumann H B, Schneider J R and Zeidler M D 1995 *J. Non-Cryst. Solids* **188** 63
[5] Iwadatea Y, Hattori T, Nishiyama S, Fukushima K, Mochizuki Y, Misawa M and Fukunaga T 1999 *J. Phys. Chem. Solids* **60** 1447
[6] Shirao K, Hattori T, Nishiyama S, Fukushima K, Iwadate Y, Misawa M, Fukunaga T, Nakazawa T, and Noda K 1999 *J. Phys. Chem. Solids* **60** 1453
[7] McGreevy R L and Pusztai L 1998 *Mol. Simul.* **1** 359
[8] Special Feature section 1992 *Neutron News* **3** 29
[9] Eungho A, Williams G A, Taylor P C, Georgiev D G, Boolchand P, Schwickert B E and Cappelletti R L 2002 *J. Non-Cryst. Solids* **299–302** 958
[10] Bureau B, Troles J, Floch M, Smektala F and Lucas J 2003 *J. Non-Cryst. Solids* **326-327** 58
[11] Bureau B, Troles J, Floch M, Smektala F, Silly G, Lucas J 2003 *Solid State Sci.* **5** 2194
[12] Mauro J C and Varshneya A K 2007 *J. Non-Cryst. Solids* **353** 1226
[13] Kaban I, Hoyer W, Petkova T, Petkov P, Beuneu B, Schöps A and Webb M A 2007 *J. Ovonic Res.* **3** 67
[14] Balboul M R, Fouad S S, Fayek S A and El-Bana M S 2008 *J. Alloys&Compounds* **460** 570
[15] Jóvári P, Bureau B, Kabanc I, Nazabal V, Beuneu B and Rütt U 2009 *J. Alloys&Compounds* **488** 39

# Nanoindentation-induced pile-up in hydrogenated amorphous silicon

**B Pantchev[1], P Danesh[1], J Wiezorek[2] and B Schmidt[3]**

[1]Institute of Solid State Physics, Bulgarian Academy of Sciences, Tzarigradsko Chaussee 72, 1784 Sofia, Bulgaria
[2]Department of Mechanical Engineering and Materials Science, Swanson School of Engineering, University of Pittsburgh, 848 Benedum Hall, Pittsburgh, PA 15261, USA
[3]Research Center Dresden-Rossendorf Inc., Institute of Ion Beam Physics and Materials Research, D-01314 Dresden, Germany

E mail:pantchev@issp.bas.bg

**Abstract**. Nanoindentation-induced material extrusion around the nanoindent (pile-up) leads to an overestimation of elastic modulus, E, and nanohardness, H, when the test results are evaluated using the Oliver and Pharr method. Factors affecting the pile-up during testing are residual stresses in film and ratio of film and substrate mechanical properties. Nanoindentation of hydrogenated amorphous silicon (a-Si:H) films has been carried out with the aim to study the effect of residual compressive stress on the pile-up in this material. To distinguish the contribution of compressive stress to the appearance of pile-up ion implantation has been used as a tool, which reduces the compressive stress in a-Si:H. Scanning probe microscope has been used for the imaging of the indent and evaluation of the pile-up. The values of E and H have been obtained from the experimental load-displacement curves using depth profiling with Berkovich tip, which has created negligible pile-up. A sharper cube corner tip has been used to study the pile-up. It has been established that pile-up is determined by the material plasticity, when the compressive stress is below 200 MPa. The contribution of mechanical stress to the pile-up is essential for the stress as high, as about 500 MPa.

## 1. Introduction

The use of thin films and materials in small volumes in photonics, micro- and nanoelectronics and micro- and nano-electromechanical systems has created a strong interest in mechanical properties of thin films. The requirements for mechanical reliability and durability become more stringent with increasing the complexity of the systems and are particularly important for the devices used as sensors for environmental control and chemical and biological application.

Thin films of hydrogenated amorphous silicon (a-Si:H) have been used in some application fields as alternative to crystalline silicon, due to the relative ease and low cost preparation. Moreover, a-Si:H can be prepared at low temperatures, rendering it attractive for integration with biomaterials and biomedical applications. Thus, the engineering and control of the mechanical properties of a-Si:H films using appropriately varied film deposition conditions becomes an important issue.

Nanoindentation testing is a technique which can be used for assessment of the elastic and plastic properties of thin films. Using continuously recording indentation technique the values of elastic (Young's) modulus, E, and nanohardness, H, can be obtained from the experimental load-displacement curves, which give the value of maximal load and the contact stiffness.

The value of projected contact area, A, can be calculated from the contact depth using the relationship, which is given by the calibrated area function. The calibration is usually conducted by the indentation of a standard sample with known mechanical parameters. In case when the mechanical properties of the studied material and the standard sample are different there might be error in the calculated contact area. For materials, which undergo plastic deformation there is pile-up at the edge of the contact with the indenter, which means that the actual contact area is larger, than the value calculated from the calibrated area function. Therefore, the employment of the Oliver and Pharr method in the case of pile-up leads to erroneous assessment of mechanical properties, and namely to an overestimation of E and H.

The primary material properties that are associated with the appearance of pile-up and determine its degree are residual stress and plasticity. It has been shown that pile-up takes place in thin films with compressive stress [2, 3]. Finite element modelling (FEM) has been used to examine the role of plasticity [4, 5] and it has been established that pile-up is significant in materials with a ratio of yield stress (the stress at which a material begins to deform plastically), Y, to elastic modulus, E, below 0.5. The lower Y/E, the more material is extruded by the indentation. In addition, it has been shown that the ratio of film and substrate hardness may also have an impact on pile-up, which typically takes place in the case of soft films on hard substrates [6-9]. As the most reported results are related to hard dielectric or soft metal films, the current knowledge is deficient on the experimental data for the pile-up induced by nanoindentaion in thin semiconductor films. Moreover, information is lacking about the relative contribution of compressive stress and plasticity to pile-up.

The aim of the present work is to establish the degree of sensitivity of pile-up to the mechanical stress. To distinguish the contributions of compressive stress and Y/E to the appearance of pile-up ion implantation has been used as a tool, which reduces the compressive stress in a-Si:H, while keeping Y/E nearly constant. The values of yield stress and elastic modulus are established using Berkovich tip, which induces negligible pile-up and the employment of the Oliver and Pharr method is legitimate. A sharp cube corner tip is used for pile-up creation and evaluation.

## 2. Experimental procedure

The a-Si:H films were prepared by plasma-enhanced chemical vapour deposition (PECVD) using 10% silane diluted in hydrogen. The deposition experiments were carried out at a substrate temperature of 270°C and a gas pressure of 1 Torr. The film thickness was about 400 nm. Corning 7059 glass (elastic modulus 67.6 GPa, hardness 7.2 GPa) was used for substrates. Glass bars $0.3\times3\times30$ mm$^3$ were used as substrates for mechanical stress measurements. The hydrogen concentration in the material was about 15 at.%, determined by nuclear reaction analysis. Raman scattering spectroscopy indicated that the a-Si:H films were amorphous.

Si ion implantation was performed at room temperature without taking measures for sample cooling. The ion beam was normal to the surface. The implantations were carried out using Si ion energies of 160 keV. The implantation doses were in the range of $1.0\times10^{13}$cm$^{-2}$ - $1.0\times10^{15}$cm$^{-2}$.

The value of mechanical stress was determined by taking the difference in the substrate curvature with the film intact and after the film was removed by etching. The radius of curvature was obtained by measuring the change in angle of a He-Ne laser beam, reflected from the sample when the latter was moved laterally. The values of total stress were calculated using constants given in [10] and the expression:

$$\sigma_{tot} = E_s d_s^2 R_{eff}^{-1}/[6(1-\nu_s)d_f] ,$$

where $\sigma_{tot}$ is negative for compressive stress, $E_s$ , $\nu_s$ and $d_s$ are the Young's modulus, Poisson's ratio and thickness of substrate, respectively; $d_f$ is the film thickness; $R_{eff}^{-1} = (R^{-1} - R_o^{-1})$. R and $R_o$ are the radii of substrate curvature with and without the film, respectively.

The intrinsic stress was obtained by abstracting the thermal stress $\sigma_{th}$ from the total stress $\sigma_{tot}$.

$$\sigma_{th} = (\alpha_s - \alpha_s)(T_d - T_m)E_f /(1- \nu_f) ,$$

where $E_f$ and $\nu_f$ are the Young's modulus and Poisson 's ratio of the film, $\alpha_s$ and $\alpha_f$ are the thermal coefficients of expansion for the substrate and the film, and $T_d$ and $T_m$ are the deposition and measurements temperatures, respectively.The stress in the studied a-Si:H films was compressive.

The present study utilized a TI 900 Triboindenter, Hysitron Inc. Nanoindentation testing was performed in quasi-static mode. The values of hardness and elastic modulus were established using diamond indenter with Berkovich tip (radius 170 nm). The maximum loads were in the range of 0.3 – 8.1 mN. The loading rates were in the range of 0.03 - 0.81 mN/s and cycle time was kept constant. The cycles were without hold time and with equal loading and unloading periods. For statistics the indentations tests were repeated 6 to 10 times in different locations of the samples. The standard deviations in the measurements were about 1.5 - 4 % of the respective means. The probe area function for Berkovich tip was regularly calibrated using a standard fused quartz sample, provided by Hysitron Inc. In addition, the projected contact area, A, was directly evaluated using scanning probe microscope (SPM). The difference between the values of E and H obtained using the calculated and measured values of A was within the mentioned statistical error.

SPM integrated in the Triboindenter system was used for surface imaging after the indentation. The imaging was carried out with a tip velocity of 2 - 6 μm/s and a contact force setpoint of 1 μN. Cube corner tip with radius of about 55 nm was used for pile-up creation and imaging.

## 3. Experimental results and discussion

Figure 1 shows the SPM images of the indentations with Berkovich tip and cube corner tip for the as-deposited a-Si:H. The contact impression on a-Si:H obtained with Berkovich tip exhibits negligible

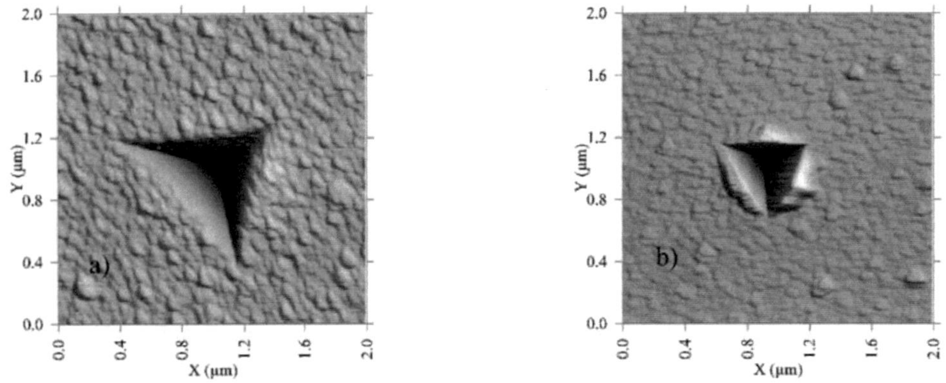


**Figure 1.** SPM images of the indentations with Berkovich (a) tip and cube corner tip ( b) for the as-deposited a-Si:H.

pile-up, which suggests that the respective P-h curves can be used for elastic modulus and hardness evaluations following the approach of Oliver and Pharr [1]. This approach has been verified by direct measuring the indents area, using the SPM images of the indentations with various maximal loads.

Figure 1(b) shows that a large pile-up of a-Si:H appears for indentations by cube corner tip. Since the ratio of film and substrate hardness have an impact on pile-up in the case of soft films on hard substrates, to avoid the influence of the substrate, Corning 7059 glass was chosen as substrate, because its hardness of 7.2 GPa is lower than the hardness of 11-13 GPa [12-14] characteristic for silicon.

The data obtained from the P-h curves for thin films on substrates are composite values of elastic modulus, $E_{rc}$, and hardness, $H_c$, i.e. characteristic of the entire film/substrate system. The values of E and H of the a-Si:H films have been established using depth profiling of $E_{rc}$ and $H_c$, which

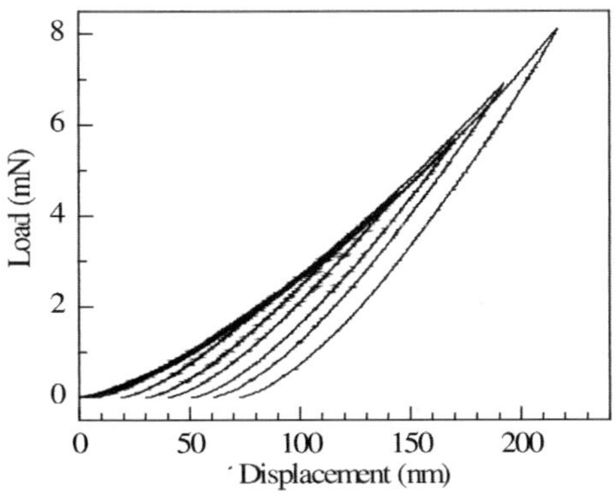


**Figure 2.** P-h curves of a series of indentations with different maximal loads up to 8.1 mN.

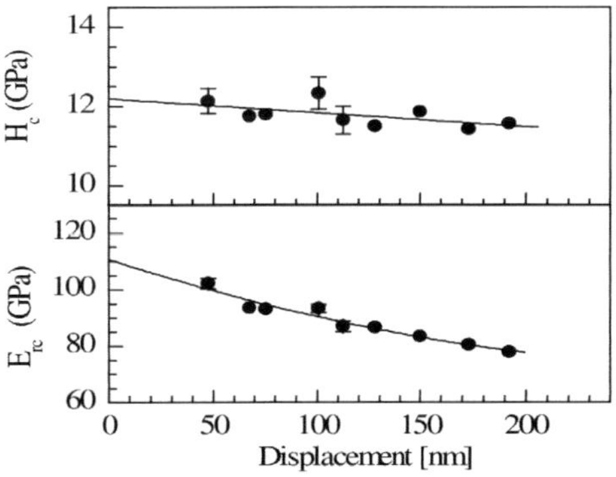


**Figure 3.** Depth profiles of $E_{rc}$ and $H_c$.

comprises a series of indentation tests with different maximal loads to achieve different maximal penetrations into the film (figure 2). Using this approach the dependencies of the composite values of elastic modulus $E_{rc}$ and hardness $H_c$ on the maximal displacement have been obtained. The extrapolation of these curves to the zero displacement [11], using second order polynomial fit, provides the values of E and H for the film (figure 3). The yield stress, Y, was estimated using the obtained value of H, and the eq. (31), page 112 in the Ref. [5]. The relationship is H= 2.3Y.

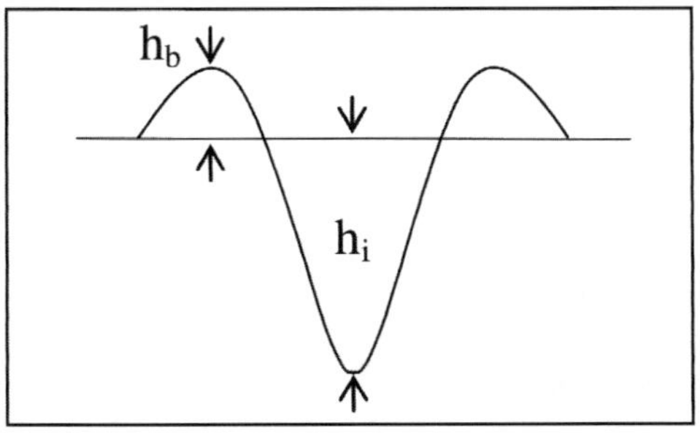


**Figure 4.** A schematic view of the cross-section of an indent.

Figure 4 shows a schematic view of the cross-section of an indent and is used for the definition of the variables $h_b$ and $h_i$. The ratio of the mean height of the bows, $h_b$, to the depth of the impression, $h_i$, has been used for quantitative estimation of the effect of ion implantation on pile-up.

Figure 5 (a, b) shows the XZ plane cross-sections of the SPM images of the indentations with cube corner tip with equal maximal load of 0.6 mN. For the as-deposited a-Si:H film there is a significant vertical displacement of material at the edges of the contact impression. The indent in the implanted film is deeper in comparison with the as-deposited film, which reflects the increase in plasticity. However, the bows around the contact are lower in the implanted film. Both changes lead to a lower value of $h_b/h_i$ (see table 1), i.e. ion implantation leads to a decrease in the pile-up.

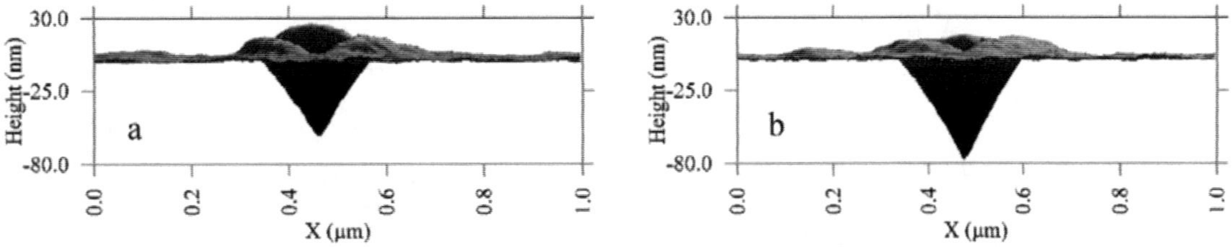


**Figure 5.** The XZ plane cross-sections of the SPM images of the indentations with cube corner tip of as-deposited (a) and implanted a-Si:H films.

Table 1 summarizes the effect of ion implantation on the mechanical properties, compressive stress and pile-up. E is calculated taking the value of 0.22 for the Poisson's ratio [15]. The elastic modulus of 119 GPa of the as-prepared a-Si:H is lower than the elastic modulus reported for crystalline silicon (c-Si), 160-180 GPa [12, 13] and closer to that of the hydrogen-free silicon layers (a-Si) amorphised by high dose ion implantation, 127 - 145 GPa [12].The obtained value of hardness of 12.3 GPa for the as-deposited a-Si:H film is comparable to the hardness of c-Si (11.4 – 13.0 GPa) [12, 13, 14] and is higher, than the hardness of hydrogen-free a-Si prepared by ion implantation-induced amorphization of c-Si or by a deposition technique (9.0 - 11.5 GPa) [12, 16, 17]. The observed change in the elastic properties of the implanted a-Si:H is similar to the effect of ion implantation on the mechanical properties of c-Si [18] and can be associated with the induced increase in the network disorder [19]. It can be suggested that the decrease in E is related to the structural change in the implanted a-Si:H, such as the implantation-induced increase in the Si-Si bond angle deviations.

**Table 1**. Elastic modulus, E, indentation hardness, H, yield stress to elastic modulus ratio, Y/E, compressive stress, $\sigma$, and pile-up parameter, $h_b/h_i$ of the as-deposited and ion implanted a-Si:H.

| Ion dose ($cm^{-2}$) | none | $1x10^{13}$ | $1x10^{14}$ | $1x10^{15}$ |
|---|---|---|---|---|
| E (GPa) | 119 ± 3 | 115 ± 3 | 108 ± 3 | 101 ± 3 |
| H (GPa) | 12.3 ± 0.35 | 11.6 ± 0.35 | 10.1 ± 0.35 | 9.4 ± 0.35 |
| Y/E | 0.045 ± 0.002 | 0.044 ± 0.002 | 0.041 ± 0.002 | 0.041 ± 0.002 |
| $\sigma$ (MPa) | 766 ± 20 | 658 ± 20 | 138 ± 10 | 31 ± 5 |
| $h_b/h_i$ | 0.33 ± 0.02 | 0.25 ± 0.01 | 0.20 ± 0.01 | 0.19 ± 0.01 |

Si ion implantation of a-Si:H leads to a decrease in hardness to the values typical for the hydrogen-free a-Si (Table 1). This change is apparently related to the implantation-induced damage and reflects the mechanism of plastic deformation of a-Si:H. Two mechanisms of plastic deformation are reported for silicon, namely, phase transformation and plastic flow [16, 19-21]. The main indicator for the operation of phase transformation during nanoindentation is the appearance of the discontinuities (kinks or pop-outs) in the P-h curves. Their absence in the obtained P-h curves suggests that the nanomechanical testing performed in the present study does not induce phase transformation, i.e. the plastic deformation of a-Si:H is accomplished by plastic flow. Plastic flow is generally considered to be associated with the presence of a high density of dangling bonds and any change in this density is expected to affect the material hardness. Ion implantation leads to an increase in the density of dangling bonds by causing atom displacements in the target and this is the origin of the decrease in the nanohardness of a-Si:H.

The decrease in both E and H due to the ion implantation, keeps the ratio Y/E almost constant for the ion doses used in this work. Having in mind the measurement accuracy, the mean value of Y/E of about 0.43 can be considered as typical for both as-deposited and implanted a-Si:H film. This makes it possible to distinguish the contribution of compressive stress to the pile-up.

The decrease in the compressive stress due to implantation is substantial (table 1). The compressive stress in a-Si:H is associated with the presence of clustered hydrogen in the silicon matrix. The ion implantation leads to a decrease in the relative amount of clustered hydrogen and in this way reduces the compressive stress [22]. The residual stress becomes negligible at the highest Si ion dose of $1\times10^{15}$ $cm^{-2}$. On the other hand, the pile-up characteristic $h_b/h_i$ decreases to 0.20 for the ion dose of $1\times10^{14}$ $cm^{-2}$ and further increase in the ion dose leads to a negligible change in $h_b/h_i$ , i.e. there is a saturation trend. It can be suggested that the value of $h_b/h_i$ of about 0.20 is related to the pile-up associated with the material plasticity and corresponds to the value of Y/E of about 0.43. The obtained results point out that for a-Si:H the effect of the residual compressive stress is essential for strongly stressed films ($\sigma \geq 500$ MPa) and is additive to the pile-up related to the plasticity. Since the value of compressive stress in a-Si:H depends strongly on the deposition conditions [23] the comparison of nanomechanical properties of various samples cannot be straightforward because of the different degree of pile-up.

## 4. Conclusion

The sensitivity of pile-up to the mechanical stress in a-Si:H films has been studied. Ion implantation was used as a tool to decrease the compressive stress in the films. The values of E and H decreased with the ion dose, while the value of Y/E remained nearly constant. Using this effect of ion implantation it was possible to distinguish the contribution of mechanical stress to pile-up. It has been established that for a-Si:H the impact of the residual compressive stress on the pile-up is essential for strongly stressed films and is additive to the pile-up related to the plasticity.

**Acknowledgments**
This study was carried out with the support of AIM - Center for Application of Ion Beams to Materials Research under RITA Contract No. 025646, and of Institute of Solid State Physics, Bulgarian Academy of Sciences, Contract No BK9-2. The authors acknowledge using facilities in the Materials Micro-Characterization Laboratory of the Department of Mechanical Engineering and Materials Science in the Swanson School of Engineering at the University of Pittsburgh.

## References

[1] Oliver W C and Pharr G M 1992 *J. Mater. Res.* **7** 1564
[2] Tsui T Y, Oliver W S and Pharr G M 1996 *J. Mater. Res* **11** 752
[3] Bolshakov A, Oliver W C and Pharr G M 1996 *J. Mater. Sci.* **11** 760
[4] Oliver W C and Pharr G M 2004 *J. Mater. Res.* **19** 3
[5] Cheng Y T and Cheng C M 2004 *Mater. Sci. Eng.* R **44** 91
[6] Bolshakov A and Pharr G M 1998 *J. Mater. Res.* **13** 1049
[7] Tsui T Y, Vlassak J and Nix W D 1999 *J. Mater. Res.* **14** 2196
[8] Tsui T Y, Vlassak J and Nix W D 1999 *J. Mater. Res.* **14** 2204
[9] Kramer D E, Volinsky A A, Moody N R and Gerberich W W 2001 *J. Mater. Res.* **16** 3150
[10] Ozawa K, Takagi N and Asama K 1983 *Jpn. J. Appl. Phys.* **22** 767
[11] Fischer-Cripps A C 2006 *Surf. Coat. Technol.* **200** 4153
[12] Follstaedt D M, Knapp J A and Myers S M 2004 *J. Mater. Res.* **19** 338
[13] Bhushan B 2004 *Handbook of Nanotechnology* (Berlin: Springer Verlag) pp 770-775
[14] Bhushan B and Li X 1997 *J. Mater. Res.* **12** 59
[15] R. Hull 1999 *Properties of crystalline Silicon* (London: INSPEC) p 102.
[16] Kulikovsky V, Vorlicek V, Bohac P, Stranyanek M, Ctvrtlik R and Kurdyumov A 2008 *Thin Solid Films* 516 5368
[17] Williams J S, Haberl B and Bradby J E 2005 *Mater. Res. Soc. Symp. Proc.* **841** R10.31/T6.3.1
[18] Szabadi M, Hess P, Kellock A J, Coufal H and Baglin J E E 1998 *Phys.Rev.* B **58** 8941
[19] Pantchev B, Danesh P, Savatinova I, Liarokapis E, Schmidt B and Grambole D 2001 *J. Phys. D:Appl. Phys.* **34** 2589
[20] Haberl B, Bradby J E, Ruffel S, Williams J S and Munroe P 2006 *J. Appl. Phys.* **100** 013520
[21] Ruffel S, Bradby J E, Williams J S, Munoz-Paniagua D, Tadayyon S, Coatsworth L L and Norton P R 2009 *Nanotechnology* **20** 135603
[22] Pantchev B, Danesh P and Schmidt B 2005 *Appl. Phys. Lett.* **87** 023014
[23] Danesh P, Pantchev B, Grambole D and Schmidt B 2002 *Appl. Phys. Lett.* **80** 2463

# Ion-beam induced hydrogen redistribution in a-Si:H-based triple layer structures

**B Pantchev[1], P Danesh[1], B Schmidt[2], D Grambole[2] and L Bischoff[2]**

[1]Institute of Solid State Physics, Bulgarian Academy of Sciences, Tzarigradsko Chaussee 72, 1784 Sofia, Bulgaria
[2]Research Center Dresden-Rossendorf Inc., Institute of Ion Beam Physics and Materials Research, D-01314 Dresden, Germany

E-mail: pantchev@issp.bas.bg

**Abstract**. Ion-beam induced hydrogen migration has been studied in triple layer structures, a-Si/a-Si:H/a-Si and $SiO_2$/a-Si:H/$SiO_2$. Nuclear reaction analysis has been used for simultaneous irradiation with MeV ions and measurement of hydrogen distribution in the structures. It has been established that there is no hydrogen loss from $SiO_2$/a-Si:H/$SiO_2$ structures, but an asymmetric redistribution due to hydrogen penetration into the bottom $SiO_2$ layer. Hydrogen loss has been observed from the a-Si/a-Si:H/a-Si structures. The inspection of the surface of these samples by means of atomic force microscope has shown that it proceeds by bubble formation and blistering at the inner interface. The observed ion-beam induced selective penetration of hydrogen into the underlying material can be used as a tool for preparation of microcavity and microchannel arrays.

## 1. Introduction

An impact of ionizing irradiation on the electronic and structural properties of hydrogenated amorphous silicon (a-Si:H) has been observed in recent studies. It has been established that electron irradiation induces structural defects in the films and that the defect creation is not initiated by recombination events, but by energy transfer during collisions [1,2]. The impact of ionizing irradiation on the hydrogen content and on its bonding configuration in a-Si:H has attracted less attention. High energy irradiation may induce hydrogen migration with different consequences. MeV ion-beam induced loss of hydrogen from organic and inorganic materials has often been observed and is usually associated with its out-diffusion through the sample surface into the vacuum environment [3-7]. Studies have reported about hydrogen accumulation near the interfaces [8] and bubble formation inside the films [9]. It has also been shown that hydrogen thermally evolved from a-Si:H film can diffuse into and be retained within the substrate [10].

Irradiation-induced hydrogen release from its bonded sites is related to the electronic excitation of the solid. For MeV ions irradiating sub-micrometer thick films this mechanism prevails the nuclear interaction between the ions and the target atoms. In the present work, MeV ion-beam induced hydrogen redistribution has been studied in a-Si:H, which has been sandwiched between thin hydrogen-free amorphous silicon (a-Si) or $SiO_2$ layers. It is shown that in these layer stacks ion-beam induced hydrogen migration strongly depends on the interface properties. Moreover, a selective penetration of hydrogen into the underlying material has been established, which could be used as a tool for microcavity and microchannel engineering.

## 2. Experimental procedure

Figure 1 shows schematically the configuration of the studied samples. The layer stacks consisted of an a-Si:H layer sandwiched between two hydrogen-free a-Si layers or between two $SiO_2$ layers. The a-Si layers were prepared by DC magnetron sputtering in argon atmosphere at a gas pressure of 6.7 Pa and a power of 250 W applied to a silicon target. The $SiO_2$ layers were deposited by RF magnetron sputtering from a fused silica target using argon/oxygen mixture, at a total pressure of 0.67 Pa and a power of 1000 W. The a-Si:H layer was deposited by plasma enhanced CVD using 10% silane diluted in hydrogen, at a pressure of 133 Pa and a power density of 100 mW/cm$^2$. After each deposition run the samples were exposed to air because of using different chambers for the layer deposition. This led to the absorption of the water molecules on the layer surface, which was particularly detectable for $SiO_2$ due to its high sensitivity to humidity. All layers constituting the structure were deposited at room temperature in order to increase the hydrogen concentration in a-Si:H and to reduce as much as possible any asymmetry in the structure properties. The film thicknesses were 100 nm for a-Si and $SiO_2$, and 50 nm for a-Si:H. The substrates were crystalline (100) Si wafers.

Hydrogen concentration has been measured at the 5 MV tandem accelerator of the Forschungszentrum Rossendorf using the following nuclear reaction:

$$^{15}N + {}^{1}H \rightarrow {}^{12}C + {}^{4}He + \gamma\ (4.43\ MeV)$$

This reaction has a narrow resonance at the energy $E_{res}$ of 6.385 MeV. By increasing the energy of $^{15}N$ ions above $E_{res}$, the ions reach the resonance energy at a certain depth of the sample due to energy loss in the material. Therefore, hydrogen depth profiles can be determined by measuring the yield of the 4.43 MeV $\gamma$ -rays as function of the energy of the $^{15}N$ ions. The $\gamma$ -rays were detected by a 4” × 4” BGO detector. The $^{15}N$ ion beam spot had an area of 2.5 mm$^2$. The depth resolution was about 8 nm. To estimate the effect of the $^{15}N$ ion irradiation on hydrogen depth distribution, the measurements were carried out on a different sample spot for each ion energy, i.e. at each specified depth. The ion fluence was varied from $6 \times 10^{13}$ to $1.7 \times 10^{15}$ ion cm$^{-2}$.

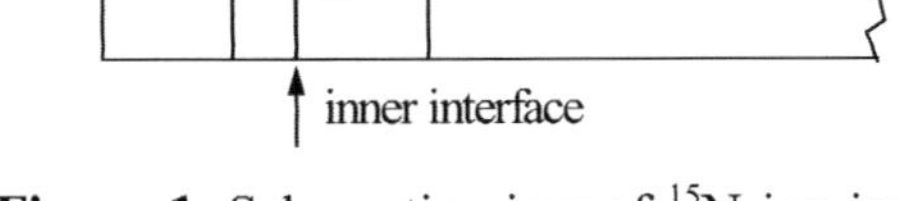


**Figure 1.** Schematic view of $^{15}N$ ion irradiation of the a-Si/a-Si:H/a-Si and $SiO_2$/a-Si:H/$SiO_2$ layer stacks.

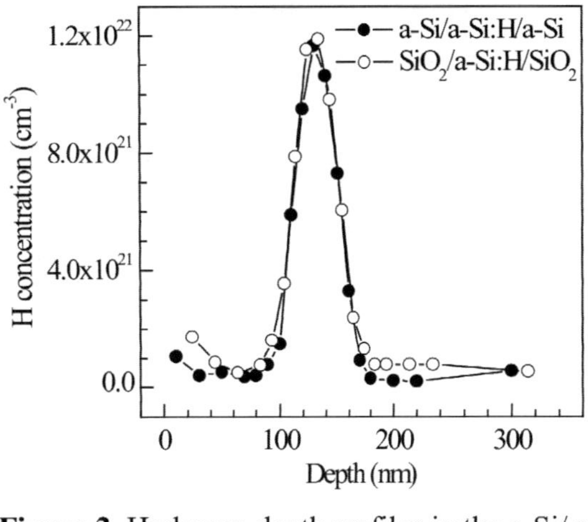


**Figure 2.** Hydrogen depth profiles in the a-Si/ a-Si:H/a-Si and $SiO_2$/a-Si:H/$SiO_2$ layer stacks irradiated with the ion fluence of 6x10$^{13}$ ion/cm$^2$.

## 3. Experimental results and discussion

Compilation of the NRA measurements carried out with different $^{15}N$ ion energies provides a hydrogen depth profile for the entire structure. It is assumed that the hydrogen distribution measured with the lowest ion fluence of $6x10^{13}$ ion/cm$^2$ is characteristic for the as-prepared structures. Figure 2 shows the initial hydrogen distribution in $SiO_2$/a-Si:H/$SiO_2$ and a-Si/a-Si:H/a-Si samples. Note that the hydrogen depth distribution in the as-deposited a-Si:H is uniform [11] and in the triple layer structure studied here it should appear as a 50 nm wide bar in the middle of the layer stack. The Gaussian form and hydrogen spreading outside the a-Si:H layer (beyond the 100 - 150 nm region) is related to the unavoidable ion straggling effect. Near the interfaces there is a difference between the hydrogen concentrations in the sputtered $SiO_2$ and a-Si layers. This difference is related to a higher water molecule absorption by the open $SiO_2$ lattice structure, as compared to a-Si. The effect is most evident in the region close to the top surface of the sandwich structures and reflects the sensitivity of the former to the air humidity. Hydrogen on the surface is not stable and its concentration decreases strongly with the irradiation.

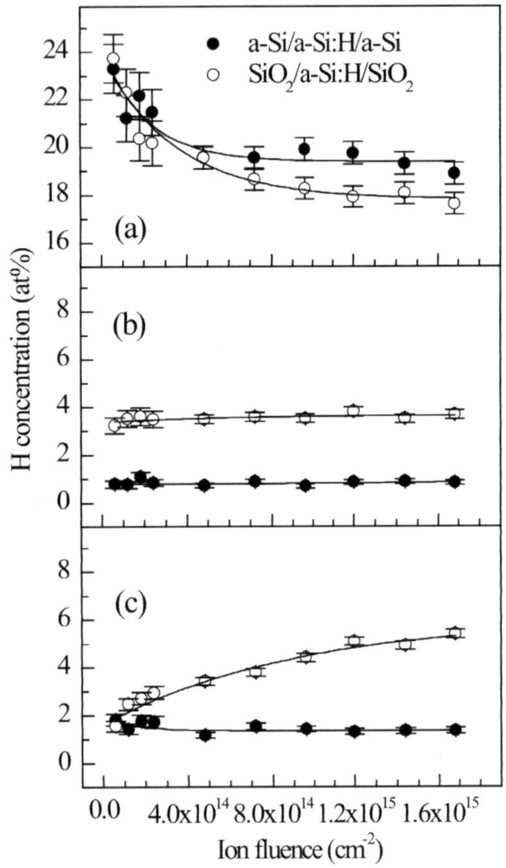


**Figure 3.** Effect of $^{15}N$ ion fluence on the hydrogen concentration in a-Si/a-Si:H/a-Si and $SiO_2$/a-Si:H/$SiO_2$ layer stacks: a) in the middle of the a-Si:H layer; b) around the outer interface; c) around the inner interface.

It has been presumed that if there is any ion-beam induced hydrogen loss from a-Si:H film it should proceed through both - outer and inner - interfaces, because of the symmetric structure and the seemingly similar properties of the interfaces. Therefore, it has been expected that the increase in ion fluence would lead to a gradual decrease in the hydrogen concentration inside the a-Si:H layer and a corresponding symmetric increase in the regions outside a-Si:H, close to the interfaces. Instead, a different behaviour has been established for the outer and inner interfaces. Figure 3 shows the variation in the hydrogen concentration with increasing ion fluence at different depths in the triple layer structures. The amount of hydrogen measured outside the a-Si:H layer is partially related to the straggling effect, but the trends are associated with the actual changes in the hydrogen concentrations and reflect its irradiation-induced dynamics.

Figure 3 (a) shows the impact of the ion fluence on the hydrogen concentration in the middle of the a-Si:H layer. As expected, the MeV $^{15}N^{++}$ irradiation leads to an essential gradual decrease in the hydrogen concentration inside the a-Si:H layer, i.e. there is irradiation-induced hydrogen out-diffusion from a-Si:H. Figure 3 (b) shows the impact of ion fluence on the hydrogen concentration at the outer interface, from the side of the top layers (depth point 80 nm). At both interfaces, a-Si/a-Si:H and $SiO_2$/a-Si:H, the hydrogen concentration is stable and does not change under the irradiation. A higher concentration of hydrogen at the $SiO_2$/a-Si:H interface is related to the water contaminated $SiO_2$. Thus, it can be suggested that hydrogen does not penetrate from a-Si:H layer into the top layers. Figure 3 (c) shows the impact of ion fluence on hydrogen concentration at the inner interface, from the side of the bottom layers (depth point 170 nm). There is a well detectable increase in the hydrogen concentration for the a-Si:H/$SiO_2$ interface, whereas for a-Si:H/a-Si interface the hydrogen concentration remains constant.

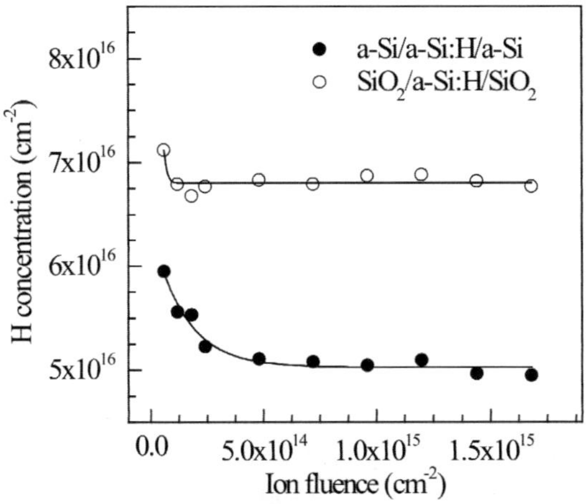


**Figure 4.** Effect of $^{15}N$ ion fluence on the hydrogen content in the entire in a-Si/a-Si:H/ a-Si and $SiO_2$/a-Si:H/$SiO_2$ layer stacks.

Figure 4 shows the irradiation induced changes in the hydrogen amount for the whole structures. The amount of hydrogen has been determined by the integration of the depth profiles over the entire thickness of the layer stack. This estimation takes into account all hydrogen, including the water molecules absorbed on the surface. In the $SiO_2$/a-Si:H/$SiO_2$ samples the hydrogen concentration decreases after the first measurements and then remains stable. The initial decrease is related to the hydrogen desorption from the water molecules adsorbed on the sample surface. Bearing in mind the data presented in figure 3 this suggests that hydrogen loss only proceeds from the a-Si:H layer and not from the whole sample. In the a-Si/a-Si:H/a-Si samples hydrogen concentration decreases continuously, which suggests that hydrogen leaves the sample and not only the a-Si:H layer.

Figure 5 shows the hydrogen depth profiles in both kinds of structures before and after the irradiation with highest fluence of $1.7x10^{15}$ ion/$cm^2$. It is seen that the ion-beam induced variation in the hydrogen distribution is asymmetric and takes place mainly in the middle of the a-Si:H layer and in the regions close to the inner interface. Final hydrogen depth distribution depends on the underlying layers. In the $SiO_2$/a-Si:H/$SiO_2$ structures hydrogen leaves a-Si:H layer by penetrating into the bottom $SiO_2$ layer. In the a-Si/a-Si:H/a-Si structures there is a decrease in the hydrogen concentration in the region at the inner interface. Specific for latter structures is the appearance of blisters (figure 6). The measurement of the depth of the exfoliated blisters using AFM gives the value of 150 nm, i.e. the sum of a-Si and a-Si:H thicknesses (figure 7). Thus, it can be suggested that the irradiation of a-Si/a-Si:H/a-Si leads to hydrogen accumulation and bubble formation at the inner interface. The hydrogen loss in these structures is associated with the delamination of the a-Si:H layer.

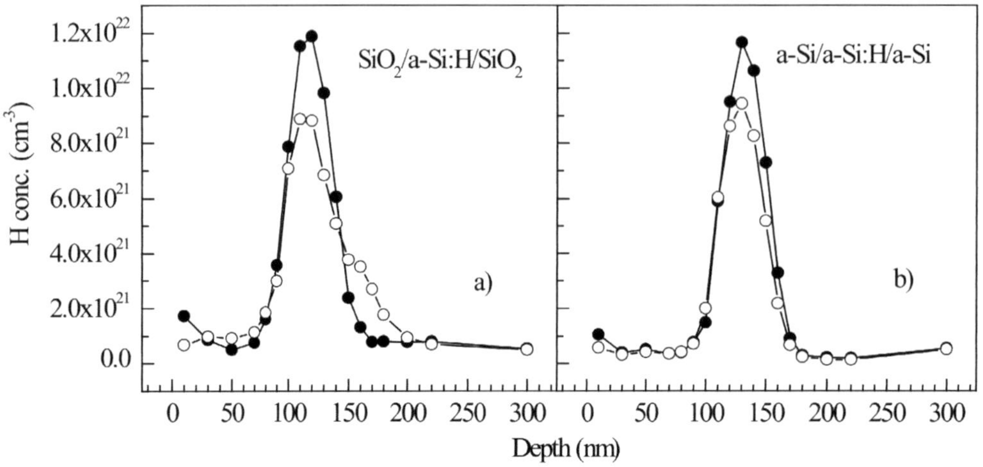


**Figure 5.** Hydrogen depth profiles in the layer stacks irradiated with the ion fluences of 6x10$^{13}$ ion/cm$^2$ and and 1.7x10$^{15}$ ion/cm$^2$: (a) a-Si/a-Si:H/a-Si; (b) $SiO_2$/a-Si:H/$SiO_2$.

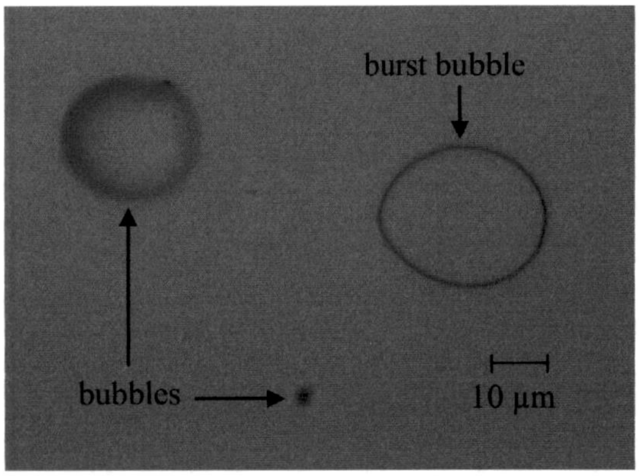


**Figiure 6.** Optical microscope image of the surface of irradiated a-Si/a-Si:H/a-Si layer stack.

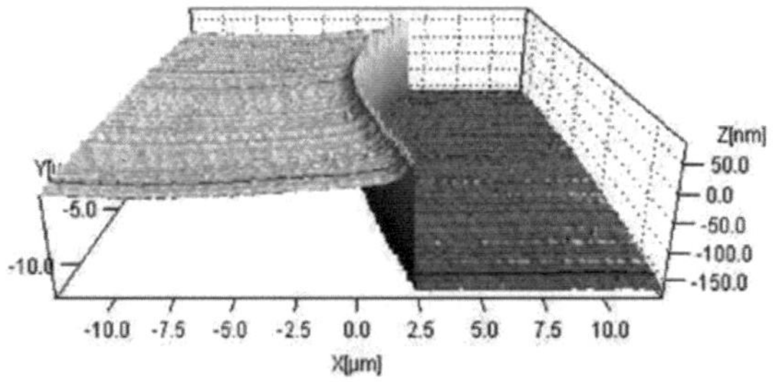


**Figure 7.** AFM image of an exfoliated blister on the surface of irradiated a-Si/a-Si:H/a-Si layer stack.

The obtained results suggest that hydrogen leaves thin a-Si:H layer in both kinds of structures through the inner interface, either by blistering and delamination, or by diffusion into the bottom layer. The hydrogen loss through the surface might also proceed, but it seems to have a minor contribution, as there is not any essential ion-beam induced change in the hydrogen concentration in the top layers and at the outer interface. The irradiation-induced decrease in the hydrogen concentration should be considered as related to the diffusion of the liberated hydrogen atoms. Hydrogen diffusion in amorphous silicon is a complex phenomenon. It strongly depends on the structure of the amorphous network, which defines the hydrogen transport path and the hydrogen chemical potential [12-14]. Since the release of hydrogen from the bonded sites is the same throughout the a-Si:H layer, the observed asymmetric redistribution of hydrogen should be ascribed to a difference in the structural properties of the inner and outer interfaces. Although the layer stacks seemed to be symmetric there might be a hidden asymmetry due to the reverse sequences of the argon/oxygen and silane/hydrogen plasma. Actually, the initial stage of the top layer deposition is associated with the ion bombardment of a-Si:H and may lead to an increase in the density of the structural defects at the outer interface. It results in a local increase in the energy difference between the hydrogen transport path and the hydrogen chemical potential. Due to the exponential dependence of the diffusion coefficient on this energy parameter even a small increase of the latter can lead to a considerable decrease in the hydrogen transport at the outer interface. Thus, the main stream of the liberated hydrogen atoms is through the inner interface.

The following hydrogen migration depends on the material properties of the bottom layer and is selective regarding the a-Si and $SiO_2$ layers. This fact could be used as a tool for microcavity and microchannel engineering. An approach for the preparation of a microchannel array is schematically shown in figure 8. A structure consisting of a-Si:H, a-Si and $SiO_2$ layers can be prepared with appropriate dimensions using conventional photolithographic methods (figure 8a). Under the impact of the MeV ions the part of a-Si:H layer on $SiO_2$ layer remains intact, due to the hydrogen penetration into the bottom layer. Hydrogen accumulation and bubbles formation proceeds on the inner a-Si:H/a-Si interface (figure 8b). Further irradiation leads to the delamination of a-Si:H layer in the

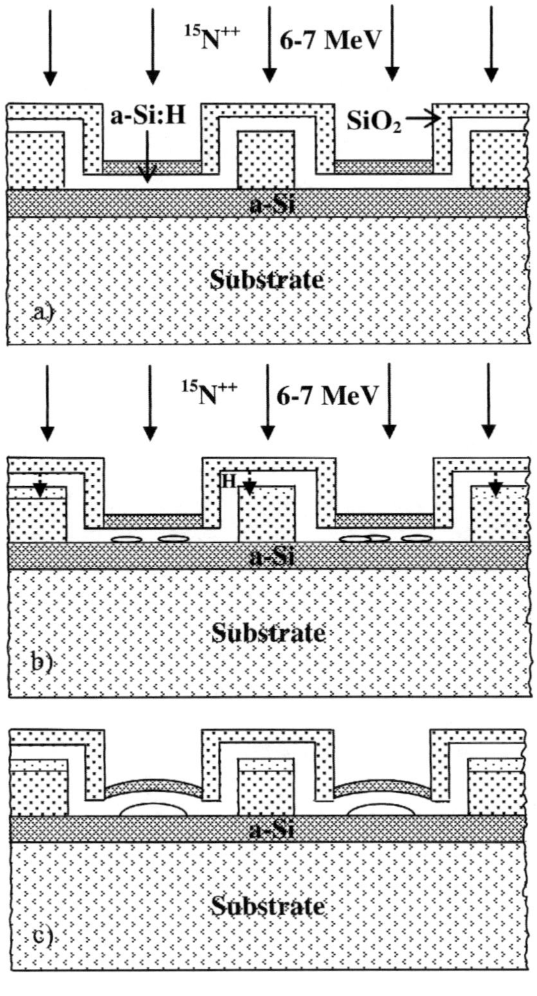


**Figure 8.** A schematic development of a microchannel array under the action of MeV ion-beam irradiation.

region over the a-Si layer (figure 8c). Note, the delaminated parts remain buckled due to the compressive stress in a-Si:H. The delamination process might be controlled by ion fluence and hydrogen concentration in the as-prepared a-Si:H. Buckling process might also depend on the thickness ratio of the layers, as well as on the mechanical stress in a-Si:H.

## 4. Conclusion

It is shown that a MeV ion beam induces hydrogen diffusion in the a-Si/a-Si:H/a-Si and $SiO_2$/a-Si:H/$SiO_2$ samples, which is sensitive to the interface properties. Hydrogen loss from the samples proceeds mainly through the inner interface. Depending on the material below the a-Si:H layer, there is either hydrogen diffusion through the interface or bubble formation and a-Si:H layer delamination.

## Acknowledgments

This study was carried out with the support of AIM - Center for Application of Ion Beams to Materials Research under RITA Contract No. 025646, and of Institute of Solid State Physics, Bulgarian Academy of Sciences, Contract No BK9-2.

## References

[1] van Swaaij R A C M M and Klaver A 2008 *J. Non-Cryst. Solids* **354** 2464
[2] Liao N M, Li W, Jiang Y D, Wu Z M, Li S B, Liu Z, Li Z, Jin X and Chen Y X 2008 *J. Phys. D: Appl. Phys.* **41** 205412
[3] Baptista D L, Garcia Irene T S and Zawislak F C 2004 *NIM* B **219-220** 846
[4] De Jong M P, Mass A J H, Van Ijzendoorn L J, Klein S S and De Voigt M J A 1997 *J. Appl. Phys.* **82** 1058
[5] Arnoldbik W M, Tomozeiu N and Habraken F H P M 2002 *NIM* B **190** 433
[6] Brockhoff A M, Arnoldbik W M and Habraken F H P M 2002 *NIM* B **190** 226
[7] Arnoldbik W M and Habraken F H P M 2007 *NIM* B **256** 300
[8] Ecker K H, Krause J, Weidenger A, Weise H P and Maser K 2000 *NIM* B **161-163** 682
[9] Ullersma E H C , Ullersma P and Habraken F H P M 2000 *Phys. Rev.* B **61** 10133
[10] Jones D I, Gibson R A, LeComber P G and Spear W E 1979 *Solar Energy Mat.* **2** 93
[11] Danesh P, Pantchev B, Grambole D and Schmidt B 2001 *J. Appl. Phys.* **90** 3065
[12] Jackson W B and Tsai C C 1992 *Phys. Rev.* B **45** 6564
[13] Beyer W 1997 *phys. stat. sol.* (a) **150** 53
[14] Beyer W 2001 *Mat. Res. Soc. Symp. Proc.* **664** A13.1.1

# Structure-related strain and stress in thin hydrogenated microcrystalline silicon films

**K Christova[1], S Alexandrova[2], A Abramov[3], E Valcheva[4], B Ranguelov[5], C Longeaud[6], S Reynolds[7], and P Roca i Cabarrocas[3]**

[1]Institute of Solid State Physics, Bulgarian Academy of Sciences, 72 Tzarigradsko Chaussee Blvd., 1784 Sofia, Bulgaria.
[2]Department of Applied Physics, Technical University, 8 Kl. Ohridski Blvd., 1797 Sofia, Bulgaria
[3]Laboratoire de Physique des Interfaces et des Couche Minces (LPICM), Ecole Polytechnique, Palaiseau 91128, France.
[4]Faculty of Physics, Sofia University, 5 J. Bourchier Blvd., 1164 Sofia, Bulgaria.
[5]Institute of Physical Chemistry, Bulgarian Academy of Sciences, Acad. G. Bonchev Str., bl. 11, 1113 Sofia, Bulgaria
[6]Laboratoire de Génie Electrique de Paris, UMR 8507 – CNRS – Supélec, Universités de Paris VI et XI, Plateau de Moulon, 11 J.Curie, 91192 Gif sur Yvette cedex, France.
[7]Division of Electronic Engineering and Physics, University of Dundee, Nethergate, Dundee DD1 4HN, UK

E-mail: kkp@issp.bas.bg

**Abstract.** The stress/strain relation for hydrogenated microcrystalline silicon (μc-Si:H) films in the thickness range 10 to 200 nm was studied. It was found from wafer curvature measurements that all deposited films exhibit compressive intrinsic stress, which decreases with film thickness. This finding is in agreement with the stress level seen from the shifts in Raman spectra. The strain was approached through the Raman Mechanical Coefficient for the thin films and was estimated to be $-1.4\times10^{-5}$ ($cm^{-1}$/MPa). The Raman spectra indicated highly crystalline films. The deconvolution of the spectra into Lorentzian components revealed inclusion of defective nanocrystallites and amorphous phase as well. The fractions of the different constituents were estimated. The contribution of the defective nanocrystallites to the overall stress in the films has been interpreted.

## 1. Introduction

The mechanical characterization of thin films is of great importance for their use since device functionality and reliability depend strongly on mechanical parameters. In the field of microcrystalline silicon films a range of diagnostic and deposition techniques has been applied to achieve further enhanced transport properties [1-4]. The relation between these properties and mechanical properties of very thin films of microcrystalline Si remains to be at issue. The results published by J. Trask et al. show that one of the ways to reach better properties is by creating films with larger grains [5] as it is found in this study. The grain and column boundaries play a particularly important role in determining electronic properties of the film, since most of the electrically-active defects are located there [6].

However, the nucleation process is a complicated one and thus, the film structure is sensitive to several factors. Apart from the process parameters, the film structure is a function of film thickness as well. It is known that the crystallinity increases with the film thickness, especially when the deposited films of μc-Si:H are obtained through PECVD method [5]. Some authors characterized microcrystalline silicon material using Raman spectroscopy and found that the optimized nc-Si:H material does not show an incubation layer [7]. In our study it is suggested that growth does pass through this stage. Bearing all this in mind, and that the microstructure of silicon films is physically related to stress/strain characteristics, in this paper attention is drawn to correlate the Raman stress induced shifts with the stress level as *independently* determined from wafer curvature measurements. The thickness dependence of the stress, of the Raman peak position, of the linewidth of this peak, and film morphology is studied. Here stress/strain characterization is implemented for the *very early stages* of deposition, since studies in the above mentioned context are still lacking. The results of this study will be used for an easier characterization of microcrystalline silicon thin films via stress/strain apprehension which is essential for reliability of device performance. The knowledge of structure related stress and strain could be applied for elaboration of technology for fabrication of films with desired parameters.

## 2. Experimental

Hydrogenated microcrystalline silicon thin films were deposited by RF (13.56 MHz) plasma enhanced chemical vapour deposition (RF PECVD) in a capacitively coupled (CCP) reactor on (111) silicon substrates. Before film deposition the Si substrates were coated with a thin layer of a-Si:N:H to avoid epitaxial film growth. Substrate thickness, $d_s$, was 380±25μm, and film thickness, $d_f$ was varied between 10 and 200 nm. The film thickness was estimated by modelling of ellipsometric spectra. This range is intentionally below that normally used in solar cell applications in order to reveal the stress condition in the early stages of deposition. A process gas mixture of tetrafluoride, $SiF_4$, $H_2$ and Ar was used for the deposition of the μc-Si:F:H films. The films were prepared, maintaining a constant gas pressure of 3 Torr in the deposition chamber. Since the film properties are very sensitive to the plasma process, low substrate temperature of 200 °C was chosen in this experiment.

Stress values, $\sigma_f$ were determined by the classical method of wafer bending and elastic parameters, Young modulus, E and Poison's ratio, ν, were calculated using Stoney's formula (1) [8]. The last term in parentheses in (1) represents the deformation $\varepsilon$ of the film-substrate system; $R_{f+s}$ and $R_0$ are the radii

$$\sigma_f = \int_0^{d_f} \sigma dz = -\frac{1}{6}\left(\frac{E}{1-\nu}\right)_s d_s^2 \left(\frac{1}{R_{f+s}} - \frac{1}{R_0}\right) \qquad (1)$$

of curvature of Si substrate with film and without film, respectively. The error in the radii determined by curvature measurements was 10%. The thermal component of the stress is not considered, since it is equal for all samples and hence does not influence the stress evaluation with film thickness. Thus, the stress calculated by (1) characterises the intrinsic stress for the film bulk.

Raman spectra were obtained by LabRAM HR micro-Raman spectrometer equipped with CCD detector and laser excitation wavelengths of 632 and 514 nm. Objective with a magnification of 100x was used. The peak position accuracy is below 0.02 $cm^{-1}$. Usually shorter wavelength is used to increase the Raman signal and to decrease the signal arising from the substrate. The penetration depth, i e. the probed volume of the film, is determined also by the wavelength. Taking into account the absorption coefficients of μc-Si:F:H films [9,10] we can assume that 632 nm beam probes the whole film volume, so that the Raman stress results can be correlated with the stress results taken from the curvature measurements. Using lower wavelengths can substantially decrease penetration depth [11], for μc-Si:F:H films the decrease amounts to 3 times [4]. Therefore, Raman scattering at 514 nm and confocal experiment are applied to reduce the depth penetration and to better resolve the different phase fractions (crystalline/amorphous) in the films.

The morphology of the films is observed by Scanning electron microscope – JEOL JSM 5300.

## 3. Results and discussions

The properties of the thin films are largely dependent on the state of stress/strain in the film. Strain in a thin film can be caused by several factors, including: lattice mismatch between film and substrate, thermal expansion coefficient differences, and different phase inclusions during growth process. The approach taken for the work presented in this paper was to examine the film properties as a function of film thickness, since such dependence gives an opportunity to discuss stress/strain relation.

We extract the effects of strain from stress dependence on the film thickness. Figure 1 shows this dependence. It also shows that the type of the stress is compressive for all films studied. The evolution character of stress follows the change in the Raman peak position (figure 2) for excitation wavelength of 632 nm as a function of the film thickness. As the film thickness increases, the Raman peak is moved towards lower frequencies, but still higher than the values for unstressed bulk crystal Si, which is 520.7 $cm^{-1}$. From the point of view of stress characterization, this means that development of compressive stress occurs [12], as in fact found from the curvature measurements shown in figure 1. From the structural point of view, the presence of peak at about 521 $cm^{-1}$ (figure 2) corresponds to films that are mostly crystalline. Another result ensues from the peaks position shift with film thickness, and it is that an evolution of film microstructure during growth is obtained, as will be discussed further.

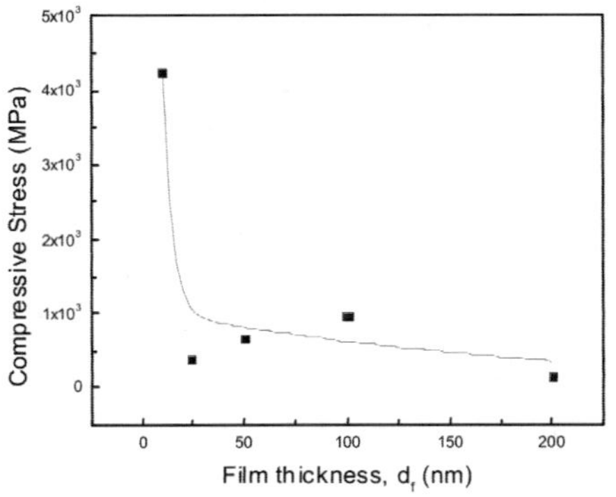


**Figure 1.** Mechanical stress values at different film thickness. Line is a guide to the eye.

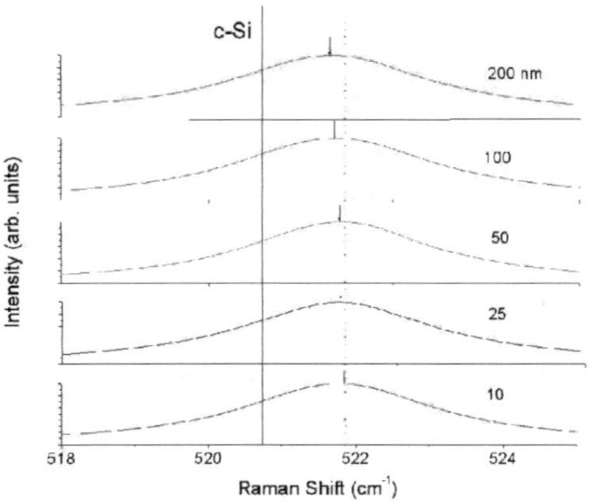


**Figure 2.** Raman shifts measured at 632 nm excitation wavelength for all film thicknesses.

Another proof in confirmation of the crystallinity of films structure is the observed (not here) constant with film thickness value of 3 $cm^{-1}$ for the full width at half maximum (FWHM). The value of this quality for crystalline Si is given to be 1.22 $cm^{-1}$ in [13]. This result is in agreement with published by Faraci et al. who found FWHM = 3 $cm^{-1}$, which they ascribed to a great number of nanocrystallites of about 100 nm size in the film [14]. Furthermore, the observed crystallites by SEM in this study are also of 100 nm size (figure 3). Thus, constant halfwidth and constant grain size are manifestation of (nano)crystallinity observed by two different analysis.

Figures 4a and 4b show the deconvoluted Raman spectra measured at 514 nm excitation wavelength for μc-Si:F:H films with thickness of 10 and 200 nm, respectively. For the thinner films the contribution of the c-Si substrate is considered, as shown in figure 4a. In thickness coordinates, the Raman peak position is shown at figure 5 for all film thicknesses studied. Together with the 521 $cm^{-1}$ peak characteristic for crystalline films two other features are observed, at around 480 $cm^{-1}$ and

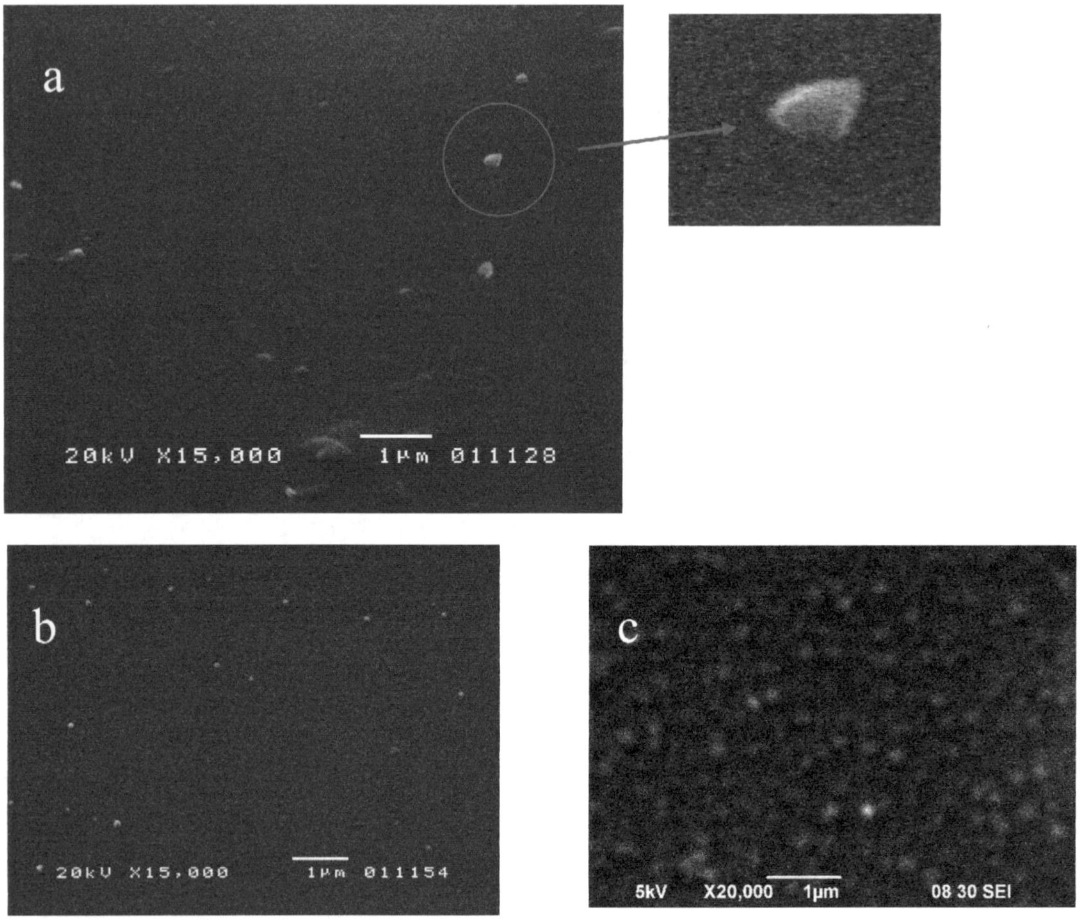


**Figure 3.** SEM images for films of 10 nm (a), 50 nm (b) and 200 nm (c) thickness.

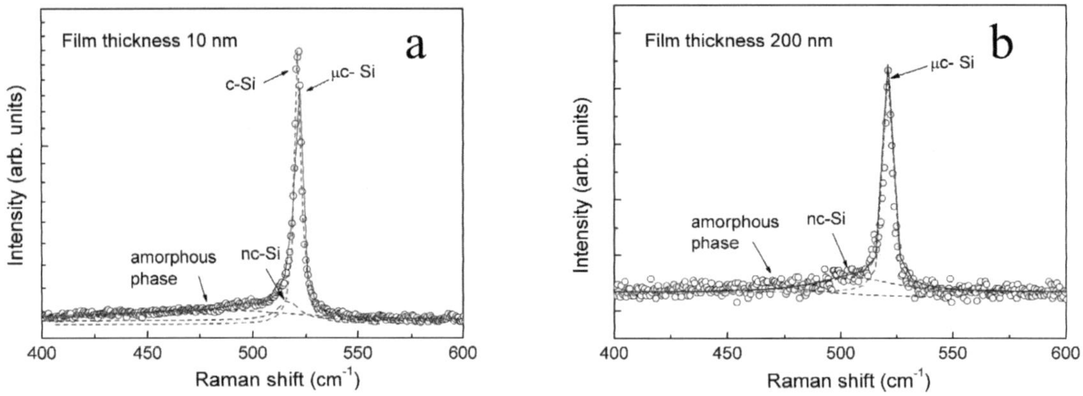


**Figure 4.** Deconvoluted Raman spectra measured at 514 nm excitation wavelength for µc-Si:F:H films of 10 nm (a) and 200 nm (b) film thickness.

515 $cm^{-1}$. The peak frequency of the Raman bands was determined from Lorentzian fits of the data. The shoulder at 480 $cm^{-1}$ indicates the presence of an amorphous fraction in the film. The low-

wavenumber peak at around 515 $cm^{-1}$ is usually assigned to a phase of defective nanocrystallites, which belongs to crystalline part of the nanocrystals, as determined by Raman emission cross-section

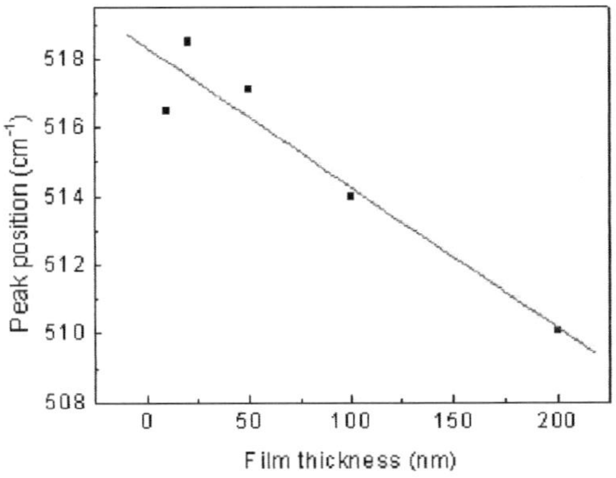


**Figure 5.** Raman shifts for films with different thickness. Excitation wavelength is 514 nm. Line is a guide to the eye.

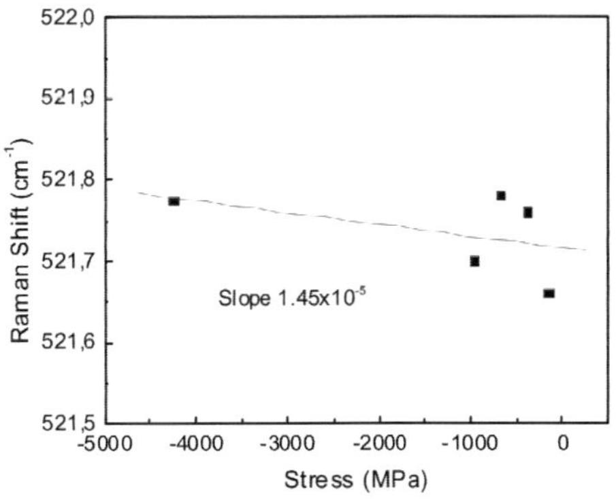


**Figure 6.** Raman shift corresponding to the measured stress (from figure 1) of μc-Si:F:H films of different thickness. The straight line is a fit assuming a linear dependence between them.

ratio in [4]. The presence of peak at 515 $cm^{-1}$ can be also attributed to smaller than 10 nm size of the grains as found recently [9] by Raman profiles that used a bimodal size distribution of crystallites. From modelling of Raman spectra, some other authors estimated that the peak at about 515 $cm^{-1}$ corresponds to grain size of about 3 nm [14]. Similar result has been obtained by means of Raman scattering for nanocrystalline silicon films [15]. According to the positions of the crystalline peak, the mean crystallite size was calculated there as being 3.5 nm. Based on these considerations, a suggestion could be drawn that namely defective nanocrystallites are smaller sized nanocrystallites in this study. However, SEM analysis was not able to show them (figure 3).

As mentioned above, the film microstructure is changed as the film thickness changes and as a result, the stress as well. To support this suggestion, amorphous/crystalline fraction analysis using integrated Raman intensities of both phases was conducted. The results are given in table 1. As it is seen, a structural transition from 10 to 200 nm films is observed. The amount of the total crystalline fraction ($X_c$+$X_{dc}$), as extracted from figures 4a and 4b, increases mainly because of the formation of defective or small size crystallites ($X_{dc}$) on the expense of the amorphous phase ($X_a$). The crystalline fraction of large size crystallites ($X_c$) remains almost constant. A rough estimate from SEM images, figure 3, gives a value of the large crystallites of about 100 nm grain size. The increased amount of $X_{dc}$ implies higher number of crystallites and greater inter-grain boundaries area. Different models have been developed to explain the observed Raman shift in microcrystalline semiconductors [16,17], which point out that the shift arises primarily due to presence of compressive stress and strongly depends on the size of the crystallites. With stress data obtained here, this would mean that defective crystallites with their greater boundaries contribute to relaxation of the stress with film thickness increasing. Considering the growth process and structure, the lowered intrinsic stress by nearly one order of magnitude compared to the stress value of 10 nm film, is a sign of completely built, finally grown film where the intrinsic stress is relieved with the thickness (200 nm film), figure 1.

The results in table 1 show that film growth proceeds through formation of amorphous and crystalline phases, which is in agreement with published in [18-20]. Since the largest part of the intrinsic stress, present at room temperature, has its origin in the growth process a structure-related discussion of the highest stress and its evolution follows. By measuring the radii of curvature before

**Table 1.** Amorphous ($X_a$) and crystalline fractions ($X_c$, $X_{dc}$), extracted from integrated Raman intensities of figure 4.

| film thickness [nm] | $X_a$ % | $X_{dc}$ % | $X_c$ % |
|---|---|---|---|
| 10 | 43.5 | 9.4 | 47.1 |
| 200 | 31.6 | 21.8 | 46.6 |

and after deposition of the μc-Si:F:H films the nature of intrinsic stress is *directly* determined in this study as compressive. A number of authors [21-23] have also found compressive stress but in rather thicker μc-Si:H films (1 μm) deposited on a glass substrate [23] or using silane-hydrogen gas mixture [21]. In the latter study, a maximum value of 1 GPa compressive stress has been measured for samples grown just before the transition to microcrystalline material. In this sense, the highest stress level of about 4 GPa for 10 nm thick film is reasonable, taking in view the different deposition conditions used in this study. To explain the stress nature *independently* of our Raman analysis we note that plasma bombardment induces compressive stress [24]. During the plasma bombardment there are a few components that take part in the film formation. The hydrogen plasma exposure results in a hydrogen-rich subsurface layer [25]. One reason for high stress value is that it is generated by unbonded hydrogen. The film is in a state in which $SiH_x$ complexes [26] are formed by H insertion into strained Si-Si bonds increasing intrinsic stress [18]. The presence of F atoms enhances this effect and μc-Si:F:H nucleation is initiated. On the other hand, according to [27], in the low temperature PECVD process, films of about 10 nm thickness are at the phase transition of amorphous to microcrystalline film structure. It is referred to as incubation phase [28] but the growth of μc-Si is produced on glass substrate, while in our case, a-Si:N:H sublayer is deposited on the Si substrate surface. This allows us to apply the same approach in analyzing the growth process. A very thin layer is sufficient for this phase transformation to occur as we observe from the Raman study (figure 4a and table 1) that both amorphous and crystalline phases are present simultaneously in the 10 nm film. So, the results about intrinsic stress, shown in figure 1, have led in suggesting that the highest stress favours the initial growth of crystallites. As SEM image of 10 nm film shows (figure 3), the material is at the transition between amorphous/microcrystalline states for which some isolated two-dimensional crystalline fragments on amorphous continuous film background are in nucleation.

Stress data (figure 1), the gradual drop in Raman shift (for both, peak at 521 and 515 $cm^{-1}$, figure 5) and SEM images with keeping same grain size (figure 3) would lead to the suggestion that μc-Si:F:H films of 50 nm thickness could be classified as being still in incubation phase, following [28] as well. Other authors cite incubation phase extending up to 60 nm [29]. The growth conditions are influenced by both substrate temperature and hydrogen dilution ratio [27]. Structural bonds in this study are less strained, while at 10 nm the film bulk is not enough for stress relaxation and requires much more deformation of the bonds. The finding of grain size of 100 nm in the film construction is consistent with peak found at about 521 $cm^{-1}$ (figure 4a, b). Nanocrystallites of order of 100 nm were inferred by other authors from modelling of Raman spectra [14]. The reason for larger grain size presence is in films deposition accomplished from $SiF_4$, compare to the deposition using $SiH_4$ as reported in [30]. The crystallite size of 16.8 nm has been estimated for films deposited from $SiH_4$ at the same temperature, 200 °C, as it is in this study but for 980 nm film thickness [23].

As it was mentioned above, a correlation between macroscopic mechanical strain or stress and the local information of the film accessed by Raman Spectroscopy can be found. The reason why such a correlation exists is the common dependence of the Raman signal and the mechanical behaviour on the micro/nanostructure of the films, which is Hooke's law or Stoney's formula (1), but in different terms. Recently, Gouadec and Colomban generalized it, giving the following relation in the case of uniaxial stress [31]:

$$\Delta\,\omega = S^{\varepsilon/\sigma}\,.\,\sigma_f, \tag{2}$$

where, $\Delta\omega$ is the Raman mode shift relative to the unstressed position; $S^{\varepsilon/\sigma}$ is the Raman Mechanical Coefficient (RMC), $\varepsilon$ is the deformation or strain, and $\sigma_f$ is the measured film stress in MPa. The physical interpretation of the proportionality coefficient $S^{\varepsilon/\sigma}$ is that it combines elastic modulus of the material to the material constants which define the lattice deformation (Raman frequency shift). A linear relationship between the film internal stress calculated by equation (1) and the Raman shift has been accepted by other authors [11,12]. Assuming the same property for our films gives the RMC value of $-1.4\text{x}10^{-5}$ $cm^{-1}$/MPa for the slope $S^{\varepsilon/\sigma}$ (figure 6).

Most of the authors who estimate the stress from Raman shift for different Si materials - from amorphous to nanocrystalline, usually use the same RMC value which is that for c-Si which is - $5.4\text{x}10^{-3}$ $cm^{-1}$/MPa [32] or its reciprocal value -250 cm.MPa [11,12,23,33]. However, RMC is strictly specific for each material. It should be emphasized that in the present study RMC for μc-Si:F:H is *independently determined* having the mechanical stress from the curvature measurements. Our RMC value is much lower than for crystalline silicon. In our opinion this result is due to the specific structure of μc-Si:F:H as evident from our SEM and Raman results. As pointed out recently [31] much lower RMC could be related to stress relaxation at grain boundaries and at interphases for nanophase materials.

Actually, the importance of our experiments is also to distinguish between the nature of the stress which accompanies film growth and structural changes in the film (as a reason this stress to appear) and the nature of the externally applied stress, for example. Strong variation of film structure with film thickness (depending on details of the deposition conditions) reflects the evaluated stress. In this study, films are deposited on crystalline Si substrates and not on a glass substrate. The curvature measurement yields the integral value of the film stress, while the Raman shifts are related to the local strain over smaller area of the film. Thus, the RMC contains implications of the physical relevance of the methods for stress determination, namely interference measurements of sample curvature and Raman spectroscopy.

## 4. Conclusion

Our experimental results show that the *very thin films* of μc-Si:F:H (up to 200 nm) grow with compressive intrinsic stress as found by the wafer curvature and Raman study. Curvature measurements allow us to *experimentally* determine the *accurate* intrinsic stress in the films and to correlate it to Raman shifts. A combined study of the internal stress and Raman shift of the crystalline mode, both measured on the same microcrystalline films, lead us to the value of the RMC material constant for microcrystalline silicon thin films deposited from $SiF_4$. The constant was estimated to be - *$1.4\text{x}10^{-5}$ ($cm^{-1}$/MPa)*. This constant characterizes the film properties resulting from the specific growth process. Besides, from our stress measurements we propose stress related growth of microcrystalline thin films. At the early stage of deposition crystallization is favoured by high stress that develops in the thin films. The mechanical stress is further reduced by an increase of the crystalline fraction linked to incorporation or creation of small size crystallites, the stress being released at their boundaries. We consider the results as giving some light to understand the development of the crystallinity of thin Si films on the base of parallel exploration of stress, Raman spectroscopy and morphology.

## References

[1] Sirenko A A, Fox J R, Akimov I A, Xi X X, Ruvimov S and Liliental-Weber Z 2000 *Solid State Comm.* **113** 553

[2] Ye D-X, Zhao Y-P, Yang G-R, Zhao Y-G, Wang G-C and Lu T-M 2002 *Nanotechnol.* **13** 615

[3] Pan B, Qian K, Xie H and Asundi A 2009 *Meas. Sci. Technol.* **20** 062001

[4] Vallat-Sauvain E, Droz C, Meillaud F, Bailat J, Shah A and Ballif C 2006 *J. Non-Cryst. Solids* **352** 1200

[5] Trask J, Cui L, Wagner A and Kortshagen U 2010 *Materials Research Society Spring Meeting* (San Francisco, CA, 5-9 April 2010) p 131
[6] Su T, Ju Tong, Yan Baojie, Yang Jeffery, Guha S and Craig Taylor P 2008 *J. Non-Cryst. Solids* **354** 2231
[7] Yan Baojie, Yue Guozhen, Yan Yanfa, Jiang Chun-Sheng, Teplin C W, Yang Jeffery and Guha S 2008 *Materials Research Society Spring Meeting* (San Francisco, CA, 24-28 March 2008) p 1955
[8] Stoney G 1909 *Proc. Roy. Soc. London Ser.* A **82** 172
[9] Ram S, Islam Md, Kumar S and Roca i Cabarrocas P 2009 *Mat. Sci. Eng.* B **159-160** 34
[10] Poruba A, Fejfar A, Remes Z, Springer J, Vanecek M, Kocka J, Meier J, Torres P and Shah A 2000 *J. Appl. Phys.* **88** 148
[11] De Wolf I 1996 *Semicond. Sci. Technol.* **11** 139
[12] De Wolf I 2003 *Spectroscopy Europe* **15** 6
[13] Weber W 2000 *Springer Series in Materials Science* vol 42, ed R Merlin (Berlin: Springer) p 81
[14] Faraci G, Gibilisco S, Russo P and Pennisi A R 2006 *Phys. Rev.* B **73** 033307
[15] Wu X. L, Siu G G, Tong S, Liu X N, Yan F, Jiang S S, Zhang X K and Feng D 1996 *Appl. Phys. Lett.,* **69** 523
[16] Fujii M, Kanzawa Y, Hayashi S and Yamamoto K 1996-II *Phys. Rev.* B **54** R8373
[17] Ager III J, Veirs D and Rosenblatt G 1991-I *Phys. Rev.* B **43** 6491
[18] Fujiwara H, Kondo M and Matsuda A 2004 *J. Non-Cryst. Solids* **338-340** 97
[19] Fontcuberta i Morral A, Bertomeu J and Roca i Cabarrocas P 2000 *Mat. Sci.Eng.* B **69-70** 559
[20] Müllerová J, Šutta P, van Elzakker G, Zeman M and Mikula M 2008 *Appl. Surf. Sci.* **254** 3690
[21] Kroll U, Meier J, Shah A, Mikhailov S and Weber J 1996 *J. Appl. Phys.* **80** 4971
[22] Stratakis E, Spanakis E, Tzanetakis P, Fritzsche H, Guha S and Yang J 2002 *Appl. Phys. Lett.* **80** 1734
[23] Paillard V, Puech P and Roca i Cabarrocas P 2002 *J. Non-Cryst. Solids* **299-302** 280
[24] Ohring M 1992 *The materials science of thin films* (San Diego, CA: Academic Press) p 429
[25] Fontcuberta i Morral A and Roca i Cabarrocas P 2006 *Eur. Phys. J. Appl. Phys.* **35** 165
[26] Hamers E A G, Fontcuberta i Morral A, Niikura C, Brenot R and Roca i Cabarrocas P 2000 *J. Appl. Phys.* **88** 3674
[27] Collins R. W, ferlauto A S, Ferreira G M, Chen Chi, Koh J, Koval R J, Lee Y, Pearce J M and Wronski C R 2003 *Solar Energy Mat. & Solar Cells* **78** 143
[28] Roca i Cabarrocas P 2000 *J. Non-Cryst. Solids* **266-269** 31
[29] Li Juan, Wu C, Meng Z, Liu J, Zhao S, Xiong S and Zhang L 2005 *Chinese J. Semicond.* **26** ISSN: 0253-4177
[30] Djeridane Y, Abramov A and Roca i Cabarrocas P 2007 *Thin Solid Films* **515** 7451
[31] Gouadec G and Colomban P 2007 *Progress in Cryst. Growth & Character. Mat.* **53** 1
[32] Anastassakis E, Cantarero A and Cardona M 1990 *Phys. Rev.* B **41** 7529
[33] Lee H C, Hong S P, Kang S K and Yeom G Y 2009 *Thin Solid Films* **517** 4100

# Stress in $LiNbO_3$ proton-exchanged waveguide layers

**K Christova, M Kuneva and S Tonchev**

Institute of Solid State Physics, Bulgarian Academy of Sciences
72 Tzarigradsko Chaussee Blvd., Sofia 1784, Bulgaria

E-mail: kuneva@issp.bas.bg

**Abstract.** Penetration of hydrogen ions in the crystal lattice of $LiNbO_3$ certainly causes stress in the proton-exchanged region, especially when the level of doping is high. In the present study, a correlation between stress and the level of doping (respectively, between stress and the phase composition of the measured layer) has been obtained. Proton-exchanged waveguide layers in X, Y and Z-cut $LiNbO_3$ have been produced at different technological conditions. These layers have been studied by spectroscopic methods for analyzing their phase composition and by the optical integral method for stress measurements. The intrinsic stress which contributes to the better characterization of such optical waveguides was calculated. An attempt to indicate stress correlation to the relative quota of the different phases in the proton exchanged layer is made.

## 1. Introduction

The easy and fast obtaining of optical waveguides in the electro-optical crystal $LiNbO_3$ by proton exchange (PE) [1] has motivated the attempts to adjust the technology for producing high quality waveguides. The surface layer modified by hydrogen diffusion has a large extraordinary index change ($\Delta n_e$) and thus, a strong waveguide and polarizing effect. The $Li_{1-x}H_xNbO_3$ layer formed by Li-H ion exchange shows complex phase behavior (up to seven different crystallographic phases - $\alpha$, $\kappa_1$,$\kappa_2$, $\beta_1$, $\beta_2$, $\beta_3$ and $\beta_4$) depending on the hydrogen concentration (value of x) [2] and PE layers may show significant optical losses, instabilities and decrease of the electro-optical coefficient. The number and the type of different phase modifications in crystal solid solutions $Li_{1-x}H_xNbO_3$, what actually is the proton-exchanged lithium niobate, are determined by the crystallographic orientation and the rate of substitution of $Li^+$ by $H^+$ (x). Each phase originates as an individual sub-layer undergoing phase jumps with gradual change of x. In every single one of them $\Delta n_e$ is a linear function of proton concentration. Within a phase transition, the value of $\Delta n_e$ and/or of the deformations perpendicular to the surface change by leap.

Infrared spectroscopy has proved its ability to characterize PE layers in $LiNbO_3$ since the Li-H substitution dramatically increases the presence of OH groups in the exchanged layer. IR absorption in the range of OH stretching vibrations (2700-3700 $cm^{-1}$) has been widely used for estimation of the phase composition [3, 4, 5, 6]. The spectra show several bands (at 3488 $cm^{-1}$, 3500 $cm^{-1}$, 3512 $cm^{-1}$, 3250-3280 $cm^{-1}$) which could be attributed to different phases forming the PE layers [3].

An optical integral method contributes to the characterization of the waveguide film from mechanical stress (MS) point of view. Since the MS is a function of the film structure, by its change due to different film composition, an additional picture for the building units of the film could be gained. Up to now the stress characterization of PE $LiNbO_3$ layers is poorly studied. The purpose of this study is to explore the relation between phase composition and mechanical stress. In this sense proton exchanged (PE) waveguide layers in X-, Y- and Z-cut $LiNbO_3$ have been studied by spectroscopic methods for analyzing their phase composition and by the optical integral method for stress measurements.

## 2. Experimental

Proton-exchanged waveguide layers were obtained at the technological conditions described in table 1.

Mode spectra were measured with resolution of ~$10^{-4}$ at $\lambda$=633 nm by using prism coupling, and the inverse WKB method [7] was used for refractive index profile reconstruction. In cases of single-mode waveguides, the depth and surface index change were calculated by sequential iterations when assuming a steplike profile.

**Table 1.** Technological and waveguide parameters of the proton-exchanged waveguides: (T – temperature, t – duration of the PE process, $T_a$ and $t_a$ – temperature and duration of the annealing, M - number of waveguide modes at $\lambda$ = 633 nm, $\Delta n_e$ – extraordinary refractive index change, d – waveguide depth)

| Sample | Sample environment | T (°C) | t (h) | M | $\Delta n_e$ | d (μm) | Phase composition (Spectroscopic evidence) | |
|---|---|---|---|---|---|---|---|---|
| | | | | | | | Mode | IR |
| Z-1 | $LiHSO_4$ (vapors) | 250 | 3.50 | 6 | 0.1491 | 2.66 | $\beta_1$,$\beta_3$,$\beta_4$ | $\beta_1$,$\beta_3$,$\beta_4$, $\kappa_1/\kappa_2$ (α) |
| Z-2 | $LiHSO_4$ (vapors) | 160 | 2.00 | 1* | 0.1529 | 0.18 | $\beta_i$ (i=1-4) | $\beta_i$ (i=1-4), $\kappa_1/\kappa_2$ (α) |
| Z-3 | $LiHSO_4$ (vapors) | 160 | 4.00 | 1 | 0.1480 | 0.28 | $\beta_1$,$\beta_3$ | $\beta_1$,$\beta_3$,($\beta_4$), $\kappa_1/\kappa_2$ (α) |
| Z-4 | $LiHSO_4$ (melt) | 160 | 7.00 | 1 | 0.1463 | 0.33 | $\beta_1$,$\beta_3$ | $\beta_1$,$\beta_3$, $\kappa_1/\kappa_2$ (α) |
| Z-7 | $LiHSO_4$ (vapors) | 180 | 5.00 | 1 | 0.1384 | 0.54 | $\beta_1$,$\beta_3$ | $\beta_1$,$\beta_3$,($\beta_4$), $\kappa_1/\kappa_2$ (α) |
| ZN-1 | $NH_4HSO_4$ (melt) | 230 | 3.33 | 9 | 0.1508 | 2.47 | $\beta_1$,$\beta_3$,$\beta_4$ | $\beta_1$,$\beta_3$,$\beta_4$, $\kappa_1/\kappa_2$ (α) |
| ZN-5 | $LiHSO_4$ (melt) | 175 | 1.50 | 1 | 0.1222 | 0.62 | $\beta_1$,$\beta_2$, ,$\beta_3$ | $\beta_1$, $\beta_2$, $\beta_3$, $\kappa_1/\kappa_2$ (α) |
| X-2 | $LiHSO_4$ (vapors) | 160 | 2.00 | 1 | 0.0999 | 0.37 | $\beta_1$,$\beta_2$ | $\beta_1$, $\beta_3$, $\kappa_1/\kappa_2$ (α) |
| X-3 | $LiHSO_4$ (vapors) | 160 | 4.00 | 1 | 0.1234 | 0.52 | $\beta_1$,$\beta_3$ | $\beta_1$, $\beta_3$, $\kappa_1/\kappa_2$ (α) |
| X-4 | $LiHSO_4$ (vapors) | 160 | 7.00 | 1 | 0.1090 | 0.70 | $\beta_1$,$\beta_2$ | $\beta_1$, $\beta_3$, $\kappa_1/\kappa_2$ (α) |
| Y-1 | 1)Benzoic acid<br>2) Air<br>3)Benzoic acid | 230<br>350<br>230 | 0.50<br>1.00<br>0.17 | 3 | 0.1360 | 2.75 | $\beta_i$(i=1-3)<br>$k_1/k_2$ (α) | $\beta_i$(i=1-3)<br>$\kappa_1/\kappa_2$ (α) |
| Y-2 | Benzoic acid + 0.5% Lithium benzoate | 230 | 3.00 | 3 | 0.1134 | 1.56 | $\beta_1$,<br>$k_1/k_2$ | $\beta_1$,<br>$\kappa_1/\kappa_2$ (α) |
| Y-3 | Benzoic acid + 1% Lithium benzoate | 230 | 7.00 | 4 | 0.1113 | 2.85 | $\beta_1$,<br>$k_1/k_2$ | α, $\beta_1$,<br>$\kappa_1/\kappa_2$ (α) |
| Y-4 | 1) Cinnamic acid vapors<br>2)Cinnamic acid melt | 300<br>230 | 0.50<br>0.08 | 3 | 0.1410 | 1.52 | $\beta_i$(i=1-3)<br>$k_1/k_2$ (α) | (α),$\beta_i$(i=1-3)<br>$\kappa_1/\kappa_2$ (α) |

* at $\lambda$ = 458 nm

Journal of Physics: Conference Series **253** (2010) 012057 doi:10.1088/1742-6596/253/1/012057

IR absorption spectra were recorded in the wavenumber range of 2700-3700 $cm^{-1}$ of OH stretching modes with FTIR spectrometer (Brucker LFS-113 V) with a resolution of 1 $cm^{-1}$. A Gaussian-Lorentzian decomposition procedure was applied to them. Figure 1 illustrates the distribution of peak frequencies and integrated intensities of the spectral components. Since the component due to the substrate α-phase was extracted from the spectra after the deconvolution, only the components of the layers' spectra are present in the histograms. The results allow the relative quota of each band to be calculated in % (table 2).

**Table 2**. The relative quota of the waveguide layer IR spectral bands (corresponding phase quota respectively) compared to the stress measured.

| SAMPLE | IR BAND INTENSITY [%] | | | | $\sigma$ x$10^9$ [N/$m^2$] |
|---|---|---|---|---|---|
| | 3280 $cm^{-1}$ | 3480 $cm^{-1}$ | 3500 $cm^{-1}$ | 3512 $cm^{-1}$ | |
| Z-1 | 16.39 | 9.12 | 55.83 | 18.77 | -0.95 |
| Z-2 | 27.33 | 11.65 | 25.54 | 35.39 | -19.83 |
| Z-3 | 20.74 | 22.00 | 57.14 | - | -4.21 |
| Z-4 | 18.48 | 21.75 | 59.89 | - | -6.43 |
| Z-7 | 17.42 | 57.96 | 13.91 | 10.68 | -2.17 |
| ZN-1 | 23.32 | 6.69 | 57.22 | 11.92 | -2.45 |
| ZN-5 | 30.66 | 16.94 | 52.30 | - | -1.17 |
| X-2 | 48.55 | 36.30 | 15.12 | - | -12.3 |
| X-3 | 48.95 | 33.95 | 17.00 | - | -0.45 |
| X-4 | 61.72 | 22.95 | 15.31 | - | 1.15 |
| Y-1 | 0.20 | 58.11 | 23.89 | 17.84 | -0.01 |
| Y-2 | 0.05 | 26.60 | 71.48 | 1.85 | -0.54 |
| Y-3 | 0.08 | 15.53 | 69.44 | 1.76 | -0.29 |
| Y-4 | 0.09 | 37.78 | 38.81 | 23.23 | -0.27 |

The quantitative calculations of layer stress have been performed by an ex situ optical interference technique [8, 9]. The role of the substrate is assumed to be of lithium niobate with no waveguide on it for each orientation – X, Y and Z. Since the deformation of the waveguide follows that of the corresponding cut substrates, the elastic parameters, Young modulus and Poison's ratio are taken for the substrates. The change in the interference fringe patterns between the substrate and an optical flat is used to measure the deformation of the layer-substrate system. The radius of curvature is actually the substrate radius. The error in the radii determined by curvature measurements was 10%. Since vapor temperature during the process of proton exchange is low, the thermal component of the stress is negligible. Thus, the calculated stress is the intrinsic stress for the waveguide layers. It has to be noted that this method for determining stress gives an integral value over the waveguide bulk. Besides, the nature of the stress is determined by its sign. If the sign is minus, the stress is compressive; the sign plus corresponds to tensile stress.

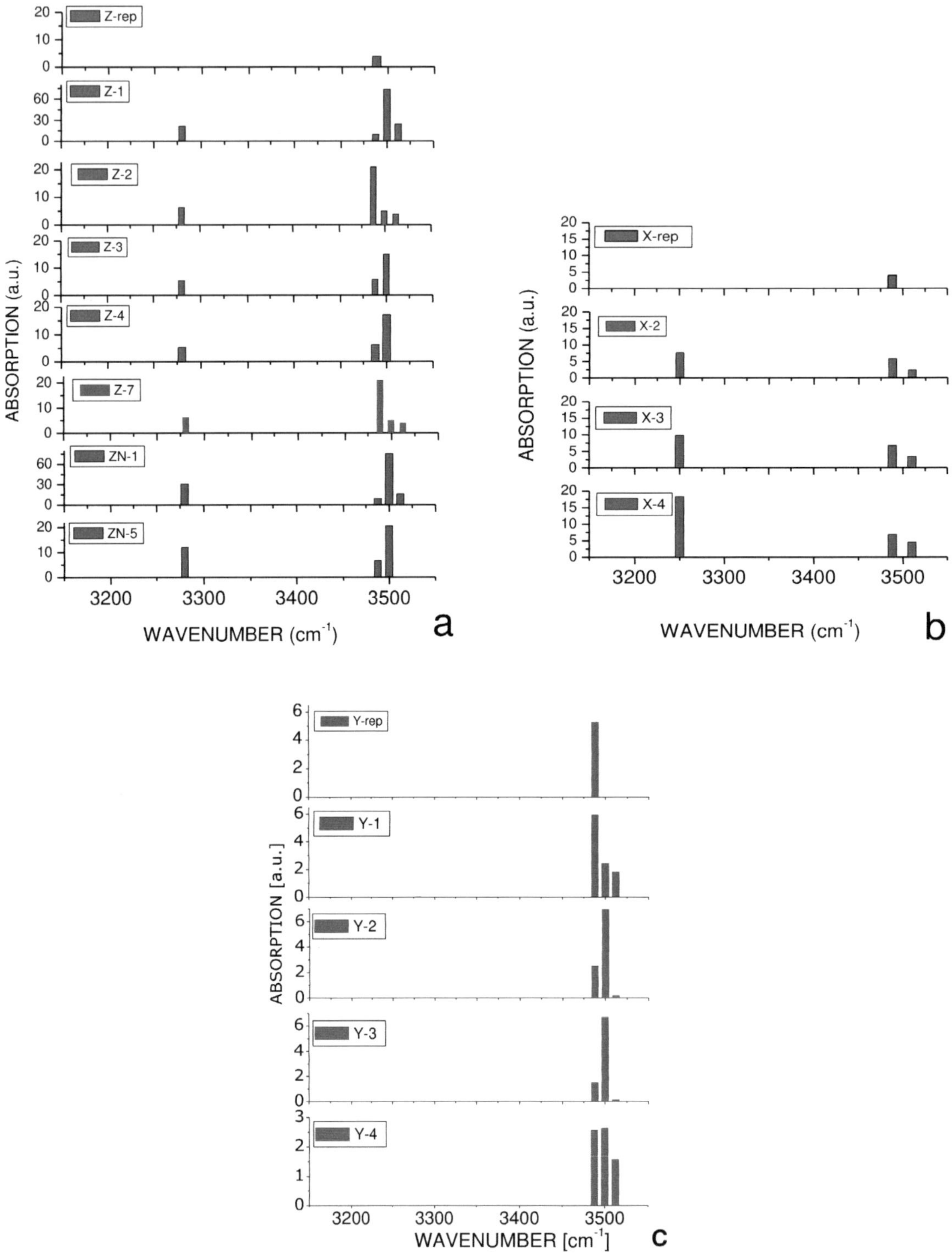


**Figure 1.** Schematic presentation of frequency position and intensity of the components in spectra of Z-cut (a), X-cut (b) and Y-cut (c) samples.

Journal of Physics: Conference Series **253** (2010) 012057 doi:10.1088/1742-6596/253/1/012057

## 3. Discussion

### *3.1. M-line spectroscopy*

The values of extraordinary refractive index change $\Delta n_e$=0.15 for Z-1 and ZN-1 and its gradual change to 0.12 on the optical profiles of the waveguide lead to the suggestion that these waveguides should contain the $\beta_4$-phase which is characterized by a higher level of proton doping compared to monocrystalline $H_xLi_{1-x}NbO_3$ ($0.65 \le x \le 0.7$) [2]. The $\beta_4$-phase can exist only in Z-cut proton-exchanged lithium niobate and only together with the $\beta_3$-and $\beta_1$-phases of lower value of x: $0.53 \le x \le 0.65$ for the $\beta_3$-phase and $0.44 \le x \le 052$ for the $\beta_1$-phase. Thus, the waveguides Z-1 and ZN-1 consist of at least three $\beta_i$-phases: $\beta_1$, $\beta_3$ and $\beta_4$. The other Z-cut single-mode waveguides should be in $\beta_1$, $\beta_1+\beta_2$ or $\beta_1+\beta_3+\beta_4$ – phase since the value of $\Delta n_e$ exceeds 0.12, indicating that $\beta_i$-phase or phases have been formed during the PE process [2].

The same considerations applied to the X- and Y-cut samples lead to the conclusion that X-3 contains a $\beta_3$-phase [2], which can be found only together with a $\beta_1$-phase. The $\beta_4$-phase could not be formed in X-cut $LiNbO_3$ [3].The lower values of $\Delta n_e$ for X-2 and X-4 mean that these layers could contain $\beta_1$ or $\beta_1+\beta_2$ phases. The shape of the optical profile could not be followed since the waveguides are single-mode. The optical profiles of Y-1 and Y-4 and the high surface refractive index change are characteristic for waveguides consisting of highly protonated $\beta_i$-phases (i=1-3) ($0.56 \le x \le 0.65$) and a layer of $\alpha$-phase ($0 \le x \le 0.12$) or $\kappa_1/\kappa_2$ phases ($0.12 \le x \le 0.56$) between them and the substrate. The optical parameters of Y-2 and Y-3 correspond to the coexistence of $\beta_1$ and $\kappa_1/\kappa_2$ phases. Y-2 has a steplike optical profile with a maximal index change of $\Delta n_e$=0.11. Thus, that waveguide should contain the $\beta_1$-phase - the only one which could be obtained by low-temperature direct proton exchange in Y-cut $LiNbO_3$ [10]. The profile shape of Y-4 shows [11] that the layer should consist of a $\beta_1$-phase together with $\kappa_1/\kappa_2$ or $\alpha$-phase which is due to the higher percent of lithium benzoate in the melt compared to the case of Y-2.

### *3.2. IR-spectra*

As it was discussed earlier [3, 4], a band centered at 3488 $cm^{-1}$ in the IR spectra is attributed to the $\alpha$-phase. All other bands could be assigned to the phases formed by PE: the broad band centered at 3250-3280 $cm^{-1}$ - to the $\beta_2$, $\beta_3$ and $\beta_4$ phases, the band at 3500 $cm^{-1}$ – to the $\beta_1$, $\kappa_1$ and $\kappa_2$ produce peaks between 3488 $cm^{-1}$ and 3500 $cm^{-1}$ (intermediate values between the $\alpha$-phase and the $\beta_1$-phase) and are spectroscopically undistinguishable. For the strongly protonated samples, a fourth additional component emerges at about 3512 $cm^{-1}$ which could be attributed to phases with higher degree of doping (the $\beta_4$ only in case of Z-cut samples). Thereby the evolution of the bands assigned to different phases versus the variations in the technological parameters could be followed. Thus, the histograms show that all samples consist of $\alpha/\kappa_1/\kappa_2$, $\beta_1$ and $\beta_4$ phases, but the technological regime chosen reflects on the thickness of sublayers formed by these phases. The contribution of the $\beta_4$-phase is stronger in samples with higher $\Delta n_e$, i.e. higher level of proton doping. The contribution of $\alpha/\kappa_1/\kappa_2$ is stronger in samples obtained at regime including annealing (Y-1) or annealing-like step (Y-4). The contribution of $\beta_1$ is stronger for samples obtained by direct proton exchange (Y-2 and Y-3). It should also be noted that all the peaks except of a large shoulder at 3280 $cm^{-1}$ are polarized showing that OH bonds lay in the oxygen planes of the crystal. The unpolarized band indicates the presence of hydrogen in interstitial positions out of the oxygen planes, which usually means a higher level of doping.

### *3.3. Stress measurements*

All but one samples studied show compressive stress regardless of crystal orientation. Such a behavior is not surprising. The network of bonds is distorted by the substitution of $H^+$ for $Li^+$ in the $LiNbO_3$ lattice. The ionic radius of $Li^+$ (1.55 Å) is higher than that of $H^+$ (0.46 Å). For all crystal orientations of the substrate, balancing forces are required to sustain equilibrium acting in compression to adjust the structure to the smaller-sized component. One of the samples (X-4) shows tensile stress. According to some authors, PE causes a little larger lattice constant of the layer $Li_{1-x}H_xNbO_3$, which results in stress generation [12].

It could be seen that the most stressed is the layer with the largest quota of the 3512 $cm^{-1}$ band (Z-2) and the least stressed is the Y-1 waveguide, where the band at 3480 $cm^{-1}$ ($\alpha$ or $\kappa_i$ phases) has largest quota. X-2 is stressed higher than other samples although it does not have the component at 3512 $cm^{-1}$. It seems that the main role in mechanical stress here is played by the quota of low-protonated phases (alpha or $\kappa_i$) – it is lower than the quota of the 3280 $cm^{-1}$ shoulder, which usually is related to interstitial hydrogen, the situation when the lattice is sufficiently disturbed. The only sample with positive stress is X-4, which has the largest shoulder centered at 3280 $cm^{-1}$ indicating the stronger presence of interstitial hydrogen. From technological point of view, it is interesting to note that the most strained layers have been obtained in two of the cases when PE was performed in vapors (Z-1 and X-2). It is quite possible (especially at higher temperatures, i.e. 250°C) that the kinetic energy of the substituting element transforms into energy of lattice deformation. In spite of the lower temperature (X-2), the larger diffusion coefficient for X-cut lithium niobate leads to the same effect. A longer diffusion when the temperature and proton source are the same decreases the MS (samples X-3 an X-4), probably due to the simultaneous process of annealing when a relaxation redistribution of the diffused protons takes place. Such low-temperature annealing (160°C) does not cause phase transitions and transformations, but probably causes homogenization of the existing phases, this way decreasing the MS in the layer.

As it could be expected, the lowest values of the magnitude of intrinsic stress were obtained with Y-cut samples. This is due to the optimization of technological conditions for obtaining of deep waveguides with high extraordinary index change in so oriented crystals. It is well known that in case of Y-cut substrates, PE leads to a fast damage of their surface exactly because of the higher MS arising under longer diffusion. Low values of MS in our case correlate with the absence of shoulders in the IR absorption spectra of these layers (figure 1-c). The appearance of 3512 $cm^{-1}$ component means only a high value of x (as it could be seen from the optical measurements), but not the presence of interstitial hydrogen. The shoulder at 3280 $cm^{-1}$ suggests both high value of x ($H^+$ at $Li^+$ positions) and interstitial hydrogen (figure 1-a, b). At high values of x, the lattice should be compressed, but the interstitial hydrogen stretches the lattice, so when the MS is positive, we could suggest that the balance is disturbed in favor of the interstitial instead of the substituting one.

Since LN is an anisotropic crystal, it should be expected that the stress is different along different directions, i.e. the bending should be different along each of the crystallographic directions. It is important to note that no elliptic interference fringes were observed during stress measurements. This means that the layer has lost its anisotropy which indeed corresponds to the total loss of electro-optical properties at high level of doping. Such kind of quasi-amorphization (paraelectric state) has been observed when Raman spectroscopy was applied for waveguide characterization showing that all $\beta_i$-phases (i=1-4) are paraelectric [13]. Actually, it is well known that at higher values of x the lattice changes from perovskite to cubic one [14].

**Conclusions**

The phase composition of proton-exchanged waveguides was studied by mode spectroscopy, IR absorption spectroscopy and mechanical stress measurements. The methods used independently gave information for the phase composition of the waveguide layer of multiphase proton-exchanged waveguides in $LiNbO_3$, and for the relative contribution of each of them.

It is demonstrated that a correlation exists between mechanical stress and the phase composition/hydrogen presence in the proton-exchanged layer as well as that the stress depends substantially on the technological regimes for obtaining waveguides. It has been shown that highly protonated phases and deep waveguides with very low mechanical stress could be obtained in Y-cut $LiNbO_3$ at specific technological conditions.

The substitution-interstitial proton ratio could be estimated from the sign and the magnitude of the mechanical stress calculated. There should be suitable fabrication conditions for obtaining a phase composition corresponding to a minimal stress in the waveguide layer.

The measurements confirm the paraelectric state of highly protonated ($\beta_i$, i=1-4) phases since such layers had lost their anisotropy.

**References**

[1] Jackel J L and Rice C E 1982 *Appl. Phys. Lett.* **41** 607-8
[2] Korkishko Yu N and Fedorov V A 1996 *Proc. SPIE* **2700** 186-95
[3] Korkishko Yu N and Fedorov V A 1996 *IEEE J. Select. Top. Quant. El.* 1-10
[4] Rams J and Cabrera M 1999 *J. Opt. Soc. Am.* B **16** 401
[5] Kuneva M, Tonchev S, Sendova-Vasileva M, Dimova-Malinovska D and Atanasov P 2002 *Sensors and Actuators* A **99** 154-9
[6] Kuneva M, Tonchev S and Atanasov P 2005 *Mater. Sci. Eng.* B **118** 301-5
[7] White J M and Heidrich P E 1976 *Appl. Opt.* **15** 151-5
[8] Christova K 1993 *PhD Thesis* ISSP BAS, Sofia, Bulgaria
[9] Skordeva E, Christova K, Tzolov M and Dimitrova Z 1998 *Appl. Phys.* A **66** 103
[10] Korkishko Yu N, Fedorov V A, Nosikov V V, Kostritskii S M and De Micheli M P 1997 *Proc. SPIE* **2997** 188-200
[11] Kuneva M, Tonchev S and Bozhkov B 2009 *J.Opt. Adv. Mater.* **11** 1529-32
[12] Fu I L, He M X, Yan Y F 2006 *Guang Pu Xue Yu Guang Pu Fen Xi* No. 12 2346-9
[13] Wu X-L, Zhang M-S and Feng D 1996 *Phys. Stat.Sol. (a)* **153** 233
[14] Rice C E 1986 *J. Sol. St. Chem.* **64** 188-99

# Experimental study of the electrically induced optical birefringence in suspension of densely packed giant lipid vesicles

**K Antonova[1,3], V Vitkova[1], G Popkirov[2], T Tenev[1], I Miloushev[1] and I Bivas[1]**
[1] G. Nadjakov Institute of Solid State Physics, Bulgarian Academy of Sciences, Sofia, Bulgaria
[2] Central Laboratory of Solar Energy and New Energy Sources, Bulgarian Academy of Sciences, Sofia, Bulgaria

E-mail: krasa@issp.bas.bg

**Abstract.** Optical birefringence $\Delta n \sim -10^{-7}$, induced in suspension of giant lipid vesicles by high-frequency (500 kHz) AC electric field was measured. The liquid phases inside and outside the vesicles were iso-osmolar solutions of sucrose and glucose (0.22 mol/L). The two solutions have different optical densities and different specific weights. The estimated volume fraction of the vesicles in the studied suspension was 48 v%. The amplitude of the applied electric field was varied in the interval 5 ÷ 30 V/mm. The observed effect was supposed to be a result from the electro-deformation of the giant vesicles leading to an optical anisotropy. The measured value of $\Delta n$ in our case is in the same range as for suspensions of large unilamellar vesicles with considerably higher lipid concentrations, as reported in the literature.

## 1. Introduction

Upon application of a high frequency AC electric field spherical vesicles change their shape to one of a prolate rotational ellipsoid with its short axis of symmetry parallel to the electric field. Such deformations have been observed in isolated vesicles [1] as well as in suspensions of giant (~20 μm) vesicles [2] and have been supposed to proceed in a suspension of small (~100 nm) vesicles [3]. The question whether the system of deformed vesicles implies an optical anisotropy has been discussed in a few publications [4-6] considering undetectable shape-transformation in vesicular suspensions of small dimension objects ($d$ ~ 100 nm). Very weak optical anisotropy of the order of $\Delta n \sim 10^{-7}$ has been measured [3] and attributed to the vesicle deformation (anisotropy of the form) and to the possible vesicle pearling into relatively long chains due to the interaction of the electric field with the vesicles.

In the present work we carried out measurements of the optical response of a suspension of giant unilamellar vesicles subjected to a high frequency AC electric field. The process of the vesicle deformation was clearly visible by a microscope camera. So far, there are no publications on electro-optics for this kind of vesicular suspensions. In our experiments we applied a technique to detect the very weak effect of the field-induced optical anisotropy, as described in [7].

[3] To whom any correspondence should be addressed.

## 2. Materials and methods

### *2.1. Preparation of giant vesicles suspension*

In our experiments giant unilamellar vesicles (GUV) were prepared by electroformation [8] from 1-stearoyl-2-oleoyl-sn-glycero-3-phospho-choline (SOPC), cat. No 850467 (Avanti Polar Lipids Inc., AL, USA). Double-distilled water was obtained from a quartz distiller. Sucrose (cat. No S7903, Sigma Ultra ®) was purchased from Sigma-Aldrich Chemie (Gremany) and D(+)-Glucose monohydrate (cat. No 108346) – from Merck (Germany). Methanol and chloroform ("for analysis" grade) were purchased from Fluka Inc. (Germany). For GUV electroformation lipid depositions were made by the careful spreading of 50 μL of SOPC solution with concentration of 1 g/L in chloroform-methanol (9:1 volume parts) organic solvent on each ITO-coated glass electrode. We used freshly prepared organic solution of the lipid (previously lyophilized and kept under vacuum at - 20°C). After the complete drying of the lipid deposits for at least 2 hours under vacuum, the electroformation cell was assembled and completely filled with 0.2 mol/L of sucrose solution avoiding air bubbles. Then AC electric field with low frequency (10 Hz) was applied and its amplitude was successively increased up to $E_{pp}$= 0.3 V/mm (where $E_{pp}$ denotes the peak-to-peak amplitude of the electric field applied). Usually, high yield of unilamellar vesicles without any visible defects was obtained after four hours accordant with the observations of other authors [9]. The initial vesicle suspension was diluted 20 times in 0.22 mol/L of glucose and left to sediment in a glass flask with conical bottom. The glucose concentration in the water solution was taken slightly higher than the sucrose one in order to obtain flaccid vesicles after the equilibration of the partial pressures of the two sugars. The difference in densities for the two sugar solutions was $\Delta\rho = 10^{-2}\,\text{kg/m}^3$ and of the refractive indexes $\Delta n = 3\cdot 10^{-3}$. In this way, the relative weight of vesicles, which are slightly heavier than the suspending solution, allowed to concentrate the vesicles at the lowest conical part of the flask. After their complete sedimentation (approx. 6 hours) the supernatant was removed and the sediment of vesicles was collected and ready for manipulation. The precipitation continued in the sample cell. As a result, closely packed vesicular suspension (lower part of the cell) in contact with free of vesicles water solution of glucose (upper part of the cell) was obtained. Volume fraction of vesicles in the initial suspension filling the cell was determined to be 48% via vesicle counting performed on an inverted phase contrast microscope Axiovert 100 (Zeiss, Germany) with a dry x63 objective. The size distribution curve of the vesicle diameters had its main maximum at 9.7 μm and its width was 3.0 μm.

### *2.2. Experimental equipment and measurements of the electro-induced optical anisotropy*

The technique for detection of very weak electro-induced optical anisotropy [7] is based on a measurement of the optical path difference between two directions in the cell (parallel and perpendicular to the electric field **E**) as a function of the applied voltage. In figure 1 the experimental set-up is represented schematically.

A polarizing microscope was used. The light from the incandescent lamp passed through an interference filter ($\lambda$ = 515 nm, $\Delta\lambda$ = 2.5 nm). The symmetry axis of the sample cell (coinciding with the electric field direction) was placed under 45 degrees towards both the crossed polarizer and analyzer, thus introducing a path difference of $\Delta l(U) = l_0 - l(U)$, where $U$ is the applied voltage and $l_0$ is the initial path difference at $U = 0$ V. A Bräce-Köhler compensator giving a variable retardation $-l_r$ enabled a total phase shift variation $\Delta l_t(U) = l(U) - l_r$. The transmitted light intensity $I(U) = I_0 \sin^2(\pi l_t(U)/\lambda)$ was detected by a photo-multiplier (PM). Using the compensator, $I_0$ was calibrated between the minimum (at $l_0$) and maximum (at $l_t = \lambda/2$) transmission, then the compensator was fixed in position $l_0 - l_r = \lambda/4$ to obtain a linear response of the transmitted intensity change $\Delta I(U) \cong I_0 \pi \Delta l_t(U)/\lambda$. To measure $\Delta I(U)$ avoiding any parasitic ionic effects in the

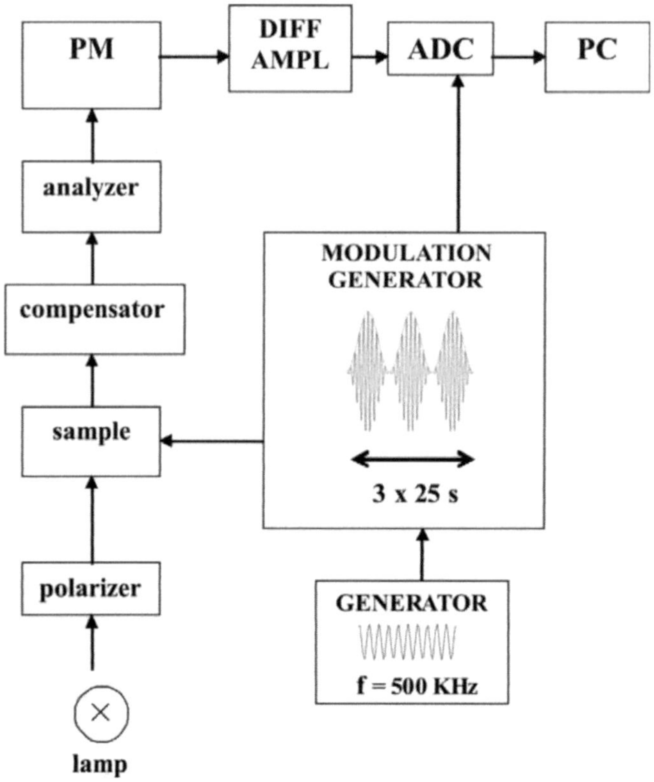


**Figure 1.** Experimental set-up

suspension, sine voltage (500 kHz) with an amplitude from $U = 5$ V up to $U = 30$ V was applied to the cell.

The amplitude of the applied voltage was modulated to almost 100% by a low frequency (0.04 Hz) sine signal from a second generator. A differential amplifier was used to subtract the DC background from the PM output signal before further amplification. Finally, both the modulation signal (used as reference) and of the optical response signal (output of the differential amplifier) were transmitted to a PC by analogue-to-digital converters (ADC) after proper filtering to avoid aliasing effects.

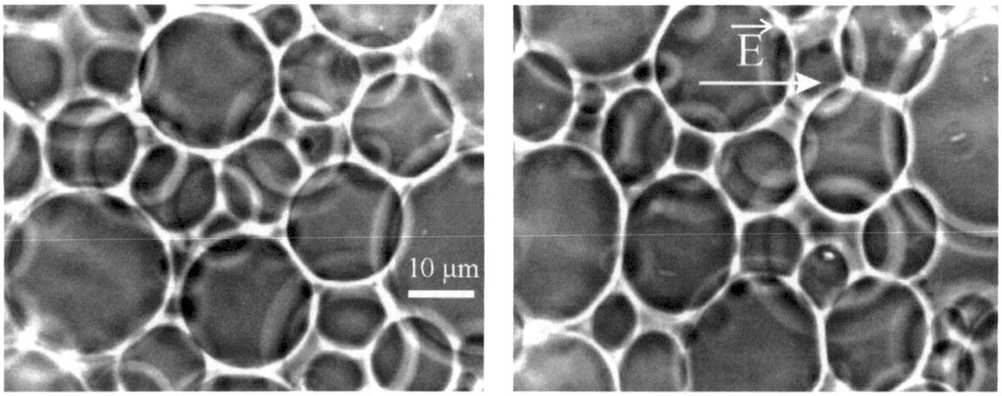


**Figure 2.** Photographs of the vesicular suspension at: a) U = 0 V; b) U = 30 V.

The homogeneity of the field **E** was achieved by the construction of the experimental cell containing the suspension. A pair of ITO electrodes was evaporated at a distance $a$ = 1 mm on each of the two glasses of the cell. An O-ring (with diameter 8 mm and cross-section 0.5mm) was used as spacer. The voltage was applied to the respective "left" and "right" electrodes on the two glasses. Thus a homogeneous lateral electric field was exerted on the vesicular suspension. Its amplitude E was assumed to be E=U/a.

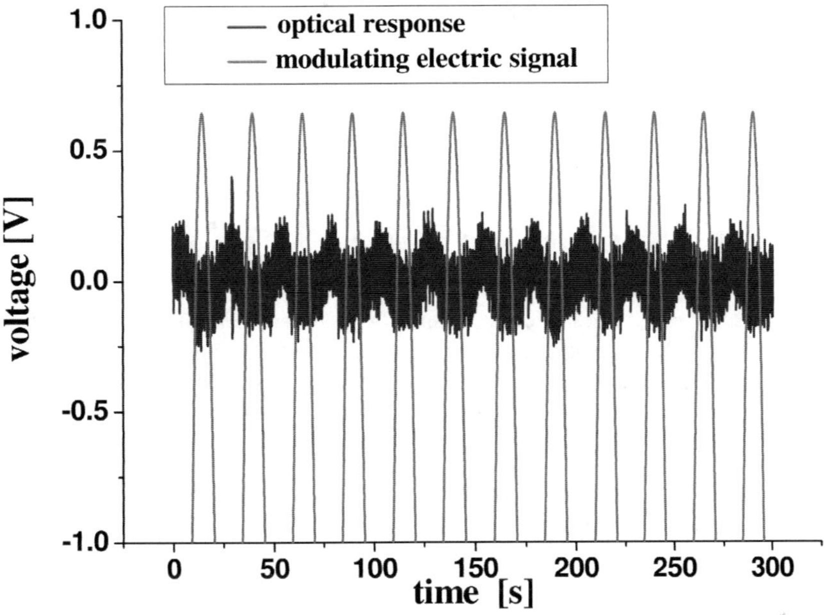


**Figure 3.** Part of the modulating signal (red) and the optical response signal (blue) shown in the time domain.

**3. Results and discussion**

Photographs of the vesicular suspension at $U$ = 0 V and at voltage $U$ = 30 V are presented in figure 2. It can be clearly seen that upon electric field the vesicles deformed into oblate rotational ellipsoids with their short axis of symmetry parallel to the electric vector **E**.

Figure 3 shows an example of the part of the modulation signal and the optical response signal as measured for 300 s. The measured magnitude of the induced optical anisotropy changed with the frequency of the modulating signal. There was slight retardation of the optical response of the suspension with respect to the electric signal possibly due to viscous effects. The period of the low frequency modulation signal was chosen long enough (25 s) to assure that the viscous effects are negligible.

The optical response of the system was decomposed into its Fourier series and the amplitude of the term corresponding to the modulating frequency was determined. A second harmonics of the optical signal was also observed (see figure 4), possibly due to the strong nonlinearity of the induced anisotropy dependence on the electric field.

The dependence of the optical anisotropy, determined from the Fourier amplitudes corresponding to the frequency of the modulating signal, on the applied field intensity is shown in figure 5. At low values of the field, the optical anisotropy $\varphi_e - \varphi_0 = \Delta I / (I_0 G)$ (where $G$ is the gain of the differential

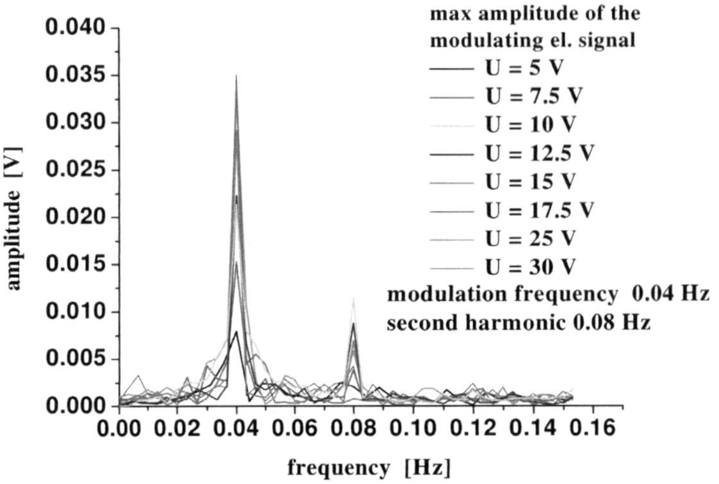


**Figure 4.** Fourier spectra of the optical response of the vesicular suspension to applied electric field. A peak corresponding to the frequency of the modulating electric signal (f=0.04 Hz) is clearly seen. Second harmonic appears as well, possibly due to the strong nonlinearity of the induced anisotropy dependence on the electric field.

amplifier) is proportional to the square of the field. At high enough fields, the anisotropy attains some maximal value and does not depend on the field any more.

One possible explanation of this phenomenon is that the eccentricity of the ellipsoid approaches its maximal value determined by the excess area of the spherical vesicle [10].

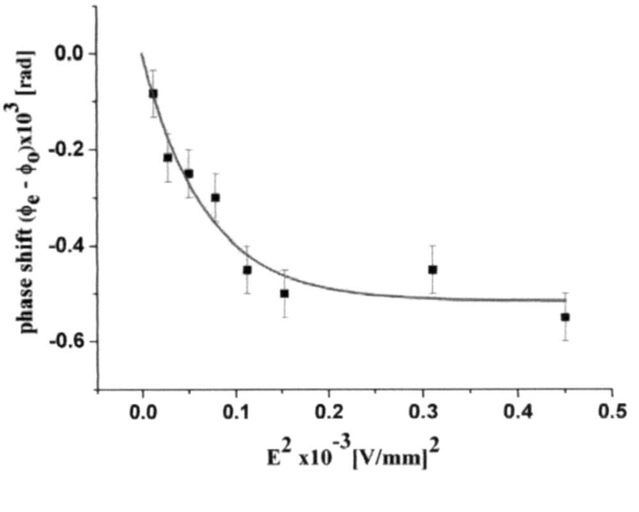


**Figure 5.** Dependence of the induced anisotropy on the squared effective electric field amplitude. The experimental data are fitted reliably with a function $f(E)=f_0\{1-exp[-(E/E_0)^2]\}$. The values of the parameters $f_0$ and $E_0$ calculated from the experimental data are $f_0$ = - 0.52 rad and $E_0$ = 8.2 V/mm.

In order to verify if there is any additional contribution to the anisotropy coming from the possible induced electric reorientation of the solution molecules (water, glucose), reference measurements with distilled water and water solution of glucose (0.22 mol/L) were carried out. No optical response was registered during both the measurements in the frames of the sensitivity of our method ($\Delta n \sim 10^{-8}$).

If the total cell thickness $d$ is assumed to contribute to the measured optical anisotropy, the maximum effective optical birefringence is estimated to be $\Delta n = \lambda(\varphi_e - \varphi_0)/(\pi d) \cong -10^{-7}$, which is in the same range as for suspensions of large unilamellar vesicles with considerably higher lipid concentrations, as reported in [3].

**Acknowledgments**

This study was supported by the National Science Fund of the Ministry of Education, Youth and Science of Republic of Bulgaria (project HTC01-121).

**References**

[1] Aranda S, Riske K, Lipowsky R and Dimova R 2008 *Biophys. J.* **95** L19
[2] Antonova K, Vitkova V and Mitov M D 2010 *EPL* **89** 38004
[3] Asgharian N, Schelly Z A 1999 *Biochim. et Biophys. Acta* **1418** 295
[4] Schelly Z A 2002 *Colloids and Surfaces A: Physicochem. and Eng. Aspects* **209** 305
[5] Richterova M, Lisy V 2003 *Laser Physics* **13** 1301
[6] Peikov V, Schelly Z A 2004 *J. Phys. Chem. B* **108** 9357
[7] Lamarque-Forget S, Martinot-Lagarde Ph, Dozov I 2001 *Jpn. J. Appl. Phys.* **40** L349
[8] Angelova M and Dimitrov D S 1988 *Progr. in Colloid and Polymer Sci.* **76** 59
[9] Politano T J, Froude V E, Jing B and Zhu Y 2010 *Colloids and Surfaces B: Biointerfaces* **79** 75
[10] Faucon J-F, Mitov M D, Meleard P, Bivas I, Bothorel P 1989 *J. Physique* **50** 2389

Journal of Physics: Conference Series **253** (2010) 012059 doi:10.1088/1742-6596/253/1/012059

# Electrical resistivity of the liquid phase of vesicular suspensions prepared by different methods

**V Vitkova[1,4], K Antonova[1], G Popkirov[2], M D Mitov[1], Yu A Ermakov[3] and I Bivas[1]**

[1] G. Nadjakov Institute of Solid State Physics, Bulgarian Academy of Sciences, Sofia, Bulgaria

[2] Central Laboratory of Solar Energy and New Energy Sources, Bulgarian Academy of Sciences, Sofia, Bulgaria

[3] Frumkin Institute of Physical Chemistry and Electrochemistry, Russian Academy of Sciences, Moscow, Russia

E-mail: victoria@issp.bas.bg

**Abstract.** Giant lipid vesicles are obtained mainly by two methods of formation: (i) electroformation and (ii) gentle hydration (spontaneous swelling). Very often the electoformation is carried out in experimental cells consisting of indium-tin oxide (ITO) coated plates as electrodes and various polymer spacers. In the present work, the influence of the ITO coatings and the polymer spacers on the electrical resistivity of the liquid medium of electroformed vesicle suspensions is examined by means of electrochemical impedance spectroscopy (EIS). Our study is intended to point out possible implications of the electroformation method, especially in cases when phenomena, related to electric properties of the vesicle membranes, are investigated.

## 1. Introduction

The experimental research in membrane biophysics widely uses giant unilamellar lipid vesicles (GUVs) as a simple model for studying the physical properties of biological membranes [1]. Together with the classical method of spontaneous swelling, the electroformation method [2] is widely applied for GUV formation. Traditionally conceived to obtain vesicles in ion-free conditions, it has been developed recently for preparation of giant lipid vesicles in physiological conditions [3,4] i.e. upon conditions of high ionic strength of the aqueous medium. At present it is widely accepted that GUV electroformation is a method of choice to produce giant vesicles also from native cell membranes assuring the preservation of the membrane compositional asymmetry and that it can be adapted to various membrane systems including synthetic lipid mixtures, natural lipid extracts, and bilayers containing membrane proteins [3,5]. It has been shown recently in the literature that the electric fields applied during electroformation can alter the chemical structure of the lipid molecules by peroxidizing lipid acyl chains, thereby modifying the phospholipid composition and material properties of the synthesized vesicles, these effects being more pronounced for poly unsaturated acyl chains [6].

There are two general types of electroformation cells most commonly used. In the first one, the electrodes are two parallel platinum wires with diameter of several hundreds of micrometers placed

[4] To whom any correspondence should be addressed.

several millimetres apart. In the second type of electroformation cells, two indium-tin dioxide-coated (ITO) plates, separated by a non-conductive spacer are used as electrodes. A possible configuration of such kind of electroformation cell is shown in figure 1. An inert material has to be used as a spacer between the two ITO-plates in order to assure stable quality of the aqueous phase during electroformation. The silicon-based two-component organic polymer polydimethylsiloxane (PDMS) [7] has gained increasing popularity and wide application in the biophysical experimental research and especially in microfluidics (soft lithography).

Good control of the composition of the suspending medium and lack of contamination of unknown origin and concentration are of major significance for the quality of vesicular suspensions. Especially ionic impurities have to be avoided when phenomena, depending on the electrolyte concentration in the aqueous phase, are being studied.

In this respect, it is important to investigate the contribution of all materials used to construct the electroformation cell on the electrochemical properties of vesicle suspensions and this constitutes the aim of the present study. The electrical resistivity of the liquid medium of the vesicular suspension is very sensitive to the presence of ionic impurities even with extremely low concentration. This is the reason why the experimental measurement of this quantity permits to verify whether the ITO coatings and the plasic spacers of the electroformation cell, as well as the lipid itself contribute to the contamination of this medium.

## 2. Materials and methods

Double-distilled water was obtained in a quartz distiller without ion-exchange filter at the outlet. Sucrose (cat. num. S7903, Sigma Ultra ®) was purchased from Sigma-Aldrich Chemie (Gremany) and D(+)-Glucose monohydrate (cat. num. 108346) – from Merck (Germany). Polydimethylsiloxane (PDMS, Sylgard 184 silicone elastomer kit) was provided by Dow Corning GmbH (Germany), and manipulated and prepared according to the manufacturer's prescriptions with base to curing agent ratio from 10:1 to 8:1.

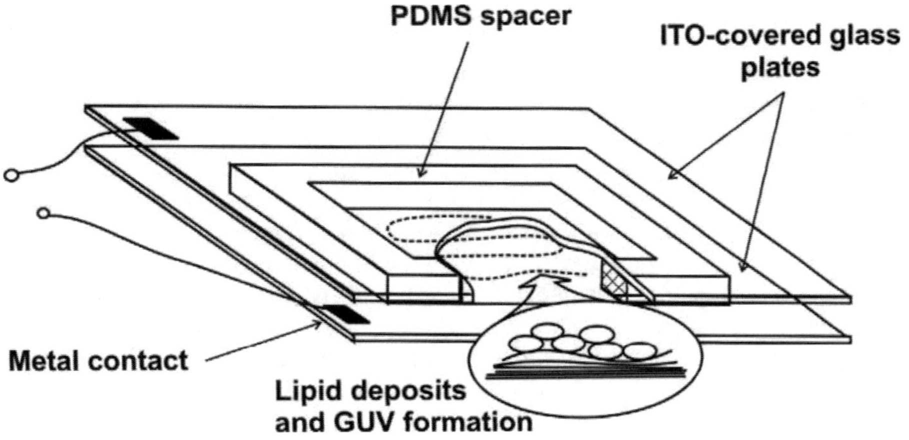


**Figure 1.** Experimental cell for electroformation – two ITO-coated plates, separated by a silicone (polydimethylsiloxane) spacer. The metal contacts between the conductive plates and the electrical wires are situated outside the cell. The used cell had thickness ~3.5 mm and area ~ 10 $cm^2$

### *2.1. Preparation of lipid coatings*

Double-distilled water, equilibrated in air, was used to prepare all the experimental solutions. Methanol and chloroform ("for analysis" grade) were purchased from Fluka Inc. (Germany). In our experiments the lipid 1-stearoyl-2-oleoyl-sn-glycero-3-phospho-choline (SOPC), cat. N° 850467 (Avanti Polar Lipids Inc., AL, USA) was used.

A typical electroformation cell is shown in figure 1. It consists of two ITO-coated plates, separated by a silicone (polydimethylsiloxane) spacer. Lipid depositions were made by the careful spreading of 50 μL of SOPC solution with concentration of 1 g/L in chloroform-methanol (9:1 volume parts) organic solvent on the ITO-side of each ITO-coated glass plates. In our experiments freshly prepared organic solution of the lipid (previously lyophilized and kept under vacuum at -20°C) was used, unless explicitly indicated. The lipid was held for at least 2 hours under vacuum until completely dry.

### *2.2. Electrochemical impedance spectroscopy*

Electrochemical impedance spectroscopy (EIS) measurements were performed with an FFT impedance spectrometer, as described earlier [8]. A broad-band voltage signal, with a small amplitude (10 mV), representing a sum of ca. 40 selected sine waves, was applied as perturbation. The measurements were performed in a two-electrode electrochemical cell configuration, either the described above electroformation cell, or a small glass cell with two Pt-electrodes each of 25 $mm^2$ surface area placed 1 cm apart. A fast potentiostat serves to interface the cell.

The impedance data quality was validated by comparing the power spectra of the perturbation voltage and the response current for each measurement, as described earlier [9].

An EIS measurement gives information about the spectrum of the sample impedance. For our analysis we use the value of the total (not the specific one) active resistance of the medium.

## 3. Results and discussion

Usually, vesicular suspensions are prepared in the manner that vesicular membranes enclose sucrose solution and the suspending medium is a glucose solution with concentration, assuring in equilibrium iso-osmolar conditions across the vesicle membrane. In order to verify if the presence of glucose (and respective ionic impurities) influences the electrical conductance of the aqueous phase we measured the resistance, $R$, of 0.2 mol/L aqueous solution of glucose and compared it with the value obtained for double-distilled water (Table 1). In the frames of the experimental precision the difference between the values of $R$ in the two cases is negligible. It can be concluded that the presence of 0.2 mol/L of sugar in the water does not influence its resistivity markedly.

In our recent study [10], a strong dependence of the electrodeformation of giant vesicles in alternating electric fields on the method the vesicles were prepared (gentle hydration, electroformation in various cells) was obtained and was ascribed as a result of an uncontrolled presence of ionic impurities in the aqueous solution. These results were confirmed by our measurements of the resistivity of vesicle suspensions (not shown) – higher resistance for suspensions, which were spontaneously swollen and lower for suspensions, electroformed in different electroformation (EF) cells using various materials as spacers. In order to reveal the sources of these differences we tried to answer the practical question if there is any contribution of the materials participating in the construction of the EF chambers, namely the ITO coating and the PDMS used to prepare the spacers.

**Table 1.** Active resistance measured by EIS at 219 Hz

| Liquid phase | R [MΩ] |
|---|---|
| Double distilled water | 0.509 |
| 0.2 mol/L Glucose solution | 0.519 |

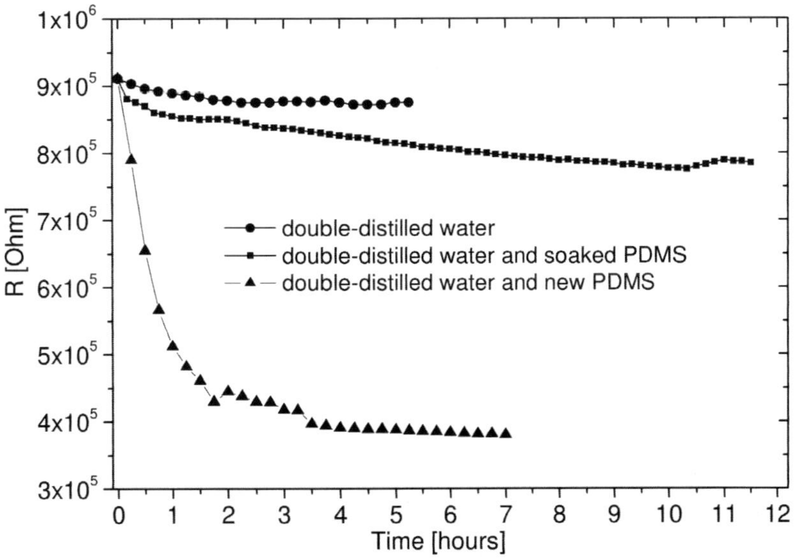


**Figure 2.** Resistance (measured by EIS; values taken at 219 Hz) of double-distilled water (●), double-distilled water with new PDMS submerged in it (▲) and double-distilled water with PDMS previously soaked for 72 hours (■). All the measurements are performed with platinum electrodes (5x5 $mm^2$) in closed glass cell.

Two types of EIS measurements were performed. The first configuration of the experimental cell consisted of a glass cell with two platinum electrodes (each of them with dimensions 5x5 $mm^2$). Small pieces of freshly prepared or soaked PDMS were added and the evolution with time of the resistance of the aqueous phase in the glass cell in presence of PDMS was measured. The results are presented in figure 2. Small fluctuations in the resistance, obtained for distilled water, are due to the temperature fluctuations during measurements. It is important to underline that before measurements double-distilled water was kept in contact with air for at least a day, i.e. carbon dioxide concentration in the water was in equilibrium with air and remained constant during measurements [11,12]. Measurable influence of sodium ions dissolved from the glass walls on the water conductivity was rejected after a control measurement of the water resistance in a quartz cell. The decrease (more than two times) of the resistance in the case of immersed freshly prepared PDMS is possibly due to emission of residual products after the polymerisation of the two-component silicone polymer PDMS. This was confirmed by EIS measurements performed after soaking the PDMS pieces for 72 hours in double-distilled water, which has been regularly changed every few hours. In this case the decrease of the water resistance in the glass cell was considerably weaker than the change measured in presence of fresh polymer, cf figure 2. This result as well as measurements with different base-to-curing agent ratios (see above) show that after the polymerisation of PDMS there are residual ionic products considerably increasing the water conductance, which can be easily washed up after long enough soaking of the fresh polymerised PDMS spacers.

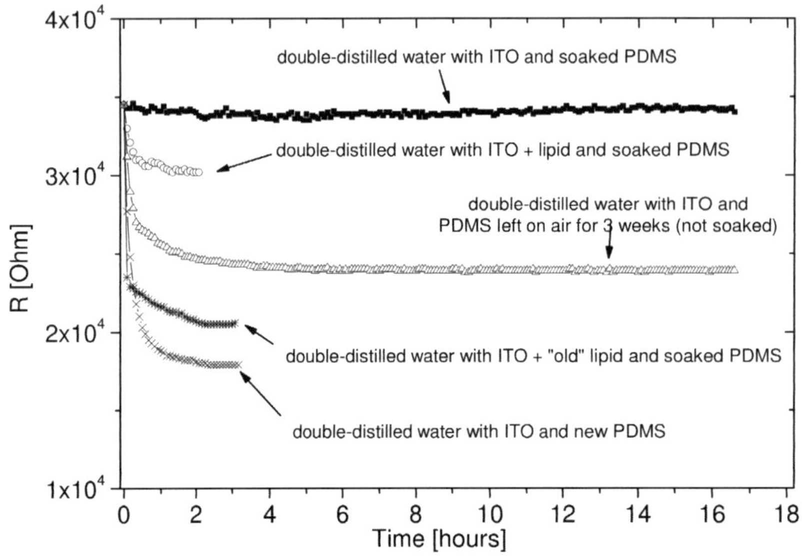


**Figure 3.** Resistance of the electroformation cell filled up with: pure water, ITO and soaked PDMS (■); water, ITO, lipid and soaked PDMS (○); water ITO, lipid and PDMS (left on air for three weeks) and not soaked (△); water, ITO, "old" lipid and soaked PDMS (✳); water, ITO and PDMS not soaked (×). Measurements performed by EIS.

These observations were confirmed in the second experimental series, where resistance measurements were performed directly in the EF cell as shown in figure 3. As known from the literature deionised water equilibrated with air (equilibrium carbon dioxide concentration reached) has specific resistance in the order of 1 MΩ.cm [12]. Our control measurement of the resistance of the EF cell filled with double distilled water and with PDMS spacer, which had been preliminarily soaked and washed, permitted to calculate the electrical resistivity of the water using the cell dimensions. Our estimation gave quite similar value to the value published in [12]. This result is an indication for the lack of emission of any undesirable ionic impurities from the ITO during the time of measurement. In order to verify if the ionic purity of commercially available lipids is detectable, we performed the same measurements but with lipid depositions on the ITO-side of each ITO-plate. The lipid depositions were prepared identically to the preparation of the lipid depositions according to the protocol described in the previous section. As it can be seen in figure 3, the presence of the new lipid film in the EF cell reduces the resistance less than 15%. This decrease is obtained to be much more pronounced if the lipid depositions were made from an "old" lipid solution (prepared a week before and kept at -20°C). This result indicates the presence of ionic impurities in the lipid stock solution due to its chemical degradation and leads us to the conclusion to recommend the usage of only freshly prepared solutions of lipids in chloroform. In the literature it has been already warned of the effects of micromolar concentrations of ion impurities, contained in the commercial lipids, if special efforts for their purification are not exerted [13].

The most considerable reduction (~ 50 %) of $R$ of the EF chamber was measured when a freshly polymerized spacer from PDMS was used (figure 3). The decrease of $R$ when the spacer has been prepared about three weeks before and meanwhile left on air was less but still important (~ 30 %).

Usually, the overall properties of vesicular suspensions obtained via electroformation in electroformation cells composed of ITO-plates and polymeric spacers are implicitly supposed to be equivalent to those of vesicles spontaneously swollen. Our experimental study is intended to draw attention to the fact that in the particular cases, described above, the use of the electroformation is susceptible to alter the vesicle properties due to the contamination of the suspending medium with uncontrolled ionic impurities, which could influence the conditions of electroformation and/or be adsorbed on the membranes or change the vesicle behaviour in certain conditions [10]. We believe that our results could be of use not only for the electroforming biophysicists but also for this part of the biophysical community working with microfuidic devices on the basis of polydimethylsiloxane.

## Acknowledgments

This study was supported by the National Science Fund of the Ministry of Education, Youth and Science of Republic of Bulgaria (project HTC01-121) and ISSP-BAS (project VK-09-05).

## References

[1] *Giant Vesicles* 2000 ed Luisi P and Walde P (Chichester: John Willey & Sons Ltd)
[2] Angelova M and Dimitrov D S 1988 *Progr. in Colloid and Polymer Sci.* **76** 59
[3] Meleard P, Bagatolli L, and Pott T 2009 *Methods in Enzymology* **465** 161
[4] Pavlic J I, Genova J, Zheliaskova A, Iglic A, Mitov M D 2010 *C R Acad Bulg Sci* **63** 497
[5] Montes L, Ahyayauch H, Ibarguren M, Sot J, Alonso A, Bagatolli L and Goñi F 2010 *Methods in Molecular biology* **606** 105
[6] Zhou Y, Berry C K, Storer P A and Raphael R M 2007 *Biomaterials* **28** 1298
[7] Mc Donald J and Whitesides G 2002 *Accounts of Chemical Research* **35** 491
[8] Popkirov G and Schindler R 1992 *Rev. Sci. Instrum.* **63** 5366
[9] Popkirov G and Schindler R 1993 *Electrochimica Acta* **38** 861
[10] Antonova K, Vitkova V and Mitov M D 2010 *EPL* **89** 38004
[11] Francis M J 2008 *J. Phys. Chem. C* **112** 14563
[12] Pashley R, Rzechowicz M, Pashley M L and Francis M J 2005 *J. Phys. Chem. B* **109** 1231
[13] Ermakov Y 2000 *Colloid J.* **62** 437

Journal of Physics: Conference Series **253** (2010) 012060 doi:10.1088/1742-6596/253/1/012060

# Conoscopic evidence of the UV light-induced flexoelectric effect in homeotropic layers of nematic liquid crystal doped with azobenzene derivatives

**Y G Marinov[1], G B Hadjichristov[1], A G Petrov[1], S Sridevi[2], U S Hiremath[2], C V Yelamaggad[2] and S K Prasad[2]**

[1] Institute of Solid State Physics, Bulgarian Academy of Sciences, 72 Tzarigradsko chaussee, Sofia 1784, Bulgaria

[2] Centre for Liquid Crystal Research, Jalahalli, Bangalore 560 013, India

Corresponding author's e-mail: ymarinov@issp.bas.bg

**Abstract**. A digitalized version of the standard method of conoscopy was employed to register the bend deformation of molecular orientation in homeotropic nematic layers caused by an in-plane applied DC electric field, and influenced by UV light illumination. Two guest-host systems prepared by mixing of a nematic liquid crystal and an azobenzene-containing photochromic liquid crystalline material featuring a longitudinal molecular asymmetry, were studied. Upon continuous UV irradiation, a photo-isomerization of the photochromic molecules occurs resulting in an enhanced flexoelectric response of the guest-host mixtures. The dependence of the photoflexoeffect on the field strength and UV light intensity was also examined.

## 1. Introduction

Flexoelectricity signifies macroscopic spontaneous polarization developed in orientationally deformed liquid crystals (LCs). It was first hypothesized in 1969 [1] for nematics with non-centrosymmetric wedge-like (or banana-like) dipolar molecules, under splay (or bend) deformation. As later established [2], quadrupole asymmetry of mesogenic molecules also leads to flexopolarization through curvature strains, making the effect universal for all LCs. In the inverse flexoeffect, curvature stresses are induced or modified by an external electric field. The flexoelectric deformations strongly depend on the boundary conditions. Of great importance are the weak anchoring and surface polarization within a few nanometers of the LC-glass interface. An interesting kind of converse flexoeffect is the flexoelectric bending in a homeotropic nematic layer. This effect takes place when a homeotropic layer is subjected to an electric field orthogonal to the nematic director [3].

Photoflexoelectric effects in photosensitive LCs and other soft materials are also known [4-6]. In particular, UV light-induced change of flexoelectric deformations has been reported for bulk thermotropic LC materials [6] and for guest-host dye-nematic mixtures [7,8]. Theory was developed for analysis of flexoelectricity in this case [8].

Conoscopy is a convenient and powerful optical technique to verify homeotropic anchoring and determine the bending of the director field in homeotropic nematic LC structures. Recently this method was successfully utilized in studying mixtures of azo dyes and LCs [9-11]. Moreover, photoinduced modification by photo-isomerization can be easily distinguished by this method. Here we report experimental results on UV light-induced photoflexoelectric effect in homeotropic nematic layer subjected to an electric field orthogonal to the director. The effect was induced by UV light illumination at 365 nm and studied by conoscopy. The influence of the magnitude of both the in-plane applied DC electric field and the intensity of the irradiated UV light on the photoflexo-response of two guest-host mixtures, was examined.

## 2. Experimental

Mixtures composed of the nematic LC 4-butyl-cyclohexane carboxylic acid 4-pentyloxy-phenyl ester (CM80 for short) and the photochromic azobenzene derivatives, 4-(4'-ethoxyphenylazo)phenyl hexanoate and 4-hexyloxybenzoloxy-4'-cyanoazobenzene, denoted hereafter as EPH and UI-147, respectively, were the subject of the present study. The chemical structures of these compounds are given in figure 1. The substances were synthesized in the Centre of Liquid Crystal Research in Bangalore, India. They were specially chosen to realize a high degree of photosensitivity as well as photoflexoelectricity caused by photo-isomerization of the azobenzene dye dopants. The LC CM80 was used as a host matrix since the compound is not UV sensitive, and exhibits negative dielectric anisotropy. The two photochromic azobenzene liquid crystalline compounds were selected considering their longitudinal molecular asymmetry. As reported elsewhere [6], this feature is essential for the flexoelectric properties of the guest-host systems studied here.

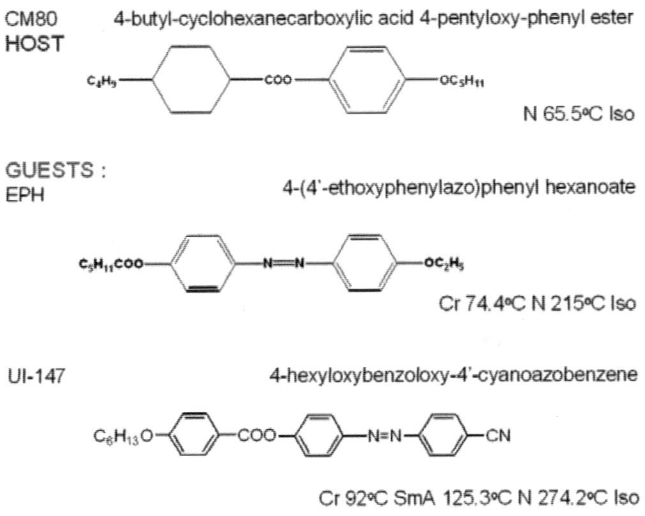


**Figure 1.** Molecular structure of the LC CM80 and the azobenzene compounds used as dopants. The clearing points of the mesogens are also given.

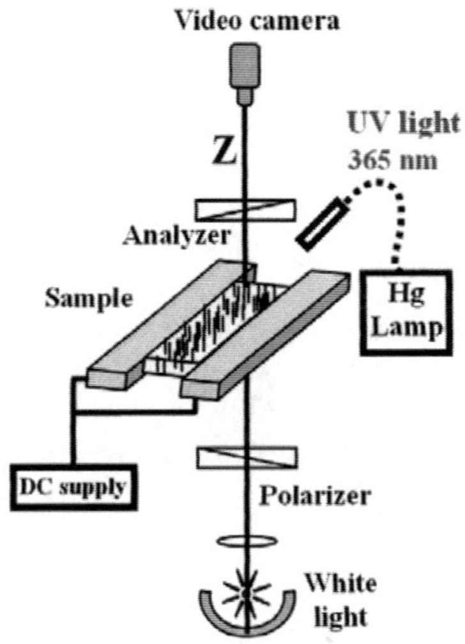


**Figure 2**. Experimental set-up for the conoscopic measurements.

Both azobenzene dyes (guests) were well soluble in the CM80 matrice (host). The mixtures were prepared at a concentration of the azobenzene guest equal to 1 wt.% in the LC CM80. The sample cell gap was fixed with 100 μm thick stainless steel spacers, also serving as electrodes. The distance between the electrodes was 2 mm. The inside surfaces of the glass plates of the cells were covered with a monolayer of silane to orient the liquid crystalline molecules homeotropically. For this purpose, a 0.032 % solution of ODS-E silane (Aldrich) in distilled water was used. By capillarity, the cells were filled with the mixtures in their isotropic phase. After filling, the samples were cooled slowly to the nematic phase to achieve a good alignment, the quality of which was checked by using a polarizing microscope (Leitz DMRXP).

To induce photoisomerization, the samples were irradiated with non-polarized UV light ŧ by using the 365 nm line of a UV spot light source (Hamamatsu LC5 type L8333-01) equipped with a quartz light guide. The distance between the output end of the light guide and the cell was 25 mm. The power of the UV light incident on the cell was measured by Hamamatsu C6080-03 UV power meter positioned at the irradiated spot in the place of the sample. The reduced (70% at $\lambda$ = 365 nm) transmission of the glass plates used for making the cells, was taken into account to arrive at the actual value of the UV light intensity (power density) absorbed by the nematic layer. A schematic diagram of the experimental setup is presented in figure 2.

For the conoscopy studies an in-plane (horizontal in the geometry shown in figure 2) electric field was applied through the in-plane electrodes to the nematic layer kept between two crossed polarizers The electric field was directed at $45^{o}$ to the polarizers. DC electric field was employed for static flexoelectric measurements. The applied DC voltage from a Keithley Electrometer (Model 6517A) created a static bend deformation that is further influenced by UV illumination.

Our digitalized version of conoscopy operates as follows: The conoscopic images of the LC azo-dye mixtures were acquired by a digital CCD camera (Optronics DEI-750D CE Digitial Output Model S60675 PAL) with a resolution of 752 (H) x 582 (V) picture elements. The pictures were taken varying the voltage in the range 20 V – 80 V, and the intensity of UV light from 1.5 mW/cm$^2$ to 2.25 mW/cm$^2$. By means of a graphical software (Image Tool for Windows, University of Texas Health Science Center), the intensity profiles across images were obtained in a direction normal to the electric field applied. In this case, both initial and final coordinates of the segment through the conoscopic isogyres were kept constant. The separation between the isogyres was determined within an experimental error of ± 1 pixel. The experiments were carried out at room temperature (25 $^{o}$C).

## 3. Results and discussion

Conoscopy is essentially an optical method that maps various directions through the LC layer by corresponding points of the conoscopic image. It employs an observation in strongly convergent light of the condenser, by using the microscopic objective lens, and a short focal length Bertrand lens. The conoscopic figure depends on the LC orientation with respect to the microscope stage. A Maltese extinction cross is commonly observed for uniaxial nematics oriented perpendicular to the stage. In such an orientation, a flexoelectric bend deformation [3] developed in the nematic layer under electric field applied in-plane to the layer, can induce a second macroscopic optical axis. As a result, a splitting of the extinction cross into two isogyres takes place. The changes in isogyres' separation induced by variation in director field bending owing to DC electric field applied can be measured as a function of the magnitude $E$ of this electric field.

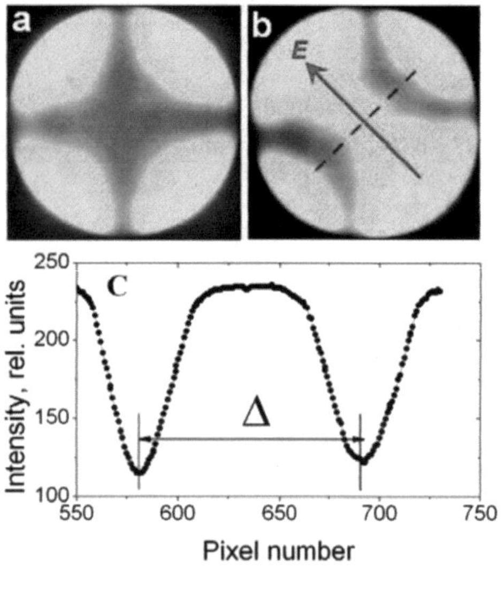


**Figure 3.** Conoscopy images illustrating the effect of flexoelectric bending in a homeotropic layer: (a) without electric field; (b) the layer is subjected to an electric field $E$ orthogonal to the director; (c) the intensity profile measured along the dashed line in picture (b).

Under orthoscopic illumination (parallel beam), in the initial field-off state the light is extinguished between crossed polarizers P ⊥ A at all angles of rotation of the samples and in the convergent beam optics of conoscopy a uniaxial cross is displayed (figure 3(a)). Applying a small in-plane DC field *E* and between diagonally crossed polarizers P(45°)/A(135°), a uniform birefringence with maximum transmission is observed. Correspondingly, the conoscopic observation (figure 3(b)) shows a splitting of the isogyres (Δ) that is a measure of the effect of flexoelectric bending. As an example, figure 3(c) presents a graph of the intensity profile measured in this case.

Figure 4 reports conoscopic interference images of homeotropic nematic layers of the two examined mixtures subjected to DC electric field and UV light illumination. The splitting of the extinction crosses due to variation of the flexoelectric bending of the nematic layer with the applied electric field, is clearly seen in the pictures. It should be noted that no hydrodynamic motion is seen up to 100 V (the maximum voltage applied).

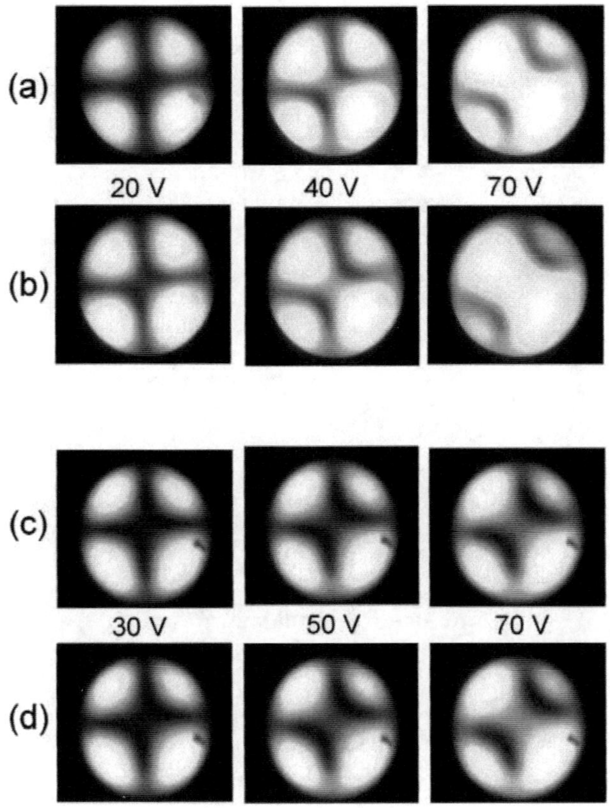


**Figure 4.** Conoscopic images of the mixtures of CM80 with 1 wt.% EPH (a,b) and CM80 with 1 wt.% UI-147 (c,d) upon increasing electric field: (a,c) no UV light illumination; (b,d) upon UV light ($\lambda$ = 365 nm) with intensity 1.9 mW/cm$^2$.

Figure 5 displays, for the two mixtures, the separation Δ between the two isogyres as obtained from the intensity profiles measured from micrographs taken in a sequence varying the applied voltage of the electric field. It is seen that Δ increases (progressive splitting of the extinction cross) upon voltage increase. This is in accordance with the theory of direct flexoelectric effect [12]. According to [3], for weak anchoring the produced curvature (the first derivative of electric field-dependent angle θ of the nematic director toward Z) is described by:

$$\frac{\partial\theta}{\partial z} = \frac{(e_{3x} + m_p)E}{K_{33}}, \tag{1}$$

where $e_{3x}$ is the bend flexoelectric coefficient, $m_p$ is the surface polarization and $K_{33}$ is the bend elastic constant of the nematic. In the case of identical anchoring at both boundaries of the nematic layer, this expression implies that the curvature is uniform bend deformation proportional to $E$. As a result, the obtained bend deformation of the molecular director field produces a macroscopic optical axis in the samples along $E$. Stronger the field, more pronounced will be the optical axis. Thus, our observations are consistent with analogous conoscopy investigation of piled uniaxial crystal [13,14].

In angular units, the splitting Δ should approximate (be less or equal to) the difference of the two boundary angles, i.e. the quantity given by Eq. (1) multiplied by the layer thickness. This difference is a measure of the field-induced birefringence of the layer. If we imagine the layer as splitted in two horizontal parts which are homogeneously tilted in different directions, the above mentioned difference equals the angle between the optical axes of the two parts. Therefore, the isogyres' separation Δ should be linearly related to $E$. Indeed, such a behavior is present for the experimental data in figure 5. Also, both dependencies have to include the zero point (more clearly seen in figure 5(a)).

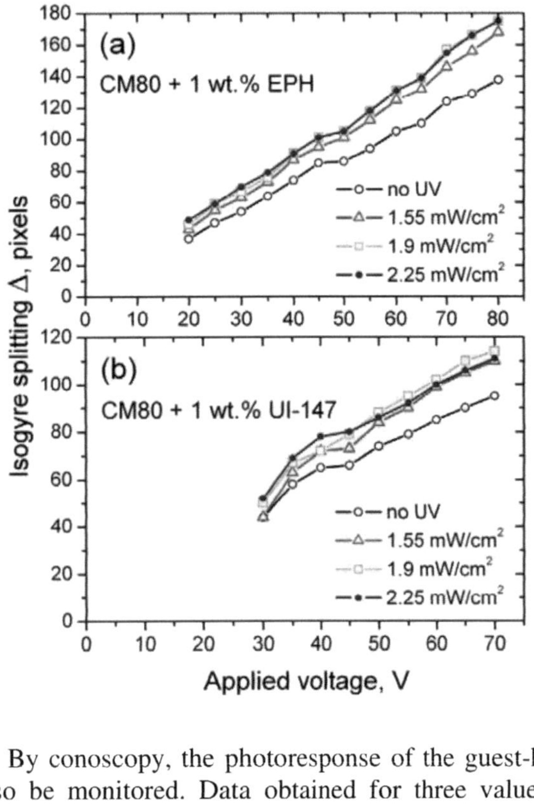


**Figure 5.** Separation of the isogyres (Δ) for the mixtures of CM80 with 1 wt.% EPH (a) and CM80 with 1 wt.% UI-147 (b) as a function of the applied DC voltage, upon UV light illumination (λ = 365 nm) at three values of the light intensity.

By conoscopy, the photoresponse of the guest-host mixtures upon continuous UV irradiation can also be monitored. Data obtained for three values of the UV light intensity (figure 5) shows an increase in Δ with increasing UV light intensity up to 2.25 mW/cm$^2$. A saturation of the flexoelectric bending effect takes place at this UV light intensity. Even at higher UV light power density, the observed conoscopy figures are still symmetric. This in turn means that the samples are dielectrically stable. It is of importance to note that the mixtures have a negative dielectric anisotropy under considered experimental conditions. Regarding Eq. (1) the enhancement of the curvature of the photoinduced nematic bend deformation is related to a possible modification of the bend flexocoefficient $e_{3x}$. or surface polarization $m_p$. Analysis of this effect is planned as a future work.

The conoscopic response of the two mixtures clearly shows an enhancement of flexoelectric deformation by UV illumination, i.e. the presence of a photoflexoeffect, obviously due to photo-isomerization of the guest molecules of azobenzene dyes. Under UV illumination, the photochromic subsystem in the examined gust-host mixtures undergo a transformation from *trans* conformers (rod-like forms) to *cis* forms having a pronounced banana shape. Owing to this geometry change, and thereby the orientation of the guest molecules within the host matrix of LC CM80, the bend flexocoefficient $e_{3x}$ of the guest-host mixtures is also modified. This photo-modification is directly reflected in the conoscopic response of the samples. The difference in the magnitude of the splitting observed for the two mixtures could be ascribed to the different efficiency of the guest dye compounds to produce conformers.

## 4. Conclusion

Homeotropically oriented samples of two guest-host liquid crystilline materials being mixtures of a small content of photochromic azobenzene dye in a photoinactive nematic were examined by digital conoscopy upon application of a DC electric field parallel to the layers in the absence of, and under UV light illumination conditions. The results obtained by this optical technique give evidence for a UV light-induced flexo-electrical effect in the mixtures driven by *trans–cis* photoisomerization of the azobenzene dopant possessing asymmetric molecular structure.

## Acknowledgments

This study was carried out as a part of Indo-Bulgarian joint research project (No. INT/BUL/B-75/07) supported by the Department of Science and Technology, New Delhi and the Ministry of Education, Youth and Science, National Science Fund of Bulgaria, contract BIn-5/07.

## References

[1] Meyer R B 1969 *Phys. Rev. Lett.* **22** 918
[2] Prost J and Marcerou J P 1977 *J. Physique (France)* **38** 315
[3] Helfrich W 1971 *Phys. Lett. A* **35** 393
[4] Petrov A G, Shonova N, Komitov L, Naydenova S, Stoylova S, Petkova J and Spassova M 1993 *20 Years Jubilee Proceedings of Institute of Solid State Physics* (Sofia, Bulgaria: ISSP)
[5] Spassova M, Petrov A G and Fendler J H 1995 *J. Phys. Chem.* **99** 9485
[6] Petrov A G 1999 *The Lyotropic State of Matter* (New York: Gordon and Breach)
[7] Hermann D S, Rudquist P, Ichimura K, Kudo K, Komitov L and Lagerwall S T 1997 *Phys. Rev. E* **55** 2857
[8] Marinov Y, Shonova N, Blinov L M and Petrov A G 1998 *Europhys. Lett.* **41** 513
[9] Kawata Y, Takatoh K, Hasegawa M and Sakamoto M 1994 *Liq. Cryst.* **16** 1027
[10] Yeh P and Gu C 1999 *Optics of Liquid Crystal Displays* (New York: John Wiley & Sons) Chapter 4.5. Conoscopy 139
[11] Ruslim C, Komitov L, Matsuzawa Y and Ichimura K 2000 *Jpn. J. Appl. Phys.* **39** L104
[12] Petrov A G, Mitov M and Derzhanski A 1978 *J. Phys. (Paris)* **39** 273
[13] Vasilevskaya A S and Sonin A S 1979 *Sov. Phys. Solid State* **21** 114
[14] MacGregor A R 1990 *J. Opt. Soc. Am. A* **7** 337

Journal of Physics: Conference Series **253** (2010) 012061 doi:10.1088/1742-6596/253/1/012061

# Influence of single-walled carbon nanotubes (< 0.001 wt %) and/or zwitter-ionic phospholipid (SOPC) surface layer on the behaviour of the gradient flexoelectric and surface induced polarization domains arising in a homeotropic E7 (a mixture of 5CB, 7CB, 8OCB and 5CT) nematic layer

**H P Hinov[1], J I Pavlič[2,2'], Y G Marinov[1], A G Petrov[1], S Sridevi[3], P M Rafailov[1] and U Dettlaff-Weglikowska[4]**

[1]Georgi Nadjakov Institute of Solid State Physics, Bulgarian Academy of Sciences, 72 Tzarigradsko chaussee blvd, 1784 Sofia, Bulgaria
[2]Faculty of Electrical Engineering, University of Ljublana, Tržaška 25, 1000 Ljubljana, Slovenia
[2']Faculty of Health Sciences, University of Ljubljana, Zdravstvena pot 5, 1000 Ljubljana, Slovenia
[3]Centre for Liquid Crystal Research, Jalahalli, Bangalore 560 013, India
[4]Korea University, Anam-Dong, Seonbuk Gu, Seoul 136-713, Korea

Corresponding author's e-mail: hinov@issp.bas.bg

**Abstract**
The influence has been studied of single-walled carbon nanotubes with a concentration between 0.0001 and 0.001 wt % and a dried zwitter-ionic phospholipid (SOPC: l-stearoyl-2-oleoyl-sn-glycero-3-phosphatidylcholine) layer of thickness, smaller than 0.5 μm, deposited only on a half of one of the two glass plates, on the behaviour of the gradient flexoelectric and surface polarization induced domains arising in a homeotropic nematic E7 (a mixture of 5CB, 7CB, 8OCB and 5CT) layer. We have observed for the first time different polar on/off formation of the surface polarization induced domains in the region of the liquid crystal cell without surface deposited lipid SOPC layer. On the other hand, the SOPC layer strongly decreases the gradient of the electric field thus leading to less-pronounced flexoelectric domains. However, the SOPC layer does not influence the creation of surface polarization induced domains and of injection induced domains arising at voltages above 4V. Appropriate dynamic light transmitted curves have been recorded and typical microphotographs have been taken.

Journal of Physics: Conference Series **253** (2010) 012061 doi:10.1088/1742-6596/253/1/012061

## 1. Introduction

After the discovery of piezoelectricity (flexoelectricity) in nematics (Ns) in 1969 by Meyer [1], many kinds of flexoeffects have been observed and studied. Among them, flexo-dielectric walls deserve a special attention [2, 3. and the citations therein]. The cross-like domains arising in a DC electric field can be observed in Ns with a small negative dielectric anisotropy ($\Delta\varepsilon < 0$)[2] and they have been interpreted as flexo-dielectric walls arising in a non-homogeneous electric field. However, cross-like domains can also be observed in Ns with a positive dielectric anisotropy ($\Delta\varepsilon > 0$) interpreted as flexo-surface polarization-induced domains [4, 5]. In this letter, we will focus our attention on the behaviour of cross-like domains in the nematic E7 with a positive dielectric anisotropy in the presence of a very small quantity of single-walled carbon nanotubes (SWCNTs) ($10^{-4}$ to $10^{-3}$ wt %) and a thin SOPC lipid layer of a thickness below 0.5 μm, deposit on half of one of the glass plates forming the sandwich liquid crystal (LC) cell (see Figure 1). The carbon nanotubes occur in two general morphologies: single-walled (SWCNTs) and multi-walled carbon nanotubes (MWCNTs). Lynch and Patrick [6] have considered for the first time the behaviour of mixtures of LCs and carbon nanotubes. Although many authors have examined the interactions between the LCs, chiefly the nematics, and the carbon nanotubes (see, for instance, the review paper [7]) including the Freedericksz transition, the electrohydrodynamics, etc., according to our knowledge, there are no papers to study the flexoelectric properties of the mixtures. The most intensively studied mixture is E7-CNTs.We have used the same mixture. For the first time we have deposed on one of the glass plates forming a sandwich LC cell a SOPC lipid layer (formally a thick lecithin layer with a thickness smaller than 0.5 μm). We know only several papers dealing with the orienting action of one-molecular lecithin layer on the Ns [8, 9]. Sometimes, one uses a very thin lipid layer with a thickness of several tens of nanometers deposited on glass plate to create giant liposomes by electric field [10, 11].

## 2. Experimental procedure: materials, technique, performance

We have used HIPco SWCNT synthesized by decomposition of CO over Fe catalyst (the so-called gas-phase catalytic method) [7]. The SWCNTs are with average diameter of around 1 nm and length of 1 μm. The nematic used was the well-known eutectic LC mixture E7 with a positive dielectric anisotropy, (purchased from Merck Eurolab (Darmstadt, Germany) containing 51 wt % of 4-cyano-4'-n-pentyl-bipheny (5CB), 25 wt % of 4-cyano-4'-n-heptyl biphenyl (7CB), 16 wt % of 4-cyano-4'-n-oxyoctyl-biphenyl (8OCB) and 8 wt % of 4-cyano-4"-n-pentyl-p-terphenyl (5CT. The available initial concentration of SWCNT was only approximately 0.01 wt %. It was diluted further by adding more E7 and the dilution procedure has been used to achieve very dilute solution down to the interval $10^{-4}$ to $10^{-3}$ wt %. We saw under the polarizing microscope that the miscibility between the nematic E7 and SWCNTs with a concentration below $10^{-3}$ wt % is very good. Furthermore, the temperature $T_c$, showing the nematic-isotropic phase transition, decreases slightly (60.1$^0$C for the pure nematic E7 versus 59.8$^0$C for the E7/SWCNTs mixture). Further, the dielectric anisotropy of the pure nematic E7 and of the E7/SWCNTs mixture have been measured and we can conclude that the dielectric anisotropy of the nematic E7 does not considerably change at minute amounts of carbon nanotubes [12, 13]. According to our previous measurements, the traces of carbon nanotubes do not change the value of the elastic constants of bend and splay but can increase the value of the surface energy stiffness constant $W_s$ (see also [14] for the case of other nematics). The E7-SWCNTs mixture was introduced by capillary forces in a previously prepared sandwich cell made by two glass plates coated with conductive layer deposed on one side and two polymer sheets with a thickness of 100 μm. Additional two metallic folders ensured the mechanical stability of the LC cell. The SOPC lipid (l-stearoyl-2-oleoyl-sn-glycero-3-phosphatidylcholine) was deposited in the following way: the glass plates were washed by chloroform (twice). The lipid film was deposited on one glass plate, on half of its side (total volume ~ 50 μl) and dried under vacuum in exicator

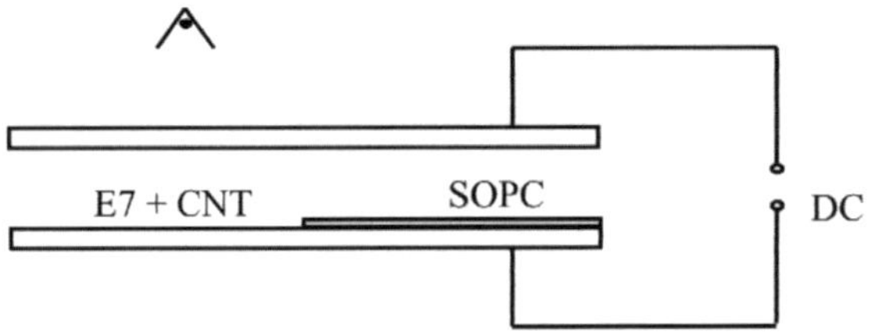


**Figure 1.** Schematic illustration of the experimental cell.

for 15 min, the cell was assembled and left overnight in air before the start of the experiment. It was performed by using polarizing microscope, photomultiplier and X-Y recorder. Simultaneously, we have made a number of microphotographs showing the details in the LC orientation.

### 3. Experimental results

The dynamic electro-optic curves shown in Figure 2 indicate that the carbon nanotubes influence the process of the formation of the gradient flexoelectric domains and of the surface polarization induced domains only in the part of the cell without SOPC lipid layer (the left-hand part of the cell shown schematically in Figure 1). More precisely, the carbon nanotubes decrease the non-homogeneity of the electric field by catching part of the ions and on reaching the electrodes they decrease the surface polarization too (see the comments made by Rahman and Lee [7]). The flexoelectric domains and the surface polarization induced domains are shown in Figure 3. The latter domains are more clear and have

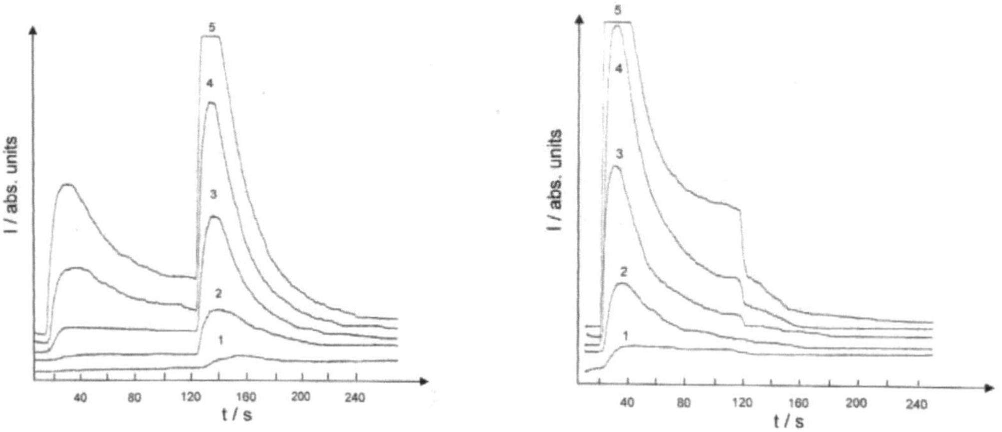


**Figure 2.** Electro-optic dynamic curves in the region of the LC/SWNTs cell without SOPC lipid layer at switch on and switch off of the applied DC voltage as follows: curves 1: 0.5 V; curves 2: 1 V; curves 3: 1.5 V; curves 4: 2 V, and curves 5: 2.5 V. The polarity of the applied DC voltage is as follows: left-hand curves: the minus (-) of the DC voltage is applied on the upper glass plate from the side of the observer; right-hand curves: the plus (+) of the DC voltage is applied on the upper glass plate. Room temperature, LC cell is with a thickness of 100 μm.

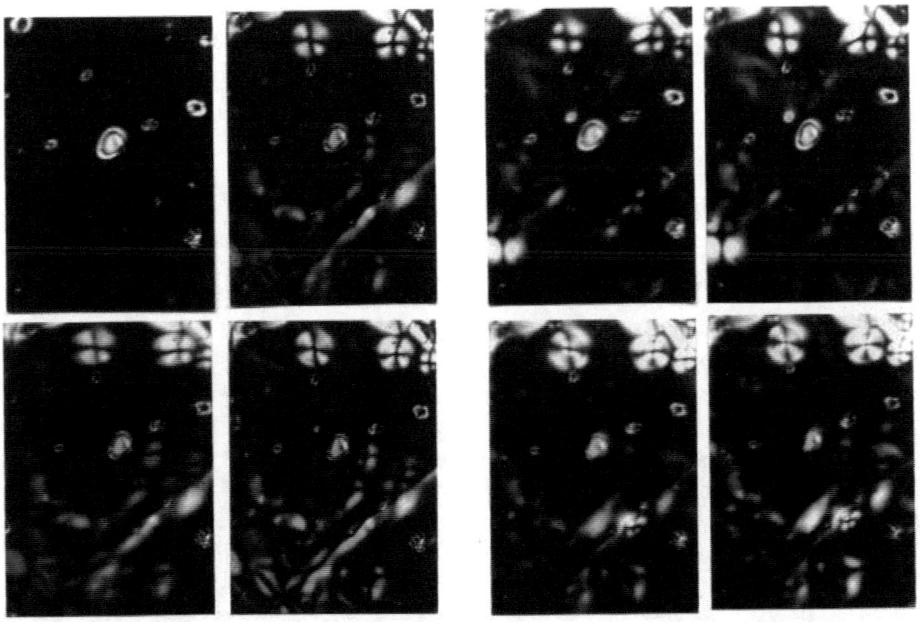

**Figure 3.** Flexoelectric domains and surface induced polarization domains in the part of the LC/SWCNTs without a SOPC lipid layer. The left-hand 4 microphotographs correspond to 0 V, 2 V, 3 V and 4 V when the minus (-) of the DC voltage is applied to the upper glass plate. The right-hand 4 microphotographs correspond to 1 V, 2 V, 3 V, and 4 V when the plus (+) of the DC voltage is applied to the upper glass plate. The exposition time was 1 or 2 minutes. Room temperature, LC cell is with a thickness of 100 μm. The long side of each photograph measures 760 μm.

cross-like form; part of the gradient flexoelectric domains are also with a circular form but they are less clear. Let us note that the polarization domains appear at the anode and they are more visible from the position of the observer. We saw an interesting effect when the (-) of the DC voltage is applied on the upper glass plate. After the switch off of the DC voltage the domains become clearer. We suggest that the domains enhance themselves due to the inverse electric field created by the accumulation of ions. All deformations arise in accordance with the results described in[2-5]. In the part of the cell with SOPC lipid layer (right-hand part of the cell shown schematically in Figure 1), we saw very weak gradient flexoelectric deformations at lower voltage of 2.5 V (see Figure 5, the first photo on the left). The dynamic electro-optic curves shown in Figure 4 point out that the carbon nanotubes do not influence the process of the formation of the surface polarization induced domains. They obviously show that the surface induced polarization domains arise at 4 V (the (-) DC voltage is applied on the upper glass plate), which at a higher voltage were transformed into injection domains. The SOPC lipid layer inhibits the formation of the gradient in the electric field along Z, the normal to the glass plates in that part of the LC cell. However, a small part of the SOPC lipid layer had been dissolved in the LC and ensured the formation of a homeotropic nematic E7 layer.

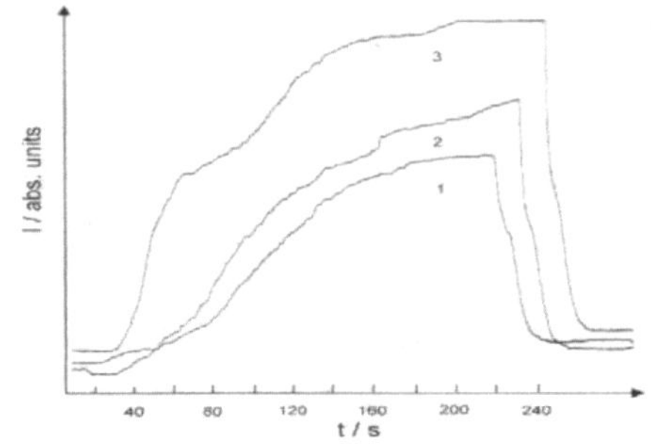


**Figure 4.** Electro-optic dynamic curves in the region of the LC/SWNTs cell with SOPC lipid layer at switch on and switch off of the applied DC voltage as follows: curve 1: 4 V; curve 2: 5 V; curve 3: 6 V. The minus (-) of the DC voltage is applied on the upper glass plate from the side of the observer. Room temperature, LC cell is with a thickness of 100 μm.

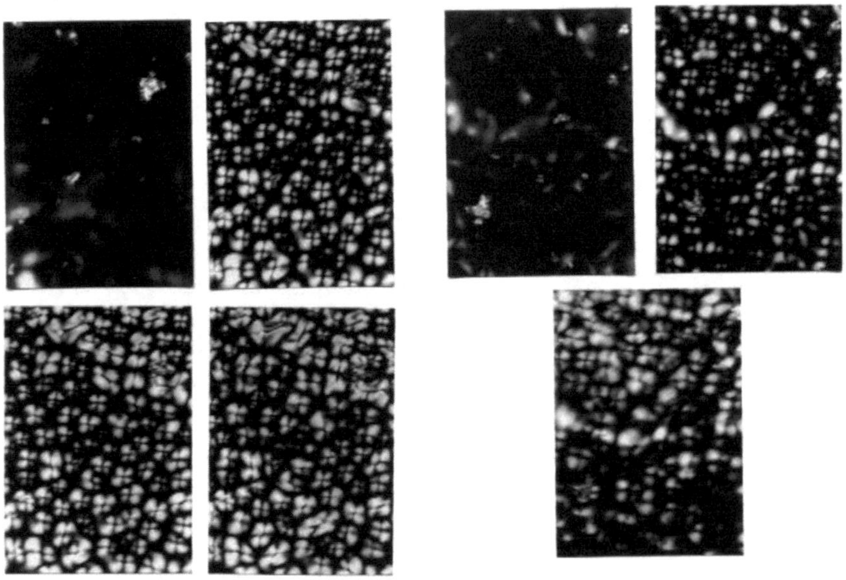

**Figure 5.** Flexoelectric deformations, surface induced polarization and injection domains in the part of the LC/SWCNTs with a SOPC lipid layer. The left-hand 4 microphotographs correspond to 2.5 V, 4V, 5 V and 6 V when the minus (-) of the DC voltage is applied to the upper glass plate. The right-hand 3 microphotographs correspond to 4 V, 5V, and 7 V when the plus (+) of the DC voltage is applied to the upper glass plate. The exposition time was 1, 2 or 3 minutes depending on the brightness of the image. Room temperature, LC cell is with a thickness of 100 μm. The long side of each photograph measures 760 μm.

Let us stress that without the SOPC lipid layer it was not possible to achieve a homeotropic orientation of the LC and one needs a special treatment of the glass plates in order to insure such an orientation. The change of the polarity of the DC voltage only raises the voltage for the corresponding pictures (see Figure 5, right-hand three photos).

## 4. Conclusions

1) In the part of the cell without SOPC lipid layer we have observed (at threshold around 1 V) the formation of both surface polarization induced domains (the more clear cross-like domains) and gradient electric field induced flexoelectric domains (the less visible deformations). The movement of the carbon nanotubes decreases the non-homogeneity of the electric field catching part of the ions, and decreases the surface polarization when the carbon nanotubes reach the electrodes. The surface polarization induced domains are more clear when the (+) of the applied DC voltage is on the upper glass plate, from the side of the observer. The change of the polarity decreases the brightness of the domains, which however significantly increases at the switch off of the voltage and then gradually decreases. In other words, we have observed for the first time an on-off polar effect.

2) In the part of the cell with the SOPC lipid layer, the non-homogeneity of the electric field is weak and the carbon nanotubes do not influence the dynamics of the domains. We have observed only surface polarization induced domains (4V or 5V depending on the sign of the polarity which were transformed into injection domains).

## Acknowledgments

This study was carried out as a part of the Indo-Bulgarian joint research project (No BIn-5/07(Bulgaria), B-71 (India)) supported by the Ministry of Education, Youth and Science, The National Science Fund of Bulgaria and the Department of Science and Technology, New Delhi, India.

## References

[1] Meyer R B 1969 *Phys. Rev. Lett.* **22** 918
[2] Hinov H P and Vistin' L K 1979 *J. Phys. (Paris)* **40** 269
[3] Hinov H P, Bivas I , Mitov M D, Shoumarov K and Marinov Y 2003 *Liq. Cryst.* **30** 1293
[4] Lavrentovich O D, Nazarenko V G, Sergan V V and Durand G 1992 *Phys. Rev. A* **45** R 6969
[5] Nazarenko V G, Klouda R and Lavrentovich O D 1998 *Phys. Rev. E* **57** R 36
[6] Lynch M D and Patrick D L 2002 *NanoLetters* **2** 1197
[7] Rahman M and Lee W 2009 *J. Phys. D: Appl. Phys.* **42** 063001
[8] Hiltrop K and Stegemeyer H 1978 *Ber. Bunsenges. Phys. Chem.* **82** 884
[9] Hiltrop K and Stegemeyer H 1981 *Ber. Bunsenges. Phys. Chem.* **85** 582
[10] Estes D J and Mayer M 2005 *Colloids & Surfaces B* **42** 115
[11] Angelova M I and Dimitrov D 1986 *Faraday Discuss. Chem. Soc.* **81** 303
[12] Scalia G, Lagerwall J P F, Schymura S, Haluska M, Giesselmann F and Roth S 2007 *Phys. Stat. Sol. (b)* **244** 4212
[13] Raynes E P, Tough R J A and Davies K A 1979 *Mol. Cryst. Liq. Cryst. Lett.* **56** 63
[14] Lu Shin-Ying and Chien Liang-Chy 2008 *Optics Express* **16** 12777

Journal of Physics: Conference Series **253** (2010) 012062 doi:10.1088/1742-6596/253/1/012062

# The electroconvection in nematic liquid crystals with short range smectic C order and negative electroconductivity anisotropy

**B Katranchev, H Naradikian, E Keskinova and M Petrov**

Institute of Solid State Physics, Bulgarian Academy of Sciences, 72 Tzarigradsko Chaussee Blvd., 1784 Sofia, Bulgaria

E-mail: mpetrov@issp.bas.bg

**Abstract**. The electroconvective instability (ECI) in nematic liquid crystals (NLCs), appearing beyond smectic C at cooling known as N with short range smecic C order, presents an unique electrooptical phenomenon due to generation and frustration of the smectic layering, which manifests inside the N temperature range. We found that the response time in this material is about an order shorter than those in the classical nematics. For the first time in NLCs were detected harmonics higher than fourth order. A twist type instability is suggested for this material.

## 1. Introduction

Anisotropic, uniaxially symmetric fluid, like nematic liquid crystals (NLCs) which have a locally preferred direction described by the director field $\boldsymbol{n}(\boldsymbol{r},t)$, with $\boldsymbol{n}^2=1$, are rich in pattern forming phenomena [1-4]. Under standard conditions, nematic LCs of positive conductive ($\sigma_a=\sigma_{||}-\sigma_{\perp}>0$), negative dielectric ($\varepsilon_a=\varepsilon_{||}-\varepsilon_{\perp}<0$) ($||$ and $\perp$ mean parallel and perpendicular to the nematic director $\boldsymbol{n}$) anisotropies and planar orientation (the director is along the $x$ direction of the LC cell's plane $xy$), the system in the presence of *d.c.* or *a.c.* electric field, applied in z direction, turns into electroconvection (EC), appearing like normal rolls (NRs) pattern with wave vector $\boldsymbol{q}$ parallel to the equilibrium director orientation $\boldsymbol{n}_o\equiv x$. This EC is well described by classical one-dimensional 1D Carr–Helfrich (CH) theory [5], but it is not applicable for the description of the more complicated 3D EC patterns. Recently a lot of papers were published [2-4], aiming to introduce more clarity in the scope of combinations of the sign and magnitude of $\varepsilon_a$ and $\sigma_a$, LC geometry (from planar to homeotropic). The EC characterizes with diversity of pattern manifestations, thus it was divided into two categories, standard and non-standard EC (sEC and nsEC) respectively depending on the possible EC driving mechanism (CH or other one still unknown). Till now a justification of the driving mechanism for the most of the nsECs does not exist.

Some experimental investigations concerning the measurement of the director's twist azimuthal angle in the xy plane and the trends of threshold field on the temperature as well as on the frequency could throw a new light in the understanding of the complicated field of the EC driving mechanisms. The nematic with short range smectic C ($S_c$) order is a suitable LC system where the twist angle variation by electric field parameters and the temperature reveals the EC mechanism in these materials i.e. the twist instability mechanism. The instability in the N with short range smectic order is strongly influenced by the generation and frustration of the smectic layers inside the N phase. Furthermore, the smectic order fluctuations in the Ns, where quasi-smecic supramolecular N complexes form undoubtedly influenced the character of the EC instability, and provoke inversion of the sign and magnitude of $\varepsilon_a$ and $\sigma_a$ as a result of the easy distribution of the charges in the layer's planes.

We will analyze the results of the 9th homologue of p,n-alkyloxybenzoic acid (9OBA), which is Ns with short range $S_C$ order built by hydrogen bonded in dimers molecules. Using both microtextural-polarization and the laser diffraction method at normal and oblique illumination, we will put into practice a real-time quantitative access to the electric field induced director field amplitude variations. Furthermore, analyzing the threshold and polarization characteristics of the patterns, their decay from EC to the ground non-excited state and the existence of the Fourier modes with frequencies harmonic to the electric field frequency, we will test the possible EC mechanism of this LC material.

## 2. Experimental results and discussion

4,n-nonyloxybenzoic acid (9OBA) display N phase between 118°C and 146°C and $S_C$ between 118°C and 92°C. The substance was filled between two parallel glass plates whose surfaces express xy plane and are coated with transparent electrodes, indium tin oxide (ITO) uni- directionally rubbed in x direction providing initial director orientation $\boldsymbol{n}_o$. The cell thickness d=12μm.

The temperature of the sample was varied with a rate of 0.1 °C $min^{-1}$ by a hot stage temperature controller Linkam TMS 90. An *ac* field $E=(0,0,E_z)$ of frequency f was applied across the sample. The EC patterns in the *xy* plane were observed by use of a video camera Hitachi and by a microscope Zeiss NU2.

The analysis of EC instability in 9OBA requires a description of the $\varepsilon_a$ and $\sigma_a$ temperature variation within the N phase. For 9OBA $\varepsilon_a$ varies between 0.03 and 0.011 at temperature variation between 146 and 118°C. The maximum value of $\varepsilon_a$ is 0.04 at 129°C and just at $T_{NC}$ the $\varepsilon_a$ sign sharply change from positive to negative. So $\varepsilon_a$ is positive but small as the average value in the N region is $\varepsilon_a \approx 0.013$ [6-8].

The $\sigma_a$ variation in the N temperature range of 9OBA is as follows: from $-0.4x10^{-11}(\Omega^{-1}cm^{-1})$ at 144°C it decrease to $-0.5x10^{-11}(\Omega^{-1}cm^{-1})$ at 118°C, i.e always negative in the nematic temperature range. An weak anomaly of $\sigma_a$ one notes at 129°C, which we mark out with $T^*$.

We emphasize that the electric behaviour of 7th and 8th homologues of p,n-alkyloxybenzoic acid differs from that of the 9th homologue, since their electroconductivity anisotropies are positive, while in 9OBA $\sigma_a<0$ in the entire N region. This could be assign to the longer molecule of 9OBA, providing easy realization of compact (co-ordination of the monomers, closed and open dimers fluctuations in space and time with trend to generate quasi-smectic layer) smectic like supramolecular complexes, still below $T_{NI}$. It is a result of the decrease of $\sigma_{||}$, with respect to $\sigma_{\perp}$, providing more easy charge distribution within the grown smectic layer. This effect depends on the tilt angle ω (the angle between the layer's normal **N** and the director), as the sign of $\sigma_a$ could be positive if $\omega \geq \pi/4$, or negative if this angle is smaller than $\pi/4$ [6]. For 9OBA $\omega \approx 34°$, while for 8OBA and 7OBA it is 55° and 50° respectively (bigger than the limit for the sign inversion). This angle dependent $\sigma_a$ sign causes the negative signs of $\sigma_a$ and provoke a EC in 9OBA (although the dimeric molecular structure) different from those of other N with short range $S_C$ order, which display positive electroconductivity. Thus the investigated here 9OBA, we can categorize as EC case where $\varepsilon_a$ is positive and small and $\sigma_a$ is negative.

We used for the diffraction a low power He-Ne laser (≈1mW) of wavelength $\lambda_L$=632,8nm. In order to study the relaxation of EC patterns the intensity of the diffracted light was monitored.

We concentrated on the EC instability below the characteristic temperature $T^*$=130°C at 12 μm for the studied here substance 9OBA, aiming to analyze this instability in the N region where the EC sets in the expected big enough quasi-smectic supramolecular clusters.

At threshold voltage 24V, 80Hz stripe domains (rolls) forming an angle ±45° with respect to $\boldsymbol{n}_o \equiv x$ are seen at 125°C. At a further overthreshold voltages the stripe domains, after fluctuations normal to the roll's axes, transform (figure 1) to a regular square pattern.

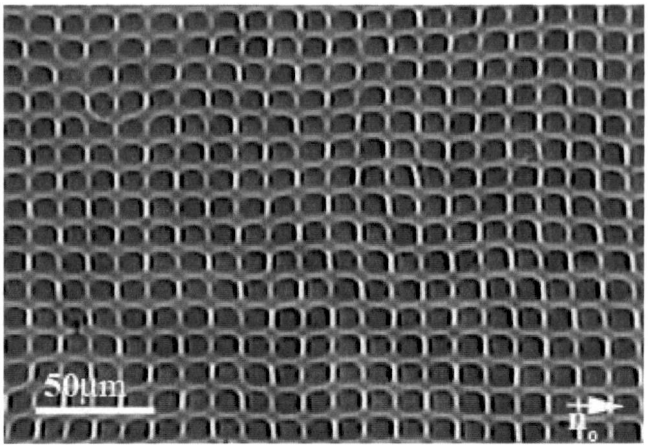


Figure 1. EC square pattern in 9OBA, d=12μm, T=125C, f=50Hz, U=26V.

At 122°C $\sigma_a$ reaches very high negative value meaning that the quasi-smectic layers are enough large to be able to ensure an easy charge distribution inside the layers. This indicates an underlined twist type instability appearing directly at threshold voltage and reveals as a regular cell-like (square form) domains arrayed in two directions including angle ≈45° with *x* direction, similar to that at 125ºC.

Both decay and response times have been scanned, extracted and measured by recording the odd and even diffraction intensities $I_{\pm1}$ and $I_{\pm2}$ at normal or oblique illumination respectively, taking into consideration that the oblique illumination is more effective one in respect to put into practice an real time access to the director amplitude at the EC onset [9-11].

The decay time is an important parameter, which is able to reveals the LC system dynamics since the EC patterns allow us to keep track of the pattern relaxation by means of the fringe intensities decay during the system transition to the equilibrium ground state.

In figure 2 we indicate, as an example, a typical 2D diffraction pattern corresponding to the EC pattern shown in figure 1. The angle of oblique illumination $\alpha$ is 10°. Increasing the voltage above threshold we indicate in figure 3(a), (b) (as an example) the response and decay curves obtained when applied different increasing over-threshold U values. As seen the response time decreases from 1000ms at 23,5V to 80ms at 25V. At the same time the decay time at 25V is ≈100ms, meaning equalization of the response and decay times, which is important for LCD technique.

Furthermore, we found for the first time, that the smectic layer generation in the N phase changes the director modulation characteristics of the EC instability, as well as the nonlinarity of the system, thus leading to inducing of the forth harmonic of the excited *ac* electric field (4f), coexisting with the second harmonic one 2f with a higher magnitude.

Using microtextural polarization analysis we estimated the amplitude of the azimuthal ***n*** modulation $\varphi_o$ at varying the over-threshold voltages by rotation the sample between crossed polarizers. If the director was chosen to be at the polarizer direction, before EC, then in the EC regime the bright regions expresses a rotation of the polarization plane of the light deviation by the ***n*** projection, while the dark regions expresses that this projection rests close to this plane. So the twist out of the tilt plane inverses the dark and bright domain lines, and the polarization direction. This is the base of the detection of the twist spatial modulation of the director (spatial variation of the effective refractive index).

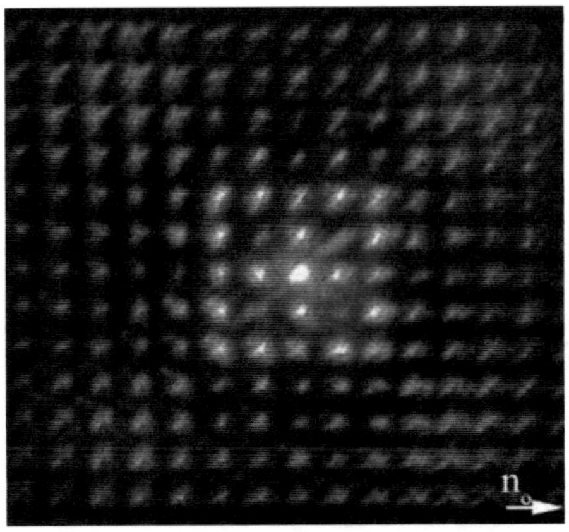


Figure 2. The 2D diffraction pattern of 9OBA. T=125°, U=26V, f=80Hz.

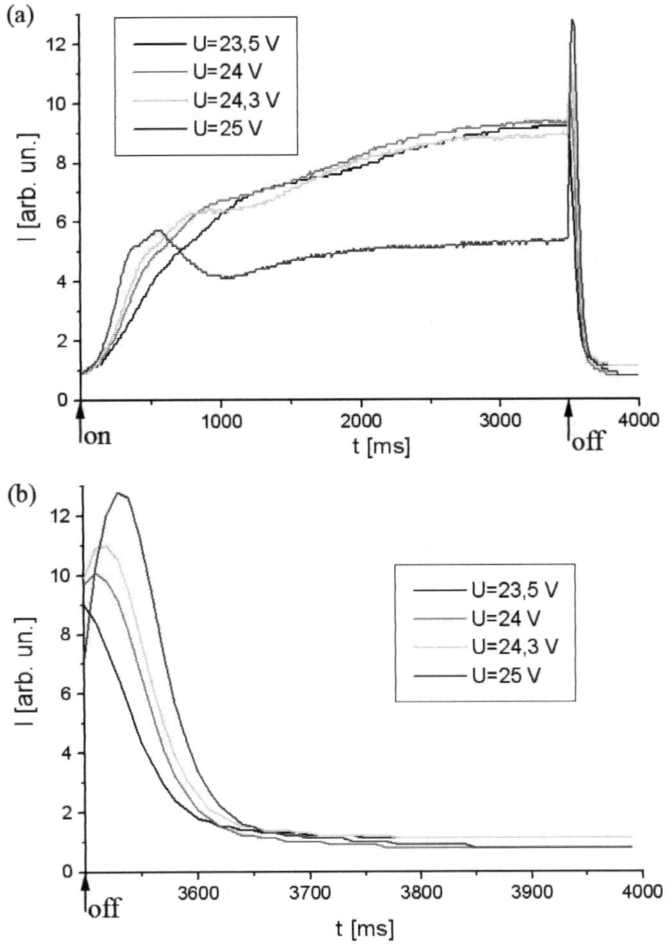


Figure 3. (a) The response times at increasing over-threshold voltages at T=122°C, f=80Hz. 'On' and 'off' on the graphics present switch on and switch off of the electric field. (b) The decay times corresponding to the response ones.

The azimuthal φ(U) dependences at 80Hz for the temperature 122° is indicated in figure 4. The azimuthal ***n*** modulation varies as $\varphi(U)=\varphi_o[(U^2-U_{th}^2)/U_{th}^2]^{1/2}$. The measured azimuthal angle variation indicates periodic twist fluctuation (periodic going out of ***n*** from the vertical *xz* plane) and is expected to be initiated from the coupling of the charge spread out in the *xy* plane and the director field, thus resulting flow field $\mathbf{v}=(v_x,v_y,v_z)$, which actually is 3D phenomenon. As a result the process of the closing of the rolls in square (figure 1) or hexagonal localized states could be due namely to such in-plane director rotation and corresponding flow in the xy plane, which acts simultaneously with that acting in *xz* plane. The distribution of the 2D far-field diffraction field intensity from the domain picture expresses two-dimensional director modulation, which is ***n*** azimuthal deflections and can be presented as $\varphi(x,y)\sim\varphi_o\cos(q_xx+q_yy)$, where $q_x$ and $q_y$ are the periodicities of the pattern. All these suggests a twist type EC instability in N with short range $S_C$ order and $\sigma_a<0$ (9OBA), resembling that of $S_C$ phase [12].

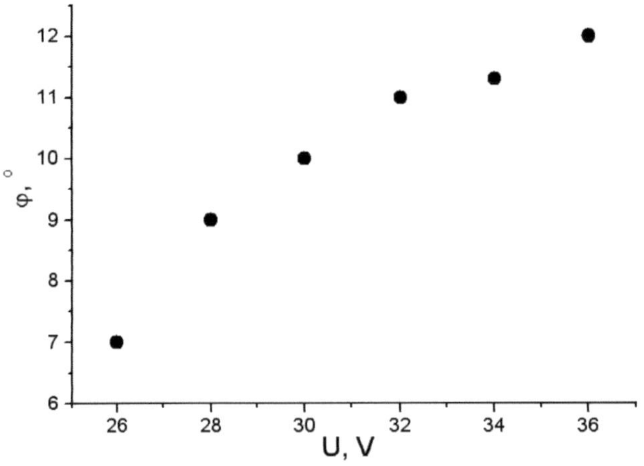


Figure 4. The azimuthal angle variation at over-threshold voltage increasing at T=122ºC, f=80Hz.

As indicated in figure 3 the decay of the EC pattern is exponential. At larger amplitude of the director we leave the linear regime and both amplitude and phase grating effects of the ***n*** modulation are important. On the other hand by oblique incidence we detected very small near threshold director deflections and in such way we have time to increase the scope of the deflection angle φ values and in turn to increase the possibility to co-ordinate the controlled switching period of time with the higher over-threshold driving voltage. Thus we keep the sharpness of the fringe.

On the base of the time-dependent director's fluctuation amplitude $\phi_o(t)$ we expresses the intensity of light diffracted by the EC domains as a Fourier expansion $I(m,t)=I_o+A_{2f}\cos2\omega t+A_{4f}\cos4\omega t+...$, where $m$ is the diffraction order and $A_{2f}$ and $A_{4f}$ are the voltage-dependent coefficients of the detected by us second and fourth harmonics. We neglect the observed in the experiment even harmonics with higher order, which will be considered in a future work.

At 122°C $\sigma_a$ reach very high negative value (as we described above) meaning that the quasi-smectic layers are enough large to be able to ensure an easy charge distribution inside the layers. This indicates underlined twist type instability at this temperature which reveals as cell-like domains arrayed in two directions including angle 45° with *x* direction.

## 3. Conclusion

In conclusion, we described the special feature of the electroconvection instability in the dimeric N phase with short range $S_C$ order of the ninth homologue of the p,n alkyloxybenzoic acid. This feature dues to the negative electroconductivity anisotropy underlined below a definite temperature ($T^*$) inside this N phase. Since the anisotropy of the dielectric constant is small and positive, we assigned the EC in this LC material to category $\varepsilon_a>0$, $\sigma_a<0$, but with specific pattern, threshold and dynamical characteristics. In the studied here material 9OBA the magnitude of the conductivity anisotropy falls always below zero. This feature due to the smaller tilt angle (34°) in 9OBA providing a negative conductivity anisotropy and EC instability closer to that in smectic C phase.

By polarization analysis of azimuthal in-plane $(x,y)$ director deflection in EC regime, we suggested twist type character of this instability, which is still not embraced by the contemporary 3D standard model, although this twist instability was discovered long ago for $S_C$. We stated also that the minimization of the free energy in the formed $S_C$ supramolecular complexes is fulfilled by a rotation of director on a cone, around the layer's normal similar to that in $S_C$ and decisively favours the twist fluctuations.

Our measurements of the response and the decay times showed that in general the response times in the region below $T^*$ of 9OBA are shorter than that in the classical Ns, as the ratio of the response and decay times is $\approx 1$. That means equalization of both response and decay times in 9OBA, which is preferable in the LC display techniques.

A fourth frequency harmonic, which is roughly an order weaker than the second one, was detected for the first time in nematic state.

## References

[1] Kramer L and Pesch W 1996 *Pattern Formation in Liquid Crystals* ed A Buka and L Kramer (New York: Springer)
[2] Buka A, Eber N, Pesch W and Kramer L 2007 *Phys. Rep.* **448** 115
[3] Pesch W, Kramer L, Eber N and Buka A 2006 *Phys. Rev.* E **73** 061705
[4] Toth-Katona T, Cauquil-Vergnes A, Eber N and Buka A 2007 *Phys. Rev.* E **75** 066210
[5] de Gennes P G and Prost J 1993 *The Physics of Liquid Crystals* second edition (New York: Oxford University Press)
[6] Rondelez F 1972 *Sol. St. Comm.* **11** 1675.
[7] Cuvatov Z, Capustin A and Trofimov A 1974 *Pisma, JETP* **19** 89
[8] Kresse H, Lucke K H, Schmidt P and Demus D 1977 *Wiss. Zeitschr. Univ. Halle* XXVI' M 147
[9] Buka A, Eber N and Pesch W July 19 2005 *Electronic-Liquid Crystal Communications*
[10] Eber N, Rozanski S A, Nemeth S, Buka A, Pesch W and Kramer L 2004 *Phys. Rev.* E **70** 061706.
[11] Zenginoglou H M and Kosmopoulos J A 1989 *Appl. Opt.* **28** 3516; 1989 *J. Opt. Soc. Am.* A **18** 573
[12] Petrov M P, Petrov A G and Pelzl G 1992 *Liq. Cryst.* **11** 865

# Does maltose influence on the elasticity of SOPC membrane?

**J Genova, A Zheliaskova and M D Mitov**

Institute of Solid State Physics, Bulgarian Academy of Sciences,
72, Tzarigradsko Chaussee Blvd., 1784 Sofia, Bulgaria.

Corresponding author's e-mail: ulia@issp.bas.bg

**Abstract**. Thermally induced shape fluctuations of giant quasi-spherical lipid vesicles are used to study the influence of the disaccharide maltose, dissolved in the aqueous solution, on the curvature elasticity $k_c$ of a lipid membrane. The influence of the carbohydrate solute is investigated throughout a considerably wide interval of concentrations. The values of the bending elastic modulus for 200 mM and 400 mM of maltose in the water solution are obtained. The data for $k_c$ in presence of maltose is compared with previously obtained results for this constant for the most popular hydrocarbons: monosaccharides glucose and fructose and disaccharides sucrose and trehalose. It is shown that the presence of maltose, dissolved in the aqueous phase surrounding the membrane does not influence on the bending elasticity with the increase of its concentration in the aqueous solution. Up to our knowledge this is the first sugar that does not show decrease of the bending elastic modulus of the lipid membrane, when present in the water surrounding it in concentration up to 400mM.

## 1. Introduction

The biological cell is the functional basic unit of life. It is the functional unit of all known living organisms. It is the smallest unit of life and is often called the building block of life. That is why the biological cell appears to be the most natural object of interest when properties and functions of living organisms are investigated. The biological cell besides its plasma membrane consists of organelles, most of which are or contain membrane structures. This is the basic motivation for the profound study of variety of physical, chemical, mechanical, rheological, etc. properties of biomembranes. Usually it is realized by the means of modeling them with much more simple and easy for theoretical description model objects and subsequently complicating the system in order to obtain model systems resembling afterwards the real membranes. The simplest models of the biological cells are the lipid vesicles (figure 1). They are closed structures formed by a model lipid membrane (a lipid bilayer). They are prepared from natural or synthetic lipids in laboratory conditions using different formation techniques.

This work is a continuation of our previously made study of the influence of carbohydrates on the elasticity of model lipid bilayers. Carbohydrates are made of 3 elements, carbon, hydrogen and oxygen. Along with proteins and fats they comprise the major components of living matter and are used for maintenance of cellular functional activities and as reserve and structural material for cells. Carbohydrates are formed by green plants in the process of photosynthesis. Humans and animal life obtain most efficiently the carbohydrates they need from the plant world. Carbohydrates provide energy for all the processes in the living cell. All life activities - digestion, blood circulation, heart activity, thinking, walking, etc. are dependent upon carbohydrates.

**Figure 1.** Scheme of a vesicle.

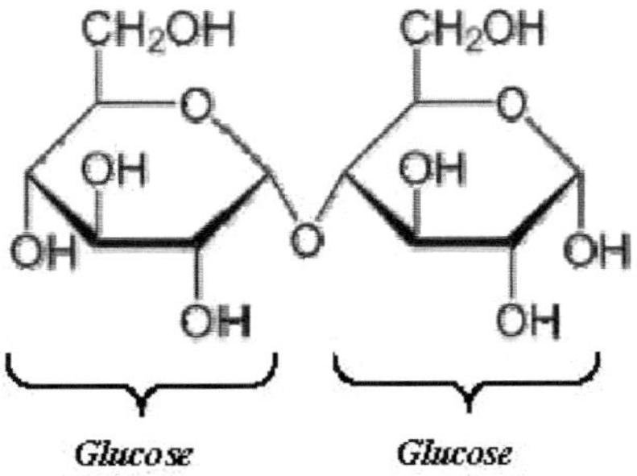


**Figure 2.** Structure of the reducing disaccharide maltose.

Carbohydrates are also known as saccharides and are classified according to numbers of single carbohydrate molecules in each chemical structure as monosaccharides, disaccharides and polysaccharides. Mono- and disaccharides can be grouped together because of their common features. They are water soluble, and they have a sweet taste and a crystalline structure, share the suffix "ose" meaning sugar and are called sugars. Monosaccharides are the only sugars that can be absorbed and utilized by the body. Disaccharides and polysaccharides are ultimately broken down into monosaccharides in the digestive process before they can be utilized by the body. Disaccharides are also divided into reducing and non reducing. Typical representatives of non reducing disaccharides are sucrose and trehalose. Typical representatives of reducing disaccharides are maltose and lactose.

Using the thermally induced shape fluctuation method we studied profoundly the influence of the most popular monosaccharides fructose and glucose and most popular no reducing disaccharides sucrose and trehalose [1, 2] and we know that all of them influence on the bending elastic modulus of the lipid membrane. The bending elastic modulus $k_c$ decreases sufficiently with the increase of the sugar concentration in the aqueous for the membrane solution.

In this work we focused on another disaccharide - maltose, which is a typical representative of reducing disaccharides (figure 2).

Maltose or malt sugar is the least common disaccharide in nature. It is present in germinating grain, in a small proportion in corn syrup, and forms on the partial hydrolysis of starch. Maltose, the disaccharide obtained by enzyme-catalyzed hydrolysis of starch, consists of two D-glucopyranoses joined by a 1,4'-beta-glycoside bond. Both maltose and cellobiose are reducing sugars because the anomeric carbons on the right-hand sugar are part of a hemiacetal. Maltose, however, is digested without difficulty and is fermented readily.

## 2. Theoretical models

The first theoretical models for the mechanical properties of lipid membranes proposed by Helfrich [3] and Evans [4] describe the elastic energy per unit area of lipid membrane, $F_c$ by the expression:

$$F_c = \frac{1}{2} k_c (c_1 + c_2 - c_0)^2 + \bar{k}_c c_1 c_2 \tag{1}$$

where: $c_1$ and $c_2$ are the membrane principal curvatures, $c_0$ is the spontaneous curvature, and $k_c$ and $\bar{k}_c$ are bending and saddle bending elastic moduli of lipid bilayer, respectively. The spontaneous curvature of a symmetric membrane in a symmetric environment vanishes, $c_o = 0$.

After the first detailed theoretical model of thermally induced shape fluctuations proposed by Milner and Safran [5], the experimental procedures, based on the analysis of thermally induces shape fluctuations of quasi-spherical vesicles are developed for the precise measurements of the bending elastic modulus [6, 7]. The fundamental expression used by the authors is [5]:

$$\left\langle \left| U_n^m(t) \right|^2 \right\rangle = \frac{k_B T}{k_c} \frac{1}{(n-1)(n+2)[\bar{\sigma} + n(n+1)]} \tag{2}$$

where $\left\langle \left| U_n^m(t) \right|^2 \right\rangle$ is the mean squared amplitude of the spherical harmonic $Y_n^m(\theta,\varphi)$, $k_B$ is the Boltzmann's constant, $T$ is the absolute temperature, $n$ is the mode number and $\bar{\sigma} = \sigma R^2 / k_c$ (or $\bar{\sigma} = \sigma R^2 / k_c + 2c_0 R + c_0^2 R^2 / 2$, if $c_0 \neq 0$) is the dimensionless membrane tension.

In fact what is measured in an experiment of fluctuating quasi-spherical giant vesicle is the equatorial cross section radius. It is shown in [6] that its time averaged angular autocorrelation function is a sum of Legendre polynomials with amplitudes $B_n$, related to the mean squared amplitudes of spherical harmonics:

$$B_n = \frac{2n+1}{4\pi} \left\langle \left| U_n^m(t) \right|^2 \right\rangle \tag{3}$$

where the factor $2n+1$ is due to the $2n+1$ different $m$-modes for a given $n$ all of them having the same mean squared amplitude and 4π comes from the different normalizations of Legendre polynomials and spherical harmonics.

In all the experimental data provided in this work stroboscopic illumination is used to remove the artifact due to the video camera integration time. Stroboscopically illuminated sample presents an instant picture of the object to the observer [8, 9].

### 3. Experimental equipment

The samples of the fluctuating giant vesicles were observed under phase contrast microscope (Axiovert 100, Zeiss, Germany, objective LD Ph2 63x NA 0.75).

The experimental equipment was improved using home-assembled stroboscopic illumination using commercial parts (xenon flash lamp L6604, external main discharge capacitor E7289-01, power supply C6096, all by Hamamatsu, Japan). The flash of the stroboscopic illumination was synchronized with the vertical pulses coming from the CCD video camera controller (C2400-60, Hamamatsu, Japan). According to the Hamamatsu data sheet, the light pulses are less than 3-4 μs long (full width at half maximum) at 2 J input energy.

The pulsed light of the stroboscopic illumination is irritating for the eyes, so the samples should be observed on an attached monitor. Due to the "sample and hold" effect of the CCD matrix the picture on the monitor is like in the case of a continuous illumination. The video signal from the camera was fed also to a frame grabber board (DT3155, Data translation, USA) mounted in a computer for a proper digitization (768 x 576 8-bit pixels). The obtained digital data was further recorded on the hard disk drive of the PC. Every second an image was acquired and recorded till the total number of images reaches a preliminary given value (about 400 or so). Nevertheless that the CCD has "square pixels" the images have to be corrected (via digital interpolation and resampling) for the difference of the scale factors in x and y directions, due to the mismatch of the CCD's pixel shift clock (in the CCD camera controller) and the pixel acquisition clock (in the frame grabber).

The value of the scale factor was determined by the ratio of the above mentioned clocks, taken from the respective data sheets and verified by x and y calibration using an object micrometer rule oriented in the respective directions.

Further details on the contour determination, mean squared amplitudes calculation and fitting procedure to determine the bending elastic modulus, $k_c$, and the dimensionless membrane tension $\bar{\sigma}$, can be found in the paper [6].

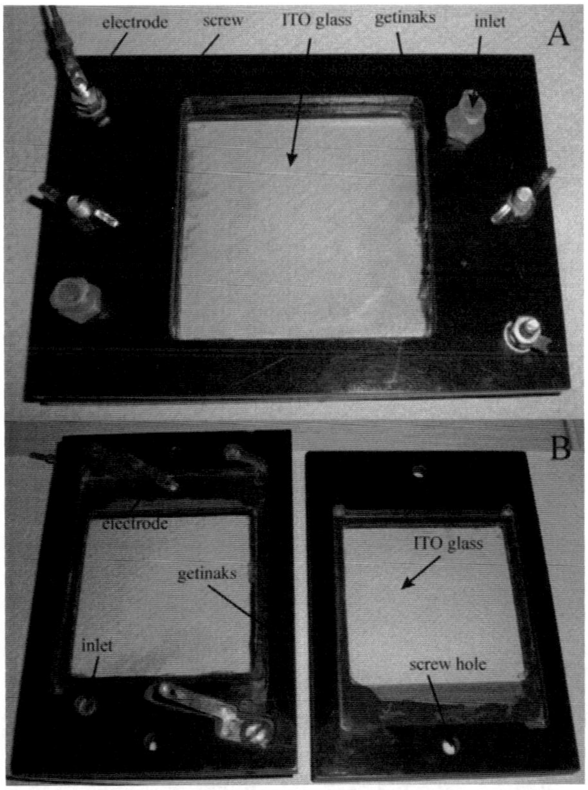


**Figure 3.** An image of the electroformation cell.

## 4. Materials and Methods

All the experiments were performed with bilayers composed of l-stearoyl-2-oleoyl-*sn*-glycero-3-phosphocholine ((C18:0/C18:1) SOPC, Avanti Polar Lipids Inc., USA). The hydrocarbons used were products with high purity > 99% (GC): D-(+)-Maltose monohydrate (Sigma - Aldrich). All the chemicals were used without any further purification.

The giant vesicles, studied in these experiments were prepared using a modification of the electroformation method. The lipid is dissolved in chloroform 1 mg/ml. A simple electroformation cell, used for all experimental procedures is shown on figure 3. The frame of the cell was made out of electro technical sheet getinaks (micarta). It is a laminated material obtained by hot pressing of paper impregnated with a thermosetting phenol-formaldehyde- or epoxy-based binder. Glasses, coated with transparent conductor, indium tin oxide (ITO; thickness of 100 ± 20 nm, resistivity of 100 Ω/□) acting as an electrode, were glued on the both sides to the cell's frame. The electrodes were 3 to 4 mm apart (d=3 or d=4 mm). Over the electrodes an ac voltage is applied, while the electroformation process is started.

A number of small drops of the lipid solution are laid on the surface of the glasses of the experimental cell in order to obtain as much as possible lipid depots for vesicle formation. Meanwhile a solution with the desired concentration is prepared in fresh double distilled water. After the entire evaporation of the solvent the experimental cell is filled with this previously made sugar solution. We established the most suitable for our experiment regime for electroformation. A low frequency (10 Hz) sinusoidal alternative voltage is applied (a step-like increase from 0.1 V PP (peak to peak) to 1.5 V PP) to the conductive glasses for about 4 hours. This procedure leads to the formation of vesicles, appropriate for our experiment. We choose giant (diameter of the order of 20-40 μm) fluctuating vesicles without any visible defects.

## 5. Results and Discussion

The analysis of thermally induced shape fluctuations of giant vesicles is used to study the influence of one typical reducing disaccharide - maltose on the elasticity of lipid membrane. The obtained experimental data for the bending elasticity modulus, $k_c$ for different concentration of the carbohydrate in the aqueous phase (0 - 400 mM) are presented in Table 1. The values for the bending elasticity modulus for every sugar concentration are calculated as a weighted average value of about 10-15 vesicles.

Comparing the obtained results we can conclude that in the frames of the error the values for the bending elastic modulus for the different concentrations of maltose in the solution are the same. So, we can claim that in such concentrations (0 – 400 mM) in the surrounding medium of the membrane the maltose does not influence on its bending elasticity.

Table 1

| **Maltose** concentration in the aqueous solution | 0 mM (pure SOPC membrane) | 200 mM | 400 mM |
|---|---|---|---|
| Bending elastic modulus of the lipid membrane $k_c$ | $(1{,}17 \pm 0{,}10) \times 10^{-19} J$ Previously obtained [1] | $(1{,}26 \pm 0{,}04) \times 10^{-19} J$ | $(1{,}23 \pm 0{,}05) \times 10^{-19} J$ |

This result is getting even more interesting in comparison with our previous results, obtained using the same method for different sugars. On the figure 4 the data for 4 different sugars are given: monosaccharides fructose and glucose and non-reducing disaccharides sucrose and trehalose [2]. For easy comparison the data for maltose is depicted on the same figure with "∘".

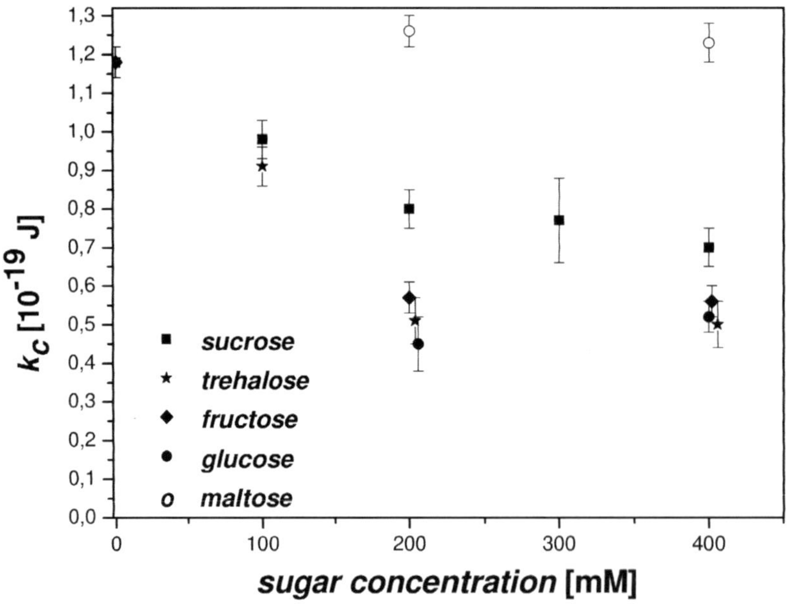


**Figure 4.** Bending elastic modulus $k_c$ of SOPC membrane as a function of the carbohydrate inclusion concentration in the aqueous phase for different types of carbohydrates: monosaccharides - glucose and fructose and disaccharides- sucrose, trehalose and maltose.

We can conclude from the received experimental data that the reducing disaccharide maltose has different behaviour in the proximity of the lipid molecules, comprising the bilayer structure than the other type of sugars: monosaccharides fructose and glucose and non-reducing disaccharides sucrose and trehalose.

**Acknowledgements**
This study was supported by Internal Grant BK-9-1 of The Institute of Solid State Physics, Bulgarian Academy of Sciences, Bulgaria.

**References**

[1] Vitkova V, Genova J, Mitov M D and Bivas I 2006 *Mol. Cryst. & Liq. Cryst.* **449** 95
[2] Genova J, Zheliaskova A and Mitov M D 2007 *J. Optoel. & Adv. Mat.* **9** 427
[3] Helfrich W Z 1973 *Naturforsch.* **28c** 693
[4] Evans E A 1973 *Biophys. J.* **13** 941
[5] Milner S T and Safran S A 1987 *Phys. Rev.* A **36** 4371
[6] Faucon J F, Mitov M D, Méléard P, Bivas I and Bothorel P 1989 *J Physique* **50** 2389
[7] Mitov M D, Faucon J F, Méléard P, Bivas I and Bothorel P 1992 *Advances in Supramolecular Chemistry*, ed G W Gokel (JAI Press Inc.: Greenwich) **93**
[8] Genova J, Vitkova V, Aladjem L, Méléard P and Mitov M D 2005 *J. Optoel. & Adv. Mat.* **7** 257
[9] Méléard P, Faucon J F, Mitov M D and Bothorel P 1992 *Europhys. Lett.* **19** 267

Journal of Physics: Conference Series **253** (2010) 012064 doi:10.1088/1742-6596/253/1/012064

# Bending elasticity of lipid membranes in presence of beta 2 glycoprotein I in the surrounding solution

**J I Pavlič[1,2], J Genova[3], A Zheliaskova[3], A Iglič[1] and M D Mitov[3]**

[1]Laboratory of Biophysics, Faculty of Electrical Engineering, University of Ljubljana,1000 Ljubljana, Slovenia;

[2]Faculty of health sciences, University of Ljubljana, 1000 Ljubljana, Slovenia

[3]Institute of Solid State Physics, Bulgarian Academy of Sciences, 72, Tzarigradsko Chaussee Blvd., 1784 Sofia, Bulgaria.

Corresponding author's e-mail: ulia@issp.bas.bg

**Abstract**. Thermally induced shape fluctuations of giant quasi-spherical lipid vesicles are used to study the bending elasticity modulus $k_c$ of a phospholipid (PHLP) membranes in presence of beta 2 glycoprotein I (β2-GPI) in the aqueous solution which surrounds the vesicle's membrane. The bending elastic modulus $k_c$ of PHLP – protein membrane was obtained for different mass concentrations of β2-GPI for pure neutral SOPC membranes and for mixed SOPC: Cardiolipin negatively charged membranes. The experimental results for the bending elastic modulus $k_c$ of the PHLP membranes does not show dependence on the concentration of β2-GPI in the range from 5.5 to 55 μg/ml, when β2-GPI is present in the aqueous solution surrounding the vesicle's membrane. Obtained results are in good agreement with predictions, based on different experiments, explaining the mechanism of binding of β2-GPI to neutral membranes.

## 1. Introduction

The lipid bilayer is a thin membrane made of two layers of lipid molecules. These membranes are flat sheets that form a continuous barrier around cells. The cell membrane of almost all living organisms and many viruses are made of lipid bilayers, as are the membranes surrounding the cell nucleus and other sub-cellular structures. Biological membranes typically include several types of lipids and integral membrane proteins. The mechanical properties of model lipid membranes containing biologically relevant proteins are of great interest, because they give valuable information on the way to more profound understanding of properties and functioning of the living cell.

β2-GPI is a soluble 50-54 kDa plasma glycoprotein, primarily synthesized in the liver. It acts as a natural anticoagulant [1, 2]. The physiological function of this protein is not fully understood, but it is known that β2-GPI is involved in variety of physiologically relevant processes such as blood coagulation, clearence of phosphatidylserine (PS) and cardiolipin (CL) containing liposomes from the circulation, membrane budding process [3] and diseases, such as the modulation of platelet-depending thrombosis, atherosclerosis and autoimmune diseases, such as antiphospholipid syndrome [4].

β2-GPI is present in human blood plasma at a concentration of 150-300 μg/ml [5]. The isoelectric point of β2-GPI is at pH between 5-7, so the protein is negatively charged at physiological pH 7.4 [6,

7]. This protein can interact with the PHLP membranes by electrostatic interactions and hydrophobic loops [6].

β2-GPI is composed of four domains, which are structurally similar, the fifth domain contains an extra C-terminal loop, which is necessary for PHLP binding [6]. Both ionic and hydrophobic interactions of the fifth domain are important for binding of β2-GPI to PHLP membranes. In contrast to other apolipoproteins, lipid-free β2-GPI does not self-associate in aqueous solution. The positively charged multilysine rich region on fifth domain interacts with negatively charged lipid head-groups and hydrophobic flexible loop putatively inserts into the PHLP layer.

The crystal structure of β2-GPI shows an overall fish-hook-like appearance with dimensions of 130 Å in vertical direction and 85 Å in horizontal direction [5], as indicated on the figure 1.

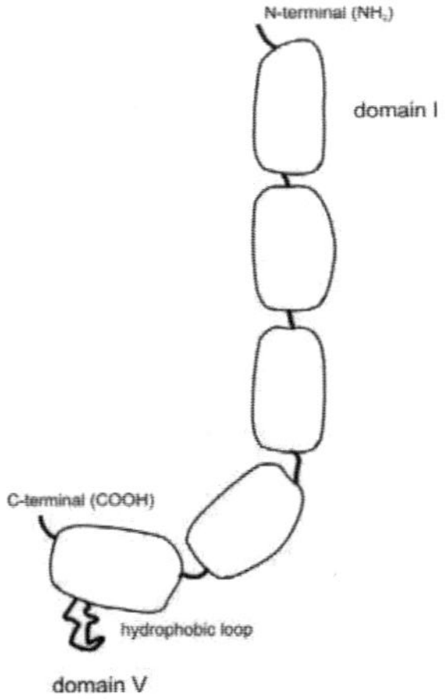


**Figure 1.** Crystal structure of β2-GPI.

The structural and biochemical data indicate a relatively simple membrane binding mechanism. Positive charges on domain V interact with the anionic PHLP head-groups and flexible loop putatively inserts into the PHLP layer; at the interface region between the acyl chains and phosphate head-groups of the PHLPs, thus anchoring itself to the membrane [5].

To better understand how β2-GPI affects the mechanical properties of model lipid membranes we can apply the experimental technique of thermally induced shape fluctuations of giant PHLP quasi-spherical vesicles.

## 2. Theory

The first theoretical models for the mechanical properties of lipid membranes proposed by Helfrich [8] and Evans [9] describe the elastic energy per unit area of lipid membrane, $F_c$ by the expression:

$$F_c = \frac{1}{2} k_c (c_1 + c_2 - c_0)^2 + \bar{k}_c c_1 c_2 \qquad (1)$$

where: $c_1$ and $c_2$ are the membrane principal curvatures, $c_0$ is the spontaneous curvature, and $k_c$ and $\bar{k}_c$ are bending and saddle bending elastic moduli of lipid bilayer, respectively. The spontaneous curvature of a symmetric membrane in a symmetric environment vanishes, $c_0 = 0$.

After the first detailed theoretical model of thermally induced shape fluctuations has been proposed by Milner and Safran [10], the experimental procedures, based on the analysis of thermally induced shape fluctuations of quasi-spherical vesicles were developed for the precise measurements of the bending elastic modulus [11, 12]. The fundamental expression of thermally induced shape fluctuations as in [10] is:

$$\left\langle \left| U_n^m(t) \right|^2 \right\rangle = \frac{k_B T}{k_c} \frac{1}{(n-1)(n+2)[\bar{\sigma} + n(n+1)]} \tag{2}$$

where $\left\langle \left| U_n^m(t) \right|^2 \right\rangle$ is the mean squared amplitude of the spherical harmonics $Y_n^m(\theta,\varphi)$, $k_B$ is the Boltzmann's constant, $T$ is the absolute temperature, $n$ is the mode number and $\bar{\sigma} = \sigma R^2 / k_c$ (or $\bar{\sigma} = \sigma R^2 / k_c + 2c_0 R + c_0^2 R^2 / 2$, if $c_0 \neq 0$) is the dimensionless membrane tension.

In fact what is measured in the experiment of fluctuating quasi-spherical giant vesicle is the equatorial cross section radius. It is shown in [11] that its time averaged angular autocorrelation function is a sum of Legendre polynomials with amplitudes $B_n$, related to the mean squared amplitudes of spherical harmonics like:

$$B_n = \frac{2n+1}{4\pi} \left\langle \left| U_n^m(t) \right|^2 \right\rangle \tag{3}$$

where the factor $2n+1$ is due to the $2n+1$ different $m$-modes for a given $n$ all of them having the same mean squared amplitude and $4\pi$ comes from the different normalizations of Legendre polynomials and spherical harmonics.

## 3. Materials and Methods

β2-GPI was isolated from pooled human plasma and was concentrated using Microcon centrifugal filter device with YM-30 Ultracel membrane [13]. Aliquots of β2-GPI in PBS (1.2 mg/ml) were stored at $-20^{\circ}$C. The final concentration of β2-GPI in the experiments was 5.5 μg/ml, 27.5 μg/ml and 55 μg/ml, which is approximately half of the free β2-GPI concentration in human plasma. Before the start of electroformation procedure, the stock suspension of 1.2 mg/ml of β2-GPI was diluted in sucrose solution to a final concentration.

For vesicle preparation we used the lipids: 1-stearoyl-2oleoyl-sn-glycero-3-choline (SOPC) and 1',3'-bis[1,2-dioleoyl-sn-glycero-3-]-sn-glycerol (cardiolipin, CL) purchased from Avanti Polar Lipids. All lipids were dissolved in chloroform in concentration of 1 mg/ml. Two different lipid mixtures were prepared: first mixture of 9:1 volume ratio of SOPC:CL (~5.5 mol% of CL) and second mixture of 4:1 volume ratio of SOPC:CL (~11.6 mol% of CL).

For electroformation we used two different cells. In figure 2 is shown the cell, made out of getinaks (micarta). Glued glasses were coated with transparent indium tin oxide conductor (ITO), which acts as an electrode. The distance between the two ITO glasses was 4 mm. Over the electrodes an ac voltage was applied, to start the electroformation process.

In figure 3 is shown a smaller electroformation cell. The tube's cap was pierced to insert four Pt electrodes. The electrodes were 3-4 mm apart. We used 4 electrodes to increase the quantity of PHLP in the whole volume and consequently to increase the number of electroformed GUVs.

Lipid mixture of SOPC and CL was put in small drops over ITO glass electrodes, and similarly, on every Pt electrode. The electroformation cell or tube cap with Pt electrodes was put in vacuum for 15-20 minutes to evaporate the organic solvent (chloroform). After the entire evaporation of the organic solvent, the cell was assembled and filled with desired aqueous solution (double distilled water or mixture of sucrose solution and phosphate buffer saline).

The cell was connected to the function generator (HP 33120A) with output termination of 50 Ω.

The preparation procedure for GUVs formation is summarized in table 1. All the experiments were done at room temperature.

**Figure 2.** Electroformation cells with ITO electrodes.

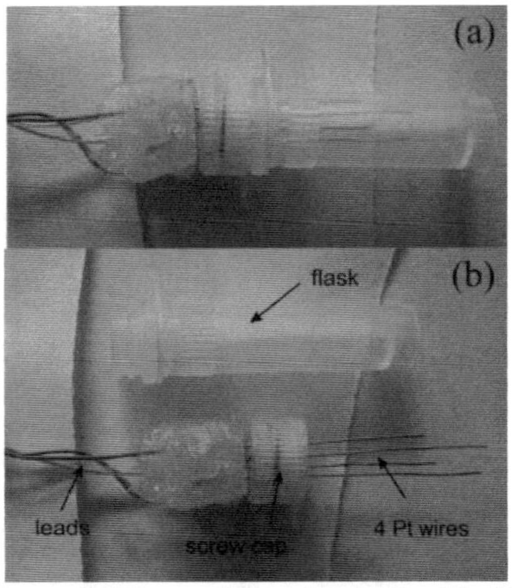


**Figure 3.** Electroformation cell with 4 Pt electrodes.

A: Assembled electroformation cell; B: Disassembled electroformation cell.

**Table 1.** Electroformation procedure for forming SOPC:CL GUVs.

| ac voltage, peak-peak [mV] | Frequency [Hz] | Time duration [min] |
|---|---|---|
| 100 | 10 | 5 |
| 200 | 10 | 5 |
| 300 | 10 | 15 |
| 600 | 10 | 60 |
| 1500 | 10 | 120 |

## 4. Results and Discussion

In the process of studying the properties of lipid vesicles an important element is their formation procedure. In order to increase the quality and quantity of lipid vesicles and to reduce the impurities in the solution, we constructed seven different modifications of electroformation cells:

A. Four platinum wires for electrodes in glass flask;
B. Four platinum wires for electrodes in plastic vial;
C. Two ITO glasses for electrodes with a rubber O-ring as a spacer;
D. Two ITO glasses for electrodes with a FFKM O-ring (Trelleborg sealing solutions) as a spacer;
E. Two ITO glasses for electrodes in getinaks frame
F. Two ITO glasses for electrodes in plexiglass frame
G. Two stripes of ITO glass for electrodes in glass flask.

In order to compare the seven different modifications of the electroformation cells we study the time dependence of their resistance. Every cell was thoroughly washed with solvents (chloroform and methanol) and distilled water. After the entire evaporation under vacuum the cells were filled with pure double distilled water and connected to a digital Lock-in amplifier (Stanford Research Systems, model SR830). A 280 Ω resistor was connected in series to the electroformation cell. At given output voltage (40 $mV_{RMS}$ @ 1 kHz sine) from the Lock-in amplifier, a voltage drop on the resistor was measured for 5 hours (approximately twice the electroformation time). The obtained results are normalized and plotted in figure 4.

The electroformation cell conductivity was measured consequently, as the voltage drop across on the 280 Ω resistor. As can be seen the conductivity of double distilled water increased with time. The increase of conductivity could come due to impurities, which are emitted into the pure water. The impurities could originate from materials used for cell assembly or from electrodes.

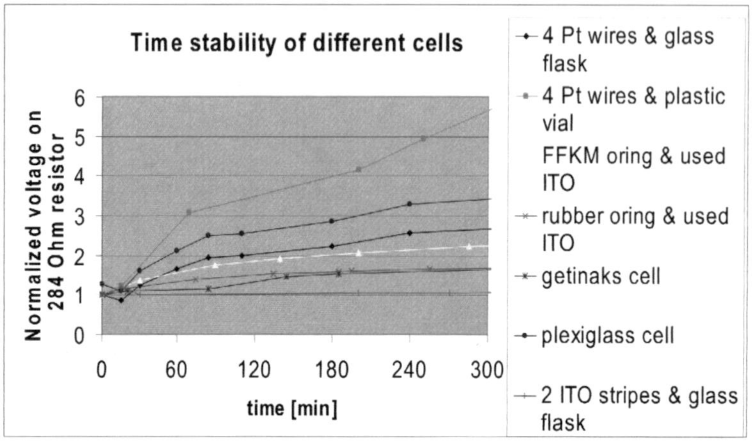


**Figure 4.** Normalized time dependence of the resistance of different cell modifications.

As it can be seen from figure 4 the most stable with time is the cell G - two stripes of ITO glass for electrodes in a glass vial, followed by the cell E- two ITO glasses for electrodes in getinaks frame and C- two ITO glasses for electrodes with a rubber O-ring as a spacer. It is also seen that the cells E and C show very similar dependencies.

The obtained experimental data give important information for the time dependence of the cell resistance of the most frequently used electroformation cells and their modifications, used by a lot of scientific groups, dealing with lipid vesicles and give motivation for usage of one cell in spite of another for every different experiment and system.

Using the method of thermally induced shape fluctuation of giant quasi spherical PHLP vesicles, we obtained the values for the bending elastic modulus $k_c$ for different systems of interest: at three different concentration of β2-GPI present in the aqueous solution 5.5 μg/ml; 27.5 μg/ml and 55 μg/ml for pure neutral SOPC membrane and mixed SOPC: CL in 9:1 volume ratio of negatively charged membrane. The obtained results for the mean values of $k_c$ for different studied systems are summarized in table 2.

For every system the mean value of the bending elastic modulus and its standard deviation is calculated from the Gaussian distributions (with their mean values and standard deviations). For calculating the mean values of $k_c$, the number of vesicles for every system was at least 7 and up to 30.

From the obtained results we can conclude that the bending elastic modulus of PHLP membranes, surrounded by glycoprotein β2-GPI does not depend on the concentration of β2-GPI in the range 5.5 – 55 μg/ml (the protein was added directly into 280 mmol/l sucrose solution used for formation of giant PHLP vesicles).

The obtained values for the bending elastic modulus $k_c$ of the studied systems are higher than the approximate value for elastic modulus of pure SOPC membranes in of 280 mmol/l sucrose solution, which is $k_c \approx (0.75 \pm 0.05) x 10^{-19} J$ [14].

In many studies the interaction of β2-GPI with liposome membranes containing negatively charged PHLPs (CL, PS, PA, PI) was shown. The interaction with neutral PHLPs (PC, PE, SM) was marginal [16, 17, 7]. But from PHLP monolayer surface pressure study [15], it was shown, that β2-GPI also interacted with neutral (DPPC) lipid monolayer, not only with negatively charged (DPPS) lipid monolayer. The interaction was indicated by an increase in initial surface pressure of lipid monolayer.

Our results indicate that there was some interaction between β2-GPI and neutral PHLP membranes. The interaction could be observed through an increase of bending elastic modulus $k_c$, obtained from thermally induced shape fluctuation method. $k_c$ does not depend much on β2-GPI concentration.

**Table 2.** The mean bending elastic modulus $k_c$ of giant vesicles for different studied systems.

| Phospholipid content | β2-GPI | Aqueous solution | Bending elastic modulus x $10^{-19}$ J | Number of studied vesicles |
|---|---|---|---|---|
| SOPC | 5,5 μg/ml | 280mM sucrose | $\langle k_c \rangle = 1{,}152 \pm 0{,}05$ | 7 |
| SOPC | 27,5 μg/ml | 280mM sucrose | $\langle k_c \rangle = 0{,}98 \pm 0{,}051$ | 8 |
| SOPC | 55,5 μg/ml | 280mM sucrose | $\langle k_c \rangle = 1{,}082 \pm 0{,}042$ | 30 |
| SOPC:CL = 9:1 | 5,5 μg/ml | 280mM sucrose | $\langle k_c \rangle = 1{,}030 \pm 0{,}05$ | 11 |
| SOPC | 0; (control) | 280mM sucrose | $\langle k_c \rangle = 1{,}002 \pm 0{,}053$ | 20 |

## Acknowledgements

This study was partially supported by Internal Grant BK-09-01 of The Institute of Solid State Physics, Bulgarian Academy of Sciences, Bulgaria and by bilateral project "Shapes and thermal shape fluctuations of lipid vesicles - relation to the properties of their membranes" from the Slovenian Research Agency.

We are very thankful for the donation of β2-GPI from dr. Mojca Frank, Department of Rheumatology, University Medical Centre Ljubljana, Ljubljana, Slovenia.

## References

[1] Schwarzenbacher R, Zeth K, Diederichs K, Gries A, Kostner G M, Laggner P and Prass R 1999 *EMBO J.* **18** 6228-6239
[2] Frank M, Sodin-Šemrl S, Irman Š, Božič B and Rozman B 2009 *Autoimmun Rev.* **9** 5-10
[3] Urbanija J, Babnik B, Frank M, Tomšič N, Rozman B, Kralj-Iglič V and Iglič A 2008 *Eur. Biophys. J.* **37** 1085-1095
[4] Gamsjaeger R, Johs A, Gries A, Gruber H J, Romanin C, Prassl R and Hinterdorfer P 2005 *Biochem. J.* **389** 665-673
[5] Bouma B, de Groot P E, van den Elsen J M H, Ravelli R B G, Schouten A, Simmelink M J A, Derksen R H W M, Kroon J and Gros P 1999 *EMBO J.* **18** 5166-5174
[6] Sodin-Šemrl S, Frank M, Ambrožič A, Pavlič J, Šuštar V, Čučnik S, Božič B, Kveder T and Rozman B 2008 *Advances in Planar Lipid Bilayers and Liposomes* vol 8 ed A Leitmannova Liu (Amsterdam: Elsevier) chapter 8 pp 243-273
[7] Wang S X, Cai G, and Su S 1999 *Biochemistry* **38** 9477-9484
[8] Helfrich W 1973 *Z. Naturforsch.* **28c** 693.
[9] Evans E A 1973 *Biophys. J.* **13** 926-940
[10] Milner S T and Safran S A 1987 *Phys. Rev. A* **36** 4371-4379
[11] Faucon J F, Mitov M D, Méléard P, Bivas I and Bothorel P 1989 *J. Phys. France* **50** 2389
[12] Mitov M D, Faucon J F, Méléard P and Bothorel P 1992 *Advances in Supramolecular Chemistry* vol 2 ed G W Gokel (Greenwich: Jai Press) p 93
[13] Frank M, Sodin-Šemrl S, Rozman B, Potočnik M and Kralj-Iglič V 2009 *Ann N Y Acad Sci.: Contemporary Challenges in Autoimmunity* **1173** 874-886
[14] Genova J, Zheliaskova A and Mitov M D 2006 *Coll. and Surf. A* **282-283** 420-422
[15] Wang S X, Cai G and SUI S 1998 *Biochem. J.* **335** 225-232
[16] Hagihara Y, Goto Y, Kato H and Yoshimura T 1995 *J Biochem.* **118** 129-136
[17] Wurm H 1984 *Int. J. Biochem.* **16** 511-515

Journal of Physics: Conference Series **253** (2010) 012065 doi:10.1088/1742-6596/253/1/012065

# Photoinduced changes in photosystem II pigments

**Atanaska S Andreeva[1], Mira C Busheva[2], Katerina V Stoitchkova[1,3] and Iren K Tzonova[2]**

[1]Sofia University, Faculty of Physics, Department of Condensed Matter Physics, 5, J. Bourchier blvd., 1164 Sofia, Bulgaria

[2]Institute of Biophysics, Bulgarian Academy of Sciences, Acad. G. Bonchev str. bl.21, 1113 Sofia, Bulgaria

E-mail: katys@phys.uni-sofia.bg

**Abstract** The photosynthetic apparatus in higher plants performs two seemingly opposing tasks: efficient harvest of sunlight, but also rapid and harmless dissipation of excess light energy as heat to avoid deleterious photodamage. In order to study this process in pigment-protein supercomplexes of photosystem II (PSII), 77 K fluorescence and room temperature resonance Raman (RR) spectroscopy were applied to investigate the changes in structure and spectral properties of the pigments in spinach PSII membranes. The high-light treatment results in a strong quenching of the fluorescence (being largest when the excitation is absorbed by carotenoids) and a red-shift of the main maximum. Decomposition of the fluorescence spectra into four bands revealed intensive quenching of F685 and F695 bands, possible bleaching of chlorophyll *a*, enhanced extent of light harvesting complexes (LHCII) aggregation and increased energy transfer to aggregated LHCII. The analysis of RR spectra revealed the predominant contribution of ß-carotene (ß-Car) upon 457.8 and 488 nm excitations and lutein (Lut) at 514.5 nm. During prolonged exposure to strong light no significant bleaching of ß-Car and weak photobleaching of Lut is observed. The results will contribute to the efforts to produce more efficient and robust solar cells when exposed to fluctuations in light intensity.

## 1. Introduction

Pigments in plants absorb almost 100% of the sunlight reaching them and transform it into other forms of energy. Using this energy the membrane bound plant pigment-protein complex of photosystem II catalyses the splitting of water into molecular oxygen via the so-called "oxygen-evolving complex" (OEC) [1-3]. PSII is surrounded by a number of trimeric light harvesting chlorophyll (Chl) *a/b* protein complexes II and some minor ones, forming the PSII–LHCII supercomplex [4]. PSII comprises core antenna proteins CP47 and CP43, reaction center subunits D1 and D2, three membrane-extrinsic subunits forming OEC and many small proteins. It binds 35 Chl *a* molecules and 12 ß-Car molecules [4,5].

The photosynthetic apparatus in higher plants performs two seemingly opposing tasks: efficient harvesting of sunlight, but also rapid and harmless dissipation of excess light energy as heat to avoid deleterious photodamage under high light intensities. In order to study this process in pigment-protein supercomplexes of PSII, 77 K steady-state emission fluorescence and room temperature (RT)

[3] To whom any correspondence should be addressed.

resonance Raman spectroscopy were applied to investigate the changes in structure and spectral properties of the major pigments in spinach PSII membrane preparations due to treatment with strong light intensity.

## 2. Materials and methods

### *2.1. Isolation of PSII submembrane particles*

PSII enriched subchloroplast fraction (BBY) was isolated from market spinach following the modified procedure [6]. BBY particles were resuspended in media containing 20mM Hepes (pH 7.5), 15 mM NaCl, 5 mM $MgCl_2$, 0.4 M sucrose and stored at -20°C. The chlorophyll content of the samples was estimated by using the method of Lichtenthaler [7].

### *2.2. High-light treatment*

The illumination of isolated PSII particles was carried out under continuous stirring at room temperature (22°C) with white light. The Chl concentration during illumination was 100 μg/ml. The membranes were illuminated on the vessel surface with intensity of 2400 $\mu E\ m^{-2}\ s^{-1}$ and 1250 $\mu E\ m^{-2}\ s^{-1}$ for Raman and fluorescence measurements, respectively. Samples for analysis were taken at different time periods (15 and 45 min) during illumination.

### *2.3. 77 K Fluorescence measurements*

The 77 K chlorophyll fluorescence emission spectra were measured using a JobinYvon JY3 spectrofluorimeter. The actinic light with low enough intensity was provided by Xenon lamp "Suprasil"- 150W, with slits width of 4 nm. The experimental spectra were corrected for the spectral sensitivity of the detection system. To normalize the emission spectra of BBY 0.5 μM fluorescein (sodium salt) was added as an internal standard to the medium. At this concentration the fluorescein did not interfere with the fluorescence emission [8].

### *2.4. Decomposition of the fluorescence emission spectra*

Detailed studies on the fluorescence of PSII complex have shown that four main bands in the 77 K PSII emission contribute to the spectrum, identified as F680, F685, F695 and F700 after their maxima positions in nm [9-13]. The first emission band, F680, is assigned to trimeric LHCII, peaking at 679 nm [10,14-16] originating from the red-most Chl *a* pigment in LHCII. F685 arises from traps of Chl *a* (absorbing at 683 nm) in D1-D2 RC complex and CP43, whereas F695 PSII emission from a 'trap' Chl *a* (absorbing at 690 nm) in CP47 [17-19]. The fourth band, F700, is characteristic for the aggregated LHCII [10-13]. It was shown [15,16] that at 77 K the spectral shape of LHCII trimers' fluorescence coincides with the fluorescence lineshape of monomeric Chl *a* in a solution [20]. The emission bands F685 and F695 in PSII fluorescence also arise from monomeric Chl *a* molecules [17-19] with different red shift in their absorption and emission maxima as compared to the maximum of trimeric LHCII, caused by various protein environment and interactions. Komura et al. [19] estimated that although the peak maximum of F685 was red-shifted, the value of its full width at half maximum (FWHM) was very close to the reported one in isolated LHCII. That is why we used the emission spectrum of trimeric LHCII for F680 and F685 spectral line shape with appropriate red shift. It was shown that the 77 K spectrum of F695 was not only red shifted as compared to the trimeric LHCII, but with FWHM about 1.5 times as broad as the trimeric LHCII one [17,19]. At critical micelle concentration (CMC) LHCsII are presented only in trimeric forms, so at this concentration the band F700 is absent and only the bands F680, F685 and F695 contribute to the total PSII spectrum. The subtraction of the sum of F680 and F685 from the total emission spectrum of PSII membranes at CMC should yield the spectral line shape of F695. To determine the scale factors and red shifts of F680 and F685 a program is made, fitting the sum of F680, F685 and F695 to the experimental PSII spectrum in the spectral range from 650 nm to 689.2 nm, where the contribution of F700 is negligibly small, using the least square method. The fitting parameters for the amplitudes of F680, F685 and F695 were the only free-running parameters. The shifts for F680 and F685 were varied in very narrow limits (0.5-1 nm) only, determined from the literature data, while the position of F695 band was fixed at the position of the spectrum obtained at CMC concentration. Considering that the experimental PSII spectrum is composed of the four bands, we obtain the fourth band F700 as a difference between the total spectrum and the sum of F680, F685 and F695.

*2.5. Resonance Raman spectra measurements*

Room temperature Resonance Raman (RR) spectra were measured using a microRaman spectrometer (Jobin-Ivon, HR 800). The excitation was provided by an argon ion laser (Innova 307, Coherent) at 457.8, 488 and 514.5 nm. The laser intensities were 4 mW and the spectral resolution was 0.5 $cm^{-1}$ for the measurements using a grating of 1800 gr/mm.

## 3. Results and discussion

*3.1. Changes in 77K fluorescence spectra*

The 77 K fluorescence emission spectra of PSII particles upon 436, 472 and 515 nm excitation are compared in figure 1 a,b,c. They are typical for PSII, having two clearly expressed maxima at 685 nm and 695 nm.

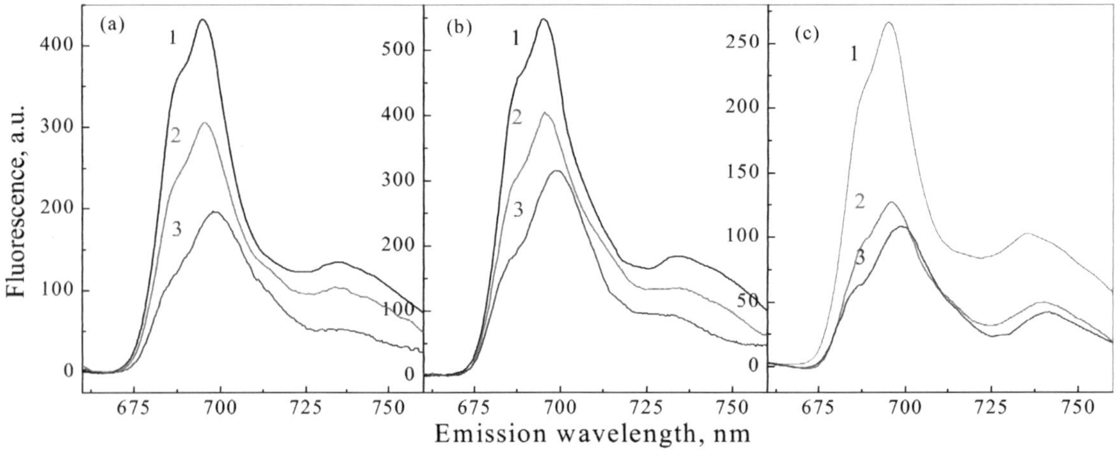


**Figure 1.** 77 K fluorescence spectra of control and light treated PSII particles: 1 - 0 min; 2 - 15 min and 3 - 45 min upon (a) 436, (b) 472 and (c) 515 nm excitation.

High-light treatment results in a strong quenching of the fluorescence and a red-shift of the main fluorescence maximum, from 695 nm in control particles to 695.4 nm and 699 nm after 15 and 45 min illumination with light intensity of 1250 μE $m^{-2}$ $s^{-1}$, respectively. Quenching is the strongest when the excitation is absorbed by carotenoids (figure 1c). The red shift suggests that the relative contribution of F695 and F700 to the whole spectrum is increased. In order to analyze in detail the contribution of the distinct emitting Chl *a* pools as well as their behavior during illumination, the spectra were decomposed as it was described in Materials and Methods. As an example of such decomposition, a typical fit is shown in figure 2.

The effect of the time of illumination on the components: F680, F685, F695, and F700 is presented in figure 3 a,b,c. All bands diminished during illumination indicating intensive quenching of the fluorescence and possible bleaching of Chl *a* and carotenoids.

The analysis of decomposed spectra shows that high-light treatment causes reduction of all fluorescence bands, mostly affecting the fluorescence of F685 and F695 bands. The increased relative contribution of aggregated LHCII to the whole spectrum and the progressive red shift of about 3 nm confirms that prolonged strong illumination could lead to increased energy transfer to the Chl *a* pools, emitting at 700 nm and enhanced extent of LHCII aggregation [16,21].

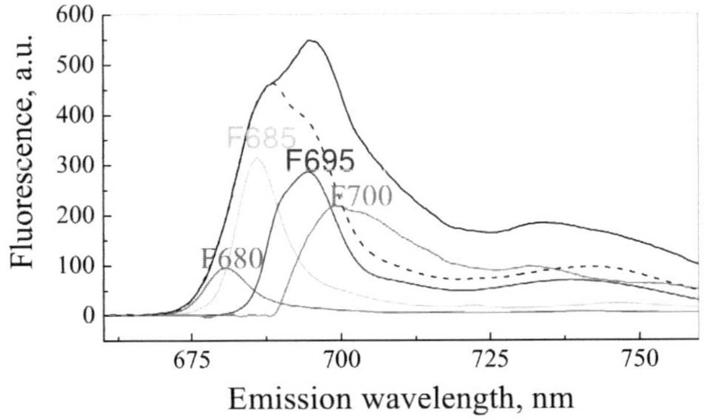


**Figure 2**. Typical decomposition of the 77 K fluorescence spectrum with four components F680, F685, F695 and F700. The sum of F680, F685 and F695 is shown with dashed line.

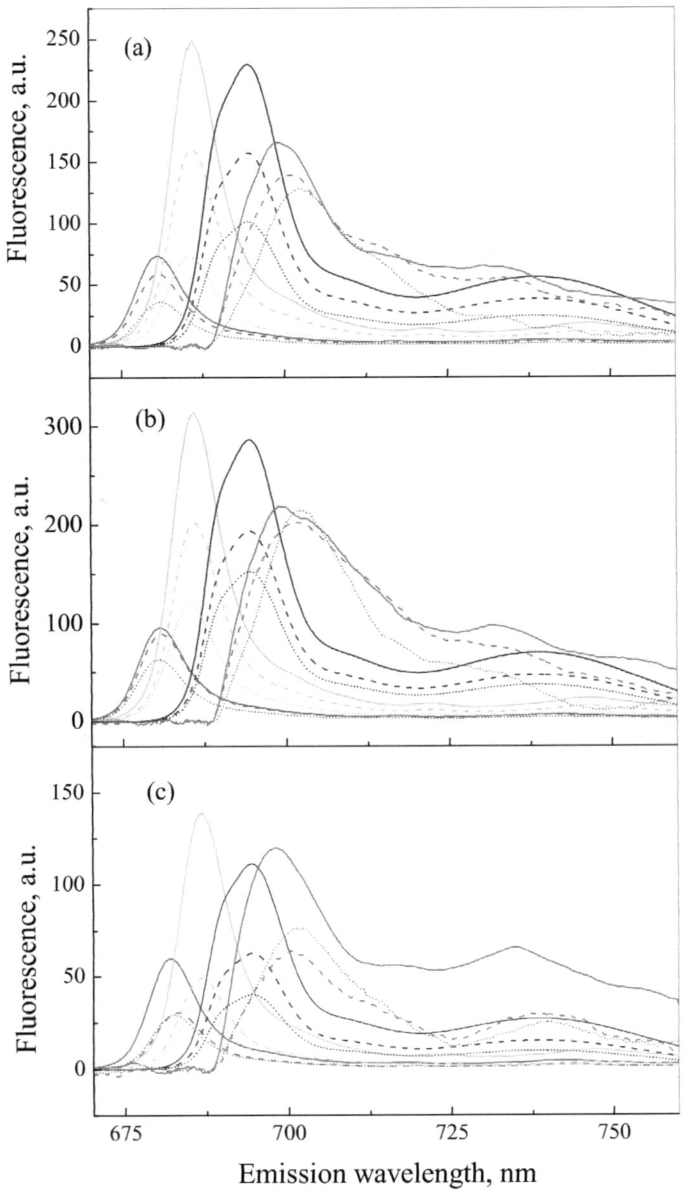


**Figure 3.** Effect of the duration of illumination: solid line - 0 min, dashed line - 15 min, dotted line - 45 min on the components: F680, F685, F695, and F700 upon (a) 436, (b) 472 and (c) 515 nm excitation.

### *3.2. Changes in the resonance Raman spectra*

We used RR scattering in an attempt to investigate possible photobleaching of carotenoid molecules presenting RR spectra of plant BBY particles at room temperature for the first time in figure 4. The spectra contained the four main bands (called from $\nu_1$ to $\nu_4$) characteristic of carotenoids [22,23]. The main carotenoid bands have been assigned [22,23] as follows: $\nu_1$ – to C=C bonds in phase stretching vibrations; $\nu_2$ – to $C_{14}$-$C_{15}$ stretches coupled to $C_{15}$-H in plane bending; $\nu_3$ – to methyl $CH_3$ in plane rocking vibrations; $\nu_4$ – to C-H out of plane bending modes coupled with $C_7$=$C_8$ torsion.

Light treatment did not cause any change in the shape of the Raman spectra. Some decrease in the intensity was observed only for the spectra upon 514.5 nm excitation comparing to the spectra upon 488 nm excitation, for which almost no change was observed.

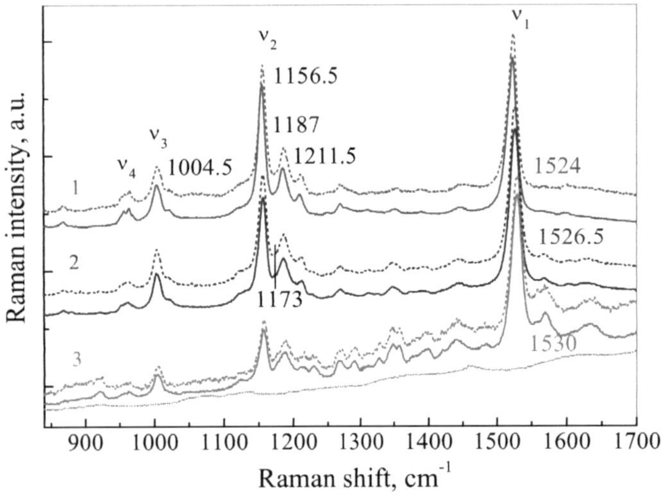


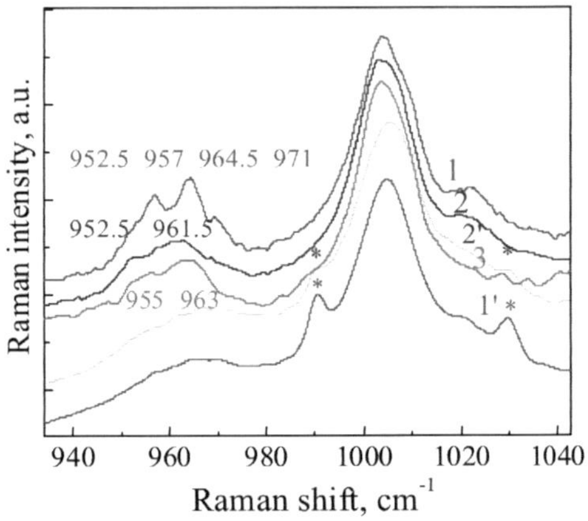


**Figure 4.** Raman spectra of control (solid line) and 45 min light treated BBY (dotted line) upon excitation: 1- 514.5 nm, 2 - 488 nm and 3 - 457.8nm. The spectra are normalized to $\nu_1$ band intensity. Spectrum of the buffer is also shown (short dotted line).

**Figure 5.** $\nu_4$ and $\nu_3$ Raman regions of BBY upon excitation: 1 - 514.5 nm, 2 - 488 nm and 3 - 457.8 nm; Lut: 1` - 514.5nm and ß-Car: 2` - 488 nm. The spectra are normalized to $\nu_3$ band intensity. Asterisks indicate Raman bands of pyridine.

In the control spectra of BBY obtained upon the different excitations the most pronounced changes were in $\nu_1$ band position and in the structure of $\nu_2$ and $\nu_4$ regions (figure 4 and figure 5). Since at RT the contribution of distinct carotenoids to RR spectra is very complex we also present RT spectra of ß-Car and Lut in pyridine (table 1). The refractive index of the solvent used is the closest to that of the protein lipid environment of the pigments. Upon 514.5 nm excitation RR spectra of the two pigments mainly differ in the $\nu_1$ band position, which is further shifted for Lut and in the $\nu_2$ region where a new band, 1173 cm$^{-1}$, exists in the spectrum of ß-Car. For the same excitation in the BBY spectrum $\nu_1$ appears only 2 cm$^{-1}$ shifted from $\nu_1$ of Lut and no traces of 1173 cm$^{-1}$ can be seen. Since such small shifts of $\nu_1$ in spectra of photosynthetic complexes are often related to different dielectric properties of the environment and pigment-protein interactions we can conclude that at 514.5 nm excitation the LHCII bound Lut RR spectrum is dominant. The great difference in the $\nu_4$ region and specifies that its structure is much more distorted when it is bound to BBY. The comparison for 488 nm excitation reveals that the predominant contribution in BBY spectrum comes most probably from ß-Car since the position of the most intensive Raman bands in $\nu_2$ and $\nu_3$ regions resembles the spectrum of ß-Car, there is a trace of 1173 cm$^{-1}$ and because the presence of neoxanthin in these complexes is low. The spectrum of BBY upon 457.8 nm mainly differs in the position of $\nu_1$ and matches the $\nu_4$ region of the spectrum upon 488 nm excitation. Since the content of ß-Car is predominant compared to other carotenoids we conclude that the signal mainly comes from ß-Car.

Journal of Physics: Conference Series **253** (2010) 012065 doi:10.1088/1742-6596/253/1/012065

The results show that there is no significant bleaching of ß-Car molecules whereas the Lut is weakly bleached. This could be explained by the shielding effect of the antenna that surrounds the core, preserving ß-Car molecules from harmful action of the high-light intensities.

**Table 1.** The main Raman lines in the spectra of Lut and ß-Car in pyridine.

| Exc. | Pigment | $\nu_1$, cm$^{-1}$ | | $\nu_2$, cm$^{-1}$ | | | $\nu_3$, cm$^{-1}$ | | | $\nu_4$, cm$^{-1}$ | |
|---|---|---|---|---|---|---|---|---|---|---|---|
| **514.5 nm** | β-Car | 1520 | 1212 | 1190 | 1173 | 1156 | 1004 | | 969 | 962 | 955 |
| | Lut | 1522 | 1212 | 1190 | | 1156 | 1004 | 972 | 966 | | 954 |
| **488 nm** | β-Car | 1524 | 1213 | 1189 | 1173 | 1156.5 | 1005.5 | 972 | 966 | 962 | 956 |
| | Lut | 1524 | 1211 | 1187 | | 1153.5 | 1002 | | 965 | | 952 |

**Acknowledgements**

This study was supported by Grants: No. DO02-167-2008 from the National Science Fund of the Ministry of Education and Science of Bulgaria and No. 050-2010 from the Scientific Research Foundation at Sofia University, Bulgaria.

**References**

[1] Mc Evoy J P and Brudvig G W 2006 *Chem. Re.* **106** 4455
[2] Messinger J and Renger G 2007 *Primary Processes of Photosynthesis: basic principles and apparatus* vol II, ed G Renger (Cambridge: Royal Society Chemistry) pp 295-352
[3] Renger G 2007 *Photosynt. Res.* **92** 407
[4] Dekker J P and Boekema E J 2005 *Biochim. Biophys. Acta* **1706** 12
[5] Guskov A, Kern J, Gabdulkhakov A, Broser M, Zouni A and Saenger W 2009 *Nature Struct. Mol. Biol.* **16** 1068
[6] Berthold D A, Babcock G T and Yocum C F 1981 *FEBS Lett.* **134** 231
[7] Lichtenthaler H K 1987 *Methods Enzymol* vol 148, eds S P Colowick, N O Kaplan, L Packer and R Donce (New York: Academic Press) pp 350–382
[8] Krause G H, Briantais J M and Vernotte C 1983 *Biochim.Biophys. Acta* **723** 169
[9] van Grondelle R, Dekker J P, Gillbro T and Sundström V 1994 *Biochim. Biophys. Acta* **1187** 1
[10] Ruban A V, Calkoen F, Kwa S L S, van Grondelle R, Horton P and Dekker J P 1997 *Biochim. Biophys. Acta* **1321** 61
[11] Andreeva A, Stoitchkova K, Busheva M and Apostolova E 2003 *J. Photochem. Photobiol.* B **70** 153
[12] Stoitchkova K, Busheva M, Apostolova E and Andreeva A 2006 *J. Photochem. Photobiol.* B **83** 11
[13] Haferkamp S, Haase W, Pascal A, van Amerongen H and Kirchhoff H 2010 *J. Biol. Chem.* **285** 17020
[14] Hemerlrijk P W, Kwa L S, van Grondelle R and Dekker J P 1992 *Biochim. Biophys. Acta* **1098** 159
[15] Palacios M A, De Weerd F, Ihalainen J A, van Grondelle R and van Amerongen H 2002 *J. Phys. Chem.* **106** 5782
[16] Andreeva A, Abarova S, Stoitchkova K and Busheva M 2009 *Eur. Biophys. J.* **38** 199
[17] Andrizhiyevskaya G E, Chojnicka A, Bautista J A, Diner B A, van Grondelle R and Dekker J P 2005 *Photosynth. Res.* **84** 173
[18] Krausz E, Hughes J L, Smith P J, Pace R J and Arskold S P 2005 *Photosynth. Res.* **84** 193
[19] Komura M, Shibata Y and Itoh S 2006 *Biochim. Biophys. Acta* **1757** 1657
[20] Boardman N K and Thorne S W (1971) *Biochim. Biophys. Acta* **46181** 222
[21] Van Oort B, van Hoek A, Ruban A V and van Amerongen H 2007 *FEBS Lett.* **581** 3528
[22] Robert B 2009 *Photosynth. Res.* **101** 147
[23] Koyama Y and Fujii R 1999 *The Photochemistry of Carotenoids* eds H Frank, A Young, G Britton and R Cogdell (Dordrecht: Kluwer Acad. Publ.) pp. 161-188

Journal of Physics: Conference Series **253** (2010) 012066 doi:10.1088/1742-6596/253/1/012066

# Integration of β-carotene molecules in small liposomes

**Atanaska Andreeva[1,3] and Antoaneta Popova[2]**

[1]Sofia University, Faculty of Physics, Department of Condensed Matter Physics, 5, J. Bourchier blvd., 1164 Sofia, Bulgaria.
[2]Institute of Biophysics, Bulgarian Academy of Sciences, Acad. G. Bonchev str. Bl 21, 1113 Sofia, Bulgaria

E.mail: andreeva@phys.uni-sofia.bg

**Abstract**. The most typical feature of carotenoids is the long polyene chain with conjugated double bonds suggesting that they can serve as conductors of electrons, acting as ''molecular wires'', important elements in the molecular electronic devices. Carotenoids are essential components of photosynthetic systems, performing different functions as light harvesting, photoprotection and electron transfer. They act also as natural antioxidants. In addition they perform structural role stabilizing the three-dimensional organization of photosynthetic membranes. Carotenoids contribute to the stability of the lipid phase, preserving the membrane integrity under potentially harmful environmental conditions. Carotenoids can be easily integrated into model membranes, facilitating the investigation of their functional roles. In carotenoid-egg phosphatidylcholine (EPC) liposomes *β*-carotene is randomly distributed in the hydrocarbon interior of the bilayer, without any preferred, well defined orientation and retains a substantial degree of mobility. Here we investigate the degree of integration of *β*-carotene in small unilamellar EPC liposomes and the changes in *β*-carotene absorption and Raman spectra due to the lipid-pigment interaction. All observed changes in *β*-carotene absorption and Raman spectra may be regarded as a result of the lipid-pigment interactions leading to the polyene geometry distortion and increasing of the environment heterogenety in the liposomes as compared to the solutions.

## 1. Introduction

Carotenoids are widespread natural molecules, which play multiple important physiological functions.The most typical feature of carotenoids is the long polyene chain with conjugated double bonds [1]. The long chain of conjugated bonds acts like a wire, allowing the electrical energy to move from one side of the molecule to the other, so called ''molecular wire'' [2]. Molecular wires, which would allow electron flow to take place between different components, are important elements in the design of molecular devices. Carotenoids are essential components of biological membranes performing multiple functions. In photosynthetic membranes they play role in light-collecting, photoprotection and electron transfer as well as stabilizing the three-dimensional integrity of bacterial and plant antenna complexes and for the assembly of functional photosystem II [3,4]. The structural role of carotenoids is probably not restricted only to the photosynthetic membranes but can be

[3] To whom any correspondence should be addressed.

Journal of Physics: Conference Series **253** (2010) 012066 doi:10.1088/1742-6596/253/1/012066

extended to the stability of the lipid phase, preserving the membrane integrity under potentially harmful environmental conditions [5].

To unravel the carotenoid contribution to the stability of the lipid phase, small unilamellar lipid vesicles containing pigment molecules have been used as a model system. Carotenoids can be easily integrated into model membranes thus making easier the investigation of the membrane thermodynamic and mechanical properties. The orientation of pigments within bilayer is dependent on the structure of the particular carotenoid and lipid bilayer composition. $\beta$-carotene molecules are randomly distributed in the hydrocarbon interior of carotenoid-egg phosphatidylcholine (EPC) liposomes without any preferred, well defined orientation and retains a substantial degree of mobility [6,7]. $\beta$-carotene, when integrated into EPC liposomes, was shown to increase motional freedom of both lipid headgroup and alkyl chains in liquid crystalline state [6].

Here we investigate the degree of integration of $\beta$-carotene in small unilamellar EPC liposomes and the changes in $\beta$-carotene absorption and Raman spectra due to the lipid-pigment interactions.

## 2. Materials and methods

### *2.1. Materials*

$\beta$-carotene, TRIS (Tris-hydroxymethylaminomethane) and pyrene were obtained from Sigma, EPC from Avanti Polar Lipids and used as purchased.

### *2.2. Liposome formation*

Lipids equivalent to 1mM (0.75mg/ml) dissolved in chlorophorm, pure or mixed with 1, 2.5 or 5 mol % $\beta$-carotene (1 mM in chlorophorm), were dried from the solvent under a gentle steam of $N_2$ to obtain a thin lipid layer on the bottom of a glass test tube. Final traces of solvent were removed form the lipid film under deep vacuum for at least 4 h followed by hydration with 50 mM TRIS buffer (pH 7.4). Small unilamellar liposomes were formed by sonication for 20 sec using ultrasound generator system. The residual, not integrated into liposomes $\beta$-carotene was removed by two steps centrifugation at 15 000xg. Supernatant contained only $\beta$-carotene-doped-EPC liposomes [8].

### *2.3. Determination of $\beta$-carotene concentration*

Concentration of $\beta$-carotene incorporated into liposomes was evaluated from absorbance spectra (350-600 nm) of ethanol extract, using molar extinction coefficient of $\beta$-carotene in ethanol at 453 nm, $141x10^3$ $M^{-1}$ $cm^{-1}$ [9]. The results were used to calculate the $\beta$-carotene incorporation yield (IY) as the ratio between concentrations of integrated and applied $\beta$-carotene.

### *2.4. Steady-state fluorescence measurements*

Membrane fluidity of pure EPC liposomes and of those containing different concentrations $\beta$-carotene was determined using the fluorescent probe pyrene that forms excimers (E) from monomers (M) by a diffusion-controlled process in fluid membranes that is directly related to membrane fluidity [10]. Pyrene was added to liposomes at concentration 4 mol % in respect to lipids from a stock ethanol solution (2 mM). Care was taken the amount of added ethanol not to exceed 0.5%. Liposome-pyrene mixture was incubated at $25^0$C for 30 min and pyrene fluorescence was recorded with a Jobin-Ivon spectrofluorimeter. Fluorescence was excited at 332 nm and registered in the region 350-550 nm. Slit widths were 4 nm. Pyrene, being apolar molecule, is completely buried within the hydrophobic region of the membrane [10]. In a monomeric form pyrene gives a rise to a fluorescent peak at 393 nm (F393) and a broad one at 470 nm, coming from the excimers (F470).

### *2.5. Resonance Raman spectra measurements*

Room temperature resonance Raman (RR) spectra were measured using a microRaman spectrometer (Jobin-Ivon, HR 800) with a grating 1800 gr/mm. The excitation was provided by an argon ion laser (Innova 307, Coherent) at 488 nm. The laser intensity was 4 mW, the spectral resolution – 0.5 $cm^{-1}$.

## 3. Results and discussion

In figure 1 are presented the chemical structures of the predominant lipid class of EPC (64% POPC – palmitoyl oleoyl phosphatidyl choline) (a) and of *β*–carotene (b). The relative length of both molecules is comparable.

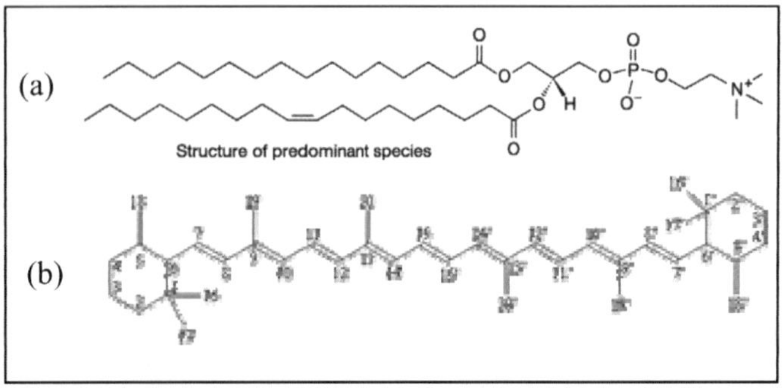


**Figure 1**. Chemical structure of predominant lipid species of EPC (a) and of *β*-carotene (b).

Liposomes were formed either from pure EPC or in the presence of *β*-carotene different concentrations - 1, 2.5 and 5 mol % in respect to lipids denoted as epc1, epc2 and 3, respectively. The absorbance spectra of *β*-carotene-doped liposomes, in TRIS buffer (figure 2a), are compared to the spectra of ethanol extracts from EPC liposomes, formed in the presence of different concentrations of *β*-carotene (figure 2b). The spectra of ethanol extracts show a typical carotene contour, with two maxima, at 454 and 476 and a shoulder at 422nm (figure 2b). In EPC liposomes spectra an additional absorbance peak is observed at 518 nm (see arrows in figure 2a).

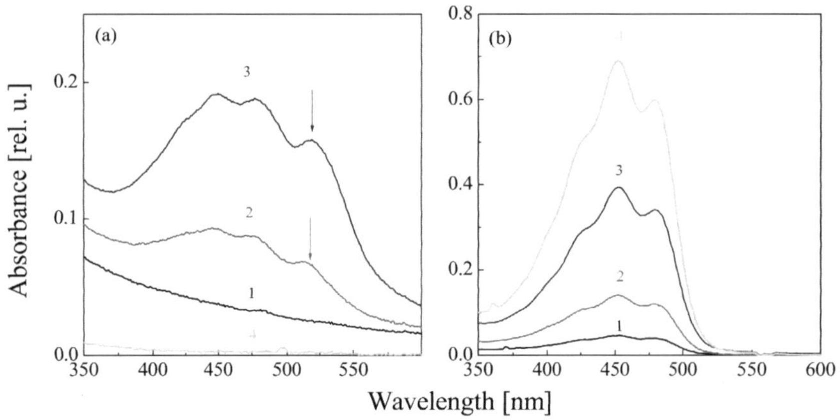


**Figure 2.** Absorbance spectra of *β*-carotene, integrated into liposomes dissolved in TRIS buffer (a) and extracted from liposomes with ethanol (b): 1 – epc1, 2 – epc2, 3 – epc3. For comparison are included the curves 4: of 22 μM *β*-carotene in TRIS buffer (a) and of 5 μM *β*-carotene in ethanol (b).

The peak at 454 nm is used for determining the concentration of *β*-carotene, integrated into small unilamellar EPC liposomes using the molar extinction coefficient of carotene in ethanol (figure 3a). As expected the concentration of *β*-carotene integrated into liposomes is linearly dependent on the concentration of applied pigment. The values of integrated and applied *β*-carotene concentrations were used to calculate the integration yield of the pigment (IY) (figure 3b). The obtained data are in a good agreement with the values obtained for IY of *β*-carotene in dipalmitoylphosphatidylcholine (DPPC) liposomes [8].

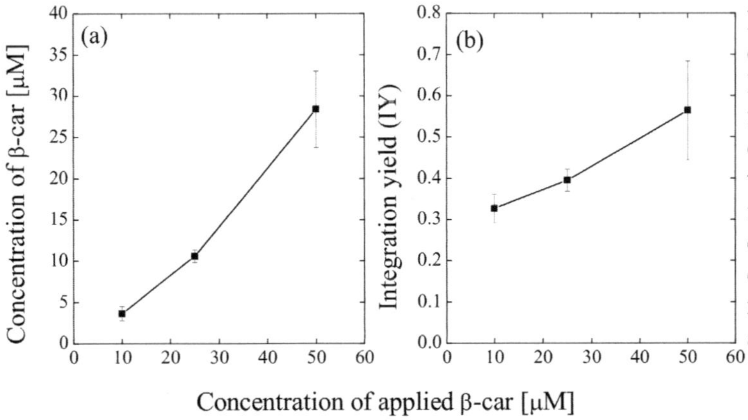


**Figure 3**. Concentration of *β*-carotene in EPC liposomes as determined from absorbance spectra of the pigment extracted with ethanol (a) and integration yield (IY), determined by the ratio concentration in liposomes/concentration of applied *β*-carotene (b).

As the observed in EPC liposomes additional absorbance peak at 518 nm is missing in the spectra of pure liposomes of the same concentration and changes in height with increase of pigment concentration, nevertheless that the concentration of liposomes is the same for all the samples (1 mM lipids), it can supposed to be due to lipid-pigment interactions. To further verify this possibility we calculated two types of ratio from the *β*-carotene spectra – A454/A476 and A454/A516, for both ethanol extract of carotene from different liposomes and of the same vesicle, dissolved in buffer. The ratio A454/A476 for the pure *β*-carotene and the ethanol extracts with different pigment concentration, dissolved in ethanol, being 1.15±0.005, 1.18±0.02, 1.17±0.01, and 1.16±0.01, respectively, is very similar, nevertheless that the calculated from the spectra concentrations of *β*-carotene are different. Both calculated ratios, for liposomes containing different concentrations of *β*-carotene and dissolved in buffer, are decreasing in value: 1.44±0, 1.39±0 and 1.03±0 for A454/A476 and 1.44±0.04, 1.39±0.04 and 1.04±0.05 for A454/A516 with increase of pigment concentration. This indicates that the observed peak at 518 nm in *β*-carotene-EPC liposomes spectra is due to the lipid-pigment interactions.

The changes in pyrene fluorescence in *β*-carotene-EPC liposomes are presented in figure 4.

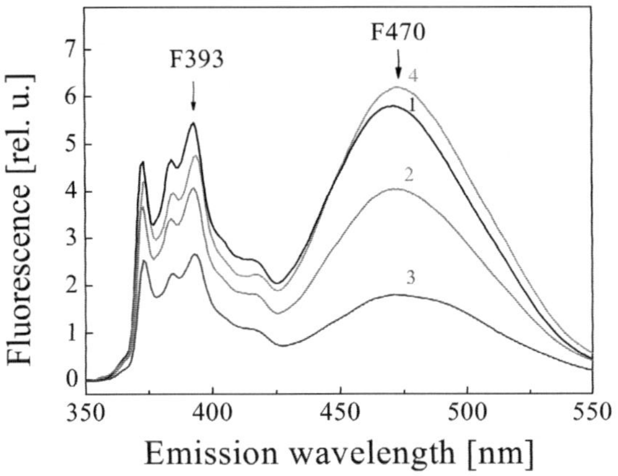


**Figure 4**. Pyrene fluorescence in EPC liposomes, in the presence of *β*-carotene: 1 – epc1, 2 – epc2, 3 – epc3 and 4 – pure. The emission maximum of excimers at 470 nm (F470) and of monomers at 393 nm (F393) are marked by arrows.

With increase of the concentration of integrated pigment the overall pyrene fluorescence is decreased. The observed decrease supposes that the pyrene fluorescence is quenched by *β*-carotene, due to an effective energy transfer from pyrene excimers to *β*-carotene. The quenching of pyrene fluorescence indicates that the pigment molecules are localized in the hydrophobic environment of the bilayer, in the vicinity of pyrene molecule, and the distance between them is not higher than 7 Å in order the supposed energy transfer to take place.

To further analyse the pigment-lipid interactions in EPC liposomes we used RR scattering. The RR spectra of $\beta$-carotene-EPC liposomes excited at 488 nm at room temperature are compared to the spectrum of $\beta$-carotene dissolved in buffer in figure 5. They manifest the characteristic for carotenoids' Raman spectrum four main frequency bands (called from $\nu_1$ to $\nu_4$). The main carotenoid bands have been assigned as follows: $\nu_1$ to in phase stretching vibrations of the C=C bonds, $\nu_2$ to C-C stretching coupled to $C_{15}$-H in plane (ip) bending, $\nu_3$ to methyl $CH_3$ ip rocking vibrations, ; $\nu_4$ – to C-H out of plane bending modes coupled with $C_7$=$C_8$ torsion [11-13].

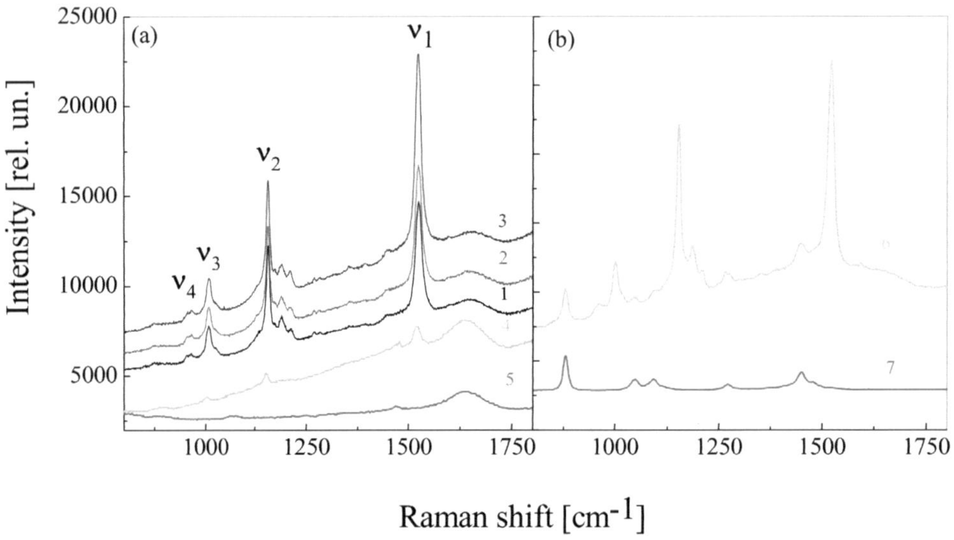


**Figure 5.** Resonance Raman spectra excited at 488 nm: (a) 1 – epc1, 2 – epc2, 3 – epc3, and 4 – $\beta$-carotene in buffer. For comparison are included spectra: (a) 5 – pure buffer and (b) 6 – $\beta$-carotene in ethanol, 7 – pure ethanol.

Differences are mainly observed in the regions of $\nu_4$ and $\nu_2$ bands. This becomes particularly clear from figure 6 where the bands were normalized in (figure 6a) to the intensity of $\nu_3$ band and in (figure 6b) to the intensity of the main $\nu_2$ band. The transitions in $\nu_4$ region are forbidden for fully planar molecules and become significant under conditions in which carotenoid undergoes molecular distortion due to the interaction with its environment [11,14]. Structured $\nu_4$ bands, indicative of out-of-plane distortions of conjugated backbone of the $\beta$-carotene molecule [11,13,14], were only observed in EPC liposomes with incorporated $\beta$-carotene in them. They exhibit two transitions at 954 and 964 $cm^{-1}$. Their intensity increases weakly with the pigment concentration, thus indicating that the polyene geometry is distorted under the pigment inclusion in liposomes. Our experimental data (figure 6b) in the *cis*-isomerization fingerprint region $\nu_2$ (1100-1300 $cm^{-1}$) [11,12,14] did not show any transition around 1134 $cm^{-1}$. It leads us to the conclusion that the presence of isomers in liposomes is not likely. We observed characteristic frequencies at 1175, 1192 and 1211 $cm^{-1}$. The 1175 $cm^{-1}$ line could be assigned to methyl rocking, where the others two to localized C-C stretching modes, according to the recently published theoretical results for $\beta$-carotene Raman-active modes [15].

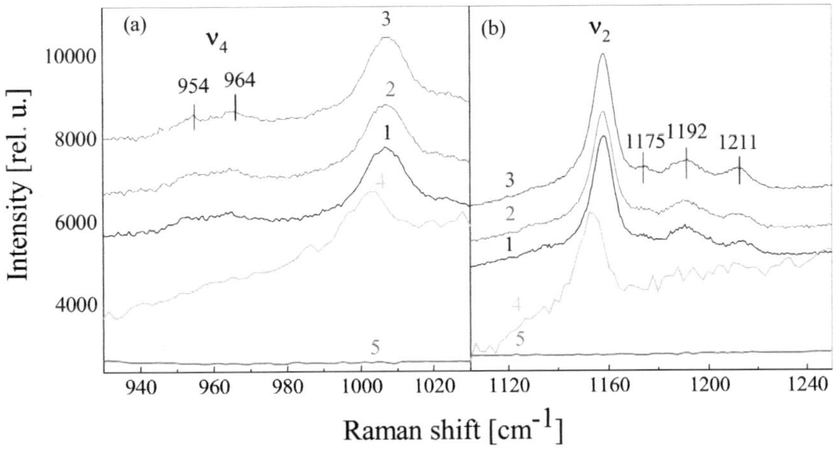


**Figure 6**. Two regions of the *β*-carotene resonance Raman spectra (shown in figure 5), where the differences are clearly seen: the $\nu_4$ region (a) and $\nu_2$ region (b). Original spectra were normalized in (a) to the intensity of $\nu_3$ band and in (b) to the intensity of $\nu_2$ band. Normalized spectra are upshifted for better visualization.

All observed changes in *β*-carotene absorption and Raman spectra may be regarded as a result of the lipid-pigment interactions leading to the polyene geometry distortion and increasing of the environment heterogenety in the liposomes as compared to the solutions.

**Acknowledgements**
This study was supported by Grants: No. DO02-167-2008 from the National Science Fund of the Ministry of Education and Science of Bulgaria and No. 050-2010 from the Scientific Research Foundation at Sofia University, Bulgaria.

**References**
[1] Fraser P D and Bramley P M 2004 *Progress in Lipid Research* **43** 228
[2] Visoly-Fisher I, Daie K, Terazono Y, Herrero C, Fungo F, Otero L, Durantini E, Silber J J, Ereno L, Gust D, Moore T A, Moore A L and Lindsay S M 2006 *PNAS* **103** 8686
[3] Moskalenko A A and Karapetyan N V 1996 *Z. Naturforsch.* **51c** 763
[4] Frank H and Cogdell R J 1993 *Carotenoids in photosynthesis*, ed A Young and G Britton (London: Chapman&Hall) pp. 253-326
[5] Havaux M 1998 *Trends Plant Sci.* **30** 147
[6] Gabrielska J and Gruszecki W I 1996 *Biochim. Biophys. Acta* **1285** 165
[7] Gruszecki W I and Strzałka K 2005 *Biochim. Biophys. Acta* **1740** 108
[8] Socaciu C, Lausch C and Diehl H A 1999 *Spectrochim. Acta A* **55** 2289
[9] http://epic.awi.de/publications/jef1997f.pdf
[10] Galla H J and Sackmann E 1974 *Biochim. Biophys. Acta* **339** 103
[11] Robert B 2009 *Photosynth. Res.* **10** 147
[12] Koyama Y and Fujii R 1999 *The Photochemistry of Carotenoids* ed H Frank, A Young , G Britton and R Cogdell (Dordrecht: Kluwer Acad. Publ.) pp. 161-188.
[13] Andreeva A and Velitchkova M 2005 *Biophys. Chem.* **114** 129.
[14] Koyama Y, Takatsuka I, Nakata M and Tasumi M 1988 *J. Raman Spectrosc.* **19** 37
[15] Tschirner N, Schenderlein M, Brose K, Schlodder E, Mroginski M A, Thomsen C and Hildebrandt P 2009 *Phys. Chem. Chem. Phys.* **11** 11471

# Dielectric function of polymer nanocomposites in small filling factor approximation

**S Sainov[1], I Vlaeva[1], T Yovcheva[2], V Dragostinova[1] and S Stavrev[3]**

[1] Institute of Optical Materials and Technologies "Acad. J. Malinovski", Bulgarian Academy of Sciences, Acad. G. Bonchev Street, Bl. 101/109, 1113 Sofia, Bulgaria
[2] Department of Experimental Physics, University of Plovdiv "P. Hilendarski", 24 Tsar Assen Street, 4000 Plovdiv, Bulgaria
[3] "Nano" S.S. Ltd, 4700 Smolyan, Bulgaria

E-mail: ssainov@optics.bas.bg

**Abstract**. The dielectric function of polymer nanocomposites is examined in a small filling factor approximation (SFFA). The linear concentration dependence on SFFA is experimentally confirmed by refractometric investigations of PVA thin films doped with titania and diamond nano-sized particles.

## 1. Introduction

Due to their interesting optical properties, ease of production, low cost, light weight, etc., organic-inorganic hybrid materials have attracted extensive scientific interest [1]. Nano-sized dopants offer ways to improve their mechanical properties together with good transparency and tunable refractive index [2]. For this purpose, high refractive index (*RI)* inorganic nanoparticles with no absorption in the visible region have been used (see [3] and references therein). The most widely used are titania ($TiO_2$) nanoparticles, that increase considerably the holographic diffraction efficiency, as was announced in [4, 5]. The technological advance in this field is very impressive: if twenty years ago $TiO_2$-containing polymer nanocomposites had a tunable *RI* up to 1.76 [6], recently, synthesis of optically transparent polyimide-$TiO_2$ thin films was reported with tunable *RI* in the range 1.583 - 2.029 [7]. Transparent $TiO_2$ – epoxy nanocomposite film with extremely high *RI* in the range 2.18 - 2.38 was prepared three years ago [8]. The investigation of the effective dielectric function $\varepsilon$ of these nanocomposite materials can elucidate the interaction between the polymer matrix and the embedded inorganic high refractive index nanoparticles. The technology of obtaining high *RI* polymer nanocomposites is a real challenge, but at the same time small *RI* changes are very important for optical applications.

In the theoretical background of the present investigation we use a small filling factor approximation (SFFA) to compare the most well known Maxwell Garnett [9], Bruggeman [10] and Odelevskii [11] mixing rules. The theoretical conclusions are confirmed experimentally by the refractive index (*RI*) measurements of polyvinyl alcohol (PVA) thin films 7 μm thick, containing different concentration of both 0 – 3 wt.% diamond and 0 – 2 wt.% $TiO_2$ nano-particles.

Journal of Physics: Conference Series **253** (2010) 012067 doi:10.1088/1742-6596/253/1/012067

## 2. Theoretical background

Following [12], we shall consider dielectric spherical particles with dielectric function $\varepsilon_i$ and diameter $d$ embedded in a polymer matrix with dielectric function $\varepsilon_m$. The filling factor is determined by the next relation:

$$f = \frac{p\pi d^3/6}{V_m + p\pi d^3/6} = \frac{V_i}{V_m + V_i}, \tag{1}$$

where $V_m$ is the volume of the matrix, $p$ is the number of particles, $V_i$ is the total particles' volume.

In the following consideration we shall use small filling factor approximation (SFFA) – $f \ll 1$.

From equation (1) using SFFA we get:

$$f = \frac{V_i}{V_m(1 + V_i/V_m)} \approx \frac{V_i}{V_m} = \frac{V_i}{V_m}\frac{\rho_i}{\rho_m}\frac{\rho_m}{\rho_i} = \frac{m_i}{M_m}\frac{\rho_m}{\rho_i} = C\frac{\rho_m}{\rho_i}, \tag{2}$$

where $C$ is the weight concentration, $M_m$, $m_i$ are the weights of the matrix and inclusions, respectively. Accordingly, $\rho_m$ and $\rho_i$ are the specific weights of the matrix and inclusions.

The average electric field $\vec{E}_e$ in the polymer matrix containing spherical nanoparticles is the average of the electric fields inside the inclusions $\vec{E}_i$ and the host polymer matrix $\vec{E}_m$:

$$\vec{E}_e = f\vec{E}_i + (1 - f)\vec{E}_m. \tag{3}$$

In the same way the average electric displacement is:

$$\vec{D}_e = \varepsilon_0\varepsilon_e\vec{E}_e = f\varepsilon_0\varepsilon_i\vec{E}_i + (1 - f)\varepsilon_0\varepsilon_m\vec{E}_m, \tag{4}$$

where $\varepsilon_0$ is the vacuum permittivity and $\varepsilon_e$ is the effective dielectric function.

When the particle's diameter $d$ is too small, the phase of the electromagnetic wave with wavelength $\lambda$ is constant when passing through the particle, i. e. $\frac{2\pi\sqrt{\varepsilon_i}d}{\lambda} \ll 1$, and one can use the quasi-static approximation [13] in order to obtain the electric field inside the inclusions:

$$\vec{E}_i = \frac{3\varepsilon_m}{(\varepsilon_i + 2\varepsilon_m)}\vec{E}_m. \tag{5}$$

From the equations (2-4), the Maxwell Garnett effective medium relation [9] is obtained:

$$\varepsilon_e = \varepsilon_m\left[1 + \frac{3f(\varepsilon_i - \varepsilon_m)}{(\varepsilon_i + 2\varepsilon_m) - f(\varepsilon_i - \varepsilon_m)}\right]. \tag{6}$$

In 1935, Bruggeman [10] proposed that inclusions are embedded in the effective medium where $\varepsilon_i = \varepsilon_m$ and obtained the following expression from the Lorentz-Lorenz relation [14]:

$$f\frac{(\varepsilon_i-\varepsilon_e)}{(\varepsilon_i+2\varepsilon_e)}+(1-f)\frac{(\varepsilon_m-\varepsilon_e)}{(\varepsilon_m+2\varepsilon_e)}=0. \tag{7}$$

From equation (6) the following quadratic equation can be obtained:

$$\begin{aligned}&\varepsilon_e^2-2A\varepsilon_e-\frac{\varepsilon_i\varepsilon_m}{2}=0,\\ &A=\frac{3f(\varepsilon_i-\varepsilon_m)+(2\varepsilon_m-\varepsilon_i)}{4}\end{aligned} \tag{8}$$

The solution of (7) yields:

$$\varepsilon_e=A+\left(A^2+\frac{\varepsilon_i\varepsilon_m}{2}\right)^{1/2}. \tag{9}$$

The same result was obtained by Odolevskii in 1951 [11].

Solving the equation (8), we can use Taylor expansion up to the first order of $f<<1$ and after some transformations we obtain:

$$\left(A^2+\frac{\varepsilon_i\varepsilon_m}{2}\right)^{1/2}=\frac{(2\varepsilon_m+\varepsilon_i)\left\{1+3f\left[(\varepsilon_i-\varepsilon_m)(2\varepsilon_m-\varepsilon_i)/(2\varepsilon_m+\varepsilon_i)^2\right]\right\}}{4}. \tag{10}$$

Finally, we obtain for the effective dielectric function:

$$\varepsilon_e=\varepsilon_m\left[1+3f\frac{(\varepsilon_i-\varepsilon_m)}{(\varepsilon_i+2\varepsilon_m)}\right]. \tag{11}$$

The same result can be achieved from equation (5), neglecting the second term in the denominator ($f<<1$). In the visible spectral region far from the absorption band, refractive index (*RI*) is a square root of the dielectric function (dielectric constant or relative permittivity) from Maxwell's electromagnetic theory ($\varepsilon^{1/2}=RI$). The relation (11) can be easily transformed into:

$$RI=\varepsilon_e^{1/2}=\varepsilon_m^{1/2}\left[1+\frac{3f}{2}\frac{(\varepsilon_i-\varepsilon_m)}{(\varepsilon_i+2\varepsilon_m)}\right]. \tag{12}$$

Despite the different approach used in the Maxwell Garnett, Bruggeman and Odelevskii's mixing theories, in the small filling factor approximation the final results coincide (equations (11) and (12)). Using SFFA, we can represent the dependence of the nanocomposite film *RI* on the particle concentration. The relative difference of refractive indices of polymer nanocomposite $n_e$ and polymer matrix $n_m$ is a linear function of the concentration in a first approximation. From (2) and (12) we have:

Journal of Physics: Conference Series **253** (2010) 012067 doi:10.1088/1742-6596/253/1/012067

$$\frac{n_e - n_m}{n_m} = const\,.\,C\,. \tag{13}$$

This can be experimentally proved by the refractometric measurements in the visible spectral region (400 nm – 800 nm). Hence, the simple relation (13) is convenient for experimental confirmation of the obtained theoretical results.

## 3. Experimental

### *3.1. Sample preparation*

The films used in the experiments contained particles of titanium dioxide with a diameter $d = 15$ nm and diamond nano-particles with a diameter of 2 nm. The nano-particles were imbedded in a matrix of polyvinyl alcohol (PVA), which is a solid colourless polymer:

$$[-CH_2 - CH(OH)_2 -]_n\,.$$

In the experiment, polyvinyl alcohol with a molecular mass 22 000 of the "Fluka" company was used, from which suspensions of 6 wt.% water solution of PVA with different quantities of titanium dioxide and diamond nano-particles were prepared.

The titanium dioxide and the diamond nano-particles are in various weight proportions relative to PVA. Their concentrations vary from 0 wt.% to 2 wt.% (titania) and from 0 wt.% to 3 wt.%, (diamond).

Glass substrates with a conductive coating of tin dioxide of the "Merck" company are used. The glass substrates are washed and dipped in a 1 %-alcohol solution of polyvinyl pyrolidone (PVP). This contributes to the improvement of the film adhesion. On a leveled table at a temperature of 20°C certain quantities of the prepared suspensions are deposited which are then dried for 24 hours at room temperature. The films obtained are 10 μm thick.

In our experiments, two types of nano-sized particles with close physical characteristics but with different diameters *d* are selected – titania and diamond. The main properties of the experimental materials are listed in table1.

**Table 1.** Physical characteristics of investigated materials

| Materials | $\varepsilon$ | $\rho$, g/cm$^3$ | *d*, nm |
|---|---|---|---|
| PVA | 2.19 | 1.20 | - |
| Titania | 5.84 | 3.51 | 15 |
| Diamond | 6.25 | 4.24 | 2 |

### *3.2. Refractive index measurements*

The refractive index is measured with a laser microrefractometer [15] at a wavelength of 632.8 nm, illustrated schematically in figure 1.

The laboratory device works on the principle of the critical angle $\varphi_c$ determination by disappearing of the diffraction pattern on the screen (6), obtained by the diffraction from the reflection metal grating (2) placed behind the sample (1) and flint glass measuring prism (4) with $RI = N$. The critical angle value when the diffraction pattern disappears is measured with a rotary table (5).

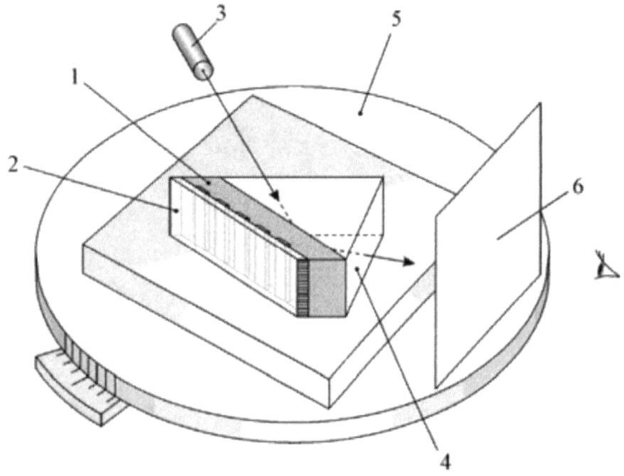


**Figure 1.** Set-up of the laser microrefractometer: 1 – sample; 2 – diffraction grating; 3 – He-Ne laser; 4 – Flint glass prism; 5 – rotary table with 1 arcmin resolution; 6 – screen.

The sample refractive index is calculated by

$$n_e = \varepsilon_e^{1/2} = N \sin\left[A - \arcsin\frac{\sin\varphi_c}{N}\right]. \tag{14}$$

The experimental uncertainty for thin film measurement [15] is $\Delta n_e \leq \pm 1.10^{-3}$.

*3.3. Results and discussion*

The surface refractive indices of the polymer nanocomposite $n_e$ and polymer matrix $n_m$ were measured at 632.8 nm with a He-Ne laser microrefractometer. The relative difference of refractive indices $(n_e - n_m)/n_m$ was calculated and the experimental uncertainty was found to be less than $\pm 3.10^{-3}$.

The experimentally obtained concentration dependence curves are plotted in figure 2 (a) – PVA with diamond nanoparticles, (b) – PVA with titania nanoparticles.

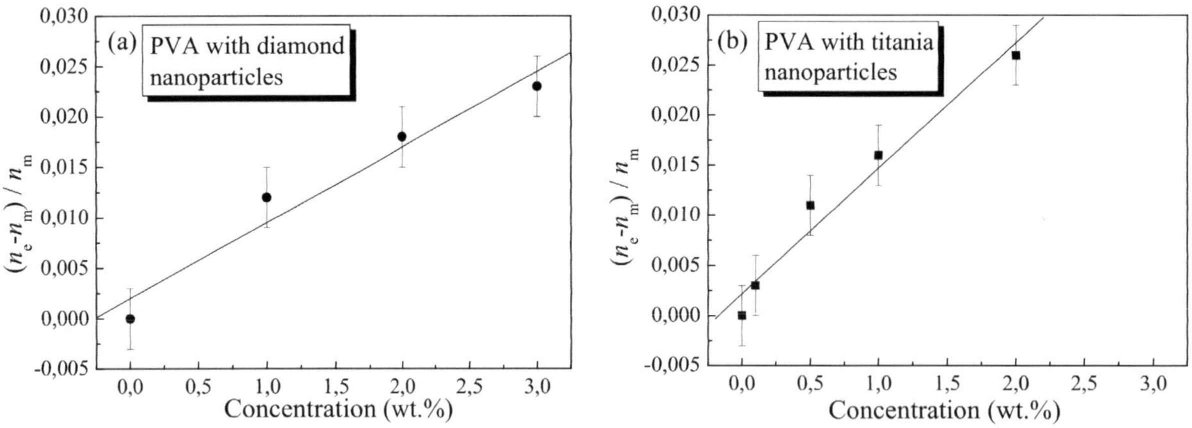


**Figure 2.** Dependence of the relative difference of refractive indices on the concentration for the PVA matrix with both diamond and titania nanoparticles at a wavelength of 632.8 nm.

Our experimental results demonstrate that for the diamond nanoparticle concentrations up to 3 wt.% and for the titania ones up to 2 wt.%, the dependence of the relative difference of refractive indices (polymer nanocomposite and polymer matrix) on the concentration is linear. Therefore, the theoretical linear dependence of the squere root of the dielectric function (refractive index in the visible spectral region) of polymer nanocomposites in small filling factor approximation on the nano-paricle concentration is confirmed in the limits of our experimental uncertainty.

## 4. Conclusion

Despite the different approach used in the Maxwell Garnett, Bruggeman and Odelevskii's mixing theories, in the small filling factor approximation the final results coincide. The linear dependence of the square root of the dielectric function (refractive index in the visible spectral region), of polymer nanocomposite film for small filling factor values, is experimentally confirmed by refractive index measurements of polyvinyl alcohol (PVA) thin films, containing either diamond or $TiO_2$ nano-particles. The different size and corresponding palarizability of the spherical nanoparticles used has no influence on the effective refractive index of PVA polymer nanocomposites. The potential optical applications of the obtained results we foresee are in thin film waveguide techniques, anti-reflective coatings, holographic optical elements, etc.

## Acknowledgements

The authors: S. Sainov, I. Vlaeva, T. Yovcheva and V. Dragostinova are grateful for the financial support under the contract DO-02/155 with the Bulgarian Ministry of Education and Science.

## References

[1] Beecroft L and Ober C K 1997 *Chem. Materials* **9** (6) 1302-07
[2] Jordan J, Jacob K, Tannenbaun R, Sharaf M and Jasiuk I 2005 *Mat. Sci. Eng.* A **393** 1-11
[3] Imai Y, Terahara A, Hakuta Y, Matsui K, Hayashi H and Ueno N 2009 *Eur. Polym. Journal* **45** 630-38
[4] Suzuki N, Tomita Y and Kojima T 2002 *Appl. Phys. Lett.* **81** (22) 4221-23
[5] Tomita Y, Suzuki N and Chikama K 2005 *Opt. Letters* **30** (8) 839-40
[6] Wang B, Wilkes C, Hedrick J, Liptak S and McGrath J 1991 *Macromol.* **24** 3449-50
[7] Chang W, Su H and Chen W 2009 *Eur. Polym. Journal* **45** 2749-59
[8] Chau J, Lin Y, Li A, Su W, Chang K, Hsu S and Li T 2007 *Mat. Lett.* **61** 2908-10
[9] Maxwell Garnett J C 1904 *Philos. Trans. R. Soc.* (London), **203** 385 – 420
[10] Bruggeman D A G 1935 *Ann. Phys.* (Leipzig), **24** 636-79
[11] Odelevskii V I 1953 *J. Techn. Phys.* **21** 667- 78
[12] Lafait J, Berthier S, Andraud C, Reillon V and Boulenguez J 2009 *CR Acad. Sci.* **10** 649-59
[13] Jackson D 1999 *Clasical electrodynamics* (New York: J Willey & Sons)
[14] Aspnes D E 1982 *Am. J. Phys.* **50** 704-9
[15] Sainov S and Dushkina N 1990 *Appl. Opt.* **29** 1406-8

Journal of Physics: Conference Series **253** (2010) 012068 doi:10.1088/1742-6596/253/1/012068

# The surface parameters modifications at nano scale for biomedical applications

**V Safonov[1], A Zykova[1,*], J Smolik[2], R Rogovska[2], N Donkov[3] and V Georgieva[4]**
[1]NSC „Kharkov Institute of Physics and Technology", Kharkov, Ukraine,
[2]Institute of Sustainable Technologies, NRC, Radom, Poland
[3]Institute of Electronics, Bulgarian Academy of Sciences, Sofia, Bulgaria
[4]Institute of Solid State, Physics Bulgarian Academy of Sciences, Sofia, Bulgaria
E-mail: zykov@bi.com.ua

**Abstract.** Functional coatings deposition is an effective way of surface modification with direct control of stoichiometry, impurity elements, functional groups and surface charges. Modified surface properties such as composition, roughness, wettability have effect on the most important processes at biomaterial interface. The aim of present study was the analysis of surface roughness and surface free energy parameters of oxide $Al_2O_3$ and $Ta_2O_5$ coatings and the possibility to separate the influence of such factors on the regularities and mechanisms of nano materials interactions with the biological objects.

## 1. Introduction

Material surface properties such as composition, roughness, topography, wettability can influence events at material interface and play a basic role at processes of dyeing, penetration, chemical absorption, biocompatibility and others. The coatings deposition is an effective way of surface modification with direct control of stoichiometry, impurity elements, functional groups and surface charges. Coatings based on Ti, Al, Zr, Ta oxides exhibit unique properties: high inductivity, density, bio- and chemically inertness, which are very important for next implant and tissue engineering applications. Cell/biomaterial interaction process and direct cell adhesion to a biological surface creates many important phenomena providing vital functions of a living organism on different levels of its organization. The effects of substrate material composition, surface chemistry and surface energy on cell adhesion and proliferation processes have been largely studied [1-4]. The interpretation of data on cell contact formation is crucially dependent on the actual area of membrane region involved in adhesion. The distance between phospholipid bilayers of membranes is in the order of 20-40 nm [5]. The characteristic size scale of cellular receptors interaction is a few 100 nm. Thus the surface relief modifications should be also controlled at a nanometer-submicronic level. The important task of cell/biomaterial interaction research is separation of different factors influenced on process such as surface chemistry, surface energy and roughness.

## 2. Materials and methods

At the present study the samples were glass substrates (Petri dishes), uncoated and oxide coated ($Al_2O_3$ (MS) magnetron sputtered and $Ta_2O_5$ (EB) e-beam evaporated) with different roughness parameters – 20, 200 and 400 nm. $Al_2O_3$ (MS) coatings deposition was performed in high vacuum

[*]To whom any correspondence should be addressed.

pumping system with the base pressure about $10^{-3}$ Pa. The main process parameters were magnetron discharge power 1-8 kW, oxygen source power 1kW, deposition rate 8 μm/hour [6]. The $Ta_2O_5$ films evaporation process was carried out at initial vacuum of $9.3\times10^{-4}$Pa, operational-mode vacuum of $4\times10^{-3}$Pa, anode current of 50mA and calculated evaporation power of 350W [7]. The deposition rate under these conditions was 28nm/min. The layer thickness and the deposition rate were controlled by a digital thin-film deposition monitor MSV-1843/H MIKI-EEV operating at 6MHz.

The coatings adhesion properties, hardness and elastic modulus, thickness were evaluated by standard methods. The surface roughness was measured by means of profilometer Hommel T-2000. surface topography was estimated by AFM methods (Quesant Instrument Corporation,USA). The structure of $Al_2O_3$ magnetron sputtered and $Ta_2O_5$ e-beam evaporated thin films were investigated by means of XPS and XRD methods. X-ray photoelectron spectroscopy was carried out using ESCALAB MkII (VG Scientific) electron spectrometer at a base pressure in the analysis chamber of $6\times10^{-8}$Pa (during the measurement $1.3\times10^{-6}$ Pa), using AlKalpha X-ray source (excitation energy hn=1486.6 eV). The instrumental resolution measured as the full width at a half maximum (FWHM) of the Ag3d5/2, photoelectron peak is 1 eV.

The contact angles were measured by means of tensiometric method in Kruss K12 Tensiometer at temperature 20° C [8]. Prior to contact angle measurements, samples were ultrasonically cleaned in acetone and deionised water and dried. The standard liquids with well-known values of surface tension, component of dispersion and polar interaction such as water, formamide, diiodo methane, ethylene glycol, α-bromo naphthalene and glycerol were used. Also the surface free energy (SFE) and its polar and dispersion components were determined by means of Owens-Wendt-Rabel-Kaelble methods [9].

## 3. Results and discussion

The coatings adhesion properties, hardness and elastic modulus were presented at Table 1. The surface roughness parameters were estimated by means of profilometer Hommel T-2000. at the range 20, 200 and 400 nm (table 2).

**Table 1.** The mechanical properties of $Al_2O_3$ magnetron sputtered and $Ta_2O_5$ deposited by EB-evaporation method coatings on glass substrates.

| Coatings type | Hardness Hv | Hardness H [Mpa] | Young Modulus [Gpa] | Adhesion [N] |
|---|---|---|---|---|
| Glass/ $Al_2O_3$ 20nm | 742.6 | 8594.8 | 203.7 | 33 |
| Glass/ $Al_2O_3$ 200nm | 747.5 | 8614.3 | 207.4 | 36 |
| Glass/ $Al_2O_3$ 400nm | 751.0 | 8105.2 | 209.8 | 37 |
| Glass/ $Ta_2O_5$ 20 nm | 129.35 | 1397.4 | 47.5 | 12.7 |
| Glass/ $Ta_2O_5$ 200 nm | 130.2 | 1412.7 | 48.8 | 12.8 |
| Glass/ $Ta_2O_5$ 400nm | 133.5 | 1452.4 | 49.3 | 13.1 |

The best mechanical properties were obtained in the case of oxide coatings $Al_2O_3$ (MS) deposited by magnetron sputtered method. X-ray diffraction profiles of $Al_2O_3$ (MS) and $Ta_2O_5$ (EB) as-deposited coatings demonstrate the amorphous nature, no peaks were observed. The high-resolution Ta 4f and 4 O 1s XPS spectra show the investigated $Ta_2O_5$ structures. Ta 4f doublets are typical for e-beam evaporated $Ta_2O_5$ and have two peaks: Ta 4f7/2 at ~ 26.2 eV and Ta 4f5/2 at the binding energy 1.9 eV higher. The Ta 4f lines of the deposited films agree well with the Ta 4f doublet representative of the Ta-0 bond in $Ta_2O_5$. The ratio Ta4f /O1s is nearly 0.37. These results clearly demonstrate that the investigated films have stoichiometric $Ta_2O_5$ composition. The O1s spectra further support this assumption. The O 1s peaks of the deposited layers are centered at binding energy of 530.6, which is consistent with reported data for $Ta_20_5$ [10]. Some additional impurities have been observed at the XPS spectra of e-beam evaporated $Ta_20_5$ (figures 1 and 2).

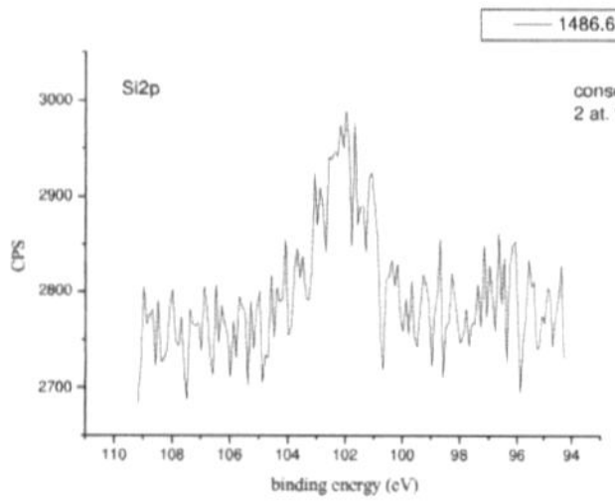


**Figure 1.** Si2p XPS spectra of $Ta_2O_5$ films deposited by EB-evaporation method.

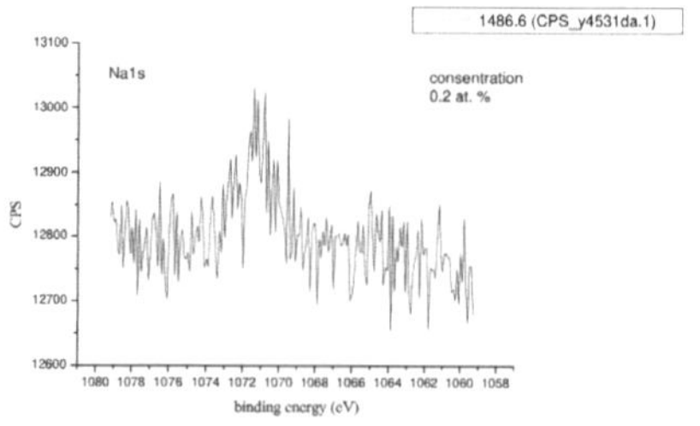


**Figure 2.** Na1s XPS spectra of $Ta_2O_5$ films deposited by EB-evaporation method.

The surface topography was observed by AFM (figure 3).

a) b) c)

**Figure 3.** Surface topography of $Al_2O_3$ (MS) coatings on glass substrates with roughness 20 (a), 200(b) and 400 (c) nm.

The standard liquids with well-known values of surface tension such as water, formamide, diiodo methane, ethylene glycol, α-bromo naphthalene and glycerol were used for tensiometric measurements. Also the surface free energy (SFE) was determined by means of Owens-Wendt-Rabel-Kaelble methods. The contact angles (water) and surface free energy calculated by means of Owens-Wendt-Rabel-Kaelble method for the uncoated glass substrates, $Al_2O_3$ and $Ta_2O_5$ oxide coated samples at various roughness parameters in the range 20, 200, 400 nm were evaluated (Table 2).

**Table 2.** The surface roughness parameters, the contact angles (water) and surface free energy for the uncoated glass substrates, $Al_2O_3$ and $Ta_2O_5$ oxide coated samples at various roughness parameters in the range 20-400 nm.

| Samples | Roughness $R_a$ (μm) | Water contact angles (°) | SFE (mN/m) by Owens-Wendt-Rabel-Kaelble method |
|---|---|---|---|
| Glass (Petri dishes) | 0.017 | 36.20±3.43 | 57.71 |
| Glass | 0.18 | 35.77±4.24 | 58.05 |
| Glass | 0.37 | 34.14±6.23 | 58.28 |
| Glass (Petri dishes) /$Al_2O_3$ MS coatings | 0.019 | 67.89±5.38 | 43.76 |
| Glass /$Al_2O_3$ | 0.21 | 64.41±6.41 | 44.97 |
| Glass /$Al_2O_3$ | 0.39 | 69.92±4.02 | 45.96 |
| Glass (Petri dishes)/ $Ta_2O_5$ EB evaporated coatings | 0.021 | 65.35±7.22 | 43.89 |
| Glass/ $Ta_2O_5$ | 0.23 | 62.41±6.43 | 44.06 |
| Glass/ $Ta_2O_5$ | 0.38 | 61.76±4.53 | 44.14 |

The data demonstrate (table 2) that for the glass substrates contact angles (water) change in the range 34-36 °, 65-68 ° for $Al_2O_3$ coated samples and 61-66 ° for $Ta_2O_5$ coatings and surface free energy calculated by means of Owens-Wendt-Rabel-Kaelble methods changes in the range 57-60 mN/m for glass substrates and 43-45 mN/m for oxide coated samples at various roughness parameters in the range 20-400 nm.

## 4. Conclusions

The best mechanical properties were obtained in the case of oxide coatings $Al_2O_3$ (MS) deposited by magnetron sputtered method (table 1). The structure and composition of coatings was examined by XRD and XPS spectra. The Ta 4f lines of the deposited films agree well with the Ta 4f doublet representative of the Ta-0 bond in $Ta_2O_5$. The results clearly demonstrate that the investigated films have stoichiometric $Ta_2O_5$ composition with some impurities (figure 1). The surface roughness and topography was observed by AFM (figure 2). The data demonstrate (table 2) that for the glass substrates contact angles (water) change in the range 34-36 °, 65-68 ° for $Al_2O_3$ coated samples and 61-66 ° for $Ta_2O_5$ coatings. The surface free energy calculated by means of Owens-Wendt-Rabel-Kaelble methods changes in the range 57-60 mN/m for glass substrates and 43-45 mN/m for coated samples at various roughness parameters in the range 20-400 nm. The results make the possibility to separate the influence of roughness and surface free energy effects on the regularities and mechanisms of nano materials interactions with the biological objects and open the perspective for a direct control of such parameters as adhesion, proliferation, differentiation of cells during their culturing.

## Acknowledgment

The study was supported by grant 4974 Targeted Research and Development Initiatives Program of Science and Technology Center in Ukraine (STCU) and also by Cooperation Program between Bulgarian Academy of Science and National Ukrainian Academy of Science.

## References

[1] Anselme K 2000 *Biomaterials* **21** 667-81
[2] Dhayal M, Cho S I, Moon J Y, Cho S J and Zykova A 2008 *J. Applied Surface Sci.* **254** 3331-8
[3] Links J, Boyan B D, Blanchard C R, Lohmann C H, Liu Y and Cochran D 1998 *Biomaterials* **19** 2219-32
[4] Xavier S P, Carvalho P S P, Beloti M M and Rosa A L 2003 *J. Dent.* **31** 173-80
[5] Hammer D A and Lauffenburger D A 1987 *J. Biophys.* **52** 475-487
[6] Zykova A, Safonov V, Luk'yanchenko V, Walkowicz J and Rogovska R 2007 *Problems of Atomic Science and Technology (Series: Plasma Physics)* **13** 200-202
[7] Todorova Z, Donkov N, Ristic Z, Bundaleski N, Petrovic S and Petkov M 2006 *J. Plasma process and polymer* **3/2** 174-178
[8] Rogowska R 2006 *Problemy Eksploatacji - Maintenance Problems* **61** 193-204
[9] Owens D K and Wendt R C 1969 *J. Appl. Polym. Sci.* **13** 1741
[10] Donkov N, Mateev E, Zykova A, Safonov V and Luk'yanchenko V 2010 *Journal of Physics: Conference Series* **223** 012030

Journal of Physics: Conference Series **253** (2010) 012069 doi:10.1088/1742-6596/253/1/012069

# Low pressure and humidity influences on the electret surface potential decay

**A Viraneva, T Yovcheva, E Gencheva and G Mekishev**

Department of Experimental Physics, University of Plovdiv, 24 Tzar Assen Str., 4000 Plovdiv, Bulgaria

E-mail: asia83@uni-plovdiv.bg

**Abstract**. In the present paper the influence of both relative humidity and low pressure on the surface potential decay of charged 20 μm polypropylene films was investigated. Sample charging was performed in a three-electrode corona discharge system. After charging the initial surface potential was measured using the method of the vibrating electrode with compensation. Two groups of tests were performed. In the first group, after charging the electrets were placed into a vacuum chamber under various low pressures (from 0.1 mbar to 1000 mbar) for 30 minutes. After that the electrets were removed from the vacuum chamber. The surface potential was measured again and the normalized surface potential was calculated. The influence of low pressure was analyzed by the equation that describes processes of desorption from the electret surface. In the second group after charging the electrets were put in desiccators at 0% and 100% relative humidity for 220 days at room temperature. After that the surface potential was measured periodically out of the desiccators. The results obtained were analyzed with the percolation 2D model. It was established that the higher values of the relative humidity led to a faster decay of the surface potential. We assume the presence of surface discharge processes.

## 1. Introduction

Electrets have been a field of investigations for many years [1-3]. They have been used in different areas of science, techniques and medicine. A number of factors, which influence the charge storage and charge transport in electrets, have been investigated [4-6]. The surface charge decay has been explained by various polarization mechanisms or charge injection and charge transport through the electrets [7]. The influence of the surface on charge relaxation has been neglected. However, there are some experimental data, which show that the surface charge relaxation should not be neglected under certain conditions in particular at different relative humidity and low pressure [8, 9]. It has been shown that corona discharge modifies the surface [10] and the surface potential decay depends on the surface charge distribution [11].

The aim of this paper is to study (investigate) the influence of the surface processes on charge decay of polypropylene corona electrets stored at different relative humidity and low pressure.

## 2. Experimental

### *2.1. Samples*

Nonmetalised samples of 20 μm thick polypropylene (PP) films manufactured by Assenova Krepost Ltd, Bulgaria were investigated. Initially the PP films were cleaned with alcohol in an ultrasonic bath

Journal of Physics: Conference Series **253** (2010) 012069 doi:10.1088/1742-6596/253/1/012069

for 4 minutes, washed in distilled water, and dried under room conditions. The 30 mm diameter samples were cut from that cleaned film and were put onto the same diameter metal pads.

### *2.2. Corona charging and surface potential measurement*

The charging of electrets in the corona discharge was carried out by means of a conventional three-electrode system consisting of grounded plate electrode, corona electrode and a grid placed between them. The distance between the corona electrode and the grid was 10 mm and the distance between the grid and the grounded plate electrode was 3 mm. The samples were charged for 1 minute under room conditions. The voltage of the corona electrode was -5 kV. The electrets surface potential was limited by the grid potential and was of the order of 500 V and 650 V.

Electrets surface potential was measured by the method of the vibrating electrode with compensation by which the estimated error was better than 5%. After charging, the initial surface potential $V_0$ was measured and the electrets were divided in two groups according to the kind of treatment – different low pressure and humidity.

### *2.3. Low pressure treatment*

After samples charging the initial electret surface potential $V_0$ was measured and then the electrets were placed into a vacuum chamber under various low pressures for 30 minutes. The pressure values in the vacuum chamber were 0.1 mbar, 1 mbar, 10 mbar, 20 mbar, 66 mbar, 132 mbar, and 1013 mbar. After that the samples with their metal pads were removed from the vacuum chamber, their surface potentials $V$ were measured again and the normalized surface potentials $V/V_0$ were calculated.

### *2.4. Humidity treatment*

To investigate the influence of relative humidity (RH) on the surface charge decay the samples were stored for 50 days in two desiccators at different RH denoted in the following manner: 1 (0% RH) and 2 (100% RH). Then the samples were charged to an initial surface potential $V_0$ and placed again in desiccators 1 and 2. The samples were marked as BA: 11, 12, 21, 22. The first figure (B) denotes the controlled RH desiccator in which the samples have been stored before charging, the second figure (A) denotes the same after charging. Periodically the surface potentials $V$ were measured again and the normalized surface potentials $V/V_0$ were calculated. The measurements were performed out of the desiccators.

## 3. Results and discussion

The conditions that propose the presence of surface discharge processes are the relative humidity and the low pressure and that is why we investigate the influence of these two factors on electrets surface potential decay.

### *3.1. Influence of low pressure on electrets surface potential decay*

Dependences of normalized surface potential $V/V_0$ on normalized pressure $p/p_0$ for PP electrets charged in a negative corona to different initial surface potential are presented in figure 1.

The symbol $V_0$ indicates the initial value of the surface potential, measured just after the sample had been charged, and $p_0$ is the atmospheric pressure. Each point in the figure 1 denotes the average value for the normalized surface potential calculated from six samples. The maximum standard deviation from the average value at 95% confidence level was 5%.

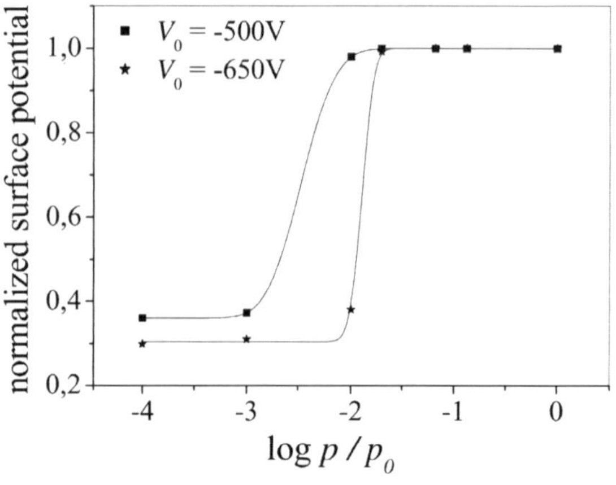


**Figure 1.** Dependence of the surface potential on pressure for PP films negatively charged to different initial surface potentials $V_0$.

The results, illustrated in figure 1 demonstrate the following peculiarities:

- Three parts are existed in each curve. At higher and lower pressures the surface potential is constant. For each curve a relatively narrow region of pressures exists where a sharp decay of surface potential occurs. Similar behaviour has also been observed in [12, 13].
- The values of the normalized surface potential at a pressure of 0.1mbar for the samples charged to a 500V initial surface potential are higher than those for the samples charged to a 650V initial surface potential.
- The higher the initial surface potential, the higher the pressure at which the sharp decay occurs. Hence, the potential sharp decay region is shifted to the lower pressures when the initial surface potential value decreases. The similar behaviour has also been observed for PET and PTFE [12, 14].

It is difficult to explain the dependence of the charge decay on pressure by means of injection and polarization models. As the pressure in the vacuum chamber decreases, desorption of ions from the electret surface is most likely to occur. As shown in [15], desorption is accompanied by surface diffusion and the following diffusion equation analyzed in [16] is proposed to describe the experimental results.

$$\frac{V}{V_0} = a + \frac{b}{2}\left(1 + erf\left(\frac{x-c}{\sqrt{2}d}\right)\right) \tag{1}$$

where $V/V_0$ is the normalized surface potential, $x = \log p / p_0$ and $a, b, c$ and $d$ are parameters. The solid line in figure 1 is the theoretical prediction and the values of the parameters are: $a$=0.3604±0.0004; $b$=0.6398±0.0004; $c$=-2.4728±0.0025; $d$=0.2542±0.0013 for initial surface potential of 500V and $a$=0.3046±0.0032; $b$=0.6954±0.0041; $c$=-1.8939±0.0077; $d$=0.0859±0.0056 for initial surface potential of 650V.

The really good agreement between the theoretical curves and the experimental data (figure 1) shows that the surface potential decays due to desorption ions from the electret surface accompanied by surface diffusion.

### *3.2. Influence of relative humidity on electrets surface potential decay*

The dependences of normalized surface potential on the time of storage for samples charged to different initial surface potentials were studied for 220 days. The curves are presented in figures 2, 3. All points are mean values from six samples and the calculated standard deviation was better than 5% with a confidence level of 90%. The surface potential time dependence curves for all samples have similar behavior.

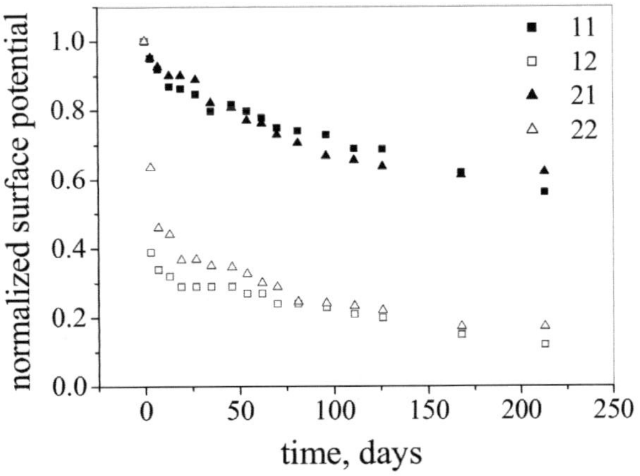


**Figure 2.** Time storage dependences of electrets surface potential for samples, charged in a negative corona to a 500 V initial surface potential.

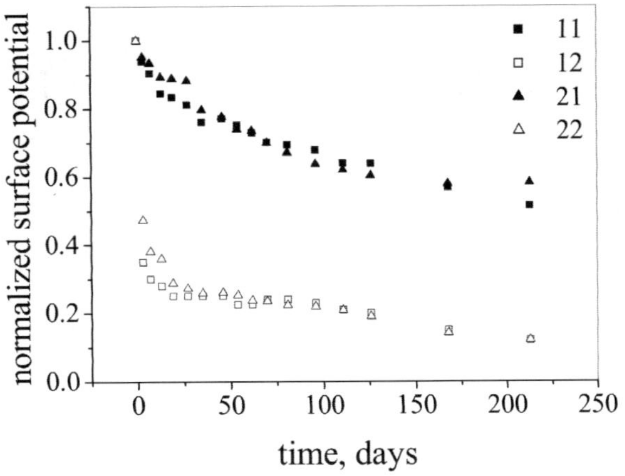


**Figure 3.** Time storage dependences of electrets surface potential for samples, charged in a negative corona to a 650 V initial surface potential.

The results presented in figures 2 and 3 show that:

- No matter whether the samples before charging are kept at a relative humidity 0% or 100% their behaviour is determined mainly from the storage conditions after charging, independently of the initial surface potential.
- In both cases of different initial surface potential (500 V and 650 V) sizable charge decay is observed if the electrets are kept after charging at the highest relative humidity 100%.
- For the samples that after charging have been stored at 0% relative humidity the established value for the normalized surface potential (after 165 days) is higher ($V/V_0$ =0.6), compared to the one for the samples stored after charging at 100% relative humidity ($V/V_0$=0.15), independently of $V_0$.

If an electret is stored at a controlled RH, adsorbed water molecules from the environment and water-molecule clusters are formed on its surface, due to moisture adsorption. The aggregation of these clusters generates a two-dimensional (2D) percolation lattice that develops with time. If the concentration of absorbed water increases, a connection of the separate clusters takes place which may lead to an unbounded cluster (UC). The specific conductivity of water is much higher than the one of the polymer samples. Consequently, UC ensures fast neutralization of electret charge and a sharp decay of the surface potential. A model that accounts for these changes is the model proposed by Kuzmin and Tairov [17, 18] based on the simultaneous utilization of the percolation theory, and the Kolmogorov's concept for the 2D kinetics of nuclei formation.

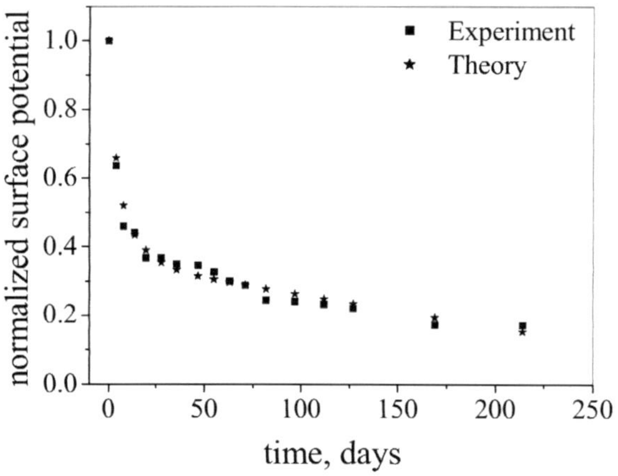


**Figure 4.** Time dependence of the surface potential for the samples 22, charged to a 500 V initial surface potential.

The surface potential time dependence curves for the samples 22 (charged to a 500 V initial surface potential) obtained from experimental studies, and the theoretical calculations, are presented in figure 4. Analogues curves were obtained for all the investigated samples. The theoretical calculations were made by computer software for nonlinear programming, using the Flexyplex software package, presented in [8].

The results obtained show a good conformity of theory with experiment. The really good agreement between the theoretical results and the experimental data (figure 4) shows that the percolation model allows to analyze electrets' surface discharge, and reveal the significant influence of the humidity level at which they have been stored. We assume that moisture adsorption is stimulated by electrization, and further cluster growth depends on the storage medium after the charging of electrets. This is extremely important, and allows one to quantitatively describe the surface charge relaxation and predict the stability of polymer electret films that function under real conditions involving various relative humidity values.

Finally, surface potential decay for electrets kept at different values for the low pressure and humidity can well be described by theoretical models that assume the availability of surface processes. Equation (1) describes linear desorption of charges which is accompanied by surface diffusion. The percolation theory presupposes surface decay of electret charges. Consequently, the influence of both low pressure and humidity on the electret surface potential decay should not be explained as a result of the injection of the surface charge into the bulk and its subsequent transport through the sample.

## 4. Conclusion

Our experimental data show that the surface charge relaxation should be the main factor determining charge decay of electrets stored under low pressure and different humidity. The used theoretical models allowed to analyze electrets surface discharge, and to reveal the significant influence of the low pressure and humidity levels at which they have been stored. This is extremely important, and should be taken into account in the study of the stability of electrets sensors and devices that are used at room temperature and function under real conditions involving various low pressure and humidity levels.

## Acknowledgements

The authors gratefully acknowledge the financial support of the National Science Fund, Bulgaria.

## References

[1] Sessler G and Gerhard-Multhaupt R 1999 *Electrets* 3rd Ed. (Laplacian: Press Morgan Hill California)
[2] Nalwa H S 1995 *Ferroelectric polymers* (Marcel Dekker Inc. New York)
[3] Hilczer B and Malecki J 1986 *Electrets* (Elsevier Amsterdam)
[4] Herous L, Nemamcha M and SAAD S 2006 *J. Engin. Appl. Sci.* **1 (4)** 329-334
[5] Gang-Jin C, Hui-ming X and Chun-feng Z 2004 *J. Zhejiang University Sci.* **5 (8)** 923-927
[6] Yovcheva T, Mekishev G and Nedev St 2005 *JOAM*, **7 (1)** 237-240
[7] Moulinie P 1999 *J. Electrostatics* **45** 265-273
[8] Yovcheva T A, Mekishev G A and Marinov A T 2004 *J. Phys.: Condens. Matter* **16** 455-464
[9] Mekishev G, Yovcheva T, Guentcheva E and Nedev St 2003 *J. Mater. Sci.: Mater. in Electronics* **14** 779-780
[10] Yovcheva T, Avramova I, Mekishev G and Marinova T 2007 *J. Electrostatics* **65** 667-671
[11] Karmazova P G and Mekishev G A 1992 *Europhys. Lett.* **19 (6)** 481-484
[12] Mekishev G A, Yovcheva T A and Viraneva A P 2007 *J. Non-Crystalline Solids* **353** 4453–4456
[13] Viraneva A P, Yovcheva T A, Gencheva E A and Mekishev G A 2010 *J. Non-Crystalline Solids* **356** 560–563
[14] Viraneva A, Yovcheva T, Gencheva E and Mekishev G, 2008 *JOAM* **10 (2)** 302 – 305
[15] Viraneva A P 2010 *PhD thesis* University of Plovdiv Bulgaria
[16] Viraneva A P, Yovcheva T A, Pisanova E S, Gencheva E A and Mekishev G A 2010 *AIP Conference Proceedings* BPU7 382-388
[17] Kuzmin Y and Tairov V 1984 *J. Tech. Phys.* **54** 964-965
[18] Kuzmin Y 1999 *Proc. 10$^{th}$ Int. Symp. on Electrets* Eds A. Konsta, A. Vassilikou-Dova, K. Vartzeli-Nikaki 55

Journal of Physics: Conference Series **253** (2010) 012070 doi:10.1088/1742-6596/253/1/012070

# X-ray photoelectron study of $Si^+$ ion implanted polymers

**T Tsvetkova[1], S Balabanov[1], L Bischoff[2], V Krastev[1], P Stefanov[3] and I Avramova[3]**

[1] Georgi Nadjakov Institute of Solid State Physics,
Bulgarian Academy of Sciences, 72 Tzarigrdasko Chaussee, 1784 Sofia, Bulgaria
[2] Forschungscentrum Dresden – Rossendorf e.V.,
P.O.Box 510119, D-01314 Dresden, Germany
[3] Institute of General and Inorganic Chemistry, Bulgarian Academy of Sciences,
Acad.G.Bonchev Str., 1113 Sofia, Bulgaria

E-mail: tania_tsvetkova@yahoo.co.uk

**Abstract**. X-ray photoelectron spectroscopy was used to characterize different polymer materials implanted with low energy $Si^+$ ions (E=30 keV, D= $1.10^{17}$ $cm^{-2}$). Two kinds of polymers were studied – ultra-high-molecular-weight poly-ethylene (UHMWPE), and poly-methyl-methacrylate (PMMA). The non-implanted polymer materials show the expected variety of chemical bonds: carbon-carbon, carbon being three- and fourfold coordinated, and carbon-oxygen in the case of PMMA samples. The X-ray photoelectron and Raman spectra show that $Si^+$ ion implantation leads to the introduction of additional disorder in the polymer material. The X-ray photoelectron spectra of the implanted polymers show that, in addition to already mentioned bonds, silicon creates new bonds with the host elements – Si-C and Si-O, together with additional Si dangling bonds as revealed by the valence band study of the implanted polymer materials.

## 1. Introduction

Polymers ultra-high-molecular weight poly-ethylene (UHMWPE) and poly-methyl-methacrylate (PMMA) find various applications. They are currently used extensively because of their structural and material properties and are especially attractive lately for their electronic properties [1]. Technologies of modifying polymer material properties by a beam of energetic ions (ion beam implantation) have found wide applications in different areas from microelectronics to medicine [2-4]. The possibilities for various application purposes of different polymer materials are essentially defined by the type and amount of the implanted ions, which often substantially modify their mechanical, optical and electrical properties [5,6]. Particularly attractive, from both fundamental view and applications related implications, are the interesting photo-luminescent properties of these materials induced by low energy $Si^+$ ion beam implantation [7,8]. The underlying $Si^+$ ion beam induced structural properties modifications, defining the optical changes and related photo-luminescent properties, are also of particular interest with view of further photo-luminescence enhancement and optimization.

In this paper, X-ray photoelectron spectroscopy (XPS) methods, combined with Raman spectroscopy measurements, have been used to characterize the $Si^+$ ion beam induced structural properties modifications of the polymer materials studied.

## 2. Experimental

Polymethyl-methacrylate (PMMA) and ultrahigh-molecular-weight poly-ethylene (UHMWPE) samples, commercially available in the form of bulk samples with thicknesses of 2 mm and shaped as squares (sides of 10 mm) and circles (diameters of 10 mm), were studied in the present work.

A Danfysik 1090 ion implanter was used for the $Si^+$ implantation at room temperature (RT). The implantation was carried out at relatively low energies (E = 30 ÷ 50 keV) and a wide range of ion doses (D = $10^{13}$ ÷ $10^{17}$ $cm^{-2}$). The beam current was kept under 2 $\mu A/cm^2$ during the implantation process, so that to maintain the target temperature below 80°C – the temperature at which the polymer material starts to decompose. A high vacuum (of the order of $10^{-6}$ Pa) was maintained in the target chamber during implantation.

The XPS studies were carried out on an ESCALAB MK II (VG Scientific) electron spectrometer with base pressure in the preparation and analysis chambers of $2\times10^{-8}$ Pa and $1\times10^{-8}$ Pa, respectively. The photoelectrons were excited using X-ray source Mg Kα (hν=1256.6 eV). The surface sensitivity was estimated as ~ 10 monolayers. A strip of gold, deposited on the surface of the sample, was used as a binding energy (BE) reference. The BE of Au $4f^{7/2}$ was assumed to be 83.8 eV. This allowed a determination of the binding energies of the C 1s, O 1s, Si 2p and Si 2s peaks with an accuracy of ±0.1 eV. The depth profiling was carried out by sputter etching at 90° using a defocused $Ar^+$ ion beam of 3 keV energy and a current density of 16 $\mu A.cm^{-2}$.

The room temperature infrared (IR) transmittance spectra were measured in the range 500 to 2000 $cm^{-1}$, using a UR-20 double-beam spectrometer. The Raman spectra were taken at room temperature using a SPEX 1403 spectrometer equipped with a photomultiplier working in a photon counting mode and a spectral resolution of 4 $cm^{-1}$. The 488 nm $Ar^+$ laser line was with a power as low as 10 mW because the thermal stability of the PMMA was only 65°C.

## 3. Results and discussion

### *3.1. X-ray photoelectron spectra*

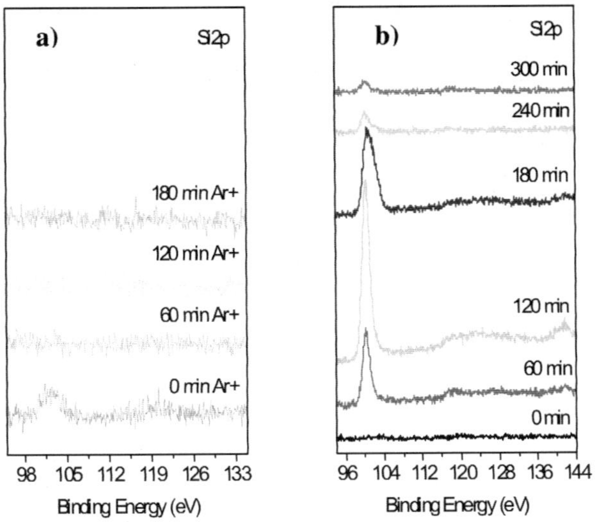


**Figure 1.** XPS spectra of Si2p electrons on the surface and after different time of $Ar^+$ sputter etching for unimplanted UHMWPE (a) and implanted UHMWPE (b).

The XPS spectra of Si2p for UHMWPE are presented in Figure 1. For the unimplanted UHMWPE (Figure 1a) no significant peaks are revealed in the spectra recorded on the surface and in depth after different intervals of argon ion sputtering.

Figure 1b shows Si2p spectra of the implanted UHMWPE. Pronounced peak at 100.3 eV characteristic of Si-C bonds in $C_2H_4/Si$ groups [9] is visible in the spectra. Maximum amplitude of that peak is found in the second line recorded after 120 min Ar ion sputtering meaning that the maximum concentration of Si ions is at some depth where they have lost their energy. After 180 min sputtering the peak is centered at 101.2 eV characteristic of silicon atoms in Si-C groups [10] due to implanted $Si^+$ ions bonded to the backbone of the polymer.

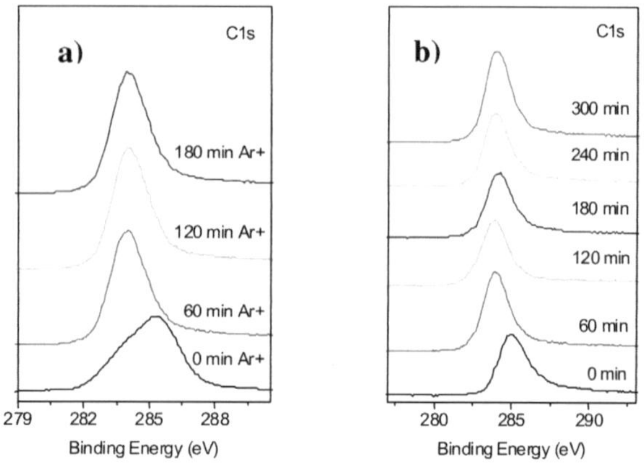


**Figure 2.** XPS spectra of C1s electrons on the surface and after different time of $Ar^+$ sputter etching for unimplanted (a) and implanted (b) UHMWPE.

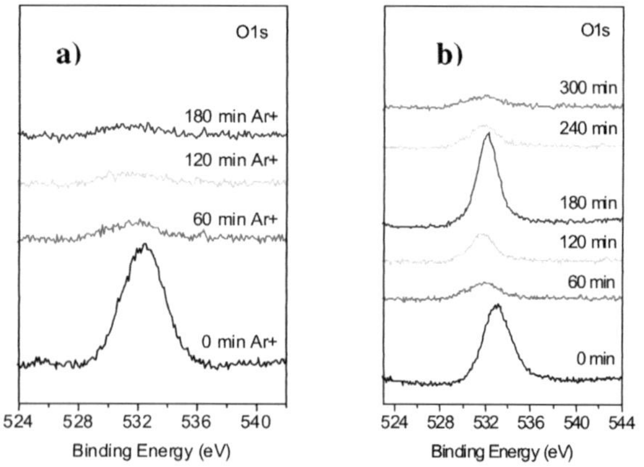


**Figure 3.** XPS spectra of O1s electrons on the surface and after different time of $Ar^+$ sputter etching for unimplanted (a) and implanted (b) UHMWPE.

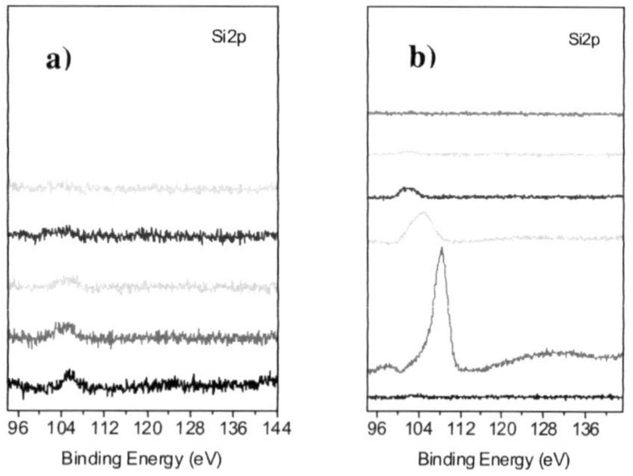


**Figure 4.** XPS spectra of Si2p electrons on the surface and after different time of $Ar^+$ sputter etching for unimplanted (a) and implanted (b) PMMA.

In Figure 2 the XPS spectra of C1s are presented for UHMWPE. Figure 2a shows C1s spectra of unimplanted UHMWPE. At the surface a large peak at 285.4 eV is visible characteristic of C atom in $-CH_2-$ groups [11] associated with a bend at 288.0 eV due to C atom in -C(O)O groups [11] which are possibly formed by oxidation. In depth the sample position of the peak shifts to lower binding energies (284.0eV) characteristic of $(-C_6H_2(CH_3)2O-)_n$ groups [12].

Figure 2b shows C1s spectra of implanted UHMWPE. At the surface a large peak at 285.0 eV is visible characteristic of C atom in -C(O)O groups[11] which are possibly formed by oxidation. In depth the position of the peak shifts to lower binding energies (283.9 eV) typical of silicon carbide [14].

The XPS spectra of O1s for UHMWPE are presented in Figure 3. In Figure 3a the O1s spectra of unimplanted UHMWPE are shown. The peak at the surface is centered at 532.5 eV characteristic of $(-CH_2)4O-)_n$ groups [13].This peak diminishes in the bulk of the sample showing surface oxidation.

Figure 3b shows O1s spectra of implanted UHMWPE. The peak recorded from the surface is centered at 533.0 eV characteristic of $(-(CH_2)4O-)_n$ groups [13].This peak due to surface oxidation diminishes after 120 min $Ar^+$ sputtering. At depth reached after 180 min a new peak at 532.2 eV appears showing $(-Si(CH_3)_2-O-)n$ groups [15]. In depth these bonds almost coincide with the maximum amplitude of Si peak in depth (Figure 1b).

The XPS spectra of Si2p for PMMA are presented in Figure 4. No significant peaks are revealed in the spectra recorded at the surface and in depth for the unimplanted PMMA (Fig.4a).

Figure 4b shows Si2p spectra of implanted PMMA. No peak is visible in a line recorded at the surface. After 60 min sputtering a large peak at 108 eV dominates. This peak is Si2p satellite at 108.4 eV [16] due to $(-Si(CH_3)_2-C_6H_4-Si(CH_3)_2-O-)_x-(Si(CH_3)_2-O-)_y$ groups. In depth of the sample recorded peaks are centered around 101.9 eV, due to $(Si(CH_3)_2-C6H_4-Si(CH_3)_2-O-)_n$ groups [16]. The peak at 104.7 eV is due to silicon oxide [17].

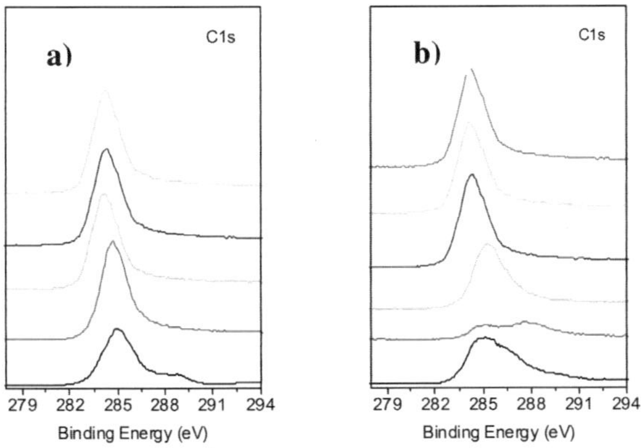


**Figure 5.** XPS spectra of C1s electrons on the surface and after different time of $Ar^+$ sputter etching for unimplanted (a) and implanted (b) PMMA.

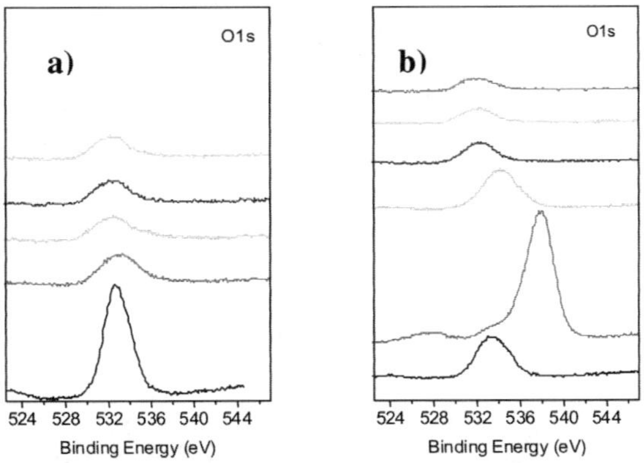


**Figure 6.** XPS spectra of O1s electrons on the surface and after different time of $Ar^+$ sputter etching for unimplanted (a) and implanted (b) PMMA.

In Figure 5 the XPS spectra of C1s are presented for PMMA. Figure 5a shows C1s spectra of unimplanted PMMA. In the line recorded at the surface the C1s peak at 285.00 eV [15] is characteristic of $CH_2C(CH_3)(C(O)OCH_2CH_2CH_2CH_3)\text{-})_n$ groups while the bend at 290.35 eV is due to $(((\text{-}CH_2CH_2\text{-})_x)\text{-}CH_2CH_2OC(O)O\text{-})_n$ groups [10]. In depth of the sample the position of the peak shifts to lower binding energies (284.1 eV) characteristic of $(\text{-}C_6H_2(CH_3)2O\text{-})_n$ groups [12].

Figure 5b shows C1s spectra of implanted PMMA. The C1s peak recorded at the surface and centered at 285.00 eV [15] is characteristic of $CH_2C(CH_3)(C(O)OCH_2CH_2CH_2CH_3)\text{-})_n$ groups while the bend at 290.35 eV is due to $(((\text{-}CH_2CH_2\text{-})_x)\text{-}CH_2CH_2OC(O)O\text{-})_n$ groups [12]. After 60 min argon ion sputtering both peaks are suppressed - with the one centered at 288 eV predominating. In depth of the sample the peaks of the spectra tend to shift to lower binding energies characteristic of silicon carbide [10].

The XPS spectra of O1s for PMMA are presented in Figure 6. In Figure 6a the O1s spectra of unimplanted PMMA are shown. The peak recorded from the surface is due to oxidation and diminishes after 120 min argon ion sputtering. In depth of the sample a new contribution with centre at 533.0 eV characteristic of $(\text{-}(CH_2)_4O\text{-})_n$ groups [13] appears.

Figure 6b shows O1s spectra of implanted PMMA. The O1s peak recorded from the surface is centered at 285.00 eV characteristic of $(\text{-}(CH_2)_4O\text{-})_n$ groups [13]. This peak is due to surface oxidation and diminishes after 60 min argon ion sputtering while a new peak at 538.8 eV appears showing presence of $(CH_3)_2CHOCH_3$ groups [13]. In depth of the sample a new peak centered at 533.0 eV characteristic of $(\text{-}(CH_2)_4O\text{-})_n$ groups [17] appears.

### *3.2. Infra-red and Raman spectra*

The IR spectra of the unimplanted and $Si^+$ implanted, at a dose of $D_3=1.10^{17}cm^{-2}$, polymethyl-methacrylate (PMMA) substrates are compared in Figure 7. As seen, the absorption of the implanted sample is higher and some features, though not intense because of the small thickness of the modified region, can be resolved. The spectra are displayed in the region from 800 to 2000 $cm^{-1}$, since in the range from 500 to 800 $cm^{-1}$ they are highly transparent. The features in these spectra are much strongly pronounced and the absorption is comparatively higher. The spectrum of the implanted PMMA substrate is also more transparent in the range1700 - 1750 $cm^{-1}$. Stretching vibrations of the C=O bonds at 1720 $cm^{-1}$ and 1780 $cm^{-1}$ [18] are reported for polyimide layers. Then, the transparency increase of our samples around these frequencies can reasonably be attributed to the breaking of C=O bonds.

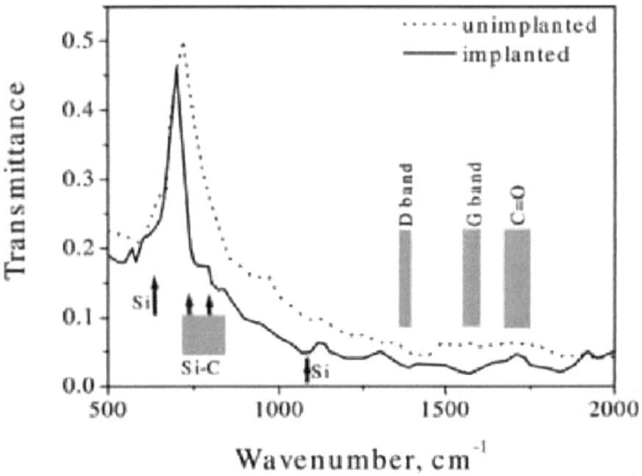


**Figure 7.** IR spectra of the unimplanted and $Si^+$ implanted at D=1.10$^{17}$cm$^{-2}$ PMMA samples.

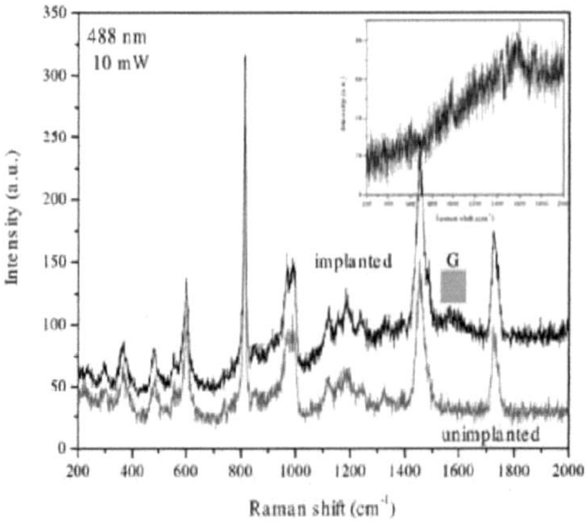


**Figure 8.** Raman spectra of the unimplanted and $Si^+$ implanted at D=1.10$^{17}$cm$^{-2}$ PMMA samples.

In Figure 8, the Raman spectra of the unimplanted and implanted with D=1.10$^{17}$cm$^{-2}$ PMMA substrates are shown, in the range from 200 to 2000 cm$^{-1}$. Because of the low scattering volume of the implanted regions, no significant difference between the spectra of unimplanted and implanted samples is seen. The only unambiguous statement concerns the increase in the scattered light intensity in the range close to 1550 cm$^{-1}$, shadowed in Figure 8. The subtraction of the unimplanted sample spectra from those of the implanted ones gave the spectrum, displayed in the inset in the same figure. It is seen that the difference in the spectra is governed by a broad band peaked at about 1550 cm$^{-1}$.

The only unambiguous statement, as it concerns the formation of new bonds, is the presence of C=C bonds. It is confirmed by the IR transmittance, as well as by the Raman spectra. The formation of graphite-like and carbon-rich amorphous structures results in the appearance of broad absorption bands in the infrared spectrum at about 1350 cm$^{-1}$ and at about 1550 cm$^{-1}$, known as D (disorder) and G (graphite) bands, respectively. The bands are usually assigned to the graphitic $sp^2$ hybridized C bonds. The G band corresponds to vibrations in the graphitic micro-particles and the D band to in-plane vibrational modes at the particle surface [19]. The experimental transmittance spectra displayed in Figure 7 show unambiguously an increase of the absorption of the implanted sample in the range 1300 and 1550 cm$^{-1}$, which is an indication of the presence of amorphous and microcrystalline graphite. It is seen, from the inset in Figure 8, that the difference in the Raman spectra of implanted and unimplanted samples is governed by a broad band peaked at about 1550 cm$^{-1}$. These spectra look like the spectrum of a network with graphitic $sp^2$ hybridized C bonds, see for example [20].

A wide variety of SiC phonon band profiles are observed, depending on the growth conditions. The frequencies of the transverse optical (TO) phonons for the different *polytypes* in which the material crystallize are close, and lie in the range 766 – 780 cm$^{-1}$ [21]. Thus, the feature at 780 cm$^{-1}$ in the spectrum of the implanted PMMA sample can be related to the SiC fundamental phonon absorption. The noticeable feature at 740 cm$^{-1}$ in Figure 7, as well as that at about 850 cm$^{-1}$, may be due to morphological factors as grains of different sizes and shapes, as well as to plasmon-phonon coupling [22]. A possible explanation of the increased absorption at 1650 – 1700 cm$^{-1}$ is the two-phonon SiC one. It is worth mentioning also that the behaviour of the spectra of the implanted PMMA samples in the range 900 – 1250 cm$^{-1}$ is very similar to that of polycarbosilane (PCS), heated at 873 K [22].

It is well known that although optical phonons in covalent semiconductors, and Si in particular, are not infrared active, these semiconductors absorb infrared radiation. The absorption is interpreted as a multiphonon one [23]. The most intense peaks in the experimental absorption spectra of Si are observed at 600 − 610 cm$^{-1}$ and at about 1100 cm$^{-1}$ [24]. Less intense two-phonon absorption is detected also, at about 700 and 850 cm$^{-1}$ [18]. Low intensity features at 610 and 1080 cm$^{-1}$ in the spectrum of the implanted PMMA sample, marked with arrows in Figure 7, then can be attributed to the absorption of unreacted Si atoms.

## 4. Conclusion

The XPS, IR and Raman spectra show that $Si^+$ ion implantation leads to the introduction of additional disorder in the polymer material. XPS spectra of the implanted polymers show that in addition to the already mentioned bonds silicon creates new bonds with the host elements atoms – Si-C and Si-O, together with additional Si dangling bonds, as revealed by the valence band study of the implanted polymer materials. The unambiguous statement, that follows from the investigations, is that the implantation of Si into PMMA leads to the formation of amorphous and nanocrystalline graphite, while the presence of SiC particles and unreacted Si atoms also looks possible.

## Acknowledgements

This work has been supported by an International Atomic Energy Agency project (No:1294/RBF). The support of Forschungszentrum Dresden – Rossendorf e.V., Germany, is also gratefully acknowledged.

## References

[1] Frank W F X, Kulisch J, Franke H, Ruch D M, Brunner S and Lessard R A 1991 *SPIE* **1559** 344
[2] Marletta G 1995 *Materials and Processes for Surface and Interface Engineering* (Dordrecht: Kluwer Academic Publishers) p 597
[3] Sloof L H, van Blaaderen A, Polman A, Hebbink G A, Klink S I, van Veggel F C J M, and Hofstraat J W 2002 *Appl. Phys. Lett.* **91** 3955
[4] Sum T C, Bettiol A A, Seng H L, Rajta I, van Kan J A and Watt F 2003 *Nucl. Instrum. & Methods* **B 210** 266
[5] Toth A, Bell T, Bertoti I, Mohai M and Zelei B 1999 *Nucl. Instr. & Methods* **B 148** 1131
[6] Evelyn A L, Ila D, Zimmerman R L, Bhat K, Poker D B, Hensley D K, Klatt C, Kalbitzer S, Just N and Drevet C 1999 *Nucl. Instrum. & Methods* **B 148** 1141
[7] Tsvetkova T, Balabanov B, Avramov L, Borisova E, Angelov I, Sinning S and Bischoff L 2009 *Vacuum* **83** S252
[8] Tsvetkova T, Balabanov S, Avramov L, Borisova E, Angelov I and Bischoff L 2010 *J. Phys.: Conf. Series* **223** 012033
[9] Stinespring C D and Wormhoudt J C 1989 *J. Appl. Phys.* **65** 1733
[10] Laoharojanaphand P and Lin T J, Stoffer J O 1990 *J. Appl. Polymer Sci.* **40** 369
[11] Clark M B Jr, Burkhardt C A and Cardella J A Jr 1989 *Macromolecules* **22** 4495
[12] Beamson G and Briggs D 1992 *High Resolution XPS of Organic Polymers: the Scienta ESCA300 Database*
[13] Bhatia Q S and Burrell M C 1991 *Polymer* **32** 1948
[14] Mizokawa Y, Geib K M and Wilmsen C W 1986 *J. Vac. Sci. Technol.* **A 4** 1696
[15] Patel N M, Dwight D W, Hedrick J L, Webster D C and McGrath J E 1988 *Macromolecules* **21** 2689
[16] Schmitt R L, Cardella J A, Magill J H and Chin R L 1987 *Polymer* **28** 1462
[17] Paparazzo E, Fanfoni M and Severini E 1992 *Appl. Surf. Sci.* **56** 866
[18] Lim H, Lee Y, Han S, Cho J and Kim K 2001 *J. Vac. Technol.* **A 19** 1490
[19] Collins R J and Fan H Y 1955 *Phys. Rev.* **93** 265
[20] György E, Mihailescu I, Baleva M, Abrashev M, Trifonova E P, Szekeres A and Perrrone A 2003 *Mat. Sci. & Eng.* **B 97** 251
[21] Freitas J A Jr. 1995 *Properties of Silicon Carbide*, ed G L Harris, *EMIS Data Reviews Series* No. **13** 29
[22] Mutschke H, Andersen A C, Clement D, Henning T and Peiter G 1999 *Astron. Astrophys.* **345** 187
[23] Lax M and Burstein E 1955 *Phys. Rev.* **97** 39
[24] Nakayama N, Tsuehiya Y, Tamada S, Kosuge K, Nagata S, Tahakiro K and Yamaguchi S 1993 *Jap. J. Appl. Phys. Lett.* **32** L1465

Journal of Physics: Conference Series **253** (2010) 012071 doi:10.1088/1742-6596/253/1/012071

# Synthesis and characterization of $RBaCo_2O_{5+x}$ (R = La, Nd, Gd, Y and Ho)

**N D Todorov[1], M V Abrashev[1], V G Ivanov[1], G V Avdeev[2] and S C Russev[1]**

[1] Faculty of Physics, University of Sofia, 1164 Sofia, Bulgaria
[2] Institute of Physical Chemistry, Bulgarian Academy of Sciences, 1184 Sofia, Bulgaria

E-mail: neno.todorov@gmail.com

**Abstract.** Ceramics of double perovskite $RBaCo_2O_{5+x}$ (R = La, Nd, Gd, Y and Ho) were prepared in air at different annealing conditions. The samples were characterized by optical microscopy, X-ray powder diffraction, scanning electron microscopy (SEM), energy-dispersive X-ray analysis (EDX), and micro-Raman spectroscopy. In the case of $RBaCo_2O_{5+x}$ (R = Gd, Y and Ho) the optical anisotropy of the tetragonal-like microcrystals observed is sufficient to determine their orientation and to obtain their polarized Raman spectra from (001)- and (100)-type oriented crystal surfaces. Due to the fact that some of the impurities observed in the ceramics have Raman spectra with intensity much larger compared to that of the main phase, special attention was paid to identify their Raman spectra. The origin of the lines observed in the polarized Raman spectra of $RBaCo_2O_{5+x}$ phase (R = Gd, Y and Ho) is discussed.

## 1. Introduction

The perovskite-like oxides and their superstructures attract the interest of the researchers for several decades because of the observation of a variety of physical phenomena in them as well their diverse physical properties. It is sufficient to note the first observations of structural second order transitions in titanates and aluminates, the discovery of high-temperature superconductivity in layered cuprates and colossal magnetoresistivity in manganites. During the last years the interest to the cobaltites with perovskite-like structure (i.e. compounds with chemical formula $(R,A)CoO_{3-\delta}$, where R is rare earth metal and A is alkaline earth metal) is also increased due to their rich structural, transport and magnetic phase diagrams [1] as well to their potential use as solid oxide fuel cell cathodes and membranes by reason of the high mobility of oxygen ions in them [2, 3]. The compounds under interest here are the double perovskites $RBaCo_2O_{5+x}$, $0 \leq x \leq 1$. Depending on the type of the rare earth and the annealing conditions (temperature and gas content atmosphere) several ordered structures with different oxygen content, different symmetry and unit cell can be prepared (see figure 1) [4]. Whereas the end members (for $x = 0$, 1) have simple tetragonal structure, the ordering of the oxygen vacancies in the compound with $x = 0.5$ doubles the unit cell and lowers the symmetry to orthorhombic. Tuning the oxygen content, coincident metal-insulator and ferromagnetic-paramagnetic phase transition near the room temperature can be observed [1]. Whereas large number of experimental methods were used to investigate their structural, transport and magnetic properties, there is only one paper on Raman spectroscopy of these materials [6]. Raman spectroscopy is an efficient tool for study

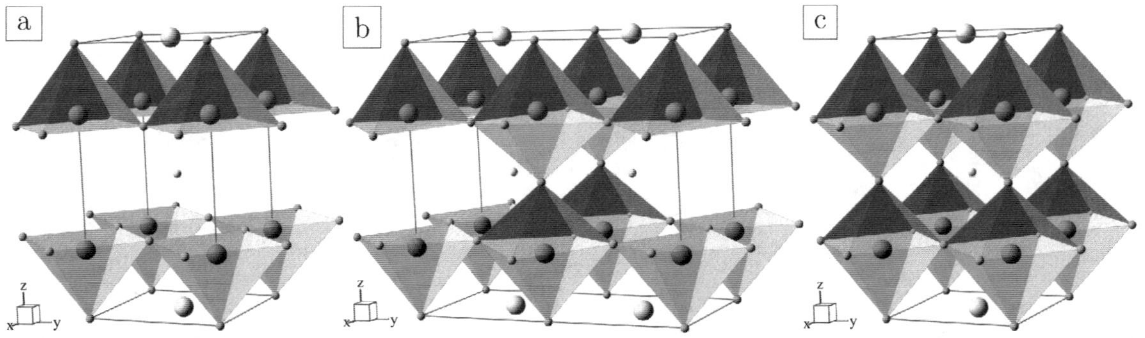


**Figure 1.** (Color online) Schematic drawing (Ref. [5]) of the $RBaCo_2O_{5+x}$ unit cell: (a) $x = 0$, (b) $x = 0.5$ and (c) $x = 1$.

of the crystal structure and their changes. In this paper we study the possibilities of this method for investigation of these type of materials.

## 2. Experimental

Samples of $RBaCo_2O_{5+x}$ (R = La, Nd, Gd, Y and Ho) were prepared by a standard solid-state reaction method. The starting materials of $R_2O_3$, $BaCO_3$, and $Co_3O_4$ were weighed in the stoichiometric ratio of $RBaCo_2O_{5+x}$. The powders were mixed in a ball mill and calcinated at 1000 and 1050° C for 48 h with intermediate grindings. After that the powders were pressed into tablets with diameter of 12 mm and sintered again in air at 1050° C for additional 24 h. Finally half of the samples were cooled in the furnace to 400° C by employing a cooling rate of 0.5° C/min. The other half of the samples was rapidly quenched from 950° C to room temperature. The tablets were polished using diamond pastes and then ultrasonically cleaned.

As prepared samples were observed under optical microscope Olympus BX41 equipped with ×100 objective as a part of LabRAM HR800 micro-Raman spectrometer used for the Raman measurements. The Raman spectra were taken in back-scattering configuration, as the excitation source was He-Ne laser (633 nm) with a laser beam spot size of $1 - 2\,\mu$m. To prevent overheating of the samples the incident laser power was under 1 mW. The elemental analysis of the synthesized samples was carried out using scanning electron microscopy (SEM; Tescan LYRA) equipped with energy-dispersive X-ray spectrometer (EDX; Bruker). The crystalline phases were determined by X-ray powder diffraction (XRD) analysis using a Philips PW 1050 diffractometer with Cu K$\alpha$ radiation and $2\theta$ in the range $20° - 90°$ with step 0.03° and an acquisition time for each step of 1 s. The diffraction patterns were refined with the PowderCell [7] software.

## 3. Results and discussion

The polycrystalline samples were analyzed initially by optical microscope. The observation of the ceramics illuminated with linearly polarized light show that for R = La and Nd, the orientation of the microcrystals is indistinguishable and the samples surface seems to be optically isotropic. For the rest samples optically anisotropic tetragonal-like microcrystals in rectangular shape can clearly be seen as shown in figure 2.

The X-ray powder diffraction analysis let us to determine the crystal phase purity of the $RBaCo_2O_{5+x}$ samples using a software algorithm. By variation of the phase dependent parameters, calculated powder diffraction patterns can be fitted to the experimental data. After the refinement the samples were indexed on cubic for R = La, orthorhombic for R = Nd, Gd and tetragonal for R = Y, Ho unit cells, which is consistent with the reports in the literature [3, 4, 8]. From

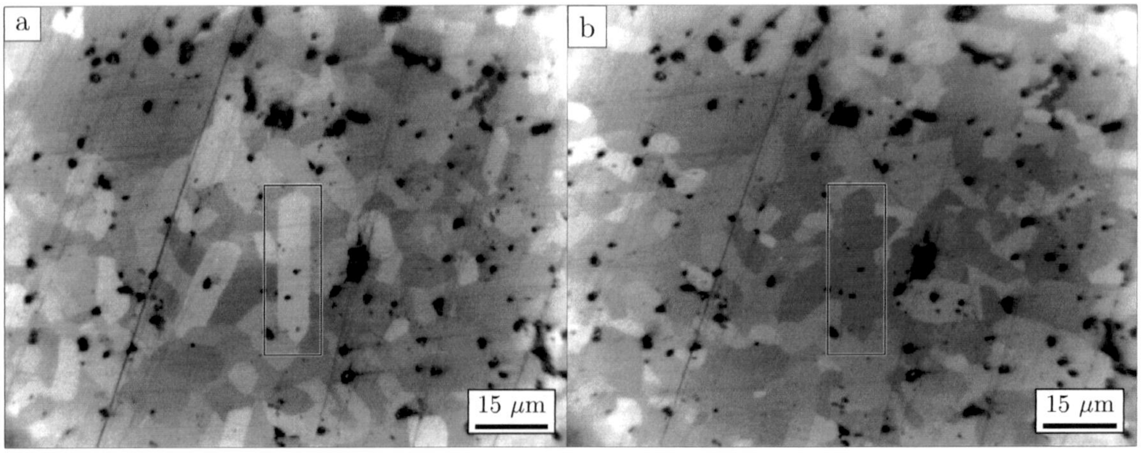


**Figure 2.** Optical images of (a) $YBaCo_2O_{5+x}$ illuminated with linearly polarized white light in the plane perpendicular to the $z$ axis (the longest side) of the delineated crystal; (b) the same crystal but the polarization of the light is in direction parallel to the $z$ axis (magnification ×100).

figure 3 can be seen that except the main phase, a perovskite-like $RBa_2Co_3O_{9-x}$ (R = La and Nd) was observed, too. These secondary phases were confirmed also by the EDX analysis. Minor impurities also exist in the Y- and Ho- samples. In figure 4 the dependence of the lattice

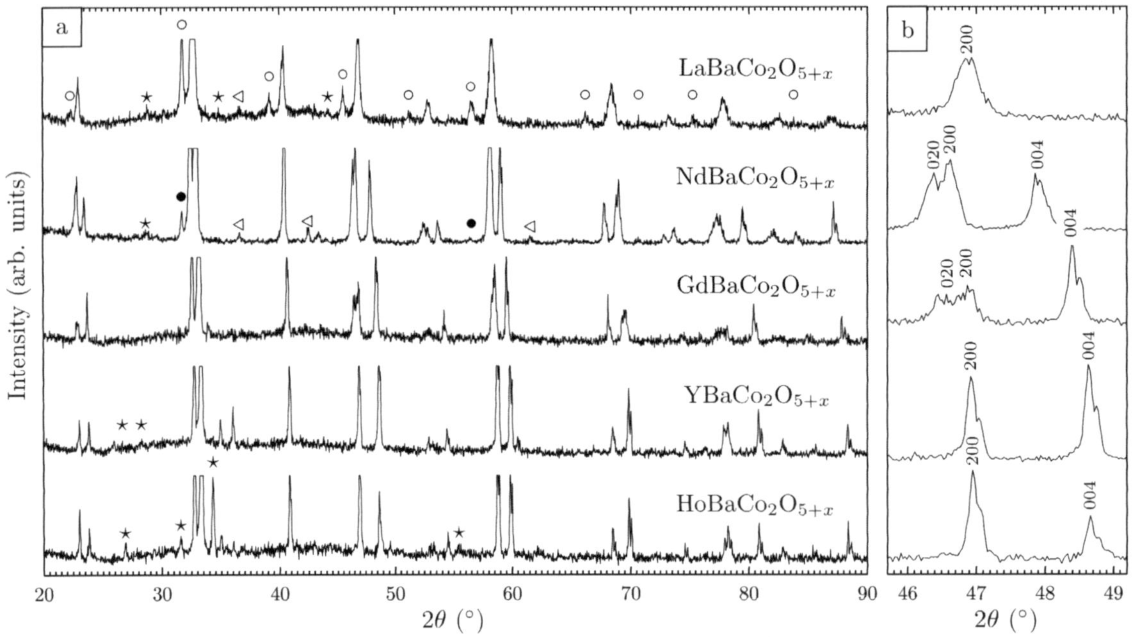


**Figure 3.** XRD patterns of $RBaCo_2O_{5+x}$ (R = La, Nd, Gd, Y and Ho). Unidentified impurities are marked with ⋆. CoO is marked with ◁. $RBa_2Co_3O_{9-x}$ is marked with ○ for R = La and • for R = Nd.

parameters and the unit cell volume (normalized to the ideal perovskite unit cell) on the ionic radius of the rare earth element were plotted. Comparing the calculated lattice parameters and the volume of the unit cell with the ones published in the literature we can evaluate the oxygen content in the $RBaCo_2O_{5+x}$ phase as follows: $x \approx 0.2$ for R = Ho and Y [9, 10], $x \approx 0.6$ for R = Gd [1], $x \approx 0.65$ for R = Nd [4, 11]. Using the empirically known dependence of the oxygen content on the type of the rare earth at equal annealing conditions, the oxygen content of La-sample also can be evaluated as $x > 0.7$ [4, 12]. Shortly, we can conclude that the main phase in Y- and Ho- samples is the tetragonal oxygen-poor $RBaCo_2O_{5.2}$, the one in the Gd- and Nd-samples is the orthorhombic $RBaCo_2O_{5.6}$, whereas the La-sample contains oxygen-rich disordered cubic perovskite (La,Ba)$CoO_{3-\delta}$.

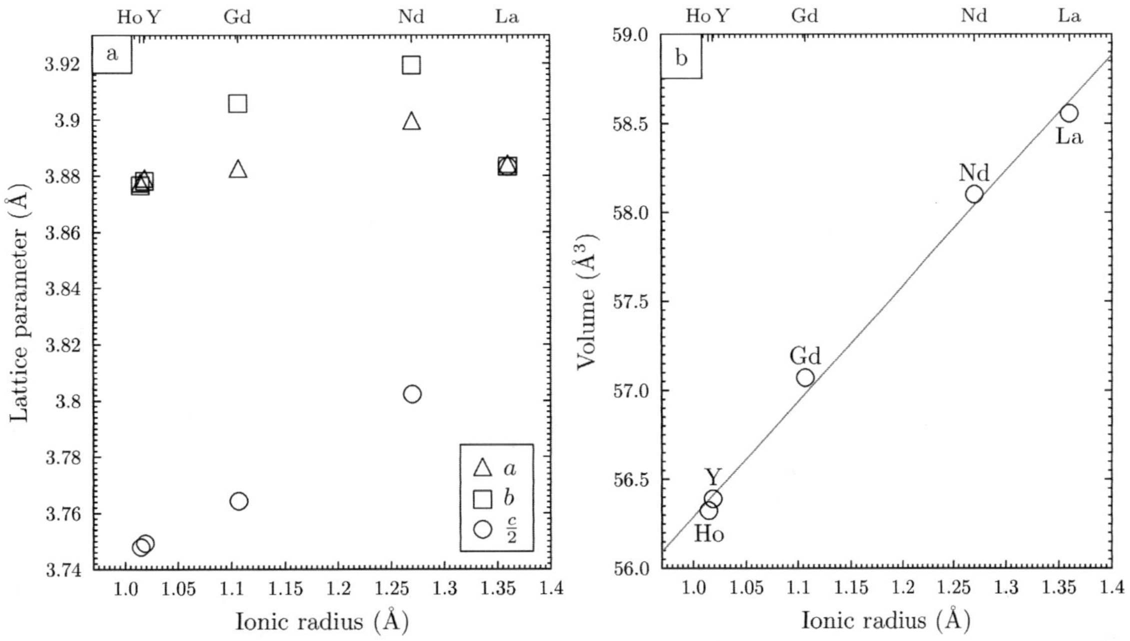


**Figure 4.** The dependence of the lattice parameters of $RBaCo_2O_{5+x}$ and the unit cell volume on the ionic radius of the rare earth element (radius of: $La^{3+}$ and $Nd^{3+}$ is in 12-fold coordination; $Gd^{3+}$ is in 9-fold coordination; $Y^{3+}$ and $Ho^{3+}$ is in 8-fold coordination [13]).

As all atoms in the ideal stoichiometric cubic perovskite occupy sites at center of symmetry, there are no Raman-active modes in it. By this reason the observed Raman lines in the spectra of perovskite-like materials can originate by activated due to an ordering of oxygen vacancies and/or by ordering of different atoms otherwise occupying equivalent sites in the ideal perovskite. In this case the symmetry of the structure lowers and the unit cell increases. As a rule for all layered perovskites, however, the intensity of the lines in Raman spectra is usually in orders lower than the spectra of starting materials. So, in our case the Raman spectra of the main phase easily can be masked by the spectra of some impurity phase, existing in negligible concentrations in the ceramics. By this reason we have carefully measured the Raman spectra of all starting materials. Some of their spectra are presented in figure 5. It is seen that the intensity of the Raman spectra of the starting materials (and the possible impurity phases) is much stronger than the spectra of the main phase. Nevertheless, in the spectra of the main phase all lines observed are intrinsic for it.

Now we can analyze the origin of the lines observed in the spectra of the tetragonal oxygen-deficient $YBaCo_2O_{5.2}$. From the selection rules of the lines, their symmetry can be determined as $E_g$ for the lines at 146 $cm^{-1}$ and near 521 $cm^{-1}$, and $A_{1g}$ for the lines at 226 $cm^{-1}$ and 455 $cm^{-1}$.

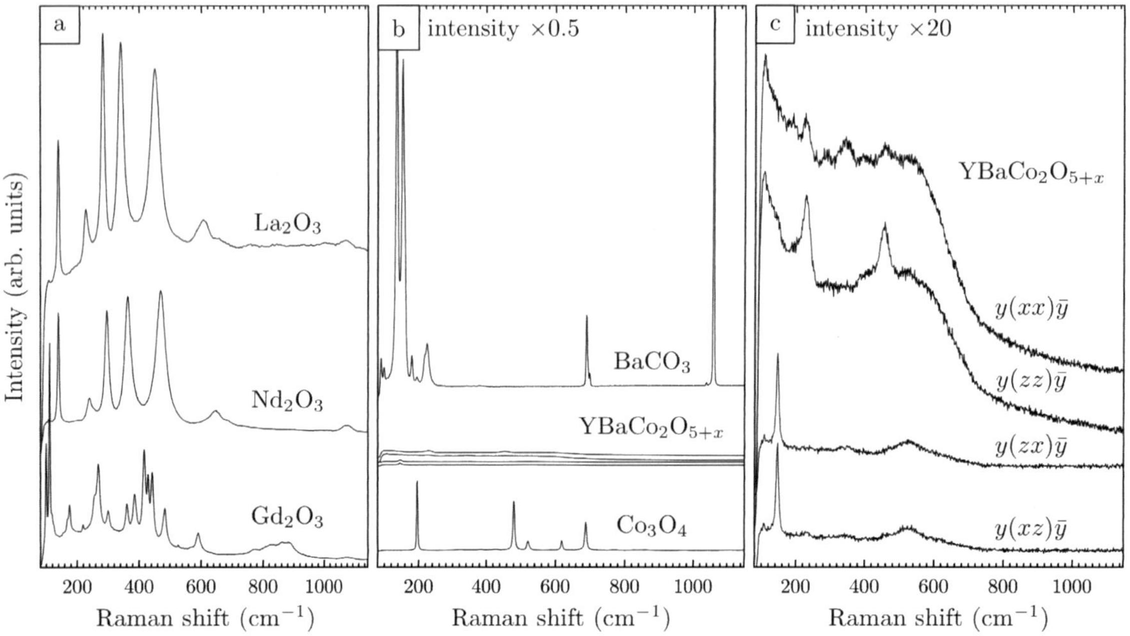


**Figure 5.** Raman spectra of the starting compounds $R_2O_3$ (R = La, Nd, Gd) (a), $BaCO_3$ and $Co_3O_4$ (b) and $YBaCo_2O_{5+x}$ (c).

From additional measurements from a (001)-oriented micro-crystal the line at 344 $cm^{-1}$ appears to be with $B_{1g}$ symmetry. Based on the lattice dynamical calculations done for the similar tetragonal $NdBaCo_2O_5$ [14], the $A_{1g}$ line at 455 $cm^{-1}$ and the $B_{1g}$ line at 344 $cm^{-1}$ can be assigned to the in-phase and out-of phase plane oxygens (both vibrations are along $z$ tetragonal axis). Lines with this symmetry and at these frequencies are typical for all layered perovskites, containing two M-O planes in the unit cell. Since the Ba, Y and O-apex atoms are in sites at center of symmetry, they cannot take part in Raman-active modes. Therefore the other lines observed can originate only by Co and O-plane vibrations. Comparing the frequencies of the observed $E_g$ line at 146 $cm^{-1}$ and $A_{1g}$ line at 226 $cm^{-1}$ with the calculated ones in [14], they can be assigned to the Co-vibrations in $xy$-plane and along $z$-direction, respectively. The origin of the broad $E_g$ line near 521 $cm^{-1}$ can be not a Raman-active mode, but rather a defect mode reflecting the oxygen non-stoichiometry. The polarized Raman spectra obtained by (100)- and (001)- oriented micro-crystals from the Ho-ceramics (not presented here) look qualitatively in the same manner. Therefore we can conclude that these spectra are typical for the tetragonal oxygen-deficient $RBaCo_2O_{5+x}$ structure.

## 4. Conclusions

The samples of double-layered perovskite $RBaCo_2O_{5+x}$ are synthesized and characterized. From the dependence of the crystal lattice parameters, the oxygen content is evaluated. Polarized micro-Raman spectra from (100)- and (001)- oriented microcrystal surfaces are obtained and analyzed. The lines observed in them are discussed and their origin is tentatively determined.

## Acknowledgments

This work was supported in part by grant No. 098/2010 of the University of Sofia as well grants No. DO02-167/2008 and No. DO02-56/2008 of the Bulgarian National Scientific Fund.

## References

[1] Taskin A A, Lavrov A N and Ando Y 2005 *Physical Review B* **71** 134414
[2] Tarancon A, Morata A, Dezanneau G, Skinner S J, Kilner J A, Hernandez-Ramirez F, Reiro F and Morante J R 2007 *Journal of Power Sources* **174** 255–263
[3] Zhang K, Ge L, Ran R, Shao Z and Liu S 2008 *Acta Materialia* **56** 4876–4889
[4] Pralong V, Caignaert V, Hebert S, Maignan A and Raveau B 2006 *Solid State Ionics* **177** 1879–1881
[5] Ozawa T C and Kang S J 2004 *Journal of Applied Crystallography* **37** 679
[6] Pashkevich Y G, Gnezdilov V P, Lemmens P, Choi K Y, Lamonova K V, Gusev A A, Guntherodt G, Barilo S N, Shiryaev S V and Bychkov G L 2004 *Spectroscopy of Emerging Materials, E. C. Faulques et.al. (eds.), Kluwer Academic Publishers* 195–204
[7] Kraus W and Nolze G 1996 *Journal of Applied Crystallography* **29** 301–303
[8] Li N, Lu Z, Wei B, Huang X, Chen K, Zhang Y and Su W 2008 *Journal of Alloys and Compounds* **454** 274–279
[9] Malavasi L, Diaz-Fernandez Y, Mozzati M C and Ritter C 2008 *Solid State Communications* **148** 87–90
[10] Akahoshi D and Ueda Y 2001 *Journal of Solid State Chemistry* **156** 355–363
[11] Burley J C, Mitchell J F, Short S, Miller D and Tang Y 2003 *Journal of Solid State Chemistry* **170** 339–350
[12] Rautama E L and Karppinen M 2010 *Journal of Solid State Chemistry* **183** 1102–1107
[13] Shannon R D 1976 *Acta Crystallographica Section A* **32** 751–767
[14] Todorov N D, Abrashev M V, Ivanov V G and Vlakhov E 2010 *AIP Conference Proceedings* **1203** 1003–1006

# A miniature capacitance dilatometer for magnetostriction and thermal expansion measurements

**Iliya Radulov, Leszek Spalek and Zacharias Viskadourakis**

Institute of Electronic Structure and Laser, Foundation for Research and Technology - Hellas, 100 N. Plastira Av. Vassilika Vouton GR-70013 Heraklion, Crete, Greece

E-mail: radulov@iesl.forth.gr

**Abstract.** In the present article we want to report about our miniature capacitance dilatometer developed for magnetostriction and thermal expansion measurements. The dilatometer is foreseen to be used on millimeter-sized single crystalline samples in the temperature range 2 - 320K. The dilatometer main body is produced of brass and is mounted in low vacuum. The capacitance was measured with a digital, self-balancing, three terminal, commercial capacitance bridge operating in the frequency range 50 Hz - 20 kHz. The temperature was determined using commercial resistive thermometers Cernox BG 1030 and Cryo-con Model 44 Temperature controller.

## 1. Introduction

To understand the lattice dynamics of the solids, where the motion of atoms always responds to cooling or heating in a way that is consistent with the symmetry of the space group to which they belong, is one of the basic aims of solid state physics. When the atoms move, the electronic structure of the solid changes, leading to changes in different physical properties, such as elastic coupling, thermal expansion (TE) or specific heat. The coupling between the magnetic and elastic regimes in a magnetostrictive material is known as Joule magnetostriction (MS). The MS is a property of the ferromagnetic materials and causes them to change shape when subjected to a magnetic field. Knowing the TE of the solids we can obtain useful information about their electronic and photonic properties, their specific heat (Gruneisen formalism), propose a thermodynamic treatment of their phase transitions (Ehrenfest relations) etc. The idea for using capacitor as a detection device was developed in 1920 by R. Whiddington [1]. His ultra-micrometer was the first capacitance device used to measure very small distances changes. In 1961 White [2] has developed his capacitance dilatometer. White's design principles of absolute and relative dilatometers were adopted and improved by number of different authors. For MS measurements, initially the strain gauges method [3] was predominately used. Green, Chanddrasekhar [4]and Fawcett[10] were among the first, that used capacitance dilatometer for MS measurements. Based on this we decided to concentrate our efforts to develop CD for both TE and MS measurements.

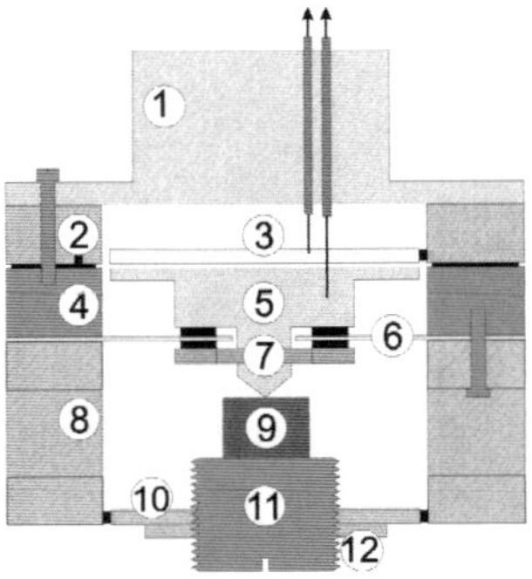


**Figure 1.** A schematic view of the capacitance dilatometer.

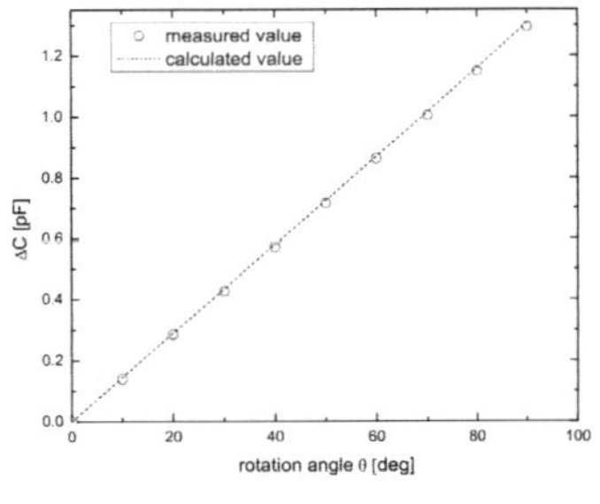


**Figure 2.** Dilatometer calibration data: response linearity.

## 2. Apparatus

A schematic draw of our capacitance dilatometer is shown on figure 1. The measuring cell is made from brass. In order to reduce the internal strains from the machining process all its parts were cleaned with 10% dilute nitric acid annealed at 200° C for about 6 h in nitrogen atmosphere .

### *2.1. Thermometry and temperature controller*

The CD is foreseen for usage with MagLab Exa system fitted with a 9T SC magnet and is capable of measurements in the range 2 ÷ 320 K. In order to ensure highest thermal stability of the measuring cell, the CD is placed in a closed stainless steel tube, evacuated to rotary pressure of about $10^{-3}$ Torr and then flooded with $^4$He exchange gas to a pressure of about 1 Torr. Sofar the thermal equilibrium and not the stability of the capacitance measuring circuit limits the resolution of the method, the most important issue is to ensure the thermal equilibrium of the whole dilatometer. For example at room temperature a temperature difference of 0.01 K causes a length change of several nm in the cell. Assuming the effective length of the cell ≈ 40 mm and typical samples length 3-5 mm we have an inaccuracy of $\alpha$ of about 10 %. To obtain an accuracy of 1% we had to stabilize the temperature of the dilatometer to within 1 mK at 300 K. To measure the sample temperature we use a Cernox BG 1030 thin film resistance sensor. The sensor tightly fitted with Apiezon N grease on the upper surface of the fixed capacitor plate (3). For temperature reading and heater control we use a Cryo-con Model 44 Temperature controller.

### *2.2. Capacitance system*

The CD represents cylindrical cell containing a flat capacitor, build by two parallel plates surrounded by three frame rings (FR), used as electrical shielding and construction frame. The upper capacitor plate (3) is fixed to the upper FR (2) with Stycast. The gluing process includes 24 h curing at room temperature followed by 24 h at about 80° C. To ensure that ring and plate are coaxial 0.2 mm thick fishing line is slipped into the gap between them. To ensure that their lower surfaces remain also coplanar both parts are placed on a flat glass plate. Three stainless steel screws passing through the connecting flange and upper FR before threading into the middle FR. The second capacitor plate (5) is bolted on the spring (6) using the nut (7). The spring is made of 0.125 mm thick Be-Cu foil and is insulated between two 0.5 mm thick Teflon washers. The spring is positioned between the middle (4) and the lower (8) FRs as shown in figure 1; the lower FR and the spring are bolted with three stainless steel screws into the middle FR. To hold the spring and lower capacitor plate concentric during the mounting we use again fishing line. For the purpose of capacitor gap evacuation and exchange gas circulation the spring has

Journal of Physics: Conference Series **253** (2010) 012072 doi:10.1088/1742-6596/253/1/012072

six small holes. The distance between the capacitor plates can be adjusted by use of separator washers placed between the upper and middle FRs or varying the thickness of the Teflon washers used with the Be-Cu spring. In order to ensure better sensitivity the initial capacitance has to be ca. 10÷15 pF. The capacitance C is measured with a digital, self-balancing, three terminal, commercial Andeen-Hagerling capacitance bridge AH2700A. According our test results for best performance the bridge has to be operated with frequency between 0.6 and 2.5 kHz (typical 1 kHz), excitation voltage 15 V rms and average over 5÷8 seconds. The connection between the capacitor plates of the dilatometer is performed first through an Ultra Miniature Coaxial Cable - Type SS (Lakeshore). Before fixing the capacitor plates to the FRs we have soldered to them short wires (from the coaxial cable center conductor). These wires are used as electrical contact points for the coaxial cables. To ground the coaxial cables shields at the dilatometer body, the shields are soldered to a small copper foil tabs, fixed between the FRs. The connection to the bridge, outside the cryostat, is performed via double-shielded BNC cables. By careful construction of the capacitance measuring circuit (shielding, avoiding of ground loops etc) the maximum sensitivity could be equal to that of the bridge 1 x $10^{-6}$ pF, which corresponds with C≈ 15 pF to a length change of 0.2 nm.

### *2.3. Sample adjustment system*

The sample adjustment system contains a sample holder screw (11), screwed into lower flange (10). The lower flange (10) is attached to the lower FR (8) using Stycast and fishing line as described above. After the sample and sample holder screw (11) are positioned appropriately they have to be fixed in place with the lock-nut (12). If necessary different sample holder screws with different lengths can be used to accommodate different sample sizes.

## 3. Test and results

The first test we have performed on our dilatometer was to check it sensitivity and linearity of the response. These tests were performed at room temperature, outside the cryostat. To avoid any kind of external disturbances the capacitance cell was protected by grounded aluminum cover. As test sample we use a cylindrical shaped polycrystalline Ni 3 mm length and 2 mm diameter. In order to measure the exact angle to capacitance dependence an azimuth disk was mounted on the lower flange and a pointer on the sample holder screw. Rotating in small steps the sample holder screw we have received the $C(\theta)$ dependence for our dilatometer. The result plotted as $\Delta C$ vs $\theta$ is shown in figure 2. In a capacitive dilatometer the dilation $\Delta L$ of a sample of length $L$ manifests as a change in the gap $d$ between a pair of capacitor plates. For an ideal parallel-plate capacitive dilatometer in vacuum the relationship between the measured capacitance $C$ and $d$ is simply $C = \varepsilon_0 \frac{S}{d}$, where $\varepsilon_0$=8.854 19 pF/m is the permittivity of free space and $S$ is the area of the capacitor plates. The dependence between the capacitor gap $d$ and the angle $\theta$ is $d = d_0 - k\theta$, where $d_0$ is the at which the gap of unloaded dilatometer and the constant $k$ is related to the thread pitch on the sample holder screw. We can rewrite the equation as $C = \varepsilon_0 \frac{S}{d_0 - k\theta}$, thus, a plot of $\Delta C$ vs $\theta$ should be a straight line whose slope yields the effective area of the capacitor plates, more detailed discussion about the capacitive edge and tilting effects is given in Pott and Schefzyk [7]. In general to determine the temperature dependent coefficient of linear TE or the field dependent MS of a sample we need to know not only the dependence with the sample, but also the dependence with the reference he material installed in the dilatometer. As discussed above the capacitance of the dilatometer is sensitive to thermal gradients, therefore it is recommended to use reference sample from the same material as the capacitance cell. Then the coefficient of linear TE is $\lambda = \lambda_{cell+unknownsample} - \lambda_{cell+ref.sample} + \lambda_{ref.sample}$. As described above our capacitance dilatometer can be used for MS and TE measurements. In general MS measurements have to be done in slow changing magnetic (recommended sweep rate 0.1÷0.2 T/min) and at measurements the temperature sweep rate strongly depends from the temperature

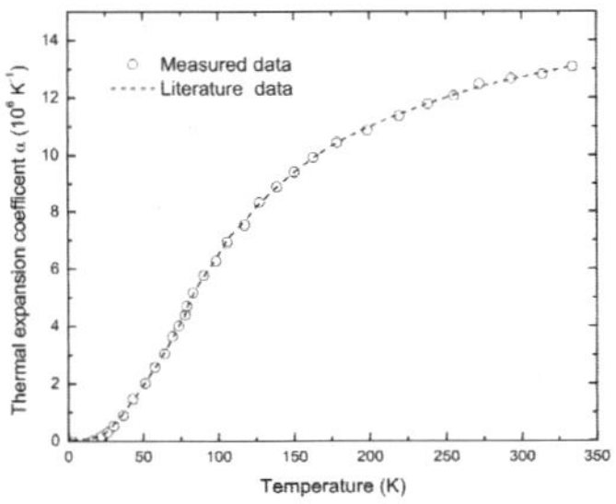


**Figure 3.** The thermal expansion of polycrystalline nickel.

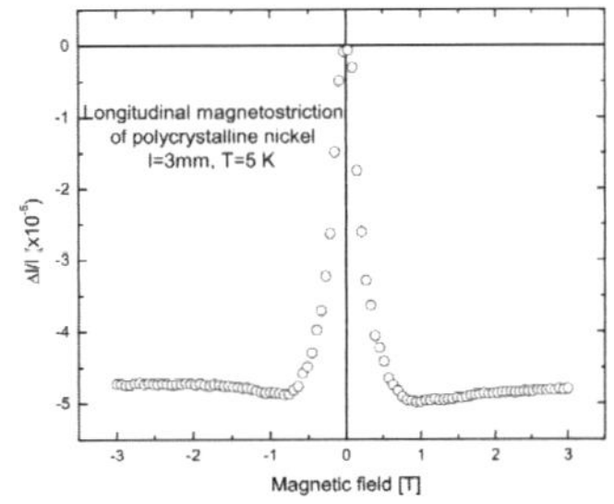


**Figure 4.** Longitudinal magnetostriction of polycrystalline nickel.

range, starting with 0.1 K/min or less for the range 2÷10 K and reaching 0.5 K/min or less for the range 180÷330 K. Using our dilatometer with the MagLab Exa system we have to consider that its temperature stabilization is better by warming and and generally data collected during warming are less noisy than during cooling. In case the temperature/ magnetic field was changed to fast we can observe "tails" (the capacitance continues change after the final temperature/ magnetic field is reached). Sometimes a small "steps" can appear in the temperature/ magnetic field dependence. We suppose these peculiarities can be caused by sudden relaxation of small strains in the CD or in the sample. In generally there are no special requirements to the samples shape. Special attention is needed in case of measuring monocrystalline samples with high magnetic anisotropy. Varying the sample orientation with respect to the applied magnetic field can cause a torque if its magnetic moment is not parallel to the field. Any motion of the sample in response to this torque will contribute to the measured capacitance change. According Birss [6] the specimen should have cylindrical symmetry, and measurements must be made as the magnetic field is rotated about the axis of symmetry. Furthermore, the specimen shape should ideally be an ellipsoid. In practice, thin discs are assumed to approximate sufficiently well to the shape of an oblate spheroid. Finally we have measured the TE and MS of a high-purity sample of polycrystalline nickel. The TE and MS results are shown in figures 3 and 4 respectively. On figure 3 the data received with our dilatometer is represented by open circles and dashed line represents the TE values for Ni published in the literature [8, 9]. The average absolute value of the fractional deviation of our nickel measurements from those in the literature is about 1.5% for the whole temperature range 2÷320 K. The longitudinal MS of a cylindrical sample of high purity, polycrystalline nickel is shown in figure 4, a pronounced negative Joule MS and a slightly positive forced MS are observed. Both of these features are consistent with published results [10, 11].

## References

[1] Whiddington R 1920 *Philosophical Magazine* Series 6 **40** 634
[2] White G K 1961 *Cryogenics* **1** 151
[3] Goldman J E 1947 *Phys. Rev.* **72** 529
[4] Green B A and Chandrasekhar B S 1963 *Phys. Rev. Lett* **11** 331
[5] Fawcett E 1970 *Phys. Rev. B* **72** 3887
[6] Birss R R, Keeler G J, Pearson P and Potton R J 1978 *J. Phys. E: Sci. Instrum.* **11** 928
[7] Pott R and Schefzyk R. 1983 *J. Phys. E* **16** 445
[8] White G K 1965 *Proc. Phys. Soc. London* **86** 159
[9] Clark A F 1968 *Cryogenics* **8** 231
[10] Fawcett E and White G K 1967 *J. Appl. Phys.* **38** 1320
[11] Nizhankovskii V I 2006 *Eur. Phys. J. B* **53** 1

Journal of Physics: Conference Series **253** (2010) 012073 doi:10.1088/1742-6596/253/1/012073

# Magnetic properties of some rare-earth nanostuctured aluminates

**V Lovchinov[1], D Petrov[2], P Simeonova[1] and B Angelov[2]**
[1] G. Nadjakov Institute of Solid State Physics, Laboratory of Environmental Physics, 1784 Sofia, Tzarigradsko Chaussee Blvd. 72, Bulgaria
[2] Department of Inorganic and Physical Chemistry, University of Food Technologies, 4002 Plovdiv, Maritsa Blvd. 26, Bulgaria

E-mail: lovcinov@issp.bas.bg

**Abstract.** Nanocrystalline single-phase $RAlO_3$ (R = Nd, Sm, Eu, Dy, Gd) has been prepared by modified Pechini's method. Malic acid has been used for the first time as a new complexing agent in the sol-gel process. It has facilited a low temperature synthesis of the compound. The characterization of the nanoparticles has been carried out by different methods. Using Physical Property Measurement System (PPMS-9 QD) the temperature and magnetic dependency of the susceptibility and magnetization of the nanostuctured aluminates were measured. The obtained results were compared with the existing ones for the single crystals and powder specimens of the same aluminates. The differences observed have been discussed in the framework of the molecular field theory for a two-sublattice system.

## 1. Introduction

The lantanides and their alloys are of current interest due to their useful magnetic properties. Especially rare-earth aluminates of the general formula $RAlO_3$ form an important series for magnetic studies, since the trivalent rare-earth ions R lies on an approximately simple cubic lattice. In this case, theoretical calculations are reduced to simple terms. On the other hand, the extensive search for new materials used as ferrofluids or magnetic refrigerants, in high-density magnetic storage, in high-frequency electronics etc. makes rare-earth aluminates promising compounds. Up to now there are many studies dealing mainly with single crystals and powder specimens of rare-earth aluminates [1-6]. It is well known that nanoscaled, nanostructured or nanocomposited materials exhibit novel physical properties as compared to those of their bulk counterparts. Recently, our group has modified Pechini's method using for the first time malic acid as a new compexing agent in the sol-gel process [7-9]. This gives us a possibility to prepare nanocrystalline single-phase aluminates.

The aim of the present study is to report preliminary results for the magnetic properties of some rare-earth aluminates prepared as nanoparticles. The results obtained were compared to previous ones for single crystals and powder specimens of the same rare-earth aluminates.

## 2. Experimental procedures

$RAlO_3$ (R = Nd, Sm, Eu, Dy, Gd) nanoparticles were synthesized by the sol-gel process using malic acid and 1, 2-ethanediol. The gel was obtained with stoichiometric amounts of the corresponding rare-earth oxides $R_2O_3$ (R = Nd, Sm, Eu, Dy, Gd), $Al(NO_3)_3.9H_2O$ and malic acid (hydroxybutanedioic, $HOOCCH_2CH(OH)COOH$). The details of the preparation procedure are given elsewhere [7-9].

X-ray diffraction patterns, TEM and SEM investigations reveal that corresponding monoaluminates exist in a single phase as nearly spherical particles of close size, coexisting with agglomerates. The size of the particles varies from 20 to 90 nm depending on the rare-earth ion. More details about XRD, TEM, SEM and X-ray photoelectron spectra are given in our earlier works on above mentioned sol-gel prepared monoaluminates [7-9].

In the present study we report preliminary results for the magnetic susceptibility and magnetization of the nanocrystalline single-phase aluminates $RAlO_3$.

The temperature dependences of the AC susceptibility and DC magnetizations of the $RAlO_3$ powder samples were measured with Physical Property Measurement System Quantum Design (PPMS-9 QD). The system ensures stepwise change of the temperature from 2 to 300 K and of the magnetic field from 0 to 9 Teslas. The accuracy is 0,2 % or $10^{-8}$ emu for the magnetic measurements and 0,01K for the temperature ones. In this work the measurements of the DC magnetization of the samples was carried out in two modes-ZFC (Zero Field Cooled) and FC (Field Cooled) at field of $H_{DC}$ = 1 kOe for DC measurements and $H_{AC}$ =10 Oe and f =1 kHz for AC measurements. For the samples $DyAlO_3$ and $GdAlO_3$ magnetization vs. magnetic field (M (H) loops) were measured at a temperature of 2K.

## 3. Results and discussion

In figures 1, 2 and 3 the temperature dependencies of the DC magnetization of all five aluminates are presented. All measurements were carried out in two modes (ZFC and FC). Both real M' and imaginary M'' parts of the magnetization were measured also.

First we examine nanostructured $NdAlO_3$, $SmAlO_3$ and $EuAlO_3$ samples (figure 1). The curves are similar to weakly magnetic systems – slow change of M with the temperature up to 50 K followed by a sharp increase in the interval between 50 and 2 K. This magnetic behaviour of the nanocrystallites corresponds to that of diluted magnetic systems in the same temperature interval. As shown in our previous work [8] for the interval between 50 and 2 K, M follows 1/T behaviour and that is why the value of M is logical to be large. It is important to note that M'' is practically equal to zero and no hysteresis effects are observed, which indicates that these compounds are paramagnetic in the whole interval in consideration (2 – 300 K).

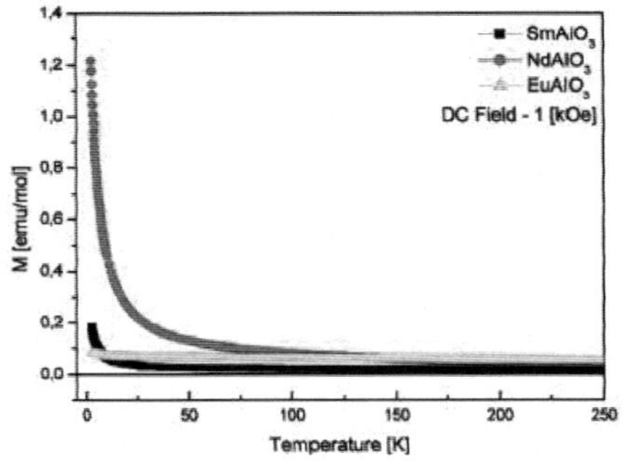


**Figure 1.** Temperature dependencies of the DC magnetization for $NdAlO_3$, $SmAlO_3$ and $EuAlO_3$.

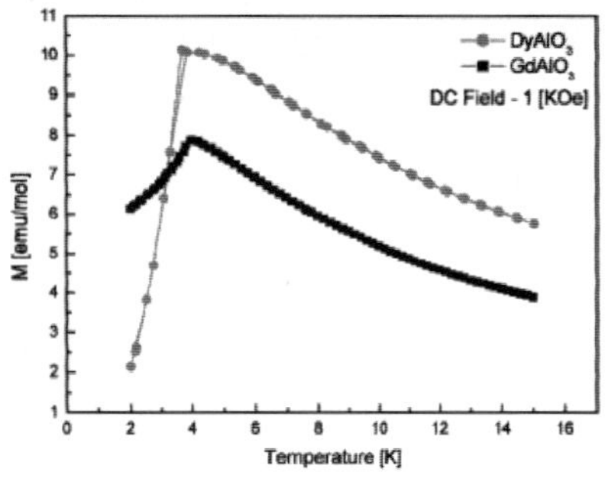


**Figure 2.** Temperature dependencies of the DC magnetization for $DyAlO_3$, and $GdAlO_3$.

Quite different is the behaviour of the other two nanostructured aluminates - $DyAlO_3$ and $GdAlO_3$ (figure 2) and (figure 3). The M(T) data indicate an antiferromagnetic ordering near 4 K.

As seen from the measurements of M(T) for the two aluminates in low field of 1 kOe (figure 2) and in high field of 20 kOe (figure 3) the type of the curves is not changed. No change is observed also in the temperature value at which antiferromagnetic ordering is found - Neel's temperature $T_N$. Changes

Journal of Physics: Conference Series **253** (2010) 012073 doi:10.1088/1742-6596/253/1/012073

are observed only for the magnitude of the magnetization as expected - proportional to the applied magnetic field. For more precise determination of the Neel's temperature $T_N$ and the paramagnetic Curie constant $\Theta$ inverse susceptibility as a function of temperature could be used to present the experimental results.

The results shown in figure 4 indicate that except Neel's temperatures for both aluminates Curie molar constants $\Theta_{Dy}$ and $\Theta_{Gd}$ could be obtained being respectively for $DyAlO_3$ - $T_{N\ Dy}$ = 3,72±0,01 [K] and $\Theta_{Dy}$ =1,2±0,1 [K], and for $GdAlO_3$ - $T_{N\ Gd}$ = 3,96 ±0,01 [K] and $\Theta_{Gd}$ = 2,1±0,1 [K].

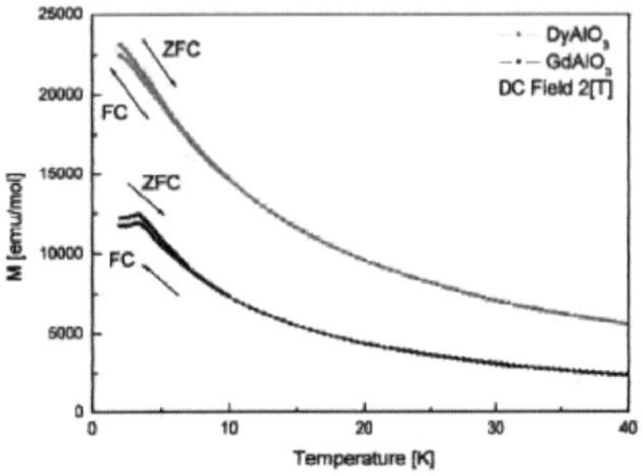


**Figure 3.** Temperature dependencies of the DC magnetization for $DyAlO_3$, and $GdAlO_3$ at high magnetic field of 2 [T]

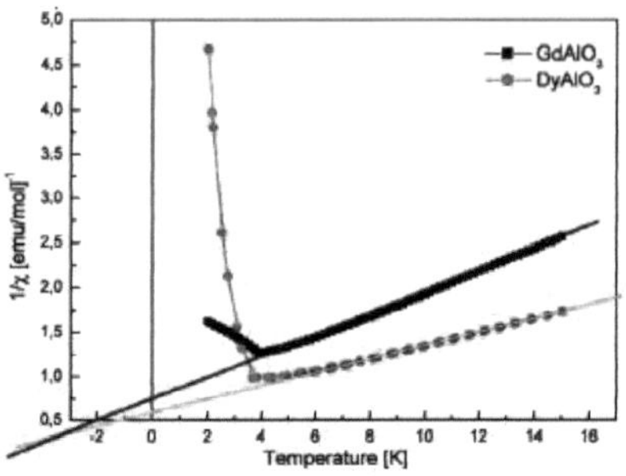


**Figure 4.** Inverse magnetic susceptibility as a function of temperatures for $DyAlO_3$, and $GdAlO_3$

Let us compare the results obtained for nanostructured aluminates with results found in the literature for single crystal samples [1-6]. The first feature is that the magnetization magnitudes of our nanostructured aluminates is by 20% lower that the same values for single crystals. For nanostructured aluminates with proven antiferromagnetic transition $DyAlO_3$ and $GdAlO_3$, this specificity is accompanied by decreased values of the molar Curie constant, e.g. for the nanostructured $DyAlO_3$ $\Theta_{Dy}$ =1,2±0,1 [our own data], whereas for the single crystal samples $\Theta_{Dy}$ = 1,5±0,1 [6]. For $GdAlO_3$ these differences are more substantial - for the nanostructured $GdAlO_3$ $\Theta_{Gd}$ = 2,1±0,1 [own data], whereas for the single crystal $\Theta_{GdAlO3}$ = 4,81±0,05 [3]. No significant differences are observed for the Neel temperatures $T_N$. According to our data $T_N$=3,72K for $DyAlO_3$ and $T_N$=3,96K for $GdAlO_3$. Reference data indicate 3,52 and 3,87K, respectively. It should be mentioned that in some studies $T_N$ values are obtained by specific heat measurements and they are closer to our own data as $\lambda$ transition at specific heat measurements reflects the bulk properties of the samples. In most theoretical studies [3,5,6] the single crystal $DyAlO_3$ and $GdAlO_3$ are considered as a two-sublattice un-axial antiferromagnet, for which the molecular field theory for a two-sublattice system is valid. In this theory each sublattice is subject to the external field H, an exchange field $H_e$ and anisotropy force.

In order to check how the external magnetic field affects the magnetization of our nanostructured systems, measurements were carried out in magnetic fields up to ±6 Tesla at temperature T=2K, i.e. far below from $T_N$.

In figure 5 the results of these measurements are presented for both $DyAlO_3$ and $GdAlO_3$. For $GdAlO_3$ a clearly expressed hysteresis in the curve M(H) is seen. In the upper left corner of the figure for clarity part of the data for the dependence M(H) are given up to fields of ±0,6 Tesla. It indicates that for $DyAlO_3$ similar hysteresis although not so clear is observed.

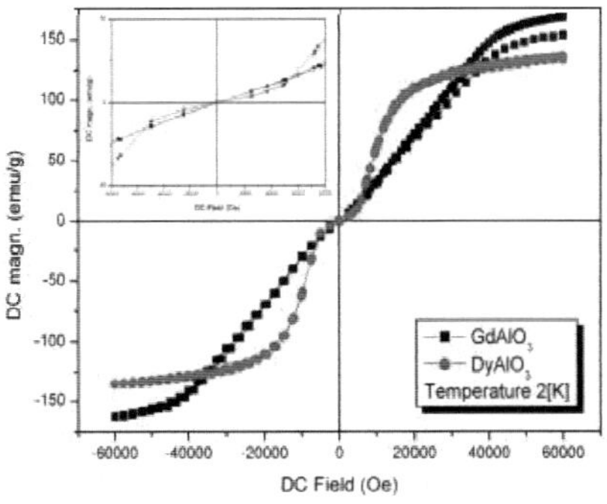


**Figure 5.** DC magnetization for $DyAlO_3$ and $GdAlO_3$ as a function of external magnetic field at temperature T=2K

This experiment shows that the nanostructured systems could also be considered as a two-sublattice unaxial antiferromagnet, for which the molecular field theory for a two-sublattice system is valid. It has to be mentioned, however, that for nanostructured systems the surface effects could play a substantial role and due to this the magnetic moments are canted. Maybe this fact is one of the reasons for lower magnetization magnitude for all nanostructured aluminates investigated.

## 4. Conclusions

Using modified Pechini method nanocrystalline single-phase compounds $RAlO_3$ (R = Nd, Sm, Eu, Dy, Gd) have been prepared.Temperature and magnetic dependencies of the magnetization and molar susceptibility of so prepared nanoparticules reveal that $NdAlO_3$, $SmAlO_3$ and $EuAlO_3$ remain paramagnetic down to the lowest temperature achieved in the experiment (2K). The temperature dependence of the magnetization for Dy and Gd nanoparticulated aluminates shows antiferomagnetic transitions at $T_N$ = 3,72K for $DyAlO_3$ and $T_N$ = 3,96K for $GdAlO_3$.

For the first time an attempt to compare the results for nanostructured aluminates to that of existing ones in the literature for monocrystalline $RAlO_3$ is made. The main conclusion from this comparison is that nanostructured aluminates shows lowest magnitude of magnetization and paramagnetic Curie constants. Probably for the nanoparticulated systems the surface effects could play a substantial role and due to the fact that magnetic moments are canted. A magnetic transition of the rare earth system to a spin glass state could be also reasonable explanation for the observed effects. In order to clarify these problems, however, additional experiments are needed.

## 5. Acknowledgements

Two of the authors (B. Angelov and D. Petrov) thank for the financial support to Project 9/10 – H of the University of Food Technologies. The rest of the authors (V. Lovchinov and P. Simeonova) would like to express their sincere gratitude to Project DO-02-352 (National Science Fund) for the financial support. Thanks are also due to Prof. K. Krezhov for the fruitful discussions and valuable suggestions.

## 6. References

[1] Holmes L, Sherwood R and VanUiterd L G 1969 *Phys.Rev.* **178** 576
[2] Bartolome F, Bartolome J and Blasco J 1996 *J.Magn.Magn.Mater.* **157/158** 491
[3] Cashion J D, Cooke A H, Thorp T L and Wells M R 1970 *Proc.Roy.Lond.*A **318** 473
[4] Holmes L M, VanUiterd L G, Hecker R R and Hull G W 1972 *Phys.Rev.*B **5** 138
[5] Cashion J D, Cooke A H, Hawkes J F B, Leask M J M, Thorp T L and Wells M R 1968 *J.Appl.Phys.* **39** 1360

[6] Kolmakova N P, Krynetskii I B, Lukina M M and Mukhin A A 1990 *Phys.Status Solidi* B **159** 845
[7] Petrov D and Angelov B 2009 *J. Sol-Gel Science and Technology* **31** 101
[8] Petrov D, Angelov B and Lovchinov V 2010 *J. Rare Earths* **28** 602
[9] Petrov D, Angelov B, Lovchinov V and Simeonova P 2009 *American Institute of Physics Conference Proceeding* **1203** 1063

# Investigation of the current injection on $J_C$ of a HTS YBCO ring

**T Nurgaliev[1], E Mateev[1], B Blagoev[1, 3], L Neshkov[1], I Nedkov[1] and L Uspenskaya[2]**

[1] Institute of Electronics BAS, 72 Tsarigradsko Chaussee, 1784 Sofia, Bulgaria
[2] Institute of Solid State Physics RAS, Chernogolovka, 142432, Russia

E-mail: blago_sb@yahoo.com

**Abstract**. The effect of the current injection from the ferromagnetic $La_{0.7}Sr_{0.3}MnO_3$ (LSMO) electrode into high temperature superconducting $Y_1Ba_2Cu_3O_{7-x}$ thin film ring on the critical current of the ring was investigated at 77 K using contactless method. The structures were prepared using magnetron sputtering, conventional photolithography and wet etching procedures. The critical current was determined from the response of the HTS ring to the AC magnetic field. The experiments demonstrated a sensitivity of the critical current to the current injection process, especially in the case of LSMO current-injecting electrode. The paper presents the experimental results and discusses possible mechanisms of the observed effects.

## 1. Introduction

Spintronics involves the exploitation of the spins of the charge carriers rather than their charge. Superconducting spintronic devices use spin-dependent properties of superconductors in superconductor/ferromagnet heterostructures [1-4]. Spin-polarized quasiparticle injection from colossal magnetoresistance materials into high temperature superconductors (HTS) causes strong nonequilibrium effects and able to affect the electronic transport properties of the superconducting films [4-6]. Ferromagnetic manganites $La_{0.7}Sr_{0.3}MnO_3$ (LSMO) and high temperature superconductors (HTS) $Y_1Ba_2Cu_3O_{7-x}$ (YBCO) can be used as components of such structures [7-9]. The similar lattice constants and growth conditions of HTS YBCO with $T_C$ ~ 92 K and the half-metal manganite compound LSMO that exhibits a FM state below $T_{Curie}$ ~ 350 K make it possible to epitaxially grow superlattices and heterostructures using various techniques [7-9]. A high spin polarization ratio is characteristic for hole-doped rare earth manganites as $La_{0.7}Sr_{0.3}MnO_3$ the spin polarization of $e_g$ – electrons of which is close 100%. For this reason $La_{0.7}Sr_{0.3}MnO_3/Y_1Ba_2Cu_3O_{7-x}$ thin film heterostructures are perspective for spintronic experiments.

It was shown experimentally, that the spin injection process in LSMO/YBCO/LSMO and LCMO/YBCO/LCMO thin film structures can reduce the critical temperature of the structures by up to 15% [4]. It was demonstrated also that the suppression of the critical current density caused by injection of the spin polarized quasiparticles into YBCO thin film through thin insulating $SrTiO_3$ or metallic Au barriers exceeds the supression caused by the unpolarised quasiparticles by several times [6,10]. In spite of the great efforts of the researchers, many aspects of the physical processes in FM/HTS structures and interfaces remain not clear yet. In this paper we propose to investigate the

[3]To whom any correspondence should be addressed

spin-injection process using contactless methods and present our preliminary experimental results obtained in a structure containing HTS YBCO and FM LSMO components.

## 2. Experimental structure

The RF and DC magnetron sputtering equipments were used for in-situ growing of HTS YBCO and FM LSMO thin films and FM/HTS double-layers on $LaAlO_3$ substrates. The growing procedure and conditions of the HTS YBCO and FM LSMO films and sandwiches were described elsewhere [11,12]. Figure 1 shows a diagram of the YBCO/LSMO structure used in this study. It consists of a thin film HTS YBCO ring, placed on a dielectric substrate, and two electrodes which are used for injection of DC current into the ring. The electrode electrically contacting with the narrow part of the ring was either an HTS YBCO (sample 1) or a FM LSMO thin film strip (sample 2), while the second electrode was an HTS strip for both samples. For preparing of structure 2 first the FM LSMO thin film electrode

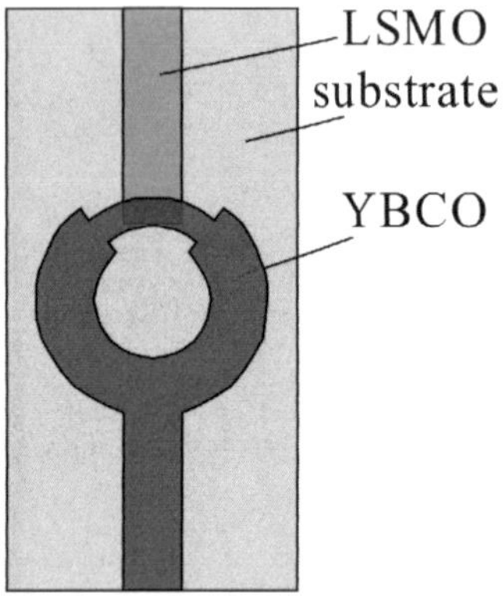


**Figure 1.** Sketch of the sample geometry chosen for contactless investigation of the current injection effect on the $J_C$ of the HTS ring.

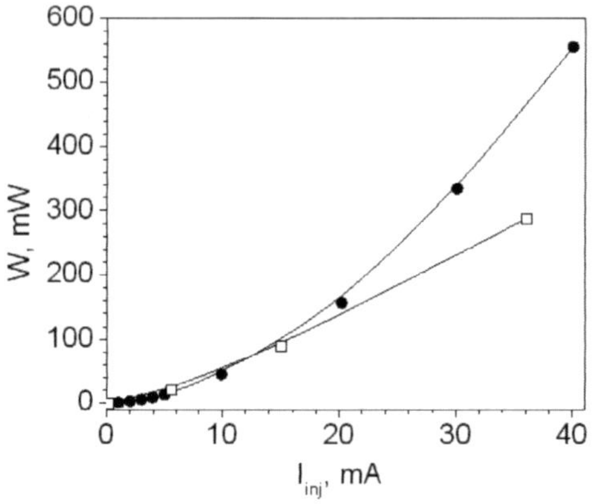


**Figure 2.** Power $W$ of the electrical losses in the sample 1 (open squares) and 2 (closed circles) at 77 K caused by the injected current $I_{inj}$.

(with the thickness of ~30 nm) was deposited by RF magnetron sputtering on 10 × 5 mm LAO substrate as a stainless steel mask was used to form a necessary configuration of the electrode during the deposition process. Afterwards the HTS YBCO thin film with the thickness of ~60 nm was grown on the top of this structure. Conventional photolithography and chemical etching were used for patterning of the HTS part of the structure and forming sample 2. Sample 1 consisted of HTS thin film components only.

Measurements were performed at 77 K in liquid nitrogen bath. AC current in the HTS ring was excited by a magnetic field of small coil, placed near the ring. Another small coil was used to register the third harmonics of the response signal, which arises when the current in the HTS ring reaches its critical value. Measurements were performed at different magnitudes of the injected DC current. If the injected DC current is not strong and the patterned structure is absolutely symmetric, it can not affect noticeably the measured value of the critical current, but can affect the parameters of the superconducting ring causing a change of the critical current. The power $W$ associated to electric losses in the samples at 77 K caused by the injected current $I_{inj}$ was small (figure 2) and was not able to affect the temperature of the samples immersed in a liquid nitrogen bath.

## 3. Results and discussion

The amplitude of the 3-d harmonics $U_{res}$ of the response signal on the amplitude $I$ of the screening current induced in the YBCO ring 1 and 2 as a function of the magnetic field of the driving coil are presented in figures 3 and 4 respectively. It can be seen that the amplitude of the third harmonics of

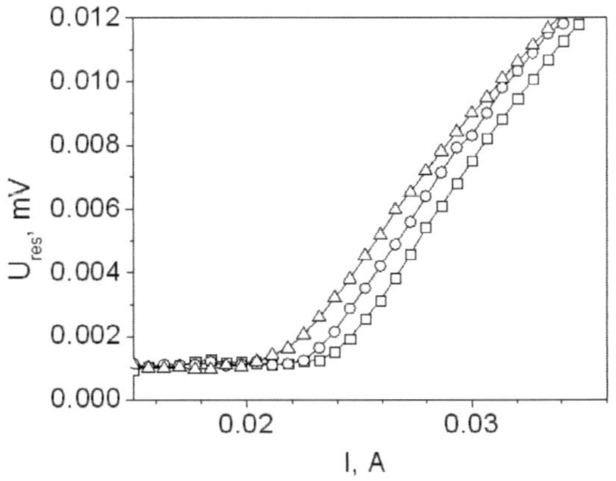


**Figure 3.** Dependence of the 3-d harmonics of the response signal on the amplitude of the screening current induced in the YBCO ring 1 (completely HTS YBCO structure) by the magnetic field of the driving coil. Injected current strength $I_{inj}$ is 0 mA (squares), 12 mA (circles) and 20 mA (triangles).

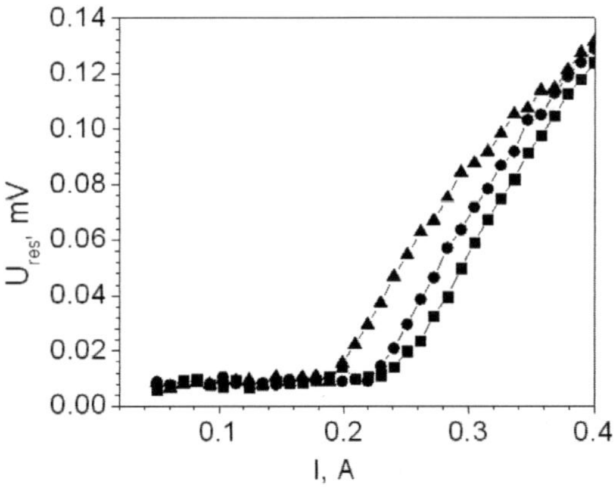


**Figure 4.** Dependence of the 3-d harmonics of the response signal on the amplitude of the screening current induced in the YBCO ring 2 (HTS YBCO structure with LSMO electrode) by the magnetic field of the driving coil. $I_{inj}$ is 0 mA (squares), 20 mA (circles) and 40 mA (triangles).

the response signal is negligible at small amplitudes of the driving current and it increases drastically when the current in the HTS ring reaches its critical value. The critical currents $I_C$ of these samples were 0.024 A and 0.23 A, respectively, in the absence of the injected current $I_{inj}$. The current injection process affects the critical current density (figures 3 and 4). The injected current $I_{inj}$ and the change of the critical current $\Delta I_C$ in each sample can be expressed in dimensionless forms as $I_{inj}/I_C$ and $\Delta I_C/I_C$, respectively. Dependence of the dimensionless critical current density on the normalized injected current strength for both samples 1 and 2 are presented in figure 5.

It can be seen that the effect of the current injection on the critical current of the HTS ring is much stronger for the sample 2 with an LSMO electrode (figure 5) than for the sample 1 with HTS electrodes. Some change of $I_C$ in completely HTS structure 1 at presence of small injected current is possibly due to presence of local unhomogeneities (i.e. weak links) of this sample which are characterized with low critical current densities and non symmetrical positions of these unhomogeneities along the sample. Change of $I_{inj}$ – polarity did not cause a noticeable change of $I_C$ for both samples. Possibly this is due partly to great area of the interface, which includes magnetic domains of LSMO with different orientations and leads to some averaging of the effect in sample 2.

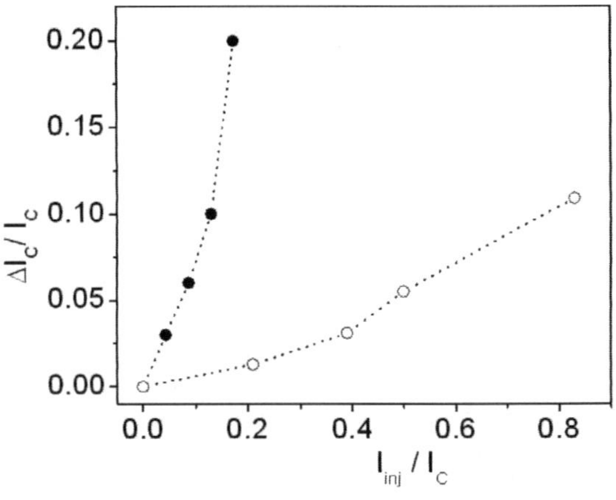


**Figure 5.** Relative change of critical current as a function of normalized injected current strengths in samples 1 (open circles) and 2 (closed circles).

There are several microscopic mechanisms of the coupling at the interface of HTS/LSMO structure [2,4,5,8,9] which affect the critical parameters of the HTS component. The proximity effect related to tunneling of Cooper pairs into the non-superconducting FM side can lead to both, a reduction and an oscillation of the critical temperature of the superconducting part of the heterostructure. On the other hand, the large exchange energy of the magnetic layer prevents Cooper pairs from tunneling into the LSMO film and this mechanism is less effective [5] for reduction of the critical parameters of HTS in the structure.

There is also an inverse proximity effect, when spin polarized quasiparticles penetrate into HTS layer on a certain length scale [1,5] of order ~10 nm [5]. According to results of [5] in HTS/manganite structure an existence of a long-range inverse magnetic proximity effect which affects the properties of YBCO films up thicknesses of more than 100 nm is possible as well. The effect is due to the anisotropy of the d-wave parameter, because the coherence length tends to approach infinity in the (110) nodal direction of the YBCO Fermi surface.

The injected current strength and polarity affect the effectiveness of the above processes and leads to modification of the critical parameters of FM/HTS structure. In the last case the changes of the electrical properties of HTS component is possible if a main part of the quasiparticles in the injected current crosses the interface without "losing" their spin-orientations. The spin-diffusion length in cuprate superconductors is of order ~ 10-20 nm [4] or more in some directions. So, a reduction of $T_C$ and $J_C$ in HTS can occur in the same length scale in the HTS component of the structure. Above mechanisms provide sensitivity of the critical parameters of HTS film to the current injection process in FM/HTS structures, as it was observed by contactless way in our experiments as well. Because of complexity of the processes in FM/HTS interfaces, additional investigations are required to explain the reflection of the effects related to the injected current polarity on the experimental results.

## 4. Conclusion

The effect of the current injection (from the LSMO electrode) into an HTS YBCO thin film ring on the critical current density of the ring was investigated at 77 K using contactless method. The structures were prepared using magnetron sputtering, conventional photolithography and wet etching procedures. The critical current density was determined from the third harmonics response of the HTS ring to the AC magnetic field. The experiments demonstrate a sensitivity of the critical current density to the current injection process to the HTS ring, especially in the case of FM LSMO current-injecting electrode, which injects spin polarized quasiparticles into the superconductor.

**Acknowledgements**

This work was supported by the European Social Fund under the project BG051PO001.3.3.04/54/2009.

## References

[1] Bergeret F S, Volkov A F and Efetov K B 2004 *Phys. Rev.* B **69** 174504
[2] Alvarez G A, Wang X L, Peleckis G, Shi D Q and Dou S X 2007*Physica* C **460-462** 438-439
[3] Moran O, Baca E and Perez F A 2008 *Microelectronics Journal* **39** 556-559
[4] Habermeier H-U, Soltan S and Albrecht J 2007 *Physica* C **460-462** 32-35
[5] Soltan S, Albrecht J and Habermeier H-U 2007 *Materials Science and Engineering* B **144** 15-18
[6] Vengalis B, Plausinaitiene V, Abrutis A and Saltyte Z 2005 *Acta Physica Polonica* A **107** 286-289
[7] Przyslupski P, Tsarou A, Dluzewski P, Paszkowicz, Minikayev R, Dybko K, Sawicki M, Dabrowski B and Kimaball C 2006 *Supercond. Sci. Technol.* **19** S38-S44
[8] Dybko K, Werner-Malento K, Aleshkevych P, Wojcik M, Sawicki M and Przyslupski P 2009 *Phys. Rev.* B **80** 144504
[9] Lin J G and Cheng S L 2006 *Physica* C **437-438** 187-189
[10] Lee K, Friedman B, Iguchi I and Cha D 2000 *Physica* C **341-348** 1611-1612

[11] Nurgaliev T, Mateev E, Blagoev B, Miteva S, Neshkov L, Strbik V, Uspenskaya L S, Benacka S, Chromik S and Nedkov I 2010 *J. Phys.: Conf. Ser.* **234** 012029

[12] Nurgaliev T, Strbik V, Miteva S, Blagoev B, Mateev E, Neshkov L, Benacka S and Chromik S 2007 *Cent. Eur. J. Phys.* **5** 6371

Journal of Physics: Conference Series **253** (2010) 012075 doi:10.1088/1742-6596/253/1/012075

# Effects induced by nuclear deformations and electron correlations on the ground-state energy of low and multiply charged helium like ions in high-temperature plasmas

**R Pavlov[1], L Mihailov[2], Ch Velchev[1], M Dimitrova-Ivanovich[1], Zh Stoyanov[2], N Chamel[3] and J Maruani[4]**

[1] Institute for Nuclear Research and Nuclear Energy, Bulgarian Academy of Sciences, 72 Tsarigradsko Chaussee, 1784 Sofia, Bulgaria, E-mail: ropavlov@inrne.acad.bg
[2] Institute of Solid State Physics, Bulgarian Academy of Sciences, 72 Tsarigradsko Chaussee, 1784 Sofia, Bulgaria, E-mail: lubomir.mihailov@gmail.com
[3] Université Libre de Bruxelles, Institut d'Astronomie et d'Astrophysique, Boulevard du Triomphe, CP226, B-1050 Brussels, Belgium, E-mail: nchamel@ulb.ac.be
[4] Laboratoire de Chimie Physique, CNRS and UPMC, 11 Rue Pierre et Marie Curie, 75005 Paris, France, E-mail: maruani@ccr.jussieu.fr

**Abstract.** The present work deals with the effects induced by nuclear deformations and electron correlations on the electron ground-state energies of low and multiply charged helium like ions. Numerical calculations of the ground state energies including mass corrections and polarization of the electronic system have been performed for such heliumoid ions with nuclear charge number from Z=2 to Z=118. A perturbation method has been developed based on a variational principle using explicitly correlated two-electron wave functions. The mass-polarization term accounting for correlation effects has been included for the first time in the minimization procedure. These effects have been found to be particularly pronounced for ions corresponding to nuclear magic numbers. The present method provides a general framework for high-precision calculations of plasma diagnostics.

## 1. Introduction

The properties and characteristics of high-temperature astrophysical and laboratory plasmas are determined by the interaction of low and multiply charged helium like ions (i.e. heliumoid ions) with electrons, ions, atoms, molecules and photons [1]. Heliumoid ions exhibit specific properties due to the unscreened long-range Coulomb field of their nucleus. High precision calculations of the electron ground-state energies of heliumoid ions requires to take into account the effects associated with nuclear deformations and electron correlations. The nuclear size effects increase steadily with increasing atomic number. The nuclear motion gives an even more significant correction to the electron energy [2,3]. This leads to two specific contributions: finite-mass effect (mass correction), electron-electron (e-e) and electron-nucleus (e-n) correlations (mass polarization). While the latter increases with increasing nuclear charge number $Z$, the former varies very little [4]. Relativity and quantum electrodynamics (QED) yield additional corrections which we do not include here.

A rich variety of resonant processes can be observed in high-temperature plasmas, due to multiply charged helium like ions [1,5]. In particular an isotope shift of the resonance lines has been recently observed [6,7] in radiative and dielectronic recombination processes. These shifts arise from changes in the charge distribution and in the nuclear mass with the neutron number $N$ [8]. This isotope shift thus indirectly reflects the variation of the mass excess with $N$.

The present work explores the contribution of both mass excess in the mass correction and mass polarization on the electron ground-state energies of heliumoid ions for the main nuclides with nuclear charge $Z$=2÷118. Effects associated with nuclear deformations are thus taken into account for the first time. A perturbation method for solving the 2-electron Schrödinger Equation (SE) has been developed based on a variational principle using Explicitly Correlated Wave Functions (ECWF) of a generalized Hylleraas type [9]. It is shown that the variational procedure leads to a generalized eigenvalue problem. This method has been applied to compute the electron ground-state energies taking into account both mass corrections and mass-polarization effects. This approach provides a general framework for including higher-order effects like for instance relativistic and QED corrections. This method is therefore very well-suited for high-precision calculations of plasma diagnostics.

## 2. Method

### *2.1. Hylleraas wave functions and Hamiltonian*

Let us consider two electrons with position vectors $\mathbf{r}_1$ and $\mathbf{r}_2$ in the Decart's coordinate system with the zero point in the nuclear centre. Introducing elliptical coordinates, $s = r_1 + r_2$, $t = r_1 - r_2$, $u = r_{12} = |\mathbf{r}_1 - \mathbf{r}_2|$, the Hylleraas-type 2-electron wave functions that we consider have the form

$$|\psi\rangle = \sum_{STU} C_{STU} |STU\rangle = \sum_{STU} C_{STU} \frac{1}{\pi\sqrt{2}} e^{-s/2} s^S t^T u^U \tag{1}$$

where $S,T,U$ are positive integers, with $T$ even as a consequence of the symmetry requirement of the spatial wave function. The overlap of two Hylleraas wave functions is therefore given by

$$\langle STU | S'T'U' \rangle = \frac{2(2\mathcal{T} + \mathcal{U} + 6)(\mathcal{S} + \mathcal{T} + \mathcal{U} + 5)!}{(\mathcal{T} + 1)(\mathcal{T} + 3)(\mathcal{T} + \mathcal{U} + 3)(\mathcal{T} + \mathcal{U} + 5)}, \ \mathcal{S}=S+S,\ \mathcal{T}=T+T',\ \mathcal{U}=U+U'. \tag{2}$$

These quantities are evaluated analytically in order to avoid numerical roundoff errors.

The non-relativistic Hamiltonian is the sum of the kinetic energy $T$, the Coulomb attractive potential $V_{en}$ due to the central charge, and the electron-electron repulsive potential $V_{ee}$ [10].

### *2.2. Scaling*

If the Hylleraas wave function (1) is subject to a coordinate scaling transformation

$$\langle \vec{r}_1 \vec{r}_2 | \Psi_\alpha \rangle = \sum_{STU} C_{STU} e^{-\alpha s/2} (\alpha s)^S (\alpha t)^T (\alpha u)^U \tag{3}$$

($\alpha$ being a scaling factor), then the overlap, kinetic, and potential matrix elements transform as

$$\langle \Psi_\alpha | \Psi_\alpha \rangle = \langle \Psi | \Psi \rangle / \alpha^6,\ \langle \Psi_\alpha | T | \Psi_\alpha \rangle = \langle \Psi | T | \Psi \rangle / \alpha^4,\ \langle \Psi_\alpha | V | \Psi_\alpha \rangle = \langle \Psi | V | \Psi \rangle / \alpha^5, \tag{4}$$

with $V = V_{en} + V_{ee}$. The expectation value of the energy thus scales as

$$\frac{\langle \Psi_\alpha | H | \Psi_\alpha \rangle}{\langle \Psi_\alpha | \Psi_\alpha \rangle} = \frac{\alpha^2 \langle \Psi | T | \Psi \rangle + \alpha \langle \Psi | V | \Psi \rangle}{\langle \Psi | \Psi \rangle} \tag{5}$$

## 3. Mass corrections

The variational calculation of the ground state energy using ECWF as trial 2-electron wave functions yields a lower bound to the exact energy. To obtain a better estimate, it is necessary to include mass correction and mass polarization. The first-order correction is given by [2,4]:

$$\varepsilon_1 = -\frac{\varepsilon}{1+\varepsilon} E_0 \approx -\varepsilon E_0 \tag{6}$$

where $E_0$ is the electron ground-state energy and $\varepsilon = m_e/M$ ($m_e$ being the electron mass and $M$ the nuclear mass). This correction is independent of the atomic state. The second-order perturbation correction given by

$$\varepsilon_2 = \varepsilon \int \nabla_1 \Psi^*(\mathbf{r}_1, \mathbf{r}_2) \nabla_2 \Psi(\mathbf{r}_1, \mathbf{r}_2) d^3 r_1 d^3 r_2 = \langle \Psi | \mathcal{E} | \Psi \rangle, \quad (7)$$

depends on the atomic state as it involves the two-electron wave function. The matrix elements of $\varepsilon_2$ can be obtained in an analogous manner as those of the kinetic energy operator and can be found in [10]. After a scaling transformation, the mass polarization (7) becomes

$$\langle \Psi_\alpha | \mathcal{E} | \Psi_\alpha \rangle = \langle \Psi | \mathcal{E} | \Psi \rangle / \alpha^4. \quad (8)$$

## 4. Variational method

The energy is minimized by varying the Hylleraas wave function with respect to both the expansion coefficients $C_{STU}$ and the scaling factor α. Variation with respect to the expansion coefficients leads to the generalized eigenvalue problem,

$$\left(\alpha^2 [T] + \alpha [V]\right) C_\alpha = E_\alpha [O] C_\alpha \quad (9)$$

where $T$, $V$, $O$ are the kinetic, potential, and overlap (symmetric) matrices respectively and $C_\alpha$ is the column vector containing the expansion coefficients. The lowest eigenvalue $E^0_\alpha$ obtained for a fixed value of α defines a function of one variable. Its minimum which can be found using standard techniques yields the best estimate of the ground-state energy.

In practice the matrices $T$, $V$ and $O$ are calculated only once. Solving the eigenvalue problem in (9) proceeds by first diagonalizing the overlap matrix $O$, with resulting eigenvalues $d_i$ and eigenvectors $Z_i$. We use Löwdin's orthogonalization. If the number of terms in the expansion (1) is too large, the set of polynomials could be overcomplete. As a consequence the overlap matrix may have very small eigenvalues which could lead to numerical instabilities. Such instabilities can be avoided by constructing the matrices $T'_{ij} = Z_i^T.T.Z_j/\sqrt{(d_i d_j)}$ and $V'_{ij} = Z_i^T.V.Z_j/\sqrt{(d_i d_j)}$ in which the small eigenvalues and their corresponding eigenvectors are omitted. Finally one has to solve the eigenvalue problem

$$\left(\alpha^2 [T'] + \alpha [V']\right) C'_\alpha = E_\alpha C'_\alpha \quad (10)$$

for various values of α in order to find the lowest energy $E^0_\alpha$. After substitution into (8) the values of the expansion coefficients $C_{STU}$ and scale parameter α, corresponding to $E^0_\alpha$, we obtain mass polarization $\varepsilon_2$ (7). In order to include $\varepsilon_2$ in the variational procedure, the operator of mass-polarization $\mathcal{E}$ is added to the Hamiltonian. Then the expectation value of the energy becomes

$$\frac{\langle \Psi_\alpha | H | \Psi_\alpha \rangle}{\langle \Psi_\alpha | \Psi_\alpha \rangle} = \frac{\alpha^2 \langle \Psi | T | \Psi \rangle + \alpha \langle \Psi | V | \Psi \rangle + \alpha^2 \langle \Psi | \mathcal{E} | \Psi \rangle}{\langle \Psi | \Psi \rangle} \quad (11)$$

The variational equations (9) and (10) correspondingly are replaced by

$$\left(\alpha^2 [T + \mathcal{E}] + \alpha [V]\right) C_\alpha = E_\alpha [O] C_\alpha \text{ and } \left(\alpha^2 [T' + \mathcal{E}'] + \alpha [V']\right) C'_\alpha = E_\alpha C'_\alpha \quad (12)$$

The nuclear masses $M$ in (6,7) are calculated using the well-known relation $M = \Delta + A - Z\, m_e + b_e$ [11], where $\Delta$ is the mass excess, $A$ is the nuclear mass number and $b_e$ is the mass equivalent of the binding energy which is neglected here. Values of $\Delta$ are taken from recent mass excess tables [12].

## 5. Results and discussion

The electron ground-state energies for the lightest heliumoid ions are indicated in Table 1. For comparison we have also shown the results obtained from calculations using different methods.

| . . . . . . . . . . . . . . . . . . . Ground state energy $E_0$ [a.u.] . . . . . . . . . . . . . . . . . . . | | | | | | | Differences between $E_0$ [14] and | | |
|---|---|---|---|---|---|---|---|---|---|
| Z | A | [2]* | [13]* | [14] | our | [15] | [16]* | our | [15] | [16]* |
| 2 | 4 | 2.9037283 | 2.9037244 | 2.9037242 | 2.9037244 | 2.9037244 | 2.8616800 | -0.0000002 | -0.0000002 | 0.0420442 |
| 3 | 7 | 7.2799088 | 7.2799088 | 7.2799132 | 7.2799134 | 7.2799134 | 7.2364152 | -0.0000002 | -0.0000002 | 0.0434980 |
| 4 | 9 | 13.655565 | 13.655565 | 13.655566 | 13.655566 | 13.655566 | 13.611299 | -0.0000002 | -0.0000002 | 0.0442666 |
| 5 | 11 | 22.030973 | 22.030972 | 22.030971 | 22.030972 | 22.030972 | 21.986234 | -0.0000006 | -0.0000006 | 0.0447365 |
| 6 | 12 | 32.406248 | 32.406247 | 32.406246 | 32.406247 | 32.406247 | 32.361193 | -0.0000006 | -0.0000006 | 0.0450532 |
| 7 | 14 | 44.781447 | 44.781445 | 44.781445 | 44.781445 | 44.781445 | 44.736164 | -0.0000001 | -0.0000002 | 0.0452811 |
| 8 | 16 | 59.156596 | 59.156595 | 59.156595 | 59.156595 | 59.156595 | 59.111143 | -0.0000001 | -0.0000001 | 0.0454523 |
| 9 | 19 | 75.531713 | 75.531712 | 75.531712 | 75.531712 | 75.531712 | 75.486126 | -0.0000003 | -0.0000004 | 0.0455856 |
| 10 | 20 | 93.906806 | 93.906807 | 93.906806 | 93.906807 | 93.906807 | 93.861113 | -0.0000005 | -0.0000005 | 0.0456925 |

**Table 1.** Absolute value of electron ground-state energies $E_0$ in atomic units (on the left panel) for heliumoid ions with $Z$=2÷10 *with corrections*: from Bethe and Salpeter's semi-empirical formula [2], from Thakkar and Koga [13], from Pekeris [14], from our calculations, from Karwowski [15] and from Hartree-Fock (HF) calculations using Froese-Fischer's code [16]. Differences between $E_0$ from Pekeris and from other calculations are shown on the right panel. Results marked with an asterisk do not include the correlation term $\varepsilon_2$.

In particular, the right panel of Table 1 shows that our results are in excellent agreement with those from Pekeris[14] which are currently regarded as being the most precise. However unlike Pekeris calculations, our calculations have been carried out not only for $Z$<10 but also for heavier ions. Complete tables for the most abundant or stable isotopes can be found in [12].

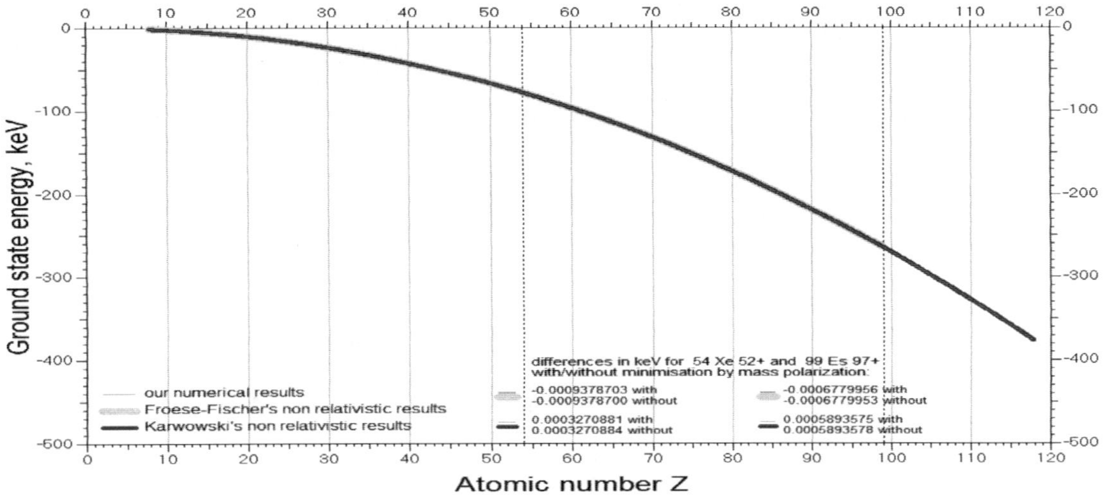


**Figure 1.** Comparison between different non-relativistic calculations of the electron ground state energy (in keV) of heliumoid ions for $Z$=2 to $Z$=118.

Figure 1 shows the electron ground-state energies as obtained from our calculations, from non-relativistic calculations of Karwowski [15] and from HF calculations using Froese–Fischer's code [16]. The curves are barely distinguishable because of the small differences between the different calculations. Figure 2 compares our results with those obtained from relativistic calculations. The Karwowski's results [17] are obtained in the scheme of relativistic Hylleraas-CI method using 2-component wave functions. HF results are relativistic, realized using the one-component scheme (Breit-Pauli) which is correct at the L-S binding level, while the two-component scheme is correct at the j-j coupling level. Figure 3 shows the effects on $\varepsilon_2$ and $E_0$ of including or not $\varepsilon_2$ in the minimization. Green lines indicate magic proton numbers. For all magic numbers up to $Z$=54, there is a peak of e-e and e-n correlations, caused by $\varepsilon_2$ (50 is an unstable magic number as observed experimentally [18]). At higher $Z$, the peaks are shifted by 2 from the magic numbers 80 and 100. Exact coincidences are expected to be found when relativistic effects are included. At $Z$>114 fluctuations in the energy disappear, which corresponds to the nuclear area of stability [18]. The peak values arise from the sphericity of magic nuclei.

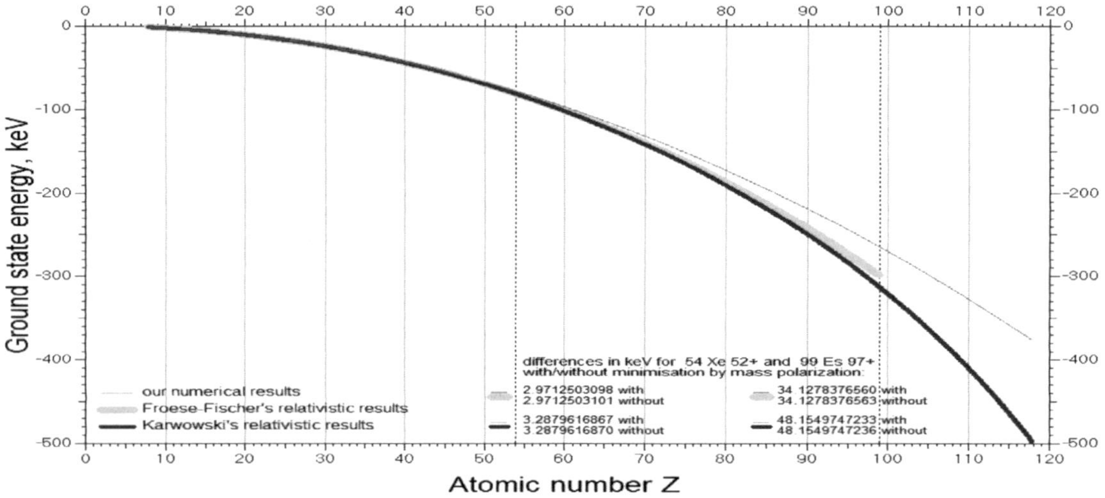


**Figure 2.** Comparison between our calculations and different relativistic calculations of the electron ground-state energy (in keV) of heliumoid ions for $Z$=2 to $Z$=118.

In conclusion, we have studied the effects of nuclear deformations and electron correlations on the electron ground-state energy of heliumoid ions for $Z$=2 to $Z$=118 using a variational-perturbation method with explicitly correlated trial wave functions of a generalized Hylleraas type. For the first time, we have included mass polarization in the minimization procedure which is found to decrease the ground-state energy. For $Z$<10, our results are in excellent agreement with Pekeris's results [14]. This demonstrates the high accuracy of our method. The direct inclusion of the mass excess in the calculation has revealed a new correlation between the ground state energy and proton magic numbers, thus showing the influence of the nuclear shape on the electronic structure. We have recently extended the calculations to all known isotopes [to be published] in order to further investigate this feature which is expected to be found in other ionic species and also in solids. Because of the restricted space available here, results will be discussed elsewhere. The explicitly correlated 2-electron wave functions that we have obtained are valid up to second-order in the perturbation theory and therefore provide a basis for calculating the higher-order relativistic corrections [2]. In future works, we will include these corrections which can be comparable to the mass polarization contribution for large $Z$. These high-precision calculations will provide reliable inputs in precise approaches for plasma diagnostics, for example in the spectral analysis of high-temperature plasmas, in deformation wave and tight-binding methods used for studying electron-ion plasma processes.

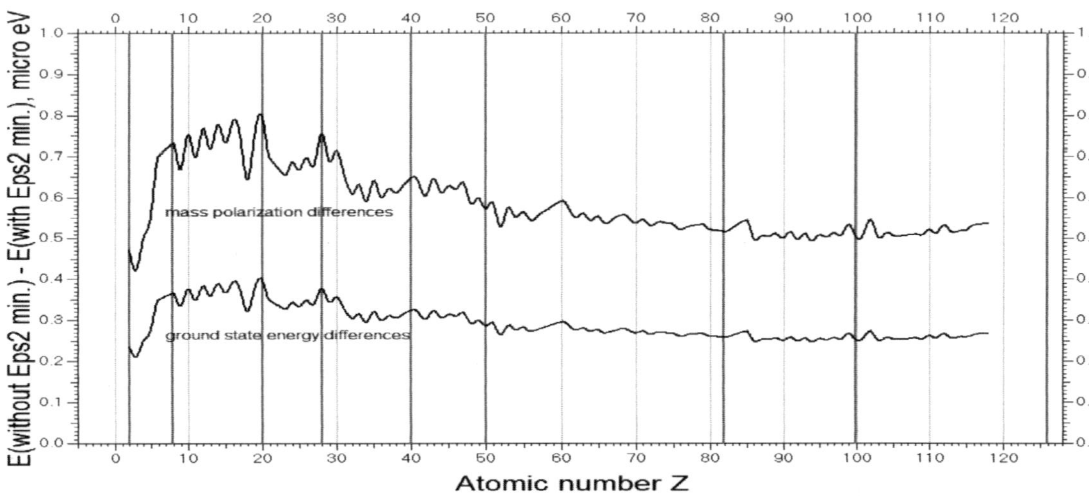


**Figure 3.** The upper (lower) curve shows the difference between $\varepsilon_2$ ($E_0$) obtained without and with including $\varepsilon_2$ in the minimization.

**Acknowledgements**
The present work was supported by the Saint-Kliment Ohridski University of Sofia, FNRS (Belgium) and the bilateral project between Wallonie-Bruxelles International and the Bulgarian Academy of Sciences. We wish to thank Prof. J. Karwowski for useful advices and for providing numerical data.

**References**

[1] Presnyakov L P, Shevelko V P and Yanev R K 1986 *Elementary processes with multiply charged ions* (Energoatomizdat Moskva, in Russian)
[2] Bethe H A and Salpeter E E 1967 *Quantum Mechanics of One- and Two-Electron Atoms* (Springer Berlin)
[3] Gombas von P 1950 *Theorie und Loesungsmethoden des Mehrteilchenproblems der Wellenmechanik* (Basel)
[4] Drake G W F 1988 *Can. J. Phys.* **66** 586
[5] Burke P G, Eissner W B, Hummer D G and Percival I C 1983 *Atoms in Astrophysics* (Plenum Press: New York and London)
[6] Brandau C, Kozhuraov C, Harman Z et al. 2008 *Phys. Rev. Lett.* **100** 073201
[7] Schiopu R et al. 2004 *Eur. Phys. J.* D**31** 21
[8] Scheid W, Palffy A, Harman Z, Kozhuharov C, Brandau C 2008 Ions *Proc. of the 27th Int. Workshop on Nuclear Theory* (23-28 June 2008, Rila, Bulgaria, Ed. Dimitrova S) 3
[9] Rychlewski J (Ed.) 2003 *Explicitly Correlated Wavefunctions in Chemistry and Physics* (Progress in Theoretical Chemistry and Physics) **13** Kluwer, Dordrecht
[10] Pavlov R, Maruani J, Mihailov L, Velchev Ch, Dimitrova-Ivanovich M, Stoianov J and Van Neck D 2008 *Proc.27th Int.Workshop on the Nuclear Theory* (23-28 June, Rila, Bulgaria) 249
[11] Kravcsov V A 1974 *Atoms Mass and Nuclei Bounding Energy* (Atom Publ., Moscow, in Russ.)
[12] Tuli J K 2005 *Nuclear Wallet Cards* National Nuclear Data Center (Upton, NY) pp 1-77
[13] Thakkar A J, Koga T, 1994 *Phys. Rev.* A **50** 854
[14] Pekeris C L 1958 *Phys. Rev.* **112** 1649; in ibid. 1959 **115** 1216; in ibid. 1962 **126** 1470
[15] Karwowski J 2009 *Private communication*
[16] Froese-Fischer C 1977 *The Hartree-Fock Method for Atoms* (Wiley, New York)
[17] Pestka G, Bylicki M, Karwowski J 2007 *J. Phys.* B*: At. Mol. Opt. Phys.* **40** 2249
[18] Eisenberg J M, Greiner W 1970 *Nuclear Theory* **1**; 1972 *Nuclear Theory* **3** (North-Holland Publishing Companny, Amsterdam – London)

Journal of Physics: Conference Series **253** (2010) 012076 doi:10.1088/1742-6596/253/1/012076

# CuBr laser beam transformations

**D N Astadjov**[1] **and S V Nakhe**[2]

[1] Metal Vapour Lasers Department, Georgi Nadjakov Institute of Solid State Physics, Bulgarian Academy of Sciences, 72 Tzarigradsko Chaussee, Sofia 1784, Bulgaria
[2] Laser Systems Engineering Division, RRCAT, P.O. CAT, Indore 452013, India

E-mail: asta55@issp.bas.bg

**Abstract**. CuBr laser beam profile transformations are studied in a beam focusing experiment. A modeling via Fourier transform is also performed with annular beams of simplified flat two-level geometry of near field: bright outer ring with a darker core. The pattern of focal beam profile i.e. far field is calculated and characterized with respect of its intensity structure. As found beam annularity has small effect on far-field intensity pattern.

## 1. Introduction

In 2000 a comprehensive reference book appeared which reviewed theory and technique, and gave basic information onto research, development and design of beam shaping systems [1]. By beam shaping the raw laser profiles which often are of random shape and not good enough for straight use, can be tailored in due way. Thus many applications of lasers can benefit from the proper redesign of beam profile. Techniques and models are abundant and they make feasible actually any profiles needed for laser applications.

Copper vapor lasers are still the most high-power and high-efficiency lasers which produce visible light ($\lambda$510nm and $\lambda$578nm) in a straightforward way without any subsequent light conversion. In some cases their profiles need tailoring as well.

To concentrate more power very frequently laser intensity is increased by focusing the radiation. The focusing causes a near-field beam profile to transforms into a far-field beam profile of intensity distribution often quite different from the initial near-field intensity distribution. Relations between near-field and far-field beam profiles are of expanding interest due to the vast employment of laser (and generally, optical) beams in science and technology. In this paper we present results of experiments as well as modeling of beam transformations observed after focusing the CuBr laser emission.

## 2. Experiment with focusing of MOPA (Master-Oscillator Power-Amplifier) CuBr laser beam

CuBr laser profile varies from annular to top-hat or Gaussian-like. In the case of MOPA CuBr laser profile variations are reported in experiments on timing and buffer gas composition [2]. While annular profiles are typical for non-hydrogen buffer or at MOPA small delays, top-hat profiles are predominant for buffer containing hydrogen or at MOPA long delays [2, 3]. This is illustrated in figures 1 and 2: in figure 1 the MOPA CuBr laser buffer has no hydrogen added and the emission is annular at all delays; in figure 2 (a hydrogen-added buffer) beam is annular at small delays only.

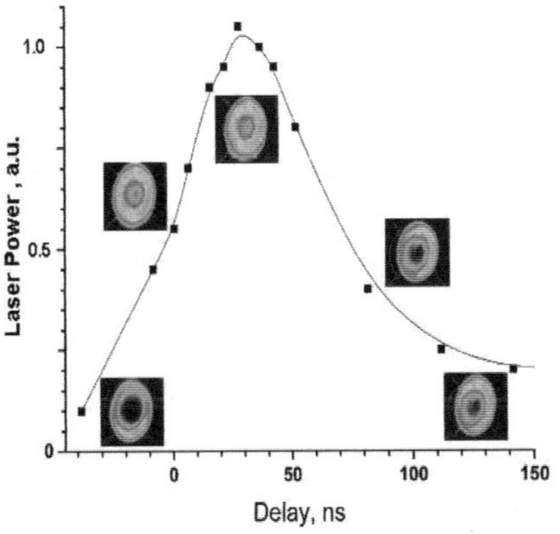


**Figure 1.** Typical near-field profiles of MOPA CuBr laser as functions of the MOPA time delay. Buffer gas is Ne of 17 torr.

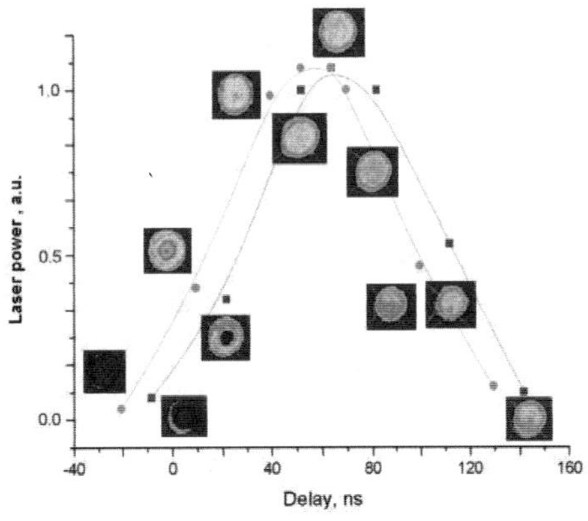


**Figure 2.** Typical near-field profiles of MOPA CuBr laser as functions of the MOPA time delay. Buffer gas is a mixture of Ne (17 torr) and $H_2$ (0.3 torr). Profile mutability is seen in two cases.

Detailed experimental pictures of far-field produced from the corresponding near-field are shown in figure 3.

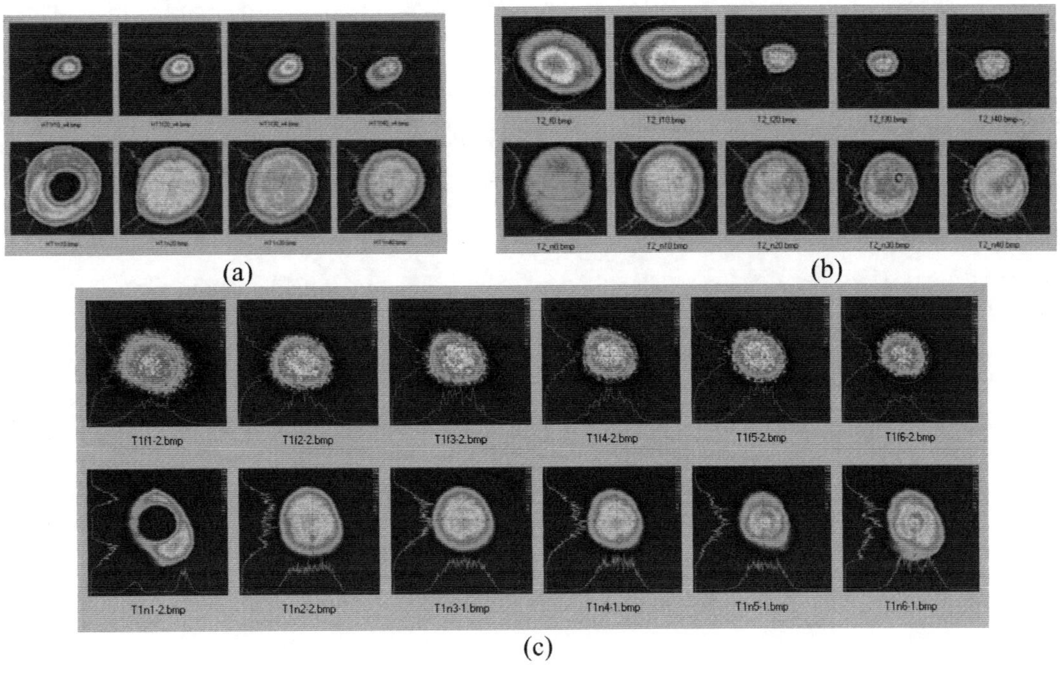


**Figure 3.** Three sets of near-field profiles (the lower traces) with the subsequent far-field profiles (the upper traces) taken experimentally with MOPA CuBr laser system.

Two-dimensional far-field beam profiles can be calculated by the two-dimensional Fourier transform of the near-field beam profiles. In real physical world, this transform is easily performed by optical lenses [4]. The focal spot profile is the far-field profile of near-field profile in front of the lens (within the accuracy of a phase coefficient). Fourier transform is a very powerful tool for optical signal processing and its most popular and simple form is the Fast Fourier Transform (FFT). Fourier transform is applicable to the complex optical field. The signal that is optically detectable is the intensity of optical field. To find the field intensity, we have to calculate the product of the optical field complex amplitude and its conjugated optical field complex amplitude [5]. It was performed via MATLAB® built-in discrete functions which resulted in discrete shape of output.

## 3. Near- and far-field parameters of flat two-level annular beam

*3.1. Near-field parameters of flat two-level annulus*

The simplest type of annulus is the normalized flat two-level annulus. We define it as a circular ring of inside concentric area of intensity, I1 (which varies from 0 to 1) and the annulus itself of intensity, I2≡1 (normalized by definition). If I1=I2, the profile is top-hat; if I1=0, the profile is 'pure' annular. Such an annulus can be described by two parameters. Its axial section is plotted in figure 4. The first parameter (referred as 'annularity'), k is k=d1/d2, where d1 and d2 are the inside and the outside diameters of annulus. The second parameter, Idip is the intensity dip of the central area:
Idip=I2-I1=1-I1.

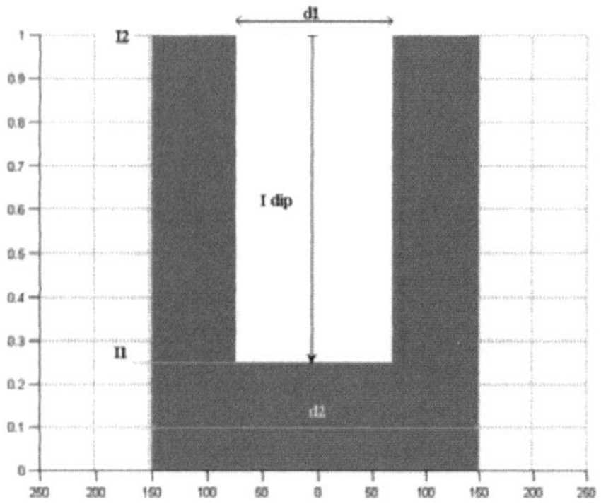


**Figure 4.** Intensity axial section of normalized two-level annulus; d1 and d2 are the inside and the outside diameters of annulus; the circular core area is of intensity, I1 (varying from 0 to 1), and I2≡1.

*3.2. Far-field parameters of flat two-level annulus*

The type of calculated pattern of far-field beam profile produced from an annular near-field beam is shown in figure 5. The profile can be described as having a prominent central peak and side peaks concentrically surrounding it (better seen in the figure inset). The appearance of side peaks comes from the discrete form of calculations we applied in the model. Actually they should smear into uninterrupted concentric rings. For FFT computations we employed MATLAB® environment.

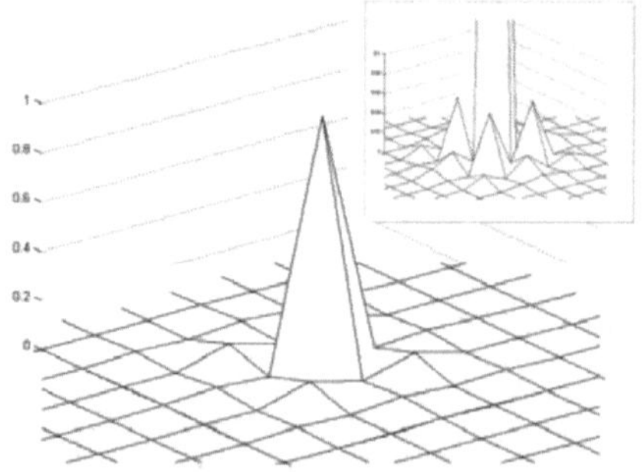

**Figure 5.** Typical calculated pattern of far-field beam profiles of a two-level annulus; the inset is x10-magnified central area.

Journal of Physics: Conference Series **253** (2010) 012076 doi:10.1088/1742-6596/253/1/012076

In figure 6 we give the patterns of far fields calculated from near fields of different values of k. As k goes up so do the side peaks. NB that in physical reality peaks are concentric rings.

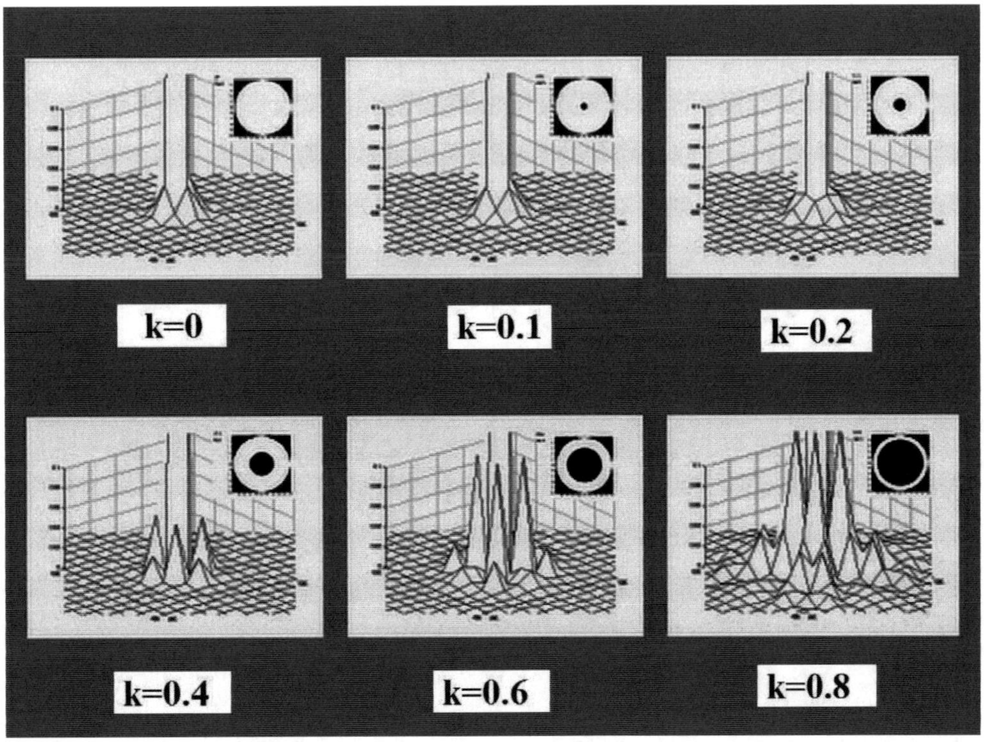


**Figure 6.** Patterns of far fields calculated from near fields of different values of annularity k. Z-axis is 10% of central peak intensity for better view of 'rings' structure.

The far-field profile parameters specified here are with regard to laser beam utilization in applications which are not susceptible to phase components of light. We pay attention to the power (energy) component of light. Nevertheless the coherency of light is a requirement for all our simulations. As a major parameter we introduce the fraction of the central peak energy to the whole energy of beam, PF0. This parameter gives a notion about energy spread within the far-field spot. The higher PF0 the lower energy spread is. In practice that means less affected surrounding (the central peak) area by light radiation. The dependence of PF0 on Idip is plotted in figure 7.

As can be seen the increase of Idip leads to the decrease the central peak energy fraction. At higher k the decrease starts at lower Idip. So at k=0.3 and Idip<0.75, the change of PF0 is less than 1.5%. While the maximum of PF0 is 0.778, for Idip=0.75 the fraction of the central peak energy PF0 is 0.767. But at k=0.8 for the same Idip of 0.75, PF0 is 0.7. The central peak energy fraction dominates the far-field energy distribution nearly over the whole range. The exception is for Idip>0.9 (and k>0.65) which occurs seldom in usual practice. Taking into account that the central peak energy is concentrated on a smaller spot area, the net impact of side energy spread diminishes furthermore.

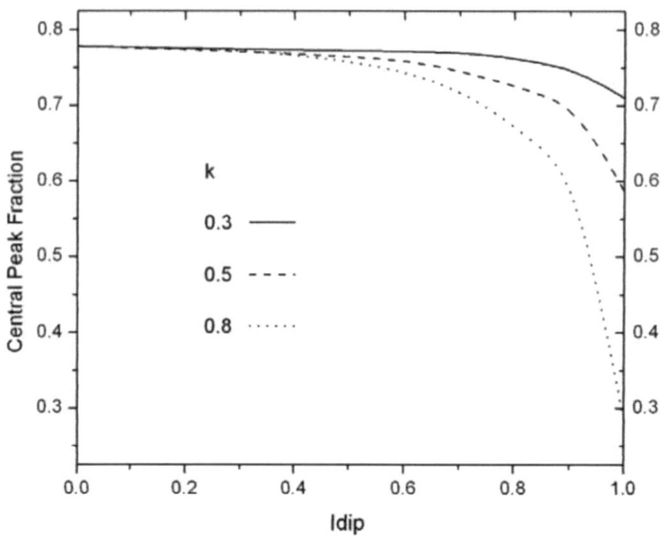


**Figure 7.** Dependence of the central peak energy fraction (PF0) on Idip (the intensity dip of the central area: Idip=1-I1) for three values of annularity, k - 0.3, 0.5 and 0.8.

For the sake of usability we performed a surface approximation of the dependence PF0 =f ( Idip, k) employing a web-based program [6]. The best approximation is given below graphically (figure 8)

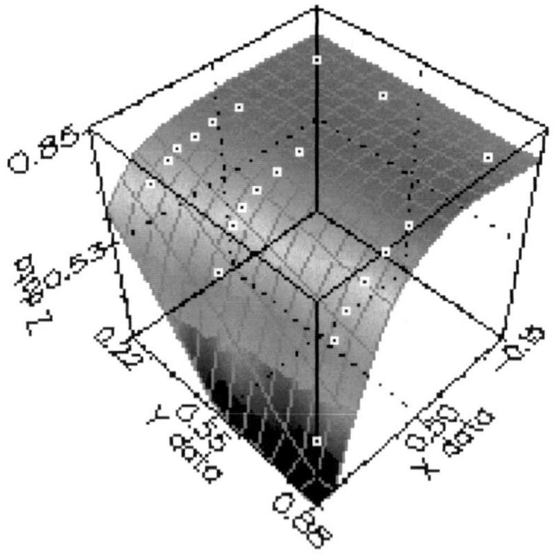

**Figure 8.** Dependence of the central peak energy fraction (PF0) on Idip (the intensity dip of the central area: Idip=1-I1) and annularity k.

Note: x ≡ Idip, y ≡ k and z ≡ PF0.

Journal of Physics: Conference Series **253** (2010) 012076 doi:10.1088/1742-6596/253/1/012076

and in letter form (equation):

$$z = a\,(\sinh(x)\,y^{1.5}) + b\,(\tan(x)\,y^{1.5}) + c\,,$$

where a = 2.0310451454356748, b = -1.9232938348115700 and c = 0.78402321357391169;

Consider that z ≡ PF0, x ≡ Idip and y ≡ k.

## 4. Conclusions

Simulations of annular beams through two-dimensional Fast Fourier Transform show that the far-field central peak energy fraction is prevailing in the focal (far-field) energy distribution virtually regardless the (near-field) shape of annular source beam. Beam annularity and beam central area intensity dip have negligible effect on the focal energy distribution. In practice that means less radiation-affected area in the vicinity of beam central peak. Taking into account that the central peak energy is confined within a smaller spot area, the net impact of side energy spread diminishes furthermore.

## Acknowledgments

The work reported here is carried out under a joint research project titled 'Indo-Bulgarian Programme of cooperation in Science & Technology: High end performance solid state power supply copper lasers for material processing.' Authors would like to acknowledge the project funding agencies. The Bulgarian part of this work was supported by Ministry of Education and Science of Bulgaria by Grant BIn3/07.

## References

[1] Dickey F M and Holswade S C 2000 *Laser beam shaping: theory and techniques (Opt. Eng. Series,* **v. 70***)* (New York: Marcel Dekker)
[2] Astadjov D N, Stoychev L and Sabotinov N V 2007 *Opt. Quant. Electron.* **39** 603-10
[3] Astadjov D N, Vuchkov N K and Sabotinov N V 1988 *IEEE J. Quantum Electronics* **24** 1927-35
[4] Goodman J W 1996 *Introduction to Fourier Optics* (New York: McGraw-Hill) pp 101-107
[5] Lee S H, Basic Principles 1981 *Optical Information Processing-Fundamentals, Topics in Applied Physics*, **48**, ed Lee S H (Berlin: Springer-Verlag) pp 2-41
[6] Phillips J R *ZunZun.com Online Curve Fitting and Surface Fitting Web Site*

Journal of Physics: Conference Series **253** (2010) 012077 doi:10.1088/1742-6596/253/1/012077

# Functionalized aerogels – new nanomaterials for energy-efficient building. Preliminary AFM, Nanoidentation and EIS studies

**G R Ivanov**[1,2], **R Tomova**[3], **S T Djambova**[1], **M Nadoliiski**[1] **and D Dimova–Malinovska**[4]

[1] Department of Physics, Faculty of Hydrotechnics, University of Architecture, Civil Engineering and Geodesy, 1 Hr. Smirnenski Blvd., 1046, Sofia, Bulgaria
[2] Advanced Technologies Ltd., compl. Yavorov, bl. 38, entr. B, 1111 Sofia, Bulgaria
[3] Institute for Optical Materials and Technologies, Bulgarian Academy of Sciences, Sofia, Bulgaria
[4] Central Laboratory of Solar Energy and New Energy Sources, Bulgarian Academy of Sciences, Sofia, Bulgaria

E-mail: george@at-equipment.com

**Abstract.** Aerogels are highly porous *nanostructured* materials with excellent thermal insulation properties. The possibility to add additional function – to functionalize the aerogels, especially to produce photovoltaic electricity, will make them an excellent candidate for energy-efficient building. Going in the direction of this midterm goal we start with the investigation of the properties of the readily available silica aerogels. Atomic Force Microscopy reveals large areas with submicrometer roughness, which allows reliable nanoidentation measurements. The average hardness was measured to be 2,2 MPa and the Young's modulus was 11 MPa, values typical for low density elastic silica aerogels. Electrochemical Impedance Spectroscopy, measured in ambient air, shows typical capacitive behaviour and the aerogel is best modelled by serially connected resistance of 37 kΩ and capacitor of 170 pF. The conductivity is interpreted in terms of proton migration, strongly dependant on air humidity.

## 1. Introduction

**Aerogels** are high surface area, highly porous *nanostructured* materials with the lowest bulk density of any known porous solid. It is derived from a gel in which the liquid component of the gel has been replaced with a gas. The result is an extremely low-density solid, with a notable effectiveness as a thermal insulator. Aerogels, which retain the free void volume of the wet gel, can be made from any material that can be processed as a gel. Outstanding properties include superior surface areas (from 100 to 1000$m^2/g$) and a bicontinuous pore–solid network in which the networked particles are 2–20 nm in diameter and covalently bonded together. The majority of pore volume in the mesoporous network resides in pores between 2 and 50 nm in size, the classical definition of mesoporosity. Aerogels are distinguished from the more commonly known xerogels by their relatively greater porosity, but more importantly by the continuity of the pore network throughout the solid, which

Journal of Physics: Conference Series **253** (2010) 012077 doi:10.1088/1742-6596/253/1/012077

facilitates diffusive mass transport at near open-medium diffusion rates. The high surface area and fast diffusive mass-transport rates within aerogels have motivated investigations into their application as catalysts, battery materials, sensor materials and supports for fuel-cell catalysts. Expressing functional materials as aerogels has yielded improved performance over analogous materials made by other means, and in some cases has revealed new mechanistic components in complex interfacial processes. In this work we review the properties and study the most widely spread and commercially available aerogels - silica aerogels.

**Silica aerogel** is a translucent material consisting of a nanostructured $SiO_2$ network [1]. These are materials with unusual properties such as high specific surface area (500–1200$m^2/g$), high porosity (80–99.8%), low density (~0.003 $g/cm^3$), high thermal insulation value (0.005W/mK), ultra low dielectric constant ($k$ = 1.0–2.0) and low index of refraction (~1.05) [2, 3, 4, 5, 6]. In fact, silica aerogels hold 15 Guinness world records, among which for smallest density solid and for best thermal insulator. Earlier production of silica aerogels included supercritical drying and expensive raw materials like tetraethoxysilane. A new process was developed [7]. The silica hydrogels were prepared with water glass as a starting material. Silation is carried out in the water phase of the hydrogel which results in a reaction induced phase separation of gel water and solvent.

Silica aerogel is a promising material for applications in building envelopes because of its high visual transmittance and its low thermal conductivity. Its thermal insulation properties are 2 times better then the mineral wool and 4 times better than Perlite. Besides its low thermal conductivity the aerogel is load bearing which makes it attractive for evacuated transparent insulation applications. Its impressive load bearing abilities are due to the dendritic microstructure, in which spherical particles of average size 2–5 nm are fused together into clusters. These clusters form a three-dimensional highly porous structure of almost fractal chains. The average size and density of the pores can be controlled during the manufacturing process.

The Parliament and Government of the European Union this year adopted a revised 2002 Directive on Energy Performance of Buildings, which requires that at the end of 2018 all new public buildings and by the end of 2020 all new buildings be of a type “nearly zero energy buildings". This definition includes both energy consumption, which implies production of the energy from renewable sources within or near building, and greenhouse gas emissions. A multi-annual plan and a strategy for energy-efficient building was developed and adopted in final form several months ago [8]. In view of this plan our midterm goal is to add additional functionality to the excellent thermal insulation properties of the aerogels – to functionalize them. More specifically, we will pursue the possibility to add the functionality of photovoltaic electric current production in the aerogels. These, probably, will be aerogels from semiconductor quantum dots, which were recently synthesized [9]. First step towards this goal is to measure some physical properties of readily available aerogels. In this paper the results from the study of the surface morphology using Atomic Force Microscope (AFM), the mechanical properties with the help of Berkovich Nanoindenter, and electrical properties by the method of Electrochemical Impedance Spectroscopy (EIS) of silica aerogel are reported.

## 2. Experimental

Monolithic hydrophilic silica aerogel in the form of cylinders 7-8 mm in height and 22 mm in diameter were purchased from Aerogel Technologies, LLC (USA). The AFM used was model Q-Scope from Ambios Technology Corporation (USA) integrated with the Nanoindenter model APEX, manufactured by CETR (USA). The AFM was equipped with 80 μm scanner (95 μm available) and was used in dynamic mode with simultaneous measurement of topography and phase contrast. The movement back and forward from the AFM to the Berkovich diamond indentation tip with radius of 70 nm was automated. Both instruments were equipped with optical microscopes and image recording. EIS was measured in ambient air with a Frequency Response Analyzer from Advanced Technologies Ltd. (Bulgaria) working in the range 0.02 Hz – 32 MHz and capable of measuring samples with resistances up to $10^{12}$ Ω. Al electrodes - 5 mm wide, were evaporated in vacuum on both sides of the aerogel and visualized with the AFM. However, the reported here EIS data was measured with electrodes directly stuck in the aerogel and stabilized with silver paste.

### 3. Results and discussion

The surface of the aerogel, measured with the AFM is shown on figure 1. The roughness of the surface at some select points was below 100 nm – smooth enough for reliable nanoidentation measurements. The evaporation of Al electrodes on the aerogel produced higher but smoother features (figure 2). On scanning at larger areas one could see the grooves of the metal discs, in which the monoliths were cast. In the optical microscopy (figure 3) it is seen that the surface has some features on a submillimeter scale. Before performing the nanoidentation tests, optically smooth surface was chosen and visualized with the AFM.

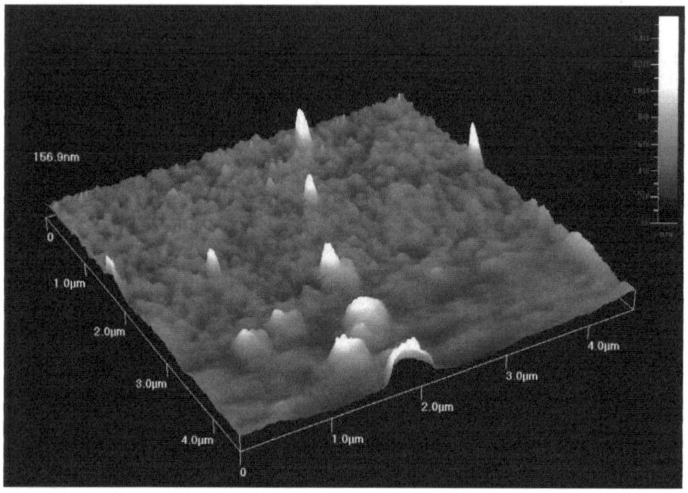


**Figure 1.** AFM image of the surface of the aerogel.

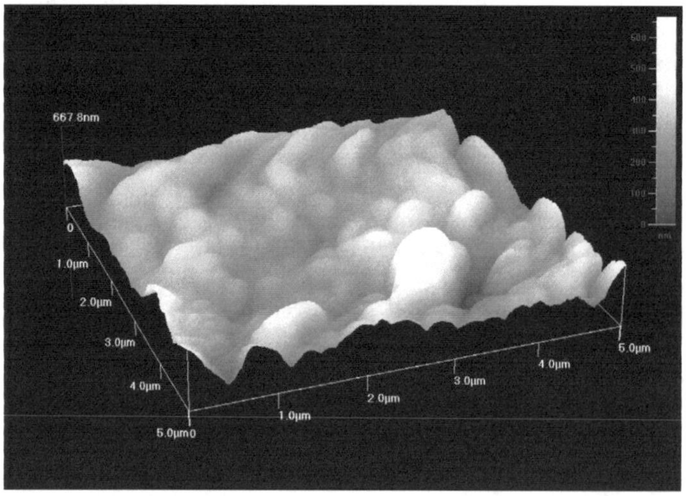


**Figure 2.** AFM image of the surface of the Al electrodes vacuum evaporated on both sides of the aerogel.

Nanoindentation permits to probe the mechanics of nanomaterials and permits direct physical property measurements of heterogeneous materials with high spatial resolution [10].The surface indents, as seen in the optical microscopes, are shown on figure 3. It should be noted that we have done a large number of indents varying both the used force and the spacing but the results for the hardness and Young's modulus were highly reproducible. One typical result is shown in figure 4. It should be pointed out that aerogels are not well characterized by hardness or the classical Oliver and Pharr theory, because homogeneous, isotropic materials are assumed here, as well as linear elasticity. Aerogels may be characterized by an extended Oliver and Pharr approach, which deals with porosity, non-linear elasticity, coatings, and we will try to access such software in the future. Nanoidentation results are shown in figure 4. In the experiment shown we use a script for measuring at 3 points and in each point we have partial loading and unloading in ¼ steps from the maximal force of 55 mN. Hardness in the experiment shown varies from 0.9 to 4.5 MPa while the Young's modulus was in the range from 2 to 20 MPa. Contact depth was in the 10 to 60 μm range. These results coincide with the data obtained for low density (0.08 – 0,.5 g/cm$^3$) samples measured previously [11], which show elastic behaviour, while the denser aerogels behave as elasto-plastic materials.

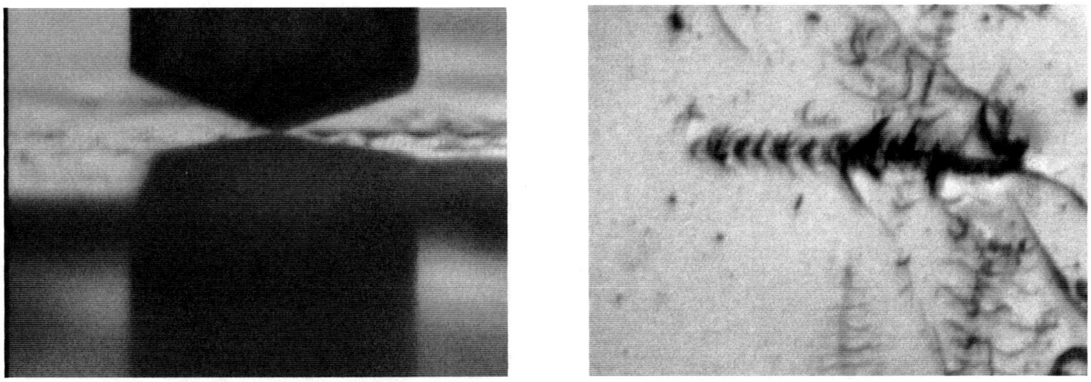

**Figure 3.** A row of indents during the indentation process (left). The Berkovich indenter and its reflection are seen. On the right, 12 indents in a row, which partly overlap due to the large force used.

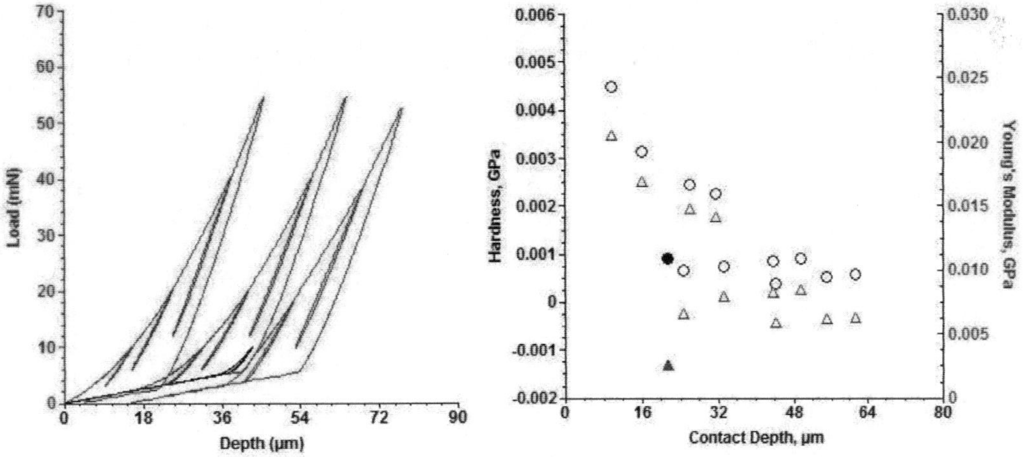


**Figure 4.** Nanoidentation results. In this experiment at 3 different locations 4 partial loadings and unloadings were performed (left). On the right are the calculated values for the hardness and Young's modulus for all 12 points. The most reliable measurement is the 1$^{st}$ one with filled symbols.

Results from the EIS measurements are shown on figure 5. The Nyquist plot is typical for pure capacitance behaviour. A modelling of the results was performed by fitting the raw data with a predefined circuitry. The best fit was obtained by serially connected resistor with value of 37 kΩ and capacitor of 170 pF. Previously a strong dependence of the resistance on relative air humidity was reported [12] for silica aerogels with relatively high density. For humidities above 80% the resistance was in the subkΩ region, while for humidities below 58 % it was in the MΩ region. These testify that the uptake of adsorbed water is a two-regime process. This mechanism of water adsorption was proposed by Dubinin and Serpinsky (D.S. theory) [13]. According to this theory, at lower relative humidity, the water forms a layer of clusters along the walls of the matrix of interconnected pores. At higher relative humidity, new water molecules start filling the remaining pore space through capillary condensation. The mobile species responsible for electrical conduction was traced out as protons and the proton conduction was highly augmented by the presence of adsorbed water. The proton migration seems to be greatly facilitated by the presence of water, suggesting that the mobility of ions is highly enhanced by the adsorbed water and the proton migration is primarily dominated by Grotthuss mechanism. In this mechanism, the proton forms a $H_3O^+$ ion and jumps to the neighboring lone pair of electrons of a water molecule.

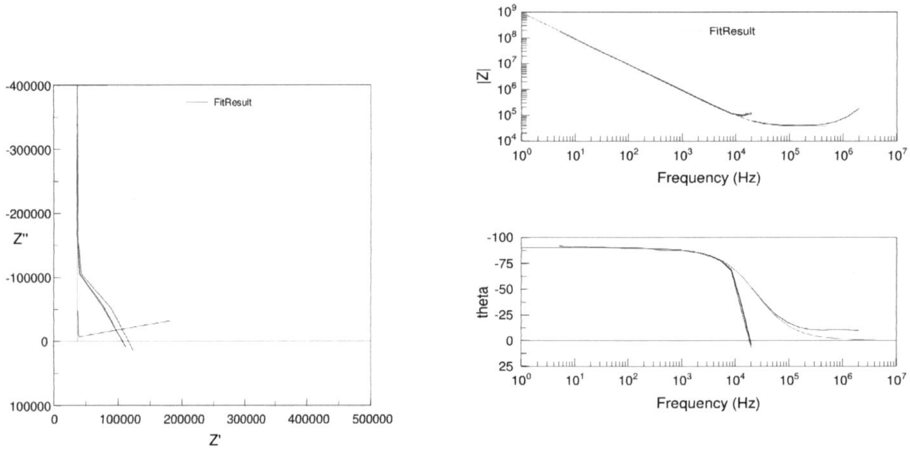


**Figure 5.** Results from the EIS measurements – Nyquist plot (left) and Bode plots.

## 4. Conclusion

Aerogels, especially if additional functionality like electricity production is added, are promising new nanomaterials in the area of energy-efficient building. Here we present preliminary studies of silica aerogels – the material that holds many records, for lowest density solid and best thermal insulator among them. We investigated in this paper the surface morphology with the help of Atomic Force Microscopy. Areas with submicron roughness are present and the evaporated Al electrodes show little increase in roughness. Mechanical properties were measured with Berkovich Nanoindenter. The average hardness was measured to be 2,2 MPa and the Young's modulus was 11 MPa, values typical for low density elastic silica aerogels. Electrochemical Impedance Spectroscopy, measured in ambient air, shows typical capacitive behaviour and the aerogel is best modelled by serially connected resistance of 37 kΩ and capacitor of 170 pF. The conductivity is interpreted in terms of proton migration, strongly dependant on air humidity.

## Acknowledgments

This work is supported under a contract with the Bulgarian National Fund "Scientific studies". Thanks are due for the Nanoidentation and AFM measurements to the Open Laboratory for Experimental Mechanics of Micro & Nanomaterials, Bulgarian Academy of Sciences, Institute of Mechanics, Department of Physical-Chemical Mechanics headed by Prof. D.Sc. R. Kotsilkova and to Dr. E. Ivanov and I. Borovanska for the help with the nanoidentation measurements. This OLEM Lab was created under NSF-BG project DO02-53. Thanks are due to Prof. D.Sc. Z. Stoynov for the helpful discussion of the EIS data.

**References**

[1] Reim M, Körner W, Manara J, Korder S, Arduini-Schuster M, Ebert H-P and Fricke J 2005 *Solar Energy* **79** 131
[2] Fricke J 1988 *J. Non-cryst. Solids* **100** 169
[3] Fricke J and Emmerling A 1992 *J. Am. Ceram. Soc.* **75** 2027
[4] Kim G S and Hyun S H 2001 *Ceramist* **4** 5
[5] Carraher Jr C E 2005 *Polym. News* **30** 386
[6] Schultz J M, Jensen K I and Kristiansen F H 2005 *Sol. Energy Mat. Sol. Cells* **89** 275
[7] Schwertfeger F, Frank D and Schmidt M 1998 *J. Non-Cryst. Solids* **225** 24
[8] E*nergy-efficient Buildings PPP. Multi-Annual Roadmap and Longer Term Strategy*, 2010 EUR 24283, European Commission
[9] Mohanan J L, Arachchige I U and Brock S L 2005 *Science* **307** 397
[10] Schuh C A Nanoindentation studies of materials 2006 *Mater. Today* **9** 32–40
[11] Moner-Girona M, Roig A, Molins E, Martınez E and Esteve J, *Appl. Phys. Lett.* 1999 **75** 653
[12] Anappara A A, Rajeshkumar S, Mukundan P, Warrier P R S, Ghosh S and Warrier K G K 2004 *Acta Materialia* **52** 369
[13] Salme I I, Baagrel A and Bandosz T J 1993 *J Phys Chem* B **103** 3877

# Study of detonation nanodiamond – plasma polymerized hexamethildisiloxan composites for medical application

**L D Pramatarova[1], N A Krasteva[2], E I Radeva[1], E V Pecheva[1], R P Dimitrova[3], T A Hikov[1], D P Mitev[4], K T Hristova[2] and G Altankov[5, 6]**
[1] Georgi Nadjakov Institute of Solid State Physics, Bulgarian Academy of Sciences, 72 Tzarigradsko Chaussee blvd., 1784 Sofia, Bulgaria,
[2] Institute for Biophysics and Biomedical Engineering, Bulgarian Academy of Science, "Acad. G. Bonchev" Str. bl.21, 1113 Sofia, Bulgaria
[3] Institute of Organic Chemistry with Phytochemistry, "Acad. G. Bonchev" Str. bl.9 Bulgarian Academy of Science, 1113 Sofia, Bulgaria
[4] Institute for Space and Solar-Terrestrial Research, Bulgarian Academy of Sciences, 72 Tzarigradsko Chaussee blvd., 1784 Sofia, Bulgaria
[5] ICREA (Institucio Catalana de Recercia i Estudias Avançats)
[6] Institute for Bioengineering of Catalonia (IBEC), Campus de Bellvitge - Feixa Llarga/Pavelló Govern 1[a], 08907 L'Hospitalet de Llobregat (Barcelona), Spain

E-mail: lpramat@issp.bas.bg

**Abstract.** The present study reports on how detonation nanodiamond (DND) - plasma poly(hexamethyldisiloxane) composites (PPHMDS) affect osteoblast cell behavior. It has been established that various modified DND nanoparticles (Ag-DND and Si-DND) can be readily integrated into virtually all polymer matrices. In particular, PPHDMS composites have been developed over the past few years because of the variety of their application as medical devices and implants. By incubation of MG-63 osteoblast-like cells on the surface of DND (Ag-DND and Si-DND) - PPHMDS composite, we tested the hypothesis that DND–based polymer composites can influence the adhesion behavior of MG-63 osteoblast-like cells. Morphological and structural characterization of DND, Ag-DND and Si-DND powders was carried out by XRD, HRTEM and EDS. For the study of the composite layers, deposited on cover glass (CG), FTIR spectroscopy has been performed in order to determine if the DND nanofiller can potentially modify the structural and chemical dynamics of the polymer matrix. The kinetic of static water contact angle of composite surfaces as a function of the as-used nanofiller DND's in polymer matrix was measured The results with MG-63 osteoblast-like cells suggest the potential of using DND-based polymer composites for application in engineering implantable scaffolds and devices.

## 1. Introduction

An important goal of materials science is the development of interfaces that integrate the functions of living cells and materials. The development of materials that serve as substrates for adherent cells is important in a range of basic and applied programs. Recently there has been a growing interest in the synthesis of composite functional materials exhibiting novel physico-chemical properties, involving integration of inorganic nanomaterials into a polymer matrix [1]. Numerous siloxan-based materials including polymerized hexamethyldisiloxane (PPHMDS) chosen for this study have been developed.

Plasma poly (hexamethyldisiloxane) (PPHMDS) has a long history of exploitation in a variety of applications, because it is easy to prepare with the well established technology of plasma polymerization [2, 3], it is non-toxic, with a very low surface tension, and it neither dissolves nor swells in a cell culture medium [4]. On the other hand, nanodiamond carbon nanostructures are of interest due to the combination of unique properties inherent to diamond and the specific surface structure of particles facilitating its fictionalization [1]. One class of nanodiamond paricles with 4 nm primary particle size is the so-called detonation nanodiamond (DND), which is synthesized by detonation of carbon-containing explosives [1, 5]. Recently, DND nanoparticles have been modified, by means of appropriate chemical reactions, and silver (Ag)- and silicon (Si)–based nanoparticles being produced. The antimicrobial effects of Ag ions are well recognized, as well as Ag nanoparticles being highly efficient against bacterial colonization and at the same time allowing adhesion and spreading of mammalian cells [2]. Production of (Si)–based nanoparticles is proposed to prevent DNDs from forming aggregates [6].

The aim of this work is the design of novel thin polymer films filled with modified DND nanoparticles. The importance of studying the interaction of osteoblast cells with polymers and polymer-based nanocomposites is their increasing use in many biological and medical applications [2, 4 and 8].

## 2. Materials and methods

### *2.1. Substrate*

Prior to deposition of the films, the cover glass substrates (of diameter 12 and 16 mm) were cleaned by standard wet chemical procedure and for 5 min in a reactor chamber by plasma at 133 Pa air pressure and 0.08 mA/cm$^2$ current density.

### *2.2. Surface modification of DND nanoparticles*

DND particles were synthesized by detonation of trinitrotoluene and hexogen at high pressure and high temperature [5]. A method to produce DNDs with a uniform metal particle distribution over the DND aggregate surface has recently been proposed [9]. The method is based on the interaction of metal salts (in our case $AgNO_3$) with the functional carboxyl groups (-COOH) present on the surface of the aggregates in DND suspensions [6]. For the production of the second group of DND (Si-DND), the method of silanization was applied.

### *2.3. Deposition of DND (Ag-DND and Si-DND) - PPHMDS composite*

For plasma polymerization, the plasma polymerization equipment was used as describe in [8]. The monomer used in this experiment was a guaranteed grade of hexamethildisiloxan (HMDS) [$(CH_3)_3$-Si-O-Si-$(CH_3)_3$] – >99% Merck (Germany). The CG substrates were placed on a Teflon plate between the electrodes. The plasma polymerization conditions were controlled by two parameters: the monomer flow rate and the current density of the glow discharge [3]. The deposition rates of plasma-polymerized (PP) films were determined by the quartz crystal microbalance technique. Using the several polymerization conditions for HMDS, it was found that the deposition rates were proportional to the polymerization time [10]. For deposition of the composites on the CG substrate, plasma polymerization equipment was used [3, 10] and the following technological regime was applied: DND particles were added to the HMDS monomer in the concentration of 0.01 g/100 ml and prior to deposition by ultrasonic shaking for 15 min. After that the container with the suspension was placed on the magnet stirrer and continuously stirred at 25$^0$C and 275 r.p.m. Plasma polymerization was performed for 20 minutes at 0.08 mA/cm$^2$ current density at about 133 Pa gas phase pressure. The thickness of the composite layers was approximately 200 nm as determined by a Dektak Stylus Profile System (Veeco Instruments Inc., Woodbury, NY, USA). After the deposition, the samples were washed under a flow of Milli-Q pure water and dried in air. Three groups of samples were thus prepared named in the text as 6-DND_PPHMDS, Ag-DND_PPHMDS and Si-DND_PPHMDS. As controls PPHMDS deposed on CG substrate were also prepared.

### *2.4. FTIR spectroscopy measurement*

The surface functional groups of the composite layers were studied by FTIR spectra obtained on a Brucker-Vector 22 FTIR-spectrometer by averaging 128 scans. All spectra were recorded at room temperature using standard instrument settings. The assignment of the absorption bands is based on experience with molecular organic compounds and literature data. The frequencies quoted are believed to be within 2 or 3 $cm^{-1}$ of the true values.

### *2.5. Cell culture and cell adhesion assay*

Human osteoblast-like MG-63 cell line (purchased from ATCC, USA) was used as a model system to examine the effects of surface coatings on osteoblast adhesion, spreading and overall morphology, as well as on FN matrix organization. Cells were maintained in Dulbecco´s modified Eagle medium (Gibco, 11960-044) supplemented with 10% FBS, 1% penicillin/streptomycin, 2 mM L-glutamine and 1 mM sodium pyruvate (Gibco, 11360-039), in a humidified atmosphere of 5% $CO_2$ in air. The culture medium was exchanged every second day. Upon reaching confluence the cells were detached with 0,05% trypsin-EDTA (Gibco, 25200-072), inactivated with FBS after detachment (approx. 5 min), and the cells being re-cultured or used for the experiments. To investigate the overall morphology and initial cell adhesion and spreading, 2 x $10^4$ cells/well were seeded to each surface in 2.0 ml serum free medium. One set of samples has been pre-coated with proteins as described above while another one was studied as plain controls. After 2 h of incubation, the cells were stained for actin and vinculin, using specific antibodies (phalloidin and mouse anti-human vinculin, respectively). Focal adhesions formed after 2 h of incubation of MG-63 cells on the samples were visualized using specific mouse monoclonal antibody (Sigma) against vinculin (1:800 dilution) in 1% bovine serum albumin (BSA). This step was followed by goat anti-mouse Cy3 conjugated secondary antibody (Jackson ImmunoResearch). After incubation with the antibodies, the samples were mounted and viewed on an inverted fluorescent microscope and at least three representative images were obtained

## 3. Results and discussion

### *3.1. Surface contact angle measurements*

Figure 1 shows the alteration of the statistic contact angle of the PPHMDS and composite films DNDs(6-DND, Ag-DND, Si-DND) studied as a function of time. The contact angle of the polymer (PPHMDS on CG) films decreased when modified DND (Si-DND, Ag-DND) were used. Decreasing

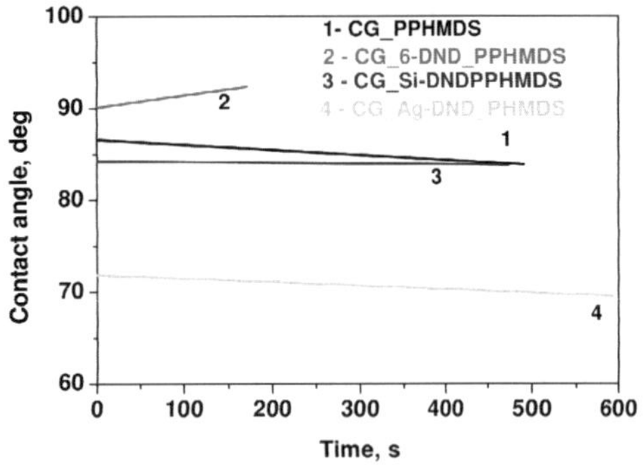


**Figure 1.** Statistic contact angle of PPHMDS and composite films DNDs(6-DND, Si-DND, Ag-DND _PPHMDS as a function of time

of the contact angle means that the surface of the grown polymer film becomes more hydrophilic. The increases in the case of the 6-DND nanofiller are explained with the increased amount of oxygenated residues on the surface of the nano-particles [11]. When Ag-cations are introduced on the DND surface, the total polymer film surface becomes even more hydrophilic due to the covalent bonds formed between the Ag cations and Si-O residues.

### 3.2. *FTIR spectroscopy measurements of composites*

On figure 2 and 3, the FTIR spectra of the DND powders and composite layers are presented respectively. One can note the significant differences in the surface functionality between the used initial DND powders and the prepared polymer layers on the CG substrate. The spectrum of plasma grown composites on the CG substrate possessed the characteristic adsorption bands of polymer methyl groups as stretching and deformation modes from the C–H bond in various different tetrahedral arrangements of the $CH_3$ groups connected to the silicon atoms as (Si-$(CH_3)_3$, Si-$(CH_3)_2$, Si-$(CH_2)$n-Si) with a maxima between 400 - 600 and 950–900 $cm^{-1}$; the Si–O rocking mode of the Si–O–Si bond at 460 $cm^{-1}$ and an asymmetric stretching vibration of the Si–O–Si group with a maximum between 1130–1040 $cm^{-1}$; a stretching vibration of the Si–C bond with a maximum at 780 $cm^{-1}$ from the Si–CH3 group and carbonyl and hydroxyl groups (–OH, >C=O) in the region 1500-1700 $cm^{-1}$. In the

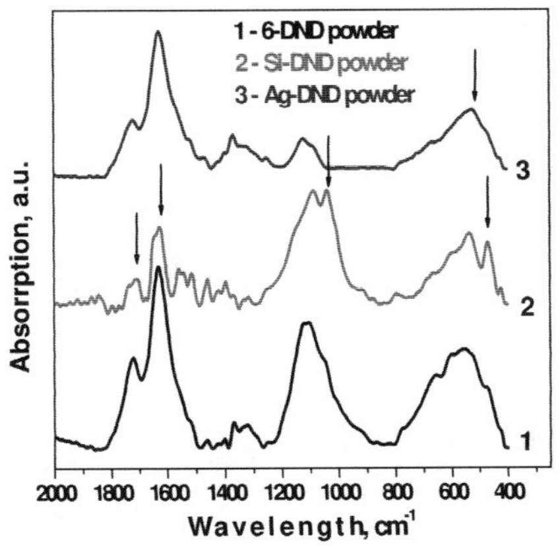


**Figure 2.** FTIR spectra of the DND powders: 6-DND (1); Si-DND (2) and Ag-DND (3).

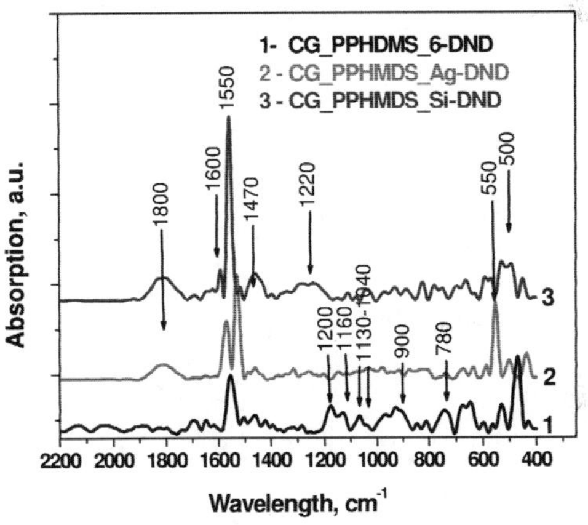


**Figure 3.** FTIR spectra of the composites: CG_6-D-PPHMDS (1); CG_Ag-DND_PPHMDS (2); CG_Si-DND_PPHMDS (3).

spectrum,the characteristic peaks for the sp bonded carbon of the DND 1550 $cm^{-1}$, a peak due to nanocrystalline diamond at approximately 1200 $cm^{-1}$ and $sp^2$ bonded carbon at 1160 $cm^{-1}$ can be clearly observed. So, the surface of the CG_DND_PPHDMS is characterized with many functional groups of the type–OH, >C=O, Si–O–Si, Si–O–C and different tetrahedral $CH_3$ which provide the surface with both hydrophobic and hydrophilic centers. The appearance of the pronounced peak at 550 $cm^{-1}$ in the spectrum of the CG_Ag-DND_PPHMDS indicates the existence of Ag cations. The splitting of the peak at 1550 $cm^{-1}$ into two peaks and the appearance of a broad band at 1800 $cm^{-1}$ confirmed the location of Ag cations at the grain boundaries of the diamond nanocparticles due to interaction of the silver ions either with the polymer Si-OH groups, or with the diamond functional groups (–OH, >C=O, Si–O–C). In the case of CG_Si-DND_PPHMDS, definite peaks at around 500, 1200-1300, 1470 and 1600 $cm^{-1}$ point towards an increased content of amorphous carbon in the grown layer due to the silanization of the DND surface.

*3.3. Cell experiments*

*3.3.1. FTIR spectra of coated with FN composites* In the present study we employed the FTIR measurements to evaluate the structural changes of composite surfaces after their covering with FN. The FTIR absorption spectra, shown in figure 5, demonstrated that both composite surfaces Si-DND_ PPHMDS and Ag-DND_ PPHMDS interact with the protein (FN). In both cases, a significant increase of the 1800 $cm^{-1}$ peaks and rearrangement of the peak's intensity in the region 1400 - 1650 $cm^{-1}$ occur in comparison with the spectra without FN shown in figure 3. This observation is due to an enhanced reaction between the FN molecule and the existing functional groups on the composite surfaces and it's irreversibly adsorption on the surface of the material. In figure 4 the protein easy interaction of the initial CG substrate covered with FN for 30 minutes is demonstrated. After 24 hours of intensive washing in DI water the same samples were measured again. Both the FTIR spectra slightly differ, which confirm that an irreversible adsorption occurs even at the beginning of the CG treatment with the FN solution.

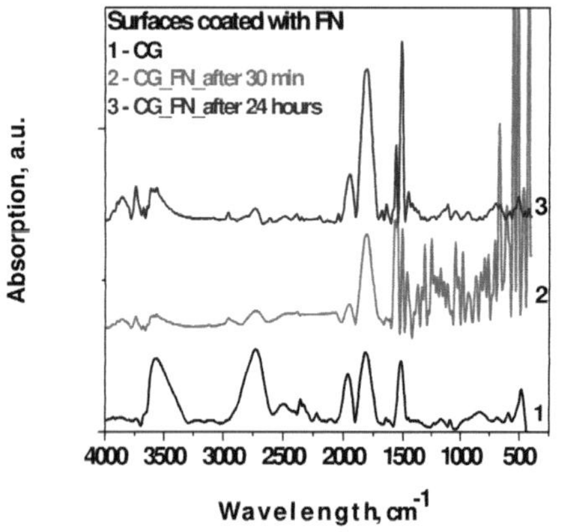


**Figure 4.** FTIR spectra of initial CG substrate (1); CG_FN after 30 min. treatment with FN (2) ; the same CG_FN sample measured again after 24 hours.

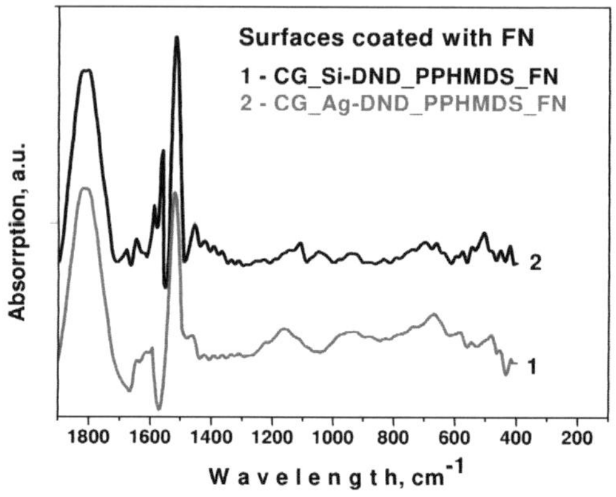


**Figure 5.** FTIR spectra of CG_Si-DND_PPHMDS (1) and CG_Ag-DND_PPHMDS (2). composite surfaces, coated with FN.

*3.3.2. Cytotoxity test* To estimate the cytotoxity of DND materials we incubated a suspension of DND (6-DND) and DND-modified particles (Si-DND and Ag-DND) with osteoblast-like cells (MG63) for 24 hours in a serum-containing medium.

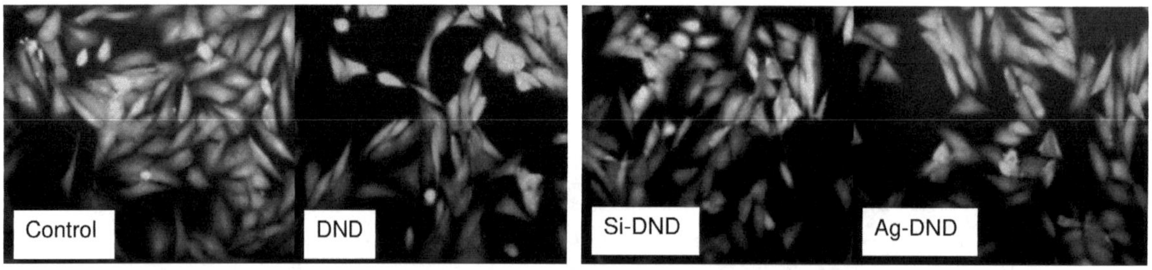


**Figure 6.** Overall morphology of MG63 osteoblasts cells incubated for 24 hours on plain cover glass slips with suspension of 6- DND and modified Si- and Ag-DND nanoparticles respectively; magnification 20x.

At the end of incubation the overall cell morphology was visualized by fluorescent diacetate (FDA) - a dye agent that stains only vital cells (figure 6). Our results showed some decrease in the number of viable cells in all the tested samples, more pronounced in cells cultured with pure 6-DND particles while in cell cultures incubated with Si- and Ag-DND particles the reduction in cell numbers was lower. Nevertheless, the attached cells were well spread, even better than the control samples, suggesting unaffected cell functions. Therefore, DND powders seem to be promising materials for some applications with cells. However, to learn more about how they affect the specific cell adhesion and functions, more detailed investigations have to be performed.

*3.3.3. Cells experiments with MG-63 osteoblast cells cultured on the composite surfaces* The purpose of the presented study was to evaluate the protein adsorption and cell adhesion on the PPHMDS-DNDs composites

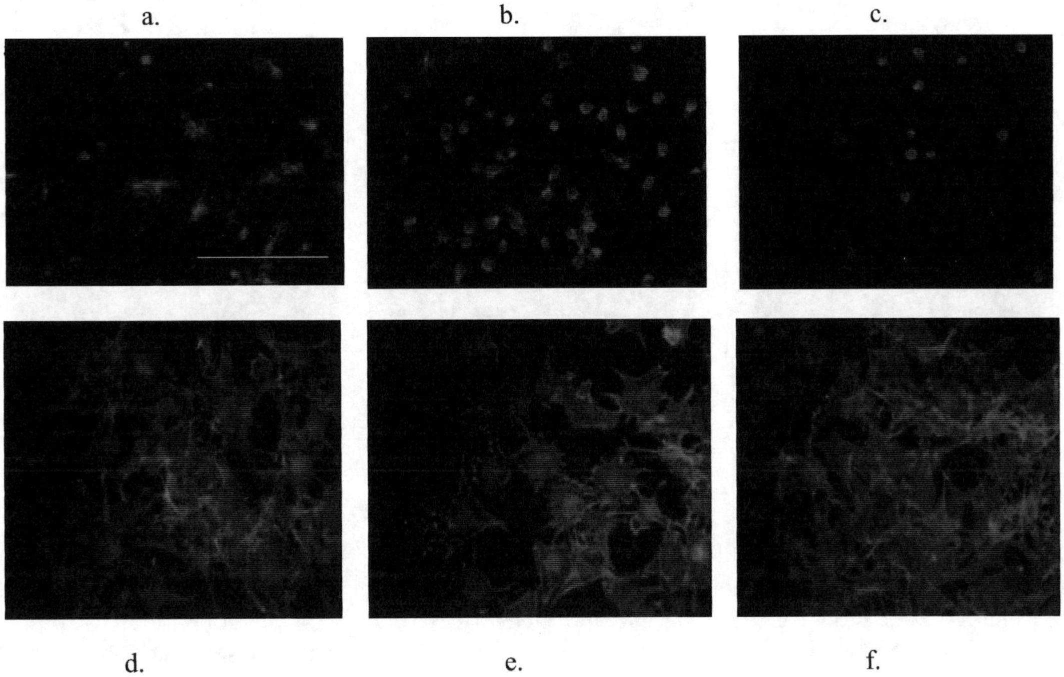


**Figure 7.** Actin staining of MG63 osteoblast-like cells, cultured for 2 hours on plain surfaces (a) CG-control; (b) Si- DND_PPHMDS (c) Ag-DND_PPHMDS; and coated with FN surfaces (d), (e) and (f) respectively; magnification 20 x.

The results from actin staining, shown in Fig. 7 revealed that the FN coating improved significantly cell adhesion and spreading (d,e,f) in contrast to the uncoated (plain) composite surfaces (a,b,c), suggesting that on the composite surfaces the FN most probably adsorbs in conformation that is more suitable for the cells. Vinculin staining of MG63 ostoblast-like cells, cultured for 2h on the composite surface cover glass controls revealed that vinculin was also better expressed on composite surfaces. It was reported [12, 13] that PPHMDS is an hydrophobic material and incubated MG63 cells are rounder, while on the PPHMDS, filled with DNDs, the cell spreading improved (figure 6 and 7- d, e, f). On the other hand, hidrophilization of the composite surfaces results in a slight decrease in spontaneous cell adhesion (Fig. 7 and 8-a, b, c). Previously we have reported that plasma polymerization of Hexamethildisiloxan (HMDS) under certain conditions [10, 12]; followed by $NH_3$ plasma treatment was effective for decreasing the PPHMDS surface contact angle and effective for protein adsorption as well as cell proliferation. In the present study we demonstrated that DND-based polymers are most effective for protein and cells attachment. The plasma polymerization of

DNDs_PPHMDS composites posses a potential for surface modification of various implant materials. The adsorption behavior of cell adhesive proteins, particularly fibronectin (FN), on the surface of novel composites (DND-based polymer) subsequently treated with $NH_3$ plasma are in progress.

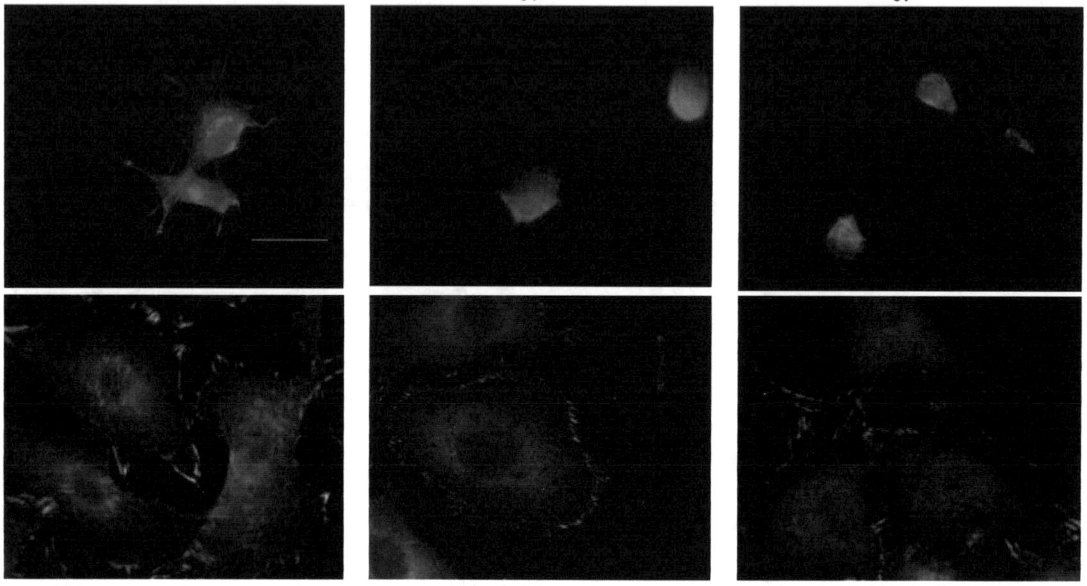


**Figure 8.** Vinculin staining of MG63 osteoblast-like cells cultured for 2 hours on plain surfaces (a) CG-control; (b) Si- DND_PPHMDS (c) Ag-DND_PPHMDS; and coated with FN surfaces (d), (e) and (f) respectively; magnification 63 x

## 4. Conclusion

Novel composites have been produced using modified DND (Ag-DND, and Si-DND) nanoparticles as fillers in PPHMDS layers and their biological response using the MG63 osteoblast cells model has been tested. It was shown that by varying the DND filler, the chemical property of hydrophobic PPHMDS films can be modified. This study demonstrates that via appropriate material design it is possible to generate surfaces allowing considerably better attachment of the osteoblast cells.

**Acknowledgments**

This work was supported by two National projects: with the Bulgarian Ministry of Education and Science (TK-X-1708/2007) and the Agency of Innovation of Bulgaria (NIF 02-54/2007). ANNA project (№ 026134(RI3), Research Infrastructure Action under the FP6 "European Integrated Activity of Excellence and Networking for Nano and Micro-Electronics Analysis") are also acknowledged.

**References**

[1] Shenderova O A, Zhirnov V V and Brenner D W 2002 *Solid State Mater.Sci.* **27** 227
[2] Vasilev K et all 2010 *Nano Letter.* **10** 202-7
[3] Radeva E, Tsankov D, Bobev K and Spassov L 1993 *J. Appl. Polym. Sci.* **50** 165
[4] Min-HsienWu 2009 *Surf. Interface Anal.* **41** 11–16
[5] Stavrev S, Lazarov S, Stoev Kh, Markov L and Ivanov V 1994 *US patent* No 5353708
[6] Baidakova M and Vul A 2007 *J. Phys. D: Appl. Phys.* **40** 6300-6311
[7] Nicholas D. Evans et all 2009 *European Cells and Materials* **18** 1-14
[8] Hikov T et all 2009 *Jour Optoelecttron. Adv. Mater.* **11** 1327 – 30
[9] Aleksenski A E, Yagovkina M A and Vul A Ya 2004 *Phys. Solid State* **46** 685-6
[10] Radeva E, Pramatarova L, Dimitrova R, Popova S, Hikov T, Fingarova D and Pecheva E *J. Optoel.Adv. Mater.* 2009 **11** 1432-35
[11] Kurosawa Sh et all *Thin Sold Films* 2006 506-7 176-9
[12] Krasteva N, K. Hristova, Radeva E, Pecheva E, Dimitrova R and Pramatarova L 2009 *American Institute of Physics* **CP1203** 688-693
[13] Hayakawa T, Yoshinari M and Nemoto K 2006 *J Oral Tissue Engin* **3** 153-160

Journal of Physics: Conference Series **253** (2010) 012079 doi:10.1088/1742-6596/253/1/012079

# Initial biocompatibility of plasma polymerized hexamethyldisiloxane films with different wettability

**N A Krasteva[1], G Toromanov[2], K T Hristova[1], E I Radeva[3], E V Pecheva[3], R P Dimitrova[4], G P Altankov[2,5] and L D Pramatarova[3]**

[1] Institute of Biophysics and Biomedical Engineering, Bulgarian Academy of Science, "Acad. G. Bonchev" Str. bl.21, 1113 Sofia, Bulgaria
[2] Institute of Bioengineering of Catalonia (IBEC), Campus de Bellvitge-Feixa Llarga/Pavelló Govern, 08907 L'Hospitalet de Llobregat, Barcelona, Spain
[3] Georgi Nadjakov Institute of Solid State Physics, Bulgarian Academy of Sciences, 72 Tzarigradsko Chaussee blvd., 1784 Sofia, Bulgaria,
[4] Institute of Organic Chemistry with Phytochemistry, "Acad. G. Bonchev" Str. bl.9 Bulgarian Academy of Science, 1113 Sofia, Bulgaria
[5]ICREA (Institution Catalan for Research and Advanced Studies)

E-mail: nataly@bio21.bas.bg

**Abstract.** Understanding the relationships between material surface properties, behaviour of adsorbed proteins and cellular responses is essential to design optimal material surfaces for tissue engineering. In this study we modify thin layers of plasma polymerized hexamethyldisiloxane (PPHMDS) by ammonia treatment in order to increase surface wettability and the corresponding biological response. The physico-chemical properties of the polymer films were characterized by contact angle (CA) measurements and Fourier Transform Infrared Spectroscopy (FTIR) analysis.Human umbilical vein endothelial cells (HUVEC) were used as model system for the initial biocompatibility studies following their behavior upon preadsorption of polymer films with three adhesive proteins: fibronectin (FN), fibrinogen (FG) and vitronectin (VN). Adhesive interaction of HUVEC was evaluated after 2 hours by analyzing the overall cell morphology, and the organization of focal adhesion contacts and actin cytoskeleton. We have found similar good cellular response on FN and FG coated polymer films, with better pronounced vinculin expression on FN samples while. Conversely, on VN coated surfaces the wettability influenced significantly initial celular interaction spreading. The results obtained suggested that ammonia plasma treatment can modulate the biological activity of the adsorbed protein s on PPHMDS surfaces and thus to influence the interaction with endothelial cells.

## 1. Introduction

The initial interaction of cells with biomaterial surfaces is of fundamental for their biocompatibility and contributes to the clinical success of implants. It is well documented that cells interact with biomaterials not directly but via a layer of adsorbed plasma proteins [1, 2]. When a biomaterial comes into contact with a biological system proteins spontaneously adsorb onto the surface within a time frame of seconds to minutes [3]. The resulting surface-bound protein layer mediates the subsequent cell attachment through interactions with cell surface receptors [4]. Both, protein adsorption and cell adhesion are strongly affected by the surfaces characteristics including the surface chemistry, topography as well as surface wettability. A lot of research has been undertaken to study the effects of the surface wettability on the cellular responses and even a certain optimal wettability has been described, however, often these studies yield contradictory results [5, 6]. The reason for this is that the surface parameters cannot be considered independently; each material with defined surface wettability may have different chemistry or topography. Therefore, to better understand the effect of this single parameter in this study we employed a wide range of wettability with a single material, thus without change in surface topography or chemistry.

Plasma polymers are attractive candidates for biomaterials modification because their surface hydrophilic can be easily modified. Plasma polymerization is a convenient and versatile solvent–free technology, providing fast and one-step synthesis. Many monomers which do not polymerized by normal (chemical) methods can be polymerized by plasma because this is a process in which the monomer is subjected to successive fragmentation and recombination, and thus forms a polymer [7]. Therefore, the structure and properties of the plasma polymers substantially differs from conventional polymers. Moreover, varying with the process parameters of plasma polymerization from one monomer could be obtained different polymers, including such with desirable structure and properties. Resulted polymers usually have highly irregular three-dimensional structure and cross-linked network which determines properties such as chemical resistance, mechanical toughness, thermal stability, and good adhesion to any substrate [8]. All these advantages have led to wide application of plasma polymer, for example as protective coatings [9], membranes [10] sensors [11-13] and biomaterial films [14, 15]. In particular, plasma polymerization using organosilicon monomers, such as hexamethyldisiloxane (HMDS), has attracted great interest because of their high deposition rates and ease ability of varying the deposition conditions to control their structure and properties, biocompatibility and low toxicity [16-18]. The relatively high hydrophobicity of HMDS has been used to prevent coagulation and inflammatory related problems in cardiovascular systems. In order to improve cell adhesion however polymer surface should be render hydrophilic. This could be realized by means of surface treatments using various gases such as Ar, $O_2$, $N_2$, $SO_2$. Indeed, several studies have showed that $O_2$-plasma treatment of polymer surfaces was able to control hydrophilicity/hydrophobicity and to introduce various functional groups [19]. We anticipate that ammonia plasma-surface modification could also be useful to influence the surface wettability, and thus affecting protein adsorption and cell behavior.

Here, we report on a simple method for modulation of surface wettability of plasma polymerized hexamethyldisiloxane (PPHMDS) films and how it affects the endothelial cells interaction. Varying with the processing parameters (current density of the plasma, monomer flow rate and time of deposition) of both, plasma polymerisation and plasma modification, we found a technological regime for deposition of ultrathin film from HMDS with lower wettability of about $96^0$, that could be rapidly hydrophilised by a single step of ammonia plasma treatment. To investigate the biological response of thus modified surfaces we have prepared model samples ranging significantly in their wettability and follow the initial cellular interaction of human umbilical vein endothelial cells (HUVEC).

## 2. Materials and methods

### *2.1. Preparation of polymer films*

Prior to polymer films deposition, glass coverslips (CG) were cleaned by repeated washing in double distillated (DI) water, followed by ultrasonic treatment in acetone (99.99% purity) for 15 min, rinsing with boiling isopropanol (99.99% purity) for 10 min and then rinsing again with DI water (about 10 times). Finally, the samples were dried under a stream of pure nitrogen (99.9 %) and loaded into the reactor chamber. The monomer in this experiment was HMDS $[(CH_3\ )_3\text{-Si-O-Si-}(CH_3)_3]$ – >99% Merck (Germany) used without further purification. Plasma deposition reactor was described in details elsewhere [20]. Briefly, thin polymer films were obtained by plasma polymerization of HMDS monomer in vacuum chamber for 10 min under following technological conditions: 0.13 mA/cm$^2$ current density and 10 l/h monomer flow rate. By using several polymerization regimes it was found that the deposition rates were proportional to the polymerization time. The thickness of the polymer layers was approximately 160 nm determined by a Dektak Stylus Profile System (Vecco Instruments Inc., Woodbury, NY, USA). In order to obtain polymer surfaces with different wettability PPHMDS films were further modified by ammonia plasma as the duration of the modification varied from 1 to

10 min. The process of ammonia plasma treatment was carried out at the same conditions as polymerization process. For cell experiments, the materials were sterilized in 70% ethanol for 10 min and then rinsed intensively with sterile PBS.

*2.2. Contact angle measurement*
Static contact angle was measured on treated polymer films by sessile drop method using Easy Drop Shape Analysis System. For each treatment two samples were tested and three droplets per sample were measured.

*2.3. FTIR spectroscopy measurement*
FTIR spectra of PPHMDS films were registered by Brucker-Vector 22 FTIR spectrometer in the range of 400 to 4000 $cm^{-1}$ wavelength. On the average of 64 scans of the polymer films were done with a resolution of 4 $cm^{-1}$ using OPUS software. All spectra were recorded at room temperature (RT) under standard instrument settings. The assignment of the absorption bands is based on experience with organic compounds and the literature data. The quoted wavelengths are believed to be within 2 or 3 cm-1 of the true values.

*2.4. Protein coating*
Before cell seeding one set of polymer films had been pre-coated with fibronectin (FN) at a concentration of 20 µg/ml in phosphate-buffered saline (PBS) pH 7.4; the other set - with fibrinogen (FG) at concentration 50 µg/ml in PBS and the third set of materials - with vitronectin (VN) at concentration 6 µg/ml in PBS. Pre-coating of the proteins was conducted for 30 min at RT and then the samples were rinsed twice with PBS.

*2.5. Cell culture*
Human Umbilical Vein Endothelial Cells (HUVEC) purchased from PromoCell (Heidelberg, Germany) was used as a model system to examine the effects of surface hydrophylicity on biological activity of three types of adhesive proteins, including fibronectin (FN), fibrinogen (FG) and vitronectin (VN). Cells were cultured in Endothelial Cell Growth Medium (PromoCell) supplemented with Supplement Mix (PromoCell), containing 0.4% ECGS/H; 2% Fetal Calf Serum, 1 ng/ml Epidermal Growth Factor, 1 µg/ml hydrocortisone and 1 ng/ml basic fibroblast factor. The culture medium was exchanged every second day. Before the experiments the confluent cells were harvested with 0,05% Trypsin/EDTA (Invitrogen, USA), subsequently washed with culture medium and finally the cells were resuspended in serum-free medium and seeded at a concentration of 50 000 cells/ml onto PHMDS films, individually placed in tissue culture plates.

*2.6. Immunofluorescence for actin and vinculin*
To study the overall cell morphology and cytoskeleton of adhering HUVEC the cells were stained for actin. For that purpose, $10^5$ cells/ well were seeded in 24-well tissue culture (TC) plates (Costar, Corning, Lowell, MA, USA), containing the polymer films, pre-coated with adhesive proteins, as described above, for 2h in serum-free medium. At the end of incubation, the cells were fixed with 4% paraformaldehyde for 10 min, permeabilized with 0.5% Triton X-1000 for 5 min and saturated with 1% albumin in PBS for 15 min. Subsequently, to visualize actin cytoskeleton the samples were incubated for 30 min with 20 µg/ml AlexaFluor 488 phalloidin (Molecular Probes, Eugene, OR) in PBS, washed several times with PBS and finally mounted in mounting media (Mowiol) (Polysciences, Warrington, PA, USA).
To visualize focal adhesion contacts, fixed, permeabilized and saturated cells were incubated for 30 min with monoclonal anti-vinculin antibody (Sigma, Cat No V9131), dissolved in 1% albumin in PBS, followed by incubation with Cy 3-conjugated Affini-Pure Goat Anti-Mouse IgG (H+L) (Jackson ImmunoResearch, Newmarket, Suffolk, UK) as a secondary antibody. After intensive washing the samples were finally mounted in Mowiol and were viewed and photographed on an inverted

fluorescent microscope (Nikon Eclipse E800; Nikon, Tokyo, Japan) where at least 3 representative images were acquired. As a positive control glass coverslips (Menzel, 15 mm in diameter) have been used.

## 3. Results and discussion

### *3.1. Surface contact angle measurements*

In order to obtain information about the change in surface wettability of PPHMDS films with the time of ammonia plasma treatment, we measure the water contact angle (WCA). The results are shown on Figure 1. As can be seen from the graphic, the WCA of PPHMDS surfaces decreased rapidly with increasing of ammonia plasma duration up to $5^{th}$ min meaning that the surface of polymers become more hydrophilic in a result of increased amount of amino groups generated on polymer surface. After 5th min however the process slowed down and from $5^{th}$ to the $10^{th}$ min only a slight decrease (less than 5 degrees) was observed. When ammonia plasma treatment exceeded 10 min, WCA reached a plateau (around 35°) and did not change anymore with the time, suggesting a saturation of the treated surfaces with ammonia groups. Therefore, we concluded that by controlling the duration of ammonia plasma treatment we can control the surface hydrophilicity of the materials.

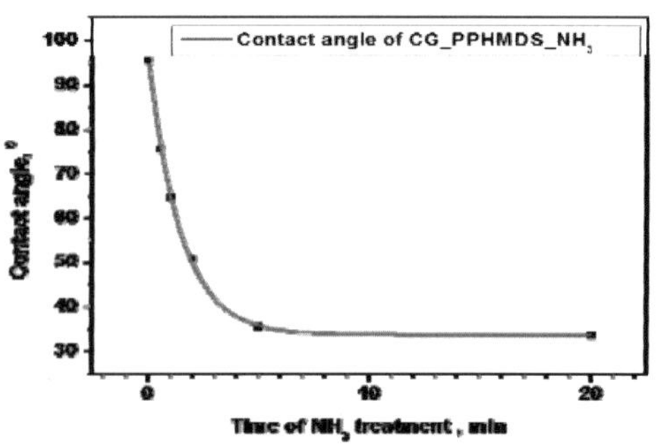


**Figure 1.** Kinetic of contact angle of PPHMDS films as a function of time of ammonia plasma treatment

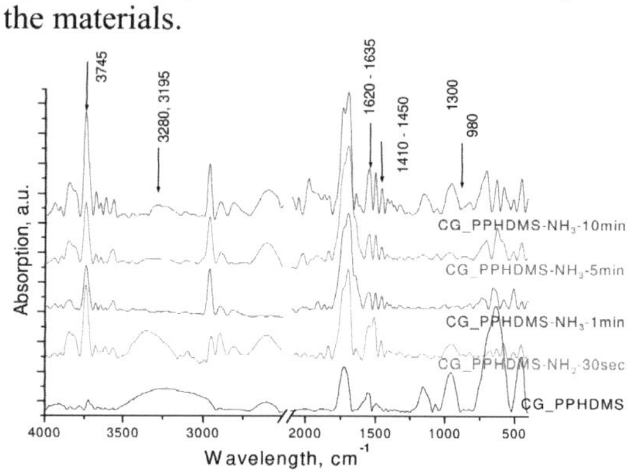


**Figure 2.** FTIR spectra of PPHMDS films, deposited on glas coverslips, untreated (in black) and treated with ammonia plasma 30 s (in red), 1 min (in blue) and 10 min (in violet)

### *3.2. FTIR spectroscopy measurements of PPHMD polymer films*

The FTIR measurements were performed to compare the surface functional groups of PPHMDS layers before and after ammonia plasma modification (Figure 2). The FTIR spectra demonstrated significant changes in the surface functionality after ammonia plasma treatment. The comparison between the spectrum of untreated PPHMDS (black line) and the spectrum of treated with ammonia for 30 sec PPHMDS (red line) pointed to an existence of $NH^{4+}$ cations (new bands in the regions of 1410-1450 $cm^{-1}$; and 1620-1635 $cm^{-1}$ ) and $NH_3$ molecules (new bands at around 3745, 3280, 3195 and 990 $cm^{-1}$ ). An interaction between the obtained in the plasma ammonium groups with both, Si-OH and OH-groups in the polymer structure occurs as can be seen from the FTIR deconvoluted spectra in the characteristic for the polymer chain range 1200-800 $cm^{-1}$. With increasing the time of ammonia plasma treatment from 1 to 10 min, ammonia molecules were grafted onto the surface of PPHMDS until the filling of all available polymer centres was achieved. A small shift and intensification of the bands around 1320, 1410-1450, 1630 and 1800 $cm^{-1}$ was observed (Figure 2 lines blue, olive and violet). Another notable change was the strong decrease in intensity of the bands for methyl groups (around 600 $cm^{-1}$) as well as broadening and intensification of the bands around 3750 $cm^{-1}$ that determined the amount of grafted ammonia. Due to the reduction of the relative part of alkyl groups in the ammonia treated polymer, a decrease of the hydrophobic and an increase of the hydrophilic character of the surface were obtained. It is worth to note that in parallel experiments using KBr as substrates only slight differences were observed in all FTIR spectra. This observation confirms the importance of the CG substrate for the properties of plasma formed layer and its surface functionality. It also shows that by ammonia plasma modification, layers with novel functional groups and importantly with strongly hydrophilic properties could be obtained on CG.

### *3.3. Cellular interaction*

Other main goal of this study was to evaluate the biological activity of thus modified PPHMDS polymer surfaces using three types of adhesive proteins e.g. FN, FG and VN, upon their adsorption to better understand how their activity is affected by the surface wettability modification. Previously, we have reported that ammonia plasma treatment of PPHMDS layers improve the proliferation and function of osteoblast cells [21]. In the present study we have used HUVE cells for the biological investigations because endothelial cells are more sensitive to any changes in substrate properties and because we wanted to understand if these modifications could affect endothelisation potential of PPHMDS layers upon implantation. As a first indicator of bioactivity, cell adhesion was investigated by monitoring the overall cell morphology and specific adhesive structures such as focal adhesion contacts, as well as the development of actin cytoskeleton. The experiments revealed that all studied PPHMDS layers adsorb FN and FG in a way (e.g. conformation) that support cellular interaction, judged by the high number of attached cells and their good spreading (Figure 3). Actin stress fibers also were well developed on both surfaces as found at higher magnification (data not shown).

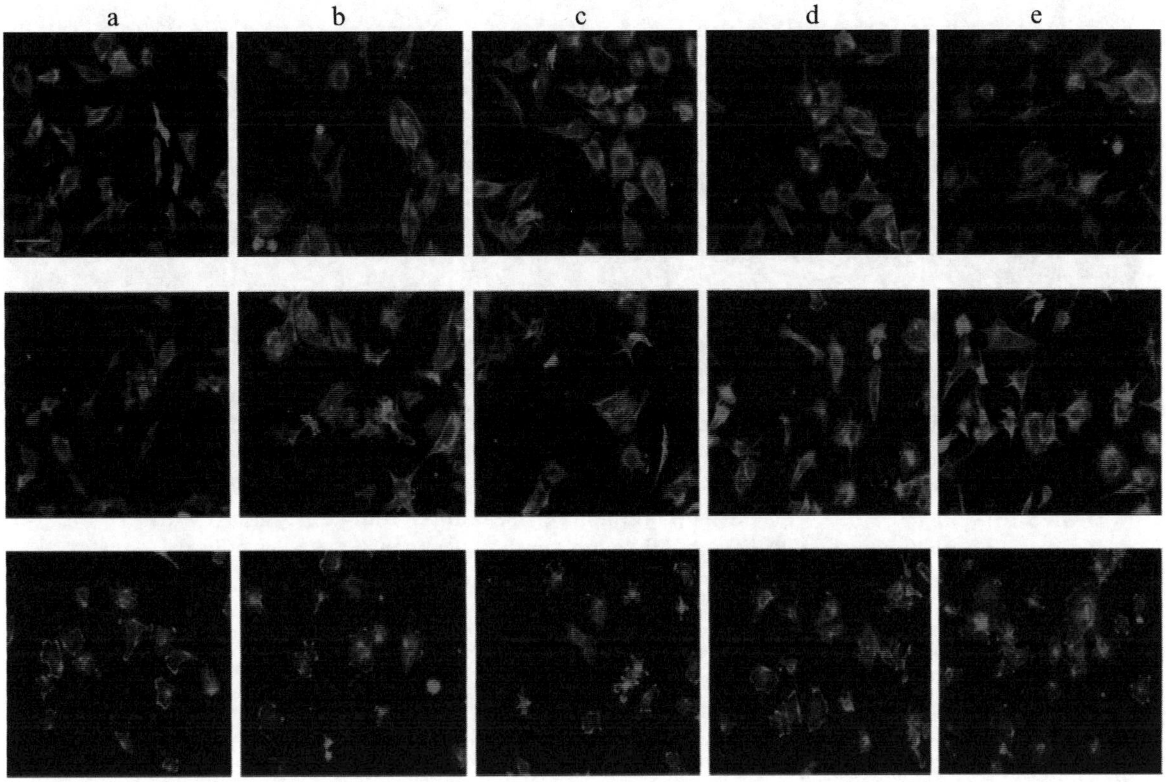


**Figure 3.** Overall morphology of HUVEC, cultured for 2 hours on plain glass coverslips (a), untreated PPHMDS (b) and PPHMDS, treated with ammonia plasma for 1 min (c), 5 min (d) and 10 min (e); pre-coated with FN (upper panel), FG (middle panel) and VN (lower panel); bar 100μm

On VN-coated surfaces, however, the cells represented rather stelate morphology with actin distributed predominantly at cell borders, suggesting that VN is worst recognized by the cells. However, the increase of wettability did not change significantly cell spreading on FN and FG while on VN-coated polymer films cell spreading was significantly improved. Thus the observed differences in the adhesion behavior of HUVEC suggested that PPHMDS layers adsorbed the adhesive proteins in a different way, resulting on more or less availability of their adhesive sequences and thus affect the cell-material interactions.

Results from vinculin staining (Figure 4) showed that it was best expressed and organized in focal adhesion contacts on FN-coated materials in comparison to FG- and VN-coated samples. However, any significant differences between ammonia treated (up to the 5 min) and non-treated materials were not observed. With increasing time of NH3 treatment however, (more than 5 min) the focal adhesions formation decreased. They become smaller (spot-like) and weaker expressed. On FG- and VN-coated materials the optimum seems to be on 5 min treated samples.

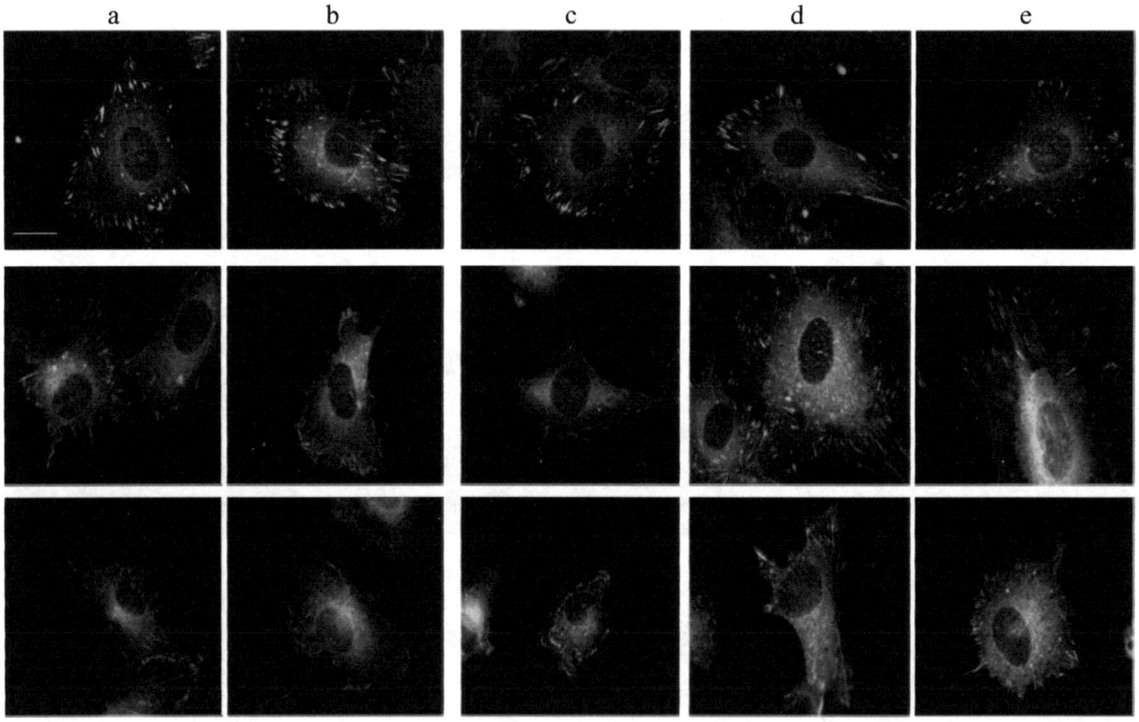


**Figure 4.** Vinculin expression in HUVEC, cultured for 2 hours on plain glass coverslips (a), untreated PPHMDS (b) and PPHMDS, treated with ammonia plasma for 1 min (c), 5 min (d) and 10 min (e); pre-coated with FN (upper panel), FG (middle panel) and VN (lower panel); bar 100μm

## Conclusion

Collectively, our results showed that surface wettability of PPHMDSO polymer films could be easy controlled by altering the duration of ammonia plasma treatment. Cell adhesion studies however revealed that the biological activity depends on the adhesive protein involved in the cellular interaction that should be taken in account in the design of new biomaterial surfaces.

**Acknowledgments**
This work was supported by three National projects: with the Bulgarian Ministry of Education and Science (TK-X-1708/2007 and HT-4-03/23.11.2004) and the Agency of Innovation of Bulgaria (NIF 02-54/2007). The financial support of Spanish Ministry for Research and Education: grant MAT2009-14440-C02-02 (to GA) is also acknowledged, as well as, the ANNA project (№ 026134(RI3) provided to IBF from the FP6 Research Infrastructure Action "European Integrated Activity of Excellence and Networking for Nano and Micro-Electronics Analysis".

**References**
[1] Andrade J D and Hlady V 1986 *Progress in Surface Science* **79:**1-64
[2] Groth T and Altankov G 1998 In: *New Biomedical Materials*, eds P I Harris and D Chapman (IOS Press, Amsterdam), p 12-23
[3] Grinnell F 1987 *Ann. NY Acad. Sci.* **56** 280-290
[4] Hynes R O 1992 *Cell* **69** 11-25
[5] Van Wachem P B, Beugeling T, Feijen J, Bantjes A, Detmers J P and Van Aken W G 1985 *Biomaterials* **6** 403-408
[6] Lee J H and Lee H B 1993 *J. Biomater. Sci. Polymer. Edn* **4** 467-481
[7] Yasuda H 1984 *J. Membrane Sci.* **18** 273-284
[8] Hamman C and Kampfrath G 1984 *Vacuum* **34 (12)** 1053-1059
[9] Sacher E, Klemberg-Sapieha J E, Schreiber H P and Wetheimer M R 1984 *J. Appl. Polym. Sci., Appl. Polym. Symp.* **38** 163-171
[10] Sakata J and Yamamoto M 1988 *J. Appl. Polym. Sci.* **42** 339-356
[11] Radeva E 1997 *Sensors and Actuators* **B44** 275-278
[12] Georgieva V, Radeva E and Spassov L 2000 *Vacuum* **5** 315-320
[13] Avramov I D, Kurosava S, Rapp M, Krawczak P and Radeva E 2002 *IEEE Sensors journal* **2(3)** 150-159
[14] Bouaidat S, Winther-Jensen B, Flygenring Christensen S and Jonsmann J 2004 *Sensors and Actuators A* **110** 390-394
[15] Kase Y and Muguruma H 2004 *Analytical Science* **20** 1143-1145
[16] Radeva E, Pramatarova L, Pecheva E, Hikov T, Iacob E, Vanzetti L, Dimitrova R, Krasteva N, Spassov T and Fingarova D 2010 *AIP Conf. Proc.* **CP1203**, 949-954
[17] Loughran M, Tsai S, Yokoyama K and Karube I 2003 *Curr. Appl. Phys.* **3** 495-499
[18] Muguruma H, Kase Y and Matsumura K 2006 *J. Phys. Chem. B* **110** 26033-26039
[19] Wei J, Yoshinari M, Takemoto S, Hattori M, Kawada E, Liu B and Oda Y 2004 *J Biomed Mater Res. Part B, Applied Biomater* **81 (1)** 66-75
[20] Radeva E, Tsankov D, Bobev K and Spassov L 1993 *J. Appl. Polym. Sci.* **50** 165-171
[21] Krasteva N et all 2010 *AIP Conf. Proc.* **CP1203** 688-693

# Dispersion of optical activity of magnesium sulfite hexahydrate single crystals

**T Dimov[1], Zh Bunzarov[2], I Iliev[1], P Petkova[1] and Y Tzoukrovski[2]**
[1] Department of Natural Sciences, Shumen University "Konstantin Preslavsky", 115 Universitetska street, 9712 Shumen, Bulgaria
[2] Physics Department, Sofia University "St. Kliment Ohridski", 1126 Sofia, Bulgaria

E-mail: dimov@shu-bg.net

**Abstract.** The magnesium sulfite hexahydrate ($MgSO_3.6H_2O$) crystals are unique because they are the only representative (with sodium periodate) of the crystallographic class $C_3$ (without a center of symmetry). The crystal symmetry suggests presence of nonlinearity, piezo- and pyro-electric properties and gyrotropy as well. Single crystals of $MgSO_3.6H_2O$ (pure and doped with Ni, Co and Zn) for the time being are grown only by the original method developed in the Laboratory for Crystal growth at the Faculty of Physics in Sofia University. The first results of optical activity of pure $MgSO_3.6H_2O$ and Zn doped $MgSO_3.6H_2O$ crystals are described and analyzed in a wide spectral range. The optical activity manifests itself in the direction (0001) as a rotation of the polarization plane.

## 1. Introduction

The magnesium sulfite hexahydrate ($MgSO_3.6H_2O$) single crystals are representative of the symmetry $C_3$ [1,2]. They are uniaxial and posses optical anisotropy in the directions different from (0001) [3].

The crystals of $MgSO_3.6H_2O$ have been produced so far with sizes up to 1.5 mm and that is the main reason that they have not been investigated properly. There are only some data in the scientific papers about their physical properties, for instance, the crystallographic parameters for pure $MgSO_3.6H_2O$ have been determined [1,4,5]. A significant piezo activity for small crystal powders has been found [1]. There is no information available for doped $MgSO_3.6H_2O$ crystals.

The first $MgSO_3.6H_2O$ single crystals (pure and doped) with sizes 40 – 50 mm have been grown by the original method developed in the Laboratory for crystal growth at the Physics Department at Sofia University [2]. It is possible to carry out experiments on samples with mentioned sizes for detailed determination of the physical characteristics of $MgSO_3.6H_2O$ and as well as of the dopand influence on its properties [6,7].

The class $C_3$ symmetry suggests the presence of gyrotropy properties of $MgSO_3.6H_2O$ crystals [8]. The present paper reports the observations of the natural optical activity in $MgSO_3.6H_2O$ and $MgSO_3.6H_2O$: Zn. It manifests itself as a rotation of the polarization plane along the optical c–axis, i.e. in the crystallographic direction (0001).

## 2. Experimental results

*2.1. Optical rotation spectra of $MgSO_3.6H_2O$*

The investigated samples of $MgSO_3.6H_2O$ are about 5 mm thick and possess double polished faces in the direction (0001). The spectrum of the optical rotating angle of $MgSO_3.6H_2O$ has been measured from 250 nm to 800 nm using SPM-2 monochromator and Glan-Taylor prisms. The absorption spectra of pure $MgSO_3.6H_2O$ present a wide structureless platform in that spectral range where the absorption coefficient is very low ($\alpha \sim 0.2$ cm$^{-1}$) [3]. The absorption coefficient starts to increase at $\lambda < 250$ nm [3]. The crystal water influence begins at $\lambda > 800$ nm [9].

The rotating angles have been measured precisely using $\varphi$ – modulation method [10]. The dependence of the rotating angle on light wave length for pure $MgSO_3.6H_2O$ is presented in figure 1. The values of the optical rotating angle increase monotonously as the light wave lengths shorten.

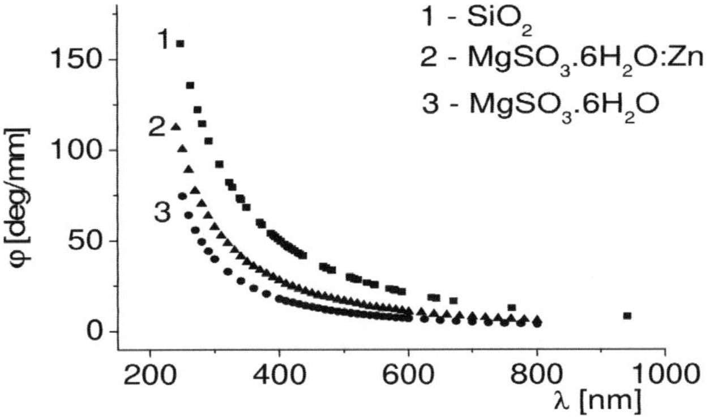


**Figure 1.** Spectra of optical rotation angle of $MgSO_3.6H_2O$, $MgSO_3.6H_2O$: Zn and $SiO_2$ [11].

*2.2. Optical rotation spectra of $MgSO_3.6H_2O$:Zn*

The spectrum of the optical rotating angle of $MgSO_3.6H_2O$: Zn has been measured from 250 nm to 800 nm using $\varphi$ – modulation method [10]. The absorption of $MgSO_3.6H_2O$: Zn is also as of the pure $MgSO_3.6H_2O$ neglective smaller in the mentioned spectral range. The dependence of the rotating angle on light wave length of Zn doped $MgSO_3.6H_2O$ is presented in figure 1. The values of the optical rotating angle increase monotonously in the transmittance range (250 – 800 nm) to the short light wave lengths region.

Eventually, the optical rotation of $MgSO_3.6H_2O$: Zn is a little larger than the optical rotation of pure $MgSO_3.6H_2O$ in the investigated spectral range. The $MgSO_3.6H_2O$ and $MgSO_3.6H_2O$: Zn crystals appear as right rotating materials.

The data of the quartz rotation angle have been taken from the literature [11] and added in figure 1. Therefore the spectrum of the quartz optical activity must be considered as a reference spectrum.

## 3. Discussion

It has been established, that the optical rotation of $MgSO_3.6H_2O$ and $MgSO_3.6H_2O$: Zn is comparable, but smaller than the optical rotation of $SiO_2$ in the investigated spectral range. The smoothly increasing of the rotation angle with shortening wave lengths in the investigated range (250 – 800 nm) can be explained by the influence of the fundamental absorption edge.

The gyration tensor of the symmetry $C_3$ can be written as [11]:

$$\begin{vmatrix} g_{11} & 0 & 0 \\ 0 & g_{11} & 0 \\ 0 & 0 & g_{33} \end{vmatrix}$$

The $MgSO_3.6H_2O$ gyrotropy must appear in accordance with the g-tensor as rotation of the polarization plane when the light propagates in the direction of the optical axis c of the crystal [8]. The polarization plane rotation of $MgSO_3.6H_2O$ and $MgSO_3.6H_2O$: Zn can be described as a circular birefringence in (0001) direction [12,13]. The gyrotropy can be observed as elliptical birefringence when the light propagates in directions, different from (0001) because the $MgSO_3.6H_2O$ crystals are anisotropic [14].

The unit cell of $MgSO_3.6H_2O$ contains two structure complexes [4,5]: the octahedron $Mg(H_2O)_6^{2+}$, where the $Mg^{2+}$ ion is situated in water surroundings and the pyramidal ion $SO_3^{2-}$ (figure 2). We assume that the structure component of the optical activity of $MgSO_3.6H_2O$ can be determined by the relationship between the mentioned complexes. The phenomenological gyrotropy theory provides that it is possible to stimulate optical activity at the symmetry $C_3$, which is connected with the acoustic modes of the crystal lattice vibrations. This optical activity must be considered jointly with the gyrotropy due to the electron transitions [11,15].

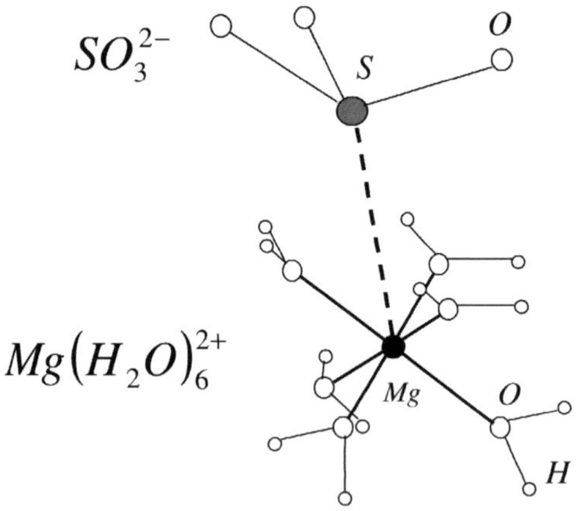


**Figure 2.** Structure complexes in the unit cell of $MgSO_3.6H_2O$: $Mg(H_2O)_6^{2+}$ octahedron and $SO_3^{2-}$ trigonal pyramid.

The contributions to the gyrotropy caused by the structure complexes in $MgSO_3.6H_2O$ can be determined mostly by the trigonal pyramid $SO_3^{2-}$, because the ideal octahedron complex with the $Mg^{2+}$ ion has a center of symmetry and therefore it does not influence the optical activity.

The increasing of the rotating angle in Zn presence in the $MgSO_3.6H_2O$ crystal lattice can be explained as follows: the $Zn^{2+}$ ion replaces the $Mg^{2+}$ and causes the octahedron deformation. One contribution to the optical activity change of the Zn doped crystal can be conditioned by the symmetry decrease of the octahedron complex [11]. The other contribution to the increase of the rotating can be caused by the influence of the Zn impurity on the relationship between the structure complexes in the unit cell.

The spectral dependence of rotating angle for optically active crystals has a complicated character. Therefore the rotating angle as a function of wave length (i.e. (λ)) is not convenient for analysis. This is the reason for the application of Kizel's method [11] for the transformation of $\varphi(\lambda)$ to $1/\varphi = f(\lambda^2)$. This transformation uses modified Drude formula:

$$1/\varphi = A\lambda^2 + B \quad (1)$$

This equation describes a linear function of $\lambda^2$ in the transmitting spectral range. A and B are constants in (1). The constant A depends on matrix elements of interband transitions; B is determined by wave length $\lambda_0$, which corresponds to the frequency $\omega_0$ of an effective harmonic oscillator.

The fundamental property of optical and magneto-optical activity [11] characterizes by the Drude dependence, because it is valid in the spectral ranges where the absorption is negligibly small and the gyrotropic parameter is real. Drude formula may be used as a criterion for determination of deviations from the linear dependence, caused by the presence of impurity atoms and structural defects of the crystal lattice as well.

Formula (1) is used for analyses of the optical rotation spectra of $MgSO_3.6H_2O$, $MgSO_3.6H_2O$: Zn and $SiO_2$. It proves definitely that the dependence $1/\varphi = f(\lambda^2)$ is linear for $MgSO_3.6H_2O$, $MgSO_3.6H_2O$: Zn and $SiO_2$ in all of the investigated spectral range (figure 3). This indicates that Drude formula is valid also for optical activity crystals in the transmittance region.

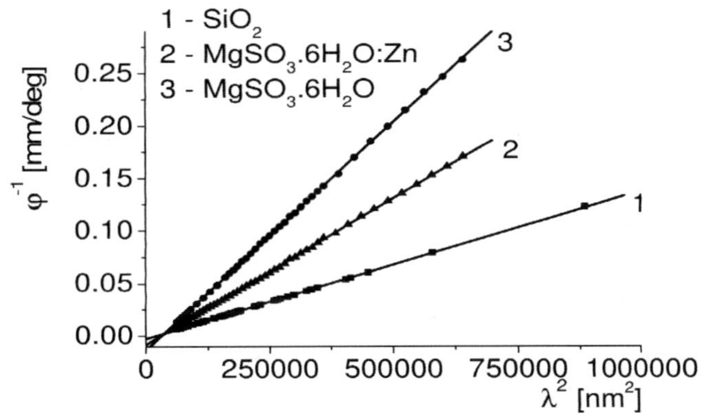


**Figure 3.** Linear dependence $1/\varphi = f(\lambda^2)$ for $MgSO_3.6H_2O$, $MgSO_3.6H_2O$: Zn and $SiO_2$ in accordance with Drude formula.

## 4. Conclusions

The presence of natural optical activity in magnesium sulfite hexahydrate ($MgSO_3.6H_2O$), pure and doped with Zn has been found experimentally as polarization plane rotation. The spectra of the optical rotation angle have been measured in the direction of the optical axis (0001) in a wide spectral range (250 – 800 nm). Zn as dopand in the crystal lattice causes the optical activity increasing of pure $MgSO_3.6H_2O$. The optical rotation of $MgSO_3.6H_2O$ and $MgSO_3.6H_2O$: Zn is compared with the optical activity of quartz and their optical activities are smaller, but comparable with quartz optical activity in the investigated spectral region. The linear dependence of Drude formula is fulfilled in the spectral region from 250 nm to 950 nm for the quartz and from 250 to 800 nm for $MgSO_3.6H_2O$ and $MgSO_3.6H_2O$: Zn.

## Acknowledgments

Financial support by Bulgarian National Fund project TK-X-1714/07 is gratefully acknowledged.

**References**

[1] Klasens H A, Perdok W and Terpstra P 1935 *Rec. Trav. Chim. Pays Bas* **54** 728-732
[2] Bunzarov Z, Saltiel S M, Kovachev T, Ljutov L and Russev D 1996 *Proceedings of SPIE* **3052** 197-200
[3] Bunzarov Zh, Petkova P, Dimov T and Iliev I 2004 *Coll. Papers Physics* Shumen University **1** 39-42
[4] Flack H 1973 *Acta Cryst.* B **29** 656-658
[5] Andersen L and Lindqvist O 1984 *Acta Cryst.* C **40** 584-586
[6] Petkova P, Bunzarov Zh, Iliev I and Dimov T 2007 *American Institute of Physics Conference Proceedings* **899** 329-330
[7] Petkova P, Bunzarov Zh, Iliev I and Dimov T *ROCAM'09* 2009 Brasov Romania (in press)
[8] Wainshtein B, Chernov A and Shuvalov L 1981 *Modern crystallography* vol 4 Moscow (in Russian) p 363
[9] Bunzarov Zh, Dimov T, Petkova P and Iliev I 2009 *American Institute of Physics Conference Proceedings* **1162** 210-214
[10] Petkova P, Dimov T and Iliev I 2007 *Journal of Physics: Conf. Series* **92** (012069) 1-4
[11] Kizel V and Burkov V 1980 *Gyrotropy of crystals* Moscow (in Russian) 277
[12] Konstantinova A, Ivanov N and Grechushnikov B 1969 *Crystallography* vol 14 (in Russian) p 283
[13] Velluz L, Legrand M and Grosjean M 1965 *Optical Circular Dichroism* (Academic: New York) p 21
[14] Sirotin Yu I and Shaskolskaya M P 1979 *Principles of Crystal Physics* Moscow (in Russian)
[15] Kizel V A *Uspehi Fizicheskih nauk* **147** 560-583

# AUTHOR INDEX

# AUTHOR INDEX

# AUTHOR INDEX